Chemistry in the Urban Atmosphere

London, United Kingdom

6–8 April 2016

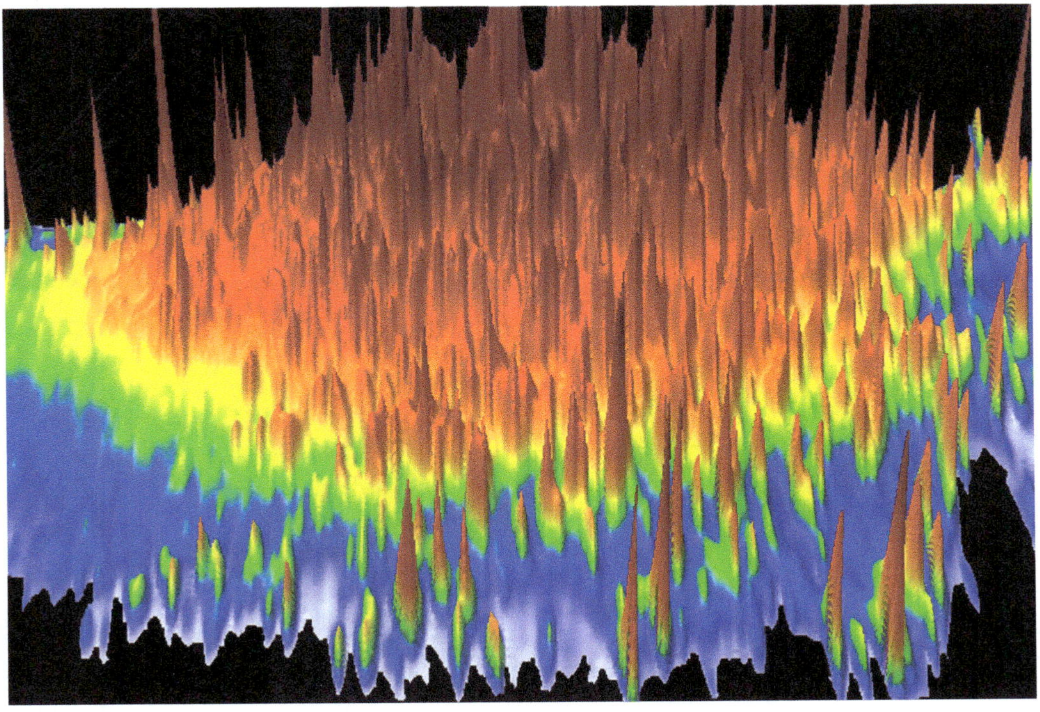

FARADAY DISCUSSIONS

Volume 189, 2016

ROYAL SOCIETY OF CHEMISTRY

The Faraday Division of the Royal Society of Chemistry, previously the Faraday Society, founded in 1903 to promote the study of sciences lying between Chemistry, Physics and Biology.

Faraday Discussions are unique international discussion meetings that focus on rapidly developing areas of chemistry and its interfaces with other scientific disciplines.

Faraday Discussions (Print ISSN 1359-6640, Electronic ISSN 1364-5498) is published 10 times a year by the Royal Society of Chemistry, Thomas Graham House, Science Park, Milton Road, Cambridge, UK CB4 0WF.

Volume 189 ISBN-13: 978-1-78262-476-9

2016 annual subscription price: print+electronic £929 US $1635; electronic only £883, US $1553. Customers in Canada will be subject to a surcharge to cover GST. Customers in the EU subscribing to the electronic version only will be charged VAT. All orders, with cheques made payable to the Royal Society of Chemistry, should be sent to RSC Order Department, Royal Society of Chemistry, Thomas Graham House, Science Park, Milton Road, Cambridge, CB4 0WF, UK.

Tel +44 (0) 1223 432398; E-mail **orders@rsc.org**

If you take an institutional subscription to any RSC journal you are entitled to free, site-wide web access to that journal. You can arrange access via Internet Protocol (IP) address at **www.rsc.org/ip**. Customers should make payments by cheque in sterling payable on a UK clearing bank or in US dollars payable on a US clearing bank.

Printed in the UK

Chemistry in the Urban Atmosphere

Faraday Discussions
www.rsc.org/faraday_d

A General Discussion on Chemistry in the Urban Atmosphere was held in London, UK on the 6th, 7th and 8th of April 2016.

CONTENTS

ISSN 1359-6640; ISBN: 978-1-78262-476-9

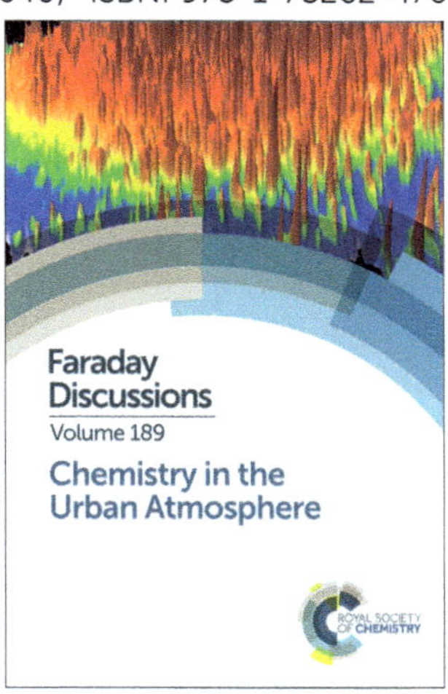

Cover
Front cover image courtesy of Dr Mohammed Salim Alam, University of Birmingham, UK.

A section from a 3D-chromatogram of the organic components sampled from urban air; each peak represents a compound, highlighting the complexity of the urban atmosphere.

INTRODUCTORY LECTURE

PAPERS AND DISCUSSIONS

Faraday Discussions

PAPER

Spiers Memorial Lecture

Introductory lecture: chemistry in the urban atmosphere

Urs Baltensperger

Received 1st April 2016, Accepted 18th April 2016
DOI: 10.1039/c6fd00065g

The urban atmosphere is characterised by a multitude of complex processes. Gaseous and particulate components are continuously emitted into the atmosphere from many different sources. These components are then dispersed in the urban atmosphere *via* turbulent mixing. Numerous chemical reactions modify the gas phase chemistry on multiple time scales, producing secondary pollutants. Through partitioning, the chemical and physical properties of the aerosol particles are also constantly changing as a consequence of dispersion and gas phase chemistry. This review presents an overview of the involved processes, focusing on the contributions presented at this conference and putting them into a broader context. Advanced methods for aerosol source apportionment are presented as well, followed by some aspects of health effects related to air pollution.

1. Introduction

Air pollution is a concern because of well-established adverse health effects: according to the World Health Organization (WHO), in 2012 about 7 million premature deaths worldwide resulted from exposure to air pollution.[1] Air pollution within cities is of special concern, because the high population density results in a potential for high total emissions, and more people suffer from the high air pollution. However, the situation is quite different for different cities: in post-industrial megacities such as Paris, successful air pollution control measures implemented in the past decades have resulted in a substantial reduction in air pollution, at least under favourable meteorological conditions.[2] Indeed, cities have the potential for more efficient per capita energy use, and lower emission factors than the national average. In contrast, in rapidly developing Asian megacities such as Beijing, air pollution is typically 10–100 times higher than in European or North American cities.[3]

Despite the improvements of the past decades, a number of air quality indicators are still exceeding the set standards, and further improvement is required. In order to develop the best strategy for this, the full complexity of atmospheric

Laboratory of Atmospheric Chemistry, Paul Scherrer Institute, 5232 Villigen, Switzerland. E-mail: urs.baltensperger@psi.ch

chemistry needs to be better understood. Our atmosphere is influenced by a wide variety of sources emitting a myriad of gas phase and particulate species. As it is impossible to identify every single compound, an adequate strategy is required, allowing us to quantify and control those species that are most harmful and/or informative, in the most efficient way possible.

When discussing air pollution in a city the concept of urban increment is often applied, a term which was originally coined by Lenschow *et al.*[4] The urban increment is defined as the increase in concentrations in the urban background relative to the rural surroundings. In an analogous way, the kerbside increment is defined as the increase in air pollution parameters at a kerbside or street site relative to the urban background (Fig. 1).

While this concept is generally accepted, the size of the urban increment is different for different regions, depending on the main emission sources in the city as well as the level of pollution in the surrounding area (see Fuzzi *et al.*[5] for a more detailed discussion). Gases and particle components that are directly emitted by the sources typically show enhanced concentrations close to the sources. This applies to NO_x but also to traffic related black carbon (BC) or hydrocarbon-like organic aerosol (*e.g.*, Mohr *et al.*[6]), as well as to species emitted by brake wear (Cu, Zr, Sb, Ba), other traffic-related elements such as Fe, or resuspended dust (Si, Ca) (Visser *et al.*[7]).

In contrast, secondary components both in the gas and aerosol phase are formed by chemical and physical transformation. Some of these secondary reactions may occur very fast. While an obvious example is the titration of NO with ozone, Gioro *et al.*[8] reported that in addition other components such as nitrate or oxidation products of anthropogenic volatile organic compounds (VOCs) can undergo relatively rapid transformation, such that under stagnant conditions they can become important locally. Other atmospheric aging processes occur on longer time scales such as several hours. As a result of this, the corresponding components are rather evenly distributed; a very well-known example of this is sulphate. Local and regional components not only differ in their chemistry but also in their size distribution: while aged secondary aerosol peaks in the accumulation mode, the size distribution of local contributions is shifted

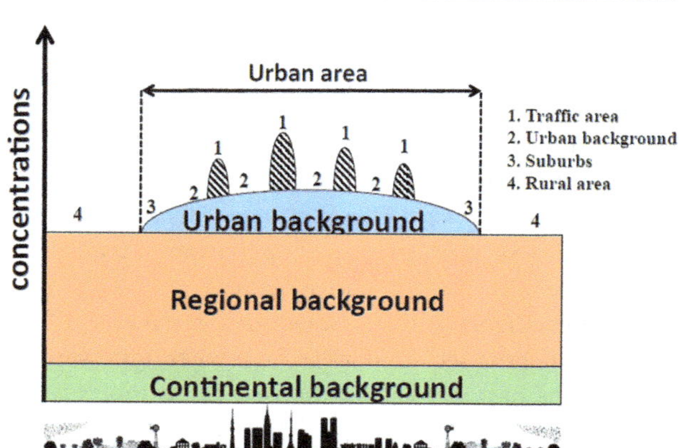

Fig. 1 The concept of regional background, urban increment and local traffic increment. Reproduced from Fuzzi *et al.*[5]

toward smaller particles with a tail in the direction of Aitken mode particles characteristic of a primary origin from exhaust emissions[8,9] and often also characterized by a larger contribution in the coarse mode.[7]

For components that have a high spatial variability a high number of instruments need to be simultaneously deployed. This is however only possible with low cost air pollution sensors, which have gained much media attention in recent years (see Lewis *et al.* (DOI: 10.1039/C5fd00201j) and references therein). For these to provide useful information, a careful evaluation of the performance of these sensors is required using real-world pollutant abundances and in the presence of a full suite of realistic co-pollutants and interferences. According to Lewis *et al.* (DOI: 10.1039/C5FD00201J) a good overall agreement was found for ozone between reference measurements and sensors, once sensor calibration values had been linked to the *in situ* reference observation. For NO and PM, general trends in atmospheric pollution were recreated by the sensors, although with notable negative and positive biases compared to the reference instruments. They conclude that the state of the art can at best be considered a 'work in progress', with promising performance for some chemicals and rather poor performance for others, and that extensive work is still required to enable useful research applications with these sensors.

2. Emission and dispersion

Emission inventories are an important tool in the top-down evaluation of air pollution. However, the Volkswagen emissions scandal revealed that these emission inventories can be substantially wrong. Both the media, as well as the general public, reacted with great surprise. However, it had already been known for several years that real-world emissions do not match the idealised test-bench emissions.[10] In 2011, scientists with the European Commission's Joint Research Centre (JRC) in Ispra, Italy, reported that average on-road emissions of tested diesel vehicles exceed allowed limits by up to 14 times.[11]

The discrepancies between bench tests and the real world were obvious even earlier than this. As an example, since 2000, the authorities of the canton of Zurich in Switzerland have employed remote sensing techniques to measure the actual emissions of CO_2, CO and HC (*via* infrared absorption) and NO (*via* UV absorption) from vehicles passing by a specific location.[12] By simultaneous detection of the license plate, the type and model year of all vehicles are retrieved. This allows for a comparison of the emission standards and real emissions for each model year (Fig. 2). The large discrepancy between actual emission standards and real emissions is obvious. In contrast, the identical measurements for gasoline cars at the same location (not shown) present emissions that are in accordance with the emission standards.

These discrepancies are confirmed by Vaughan *et al.* (DOI: 10.1039/c5fd00170f). Using the eddy-covariance technique from an aircraft flying at low altitude over London, they performed flux measurements of NO_x. They found a significant discrepancy between emission estimates and actual flux measurements, with the highest underestimation being a factor of two in central London. Using an enhanced version of the emission inventory, which is based on real-world traffic emissions, better agreement was achieved, with measured to inventory ratios of 1.1 to 1.5.

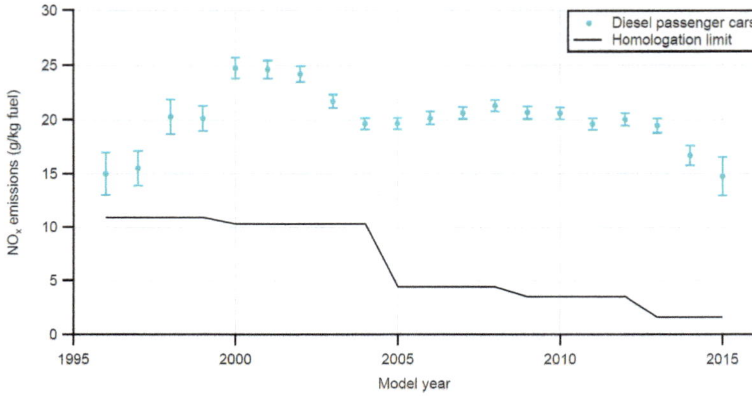

Fig. 2 Temporal evolution of real-world emissions of diesel passenger vehicles based on remote sensing at a road site in Switzerland, compared to the homologation limit of diesel passenger vehicles.[12] The actual emissions are based on NO measurements, which are transformed to NO_x values using primary NO_2 emission data based on the Handbook Emission Factors for Road Transport.[13] The discrepancy is obvious. Only for very recent model years have the real-world emissions started to decrease.

These results highlight again the critical importance of obtaining independent measurements of pollutant emission rates from vehicles during on-road driving conditions and using these data in emission inventories, rather than relying on emission data obtained during artificial test driving conditions or provided by vehicle manufacturers. At least there is a clear way forward. By 2017, the European Commission plans to establish a 'Real Driving Emissions' procedure that mandates on-road emissions tests for all types of passenger vehicles. To ensure that vehicles meet the NO_x emission limits on the road—and not just during lab conditions, which are open to manipulation—the procedure will make use of portable emissions measurement systems.[10]

In the last few years, the fraction of directly emitted NO_2 from diesel vehicles has increased due to vehicle after-treatment technologies which oxidise carbon monoxide and hydrocarbons and generate NO_2 to reduce particle emissions. Carslaw *et al.* (DOI: 10.1039/c5fd00162e) report that in London the NO_2/NO_x emission ratio increased substantially from around 5 vol% in 1996 to around 20 to 25 vol% in 2010, with an important effect on ambient NO_2 concentrations, especially close to roads. However, after 2010, this ratio decreased to about 15 vol% at the end of 2014 in both inner and outer London. The reasons for the decrease are likely related to considerable reductions in the NO_2/NO_x ratio for Euro IV/V heavy duty vehicles, which is confirmed by remote sensing emission measurements. In addition, Euro 4 and 5 diesel passenger cars show a decrease in the NO_2/NO_x ratio with increasing vehicle age, and the increased use of 'low-NO_2' SCRT systems (consisting of a continuously regenerating trap (CRT) to reduce particle emissions, and a selective catalytic reduction (SCR) system to reduce NO_x emissions) in public buses has further contributed to an NO_2 decrease. This combined decrease may have a potentially important effect on further reducing real-world NO_2/NO_x ratios from vehicles in the future.

The ability to predict NO_2 concentrations within urban street networks is important for the evaluation of strategies to reduce exposure to NO_2. However, this involves the coupling of several complex processes, *i.e.*, traffic emissions under different levels of congestion and dispersion *via* turbulent mixing, as well as chemical processes of relevance at the street-scale. Tomlin *et al.* (DOI: 10.1039/c5fd00159e) performed such a modelling study on a street canyon within the city of York, UK. Overall, the predicted NO_2 concentration at a number of in-street sites was uncertain to within approximately a factor of 2. Wall roughness length and background wind direction were found to be the most important physical parameters where the latter can be measured above-roof but is not a routine measurement. Concerning chemical processes, the short time-scale, in-street processes seemed to have the biggest effect on the NO_2 concentration, in particular the activation energy for the reaction of NO with O_3. It turned out that mitigation strategies in congested street canyon scenarios should focus on reducing primary NO_2 emissions rather than on traffic demand, while for less congested scenarios, reducing traffic demand did result in a reduction of the NO_2 concentration.

3. Gas phase chemistry

The question of emissions is tightly linked to air quality. Models are important tools to forecast how air quality evolves due to technological developments and regulatory action. A major challenge relates to the question of why air pollution has remained above acceptable limits in spite of the reduction efforts. Ozone is a prime example. Skouloudis and Rickerby (DOI: 10.1039/c5fd00189g) report that ozone concentrations did not decrease between 1995 and 2010 in their study areas and present the complexity of the required modelling approaches. As outlined in the recent review by Monks *et al.*,[14] in many places in the Northern Hemisphere mid-latitudes, ozone in some senses remains a paradox, with the high summertime peak levels decreasing but the regional background levels increasing or at least staying constant. As an example, Fig. 3 shows the situation for the UK, with

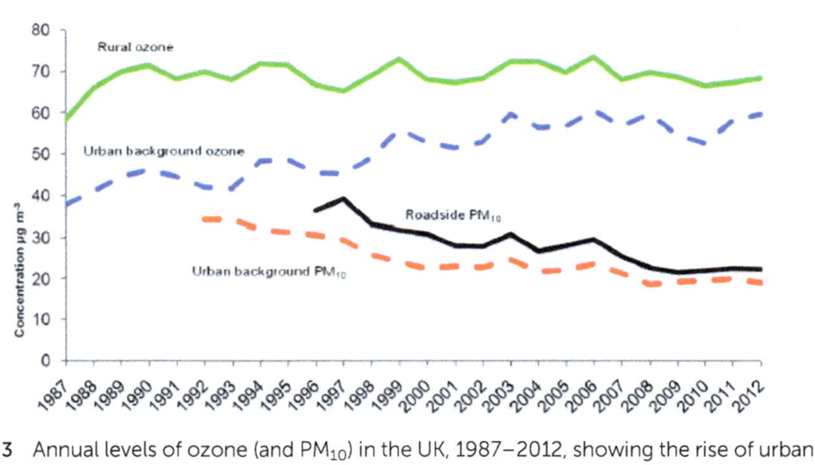

Fig. 3 Annual levels of ozone (and PM_{10}) in the UK, 1987–2012, showing the rise of urban ozone concentrations towards the rural ones and the relatively flat rural concentrations over the last decade, along with declining PM_{10} values. Reproduced from Monks *et al.*[14]

constant rural ozone concentrations, while the urban background ozone concentrations have increased, and the PM_{10} concentrations have decreased.[14] Clearly, the issues pertaining to tropospheric ozone are not "solved".[15]

In line with these observations, Ehlers *et al.* (DOI: 10.1039/c5fd00180c) report that the average number of hours per year for all survey stations in Germany exceeding the European ozone alarm value of 240 $\mu g\ m^{-3}$ has decreased since 1990, while average ozone concentrations have instead stayed constant or slightly increased in the same period. The ozone concentration also depends on VOC concentrations and reactivities. In urban areas, the major emission source of VOCs are gasoline cars during cold start while the major source of NO_x is expected to be diesel vehicles. Therefore, Ehlers *et al.* (DOI: 10.1039/c5fd00180c) explored how the change of the VOC and the NO_x emissions from traffic over the last 20 years affected the production of ozone, using VOC and NO_x reactivities rather than concentrations, and applying a box model using the master chemical mechanism (an extended version of MCM 3.1).[16] VOC concentrations decreased by a factor of 15 in the last 20 years, while the NO_x emissions stayed almost constant over this time. This has resulted in a substantially lower ozone production rate in the region where the ozone production is limited by the concentration of VOCs (corresponding to the whole urban and part of the rural area) than if the NO_x concentrations had been reduced simultaneously. If the decrease in traffic NO_x emissions started now, there would be a two-fold benefit, *i.e.*, the avoidance of high ozone production rates and high NO_2 concentrations in the cities.

While the MCM is a near-explicit chemical mechanism which describes the detailed gas-phase chemical processes involved in the tropospheric degradation of a series of emitted VOCs, it is computationally not feasible to include the full mechanism in air quality models. Malkin *et al.* (DOI: 10.1039/c5fd00218d) therefore performed studies with a box model, using MCM 3.2 (ref. 16) and compared this to two reduced chemistry schemes which are used in regional air quality models (*e.g.*, the regional chemistry transport model EMEP). By running the three box models to steady state they determined the full ozone creation potential for the investigated situations. However, this took up to seven days and resulted in ozone concentrations much higher than observed (up to 1240 ppb), indicating that the real London atmosphere was far from steady state. While the simplified models matched the *in situ* ozone predicted by the MCM box model to within 40% on most days in summer and winter, HONO had a major impact on the ozone chemistry, and was not included in the simplified models. HONO can play a dominant role in radical production, particularly in winter, and inclusion of a radical source from HONO and an accurate representation of the HONO concentrations observed in urban centres would improve the *in situ* ozone predicted by these reduced chemistry schemes.

As shown by Malkin *et al.* (DOI: 10.1039/c5fd00218d), HONO is an important source of OH, which in turn strongly influences ozone chemistry. Boundary layer levels of HONO have therefore been an active field of research in the last few decades. Numerous studies show that additional HONO sources may exceed the known production rate through the termolecular homogeneous reaction of OH radicals with NO by a factor of 10. Of these, photo-enhanced NO_2-to-HONO conversion upon organic films may be the most efficient (see Crilley *et al.* (DOI: 10.1039/c5fd00224a) and references therein). Photo-stationary steady-state analysis may help in identifying the strength of the unknown HONO source, but only if the photochemical steady-state is reached, which may require tens of minutes. Tong *et al.* (DOI: 10.1039/c5fd00163c) studied the formation mechanism of HONO in Beijing in winter and found that in urban areas the homogeneous reaction dominated, while in suburban areas the heterogeneous production rate was about ten times higher than the homogeneous one.

Despite the fact that the basic photochemistry of ozone has been known for decades it is still not possible to predict the highest ozone levels, which present the greatest health threat. Brune *et al.* (DOI: 10.1039/C5fd00204d) compared measured concentrations of OH and HO_2, and OH reactivity with a box model and found that measured HO_2 far exceeded modelled HO_2 in NO_x-rich plumes. They found evidence that the discrepancy was not due to the HO_2 measurement or to the sampling of transported plumes but instead to either emissions of unknown organic species that accompany the NO emissions or unknown photochemistry, calling for more laboratory and modelling kinetic studies. Kim *et al.* (DOI: 10.1039/c5fd00230c) measured the OH reactivity and compared it to trace gas observations. While at an urban site the total OH reactivity could in general be accounted for with the measured trace gases, this was not the case for a suburban forest area, indicating a large missing OH reactivity amounting to approximately 70% of the total. Similar findings had been reported from a boreal forest[17] and call for a careful evaluation of the whole photochemistry involved including potentially unidentified biogenic VOCs.

Besides the photochemistry of the present, the prediction of future photochemistry as a result of changes in the use of energy sources such as biofuels presents another important challenge. Dunmore *et al.* (DOI: 10.1039/C5fd00190k) found ethanol to be the most abundant VOC in London's atmosphere during a winter and a summer campaign. They identified a combination of combustion and evaporative emissions from the use of ethanol blended gasoline as the likely main emission source of ethanol and acetaldehyde, calling for adaptations of the emission inventories. Given the long life-time of ethanol, the full effects of this unaccounted emission are expected to occur downwind of the emission, such as increases in peroxyacetyl nitrate (PAN) and reduction in NO_2. Wallington *et al.* (DOI: 10.1039/C5fd00205b) present a literature review addressing the impact of ethanol blend levels (E0–E85) on emissions from gasoline light-duty vehicles in the U.S. and report on measurements with a biodiesel surrogate (butyl nonanoate). They conclude that emissions of NO_x, non-methane hydrocarbons, particulate matter (PM), and air toxics from modern gasoline and diesel vehicles are not increased by a higher biofuel content over the range in which the vehicles are designed to operate. In this context, a systems perspective (fuel and vehicle) is needed to fully understand, and optimize, the benefits of biofuels when blended into gasoline and diesel.

4. Volatility and partitioning

For a long time, particulate emissions have been treated as non-volatile. This resulted in a large overestimation of the ratio of primary to secondary organic aerosol. Robinson *et al.*[18] then introduced a change in paradigm, showing that most primary organic-particulate emissions are semi-volatile. Thus, they partially evaporate with atmospheric dilution, creating substantial amounts of low-volatility gas-phase material, which, upon oxidation, re-condenses to the particle phase.

Paciga *et al.*[19] estimated the volatility distribution of Paris winter and summer aerosol based on measured thermograms in the temperature range 20–200 °C. They termed the OA components semi-volatile (SVOCs with effective saturation concentration, C^*, values of 1, 10, and 100 μg m^{-3}), low volatility (LVOCs with C^* values of 10^{-3}, 10^{-2}, and 0.1 μg m^{-3}), and extremely low volatility (ELVOCs with $C^* \leq 10^{-4}$ μg m^{-3}). They found quite similar volatility distributions for the two seasons (Fig. 4), which they considered rather surprising, especially accounting for the corresponding uncertainties.

Gas-to-particle partitioning of organic aerosols (OA) is governed by Raoult's law, and depends on the existing mass of particles into which organic gases can dissolve. This raises the possibility of a non-linear response of particle-phase OA mass to the emissions of precursor VOCs that contribute to this partitioning mass, suggesting that reductions in VOC emission would have a more-than-proportionate benefit in lowering ambient OA concentrations. Madronich *et al.* (DOI: 10.1039/c5fd00209e) re-examined the linearity question using volatility spectra from several sources, *i.e.* chamber studies of selected aerosols, volatility inferred for aerosols sampled in two megacities, Mexico City and Paris, and an explicit chemistry model (GECKO-A). They present an overview of literature data on volatility distributions, for a variety of aerosol samples, including POA from different sources, SOA from biogenic emissions, as well as different components of atmospheric aerosols such as HOA (hydrocarbon-like organic aerosol), BBOA (biomass burning OA), COA (cooking OA), SV-OOA (semi-volatile oxygenated OA), and LV-OOA (low-volatility oxygenated OA). They used the volatility distributions to estimate the scaling of organic aerosols on VOC loading, OA ~ VOC^S, where S is a non-linear scaling or sensitivity exponent. They concluded that this sensitivity exponent was lower (on the order of 1.2) in complex urban mixtures than in aerosol from chamber experiments. Photochemistry pushes organic material from higher volatility to lower volatility, at least in the urban environment. Therefore, aerosols that have undergone intense photochemical processing (*e.g.*, OOA) show sensitivities very close to unity, while for example HOA and BBOA show higher sensitivities. It should be noted, however, that the role of

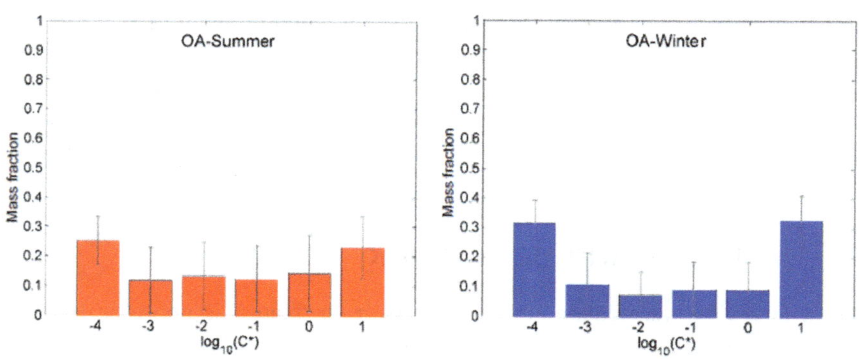

Fig. 4 Estimated volatility distributions for summer (left panel) and winter total OA (right panel). The error bars correspond to the fitting uncertainties. Reproduced from Paciga *et al.*[19]

photochemistry may reverse after longer reaction times, leading to net fragmentation (and increased volatility) rather than further functionalization.

Partitioning may also have effects on the size distribution of ultrafine particles (UFPs). The formation of UFPs in traffic exhaust is mainly driven by sulphuric acid and hydrocarbons where the latter are mostly associated with lubricating oil,[20] but they are believed to mostly contain a small non-volatile fraction (see Nikolova *et al.* (DOI: 10.1039/c5fd00164a) and references therein). After emission, coagulation results in larger UFP diameters and lower concentrations while dilution may result in smaller sizes due to partitioning. Nikolova *et al.* (DOI: 10.1039/c5fd00164a) developed a model to simulate the particle size distribution as a function of time after emission. The model accounts for the complex mixture of 17 SVOCs which were modelled as *n*-alkane proxies. They found that deposition was more important than coagulation in reducing the number concentration, but, overall, dilution was the major mechanism in altering the size distribution. The modelled nucleation-mode peak diameter at 23 nm decreased to 9 nm in 120 s. No SVOC species remained in particles in the sub-10 nm range; rather, these consisted only of non-volatile material. In order to simulate the nucleation-mode UFP observed, therefore, either non-volatile cores in the nucleation-mode particles emitted from the street canyon, or a composition consisting in part of organic compounds with volatilities much lower than those of $C_{32}H_{66}$ *n*-alkanes, are required. Otherwise one would expect an accelerated evaporation with decreasing size, due to an increasing Kelvin (curvature) effect with decreasing particle size, also known as the Nano-Köhler effect.[21]

This calls for a better characterization of these SVOCs from traffic exhaust. Alam *et al.* (DOI: 10.1039/C5fd00185d) report on SVOCs in engine exhaust, before and after a diesel oxidation catalyst (DOC). These SVOCs are typically hydrocarbons in the C15–C35 range, and are largely uncharacterised because they are unresolved by traditional gas chromatography and form a large hump in the chromatogram referred to as Unresolved Complex Mixture (UCM). Using thermal desorption coupled to two-dimensional gas-chromatography time-of-flight mass spectrometry (TD-GC × GC-ToF-MS) they were able to identify straight and branched alkanes (C11–C33), cycloalkanes (C11–C25), various cyclic aromatics, polyaromatic hydrocarbons (PAH), decalins and alkyl benzenes. The DOC reduced the concentrations of n-alkanes, alkyl-cyclohexanes and PAH only by 19 to 28%. This is much lower than claimed by the DOC technology (50–70%), most probably due to reactivity. This may have implications for SOA production.

At the moment of emission, freshly emitted primary particles are externally mixed with the pre-existing aerosol. Due to the semi-volatile nature of the emissions, this externally mixed particle population will start to evolve and converge immediately after emission. Robinson *et al.* (DOI: 10.1039/C5fd00214a) mixed two different aerosol types to investigate the 'Marcolli mixing',[22] *i.e.*, the thermodynamic mixing *via* gas-phase exchange of semi-volatile vapours under conditions where coagulation is negligible. They showed that if the populations are miscible and partitioning may occur this mixing occurs on a timescale of minutes to hours, depending on the volatility of the constituents. In contrast, non-fully miscible aerosol types such as populations of relatively polar secondary organic aerosol and relative non-polar primary organic aerosol remain distinct for hours. Oxidation evidently can suppress this miscibility barrier, but the specific thresholds for miscibility remain to be elucidated.

Zhu *et al.* (DOI: 10.1039/c5fd00175g) investigated this mixing using a size-composition resolved aerosol model (SCRAM) for a winter period in Paris and comparing the results with single particle observations based on an ATOFMS (aerosol time-of-flight mass spectrometer) obtained during the same time period. They found a consistent average value of the single particle diversity, which represents the average number of species within each particle, between simulation and measurement (2.91 and 2.79 respectively). They also calculated a mixing-state index.[23] The spatial distribution of the mixing-state index showed that particles are more externally mixed in urban regions in and around Paris, while they are more internally mixed in rural regions. This indicates that traditional aerosol models, which assume that particles are internally mixed, may be suitable for simulating rural regions, while in urbanised areas, the internal-mixing assumption does not hold.

5. Oligomers

Oligomers are closely linked to partitioning. Oligomerisation reduces the volatility of involved species considerably without further oxidation and therefore increases the partitioning of a semi-volatile species to the condensed phase, resulting in a much higher fraction found in the aerosol phase than would be expected from its volatility.[24,25]

Oligomers were first detected in ambient aerosols;[26] however, their source was unclear. In 2004 oligomers were then shown to be formed in secondary organic aerosol.[27,28] This created a wide field of research, which is summarized in the review paper by Hallquist *et al.*[24] Oligomers were found in a wide range of different types of aerosols.[29] The temporal evolution of the mass spectrometric signal indicated that oligomers continued to form in SOA over a time scale of at least 10 hours.[30] However, their mass spectrometric detection did not allow for quantification.

Even though Hoffmann *et al.*[31] had already reported dimer formation in the gas phase during the oxidation of α-pinene in a smog chamber in 1999, and Hallquist *et al.*[24] indicated the possibility of this formation process in Fig. 7 of their review article, an understanding of the formation mechanism of oligomers remained elusive. This changed with the development of new instrumentation when high-molecular weight compounds were detected in the gas phase using atmospheric pressure interface time of flight (APi-TOF) mass spectrometers. Using such instruments, Ehn *et al.*[32] found high concentrations of highly oxygenated molecules (HOMs). Note that these HOMs were originally coined ELVOCs (extremely low-volatility oxygenated compounds) – however, when referring to the species detected by an APi-TOF, the term HOMs should be used, since only a fraction of the registered species are of extremely low volatility.[33] Crounse *et al.*[34] then found that these HOMs are formed by only one attack of an oxidant, followed by internal rearrangement and further addition of O_2 in a process called autoxidation. Many species formed can further react to form covalently bound dimers[35] and larger oligomers by clustering, representing an efficient path for new particle formation.[32] Riccobono *et al.*[36] linked the temporal evolution of these HOMs (in the presence of sulfuric acid) with the concentrations of newly formed particles in the CLOUD[37] chamber and established a clear link between HOMs and new particle formation, as shown in Fig. 5. Very recently,

Kirkby *et al.* showed that this nucleation also proceeds with pure HOMs, in the absence of sulphuric acid.[38] Moreover, Bianchi *et al.* demonstrated, through measurements at the high-alpine site Jungfraujoch, Switzerland, that this mechanism does indeed occur in the ambient atmosphere.[39]

Since the first detection of oligomers, a large number of studies have investigated their chemical composition, and only a small selection of studies can be mentioned here. Ultra-high resolution mass spectrometry revealed the full complexity of the spectra, with multiple peaks for each unit mass.[40] Kourtchev *et al.*[41] reported that none of the investigated aging processes produced an observable effect on the oligomers formed from the ozonolysis of the BVOC mixture and that they were found to be equally abundant in both "fresh" and "aged" SOA. Additional separate dark ozonolysis and OH-initiated experiments with α-pinene showed that oligomers are an important group of compounds in both oxidation schemes. However, oligomers in the OH-initiated SOA were shifted towards higher masses. Despite of all these studies, a quantification of oligomers produced in the gas-phase *versus* intra-particle reactions is still missing.

6. Aerosols: characterization and source apportionment

The contribution of the various sources needs to be known in order to be able to provide recommendations on how to reduce air pollution. Concerning aerosols, the last few years have brought tremendous progress through the introduction of the aerosol mass spectrometer and sophisticated statistical analysis of the data, such as positive matrix factorization (PMF)[42,43] or the multilinear engine 2 (ME-2) where individual factors can be constrained,[44] and procedures have been developed to efficiently find the optimal solution.[45]

Pandis *et al.* (DOI: 10.1039/c5fd00212e) present an overview of particulate matter pollution in five cities (Athens and Paris in Europe, Pittsburgh and Los Angeles in the United States, and Mexico City in Central America). In all cases, reductions of emissions by industrial and transportation sources have resulted in significant improvements in air quality during the last few decades. However, these changes have resulted in the increasing importance of secondary particulate

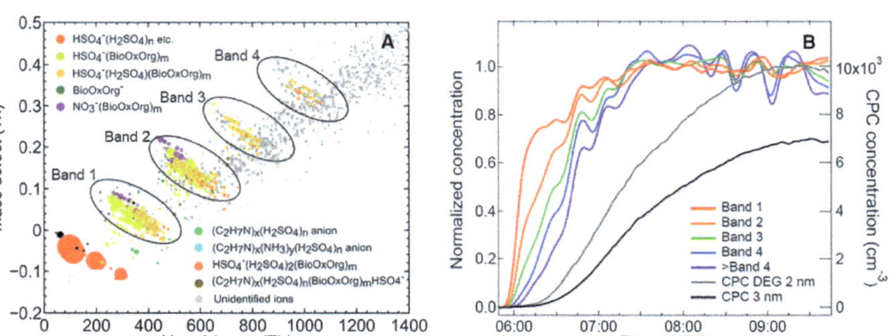

Fig. 5 (A) Mass-defect plot of negative clusters measured by the APi-TOF and (B) the time series of the normalised concentration of growing clusters grouped in bands and the time series of the concentration of newly formed particles. Reproduced from Riccobono *et al.*[36]

matter (PM) which dominates over primary in most cases. At the same time, the relative importance of long-range transport has increased. As an example, Beekmann *et al.*[2] found that about 70% of the fine PM mass was transported into the megacity Paris from upwind regions of France and continental Europe, and it is expected that this is representative of any modern megacity. Therefore, at least an area extending up to approximately 500 km from the city should be examined, together with the urban area itself, in efforts to further reduce fine PM concentrations (DOI: 10.1039/c5fd00212e). Due to the controls that have been successfully imposed on the corresponding emissions the contribution of the "usual suspects" to fine PM (local transportation and heavy industry, which together comprised from less than 20% to a maximum of 40% of the fine PM), has been decreasing. This has resulted in the increase of the relative importance of other sources, such as cooking, residential biomass burning, and nearby agricultural burning (DOI: 10.1039/c5fd00212e).

The costs associated with the operation and analysis of such aerosol mass spectrometers do not allow for the long-term campaigns that would be required *e.g.* to determine seasonal variations. With a simpler instrument, the aerosol chemical speciation monitor (ACSM)[46] such long-term operation becomes feasible.[47] However, this requires adaptation of the statistical procedures, since all PMF methods are based on the assumption of constant factor profiles. While this assumption may be reasonably close to reality for primary components, this is obviously not the case for secondary components, which are formed during atmospheric aging and are constantly evolving (Ng *et al.*).[48] Therefore, novel analysis approaches including automatic rolling features, whereby factors are allowed to evolve over time, are being developed for these applications (Canonaco *et al.*, A novel strategy for the source apportionment of long-term ACSM data based on ME-2 with SoFi: Automatic Rolling SoFi (AuRo SoFi), in preparation).

Another approach to performing source apportionment over a range of seasons is based on sampling on filters or cascade impactors (for better size resolution), followed by off-line analysis of extracts of the aerosol material. This can either be done directly by analysis of the liquid samples (DOI: 10.1039/c5fd00228a) or after re-nebulization of the solution followed by analysis with an AMS.[49,50] van Pinxteren *et al.* (DOI: 10.1039/c5fd00228a) reported on a source apportionment study performed simultaneously at four sites (traffic, traffic/residential, urban background, and regional background) in Leipzig, Eastern Germany, using size resolved sampling with cascade impactors in the years 2013 to 2015. The results of the chemical analyses were then analysed by PMF. They found that only about 50% of the PM was transported into the city, in contrast to the 70% mentioned above for Paris. The lower fraction is, however, to a great extent explained by the fact that van Pinxteren *et al.* (DOI: 10.1039/c5fd00228a) analysed PM_{10}, in contrast to the above studies which focused on fine PM (which, depending on the used instrumentation refers to PM_1 or $PM_{2.5}$). The traffic contribution at the most polluted site amounted to 30–40%, where about half of the traffic related PM_{10} was in the coarse mode. This underlines the importance of the coarse mode and points to difficulties for further reductions, as the non-exhaust emissions are difficult to address. Moreover, 30 to 40% of PM_{10} was explained by coal combustion when air masses arrived from Eastern regions during winter, again underlining the importance of regional pollution.

Carbon-14 is another powerful tool for source apportionment. Since fossil carbon does not contain any ^{14}C atoms while recent carbon contains the contemporary fraction of ^{14}C atoms, the ^{14}C measurements provide a discrimination between fossil and non-fossil carbon. Then, by assigning hydrocarbon-like organic aerosol (HOA) to fossil OA, and cooking organic aerosol (COA) and biomass burning organic aerosol (BBOA) to non-fossil OA, one can split the remaining carbon (constituting the SOA) into a fossil and a non-fossil fraction (*e.g.*, Beekmann *et al.*,[2] Huang *et al.*[51]). Minguillón *et al.* (DOI: 10.1039/c5fd00182j) used combined ACSM and carbon-14 analyses to perform a source apportionment of the organic aerosol at an urban background site in Barcelona in summer. They first used ME-2 to resolve the primary components, HOA and COA along with two different oxygenated OA components. Then they used carbon-14 for further discrimination, and found that in the majority of cases the fossil organic carbon was secondary. This result is somewhat different from the one found for Paris in summer using the same approach.[2] Further research is needed to get more insight into these fractions.

Tong *et al.* (DOI: 10.1039/C5fd00206k) analysed the composition of fine particles (diameter ≤ 2.5 μm) at an urban background site (Elms Road Observatory Site) and a road tunnel (Queensway) in Birmingham, UK, with direct infusion, nano-electrospray ionisation ultrahigh resolution mass spectrometry (UHRMS). They found a higher number of mono-aromatics and PAHs in the tunnel samples compared to the urban background site. They concluded that traffic emission can be a source for a very large number of oxidised mono-aromatics and PAHs in the urban atmosphere, which could potentially be used to better constrain vehicle exhaust emission profiles and effects of atmospheric aging on these particle emission.

It should, however, be kept in mind that biomass burning and coal combustion can also be large sources of PAHs. Elser *et al.*[52] reported extremely high PAH

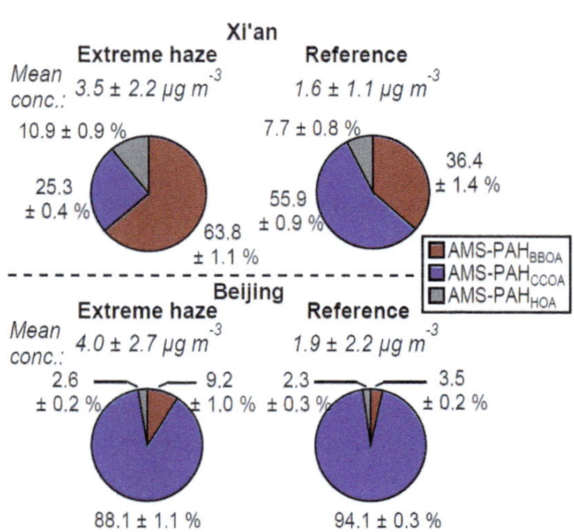

Fig. 6 Average AMS-PAH concentrations and relative contributions of the different sources to the measured AMS-PAHs for Xi'an and Beijing, each for two different periods (extreme haze and a less polluted period called reference). Errors represent the standard deviations from multiple PMF solutions. Reproduced from Elser *et al.*[52]

concentrations as measured with an aerosol mass spectrometer, with frequent peak concentrations above 10 μg m^{-3}. They attributed the majority of these PAHs to biomass burning in Xi'an and to coal combustion in Beijing (Fig. 6), confirming recent results from Beijing.[51]

Intermittent, strict control on emissions implemented in Beijing, China, during the 2015 China Victory Day Parade (V-day Parade) presented an excellent means to test the effects of short-term emission control on aerosol chemistry. Li *et al.* (DOI: 10.1039/c6fd00004e) used an ACSM and a MAAP (multiangle-absorption photometer) to determine the evolution of aerosol chemistry and, using PMF, the contributions of the various sources to PM_1. The average PM_1 mass concentration was 11.3 (±6.7) μg m^{-3} during the parade, which was 63.5% lower than before the parade. The aerosol composition also changed, but not extensively. The contribution of organics to the submicron aerosol increased from 41.3% before the control period to 53.2% during the parade, with associated reductions in the contributions of the inorganic species sulphate, nitrate, and ammonium. POA from traffic and cooking emissions decreased by 41.5% during the parade whereas secondary organic aerosol (SOA) was reduced by 59%, which is somehow against expectations. This may partly be due to changes in the meteorological conditions, even though there were no obvious changes in the local temperature, RH, wind speed and planetary boundary layer (PBL) height except for the fourth (last) day of the parade – when some precipitation occurred – the data of which were removed for the comparison.

The source apportionment of secondary organic aerosol (SOA) remains a difficult problem, due to a number of reasons. SOA is constantly evolving, such that the chemical signature varies as a function of aging time, exemplified by its degree of oxygenation. Cooking OA and biomass burning OA undergo chemical aging as well (DOI: 10.1039/c5fd00212e). Specific marker compounds which originate from a single precursor VOC are known in several cases; however, they cannot easily be used to quantify the fraction of SOA from this precursor, because depending on the actual conditions during the photooxidation (*e.g.* ratio of oxidant concentrations, VOC to NO_x ratio, *etc.*) the yield of this marker compound can vary substantially.

7. Health related effects

Numerous epidemiological studies have established a link between air pollution and adverse health effects/mortality, including many more since the famous six-cities study.[53] A major deficiency of these epidemiological studies is the fact that they are mostly based on correlations of a certain aerosol parameter, mostly $PM_{2.5}$, in this way assuming that the toxicity of all aerosol components is identical. This is obviously not the case, and health effect studies that are linked to specific source categories are urgently needed. Lelieveld *et al.*[54] used a global atmospheric chemistry model to investigate the link between premature mortality and seven emission source categories in urban and rural environments. They found that emissions from residential energy use such as heating and cooking, prevalent in India and China, have the largest impact on premature mortality globally. However, they assumed that all particles are equally toxic, and only performed a sensitivity study to account for varying toxicity. More detailed information on the health related toxicity of particles from different sources is urgently needed.

Many studies have shown a link between adverse health effects and the oxidative potential of PM (*e.g.*, Strak *et al.*[55]). It is believed that these effects are either caused by direct exposure to oxidants such as reactive oxygen species (ROS), or by the generation of ROS driven by air pollution components in human cells.[56] Shirmohammadi *et al.* (DOI: 10.1039/c5fd00166h) determined the oxidative potential as determined by the dithiothreitol (DTT) assay, of ambient $PM_{2.5}$ and $PM_{0.18}$ in Los Angeles, CA. They then linked the DTT activity with PM sources identified by a molecular marker-chemical mass balance (MM-CMB) model (*i.e.* vehicle tailpipe emissions, wood smoke, primary biogenic, secondary organic carbon, crustal material, vehicular abrasion, secondary ions and sea salt) through multilinear regression analysis. They found primary biogenic emissions, SOC and vehicular abrasion in the $PM_{2.5}$ size range, along with vehicular abrasion and vehicle tailpipe emissions in the $PM_{0.18}$ size range, to be the major sources of the DTT activity. They also compared their findings with studies from previous years. Ambient $PM_{0.18}$ showed a consistent decrease in DTT activity levels in the past decade, likely due to major reductions in tailpipe emissions in the LA basin as a result of stringent regulations on mobile sources. In the $PM_{2.5}$ size range, however, DTT activity levels showed overall a slight increase over the years, linked to non-tailpipe emissions.

Shafer *et al.* (DOI: 10.1039/c5fd00196j) applied an *in vitro* rat alveolar macrophage model and cytokine induction (TNFα) in addition to the DTT assay to examine relationships between PM-induced generation of ROS and PM composition. For this purpose they took size fractionated PM samples (>PM_7, PM_{3-7}, and PM_3) from roadside canyon, roadside motorway, and background urban sites in six European cities (Amsterdam, Frankfurt, London, Milan, Stockholm and Thessaloniki) and from three rural sites (one close to the North sea, one in Central Germany, and one on the Southern slope of the Alps in Northern Italy). They found a remarkable uniformity in DTT activity across the continent, with only a 4-fold difference across the urban sites. The DTT specific activity was greater at urban background sites (and also rural sites) than at urban canyon locations. 76% of the DTT activity was found in PM_3, reflecting both the large contribution of PM_3 to total PM mass levels and importantly the higher specific oxidative activity of PM_3 in comparison with the larger particles. Also, the soluble fraction of total DTT activity was very high (94%). In contrast to other studies they observed only poor correlations between the oxidative potential and the measured 28 elements in PM_3. They concluded that the chemical drivers of oxidative capacity are not too dissimilar between sites and that oxidative activity may not be only an urban problem. In contrast, the TNFα induction activity of the aerosol showed much larger variability.

A number of recent studies indicate that the oxidative potential of the aerosol is related to the organic fraction, and specifically to the secondary organic aerosol. Miljevic *et al.*[57] investigated the ROS concentrations of logwood burning aerosol. They found that 80–100% of the ROS concentration was removed by a thermodenuder at 250 °C, indicating that the ROS observed in this study were related to the organic fraction of the logwood burning aerosol. Platt *et al.*[58] demonstrated in chamber studies that oxidation processes produced a secondary organic aerosol from vehicle exhaust with dramatically greater ROS activity than primary emissions (Fig. 7), which finding was confirmed by a later study.[59] Further research is needed to find the best possible indicator (or combination of indicators) for

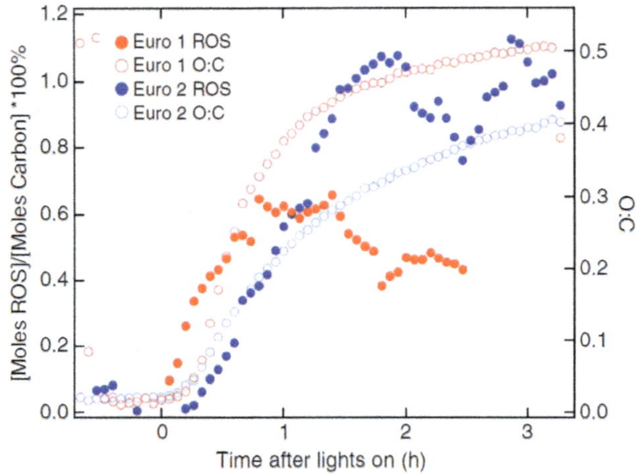

Fig. 7 Reactive oxygen species emissions from a Euro 1 (red) and Euro 2 (blue) two-stroke scooter. ROS concentration measured in moles of hydrogen peroxide equivalents is normalized to the molar organic carbon concentration per m^3 inside the smog chamber to give a percentage. The elemental O : C ratios of the organic aerosol as measured with an Aerodyne aerosol mass spectrometer are also shown. Reproduced from Platt *et al.*[58]

adverse health effects. This could involve a variety of cell cultures as done in a recent toxicity study where normal and diseased airway epithelia were exposed to aged gasoline exhaust particles,[60] or any combination of *in vitro* and *in vivo* studies.

Acknowledgements

I am grateful for the many contributions of past and present members of the Laboratory of Atmospheric Chemistry at the Paul Scherrer Institute who have made this review possible. This work is financially supported by various sources, including the Swiss National Science Foundation, the Swiss Federal Office for the Environment, and the European Commission through a variety of projects.

References

1 WHO, Review of Evidence on Health Aspects of Air Pollution – the REVIHAAP project. Final technical report, World Health Organization, Geneva, www.euro.who.int/__data/assets/pdf_file/0004/193108/REVIHAAP-Final-technical-report-final-version.pdf, 2013.

2 M. Beekmann, A. S. H. Prévôt, F. Drewnick, J. Sciare, S. N. Pandis, H. A. C. Denier van der Gon, M. Crippa, F. Freutel, L. Poulain, V. Ghersi, E. Rodriguez, S. Beirle, P. Zotter, S. L. von der Weiden-Reinmüller, M. Bressi, C. Fountoukis, H. Petetin, S. Szidat, J. Schneider, A. Rosso, I. El Haddad, A. Megaritis, Q. J. Zhang, V. Michoud, J. G. Slowik, S. Moukhtar, P. Kolmonen, A. Stohl, S. Eckhardt, A. Borbon, V. Gros, N. Marchand, J. L. Jaffrezo, A. Schwarzenboeck, A. Colomb, A. Wiedensohler, S. Borrmann, M. Lawrence, A. Baklanov and U. Baltensperger, *Atmos. Chem. Phys.*, 2015, **15**, 9577–9591.

3 M. Kulmala, *Nature*, 2015, **526**, 497–499.
4 P. Lenschow, H. Abraham, K. Kutzner, M. Lutz, J. Preuss and W. Reichenbacher, *Atmos. Environ.*, 2001, **35**, S23–S33.
5 S. Fuzzi, U. Baltensperger, K. Carslaw, S. Decesari, H. Denier van der Gon, M. C. Facchini, D. Fowler, I. Koren, B. Langford, U. Lohmann, E. Nemitz, S. Pandis, I. Riipinen, Y. Rudich, M. Schaap, J. G. Slowik, D. V. Spracklen, E. Vignati, M. Wild, M. Williams and S. Gilardoni, *Atmos. Chem. Phys.*, 2015, **15**, 8217–8299.
6 C. Mohr, R. Richter, P. F. DeCarlo, A. S. H. Prévôt and U. Baltensperger, *Atmos. Chem. Phys.*, 2011, **11**, 7465–7482.
7 S. Visser, J. G. Slowik, M. Furger, P. Zotter, N. Bukowiecki, F. Canonaco, U. Flechsig, K. Appel, D. C. Green, A. H. Tremper, D. E. Young, P. I. Williams, J. D. Allan, H. Coe, L. R. Williams, C. Mohr, L. Xu, N. L. Ng, E. Nemitz, J. F. Barlow, C. H. Halios, Z. L. Fleming, U. Baltensperger and A. S. H. Prévôt, *Atmos. Chem. Phys.*, 2015, **15**, 11291–11309.
8 C. Gioro, A. Tapparo, M. Dall'Osto, D. C. S. Beddows, J. K. Esser-Gietl, R. M. Healy and R. M. Harrison, *Environ. Sci. Technol.*, 2015, **49**, 3330–3340.
9 R. M. Harrison, D. C. S. Beddows and M. Dall'Osto, *Environ. Sci. Technol.*, 2011, **45**, 5522–5528.
10 Q. Schiermeier, Nature News, 24 September, 2015, http://www.nature.com/news/the-science-behind-the-volkswagen-emissions-scandal-1.18426, last accessed on 30 March 2016.
11 M. Weiss, P. Bonnel, R. Hummel, A. Provenza and U. Manfredi, *Environ. Sci. Technol.*, 2011, **45**, 8575–8581.
12 AWEL, Bericht und Auswertung RSD Messungen 2015, Zürich, 2015 (in German), www.awel.zh.ch/dam/baudirektion/awel/luft_asbest_elektrosmog/verkehr/rsd/dokumente/RSD_Bericht_2015.pdf, last accessed on 20 March 2016.
13 Handbook Emission Factors for Road Transport, HBEFA 3.2, 2014, http://www.hbefa.net/, last accessed on 30 March 2016.
14 P. S. Monks, A. T. Archibald, A. Colette, O. Cooper, M. Coyle, R. Derwent, D. Fowler, C. Granier, K. S. Law, G. E. Mills, D. S. Stevenson, O. Tarasova, V. Thouret, E. von Schneidemesser, R. Sommariva, O. Wild and M. L. Williams, *Atmos. Chem. Phys.*, 2015, **15**, 8889–8973.
15 C. B. B. Guerreiro, V. Foltescu and F. de Leeuw, *Atmos. Environ.*, 2014, **98**, 376–384.
16 M. E. Jenkin, S. M. Saunders, V. Wagner and M. J. Pilling, *Atmos. Chem. Phys.*, 2003, **3**, 181–193.
17 A. C. Nolscher, J. Williams, V. Sinha, T. Custer, W. Song, A. M. Johnson, R. Axinte, H. Bozem, H. Fischer, N. Pouvesle, G. Phillips, J. N. Crowley, P. Rantala, J. Rinne, M. Kulmala, D. Gonzales, J. Valverde-Canossa, A. Vogel, T. Hoffmann, H. G. Ouwersloot, J. V. G. de Arellano and J. Lelieveld, *Atmos. Chem. Phys.*, 2012, **12**, 8257–8270.
18 A. L. Robinson, N. M. Donahue, M. K. Shrivastava, E. A. Weitkamp, A. M. Sage, A. P. Grieshop, T. E. Lane, J. R. Pierce and S. N. Pandis, *Science*, 2007, **315**, 1259–1262.
19 A. Paciga, E. Karnezi, E. Kostenidou, L. Hildebrandt, M. Psichoudaki, G. J. Engelhart, B. H. Lee, M. Crippa, A. S. H. Prévôt, U. Baltensperger and S. N. Pandis, *Atmos. Chem. Phys.*, 2016, **16**, 2013–2023.

20 H. Sakurai, P. Park, P. H. McMurry, D. D. Zarling, D. B. Kittelson and P. J. Ziemann, *Environ. Sci. Technol.*, 2003, **37**, 5487–5495.

21 M. Kulmala, J. Kontkanen, H. Junninen, K. Lehtipalo, H. E. Manninen, T. Nieminen, T. Petäjä, M. Sipilä, S. Schobesberger, P. Rantala, A. Franchin, T. Jokinen, E. Järvinen, M. Äijälä, J. Kangasluoma, J. Hakala, P. P. Aalto, P. Paasonen, J. Mikkilä, J. Vanhanen, J. Aalto, H. Hakola, U. Makkonen, T. Ruuskanen, R. L. Mauldin III, J. Duplissy, H. Vehkamäki, J. Bäck, A. Kortelainen, I. Riipinen, T. Kurtén, M. V. Johnston, J. N. Smith, M. Ehn, T. F. Mentel, K. E. J. Lehtinen, A. Laaksonen, V.-M. Kerminen and D. R. Worsnop, *Science*, 2013, **339**, 943–946.

22 C. Marcolli, B. Luo, T. Peter and F. Wienhold, *Atmos. Chem. Phys.*, 2004, **4**, 2593–2599.

23 N. Riemer and M. West, *Atmos. Chem. Phys.*, 2013, **13**, 11423–11439.

24 M. Hallquist, J. C. Wenger, U. Baltensperger, Y. Rudich, D. Simpson, M. Claeys, J. Dommen, N. M. Donahue, C. George, A. H. Goldstein, J. F. Hamilton, H. Herrmann, T. Hoffmann, Y. Iinuma, M. Jang, M. Jenkin, J. L. Jimenez, A. Kiendler-Scharr, W. Maenhaut, G. McFiggans, T. F. Mentel, A. Monod, A. S. H. Prévôt, J. H. Seinfeld, J. D. Surratt, R. Szmigielski and J. Wildt, *Atmos. Chem. Phys.*, 2009, **9**, 5155–5236.

25 J. L. Jimenez, M. R. Canagaratna, N. M. Donahue, A. S. H. Prévôt, Q. Zhang, J. H. Kroll, P. F. DeCarlo, J. D. Allan, H. Coe, N. L. Ng, A. C. Aiken, K. D. Docherty, I. M. Ulbrich, A. P. Grieshop, A. L. Robinson, J. Duplissy, J. D. Smith, K. R. Wilson, V. A. Lanz, C. Hueglin, Y. L. Sun, J. Tian, A. Laaksonen, T. Raatikainen, J. Rautiainen, P. Vaattovaara, M. Ehn, M. Kulmala, J. M. Tomlinson, D. R. Collins, M. J. Cubison, E. J. Dunela, J. A. Huffman, T. B. Onasch, M. R. Alfarra, P. I. Williams, K. Bower, Y. Kondo, J. Schneider, F. Drewnick, S. Borrmann, S. Weimer, K. Demerjian, D. Salcedo, L. Cottrell, R. Griffin, A. Takami, T. Miyoshi, S. Hatakeyama, A. Shimono, J. Y. Sun, Y. M. Zhang, K. Dzepina, J. R. Kimmel, D. Sueper, J. T. Jayne, S. C. Herndon, A. M. Trimborn, L. R. Williams, E. C. Wood, C. E. Kolb, U. Baltensperger and D. R. Worsnop, *Science*, 2009, **326**, 1525–1529.

26 N. Havers, P. Burba, J. Lambert and D. Klockow, *J. Atmos. Chem.*, 1998, **29**, 45–54.

27 M. Kalberer, D. Paulsen, M. Sax, M. Steinbacher, J. Dommen, A. S. H. Prévôt, R. Fisseha, E. Weingartner, V. Frankevich, R. Zenobi and U. Baltensperger, *Science*, 2004, **303**, 1659–1662.

28 M. P. Tolocka, M. Jang, J. Ginter, F. Cox, R. Kamens and M. Johnston, *Environ. Sci. Technol.*, 2004, **38**, 1428–1434.

29 U. Baltensperger, M. Kalberer, J. Dommen, D. Paulsen, M. R. Alfarra, H. Coe, R. Fisseha, A. Gascho, M. Gysel, S. Nyeki, M. Sax, M. Steinbacher, A. S. H. Prévôt, S. Sjögren, E. Weingartner and R. Zenobi, *Faraday Discuss.*, 2005, **130**, 265–278.

30 D. S. Gross, M. E. Gälli, M. Kalberer, A. S. H. Prévôt, J. Dommen, M. R. Alfarra, J. Duplissy, K. Gaeggeler, A. Gascho, A. Metzger and U. Baltensperger, *Anal. Chem.*, 2006, **78**, 2130–2137.

31 T. Hoffmann, R. Bandur, U. Marggraf and M. Linscheid, *J. Geophys. Res.*, 1998, **103**, 25569–25578.

32 M. Ehn, J. Thornton, E. Kleist, M. Sipilä, H. Junninen, I. Pullinen, M. Springer, F. Rubach, R. Tillmann, B. Lee, F. Lopez-Hilfiker, S. Andres, I. Acir,

M. Rissanen, T. Jokinen, S. Schobesberger, J. Kangasluoma, J. Kontkanen, T. Nieminen, T. Kurten, L. Nielsen, S. Jorgensen, H. Kjaergaard, M. Canagaratna, M. Dal Maso, T. Berndt, T. Petäjä, A. Wahner, V. Kerminen, M. Kulmala, D. Worsnop, J. Wildt and T. Mentel, *Nature*, 2014, **506**, 476–479.
33 Jasmin Tröstl *et al.*, *Nature*, 2016, **533**, 527–531.
34 J. D. Crounse, L. B. Nielsen, S. Jørgensen, H. G. Kjaergaard and P. O. Wennberg, *J. Phys. Chem. Lett.*, 2013, **4**, 3513–3520.
35 T. Jokinen, M. Sipilä, S. Richters, V. M. Kerminen, P. Paasonen, F. Stratmann, D. Worsnop, M. Kulmala, M. Ehn, H. Herrmann and T. Berndt, *Angew. Chem., Int. Ed.*, 2014, **53**, 14596–14600.
36 F. Riccobono, S. Schobesberger, C. E. Scott, J. Dommen, I. K. Ortega, L. Rondo, J. Almeida, A. Amorim, F. Bianchi, M. Breitenlechner, A. David, A. Downard, E. M. Dunne, J. Duplissy, S. Ehrhart, R. C. Flagan, A. Franchin, A. Hansel, H. Junninen, M. Kajos, H. Keskinen, A. Kupc, A. Kürten, A. N. Kvashin, A. Laaksonen, K. Lehtipalo, V. Makhmutov, S. Mathot, T. Nieminen, A. Onnela, T. Petäjä, A. P. Praplan, F. D. Santos, S. Schallhart, J. H. Seinfeld, M. Sipilä, D. V. Spracklen, Y. Stozhkov, F. Stratmann, A. Tomé, G. Tsagkogeorgas, P. Vaattovaara, Y. Viisanen, A. Vrtala, P. E. Wagner, E. Weingartner, H. Wex, D. Wimmer, K. S. Carslaw, J. Curtius, N. M. Donahue, J. Kirkby, M. Kulmala, D. R. Worsnop and U. Baltensperger, *Science*, 2014, **344**, 717–721.
37 J. Kirkby, J. Curtius, J. Almeida, E. Dunne, J. Duplissy, S. Ehrhart, A. Franchin, S. Gagne, L. Ickes, A. Kurten, A. Kupc, A. Metzger, F. Riccobono, L. Rondo, S. Schobesberger, G. Tsagkogeorgas, D. Wimmer, A. Amorim, F. Bianchi, M. Breitenlechner, A. David, J. Dommen, A. Downard, M. Ehn, R. C. Flagan, S. Haider, A. Hansel, D. Hauser, W. Jud, H. Junninen, F. Kreissl, A. Kvashin, A. Laaksonen, K. Lehtipalo, J. Lima, E. R. Lovejoy, V. Makhmutov, S. Mathot, J. Mikkila, P. Minginette, S. Mogo, T. Nieminen, A. Onnela, P. Pereira, T. Petaja, R. Schnitzhofer, J. H. Seinfeld, M. Sipila, Y. Stozhkov, F. Stratmann, A. Tome, J. Vanhanen, Y. Viisanen, A. Vrtala, P. E. Wagner, H. Walther, E. Weingartner, H. Wex, P. M. Winkler, K. S. Carslaw, D. R. Worsnop, U. Baltensperger and M. Kulmala, *Nature*, 2011, **476**, 429–433.
38 Jasper Kirkby *et al.*, *Nature*, 2016, **533**, 521–526.
39 F. Bianchi *et al.*, *Science*, 2016, **352**(6289), 1109–1112.
40 A. Reinhardt, C. Emmenegger, B. Gerrits, C. Panse, J. Dommen, U. Baltensperger, R. Zenobi and M. Kalberer, *Anal. Chem.*, 2007, **79**, 4074–4082.
41 I. Kourtchev, J. F. Doussin, C. Giorio, B. Mahon, E. M. Wilson, N. Maurin, E. Pangui, D. S. Venables, J. C. Wenger and M. Kalberer, *Atmos. Chem. Phys.*, 2015, **15**, 5683–5695.
42 V. A. Lanz, M. R. Alfarra, U. Baltensperger, B. Buchmann, C. Hueglin and A. S. H. Prévôt, *Atmos. Chem. Phys.*, 2007, **7**, 1503–1522.
43 I. M. Ulbrich, M. R. Canagaratna, Q. Zhang, D. R. Worsnop and J. L. Jimenez, *Atmos. Chem. Phys.*, 2009, **9**, 2891–2918.
44 V. A. Lanz, M. R. Alfarra, U. Baltensperger, B. Buchmann, C. Hueglin, S. Szidat, M. N. Wehrli, L. Wacker, S. Weimer, A. Caseiro, H. Puxbaum and A. S. H. Prévôt, Source attribution of submicron organic aerosols during wintertime inversions by advanced factor analysis of aerosol mass spectra, *Environ. Sci. Technol.*, 2008, **42**, 214–220.

45 F. Canonaco, M. Crippa, J. G. Slowik, U. Baltensperger and A. S. H. Prévôt, *Atmos. Meas. Tech.*, 2013, **6**, 3649–3661.

46 N. L. Ng, S. C. Herndon, A. Trimborn, M. R. Canagaratna, P. L. Croteau, T. B. Onasch, D. Sueper, D. R. Worsnop, Q. Zhang, Y. L. Sun and J. T. Jayne, *Aerosol Sci. Technol.*, 2011, **45**, 780–794.

47 F. Canonaco, J. G. Slowik, U. Baltensperger and A. S. H. Prévôt, *Atmos. Chem. Phys.*, 2015, **15**, 6993–7002.

48 N. L. Ng, M. R. Canagaratna, Q. Zhang, J. L. Jimenez, J. Tian, I. M. Ulbrich, J. H. Kroll, K. S. Docherty, P. S. Chhabra, R. Bahreini, S. M. Murphy, J. H. Seinfeld, L. Hildebrandt, N. M. Donahue, P. F. DeCarlo, V. A. Lanz, A. S. H. Prévôt, E. Dinar, Y. Rudich and D. R. Worsnop, *Atmos. Chem. Phys.*, 2010, **10**, 4625–4641.

49 Y. Sun, Q. Zhang, M. Zheng, X. Ding, E. S. Edgerton and X. Wang, *Environ. Sci. Technol.*, 2011, **45**, 4854–4861.

50 K. R. Daellenbach, C. Bozzetti, A. Křepelová, F. Canonaco, R. Wolf, P. Zotter, P. Fermo, M. Crippa, J. G. Slowik, Y. Sosedova, Y. Zhang, R.-J. Huang, L. Poulain, S. Szidat, U. Baltensperger, I. El Haddad and A. S. H. Prévôt, *Atmos. Meas. Tech.*, 2016, **9**, 23–39.

51 R.-J. Huang, Y. Zhang, C. Bozzetti, K.-F. Ho, J.-J. Cao, Y. Han, K. R. Daellenbach, J. G. Slowik, S. M. Platt, F. Canonaco, P. Zotter, R. Wolf, S. M. Pieber, E. A. Bruns, M. Crippa, G. Ciarelli, A. Piazzalunga, M. Schwikowski, G. Abbaszade, J. Schnelle-Kreis, R. Zimmermann, Z. An, S. Szidat, U. Baltensperger, I. E. Haddad and A. S. H. Prevot, *Nature*, 2014, **514**, 218–222.

52 M. Elser, R.-J. Huang, R. Wolf, J. G. Slowik, Q. Wang, F. Canonaco, G. Li, C. Bozzetti, K. R. Daellenbach, Y. Huang, R. Zhang, Z. Li, J. Cao, U. Baltensperger, I. El-Haddad and A. S. H. Prévôt, *Atmos. Chem. Phys.*, 2016, **16**, 3207–3225.

53 D. W. Dockery, C. A. Pope, X. Xu, J. D. Spengler, J. H. Ware, M. E. Fay, B. G. Ferris and F. E. Speizer, *N. Engl. J. Med.*, 1993, **329**, 1753–1759.

54 J. Lelieveld, J. S. Evans, M. Fnais, D. Giannadaki and A. Pozzer, *Nature*, 2015, **525**, 367–371.

55 M. Strak, N. A. H. Janssen, K. J. Godri, I. Gosens, I. S. Mudway, F. R. Cassee, E. Lebret, F. J. Kelly, R. M. Harrison, B. Brunekreef, M. Steenhof and G. Hoek, *Environ. Health Perspect.*, 2012, **120**, 1183–1189.

56 K. Donaldson, V. Stone, P. J. A. Borm, L. A. Jimenez, P. S. Gilmour, R. P. F. Schins, A. M. Knaapen, I. Rahman, S. P. Faux, D. M. Brown and W. MacNee, *Free Radical Biol. Med.*, 2003, **34**, 1369–1382.

57 B. Miljevic, M. F. Heringa, A. Keller, N. K. Meyer, J. Good, A. Lauber, P. F. DeCarlo, K. E. Fairfull-Smith, T. Nussbaumer, H. Burtscher, A. S. H. Prévôt, U. Baltensperger, S. E. Bottle and Z. Ristovski, *Environ. Sci. Technol.*, 2010, **44**, 6601–6607.

58 S. M. Platt, I. E. Haddad, S. M. Pieber, R.-J. Huang, A. A. Zardini, M. Clairotte, R. Suarez-Bertoa, P. Barmet, L. Pfaffenberger, R. Wolf, J. G. Slowik, S. J. Fuller, M. Kalberer, R. Chirico, J. Dommen, C. Astorga, R. Zimmermann, N. Marchand, S. Hellebust, B. Temime-Roussel, U. Baltensperger and A. S. H. Prévôt, *Nat. Commun.*, 2014, **5**, 3749.

59 A. Saffari, S. Hasheminassab, D. Wang, M. M. Shafer, J. J. Schauer and C. Sioutas, *Atmos. Environ.*, 2015, **120**, 286–296.

60 L. Künzi, M. Krapf, N. Daher, J. Dommen, N. Jeannet, S. Schneider, S. Platt, J. G. Slowik, N. Baumlin, M. Salathe, A. S. H. Prévôt, M. Kalberer, C. Strähl, L. Dümbgen, C. Sioutas, U. Baltensperger and M. Geiser, *Sci. Rep.*, 2015, **4**, 11801.

Faraday Discussions

PAPER

Single-particle measurements of phase partitioning between primary and secondary organic aerosols

Ellis Shipley Robinson,[ab] Neil M. Donahue,*[a] Adam T. Ahern,[a] Qing Ye[a] and Eric Lipsky[ac]

Received 8th December 2015, Accepted 8th January 2016

DOI: 10.1039/c5fd00214a

Organic aerosols provide a measure of complexity in the urban atmosphere. This is because the aerosols start as an external mixture, with many populations from varied local sources, that all interact with each other, with background aerosols, and with condensing vapors from secondary organic aerosol formation. The externally mixed particle populations start to evolve immediately after emission because the organic molecules constituting the particles also form thermodynamic mixtures – solutions – in which a large fraction of the constituents are semi-volatile. The external mixtures are thus well out of thermodynamic equilibrium, with very different activities for many constituents, and yet also have the capacity to relax toward equilibrium *via* gas-phase exchange of semi-volatile vapors. Here we describe experiments employing quantitative single-particle mass spectrometry designed to explore the extent to which various primary organic aerosol particle populations can interact with each other or with secondary organic aerosols representative of background aerosol populations. These methods allow us to determine when these populations will and when they will not mix with each other, and then to constrain the timescales for that mixing.

1 Introduction

The urban atmosphere is incredibly complex because most cities have a wide variety of sources (and different cities can have wildly different distributions of sources). Here we shall consider organic aerosols (OA) as one measure of urban complexity because they have such a rich composition; unlike the gas-phase with individual molecules, suspended particles with diameters between 1 and 1000 nm contain between a few and more than 10^9 molecules, often from many different sources. Particles are thus a bellwether for complexity. Furthermore, urban areas

[a]*Center for Atmospheric Particle Studies, Carnegie Mellon University, 5000 Forbes Avenue, Pittsburgh, Pennsylvania 15213, USA. E-mail: nmd@cmu.edu*

[b]*Earth System Research Laboratory, National Oceanographic and Atmospheric Administration, 325 Broadway, Boulder, Colorado 80305-3337, USA*

[c]*Penn State Greater Allegheny, 4000 University Drive, McKeesport, Pennsylvania 15132, USA*

exist in a regional context. While "urban outflow" especially from mega cities is a well recognized concept, fine particles typically reside in the atmosphere for the order of 1 week;[1] because of this, the inflow to urban areas contains a rich background of (often chemically processed, or "aged") aerosols even before local emissions are considered. Thus the concept of "urban inflow" where regionally processed air masses are received by high population receptor areas may be a more meaningful concept than the urban outflow.

Organic aerosols are complex mixtures of molecules, many of which are semi-volatile at typical atmospheric concentrations (1–50 $\mu g\ m^{-3}$) and partition between the condensed organic and vapor phases. These myriad constituents form a solution, where constituent partial pressures are dictated by their mole fractions in the condensed phase.[2,3] For this reason, secondary OA (SOA) models are highly sensitive to the inclusion of primary OA (POA) as an absorbing phase.[4–6] Whether SOA and POA form a solution or not is one of the main ways in which anthropogenic emissions may influence the atmospheric burden of biogenic SOA.[7] The mass uptake of semi-volatile constituents formed in the gas phase by a pre-existing organic aerosol mass depends on the activity coefficients of those semi-volatile constituents in the absorbing medium. Thus, knowledge of the miscibility between different types of OA is important for understanding, for example, the influence of anthropogenic activity on biogenically-derived aerosols as well as predicting overall OA mass in air quality and climate models.

In the urban context, it is of great interest to know how quickly heterogeneous, externally mixed populations of organic particles will homogenize to become (more or less) internally mixed. Initially, particles will have a well-defined source, consisting almost entirely of molecules from that one source. However, at some rate, the semi-volatile constituents in different populations will mix with each other, and secondary species will condense on to and mix in to many particles. In addition, if the background aerosol consists of aged particles – even from similar upwind primary sources – the mixing process may be different than for very fresh particles from nearby sources. One example of the "urban inflow" and the effect of regional aging is the strong seasonal cycle observed for lubricating-oil tracers (hopanes) during the Pittsburgh Air Quality Study in 2001. Total organic aerosol levels in Pittsburgh were only slightly higher than the regional values during most of the study.[8] The hopane concentrations (as well as the ratios of hopanes to elemental carbon) fell to near the detection limit during the summer at a regional background site yet became comparable to an urban background site in the winter[8] despite strong evidence that the vehicular contributions to organic aerosol were relatively constant.[9] Similar evidence for strong regionality and regional photochemical aging from the TORCH experiment in England was presented by McFiggans *et al.*[10] during a 2005 Faraday Discussion, where the bulk sub-micron organic aerosol evolved from a relatively reduced, primary, vehicular mode in urban centers to a relatively oxidized accumulation mode regionally. In that study there were strong indications that organics in the vehicular mode were depleted with unusual speed, suggesting possible material transport from the primary vehicular mode into the secondary accumulation mode *via* the gas phase.

A few experimental studies have attempted to diagnose phase partitioning between different classes of atmospheric OA. Hildebrandt *et al.*[11] showed that SOA mass yields from the ozonolysis of α-pinene (the most common laboratory representation of biogenic SOA) were enhanced by the presence of SOA from toluene (a common laboratory representation of anthropogenic SOA). From this mass enhancement, they concluded that the two SOA types were miscible. Using similar data analysis techniques, Song *et al.*[12] concluded that SOA from α-pinene ozonolysis did not mix with lubricating oil OA, a surrogate for hydrophobic POA. Asa-Awuku *et al.*,[13] using chemically-resolved aerosol size distributions, similarly found that SOA from α-pinene ozonolysis remained phase-separated from motor-oil POA. On the other hand, they concluded that diesel-exhaust POA (a more oxidized, primary anthropogenic aerosol) and SOA did mix, suggesting a high variability between the character of POA model systems as it pertains to partitioning.

The above studies present a complicated picture of what the anthropogenic influence on biogenic OA might be, and for that matter how anthropogenic emissions are influenced by background OA. However, they share in common inferences about thermodynamic interactions between SOA and POA from indirect or at most size-resolved measurements, as opposed to direct measurements of the number of phases. Single-particle measurements, on the other hand, provide direct constraints on the interactions among chemically distinct aerosol populations. For example, Robinson *et al.*[14] showed that anthropogenic SOA partitions rapidly into biogenic SOA, using single-particle mass spectrometry. Here we measure phase-partitioning between POA and SOA systems, including two canonical laboratory surrogates for atmospheric OA: biogenic SOA (from α-pinene ozonolysis), and anthropogenic POA (from diesel engine exhaust, as well as from motor oil). We also perform these experiments with photochemically aged diesel-exhaust POA, demonstrating how atmospheric processing can lead to significant changes in mixing.

2 Methods

We conducted a series of aerosol population interaction experiments first described in Robinson *et al.*[14] to explore the interactions of various systems, each consisting of two different aerosol types. The key observations were single-particle mass spectrometer measurements of the particle populations using two different Aerodyne Aerosol Mass Spectrometers (AMS). All experiments presented follow the same general scheme, depicted in Fig. 1. First we prepared two aerosol populations in separate volumes. Following the preparation steps, when net condensation of organic vapors to particles of both populations had ceased, we transferred the two aerosol populations into a Teflon smog chamber (in most cases we transferred one population into the chamber where the other had been prepared). From this time onwards, vapors from one population were available for absorption into particles from the other population (and visa versa); we denote this as "time from combination" in all subsequent figures (the combination time is $t = 0$).

We use the word "interaction" to describe all aspects of the population ... interactions. The word "mixing" deserves special attention, as it has several potential meanings. Here we shall reserve "mixing" for the thermodynamic sense, in which different molecules interact in a single (organic) phase characterized by molecular activities determined by both the solute mole (or mass) fraction and the molar (or mass based) activity coefficients of the solute in the solvent. We shall use "separate" or "combined" to describe the two particle populations

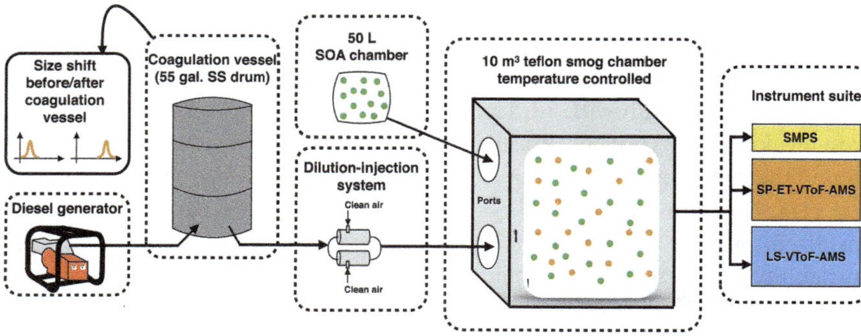

Fig. 1 Simplified experimental schematic for mixing experiments between diesel-exhaust primary organic aerosol (POA) (represented by orange) and secondary organic aerosol (SOA) from α-pinene ozonolysis (represented by green). For the aged POA experiment, we subjected fresh exhaust to OH oxidation formed in the smog chamber *via* UV photolysis of nitrous acid (HONO) gas prior to the addition of SOA. For the squalane/POA experiment, we prepared squalane aerosols directly in the smog chamber prior to the injection of the POA from the engine coagulation-vessel diluter system.

before or after combination – in the CMU smog chambers the stirring timescale (another synonym) is of the order of 5 minutes, whereas the experiments last for hours, and so the combination event is effectively instantaneous. The term "mixing state" is also common in aerosol science, and that generally means whether two aerosol populations have the same composition (they are "internally mixed") or different compositions (they are "externally mixed"). The strict definition of mixing state applies to a given particle size (for sectional models or observations) or mode (for modal representations). It is important to note that (especially in the atmosphere) particles can be "internally mixed" and yet not form a "mixture" in the thermodynamic sense we use here, as individual particles can contain more than one distinct phase (for example a polar and a non-polar organic phase).[15,16] However, we conduct these experiments under conditions where coagulation is minimal (with timescales of many days in the post-combination chamber) and so the only way that the compositions of the two particle populations can converge after combination is through thermodynamic mixing *via* gas-phase exchange of semi-volatile vapors ("Marcolli mixing"[17]). That is the process we wish to explore, and these experiments are designed so that the signal (composition convergence of two particle populations) is driven solely by that process.

2.1 Aerosol Generation

We prepared diesel exhaust POA by running a small diesel generator (Yanmar L70AE air-cooled diesel engine connected to a 4.5 kW generator). In all experiments, we ran the generator for five minutes without any applied load before directing the exhaust into a 55-gallon stainless-steel drum (labeled "coagulation vessel" in Fig. 1). We chose to run the engine with no load for two reasons: first, diesel engines tend to produce a higher ratio of organic to elemental carbon at low load,[18] which helps us maximize the mixing signal we are seeking. Second, as diesel PM is a mixture of unburned and partially-burned diesel fuel and lubricating oil, low-load emissions are skewed towards unburned fuel, which tends to be of higher volatility than lubricating oil.[19] As volatility is a key property for mixing experiments – higher volatility compounds will mix faster than lower volatility ones – a higher proportion of diesel fuel to lubricating oil in the exhaust is preferable. Given that the composition of diesel exhaust PM is a function of engine load (among many other factors such as engine temperature, fuel and lubricating-oil type, and the age and condition of the engine), it would be interesting to investigate the mixing ability for diesel POA with respect to load and other engine parameters.

After the 5 minute warmup, we filled the coagulation vessel for three minutes before shutting off the engine and sealing all ports on the vessel. This is the sole case where we allowed (even exploited) coagulation. The single-particle methods we employ are sensitive to particles larger than *c.* 150 nm diameter, and the primary POA particles from the generator were much smaller than that; consequently, we allowed a large number concentration of those primary particles to grow *via* coagulation (as a separate isolated population) before combining them with a second population. After an empirically-determined 15 minutes settling interval, we siphoned the POA from the coagulation vessel into the Teflon smog chamber using an ejector-diluter (20 : 1 dilution; Dekati Ltd., Helsinki, Finland, with additional dilution of >10 : 1 *via* introduction into the smog chamber).

We used the following procedure in all experiments where we chemically aged diesel-exhaust POA. First we allowed the diesel exhaust to rest in the smog chamber following injection for ~45 minutes. Next we generated hydroxyl radicals (OH) in the chamber *via* photolysis of nitrous acid (HONO). To inject the HONO we bubbled filtered air through a 1 : 2 (by volume) solution of 0.1 M $NaNO_2$ and 0.05 M H_2SO_4 for 15 minutes, carrying HONO into the smog chamber. After HONO injection we turned on UV lights (350 nm peak emission[20]) for 30 minutes to generate OH. Finally, we turned the UV lights off at least 30 minutes prior to the combination time in order to ensure the completion of all photochemistry prior to addition of the second aerosol population. We also prepared motor-oil POA by flash vaporization using the stainless-steel heater.

We prepared isotopically-labeled (D_{62}) squalane aerosol following the same procedure described in Robinson and co-workers.[14,16] Briefly, we applied 1 μL of D_{62}-squalane to a reservoir on the end of a stainless-steel sheath containing a hermetically-sealed conductive heater. We then inserted the vaporizer into the smog chamber and power cycled the heater for ~10 minutes until the vapor condensed (homogeneously nucleated) into pure, isotopically labeled squalane particles.

We prepared SOA *via* ozonolysis of α-pinene. In some experiments, we generated the SOA in the large Teflon smog chamber (*e.g.* Fig. 4). In these cases, we used ammonium sulfate particles (formed through atomization of an aqueous ammonium sulfate solution) as seeds for SOA condensation. After we injected ammonium sulfate, we vaporized 1.2 μL of α-pinene in a heated tube and flushed it into the chamber with a stream of hot, particle- and gas-filtered air. For these experiments we did not have a direct measurement of the initial α-pinene mixing ratio, so we assumed it was ~20 ppb based on a unit transfer efficiency and a 10 m^3 smog-chamber volume (consistent with many past experiments).[21] We then charged the chamber with excess ozone (~2 ppm), after which SOA formation commenced rapidly. Prior to the combination of the SOA particles with any

POA, we scrubbed the chamber of any excess ozone remaining after SOA formation. In control experiments, we found diesel exhaust to strongly react with ozone, presumably from the presence of small alkenes (see *e.g.* Schauer *et al.*,[22] Erickson *et al.*[23]) reacting to form hydroxyl radicals which can then react with many IVOC and SVOC compounds from diesel exhaust known to have high SOA-forming potential.[24] To rid the system of ozone, we added tetra-methyl ethylene (TME) into the chamber until the ozone monitor reading fell below the baseline levels. It should be noted that we did not add an OH scavenger (other than the TME), and thus the SOA in these experiments was additionally aged beyond first-generation SOA ozonolysis products.[25]

2.2 Instrument suite

The Carnegie Mellon University smog chamber has been described in detail elsewhere.[26] Briefly, the smog chamber used for this work is a 10 m^3 Teflon bag (Welch Fluorocarbon, Inc.; Dover, NH, USA) housed in a temperature-controlled room. We measured temperature and relative humidity using a Vaisala humidity monitor (HNP-233; Vaisala, Helsinki, Finland) and logged the data through a LabView data-acquisition computer (National Instruments; Austin, TX, USA). We measured ozone mixing ratios in the smog chamber with a UV-photometric ozone monitor (Dasibi 1008-PC; Dasibi Environmental Corp., Glendale, CA, USA). To generate ozone, we flowed molecular oxygen through a corona-discharge ozone generator upstream of the smog chamber. Prior to all experiments, we flushed the chamber with vapor-, water-, and particle-free air for at least 18 hours.

We measured the total particle number and volume concentrations using a Scanning Mobility Particle Sizer (SMPS 3936; TSI, Minneapolis, MN, USA). We measured ensemble OA mass and chemical composition using a High-Resolution Time-of-Flight Aerosol Mass Spectrometer (HR-ToF-AMS; Aerodyne Research, Inc., Billerica, MA, USA) operated in single-reflectron "V-mode." Within V-mode operation, the HR-ToF-AMS can be operated to provide two types of data: bulk mass spectra (MS) and size-resolved chemical composition using particle time-of-flight (PToF). We analyzed the ensemble OA data from the HR-ToF-AMS using the SQUIRREL (v1.51A) and PIKA (v1.12C) software packages written by Donna Sueper of Aerodyne Research, Inc. (available at http://cires1.colorado.edu/jimenez-group/ToFAMSResources/ToFSoftware/index.html). We used two single-particle mass spectrometers in this work: (1) a light-scattering-module equipped HR-ToF-AMS, which has been used previously in smog-chamber OA mixing experiments (see Robinson *et al.*, 2013,[14] Robinson *et al.*, 2015,[16] and Ye *et al.*, 2016[27]), and (2) a soot-particle AMS (SP-AMS) equipped with Event Trigger (ET) capability.

The light-scattering single-particle (LSSP) module for the HR-ToF-AMS has been described previously by Cross *et al.*[28] Briefly, LSSP data constrain single-particle composition and size. The LSSP mode differs from typical operation of the AMS by using laser light scattering from a continuous-wave laser beam (405 nm, 50 mW; LC BCL-050-405, CrystaLaser) to trigger acquisition of individual particle mass spectra for a given chopper cycle (~7 ms) as opposed to averaging all spectra (and not saving individual spectra) over longer times typical of standard operation (~1 min). Individual particle mass spectra from LSSP can be sorted into three groups: prompt, delayed, and null. Prompt particles are those with a total mass signal above a threshold of 6 ions (operationally-defined as in Liu *et al.*[29])

with signal appearing in the mass spectrometer within a narrow time window determined by the particle velocity, which in turn is measured between the chopper and the laser. Delayed particles have a mass signal above the threshold, but their signal appears in the mass spectrometer later than expected based on their measured velocity. The distinction between prompt and delayed is defined as particles whose actual arrival time is no greater than 20% (prompt) or is greater than 20% (delayed). Null particles are those that scatter laser light but whose mass signal is below threshold.

Event Triggering (ET) is an alternative single-particle acquisition mode for the AMS that does not rely on a laser scattering signal to initiate acquisition of individual mass spectra within a single chopper cycle. It relies on a new high-speed data acquisition card that can be programmed to trigger data acquisition when selected ion (m/z) signals exceed a specified threshold. The LSSP mode enables external triggering of data acquisition and also determination of particle collection efficiency *via* the ratio of null to observed particles.[28] However, single-particle acquisition with the original electronics takes approximately 160 ms, so even with unit collection efficiency the single-particle data rate is limited to 6 Hz. For bouncy particles with low collection efficiency, the LSSP data rate can get very low, as many triggered cycles are recorded as "null." Event triggering mode operates on the principle of ion thresholding for "regions of interest" (ROI) in the mass spectrum. In the case of a mixing experiment with two aerosol populations where one significantly outnumbers the other (*e.g.* $\sim$20 : 1 of type A *vs.* type B particles), the user can define an ROI to ensure collection of a statistically meaningful number of particles from each population ($\sim$1 : 1). For instance, the user could alternate between an ROI threshold that only applied to ions unique to particle type A and an ROI threshold that only applied to ions unique to type B, thus ensuring collection of data from each population. Because ET is based on signal levels, collection efficiency can not be measured but the ET data rate can remain near the theoretical limit (which is the chopper speed, 145 Hz for our case) so long as the incident particle transmission rate through the chopper exceeds that rate.

3 Results and discussion

3.1 POA – squalane mixing

The first question of interest is "how quickly will urban aerosols from different sources mix with each other?" Work in the last two decades has shown that a large fraction of diesel POA is semi volatile.[24,30] This has important implications for emissions testing.[31] Further, the gas-phase oxidation of POA emissions can dominate their behavior regionally.[32] Many diesel-POA constituents have saturation concentrations between 1 and 1000 μg m^{-3},[33] so these compounds will be present with significant fractions in both the gas phase and the particle phase under ambient conditions. Furthermore, the aerosol condensation sink in the urban atmosphere typically varies between 0.5 and 5 min^{-1},[34] so equilibration timescales in the urban atmosphere should be in the order of minutes.[35] The full exchange timescale for less volatile constituents will in turn be limited by their evaporation timescales (in the order of 1 h for species with $C^* \simeq 1$ μg m^{-3} and $d_p \simeq 200$ nm).[36] We thus designed an experiment to test whether we could observe the exchange of semi-volatile POA components between diesel POA and a surrogate second POA population (D_{62}-squalane). Squalane is completely reduced, non-polar, and liquid; it thus should be miscible with the relatively un-oxidized straight and branched alkanes comprising the diesel POA. Squalane also has a unique mass spectrum because of the isotopic labeling, which we have previously shown to be useful for separating particle populations in mixing experiments.[14,16]

Fig. 2 shows the results of this mixing experiment between D_{62}-squalane particles and diesel exhaust POA using the SP-AMS in ET mode. We first formed D_{62}-squalane particles in the 10 m^{-3} smog chamber by flash vaporization. After sufficient time for characterization, we added POA from the diesel generator to the smog chamber (as depicted in Fig. 1), initiating the start of the mixing experiment. Fig. 2a shows single-particle similarity (using D_{62}-squalane as a reference MS) *vs.* time after combination for each individual particle. The symbol color also shows the signal fraction at *m*/*z* 57 (f_{57}), which is indicative of

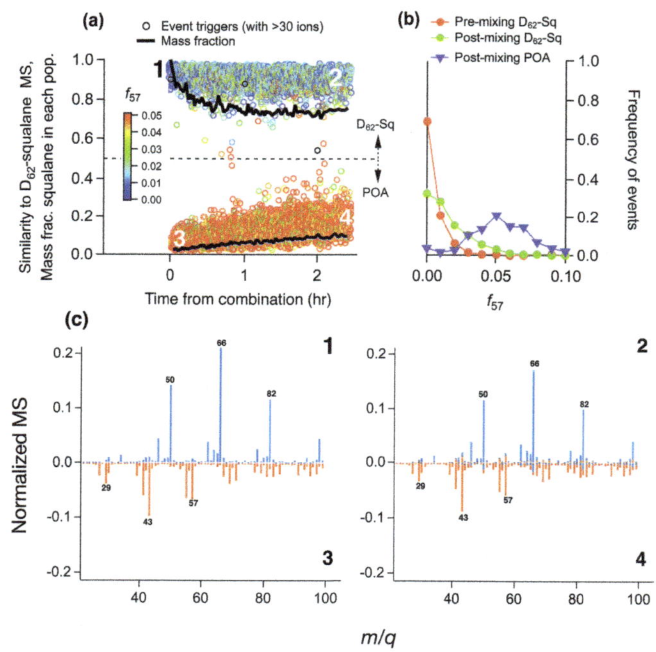

Fig. 2 (a) Aerosol mass spectrometer (AMS) event triggering (ET) time series showing single-particle mass spectrometer (MS) data for a D_{62}-squalane OA plus diesel POA mixing experiment. Each data point (each particle) is plotted based on the mass spectral similarity to a D_{62}-squalane reference. The squalane particles rapidly take up POA material, shown both in a change in the mass fraction of squalane (black line), and by the shift towards higher f_{57} values. The POA particles show a slow absorption of D_{62}-squalane material, as their MS similarity values and squalane mass fraction continuously increase for the duration of the experiment. (b) Histogram showing the distribution of f_{57}, as measured by AMS ET, in the squalane population before (red) and after (green) mixing, and in the POA population (purple) after mixing. There is a clear shift in the distribution of f_{57} in the squalane particles before and after mixing, which we attribute to absorption of POA vapors. (c) Aggregate mass spectra for the periods indicated by numbers in (a) with D_{62}-squalane signals plotted in blue and POA signals plotted in orange (facing downward for the POA population). Significant exchange of vapors between the populations is evident in spectra 2 and 4.

diesel POA content. The very high single-particle data rate for the ET is evident in the dense groups of symbols; we obtain ~100 mass spectra from each population every minute. The variability in the individual particle signals largely reflects statistical variation due to the relatively small number of ions per particle rather than truly variable composition among the sampled particles.

The ET data rate was high enough that we could quantitatively analyze the aggregate mass spectra from the two particle populations every minute. We calculate the average mass fraction of squalane within each particle population with one-minute averaging. The results shown in Fig. 2a are as expected, given the chemical similarity of the POA and squalane: both absorb semi-volatile vapors from the other aerosol population. This result is indicated by the calculated mass fractions (which show squalane entering the POA population, and *vice versa*), and is also visible as an increase in f_{57} in the squalane population. The signal at m/z 57 is largely comprised of the $C_4H_9^+$ ion. Because the squalane is fully isotopically labeled, it contains almost no native m/z 57 in its mass spectrum, which is illustrated in the histogram of single-particle data in Fig. 2b as well as the representative mass spectrum in Fig. 2c. Prior to mixing, there is very little m/z 57 in the squalane particles, shown by the red trace in Fig. 2b. What little there is comes from slight impurities of non-labeled squalane in our deuterated standard. There is a significant fraction of m/z 57, however, in the POA particles (purple trace in Fig. 2b). Following the addition of POA to the chamber, the distribution of f_{57} in the squalane population shifts to larger values, shown by the green trace in Fig. 2b, indicating absorption of POA vapors.

A key feature of the single-particle data is that we can use several measures to isolate individual populations and then aggregate the mass spectra for quantitative analysis. In this case separating the two populations is trivial. In Fig. 2c we show aggregate mass spectra for the two populations just after combination ($0 < t < 2$ min) and at the end of the experiment (from $2.0 < t < 2.5$ h), with the squalane particle mass spectra oriented upward and the diesel-POA mass spectra oriented downward. We use a short interval for the first period because the squalane particle composition is changing rapidly. Given the chemical similarity between squalane and POA, it is not surprising that we observe absorptive exchange between the two populations. Additionally, this absorption occurs on timescales that make sense given the respective volatility distribution of POA (which contains semi-volatile compounds in the $C^* = 10^0$ to 10^3 μg m^{-3} range) and the much-lower volatility of squalane ($C^0 \sim 10^{-1}$ μg m^{-3}). These timescales are illustrated by the black traces in Fig. 2a, which show rapid uptake of POA vapors by squalane, and a much slower uptake of squalane into the POA. The relevant timescales are for complete evaporation of different constituents from the original host particles. For example, D_{62}-squalane must evaporate from the squalane particles before it can mix into the diesel-POA particles, and the evaporation timescale for $C^* = 0.1$ material from 200 nm particles is about 20 h.[36] We thus see a steady rise in the D_{62}-squalane content in the diesel-POA particles up to around 10% after 2 h. On the other hand, the originally pure D_{62}-squalane particles rapidly acquire roughly 30% by mass diesel-POA constituents. This is consistent with >30% of the diesel-POA being semi volatile, with $C^* > 1$ μg m^{-3} at the loadings in the CMU smog chamber ($C_{POA} \simeq 10$ μg m^{-3}). Constituents with a volatility much higher than this would not have a high enough gas-phase activity after dilution to add a significant mass fraction to the originally pure D_{62}-squalane particles.

This experiment confirms that (1) diesel POA is comprised of compounds capable of rapid gas-phase mass transfer, and that (2) no particle-phase diffusion limitations exist that would impede any mixing process for these systems (though diffusion into the diesel POA would have to be very slow for this to impede the squalane mass transfer, which is slow in any event). It also is a clear demonstration of the value of the event-trigger mode in the AMS for examination of externally-mixed aerosol populations and to measure their changes over time.

In the urban context, it is likely that POA particles from different sources will exchange a substantial fraction of semi-volatile constituents, depending on their exact volatility distributions, within a few minutes of emission and thus become less obviously associated with a specific source. However, provided that the source also emits a fair fraction of material with lower volatility (less than or equal to that of squalane), the particles will also retain a signature of their origin, even if they are purely organic. Of course, if they also contain identifiable non-volatile components such as elemental carbon or refractory metals, that unique source signature will also remain intact.

3.2 SOA–POA mixing

After establishing that diesel exhaust POA can participate in gas-phase exchange with another aerosol population comprised of miscible material, and that this mass-transfer can be detected using our single-particle mass spectrometers, we conducted several experiments to test the mixing behavior between POA and SOA (from the ozonolysis of α-pinene). Previously, we have demonstrated that there are no mass-transfer barriers to absorption of vapors into SOA derived from α-pinene ozonolysis (α-pinene SOA);[14,27] thus, any mixing we observe between SOA and POA can be attributed to the miscibility between these two OA types, and the absence of mixing can be attributed to thermodynamic immiscibility. We show the scheme for these experiments in Fig. 1.

Motor oil (MO) has previously been used as a proxy for POA in laboratory experiments probing aging,[37] mixing,[13] and SOA-formation.[38] Motor oil also has been identified by Worton *et al.*[39] as the dominant primary organic particulate material from motor vehicles in the atmosphere. We conducted a mixing experiment using motor oil as a proxy for POA, and observed no mixing with SOA as shown in Fig. 3. For this experiment we flash vaporized motor oil into the main chamber and subsequently injected SOA formed *via* ozonolysis in a 50 L sample bag. The LSSP data in Fig. 3a show two distinct populations after mixing with little evidence for exchange of the POA marker (m/z 57). There is also a clear shoulder in the particle time of flight data shown in Fig. 3c.

In Fig. 3d we show aggregated mass spectra for the POA population before and after SOA injection (spectra 1 and 2) as well as the aggregated SOA population (spectrum 3). There is no evident change in the POA mass spectrum and no uptake of characteristic tracers such as m/z 29. However, there is some evidence for limited uptake of SOA vapors into the POA – specifically characteristic "picket fence" features for $m/z > 60$, which are absent in pure α-pinene SOA (presented below). Our interpretation is that the POA vapors have a large but finite activity coefficient in the SOA particles and thus can be taken up to a limited degree. Furthermore, the SOA vapors show no sign of partitioning into the POA particles, suggesting an even larger activity coefficient. This is consistent with theoretical

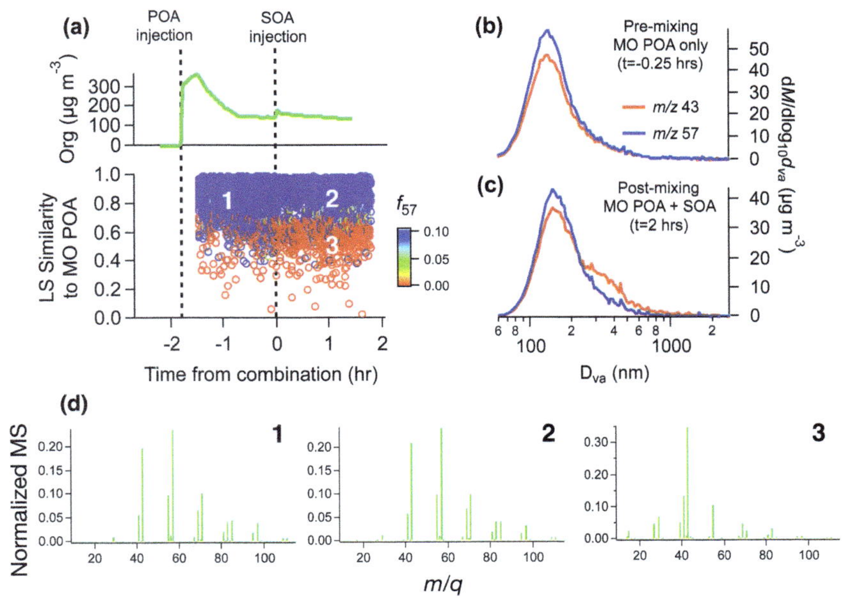

Fig. 3 Mixing experiment for motor-oil (MO) POA and α-pinene SOA. (a) Similarity of particles observed *via* AMS LSSP to a motor-oil reference spectrum based on the aggregate spectrum for $t < 0$. The POA and SOA populations remain distinct for 1.5 h. (b and c) Particle time-of-flight signals are of limited value because of the high overlap in the POA and SOA size distributions, with the SOA showing as a shoulder at 300 nm on the MO POA size distribution. (d) Aggregate mass spectra for the MO POA population before (1) and after (2) SOA injection as well as the SOA population spectrum (3). There is no evident change in the MO POA spectra, but signs of limited POA vapor uptake in the SOA spectrum given signals for $m/z > 60$, which are uncharacteristic of this SOA.

expectations[40] and suggests that if anything SOA particles are (weak) seeds for POA condensation rather than POA being effective seeds for (absorptive) SOA condensation. This also confirms that this commonly used laboratory POA surrogate, which may be the most common actual atmospheric constituent of motor-vehicle POA,[39] has very limited solubility in the SOA we used.

In Fig. 4 we show the results of a mixing experiment between SOA and fresh diesel-exhaust POA. In this experiment, we first produced SOA in the smog chamber for characterization. After scrubbing the ozone from the chamber by adding TME, we injected POA particles from the coagulation barrel into the chamber to combine the two populations ($t = 0$ h). Fig. 4a contains time sequences of total organic mass (top) and single-particle (LSSP) similarity (bottom) with POA as the reference MS. Single-particle data points are colored with their f_{57} values. From Fig. 4a, prior to $t = 0$ h we see that SOA particles have similarity values less than 0.6, and do not contain any m/z 57 in their "pure" mass spectrum. The introduction of POA to the chamber at $t = 0$ h is evident in the single-particle data, identified by high similarity and non-zero f_{57} values. Over the full course of the experiment, these two distinct populations persist.

While composition information on the individual-particle level unambiguously constraints the mixing behavior between two populations, here we can also use size-resolved bulk composition data to answer the mixing question for these

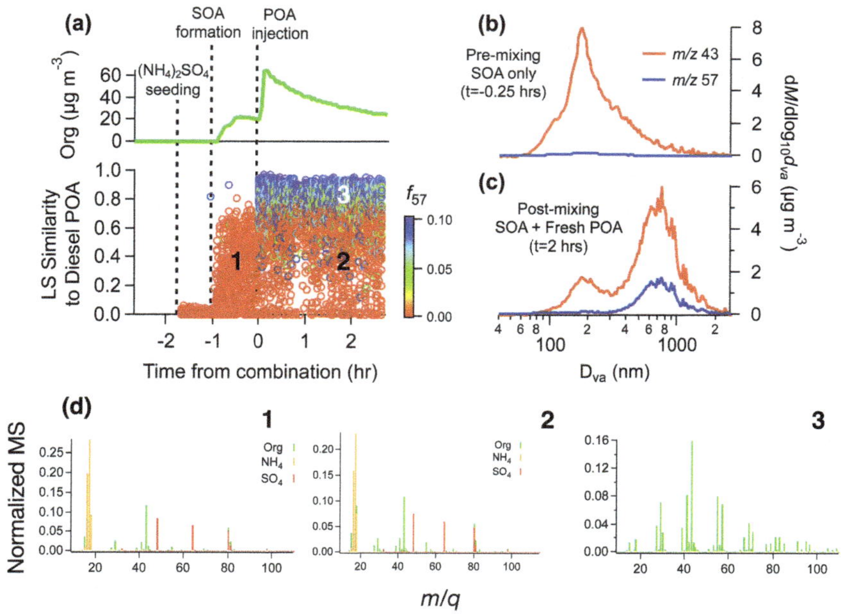

Fig. 4 (a) AMS light-scattering single-particle (LSSP) time-series showing single-particle MS data for an α-pinene SOA plus diesel POA mixing experiment, where each data point shows the mass spectral similarity to diesel POA. The color scale differentiates the two populations, as there is very little native *m*/*z* 57 in SOA, which continues to be true in the SOA mode for the duration of the mixing experiment. These data show, to first order, limited vapor exchange. Similar to the analysis method of Asa-Awuku *et al.*, (b and c) show AMS PToF distributions for an α-pinene SOA plus diesel POA mixing experiment, averaged over 15 minutes around the indicated times. Two distinct modes persist for the entire experiment, with little indication of *m*/*z* 57, a unique marker for diesel POA, absorbing into the smaller SOA mode. The distribution of *m*/*z* 43 (a fragment common to both SOA and POA) shows the size separation between the two populations in the experiment. (d) Aggregated mass spectra for SOA before (1) and after (2) POA injection and POA (3) after injection. The SOA spectra show evidence for limited absorption of POA vapors.

POA and SOA populations (this approach is fully described in Asa-Awuku *et al.*[13]). Fig. 4b and c show the mass distributions of *m*/*z* 43 and 57 before and after we added POA to the chamber, respectively. In this experiment, the SOA mass distribution is substantially smaller (200 nm mode) than the POA (700 nm mode). This modal separation allows us to look for chemical tracers from each aerosol type in each mode. As in the single-particle data in Fig. 4a, we see no *m*/*z* 57 in the SOA (smaller mode) at the end of the experiment, indicative that no diesel POA vapors were absorbed into the SOA particles.

In Fig. 4d we show the aggregate mass spectra for three separate populations: the SOA for $t < 0$ h, the SOA for $t > 0$ h, and the POA for $t > 0$ h. The POA spectrum (3) is very similar to the POA spectrum shown in Fig. 2c, suggesting little uptake of SOA vapors into the POA. However, the SOA spectra do change significantly after POA addition. Specifically, the SOA spectrum before POA injection has almost no signal (other than from the ammonium sulfate seeds, indicated in gold and red) for $m/z > 60$, while the spectrum afterwards does show features at higher masses. In this case the increase at $m/z = 43$ is also probably due to POA vapor uptake;

however, the limited *m/z* 57 uptake (especially in the p-ToF data) renders this result ambiguous. Isotopic labeling (*e.g.* ^{13}C) of the α-pinene would be useful for better separation of these spectra, but we did not have access to a labeled precursor.

Our observations replicate findings of Asa-Awuku *et al.*,[13] who also found that motor-oil POA and SOA did not mix completely. For the urban atmosphere, the implication is that at least shortly after POA emission, the POA suspension will remain externally mixed with respect to background aerosols, to the extent that the background accumulation mode particles are relatively highly oxidized.[41] However, what is less certain is how these populations will interact after some atmospheric oxidation ages the two populations.

3.3 SOA – aged POA mixing

As a final test, we explored the interaction of aged diesel-POA with α-pinene SOA. For these experiments, we injected POA from the coagulation chamber into the main smog chamber and then exposed the POA to OH radicals formed *via* HONO photolysis, as described in the Methods section. After approximately 3 h of aging, we then introduced α-pinene SOA formed *via* ozonolysis in a second 50 liter "pillow" bag, as described in Robinson *et al.*[14] though with the added step of scavenging excess ozone prior to combining the two populations. In Fig. 5 we show the bulk and single-particle (LSSP) mass spectra from one such experiment. These data confirm that diesel emissions are a potent source of secondary mass,

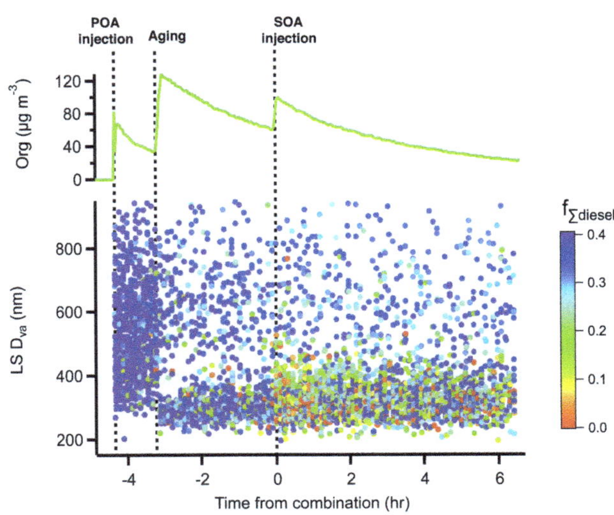

Fig. 5 AMS LSSP time-series showing single-particle d_{va} data (bottom panel) for an α-pinene SOA plus diesel POA mixing experiment, as well as total organic mass measured in bulk operating mode (top panel). In this experiment we "aged" the POA using OH radicals produced *via* HONO photolysis. This led to significant new-particle formation as well as a large increase in the total aerosol mass concentration. Due to similarities in the MS of aged POA and SOA, we plot against single-particle size instead of similarity for better visual separation. The SOA particles are coincident in size to the nucleated particles formed from POA vapors during aging, but show much lower signals for a set of POA mass tracers (shown with symbol color) and thus appear in red and green hues.

as the aging doubles or quadruples the OA mass (depending on whether one compares to the initial POA level of 60 μg m^{-3} or the level of 30 μg m^{-3} just before the onset of aging).[24,42] The addition of SOA at $t = 0$ h is also plainly evident in the total AMS mass trace as another 30 μg m^{-3} mass increase.

The LSSP data in Fig. 5 are more difficult to interpret. Here we plot the LSSP data according to the aerodynamic diameter as determined by the light-scattering particle time of flight. The symbol color is also based on the fraction of the sum of tracers unique to the diesel POA ($\sum m/z$ 39, 55, 57, 65, 67, 69, 71, 77, 79, 81, 83, 85, 91) to the total OA. Several experimental challenges are evident in Fig. 5. First, the POA aging in this case included substantial new-particle formation as we saw the emergence of a nucleation mode (or condensation onto primary particles smaller than the transmission window of the AMS aerosol lens), and these new particles grew large enough to observe with light scattering. The larger POA particles (the original population) evident before aging also grew, and so the LSSP data show two distinct modes after aging. However, the new particles dominate the number distribution and so dominate the LSSP triggering. This reduces the hit rate for the initial primary particle population and compromises the signal to noise. Second, the SOA particles are almost the same size as the new particles formed during POA aging, and so overlap in diameter space. In contrast to our experiment with motor oil and SOA, the size-resolved bulk spectra (particle time of flight) thus provide no information about possible mixing. However, the SOA particles have a low fraction of primary diesel markers in the LSSP mass spectra and are thus clearly evident (green and red hues).

From the LSSP data alone it is difficult to tell whether the aged POA aerosols (either the original, larger particles or the nucleated particles) and the SOA from α-pinene mixed with each other. Once again we can separate the populations based on various observables; here we shall use both vacuum aerodynamic diameter and the abundance of diesel tracers. We show aggregated mass spectra from six populations in Fig. 6, indicated with boxes in the inset: POA before aging; larger POA after aging ($d_{va} > 400$ nm); smaller (nucleated) particles formed from POA vapors during aging ($d_{va} < 400$ nm); larger POA after aging and SOA introduction ($d_{va} > 450$ nm); and two populations of smaller particles after SOA introduction. The last two populations are those with the aggregate fraction of diesel markers $f_{diesel} > 0.25$, presumably the population of nucleated particles, and those with the aggregate fraction of diesel markers $f_{diesel} < 0.25$, presumably the added SOA particles.

The aggregated mass spectra indicate a modest degree of mixing. The fragments at m/z 29 (CHO^+) and m/z 43 ($C_3H_7^+$ and $C_2H_3O^+$) are revealing – especially the ratio $f_{29} : f_{43}$, or 29 : 43. The relatively high signal intensity at m/z 29 in these spectra is unusual; however, one feature of the single-particle data is that bulk interference at m/z 28 and 29 (from N_2 and $^{15}N^{14}N$) is absent because the particle signals arrive many milliseconds after air. In these aggregated mass spectra the initial POA particles before and after the POA aging are almost identical, based on their mass spectra (1 and 2). However, after SOA introduction (population 4) there is a significant increase in the 29 : 43 signal, indicating that some SOA vapors did absorb into the aged POA particles, unlike both fresh POA examples presented above. The mass spectrum for the nucleated particles that formed from POA vapors during the aging step (population 3) is somewhat similar to the POA mass spectrum but there is much less signal at higher masses. This is consistent with

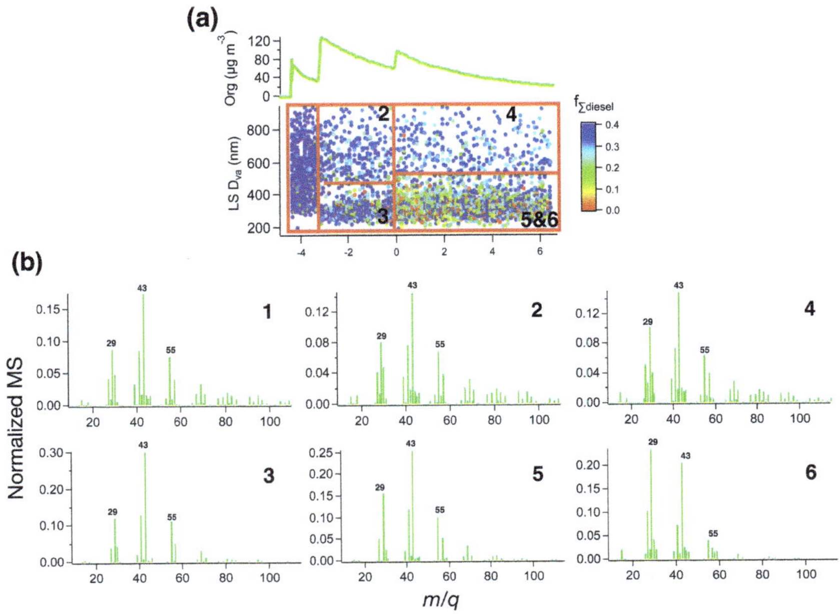

Fig. 6 (a) Selection regions for aggregated mass spectra for different particle populations from the experiment shown in Fig. 5. (b) The mass spectra. The POA mode (1, 2, 4) shows little change during aging (1–2) but significant uptake of SOA vapors after SOA injection, shown by increasing 29 : 43. The particles formed during POA aging (3) have a similar 29 : 43 to the fresh POA but the mass spectrum shows much less signal for $m/z > 60$, consistent with a somewhat more oxidized carbon backbone that is more easily fragmented after vaporization and electron ionization. Those new POA particles also show an increase in 29 : 43 after SOA injection, while the SOA particles (6) show the highest 29 : 43 and limited signal for $m/z > 60$, consistent with modest uptake of POA vapors.

a more aged – and more easily fragmented – molecular composition. Finally, the smaller particles after SOA introduction have the most notable differences. The population of nucleated particles (population 5 after SOA introduction) resembles population 4 after SOA introduction, but still with less signal at higher masses. The SOA population (population 6) has by far the highest 29 : 43, but does show some signal for $m/z > 60$, consistent with modest uptake of POA vapors. Broadly, these results suggest that the three populations (aged primary POA particles, newly-formed particles from POA vapors during the aging step, and α-pinene SOA) are partially miscible in both directions. More extensive aging of POA can generate oxidized POA with large signal at $m/z = 44$,[43] and these mixing data show that even modestly oxidized POA can begin to take up SOA vapors *via* absorption. Thus we expect POA and SOA populations to become miscible as urban plumes age, probably on a timescale of hours as shown in these experiments.

4 Conclusions

Part of the complexity of the urban atmosphere is that particle populations do not stay still. The ability of molecules within one particle population to migrate to other populations has implications for source apportionment in the source-rich urban atmosphere.[44] This gas-phase mixing also can significantly affect the overall organic mass, as the addition of absorbing aerosol mass can draw in organic vapors *via* Raoult's law.[4,5,45] Understanding what controls gas-phase exchange, especially in urban atmospheres where regional background oxidized aerosol meets fresh anthropogenic emissions, is clearly important.

Here we have shown that initially distinct, externally mixed aerosol populations will converge *via* gas-phase exchange of semi-volatile organic vapors on a timescale of minutes to hours, depending on the volatility and miscibility of the constituents. Single-particle observations of particle composition are an essential feature of the experiments we have presented. However, not all populations are fully miscible. Populations of relatively polar secondary organic aerosol and relatively non-polar primary organic aerosol remain distinct for hours even though other experiments reveal that these populations will accept (and donate) vapors to other populations if they are miscible. Oxidation evidently can suppress this miscibility barrier, but the specific thresholds for miscibility remain to be elucidated.

Acknowledgements

This research was supported by grant CHE1412309 from the National Science Foundation. The High-Resolution Aerosol Mass Spectrometer was purchased with Major Research Instrumentation funds from NSF CBET0922643 and the Wallace Research Foundation. The authors gratefully acknowledge the support of Joel Kimmel, Harald Stark, and Donna Sueper for development of the Event Triggering Single-Particle instrumentation and software used in this research. The authors would also like to thank John-Charles Baucom for his help with the design and construction of the heated vaporizer used for generating squalane and motor oil particles.

References

1 R. Jaenicke, Atmospheric Aerosols and Global Climate, *J. Aerosol Sci.*, 1980, **11**, 577–588.

2 J. F. Pankow, An absorption model of gas/particle partitioning of organic compounds in the atmosphere, *Atmos. Environ.*, 1994, **28**, 185–188.

3 N. M. Donahue, A. L. Robinson, C. O. Stanier and S. N. Pandis, Coupled Partitioning, Dilution, and Chemical Aging of Semivolatile Organics, *Environ. Sci. Technol.*, 2006, **40**, 2635–2643.

4 F. M. Bowman and A. M. Karamalegos, Estimated Effects of Composition on Secondary Organic Aerosol Mass Concentrations, *Environ. Sci. Technol.*, 2002, **36**, 2701–2707.

5 B. Koo, A. S. Ansari and S. N. Pandis, Integrated approaches to modeling the organic and inorganic atmospheric aerosol components, *Atmos. Environ.*, 2003, **37**, 4757–4768.

6 R. J. Griffin, K. Nguyen, D. Dabdub and J. H. Seinfeld, A coupled hydrophobic-hydrophilic model for predicting secondary organic aerosol formation, *J. Atmos. Chem.*, 2003, **44**, 171–190.

7 C. R. Hoyle, M. Boy, N. M. Donahue, J. L. Fry, M. Glasius, A. Guenther, A. G. Hallar, K. Huff Hartz, M. D. Petters, T. Petaja, T. Rosenoern and

A. P. Sullivan, A review of the anthropogenic influence on biogenic secondary organic aerosol, *Atmos. Chem. Phys.*, 2011, **11**, 321–343.

8 A. L. Robinson, N. M. Donahue and W. F. Rogge, Photochemical Oxidation and Changes in Molecular Composition of Organic Aerosol in the Regional Context, *J. Geophys. Res.: Atmos.*, 2006, **111**, D03302.

9 R. Subramanian, N. M. Donahue, A. Bernardo-Bricker, W. F. Rogge and A. L. Robinson, Contribution of motor vehicle emissions to organic carbon and fine particle mass in Pittsburgh, Pennsylvania: Effects of varying source profiles and seasonal trends in ambient marker concentrations, *Atmos. Environ.*, 2006, **40**, 8002–8019.

10 G. McFiggans, M. R. Alfarra, J. Allan, K. Bower, H. Coe, M. Cubison, D. Topping, P. Williams, S. Decesari, C. Facchini and S. Fuzzi, Simplification of the representation of the organic component of atmospheric particulates, *Faraday Discuss.*, 2005, **130**, 341–362.

11 L. Hildebrandt, K. M. Henry, J. H. Kroll, D. R. Worsnop, S. N. Pandis and N. M. Donahue, Evaluating the Mixing of Organic Aerosol Components Using High-Resolution Aerosol Mass Spectrometry, *Environ. Sci. Technol.*, 2011, **45**, 6329–6335.

12 C. Song, R. A. Zaveri, M. L. Alexander, J. A. Thornton, S. Madronich, J. V. Ortega, A. Zelenyuk, X.-Y. Yu, A. Laskin and D. A. Maughan, Effect of hydrophobic primary organic aerosols on secondary organic aerosol formation from ozonolysis of alpha-pinene, *Geophys. Res. Lett.*, 2007, **34**, L20803.

13 A. Asa-Awuku, M. A. Miracolo, J. H. Kroll, A. L. Robinson and N. M. Donahue, Mixing and phase partitioning of primary and secondary organic aerosols, *Geophys. Res. Lett.*, 2009, **36**, L15827.

14 E. S. Robinson, R. Saleh and N. M. Donahue, Organic Aerosol Mixing Observed by Single-Particle Mass Spectrometry, *J. Phys. Chem. A*, 2013, **117**, 13935–13945.

15 A. Zuend and J. H. Seinfeld, Modeling the gas-particle partitioning of secondary organic aerosol: the importance of liquid-liquid phase separation, *Atmos. Chem. Phys.*, 2012, **12**, 3857–3882.

16 E. S. Robinson, R. Saleh and N. M. Donahue, Probing the evaporation dynamics of mixed SOA/squalane particles using size-resolved composition and single-particle measurements, *Environ. Sci. Technol.*, 2015, **49**, 9724–9732.

17 C. Marcolli, B. Luo, T. Peter and F. Wienhold, Internal mixing of the organic aerosol by gas phase diffusion of semivolatile organic compounds, *Atmos. Chem. Phys.*, 2004, **4**, 2593–2599.

18 J. D. McDonald, M. J. Campen, K. S. Harrod, J. Seagrave, S. K. Seilkop and J. L. Mauderly, Engine-Operating Load Influences Diesel Exhaust Composition and Cardiopulmonary and Immune Responses, *Environ. Health Perspect.*, 2011, **119**, 1136–1141.

19 T. R. Dallmann, T. B. Onasch, T. W. Kirchstetter, D. R. Worton, E. C. Fortner, S. C. Herndon, E. C. Wood, J. P. Franklin, D. R. Worsnop, A. H. Goldstein and R. A. Harley, Characterization of particulate matter emissions from on-road gasoline and diesel vehicles using a soot particle aerosol mass spectrometer, *Atmos. Chem. Phys.*, 2014, **14**, 7585–7599.

20 A. A. Presto, K. E. Huff Hartz and N. M. Donahue, Secondary Organic Aerosol production from terpene ozonolysis: 1. Effect of UV radiation, *Environ. Sci. Technol.*, 2005, **39**, 7036–7045.

21 A. A. Presto and N. M. Donahue, Investigation of α-pinene + ozone Secondary Organic Aerosol Formation at Low Total Aerosol Mass, *Environ. Sci. Technol.*, 2006, **40**, 3536–3543.

22 J. J. Schauer, M. J. Kleeman, G. R. Cass and B. R. T. Simoneit, Measurement of Emissions from Air Pollution Sources. 2. C_1 through C_{30} Organic Compounds from Medium Duty Diesel Trucks, *Environ. Sci. Technol.*, 1999, **33**, 1578–1587.

23 M. H. Erickson, M. Gueneron and B. T. Jobson, Measuring long chain alkanes in diesel engine exhaust by thermal desorption PTR-MS, *Atmos. Meas. Tech.*, 2014, **7**, 225–239.

24 A. L. Robinson, N. M. Donahue, M. K. Shrivastava, A. M. Sage, E. A. Weitkamp, A. P. Grieshop, T. E. Lane, J. R. Pierce and S. N. Pandis, Rethinking Organic Aerosols: Semivolatile Emissions and Photochemical Aging, *Science*, 2007, **315**, 1259–1263.

25 L. Qi, S. Nakao and D. R. Cocker III, Aging of secondary organic aerosol from α-pinene ozonolysis: Roles of hydroxyl and nitrate radicals, *J. Air Waste Manage. Assoc.*, 2012, **62**, 1359–1369.

26 C. O. Stanier, R. K. Pathak and S. N. Pandis, Measurements of the Volatility of Aerosols from α-Pinene Ozonolysis, *Environ. Sci. Technol.*, 2007, **41**, 2756–2763.

27 P. Ye, X. Ding, Q. Ye, E. S. Robinson and N. M. Donahue, Uptake of semi-volatile secondary organic aerosol formed from α-pinene into non-volatile polyethylene glycol probe particles, *J. Phys. Chem. A*, 2016, **120**, 1459–1467.

28 E. Cross, T. Onasch, M. Canagaratna, J. Jayne, J. Kimmel, X. Yu, M. Alexander, D. Worsnop and P. Davidovits, Single particle characterization using a light scattering module coupled to a time-of-flight aerosol mass spectrometer, *Atmos. Chem. Phys.*, 2009, **9**, 7769–7793.

29 S. Liu, L. M. Russell, D. T. Sueper and T. B. Onasch, Organic particle types by single-particle measurements using a time-of-flight aerosol mass spectrometer coupled with a light scattering module, *Atmos. Meas. Tech.*, 2013, **6**, 187–197.

30 C. A. Stroud, J. Liggio, J. Zhang, M. Gordon, R. M. Staebler, P. A. Makar, J. Zhang, S.-M. Li, C. Mihele, G. Lu, D. K. Wang, J. Wentzell, J. R. Brook and G. J. Evans, Rapid organic aerosol formation downwind of a highway: Measured and model results from the FEVER study, *J. Geophys. Res.: Atmos.*, 2014, **119**, 1663–1679.

31 A. L. Robinson, A. P. Grieshop, N. M. Donahue and S. W. Hunt, Updating the conceptual model for fine particle mass emissions from combustion systems, *J. Air Waste Manage. Assoc.*, 2010, **60**, 1204–1222.

32 M. K. Shrivastava, T. E. Lane, N. M. Donahue, S. N. Pandis and A. L. Robinson, Effects of gas particle partitioning and aging of primary emissionson urban and regional organic aerosol concentrations, *J. Geophys. Res.: Atmos.*, 2008, **113**, D18301.

33 A. P. Grieshop, M. A. Miracolo, N. M. Donahue and A. L. Robinson, Constraining the Volatility Distribution and Gas-Particle Partitioning of Combustion Aerosols Using Isothermal Dilution and Thermodenuder Measurements, *Environ. Sci. Technol.*, 2009, **43**, 4750–4756.

34 I. Salma, T. Borsós, T. Weidinger, P. Aalto, T. Hussein, M. Dal Maso and M. Kulmala, Production, growth and properties of ultrafine atmospheric aerosol particles in an urban environment, *Atmos. Chem. Phys.*, 2011, **11**, 1339–1353.

35 R. Saleh, N. M. Donahue and A. L. Robinson, Time Scales for Gas-Particle Partitioning Equilibration of Secondary Organic Aerosol Formed from Alpha-Pinene Ozonolysis, *Environ. Sci. Technol.*, 2013, **47**, 5588–5594.

36 N. M. Donahue, A. L. Robinson, E. R. Trump, I. Riipinen and J. H. Kroll, in *Atmospheric and Aerosol Chemistry*, ed. V. F. McNeill and P. A. Ariya, Topics in Current Chemistry, Springer, 2014, vol. 339, Chapter Volatility and Aging of Atmospheric Organic Aerosols, pp. 97–144.

37 G. Isaacman, A. W. H. Chan, T. Nah, D. R. Worton, C. R. Ruehl, K. R. Wilson and A. H. Goldstein, Heterogeneous OH Oxidation of Motor Oil Particles Causes Selective Depletion of Branched and Less Cyclic Hydrocarbons, *Environ. Sci. Technol.*, 2012, **46**, 10632–10640.

38 M. A. Miracolo, A. A. Presto, A. T. Lambe, C. J. Hennigan, N. M. Donahue, J. H. Kroll, D. R. Worsnop and A. L. Robinson, Photo-Oxidation of Low-Volatility Organics Found in Motor Vehicle Emissions: Production and Chemical Evolution of Organic Aerosol Mass, *Environ. Sci. Technol.*, 2010, **44**, 1638–1643.

39 D. R. Worton, G. Isaacman, D. R. Gentner, T. R. Dallmann, A. W. H. Chan, C. Ruehl, T. W. Kirchstetter, K. R. Wilson, R. A. Harley and A. H. Goldstein, Lubricating Oil Dominates Primary Organic Aerosol Emissions from Motor Vehicles, *Environ. Sci. Technol.*, 2014, **48**, 3698–3706.

40 N. M. Donahue, S. A. Epstein, S. N. Pandis and A. L. Robinson, A Two-Dimensional Volatility Basis Set: 1. Organic Mixing Thermodynamics, *Atmos. Chem. Phys.*, 2011, **11**, 3303–3318.

41 J. L. Jimenez, *et al.*, Evolution of Organic Aerosols in the Atmosphere, *Science*, 2009, **326**, 1525–1529.

42 E. A. Weitkamp, A. T. Lambe, N. M. Donahue and A. L. Robinson, Laboratory Measurements of the Heterogeneous Oxidation of Condensed-Phase Organic Molecular Makers for Motor Vehicle Exhaust, *Environ. Sci. Technol.*, 2008, **42**, 7950–7956.

43 A. M. Sage, E. A. Weitkamp, A. L. Robinson and N. M. Donahue, Evolving mass spectra of the oxidized component of organic aerosol: results from aerosol mass spectrometer analyses of aged diesel emissions, *Atmos. Chem. Phys.*, 2008, **8**, 1139–1152.

44 R. C. Moffet, B. de Foy, L. T. Molina, M. J. Molina and K. A. Prather, Measurement of ambient aerosols in northern Mexico City by single particle mass spectrometry, *Atmos. Chem. Phys.*, 2008, **8**, 4499–4516.

45 K. Dzepina, R. M. Volkamer, S. Madronich, P. Tulet, I. M. Ulbrich, Q. Zhang, C. D. Cappa, P. J. Ziemann and J. L. Jimenez, Evaluation of recently-proposed secondary organic aerosol models for a case study in Mexico City, *Atmos. Chem. Phys.*, 2009, **9**, 5681–5709.

Faraday Discussions

PAPER

Molecular composition of organic aerosols at urban background and road tunnel sites using ultra-high resolution mass spectrometry

Haijie Tong,†[a] Ivan Kourtchev,[a] Pallavi Pant,[b] Ian J. Keyte,[b] Ian P. O'Connor,[c] John C. Wenger,[c] Francis D. Pope,[b] Roy M. Harrison‡[b] and Markus Kalberer*[a]

Received 2nd December 2015, Accepted 21st December 2015
DOI: 10.1039/c5fd00206k

Organic aerosol composition in the urban atmosphere is highly complex and strongly influenced by vehicular emissions which vary according to the make-up of the vehicle fleet. Normalized test measurements do not necessarily reflect real-world emission profiles and road tunnels are therefore ideal locations to characterise realistic traffic particle emissions with minimal interference from other particle sources and from atmospheric aging processes affecting their composition. In the current study, the composition of fine particles (diameter ≤2.5 μm) at an urban background site (Elms Road Observatory Site) and a road tunnel (Queensway) in Birmingham, UK, were analysed with direct infusion, nano-electrospray ionisation ultrahigh resolution mass spectrometry (UHRMS). The overall particle composition at these two sites is compared with an industrial harbour site in Cork, Ireland, with special emphasis on oxidised mono-aromatics, polycyclic aromatic hydrocarbons (PAHs) and nitro-aromatics. Different classification criteria, such as double bond equivalents, aromaticity index and aromaticity equivalent are used and compared to assess the fraction of aromatic components in the approximately one thousand oxidized organic compounds at the different sampling locations.

[a]*Centre for Atmospheric Science, University of Cambridge, Lensfield Road, Cambridge, CB2 1EW, UK. E-mail: markus.kalberer@atm.ch.cam.ac.uk*

[b]*School of Geography, Earth and Environmental Sciences, University of Birmingham, Edgbaston, Birmingham, B15 2TT, UK*

[c]*Department of Chemistry and Environmental Research Institute, University College Cork, Cork, Ireland*

† Present address: Max Planck Institute for Chemistry. Hahn-Meitner-Weg 1, 55128 Mainz, Germany.

‡ Also at: Department of Environmental Sciences/Center of Excellence in Environmental Studies, King Abdulaziz University, PO Box 80203, Jeddah, 21589, Saudi Arabia.

1. Introduction

Aerosol particles are emitted into the atmosphere by natural and anthropogenic processes and about 20–90% of submicron particles are composed of organic material.[1,2] Road traffic, and especially engine exhaust emissions, are responsible for a substantial fraction of urban submicron particles.[3] In addition to directly emitted particles, combustion engines can emit significant amounts of gaseous volatile organic compounds which can generate additional particle mass due to gas-to-particle conversion processes such as oxidation reactions, leading to less volatile organic compounds and the formation of so-called secondary organic aerosol mass (SOA).[4,5] The chemical composition of road traffic related exhaust particles is highly complex and especially the effects of SOA formation and atmospheric aging, which alter the particle composition during their entire lifetime, are not well understood. The particle composition also depends on traffic characteristics such as vehicle fleet composition and the type of fuel used, or meteorological conditions such as long range atmospheric transport.[6,7] In addition to traffic there are a large number of other natural and anthropogenic sources contributing to urban particle mass. Assessment of the most important particle sources and their contribution to the ambient particle mixture is only possible through a detailed chemical characterisation of ambient particles and the major potential sources. Inhalation of these particles has been linked to a number of adverse health effects[8,9] and carcinogenic compounds such as polycyclic aromatic hydrocarbons (PAHs) and transition metals are potentially responsible for their toxicity.[10–12]

Particles emitted by an average vehicle fleet at a specific location can be assessed ideally with samples collected in road tunnels, because they provide a real-world mixture of vehicles and driving conditions. Currently only a few studies have focused on traffic emissions in urban aerosol in the UK.[13,14] These studies utilized a range of traffic-specific organic markers, such as hopanes, steranes and PAHs, which allows source apportionment of different traffic sources. However, the huge complexity of combustion exhaust particles with thousands of organic compounds[15] warrants a broad characterisation of the organic particle composition to potentially complement existing particle source profiles.

In our current study, direct infusion negative nano-electrospray ionisation (nanoESI) ultrahigh resolution mass spectrometry (UHR-MS) was used to characterise and compare a broad range of particle components in fine particles from a road tunnel, characteristic of directly emitted vehicle exhaust particles, and an urban background site in Birmingham, UK. The results show that in both sampling locations a large number of oxidised mono-aromatic compounds and PAHs are present with significant numbers of nitrogen-containing aromatic compounds, especially in the tunnel samples. In contrast, sulfur-containing organics have greater relative abundance at the urban background site, indicating the effect of atmospheric aging processes or the presence of non-traffic sources.

2. Method

2.1 Sample collection and analysis

Sampling was conducted at the A38 Queensway Tunnel (QT) and the Elms Road Observatory Site (EROS) in Birmingham, UK, between September 11 and 21, 2012.

Five filter samples were analysed from each site (Table 1). More detailed information about the sampling sites is given elsewhere (Pant *et al.*, submitted).

Briefly, the Queensway tunnel is naturally ventilated, 545 metres long and in a North/South direction allowing 50 kilometres per hour vehicle speed. It has two lanes on each side and a 25 m^2 cross-section. The daily traffic volume through the tunnel is around 25 000 vehicles (Pant *et al.*, submitted). The aerosol samplers were placed in the emergency lay-by area at a distance of 1.5 m from the road in the southbound lane. The EROS site was chosen as an urban background site, located at 52.45° N, 1.93° W, and about 3.5 km southwest of Birmingham city centre with no significant particle sources in the immediate vicinity of the site. At both sites 24 hour integrated $PM_{2.5}$ aerosol samples were collected on pre-baked 150 mm quartz filters (Whatman QM-A) using a high-volume samplers (Digitel DHA-80) operating at a flow rate of 500 L min^{-1}.

Organic carbon (OC) and elemental carbon (EC) concentrations of all aerosol samples were determined using a Sunset Laboratory thermal/optical carbon analyser using the EUSAAR2 protocol[16] and results are shown in Table 1.

Detailed information on the aerosol sample extraction and UHRMS analysis can be found elsewhere.[17] Briefly, a part of the quartz fibre filter (35±7 cm^2) was extracted three times with 5 mL of methanol (Optima® LC/MS grade, Fisher Scientific) under ultrasonic agitation for 30 min in ice cold water. The extracts were combined, filtered through a 0.2 μm ISO-Disc™ PTFE filter (Supelco, Bellefonte, PA, USA), and reduced by volume to the same OC concentration under a gentle stream of nitrogen. The extracts were analysed using an ultrahigh resolution LTQ Orbitrap Velos mass spectrometer (Thermo Fisher, Bremen, Germany) equipped with a TriVersa Nanomate robotic nanoflow chip-based electrospray ionisation (ESI) source (Advion Biosciences, Ithaca NY, USA).[18]

The Orbitrap MS was calibrated using an Ultramark 1621 solution (Sigma-Aldrich, UK) providing mass accuracy of the instrument below 1 ppm. The instrument mass resolution was 100 000 at *m*/*z* 400. The direct infusion, negative mode nanoESI parameters were as follows: the ionization voltage and back pressure were set at −1.4 kV and 0.8 psi, respectively. The inlet temperature was 200 °C.

Table 1 Concentration (μg m^{-3}) of organic carbon (OC), elemental carbon (EC), and total carbon (TC) in samples from the Birmingham background and tunnel sites

Sample type	Sampling date	OC	EC	TC
Tunnel	11.09–12.09	13.4 ± 0.69	15.1 ± 0.78	28.6 ± 1.47
Tunnel	12.09–13.09	19.5 ± 1.00	17.1 ± 0.87	36.6 ± 1.87
Tunnel	13.09–14.09	17.3 ± 0.89	18 ± 0.92	35.3 ± 1.81
Tunnel	14.09–15.09	20.0 ± 1.02	22 ± 1.12	42 ± 2.14
Tunnel	15.09–16.09	16.9 ± 0.86	19.3 ± 0.99	36.1 ± 1.85
Background	11.09–12.09	0.41 ± 0.04	0.4 ± 0.04	0.80 ± 0.08
Background	12.09–13.09	3.12 ± 0.18	1.89 ± 0.12	5.01 ± 0.29
Background	13.09–14.09	1.25 ± 0.08	0.96 ± 0.07	2.21 ± 0.15
Background	14.09–15.09	1.42 ± 0.09	0.85 ± 0.06	2.27 ± 0.16
Background	15.09–16.09	2.05 ± 0.12	0.97 ± 0.07	3.02 ± 0.19

The mass spectral data analysis was performed using procedures described in detail elsewhere.[17] Briefly, for each sample analysis, 60–90 mass spectral scans were averaged into one mass spectrum. The high mass resolution and accuracy measurements allows for an unambiguous assignment of formulas for most peaks in the mass spectra up to a mass range of about *m*/*z* 450.[19] Molecular assignments were made using Xcalibur 2.1 software (Thermo Fisher Scientific, USA) applying the following constraints: $0<{}^{12}C\leq100$, ${}^{13}C\leq1$, ${}^{1}H\leq200$, ${}^{16}O\leq50$, ${}^{14}N\leq5$, ${}^{32}S\leq2$, ${}^{34}S\leq1$. The double bond equivalent (DBE) for each individual formula was calculated using Xcalibur 2.1 software. All molar ratios, DBE factors and chemical formulae presented in this paper refer to neutral molecules. Data processing was performed using a Mathematica 10.1 (Wolfram Research Inc., UK) code developed in-house that employed several conservative rules and constraints to identify unambiguous elemental formulas for measured accurate mass spectral peaks as discussed in more detail in our previous studies.[17,20] Briefly, the assigned formulas were checked to satisfy the nitrogen-rule. Elemental formulae containing ${}^{13}C$ or ${}^{34}S$ were checked for the presence of ${}^{12}C$ or ${}^{32}S$ counterparts respectively, and if they were not present the formula with the next largest mass error was considered. The background spectra obtained from the procedural blanks were also processed using the rules mentioned above. The formulae lists of the background spectra were subtracted from those of the ambient samples and only formulae with a sample to background ratio ≥10 were retained.[17,20]

The samples from the two Birmingham sites are compared to aerosol collected at an industrial harbour site in Cork, Ireland. The sampling and analysis of these samples are described elsewhere in detail.[18]

2.2 Interpretation of direct-injection electrospray ionisation high-resolution mass spectrometry data

Only ions found in all five tunnel or all five background samples, respectively, referred to as 'common ions' in the text, were considered in the analysis presented here. These common ions represent the compounds present in the aerosol at a specific site over several days, which ensures they represent a typical average traffic source profile and an average urban organic background profile, respectively.

Electrospray ionisation UHRMS is a highly sensitive mass spectrometry technique to analyse polar compounds, and due to its soft ionisation process analyte fragmentation is minimal, which allows characterisation of complex samples without prior chromatographic separation.[21–23] However, non-polar compounds (such as alkanes or non-oxidised aromatic compounds) cannot be ionised and measured with ESI, or only with very poor ionisation efficiencies.[24] In addition, all ESI techniques are prone to matrix effects, which cause varying ionisation efficiency (and therefore signal intensities) depending on the overall sample composition (*i.e.* matrix). By considering common ions only we minimize sample-to-sample variations due to these matrix effects and obtain mass spectra representative for the respective sampling locations.

This study aims to compare and contrast differences in the chemical composition of the approximately 1000 common ion formulae measured at the three urban sites. This reflects the chemical diversity and complexity of particles at the various sampling sites but does not allow for inferring of concentrations of compounds, due to the ionisation efficiency effects discussed above (matrix effects).

3. Results and discussion

A main emphasis of this study is to characterise the very broad and complex composition of oxidized organic compounds (formulae) in the Birmingham tunnel samples and compare them with the aerosol composition at the Birmingham urban background and the Cork harbour sites, with the goal of identifying compositional influences of traffic exhaust emissions (*i.e.* the tunnel samples) on ambient urban particle composition.

The mass spectra in Fig. 1a and b show common ions of the five samples collected at the background and the tunnel site. As described above, the common ions represent those mass spectral peaks that are present over several days at the locations and can thus be seen as robust profiles for the average traffic source at the Queensway tunnel in Birmingham (Fig. 1b) and average Birmingham urban background composition (Fig. 1a). In total about 970 and 1000 common ions were detected in the background and tunnel samples, respectively.

Fig. 1 clearly shows that formulae containing only C, H and O atoms (*i.e.* containing no nitrogen and sulfur functional groups, red peaks in Fig. 1) are the dominant formula subgroup in both the urban background and the tunnel particles with a fraction of ~40% and ~44% (384 and 439 ions), respectively. This compares to an earlier study where we observed about 60% of all ions in the CHO-subgroup at an industrial harbour site in Cork, Ireland.

N-containing formulae (CHON subgroup) make up about 28% of the background sample and 39% of the tunnel sample (see more detailed discussion below). In contrast, S-containing organics (*i.e.* CHOS and CHONS subgroups) were more prominent in the urban background samples compared to the tunnel samples. This is likely reflecting the longer atmospheric aging and heterogeneous reactions of particle components collected at the background site, with sulphate and sulfuric acid present in greater abundance in urban background particles compared to particles at the tunnel site, which represent more direct emission conditions with low sulfur content.[25,26]

There is significant overlap between a number of molecular formulae (including N and S-containing organics) at both the Birmingham tunnel and the background site (~60%), as shown in the Venn diagram (Fig. 1c), indicating that the chemical composition of particles at the background site might be significantly impacted by traffic emissions. It must be noted, however, that the online ESI-MS technique used in this study is only semi-quantitative as discussed above and therefore additional (*e.g.* chromatography) analyses would be needed to confirm these similarities on a quantitative level.

Fig. 2 shows van Krevelen plots for the common ions of the Birmingham background, the tunnel and the Cork samples, where H/C (hydrogen to carbon) *versus* O/C (oxygen to carbon) elemental ratios of all formulae detected in these samples indicate the degree of oxidation.[17,18,27,28] The color code in Fig. 2 indicates the double bond equivalent (DBE) value of each formula. The DBE is calculated as[29]

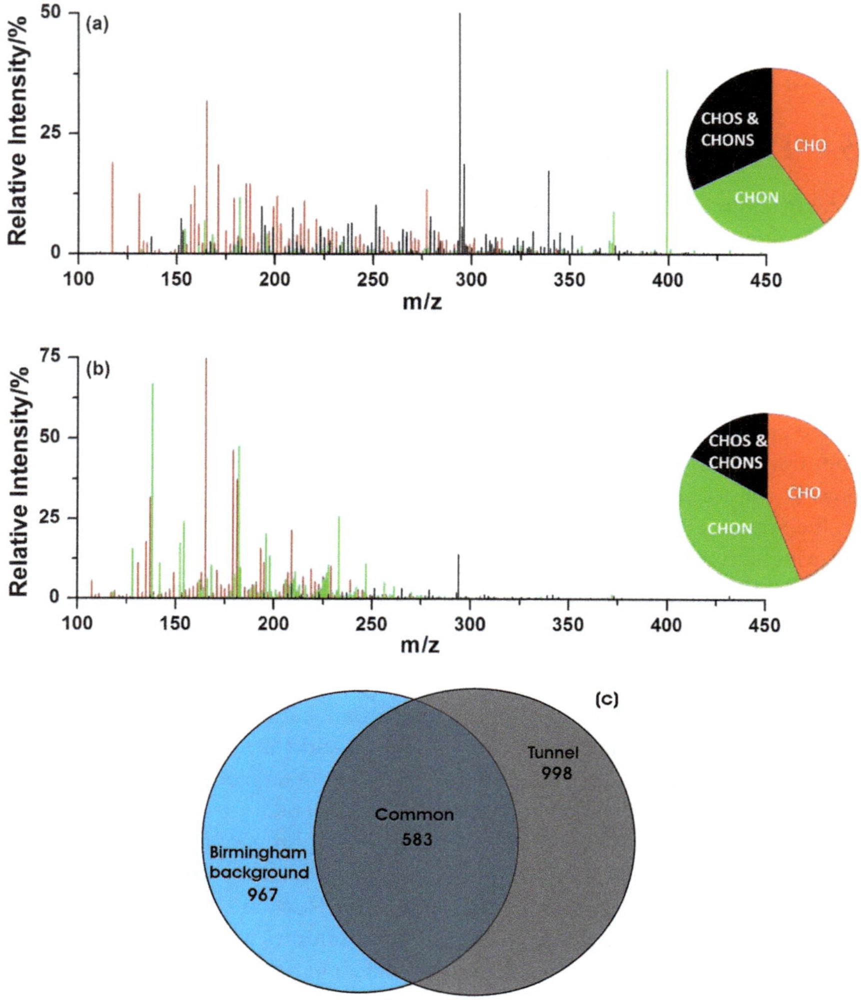

Fig. 1 Negative ionization nanoESI mass spectra of common ions in aerosol samples collected at a background (a) and tunnel (b) site in Birmingham, UK. The pie charts and color code of the mass spectra indicate the abundance of organic compounds which contain only C, H and O atoms and N- and S-containing compounds, respectively. Common ions are defined as ions which are detected in all five background or five tunnel samples analysed in this study, respectively. Thus, the common ion mass spectra represent a robust signature of the organic aerosol composition at these two sites. (c) Venn diagram showing the number of all molecular formulae identified in the Birmingham tunnel and background samples, and common formulae at both sites.

$$\mathrm{DBE} = 1 - \frac{N_\mathrm{H}}{2} + \frac{N_\mathrm{N}}{2} + \frac{N_\mathrm{S}}{2} + N_\mathrm{C} \tag{1}$$

where 'N' is the number of atoms (C, H, N, and S) in a formula of $C_cH_hO_oN_nS_s$.

The less oxidised and hydrogen-rich formulae (H/C>1.5) at all three locations are likely aliphatic compounds. Previous studies that utilised UHRMS attributed some of these molecules to fatty acids.[17,30] Fatty acids are a signature of a number

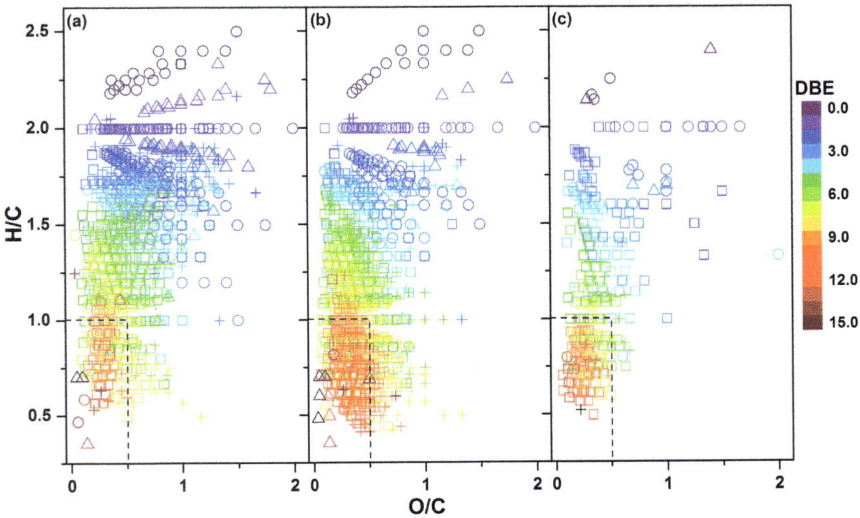

Fig. 2 Van Krevelen diagrams (HC *versus* O/C atomic ratios) including double bond equivalent (DBE) values of all common ions detected in the background (a) and tunnel (b) samples from Birmingham and from Cork harbour (c). Symbols: squares indicate CHO, crosses CHON, circles CHOS, and triangles CHONS formulae. The black dashed squares indicate compounds with low H/C ratio and high DBE values, a typical signature of mono-aromatic compounds and PAHs.

of primary natural sources, including marine biota,[31] plant waxes,[32] and combustion of biomass material.[33,34] Although less abundant than in the background site, a significant number of these aliphatic compounds are also found in the tunnel sample, suggesting that vehicle emissions are also a potential source of these saturated aliphatic compounds.

Formulae with $0<H/C\leq1.0$ and $O/C\leq0.5$ (dotted-line squares in Fig. 2a–c) dominantly have high DBE values (≥7.0), which is consistent with oxidized polycyclic aromatic hydrocarbons (PAHs; the smallest PAH, naphthalene, $C_{10}H_8$ has an H/C of 0.8 and 7 DBE). The highest number of these PAH-like compounds is found in the tunnel samples with 480 ions (48% of all peaks in the tunnel samples), compared to only 186 ions in the urban background samples (19%). The samples from Cork harbour contain 110 ions in the PAH range (38%). The comparable fraction of high-DBE formulae in the tunnel and Cork harbour samples likely reflects the traffic and industrial activity influencing particle composition at these sites.

To evaluate further whether this large number of formulae with high DBEs are indeed of an oxidized (poly-)aromatic structure, the formulae in the CHO subgroup were probed in more detail. Fig. 3a–c shows the H/C ratio of all CHO subgroup formulae as a function of their molecular weight and Fig. 3d–f the O/C ratio at all three locations. Fig. 3a (Birmingham urban background) and 3b (Birmingham tunnel) show that H/C of most of the formulae are larger than 1.0. Only a small fraction of the formulae (~21% for background and ~36% for tunnel) have $H/C<1.0$. Fig. 3 clearly shows that most formulae with a DBE >10 have molecular masses >200 Da, which is consistent with slightly oxidized

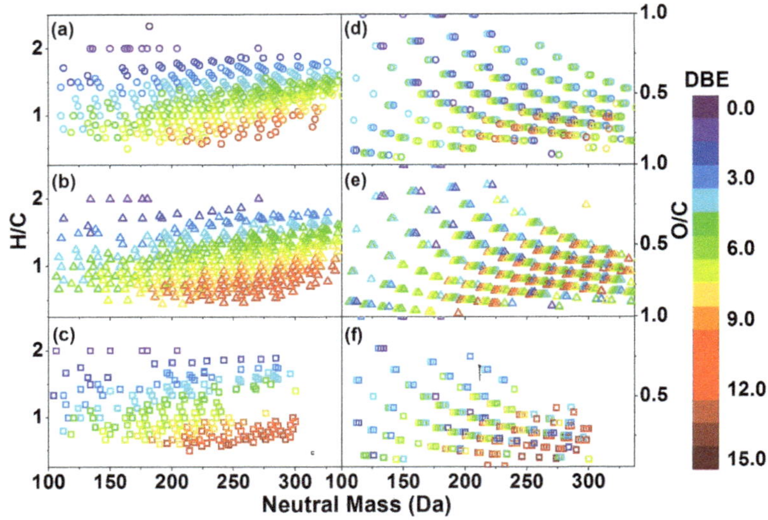

Fig. 3 H/C (a, b and c) and O/C (d, e and f) values as a function of neutral mass and DBE values of all CHO formulae from Birmingham background (a and d), tunnel (b and e) and Cork harbour (c and f) samples. The fractions of compounds with O/C>1.0 for all samples is <2% and these are not shown. The high number of formulae with DBE >7, masses >200 Da and H/C<1 supports the presence of a large number of oxidized PAHs.

3-ring PAHs (*e.g.*, dihydroxy-anthracenes) and larger polyaromatics. A similar profile of high-mass, high-DBE formulae is present in the Cork harbour samples (Fig. 3c and f).

The vast majority of the high-DBE (PAH-like) compounds have low or moderate O/C ratios with values mostly between 0.2–0.5. This is rather low compared to many other, often highly oxidized organic compounds in ambient aerosols.[18,35] For PAHs, however, these O/C ratios represent a high degree of oxidation: using again the example of anthracene ($C_{14}H_{10}$), an O/C ratio of about 0.3 would correspond to 4–5 oxygen atoms in the molecule.

DBE and molecular mass can, however, be inadequate to identify whether a specific molecular formula is potentially a (poly-)aromatic structure. Therefore, two parameters were developed in recent years, the aromaticity index (AI) and aromaticity equivalent (X_c), to characterize aromatic and poly-aromatic compounds in highly complex compound mixtures.[36,37] The AI provides a measure of the C–C double bond density by considering the contribution of heteroatoms to π-bonds, which normally has a range of 0 to 1. The AI can be calculated for each formula determined in a UHRMS spectrum and is calculated as[36]

$$\mathrm{AI} = \frac{\mathrm{DBE_{AI}}}{\mathrm{C_{AI}}} = \frac{1 + N_\mathrm{C} - N_\mathrm{O} - N_\mathrm{S} - 0.5N_\mathrm{H}}{N_\mathrm{C} - N_\mathrm{O} - N_\mathrm{S} - N_\mathrm{N}} \tag{2}$$

where the '*N*' in eqn (2)–(5) is the atom number for H, C, O, N, or S. $\mathrm{DBE_{AI}}$ is the minimum number of C–C double bonds plus rings in the molecule. The $\mathrm{DBE_{AI}}$ and $\mathrm{C_{AI}}$ were calculated through the following equations:

$$DBE_{AI} = 1 + N_C - N_O - N_S - 0.5N_H \tag{3}$$

$$C_{AI} = N_C - N_O - N_N - N_S \tag{4}$$

The AI value of >0.5 is indicative of aromatic compounds and the AI ≥0.67 of PAHs. However, for (poly-)aromatic compounds with significant alkylation, AI values drop below these thresholds and can thus be wrongly assigned as non-aromatic structures. To avoid such mis-classifications an improved "Aromaticity Equivalent" parameter X_c has been proposed.[37] X_c normally has values between 0 and 3.0 and is calculated using the following equation:[37]

$$X_c = \frac{3(DBE - mN_O - nN_S) - 2}{DBE - mN_O - nN_S} \tag{5}$$

where the parameters of 'm' and 'n' in eqn (5) account for the fraction of oxygen and sulfur atoms involved in π-bond structures of a compound and are different for different functional groups. For carboxylic acid, ester, and nitro functional groups $m = n = 0.5$.[37] For compounds with other functional groups such as aldehydes, ketones, nitroso, cyanate, alcohol, or ethers, 'm' and 'n' need to be adjusted to 1 or 0. Because electrospray ionisation, used in this study, is most sensitive to carboxylic acids, the X_c analysis below was performed assuming that $m = n = 0.5$. When the molecular formula had an odd number of oxygen or sulfur atoms, the sum $(mN_O + nN_S)$ in eqn (5) was rounded down to the closest integer as detailed elsewhere.[37]

The DBE values *versus* the number of carbon atoms as well as AI values for all CHO subgroup formulae in the Birmingham background and tunnel samples are shown in Fig. 4a. The peak intensity is indicated by the size of symbols and naphthalene ($C_{10}H_8$) and anthracene ($C_{14}H_{12}$) are shown as red crosses in Fig. 4a and b. The vast majority of compounds at the background site (circles) have very low DBE (<5) and AI (≤0.5, grey symbols) values demonstrating the low abundance of oxidised mono-aromatics and PAHs at this site. There is only a limited number of formulae with AI values of 0.53–0.67 (32 formulae) and ≥0.67 (6 formulae) in the background sample, which are the thresholds for mono-aromatics and polyaromatics, respectively. This is in contrast to the tunnel samples (Fig. 4a, triangles) where a larger number of formulae have AI values of 0.53–0.67 (65 formulae; mono-aromatics) and AI ≥0.67 (22 formulae) indicating the presence of more oxidised PAHs than at the background site.

Fig. 4b shows the same data as Fig. 4a but the color code indicates the value of X_c. Using the X_c classification there are 128 formulae in the background samples with an X_c value between 2.5 and 2.71 (indicative of mono-aromatics), significantly more than the AI classification suggests. Similarly, X_c suggests that more PAHs are present in the background sample than the AI classification, with 44 formulae having an X_c>2.71. The tunnel samples also contain a larger number of formulae consistent with mono-aromatic (173 formulae, blue and brown triangles, Fig. 4b) and poly-aromatic (106 formulae, pink and purple triangles, Fig. 4b) structures.

The considerably larger number of PAHs identified using the X_c classification compared to the AI suggests that there is a large number of alkylated PAHs which are wrongly assigned as non-aromatic by the AI classification.

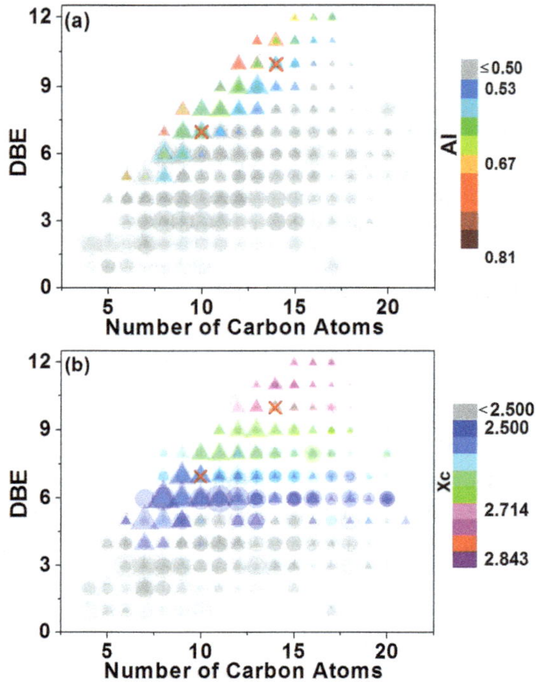

Fig. 4 Double bond equivalents (DBEs) *versus* number of carbon atoms, (a) aromaticity index (AI), and (b) aromaticity equivalent (X_c) of all CHO formulae in the Birmingham background (circles) and tunnel (triangle) samples. The red crosses indicate a two- and three-ring PAH (naphthalene ($C_{10}H_8$) and anthracene ($C_{14}H_{10}$)). The grey points indicate non-aromatic formulae with AI≤0.50 or X_c<2.500. A significant number of non-aromatic formulae with DBE >6 is clearly visible in both samples. Only formulae with DBE >9 and X_c>2.7 can be clearly assigned as PAHs. The size of the symbols reflects the relative peak intensities in the mass spectra on a logarithmic scale.

Fig. 4a and b highlight the advantage of the AI and especially of the X_c classification compared to using only DBEs as an indicator of aromaticity or PAHs. When only considering DBEs, all formulae with a DBE ≥7 have to be assumed to be potentially PAHs. However, Fig. 4 shows that there are a significant number of formulae with 7 or more DBEs (76 formulae for the background and 101 formulae for tunnel samples), which cannot be explained by polyaromatic structures using the more detailed concepts of AI or X_c (blue and green data points in Fig. 4 with DBE >7).

Overall there are more mono-aromatics at the Birmingham background and tunnel sites (using X_c), with 128 and 173 formulae, respectively, than PAHs (44 formulae at background and 106 at tunnel site). The much lower number of oxidised (poly-)aromatics at the background site is likely due to dilution affecting the background site more than the tunnel samples. Importantly, 41 of the PAHs detected in the tunnel samples were also present at the background site (out of 44 PAHs detected in the background sample) suggesting that almost all PAHs detected at the background site have a contribution from traffic exhaust sources.

Fig. 5 shows the H/C (a, b, and c) and O/C ratios (d, e, and f) *versus* the neutral mass and X_c of the Birmingham background (a and d), Birmingham tunnel (b and e), and Cork harbour (c and f) samples. This illustrates clearly the much larger number of mono-aromatic compounds (blue and green data points) compared to PAHs (pink to purple). The mass range and O/C ratio of mono-aromatics shows the presence of a high degree of oxidation (O/C up to *ca.* 0.7) and alkylation (molecular masses up to about 330 Da). In contrast, oxidised PAHs are less abundant and their lower H/C ratios (compared to mono-aromatics) with values mostly below 1 indicate a generally low degree of alkylation. The industrial harbour site in Cork has a more balanced fraction of mono-aromatics and PAHs.

The O/C ratios of organic compounds can be misleading when assessing the degree of oxidation because non-oxidative reactions (such as dehydrogenation) can change the O/C but may not affect the degree of oxidation of a compound. Thus, an alternative metric, the 'carbon oxidation state', has been applied to characterize aerosols composition[38] and can be calculated through the following equation:

$$\overline{\mathrm{OS}}_{\mathrm{C}} = -\sum_i \mathrm{OS}_i \frac{N_i}{N_{\mathrm{C}}} \tag{6}$$

where the $\overline{\mathrm{OS}}_{\mathrm{C}}$ is the average carbon oxidation state. OS_i is the oxidation state associated with element I, and N_i and N_{C} are moles of elements i and of carbon, respectively. For the application of $\overline{\mathrm{OS}}_{\mathrm{C}}$ to CHO subgroup formulae only, eqn (6) can be simplified to:

$$\overline{\mathrm{OS}}_{\mathrm{C}} \approx 2\frac{\mathrm{O}}{\mathrm{C}} - \frac{\mathrm{H}}{\mathrm{C}} \tag{7}$$

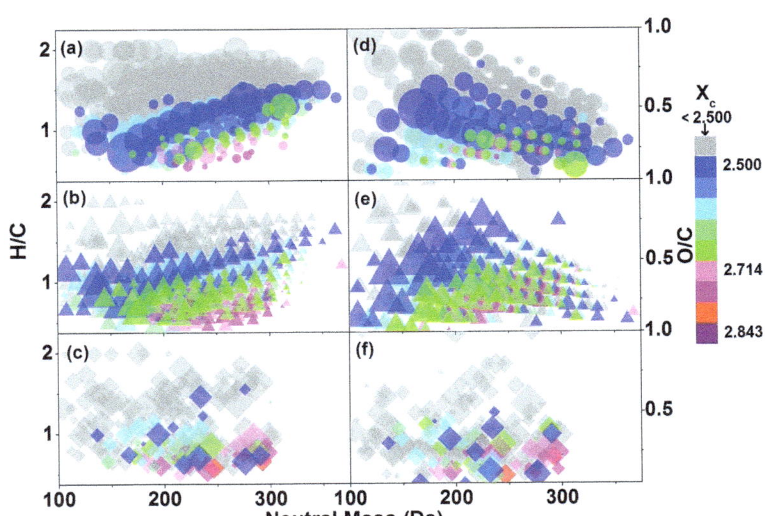

Fig. 5 H/C (a, b and c) and O/C (d, e and f) values of all CHO formulae are shown as a function of their neutral mass from Birmingham background (a and d), tunnel (b and e) and Cork harbour (c and f) samples with their X_c values color-coded. Grey data points indicate non-aromatic compounds (X_c<2.5), blue to green data (2.5 < X_c<2.7) are mono-aromatic compounds and pink to purple data (X_c>2.7) includes PAHs. The size of the symbols reflects the relative peak intensities in the mass spectra on a logarithmic scale.

where O/C and H/C are the elemental ratio of oxygen and hydrogen to carbon, respectively. The carbon oxidation states of all CHO subgroup formulae in the Birmingham background and tunnel samples are shown in Fig. 6, with their corresponding X_c as color code. Especially in the tunnel samples, aromatics and poly-aromatics are generally at the higher end of the distributions of $\overline{OS}_C$ values for a particular carbon number, and almost all PAHs (X_c>2.7) have a $\overline{OS}_C > -0.6$. Such high $\overline{OS}_C$ values are generally associated with semi- or low-volatility aerosol components.

In general, upper limit $\overline{OS}_C$ values are 1 for both Birmingham samples, which is lower compared to pure secondary organic aerosol (from chamber studies) where the $\overline{OS}_C$ goes up to 2, demonstrating the relatively fresh nature of most components in these urban particles and compared to remote locations where long-range transport and oxidation cause a higher degree of oxidation.[30]

Besides CHO-containing compounds there are a considerable number of organic compounds containing nitrogen and sulfur (Fig. 1). Fig. 7 shows three different subgroups, (CHON, CHOS, and CHONS), while CHN, CHS, and CHNS components are not included because they account for <0.5% of the total formulae in the samples. At the Birmingham background site (Fig. 7a) the CHON subgroup is the dominant compound group with about 47% of all N- or S-containing formulae. Only a relatively small number of components (81 CHON formulae, 7 CHOS formulae, and 5 CHONS formulae) have $X_c \geq 2.5$ (at the background samples) and thus can be categorized as N- or S-containing aromatic compounds. The tunnel samples (Fig. 7b) have much larger number fraction of CHON components (391 formulae, ~70%) indicating a higher degree of nitration. The higher number of aromatic structures (303 CHON formulae with $X_c \geq 2.5$) in the tunnel samples supports the presence of many nitro-(poly-)aromatics.

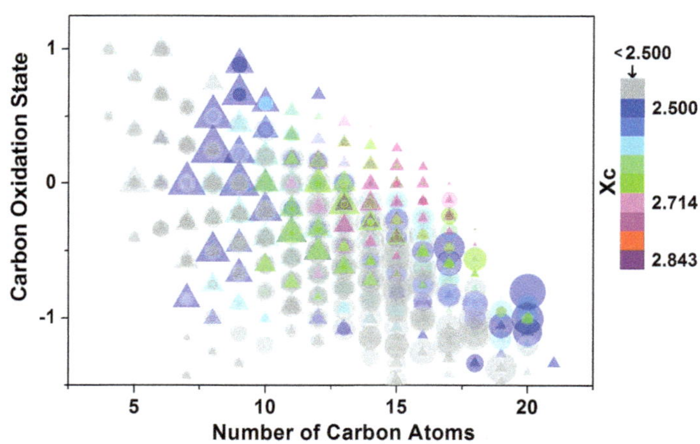

Fig. 6 Carbon oxidation state *versus* the carbon atom number of all CHO formulae in the Birmingham tunnel (triangle) and Birmingham background (circles) formulae (X_c values color-coded). Especially for the tunnel samples, the high degree of alkylation of mono-aromatics with up to 10–11 carbon atoms is clearly visible, as well as the generally lower degree of oxidation of PAHs (*i.e.* lower carbon oxidation state values) compared to mono-aromatics. The size of the symbols reflects the relative peak intensities in the mass spectra on a logarithmic scale.

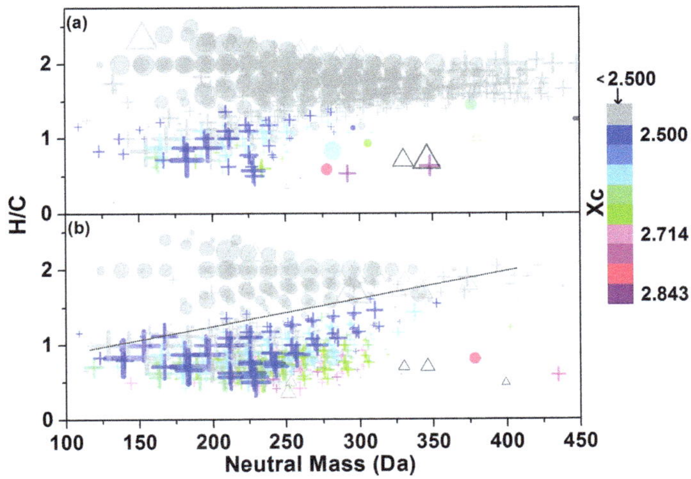

Fig. 7 H/C atomic ratios *versus* molecular mass of common ions of all CHON (crosses), CHOS (circles) and CHONS (triangles) formulae in the Birmingham background (a) and tunnel (b) samples (X_C values color-coded). Aerosol particles in the tunnel samples include significantly higher numbers of nitrogen-containing (poly-)aromatic formulae. The size of the symbols reflects the relative peak intensities in the mass spectra on a logarithmic scale.

A comparison of Fig. 3 (panels a, b, d and e) and Fig. 7 shows that aromatic compounds in the Birmingham tunnel samples with DBE $\leq$9, H/C<1, O/C<0.7, and neutral mass <300 contain a significant fraction of CHON subgroup formulae. In contrast, the aromatic compounds in the urban background samples in the same DBE, H/C, and O/C value range are mainly composed of CHO subgroup formulae and only contain a small fraction of CHON, CHOS, or CHONS components.

The most intensive N-containing compound in the mass spectra from tunnel samples was an ion at *m*/*z* 138.01988 (RI >65%) with a molecular formula of $C_6H_5NO_3$, possibly corresponding to nitrophenols (2-nitrophenol and/or 4-nitrophenol). Formulae with likely the same core-structure but with an additional one and two methyl groups (*i.e.* $C_7H_7NO_3$ and $C_8H_9NO_3$) were also detected. These compounds have been previously observed in aerosol from urban locations *e.g.*, Rome, Italy,[39] Mainz, Germany[40] and Cork, Ireland,[18] and were mainly attributed to traffic emissions. However, nitrophenols were previously also found in various other sources including decomposition of herbicides and insecticides and burning of coal and wood,[41] although in urban environments primary vehicle engine emissions are believed to be their major source.[42] Another nitrogen-containing homologous series observed with high signal intensities included molecules with the following molecular formulae: $C_6H_5NO_4$, $C_7H_7NO_4$, $C_8H_9NO_4$, $C_9H_{11}NO_4$ and $C_{10}H_{13}NO_4$. These formulae were tentatively identified as nitroaromatic compounds, *e.g.*, nitrocatechols, nitrophenols, nitroguaiacols and nitrosalicylic acids. These nitroaromatics have been recently detected in aerosol samples from urban locations, *e.g.*, Ljubljana, Slovenia[43] and rural environments *e.g.*, Saxony, Germany,[44] K-Puszta, Hungary,[45] Hyytiälä, Finland,[17] Los Angeles, USA[46] and Cork, Ireland,[18] and were mainly attributed to biomass burning sources. Although direct infusion analysis does not allow structural or isomeric molecular elucidation, our results suggest that these identified series can be associated not only with biomass burning but also with traffic emission sources. In addition, formulae with $OSc<-1$ and more than 7 carbon atoms, which is indicative of primary biomass burning aerosol,[38] were observed in the background Birmingham samples (Fig. 6). Their small fraction compared to the overall observed number of formulae (*ca.* 8%) suggests a rather modest influence of biomass burning at the Birmingham background site compared to traffic sources.

Only a very small number of known oxidized and nitrated PAHs were detected in the Birmingham background and tunnel samples, *e.g.* formulae that correspond to nitro-naphthalenes and hydroxy-nitro-naphthalenes. These formulae were measured with low signal intensities, which is likely due to their poor ionisation efficiency in negative mode ESI-MS.

Using gas chromatography-mass spectrometry, samples from the same two locations in Birmingham were analysed and several nitro-PAHs and oxidized PAHs with a carbonyl functional group were identified.[47] These detailed chromatography analyses concluded that PAHs at the background site were dominated by traffic emissions, which is supported by the analyses presented here, but that other sources such as wood burning have become increasingly important over the last two decades. This study[47] also provided strong evidence of increased levels of nitro-PAHs in the tunnel samples, likely due to increased use of diesel cars, which is also supported by the nitro-aromatics identified in the current study.

Formulae in the CHOS subgroup are more prominent in the background sample (~36% of all N or S containing formulae) than in the tunnel samples (~24%). The higher number of CHOS components at the background site could be explained by atmospheric aging reactions of organics with H_2SO_4 leading to the formation of organosulfates. The number of aromatic sulphates (using AI or X_c), however, is very small in both samples (less than 10 formulae).

The composition of the tunnel samples shows a distinct boundary between CHON and S-containing subgroups (CHOS and CHONS) (dotted line in Fig. 7b). Above this line, CHOS and CHONS subgroups are dominating whereas below the line, CHON subgroup formulae are most abundant. Thus CHOS and CHONS components have on average a larger H/C ratio than the CHON compounds. About 91% of the 168 CHOS and CHONS compounds have $\geq$4 oxygen atoms and are likely organosulfates. This clear distinction in the H/C ratio between S- and N-containing compounds is not observed in the Birmingham background samples where on average the CHON subgroup has a higher H/C ratio than in the tunnel samples.

As shown in Fig. 8a about 96% of all molecular formulae associated with mono-aromatic compounds found in background Birmingham samples were also present in tunnel samples. A slightly smaller but still very high fraction of PAH-like molecular formulae (~87%) from the background samples were also present in the tunnel aerosol, indicating that the majority of these molecules at the background site are likely to be associated with the traffic emissions.

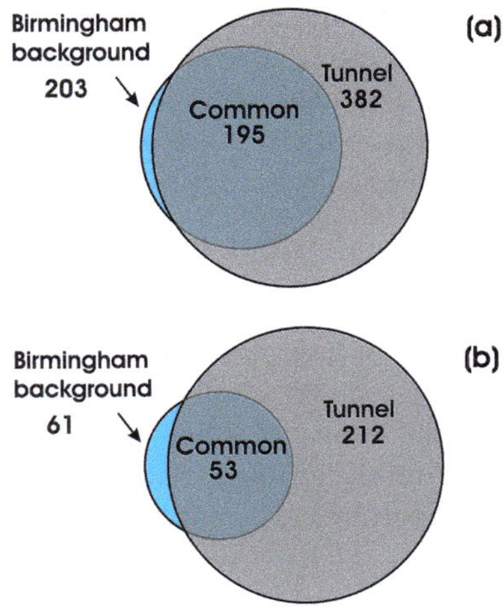

Fig. 8 Venn diagrams showing the number of molecular formulae associated with (a) mono-aromatic and (b) PAH-like species in the Birmingham tunnel and background samples, and common formulae present at both locations. The size of the circles reflects the relative ratio of molecular formulae identified in the samples.

4. Conclusions

The organic composition of ambient aerosol was analysed and compared from a road tunnel, an urban background and industrial harbour sites using direct infusion, negative mode nanoESI UHR-MS. Components of the CHO-subgroup (containing only C, H and O atoms) were most dominant at all sites. A higher number of mono-aromatics and PAHs were identified in the tunnel samples compared to the urban background sites. Double bond equivalents, aromaticity index and aromaticity equivalent approaches were compared to identify the distribution of oxidized mono-aromatics and PAHs among the approximately one thousand organic compounds (formulae) present in the mass spectra at these sites.

Nitrogen- and sulfur-containing formulae were also investigated and a higher number of N-containing mono-aromatics were observed in the tunnel samples compared to the background samples, with some nitro-aromatic compounds tentatively identified. In contrast, a higher fraction of S-containing formulae, likely organic-sulphates, were found in the urban background site, indicating the influence of atmospheric aging processes or the presence of non-traffic sources.

Using common ions from the tunnel samples integrated over five days represents a robust average organic composition profile of vehicle emissions, and our results suggest that traffic emission can be a source for a very large number of oxidised mono-aromatics and PAHs in the urban atmosphere, which could potentially be used to better characterise and constrain vehicle exhaust emission profiles and effects of atmospheric aging on these particle emissions.

Although direct-injection ESI-UHRMS analyses (performed in this study) allow the separating of 1000s of compounds in a single analysis, they cannot unambiguously identify individual compounds at a molecular level. In addition, ESI-MS is mostly sensitive to compounds with polar functional groups. Thus, combining

ESI-UHRMS data with results from other techniques, which are sensitive to less polar organic compounds and are capable of separating 1000s of compounds in a single analysis, such as two-dimensional gas chromatography-mass spectrometry, might be a promising approach to characterize comprehensively the entire organic compositional complexity in atmospheric particles on a molecular level.

Acknowledgements

We would like to thank Olalekan Popoola at University of Cambridge for helping to improve the Mathematica code for mass spectral data evaluation. We thank Salim Azzi, John Poole, Paul Lees and Amey plc in general for access to the Queensway Tunnel. PP acknowledges financial support from the University of Birmingham. Financial support by ERC grant 279405 is acknowledged.

Notes and references

1 J. Jimenez, M. Canagaratna, N. Donahue, A. Prevot, Q. Zhang, J. Kroll, P. DeCarlo, J. Allan, H. Coe and N. Ng, *Science*, 2009, **326**, 1525–1529.
2 M. Kanakidou, J. Seinfeld, S. Pandis, I. Barnes, F. Dentener, M. Facchini, R. V. Dingenen, B. Ervens, A. Nenes and C. Nielsen, *Atmos. Chem. Phys.*, 2005, **5**, 1053–1123.
3 A. Q. E. GROUP, *Fine Particulate Matter ($PM_{2.5}$) in the United Kingdom*, Department for Environment, Food and Rural Affairs, London, 2012.
4 S. Platt, I. E. Haddad, S. Pieber, R.-J. Huang, A. Zardini, M. Clairotte, R. Suarez-Bertoa, P. Barmet, L. Pfaffenberger and R. Wolf, *Nat. Commun.*, 2014, **5**, 3749.
5 D. R. Gentner, G. Isaacman, D. R. Worton, A. W. Chan, T. R. Dallmann, L. Davis, S. Liu, D. A. Day, L. M. Russell and K. R. Wilson, *Proc. Natl. Acad. Sci. U. S. A.*, 2012, **109**, 18318–18323.
6 P. Pant and R. M. Harrison, *Atmos. Environ.*, 2013, **77**, 78–97.
7 V. Franco, M. Kousoulidou, M. Muntean, L. Ntziachristos, S. Hausberger and P. Dilara, *Atmos. Environ.*, 2013, **70**, 84–97.
8 D. Brugge, J. L. Durant and C. Rioux, *Environ. Health*, 2007, **6**, 23.
9 N. Künzli, R. Kaiser, S. Medina, M. Studnicka, O. Chanel, P. Filliger, M. Herry, F. Horak, V. Puybonnieux-Texier and P. Quénel, *Lancet*, 2000, **356**, 795–801.
10 U. Pöschl and M. Shiraiwa, *Chem. Rev.*, 2015, **115**, 4440–4475.
11 I. J. Keyte, R. M. Harrison and G. Lammel, *Chem. Soc. Rev.*, 2013, **42**, 9333–9391.
12 K. Slezakova, D. Castro, C. Delerue-Matos, M. da Conceição Alvim-Ferraz, S. Morais and M. do Carmo Pereira, *Atmos. Res.*, 2013, **127**, 141–147.
13 R. M. Harrison, R. Tilling, M. a. S. C. Romero, S. Harrad and K. Jarvis, *Atmos. Environ.*, 2003, **37**, 2391–2402.
14 D. Smith and R. M. Harrison, *Atmos. Environ.*, 1996, **30**, 2513–2525.
15 C. P. Rüger, T. Miersch, T. Schwemer, M. Sklorz and R. Zimmermann, *Anal. Chem.*, 2015, **87**(13), 6493–6499.
16 F. Cavalli, M. Viana, K. Yttri, J. Genberg and J.-P. Putaud, *Atmos. Meas. Tech.*, 2010, **3**, 79–89.
17 I. Kourtchev, S. Fuller, J. Aalto, T. M. Ruuskanen, M. W. McLeod, W. Maenhaut, R. Jones, M. Kulmala and M. Kalberer, *Environ. Sci. Technol.*, 2013, **47**, 4069–4079.

18 I. Kourtchev, I. O'Connor, C. Giorio, S. Fuller, K. Kristensen, W. Maenhaut, J. Wenger, J. Sodeau, M. Glasius and M. Kalberer, *Atmos. Environ.*, 2014, **89**, 525–532.
19 N. W. Green and E. M. Perdue, *Anal. Chem.*, 2015, **87**, 5086–5094.
20 I. Kourtchev, J.-F. Doussin, C. Giorio, B. Mahon, E. Wilson, N. Maurin, E. Pangui, D. Venables, J. Wenger and M. Kalberer, *Atmos. Chem. Phys.*, 2015, **15**, 5683–5695.
21 C. M. Whitehouse, R. N. Dreyer, M. Yamashita and J. B. Fenn, *Anal. Chem.*, 1985, **57**, 675–679.
22 M. Kandiah and P. L. Urban, *Chem. Soc. Rev.*, 2013, **42**, 5299–5322.
23 S. Banerjee and S. Mazumdar, *Int. J. Anal. Chem.*, 2012, **2012**, 282574.
24 L. Molnárné Guricza and W. Schrader, *J. Mass Spectrom.*, 2015, **50**, 549–557.
25 J. D. Surratt, J. H. Kroll, T. E. Kleindienst, E. O. Edney, M. Claeys, A. Sorooshian, N. L. Ng, J. H. Offenberg, M. Lewandowski and M. Jaoui, *Environ. Sci. Technol.*, 2007, **41**, 517–527.
26 P. Schmitt-Kopplin, A. Gelencser, E. Dabek-Zlotorzynska, G. Kiss, N. Hertkorn, M. Harir, Y. Hong and I. Gebefügi, *Anal. Chem.*, 2010, **82**, 8017–8026.
27 S. A. Nizkorodov, J. Laskin and A. Laskin, *Phys. Chem. Chem. Phys.*, 2011, **13**, 3612–3629.
28 Y.-H. Lin, Z. Zhang, K. S. Docherty, H. Zhang, S. H. Budisulistiorini, C. L. Rubitschun, S. L. Shaw, E. M. Knipping, E. S. Edgerton and T. E. Kleindienst, *Environ. Sci. Technol.*, 2012, **46**, 250–258.
29 F. W. McLafferty and F. Tureček, *Interpretation of mass spectra*, University Science Books, 1993.
30 I. Kourtchev, S. Fuller, C. Giorio, R. Healy, E. Wilson, I. O'Connor, J. Wenger, M. McLeod, J. Aalto and T. Ruuskanen, *Atmos. Chem. Phys.*, 2014, **14**, 2155–2167.
31 H. Tervahattu, J. Juhanoja and K. Kupiainen, *J. Geophys. Res.*, 2002, **107**, 4319.
32 B. R. Simoneit, R. Cox and L. Standley, *Atmos. Environ.*, 1988, **22**, 983–1004.
33 D. R. Oros and B. R. Simoneit, *Appl. Geochem.*, 2001, **16**, 1545–1565.
34 D. R. Oros and B. R. Simoneit, *Appl. Geochem.*, 2001, **16**, 1513–1544.
35 M. Ehn, J. A. Thornton, E. Kleist, M. Sipilä, H. Junninen, I. Pullinen, M. Springer, F. Rubach, R. Tillmann and B. Lee, *Nature*, 2014, **506**, 476–479.
36 B. Koch and T. Dittmar, *Rapid Commun. Mass Spectrom.*, 2006, **20**, 926–932.
37 M. M. Yassine, M. Harir, E. Dabek-Zlotorzynska and P. Schmitt-Kopplin, *Rapid Commun. Mass Spectrom.*, 2014, **28**, 2445–2454.
38 J. H. Kroll, N. M. Donahue, J. L. Jimenez, S. H. Kessler, M. R. Canagaratna, K. R. Wilson, K. E. Altieri, L. R. Mazzoleni, A. S. Wozniak and H. Bluhm, *Nat. Chem.*, 2011, **3**, 133–139.
39 A. Cecinato, V. Di Palo, D. Pomata, M. C. T. Scianò and M. Possanzini, *Chemosphere*, 2005, **59**, 679–683.
40 Y. Zhang, L. Müller, R. Winterhalter, G. Moortgat, T. Hoffmann and U. Pöschl, *Atmos. Chem. Phys.*, 2010, **10**, 7859–7873.
41 W. E. Shafer and J. Schönherr, *Ecotoxicol. Environ. Saf.*, 1985, **10**, 239–252.
42 J. Tremp, P. Mattrel, S. Fingler and W. Giger, *Water, Air, Soil Pollut.*, 1993, **68**, 113–123.
43 Z. Kitanovski, I. Grgić, F. Yasmeen, M. Claeys and A. Čusak, *Rapid Commun. Mass Spectrom.*, 2012, **26**, 793–804.

44 Y. Iinuma, O. Böge, R. Gräfe and H. Herrmann, *Environ. Sci. Technol.*, 2010, **44**, 8453–8459.
45 M. Claeys, R. Vermeylen, F. Yasmeen, Y. Gómez-González, X. Chi, W. Maenhaut, T. Mészáros and I. Salma, *Environ. Chem.*, 2012, **9**, 273–284.
46 X. Zhang, Y.-H. Lin, J. D. Surratt and R. J. Weber, *Environ. Sci. Technol.*, 2013, **47**, 3685–3693.
47 I. J. Keyte, *The concentrations, behaviour and fate of polycyclic aromatic hydrocarbons (PAHs) and their oxygenated and nitrated derivatives in the urban atmosphere*, University of Birmingham, 2015.

Faraday Discussions

PAPER

The characterisation of diesel exhaust particles – composition, size distribution and partitioning

Mohammed S. Alam,[a] Soheil Zeraati-Rezaei,[b] Christopher P. Stark,[a] Zhirong Liang,[b] Hongming Xu[b] and Roy M. Harrison†*[a]

Received 20th November 2015, Accepted 12th January 2016
DOI: 10.1039/c5fd00185d

A number of major research questions remain concerning the sources and properties of road traffic generated particulate matter. A full understanding of the composition of primary vehicle exhaust aerosol and its contribution to secondary organic aerosol (SOA) formation still remains elusive, and many uncertainties exist relating to the semi-volatile component of the particles. Semi-Volatile Organic Compounds (SVOCs) are compounds which partition directly between the gas and aerosol phases under ambient conditions. The SVOCs in engine exhaust are typically hydrocarbons in the C_{15}–C_{35} range, and are largely uncharacterised because they are unresolved by traditional gas chromatography, forming a large hump in the chromatogram referred to as Unresolved Complex Mixture (UCM). In this study, thermal desorption coupled to comprehensive Two Dimensional Gas-Chromatography Time-of-Flight Mass-Spectrometry (TD-GC × GC-ToF-MS) was exploited to characterise and quantify the composition of SVOCs from the exhaust emission. Samples were collected from the exhaust of a diesel engine, sampling before and after a diesel oxidation catalyst (DOC), while testing at steady state conditions. Engine exhaust was diluted with air and collected using both filter and impaction (nano-MOUDI), to resolve total mass and size resolved mass respectively. Adsorption tubes were utilised to collect SVOCs in the gas phase and they were then analysed using thermal desorption, while particle size distribution was evaluated by sampling with a DMS500. The SVOCs were observed to contain predominantly *n*-alkanes, branched alkanes, alkyl-cycloalkanes, alkyl-benzenes, PAHs and various cyclic aromatics. Particle phase compounds identified were similar to those observed in engine lubricants, while vapour phase constituents were similar to those measured in fuels. Preliminary results are presented illustrating differences in the particle size distribution and SVOCs composition when collecting samples with and without a DOC.

[a]*Division of Environmental Health and Risk Management, School of Geography, Earth and Environmental Sciences, University of Birmingham, Edgbaston, Birmingham B15 2TT, UK. E-mail: r.m.harrison@bham.ac.uk; Fax: +44 (0)121 414 3708; Tel: +44 (0)121 414 3494*

[b]*School of Mechanical Engineering, University of Birmingham, Edgbaston, Birmingham B15 2TT, UK*

† Also at: Department of Environmental Sciences/Center of Excellence in Environmental Studies, King Abdulaziz University, PO Box 80203, Jeddah, 21589, Saudi Arabia.

The results indicate that the DOC tested is of very limited efficiency, under the studied engine operating conditions, for removal of SVOCs, especially at the upper end of the molecular weight range.

1.0 Introduction

Urban air pollution is comprised of a highly complex mixture of compounds both in the vapour and particulate phase. These complex mixtures can enter the atmosphere through various natural and anthropogenic processes such as incomplete combustion of fossil fuels, biomass burning and forest fires and may also be formed as products of gas to particle conversion through atmospheric reactions. Urban areas therefore, provide a natural focus for research on air quality as they experience high emissions from space heating and road transport associated with their high population densities.[1] Road traffic generated particulate matter (PM) has gained major interest over the last few decades owing to their associated detrimental impacts on human health.[2–4] PM emissions from road vehicles can be generated from exhaust and non-exhaust emission types; for example, exhaust tailpipe and abrasion of vehicle parts including brakes, clutch and tyres, and re-suspension of dust.[5] There are currently more than 37 million vehicles on UK roads. In 1994 the UK diesel passenger cars comprised 7.4% of the fleet and increased to 34.5% of the total by 2013.[6] This means that approximately 1 in 3 light duty vehicles in the UK use diesel engines. Furthermore, in 2014, 50% of all new registration vehicles in the UK were diesel, while 48% were petrol and 2% were alternative fuel vehicles. This increase in diesel engines has led to predictions that diesel fuel will be the number one transport fuel in the world by 2020.[7]

Primary vehicle exhaust emissions can exist in the gas phase or as liquids or solids in aerosols, where the partitioning between the phases is determined by the compound's vapour pressure, its occurrence as a pure substance or as a mixture in the condensed phase and its water solubility. Once emitted into the atmosphere the compounds with vapour pressures of $>10^{-5}$ atm, 10^{-5} to 10^{-11} atm and $<10^{-11}$ atm, at ambient temperature are expected to be volatile (in the gas phase), semi volatile and non-volatile (in the condensed phase), respectively. A semi-volatile organic compound (SVOC) includes any compound with >1% of its mass in both the condensed and vapour phases.[8] The composition of primary vehicle exhaust aerosol and its contribution to secondary organic aerosol (SOA) formation still remains elusive, where many uncertainties exist relating to the semi-volatile component of the particles.

Diesel exhaust particles consist mainly of agglomerated solid carbonaceous material and ash, together with organic and sulphur containing compounds.[9] The solid carbon is formed during combustion and is subsequently oxidised, while a small fraction of the fuel and motor/lubricating oil are not oxidised and are emitted as VOC/SVOCs in the exhaust. The fraction associated with unburned fuel and lubricating oil is known to vary with engine design and operating conditions, ranging from <10% to >90%.[9] Many studies have investigated the composition of the organic fraction of the particles, reporting targeted groups of compounds such as aliphatic-alkanes, cyclic-alkanes, polycyclic aromatic hydrocarbons (PAHs), steranes and hopanes.[10–14] Limited studies, however, focus on higher carbon number species (>C_{15}) owing to instrumental/experimental limitations.[15,16] The SVOCs in diesel exhaust emissions are typically hydrocarbons in the C_{15}–C_{25} range, and are largely uncharacterised because they are unresolved by traditional gas chromatography and form a large hump in the chromatogram referred to as Unresolved Complex Mixture (UCM). To add further complexity, the composition of exhaust aerosol is also dependent upon the sampling collection method. When collecting diesel exhaust aerosol, it is diluted and cooled, which leads to the transformation of gas phase volatile species in the tail pipe into solid and liquid PM.[17,18] Dilution and cooling processes therefore determine the relative amounts of species that adsorb and/or condense onto existing particles and nucleate to form new particles.[19–21]

The size distribution of diesel PM has been extensively studied.[9,21–27] The majority of the particle mass exists in the 100–300 nm diameter range (accumulation mode) and is where the carbonaceous agglomerates and associated adsorbed materials are found. Approximately 1–20% of the particle mass and >90% of the particle number exists in the 5–50 nm diameter range (nucleation mode) and is where the volatile organic and sulphur containing compounds that form during exhaust dilution and cooling are assumed to reside.[9] Particles with diameter > 2.5 μm (coarse mode) comprise typically 20–50% of the particle mass in the PM_{10} size range. The physico-chemical properties of particles influence their residence time in the atmosphere, and determine their surface area which governs their availability for atmospheric reactions, optical properties of particles which influence atmospheric visibility and light scattering, and human health effects.

This paper presents the use of thermal desorption coupled to comprehensive Two Dimensional Gas-Chromatography Time-of-Flight Mass-Spectrometry (TD-GC × GC-ToF-MS) to characterise and quantify the composition of SVOCs from diesel exhaust emissions, both in the gas and particulate phases. The composition was also investigated for different particle size fractions using collection by cascade impaction (nano-MOUDI), while particle number size distribution was evaluated by sampling with a DMS500.

2.0 Experimental

2.1 Engine sampling

Experiments were conducted with a light-duty diesel engine; 2.2 L, 4-cylinder inline compression ignition (CI) engine equipped with a common rail direct injection system, a variable-nozzle-turbine (VNT) turbocharger and a diesel oxidation catalyst (DOC). The open engine control unit (Open-ECU) allows full control over the engine operating parameters. A schematic of the engine testing system is illustrated in Fig. 1.

30 min samples were collected at steady state engine operating conditions at a load of 5.9 bar brake mean effective pressure (BMEP) and speed of 1800 revolutions per minute (RPM), before and after the DOC. Fuel injection pressure was maintained at 1000 bar and a pilot-main fuel injection strategy was used. Exhaust gas recirculation (EGR) percentage was fixed at 12%. Sampling was started after ensuring that the engine was fully warmed-up and thermally stable. Standard European, EN590 specifications, ultra-low sulphur diesel (ULSD, $S < 10$ ppm) and 5W-30 part synthetic engine lubricating oil were used for the experiments.

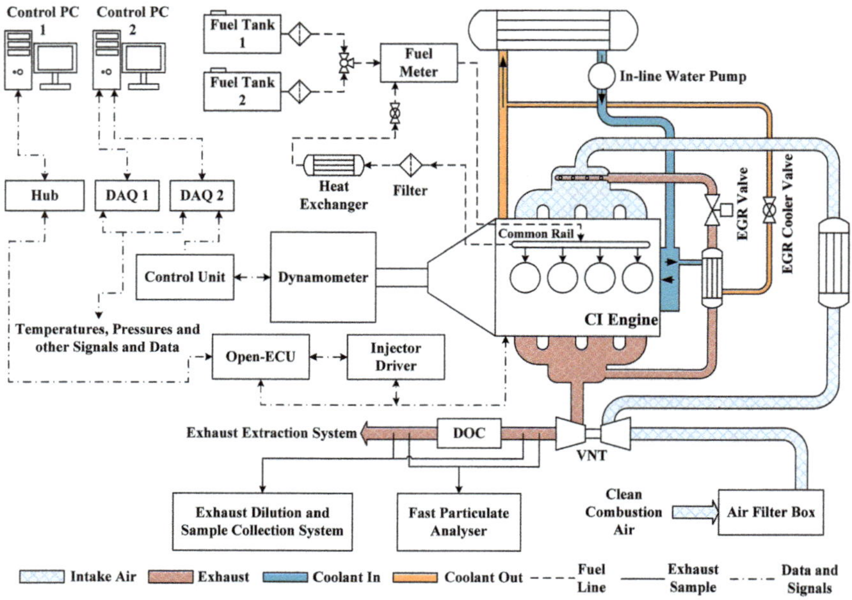

Fig. 1 Schematic of the engine testing system.

Particulate number and size distribution were measured by a Cambustion DMS500 MKII fast particulate analyser and averaged over 1 minute for each sample with 4 samples per experiment. A schematic of the in-house exhaust dilution and sample collection system is shown in Fig. 2.

The dilution system incorporates a modified TSI 3302A diluter. The diluter was modified to use cleaned compressed air by using a high flow rate moisture trap and two HEPA filters. The undiluted hot exhaust sample was delivered *via* a heated line maintained at 191 °C and subsequently mixed with the dilution air at ambient temperature. The dilution ratio (DR) was kept at 1 : 30 ± 5 confirmed by measurements of nitric oxide (NO) before and after the dilution system. The diluted sample was divided into two flow streams, (see Fig. 2, labelled 1 and 2). A flow rate of 9 L min^{-1} was fixed for stream 1, created by a vacuum pump. A calibrated rotameter and needle valve combination was used to measure the flow rate, and was removed while sampling to avoid particle losses. The second rotameter downstream of the adsorption tube was fixed to ensure the flow rates remained constant throughout the engine sampling period. A nano-Micro Orifice Uniform Deposit Impactor (nanoMOUDI) 125 instrument (MSP Copley Scientific, UK) was utilised for collecting different size particles *via* this flow stream. The nanoMOUDI has fourteen stages with nominal 50% cut-points of 10 000, 5600, 3200, 1800, 1000, 560, 320, 180, 100, 56, 32, 18, 10 and <10 nm. The sample was collected for 30 min, using polypropylene backed PTFE 47 mm filters (Whatman, Maidstone, UK). Stream 2, demonstrated a fixed flow rate of 1.8 L min^{-1}. Stainless steel thermal adsorption tubes packed with 1 cm quartz wool, 300 mg Carbograph 2TD 40/60 (Markes International) were used to collect gas phase constituents from stream 2; downstream of a polypropylene backed PTFE 47 mm filter (Whatman, Maidstone, UK) to remove any particulates. The temperatures at the sampling points were 25 ± 5 °C.

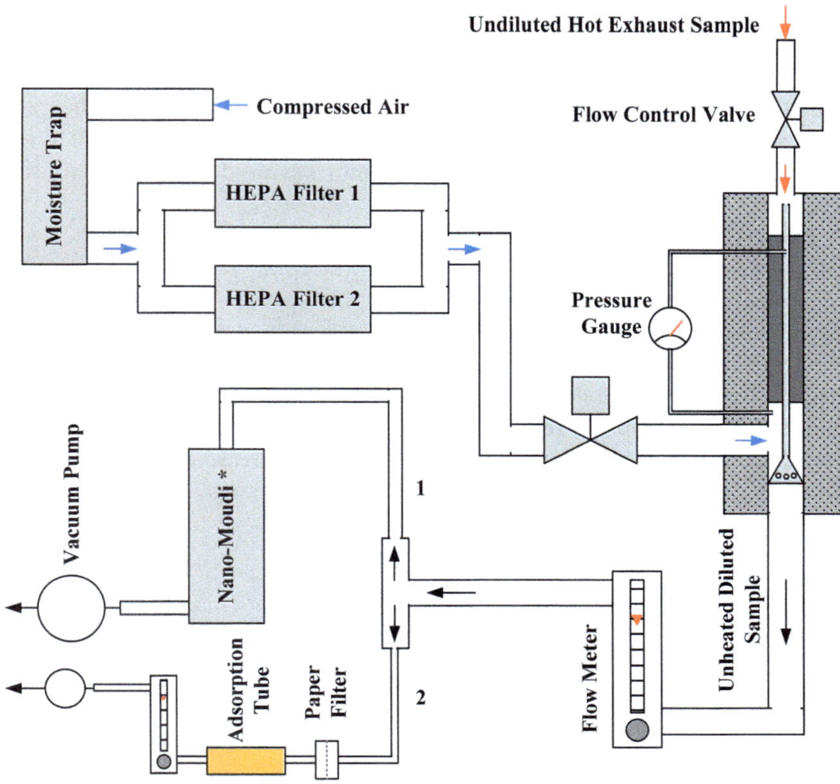

Fig. 2 Schematic of the exhaust dilution and sample collection system.

2.2 Analytical instrumentation

Adsorption tubes were desorbed using thermal desorption (Unity 2, Markes International, Llantrisant, UK) and samples were subsequently analysed using a gas chromatograph (GC, 7890A, Agilent Technologies, Wilmington, DE, USA) equipped with a Zoex ZX2 cryogenic modulator (Houston, TX, USA). The first dimension was equipped with a SGE DBX5, non-polar capillary column (30 m, 0.25 mm ID, 0.25 μm – 5% phenyl polysilphenylene-siloxane), and the second dimension column equipped with a SGE DBX50 (4.0 m, 0.1 mm ID, 0.1 μm – 50% phenyl polysilphenylene-siloxane). The GC × GC was interfaced with a Bench-ToF-Select, time-of-flight mass spectrometer (ToF-MS, Markes International, Llantrisant, UK). The scan speed was 50 Hz with a mass range of 35 to 600 *m*/*z*. Ionisation energies that were explored in this work were 14 eV and 70 eV. All data produced were processed using GC Image v2.3 (Zoex Corporation, Houston, US).

2.3 Analysis of samples

The filters were spiked with 50 μL of 1 ng μL^{-1} deuterated internal standards for quantification. Filters were immersed in dichloromethane (DCM), ultrasonicated for 20 min at 20 °C. The extract was concentrated to 50 μL under a gentle flow of nitrogen for analysis using GC × GC-ToF-MS. 1 μL of the extracted sample was injected in a split ratio 100 : 1 at 300 °C. The initial temperature of the primary oven (120 °C) was held for 2 min and then increased at 2 °C min^{-1} to 210 °C, followed by 1.5 °C min^{-1} to 325 °C. The initial temperature of the secondary oven (120 °C) was held for 2 min, and then increased at 3 °C min^{-1} to 200 °C, followed by 2 °C min^{-1} to 300 °C and a final increase of 1 °C min^{-1} to 330 °C to ensure all species passed through the column. The transfer line temperature was 330 °C and the ion source temperature was 280 °C. Helium was used as the carrier gas at a constant flow rate of 1 mL min^{-1}.

Adsorption tubes were desorbed using TD. Briefly, the tubes were spiked with 1 ng of deuterated internal standard for quantification, and desorbed onto the cold trap at 350 °C for 15 min (trap held at 20 °C). The trap was then purged onto the column in a split ratio of 102 : 1 at 350 °C and held for 4 min. The initial temperature of the primary oven (90 °C) was held for 2 min and then increased to 2 °C min^{-1} to 240 °C, followed by 3 °C min^{-1} to 310 °C and held for 5 min. The initial temperature of the secondary oven (40 °C) was held for 2 min and then increased at 3 °C min^{-1} to 250 °C, followed by an increase of 1.5 °C min^{-1} to 315 °C and held for 5 min. Helium was used as the carrier gas at a constant flow rate of 1 mL min^{-1}.

3.0 Results and discussion

A typical two dimensional separation for diesel fuel and an engine lubricant is shown in Fig. 3A and B, respectively. All compounds were separated by volatility in the *x*-axis (first dimension) and by polarity in the *y*-axis (second dimension). Peak identification was based on retention indices and mass spectral data from the (National Institute of Standards and Technology) NIST library. The GC × GC conditions for the two contour plots illustrated in Fig. 3A and B are outlined in Section 2.3, similar to the filters, and are not optimum conditions for separating diesel or motor oil samples. The purpose for this was to get an indication of the positioning of compounds in the contour plots that are present in the filters and adsorption tubes that may come from diesel fuel and/or engine oil. A detailed investigation of the separation of engine oils and diesel fuel using GC × GC-ToF-MS is given elsewhere.[28-30] Fig. 3A shows that diesel fuel contains up to C_{20} compounds which are represented in the first portion of the contour plot, whereas engine oil samples are dominated by C_{18}–C_{36} compounds forming a UCM in the latter portion of the contour plot (Fig. 3B). The presence of both these UCMs are observed in transient diesel engine exhaust filter samples using the same GC × GC operating conditions (data not shown here) indicating the importance in identifying constituents in diesel and lubricating oils, which are both precursors of engine exhaust emissions. Sutton *et al.* (2005) isolated individual hydrocarbon fractions from a UCM of biodegraded crude oil and suggested that the entire oil UCM could contain approximately 250 000 unidentified compounds.[31]

Stainless steel thermal adsorption tubes were used to collect gas phase emissions and analysed using TD-GC × GC-ToF-MS. The chromatogram for a 30 min sample collected at steady state engine operating conditions at a load of 5.9 bar BMEP and speed of 1800 RPM, before the DOC is shown in Fig. 4. The 3D chromatographic image indicates an abundance of gas phase constituents up to C_{22}, although *n*-alkanes are present up to C_{33} ($RT_1 > 60$ min, not shown in Fig. 4). The chromatogram is very similar to Fig. 1A, the diesel fuel, indicating that the

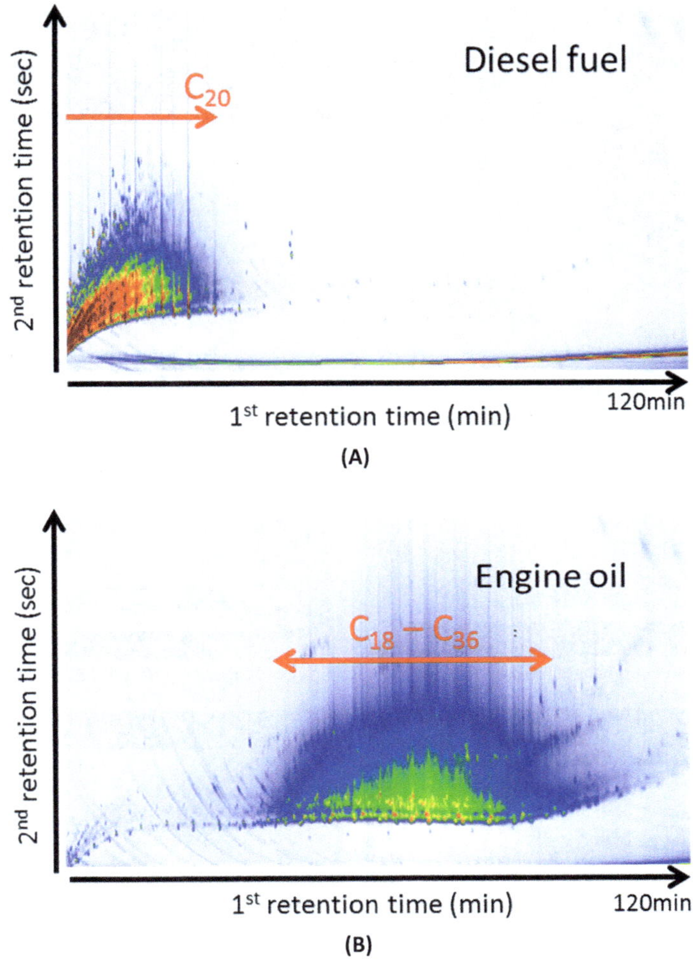

Fig. 3 A two dimensional separation (contour plot) of (A) diesel fuel, and (B) engine lubricant. *x*-Axis separation on a volatility basis; *y*-axis separation on a polarity basis. Each coloured spot represents an individual species and has a corresponding full mass spectrum. The intense red spots are major peaks, while the blue spots represent less abundant peaks.

majority of the gas phase hydrocarbon engine emission is from the diesel fuel. Recently, Dunmore *et al.* (2015) showed how diesel related hydrocarbons could dominate gas phase reactive carbon in urban environments.[32] Their study showed that approximately 60% of the winter primary hydrocarbon hydroxyl radical reactivity was from diesel related hydrocarbons and that they could contribute to up to 50% of the ozone production potential. The authors measured up to C_{13} and estimated the unmeasured diesel emissions (up to C_{22}) using fuel composition-based emission factors from Gentner *et al.* (2013).[33]

Approximately 12 000 peaks were found in the chromatography illustrated in Fig. 4A for the gas phase samples compared to 8000 for the diesel fuel (Fig. 1A). Compounds that were identified include straight and branched alkanes (C_{11}–C_{33}) (Fig. 4B), cycloalkanes (C_{11}–C_{25}) (Fig. 4C), C_9–C_{12} aromatics (Fig. 4D and E),

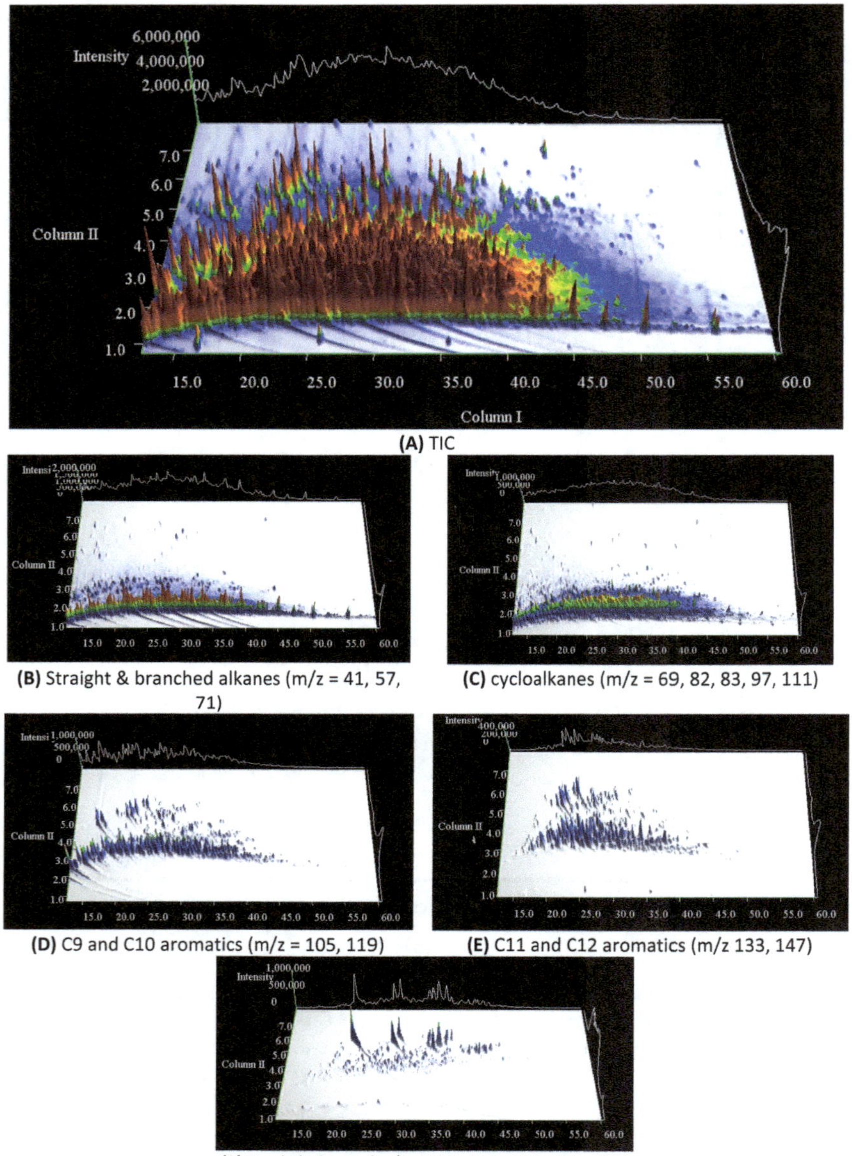

Fig. 4 3 dimensional separations of gas phase constituents collected with an adsorption tube and analysed using TD-GC × GC-ToF-MS. (A) Total ion chromatogram (B)–(F) selected ion chromatograms. The white solid line signifies 1D chromatography.

naphthalenes (Fig. 4F), as well as aldehydes, ketones, alkanoic acids, alkyl benzenes, nitriles, alkyl phenyl ketones, alkyl-biphenyls, furans, furanones, decalins, PAHs and sulphur containing compounds including benzothiazole and diphenyl sulfone. Compounds not found in the NIST library were further investigated and tentatively identified using the molecular mass of species found at the same retention times from 14 eV ionisation mass spectral data.

In addition to the qualitative data above, preliminary quantitative data for *n*-alkanes, *n*-alkylcylohexanes and PAHs were calculated in order to be used as surrogates for modelling evaporation and composition changes of traffic induced ultrafine particles.[34] These quantitative data are shown in Table 1 for total PM (gas phase, particulate and total concentrations) and compared to those data measured after the DOC. The DOC is designed to limit the amount of harmful emissions released to the environment from diesel engines. Its primary functions are the oxidation of CO, unburned hydrocarbons and NO, and it can also provide some oxidation of soluble organic fraction (SOF) components to improve soot filter performance.[35] Diesel oxidation catalysts typically reduce emissions of PM by 20–40% or more and gaseous emissions by 50–70%.[36] Table 1 shows the concentration reduction of *n*-alkanes when measuring these species after the DOC, with a total percentage decrease for C_{11} to C_{33} of 19, 28 and 24% in the vapour, particulate and the sum of the vapour and particulate phases, respectively. These data, for alkanes, suggest that the efficiency of the DOC for higher molecular weight compounds (HMW, *e.g.* >C_{28}), is less than for lower molecular weight compounds (LMW). The total reduction of the sum of vapour and particulate phases for the HMW (C_{28}–C_{32}) was <10%. The results are, however, indicative of some re-partitioning of the particulate component into the vapour phase.

The efficiency of reducing the particulate emissions for both *n*-alkylcyclohexanes and *n*-alkanes is greater than for the gaseous phase. The *n*-alkylcyclohexane compounds show a total reduction of 19, 27 and 21% after the DOC, for the vapour, particulate and the sum of the vapour and particulate phases, respectively. For C_{12}–C_{26} compounds the total (sum of vapour and particulate phase) concentration reductions were 2592 to 2050 ng m^{-3}. The concentration reductions of individual compounds in the vapour phase are few, whereas all concentrations in the particulate phase are reduced efficiently by >20% (for LMW compounds, >C_{20}) and to a lesser extent for HMW (>5% for >C_{21}). Since the DOC is expected predominantly to adsorb and oxidise vapour, the results appear to reflect very active re-partitioning between the particulate and vapour phases.

PAH concentrations illustrate a larger total reduction of 28, 26 and 27% after the DOC for the vapour, particulate and the sum of the vapour and particulate phases, respectively. This reduction is more pronounced for the species that are more likely to be in the gas phase (NAP to ANT).[37,38] The greater reduction of ANT (68%) in comparison to other PAHs in the presence of the DOC may explain why roadside concentrations of ANT are generally low in the UK in comparison to some developing countries.[39,40] For HMW PAHs (PYR to B(a)P) the efficiency of the DOC is reduced for the vapour phase but appears efficient for the particulate phase (see Table 1).

A nanoMOUDI 125 instrument (MSP Copley Scientific, UK) was utilised for collecting different size particles. Each of the fourteen stages of the nanoMOUDI with nominal 50% cut-points of 10 000, 5600, 3200, 1800, 1000, 560, 320, 180, 100, 56, 32, 18, 10 and <10 nm, were analysed using GC × GC-ToF-MS. Approximately 1500 peaks were found in the majority of each of the 13 stage two-dimensional chromatograms, using the blob detection technique performed by GC Image v2.3 (Zoex Corporation, Houston, US) software, with a UCM similar to that found in Fig. 1B, engine lubricant; consistent with other studies.[41] Compounds identified included straight and branched chain alkanes (C_{12}–C_{33}), cycloalkanes (C_{11}–C_{25}),

Table 1 Concentrations of *n*-alkanes, *n*-alkylcyclohexanes and PAHs in the vapour phase, particulate phase and sum of vapour and particulate phase, before and after the DOC, and their percentage reductions

	Vapour phase			Particulate phase			Total (v + p)		
	Before DOC	After DOC	% reduction	Before DOC	After DOC	% reduction	Before DOC	After DOC	% reduction
***n*-Alkanes (μg m^{-3})**									
C_{12}	1.07	0.87	18%	0.15	0.05	70%	1.21	0.92	24%
C_{13}	1.07	0.88	18%	0.47	0.30	37%	1.54	1.18	24%
C_{14}	0.90	0.72	20%	0.44	0.28	37%	1.34	1.00	25%
C_{15}	1.00	0.99	1%	0.50	0.27	46%	1.50	1.26	16%
C_{16}	0.62	0.67	−8%	0.51	0.38	24%	1.13	1.05	7%
C_{17}	0.34	0.46	−34%	0.32	0.14	57%	0.67	0.60	11%
C_{18}	0.24	0.20	17%	0.36	0.26	27%	0.59	0.46	23%
C_{19}	0.21	0.16	24%	0.22	0.16	27%	0.44	0.33	25%
C_{20}	0.20	0.10	48%	0.26	0.20	21%	0.46	0.31	33%
C_{21}	0.23	0.05	79%	0.31	0.25	21%	0.54	0.29	45%
C_{22}	0.29	0.06	78%	0.26	0.22	16%	0.55	0.28	49%
C_{23}	0.34	0.07	80%	0.31	0.18	41%	0.65	0.25	62%
C_{24}	0.29	0.09	69%	0.26	0.24	9%	0.56	0.33	40%
C_{25}	0.21	0.08	62%	0.48	0.37	22%	0.68	0.45	34%
C_{26}	0.19	0.10	49%	0.45	0.32	29%	0.64	0.42	35%
C_{27}	0.10	0.11	−8%	0.51	0.28	45%	0.61	0.39	36%
C_{28}	0.10	0.15	−46%	0.49	0.39	21%	0.59	0.54	9%
C_{29}	0.13	0.16	−26%	0.50	0.44	12%	0.63	0.60	4%
C_{30}	0.08	0.14	−72%	0.49	0.41	17%	0.57	0.55	4%
C_{31}	0.10	0.13	−38%	0.42	0.35	17%	0.52	0.48	7%
C_{32}	0.07	0.08	−18%	0.33	0.30	8%	0.39	0.38	4%
C_{33}	0.00	0.00	n/a	0.24	0.19	21%	0.24	0.19	21%
Total	*7.77*	*6.28*	*19%*	*8.28*	*5.97*	*28%*	*16.1*	*12.2*	*24%*
***n*-Alkyl-cyclohexane (ng m^{-3})**									
C_{12}	659.0	540.0	18%	0.0	0.0	n/a	659.0	540.0	18%
C_{13}	295.0	161.0	45%	0.0	0.0	n/a	295.0	161.0	45%
C_{14}	331.0	263.0	21%	33.4	15.1	55%	364.0	279.0	23%
C_{15}	179.0	177.0	1%	61.2	35.6	42%	240.0	212.0	12%
C_{16}	146.0	169.0	−16%	86.7	25.9	70%	233.0	195.0	16%
C_{17}	57.0	49.0	14%	55.0	41.8	24%	112.0	91.0	19%
C_{18}	29.0	35.0	−21%	27.1	12.4	54%	56.0	47.0	16%
C_{19}	12.0	25.0	−108%	63.2	36.0	43%	75.0	61.0	19%
C_{20}	11.0	15.0	−36%	52.1	41.8	20%	63.0	57.0	10%
C_{21}	33.0	17.0	48%	67.7	64.1	5%	101.0	81.0	20%
C_{22}	12.0	13.0	−8%	74.2	69.8	6%	87.0	83.0	5%
C_{23}	37.0	24.0	35%	85.0	82.1	3%	122.0	106.0	13%
C_{24}	14.0	18.0	−29%	58.6	50.0	15%	72.0	68.0	6%
C_{25}	42.0	5.0	88%	49.0	46.1	6%	91.0	51.0	44%
C_{26}	0.0	0.0	n/a	21.3	18.8	12%	21.0	19.0	10%
Total	*1857*	*1511*	*19%*	*734*	*539*	*27%*	*2592*	*2050*	*21%*
PAH (ng m^{-3})									
NAP	402.0	281.0	30%	0.0	0.0	n/a	402.0	281.0	30%
ACEY	71.0	47.5	33%	0.0	0.0	n/a	71.0	47.5	33%
ACE	42.7	32.5	24%	0.0	0.0	n/a	42.7	32.5	24%
FLU	38.8	30.6	21%	0.0	0.0	n/a	38.8	30.6	21%

Table 1 (*Contd.*)

	Vapour phase			Particulate phase			Total (v + p)		
	Before DOC	After DOC	% reduction	Before DOC	After DOC	% reduction	Before DOC	After DOC	% reduction
PHE	31.9	19.4	39%	1.5	1.2	23%	33.5	20.6	38%
ANT	38.3	12.2	68%	0.3	0.3	23%	38.6	12.4	68%
PYR	16.1	19.6	−21%	17.9	10.8	40%	34.0	30.4	11%
FLR	18.3	21.5	−17%	18.9	10.8	43%	37.1	32.3	13%
CHR	15.1	15.4	−2%	18.7	13.7	26%	33.7	29.1	14%
B(a)A	11.9	12.0	−1%	13.6	11.1	19%	25.5	23.1	10%
B(a)P	16.6	16.6	0%	8.9	7.5	16%	25.5	24.1	5%
B(ghi) P	0.0	0.0	n/a	30.2	26.5	12%	30.2	26.5	12%
Total	*702*	*508*	*28%*	*110*	*81.8*	*26%*	*812*	*590*	*27%*

alkyl benzenes, PAHs, decalins, steranes and hopanes. Sulphur containing compounds that were present in the gas phase were not observed in the particulate phase. The concentration distribution of *n*-alkane compounds throughout the nanoMOUDI is shown in Table 2. Interestingly, the largest concentrations were observed in different size fractions according to MW. For example, for LMW compounds (C_{12}–C_{16}) the largest concentrations measured were in the size fraction 1.0 μm in diameter, whereas for HMW compounds (C_{26}–C_{31}) the largest concentrations were observed in the 0.1 μm diameter particle fraction.

The bold numbers are the largest concentrations found for those carbon number species. There is a clear trend for the maximum concentration of the alkanes to be associated with smaller particles as molecular weight increases. This is probably a consequence of the Kelvin effect which increases the equilibrium vapour pressure, and hence volatility of compounds, as particle surface curvature increases with diminishing particle size. Hence the more volatile lower molecular weight compounds are destabilised, with an increased tendency to partition into the vapour phase as particle size decreases. This result is consistent with the theoretical predictions made within the FASTER project and with the results of thermodenuder experiments.[34,42] Studies have suggested that particles in the 5–100 nm diameter range consist mainly of HMW hydrocarbons associated with engine lubricant,[10,11] which is in broad agreement with the results in this study as the largest concentrations for C_{32} and C_{33} are found in the smaller size ranges (56 nm diameter) compared to other LMW alkanes (see Table 2). Furthermore, roadside aerosol studies using hygroscopic, organic and volatility Tandem Differential Mobility Analyser techniques have shown that particles in the nucleation mode have partly evaporated leaving a 10 nm non-volatile core.[41]

The sum of the total size fractions indicates that the largest concentrations found in the PM are found in the C_{23}–C_{29} region, which also corresponds to the motor oil UCM region shown in Fig. 1B. Schauer *et al.* (1999) reported that the gas phase portion of the UCM consisted of HMW SVOCs, which accounted for slightly less than half that of the total carbonyl mass emission rate, while the particle phase portion of the UCM was similar to motor oil when analysed by GC/MS,

Table 2 Concentrations (ng m^{-3}) of *n*-alkanes (C_{12}–C_{33}) in the particulate phase size fractions 10 μm to 0.01 μm measured using nano-MOUDI 125ii. The bold numbers are the largest concentrations found for those carbon number species

	Size fractions (μm)														
	<0.010	0.010	0.018	0.032	0.056	0.10	0.18	0.32	0.56	1.0	1.8	3.2	5.6	10	Total
C number	Concentrations (ng m^{-3})														
C_{12}	6.61	6.35	9.04	9.07	12.8	13.4	16.4	18.2	19.6	**20.3**	15.9	16.6	17.0	19.6	201.0
C_{13}	4.15	5.19	7.44	8.28	9.72	7.88	15.7	14.4	19.7	**19.8**	16.5	15.7	16.8	19.6	181.0
C_{14}	5.30	5.41	5.29	7.65	9.46	10.5	12.8	12.6	13.3	**19.4**	15.4	14.0	14.3	18.3	164.0
C_{15}	5.06	5.20	6.55	6.46	6.70	8.13	10.7	12.4	14.8	**21.0**	13.7	11.7	13.9	14.8	151.0
C_{16}	3.66	3.02	3.98	6.42	6.47	7.86	7.84	9.70	10.5	**13.5**	10.4	11.8	12.9	13.3	121.0
C_{17}	4.28	4.54	2.71	4.56	5.74	5.92	8.54	11.0	**14.2**	7.83	10.2	10.6	10.9	12.7	114.0
C_{18}	1.17	1.66	1.27	1.74	2.16	3.07	4.68	5.76	**10.7**	8.23	5.28	3.95	8.23	8.43	66.0
C_{19}	1.89	2.69	2.10	2.69	3.52	3.30	5.64	5.97	**9.90**	8.31	5.30	4.52	5.24	5.21	66.0
C_{20}	2.77	2.20	3.87	4.28	7.74	9.37	12.4	**16.0**	7.54	6.82	4.81	4.24	5.93	5.32	93.0
C_{21}	3.22	2.42	4.96	6.11	9.56	12.0	**20.4**	15.1	9.04	5.60	6.64	6.22	6.18	7.29	115.0
C_{22}	3.47	3.47	8.44	12.4	16.2	18.2	**24.2**	21.3	16.0	16.0	9.98	8.20	4.67	2.40	165.0
C_{23}	8.07	6.33	9.53	15.8	17.9	26.1	**30.2**	26.5	17.7	22.4	18.3	9.16	8.60	2.62	219.0
C_{24}	8.22	6.84	20.1	24.7	30.4	46.8	**53.4**	42.7	37.5	37.7	21.4	18.4	15.1	9.11	372.0
C_{25}	11.1	13.1	28.7	33.4	32.6	45.8	**64.2**	47.1	28.4	42.0	30.9	26.0	9.60	11.2	424.0
C_{26}	12.0	8.96	31.1	24.1	43.6	**53.3**	50.2	41.5	19.8	28.7	22.4	21.8	13.1	10.4	381.0
C_{27}	7.78	8.16	28.9	33.3	46.9	**60.3**	55.8	38.8	17.6	30.0	9.54	17.6	8.44	7.78	371.0
C_{28}	12.0	13.6	26.7	25.3	42.3	**51.6**	46.7	29.2	17.8	19.7	4.67	15.1	5.09	8.44	318.0
C_{29}	16.8	26.8	31.1	44.4	36.2	**47.4**	28.0	22.3	12.0	3.99	3.98	19.1	3.30	2.68	298.0
C_{30}	4.67	2.69	7.13	18.2	21.3	**30.1**	22.4	18.4	17.6	15.8	13.5	11.4	5.18	6.11	195.0
C_{31}	2.77	5.17	6.83	10.2	16.7	**18.9**	16.5	17.5	6.83	6.68	8.35	7.00	3.42	1.70	129.0
C_{32}	7.47	2.60	5.13	12.4	**18.5**	10.3	13.4	13.1	5.13	10.8	8.30	4.97	3.57	4.80	120.0
C_{33}	7.06	8.38	7.79	13.8	**16.8**	11.1	9.47	9.10	7.93	7.62	5.91	4.65	2.94	6.18	119.0
Total	140.0	145.0	259.0	325.0	413.0	501.0	**530.0**	449.0	334.0	372.0	261.0	263.0	194.0	198.0	**4384.0**

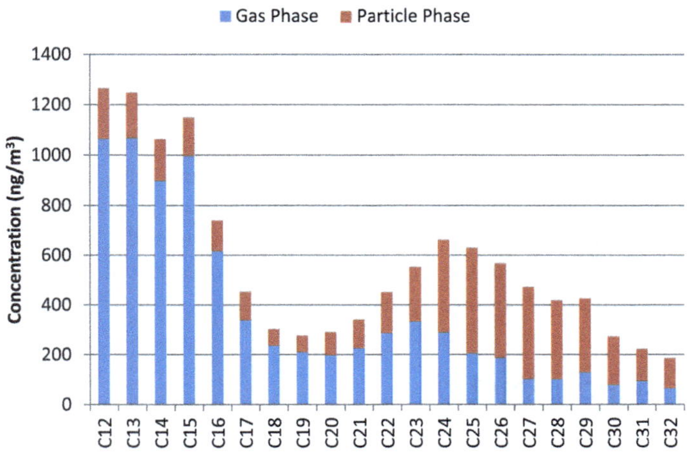

Fig. 5 Gas/particle phase distribution of *n*-alkane concentration.

consistent with the results found in this study.[41] Fig. 5 illustrates the gas/particle phase distribution of *n*-alkane concentrations. A bimodal distribution is observed giving rise to a diesel (<C_{20}) source and an engine lubricant (C_{21}–C_{27}) source. Elevated concentrations for the LMW alkanes exist in the gas phase, while HMW (>C_{24}) are predominantly in the particle phase.

4.0 Conclusion

Thermal desorption coupled to comprehensive Two Dimensional Gas-Chromatography Time-of-Flight Mass-Spectrometry (TD-GC × GC-ToF-MS) was exploited to characterise and quantify the composition of SVOCs from the exhaust emission. Samples were collected from the exhaust of a diesel engine, before and after a diesel oxidation catalyst (DOC). This preliminary dataset shows that the SVOCs were observed to contain predominantly straight and branched alkanes (C_{11}–C_{33}), cycloalkanes (C_{11}–C_{25}), various cyclic aromatics, PAHs, decalins and alkyl benzenes. Sulphur containing compounds were observed only in the gas phase, as were ketones, aldehydes and naphthalenes. Particle phase compounds identified were similar to those observed in engine lubricants, while vapour phase constituents were similar to those measured in fuels. Sampling after the DOC showed total reductions of 19, 19 and 28% for gas phase *n*-alkanes, alkyl-cyclohexanes and PAHs, respectively; although the DOC illustrated less efficiency for HMW compounds (in the gas phase). With current DOC technology claiming 50–70% reduction in emissions in the gas phase alone, this study shows that there is large variation between compound classes and compounds within a compound class, most probably due to reactivity.

Although studies have reported the size distribution and composition of fine PM emitted from diesel engines, with chemical speciation such as sulphate, ammonium, nitrate, EC and OC;[43–45] or have investigated gas and particle phase tailpipe emissions reporting VOCs, SVOCs and particle associated OC,[41] very limited studies have investigated the organic composition of diesel exhaust emissions corresponding to different size fractions. We present the *n*-alkane (C_{12}–C_{33}) concentrations measured in 13 different size fractions, showing an increase in the non-volatile concentration with decreasing size fractions. The mode peak for different alkanes was found to be dependent upon MW, with HMW concentrations peaking at 56–100 nm and LMW concentrations peaking at 1 μm. Gas and particle phase distributions of alkanes were also presented showing an overall (total) bimodal distribution, with >70% in the gas phase for compounds $\leq C_{20}$ and <30% for compounds $\geq C_{20}$. The DOC tested is of very limited efficiency (4–68% for the compounds measured); under the studied engine operating conditions, and appears to cause substantial re-partitioning of compounds from the condensed phase into vapour.

Acknowledgements

This work is part of the FASTER project, sponsored by the European Research Council (ERC). The authors would also like to thank Yasser Al Qahtani for his contribution during engine sampling.

References

1 R. M. Harrison, M. Dall'Osto, D. C. S. Beddows, A. J. Thorpe, W. J. Bloss, J. D. Allan, H. Coe, J. R. Dorsey, M. Gallagher, C. Martin, J. Whitehead, P. I. Williams, R. L. Jones, J. M. Langridge, A. K. Benton, S. M. Ball, B. Langford, C. N. Hewitt, B. Davison, D. Martin, K. F. Petersson, S. J. Henshaw, I. R. White, D. E. Shallcross, J. F. Barlow, T. Dunbar, F. Davies, E. Nemitz, G. J. Phillips, C. Helfter, C. F. Di Marco and S. Smith, *Atmos. Chem. Phys.*, 2012, **12**, 3065–3114.

2 J. L. Mauderly, *Environ. Health Perspect.*, 1994, **102**, 165–171.

3 M. Masiol, A. Hofer, S. Squizzato, R. Piazza, G. Rampazzo and B. Pavoni, *Atmos. Environ.*, 2012, **60**, 375–382.

4 J. Rissler, E. Swietlicki, A. Bengtsson, C. Boman, J. Pagels, T. Sandström, A. Blomberg and J. Löndahl, *J. Aerosol Sci.*, 2012, **48**, 18–33.

5 P. Pant and R. M. Harrison, *Atmos. Environ.*, 2013, **77**, 78–97.

6 SMMT, *The Society of Motor Manufacturers and Traders – Motor Industry Facts 2015*, 2015.

7 ExxonMobil, *The Outlook for Energy: A view to 2040*, Tech. rep., Exxon Mobil Corporation, Texas, 2014.

8 N. M. Donahue, A. L. Robinson, C. O. Stanier and S. N. Pandis, *Environ. Sci. Technol.*, 2006, **40**, 2635–2643.

9 D. B. Kittelson, *J. Aerosol Sci.*, 1998, **29**, 575–588.

10 H. Sakurai, H. J. Tobias, K. Park, D. Zarling, K. S. Docherty, D. B. Kittelson, P. H. McMurry and P. J. Ziemann, *Atmos. Environ.*, 2003, **37**, 1199–1210.

11 H. J. Tobias, D. E. Beving, P. J. Ziemann, H. Sakurai, M. Zuk, P. H. McMurry, D. Zarling, R. Waytulonis and D. B. Kittelson, *Environ. Sci. Technol.*, 2001, **35**, 2233–2243.

12 J. J. Schauer, M. J. Kleeman, G. R. Cass and B. R. T. Simoneit, *Environ. Sci. Technol.*, 1999, **33**, 1578–1587.

13 B. Zielinska, J. Sagebiel, J. D. McDonald, K. Whitney and D. R. Lawson, *J. Air Waste Manage. Assoc.*, 2004, **54**, 1138–1150.

14 L. Huang, S. V. Bohac, S. M. Chernyak and S. A. Batterman, *Atmos. Environ.*, 2015, **102**, 228–238.
15 A. W. H. Chan, G. Isaacman, K. R. Wilson, D. R. Worton, C. R. Ruehl, T. Nah, D. R. Gentner, T. R. Dallmann, T. W. Kirchstetter, R. A. Harley, J. B. Gilman, W. C. Kuster, J. A. de Gouw, J. H. Offenberg, T. E. Kleindienst, Y. H. Lin, C. L. Rubitschun, J. D. Surratt, P. L. Hayes, J. L. Jimenez and A. H. Goldstein, *J. Geophys. Res.: Atmos.*, 2013, **118**, 6783–6796.
16 G. Isaacman, K. R. Wilson, A. W. H. Chan, D. R. Worton, J. R. Kimmel, T. Nah, T. Hohaus, M. Gonin, J. H. Kroll, D. R. Worsnop and A. H. Goldstein, *Anal. Chem.*, 2012, **84**, 2335–2342.
17 B. Giechaskiel, M. Maricq, L. Ntziachristos, C. Dardiotis, X. Wang, H. Axmann, A. Bergmann and W. Schindler, *J. Aerosol Sci.*, 2014, **67**, 48–86.
18 M. Matti Maricq, *J. Aerosol Sci.*, 2007, **38**, 1079–1118.
19 E. M. Lipsky and A. L. Robinson, *Environ. Sci. Technol.*, 2006, **40**, 155–162.
20 A. L. Robinson, N. M. Donahue, M. K. Shrivastava, E. A. Weitkamp, A. M. Sage, A. P. Grieshop, T. E. Lane, J. R. Pierce and S. N. Pandis, *Science*, 2007, **315**, 1259–1262.
21 I. Abdul-Khalek, D. Kittelson and F. Brear, *SAE Tech. Pap. Ser.*, 1999, 1999-01-1142, DOI: 10.4271/1999-01-1142.
22 M. J. Kleeman, J. J. Schauer and G. R. Cass, *Environ. Sci. Technol.*, 2000, **34**, 1132–1142.
23 A. Charron and R. M. Harrison, *Atmos. Environ.*, 2003, **37**, 4109–4119.
24 J. Schneider, N. Hock, S. Weimer and S. Borrmann, *Environ. Sci. Technol.*, 2005, **39**, 6153–6161.
25 A. Domínguez-Sáez, M. Viana, C. C. Barrios, J. R. Rubio, F. Amato, M. Pujadas and X. Querol, *Environ. Sci. Technol.*, 2012, **46**, 11187–11194.
26 T. Rönkkö, A. Virtanen, K. Vaaraslahti, J. Keskinen, L. Pirjola and M. Lappi, *Atmos. Environ.*, 2006, **40**, 2893–2901.
27 L. Pirjola, M. Karl, T. Rönkkö and F. Arnold, *Atmos. Chem. Phys.*, 2015, **15**, 10435–10452.
28 M. S. Alam, C. Stark and R. M. Harrison, *Anal. Chem.*, 2016, **88**, 4211.
29 G. S. Frysinger and R. B. Gaines, *J. High Resolut. Chromatogr.*, 1999, **22**, 251–255.
30 D. R. Gentner, G. Isaacman, D. R. Worton, A. W. H. Chan, T. R. Dallmann, L. Davis, S. Liu, D. A. Day, L. M. Russell, K. R. Wilson, R. Weber, A. Guha, R. A. Harley and A. H. Goldstein, *Proc. Natl. Acad. Sci. U. S. A.*, 2012, **109**, 18318–18323.
31 P. A. Sutton, C. A. Lewis and S. J. Rowland, *Org. Geochem.*, 2005, **36**, 963–970.
32 R. E. Dunmore, J. R. Hopkins, R. T. Lidster, J. D. Lee, M. J. Evans, A. R. Rickard, A. C. Lewis and J. F. Hamilton, *Atmos. Chem. Phys.*, 2015, **15**, 9983–9996.
33 D. R. Gentner, D. R. Worton, G. Isaacman, L. C. Davis, T. R. Dallmann, E. C. Wood, S. C. Herndon, A. H. Goldstein and R. A. Harley, *Environ. Sci. Technol.*, 2013, **47**, 11837–11848.
34 I. Nikolova, A. R. MacKenzie, X. Cai, M. S. Alam and R. M. Harrison, *Faraday Discuss.*, 2016, DOI: 10.1039/C5FD00164A.
35 A. Russell and W. S. Epling, *Catal. Rev. Sci. Eng.*, 2011, **53**, 337–423.
36 T. V. Johnson, *SAE Tech. Pap. Ser.*, 2001, 2001-01-0184, DOI: 10.4271/2001-01-0184.

37 M. S. Alam, J. M. Delgado-Saborit, C. Stark and R. M. Harrison, *Atmos. Chem. Phys.*, 2014, **14**, 2467–2477.

38 I. J. Keyte, R. M. Harrison and G. Lammel, *Chem. Soc. Rev.*, 2013, **42**, 9333–9391.

39 M. S. Alam, J. M. Delgado-Saborit, C. Stark and R. M. Harrison, *Atmos. Environ.*, 2013, **77**, 24–35.

40 M. A. Alghamdi, M. S. Alam, J. Yin, C. Stark, E. Jang, R. M. Harrison, M. Shamy, M. I. Khoder and I. I. Shabbaj, *Sci. Total Environ.*, 2015, **506–507**, 401–408.

41 J. J. Schauer, M. J. Kleeman, G. R. Cass and B. R. T. Simoneit, *Environ. Sci. Technol.*, 1999, **33**, 1578–1587.

42 P. Tiitta, P. Miettinen, P. Vaattovaara, J. Joutsensaari, T. Petäjä, A. Virtanen, T. Raatikainen, P. Aalto, H. Portin, S. Romakkaniemi, H. Kokkola, K. E. J. Lehtinen, M. Kulmala and A. Laaksonen, *Atmos. Environ.*, 2010, **44**, 976–986.

43 M. J. Kleeman, J. J. Schauer and G. R. Cass, *Environ. Sci. Technol.*, 2000, **34**, 1132–1142.

44 M. A. Robert, M. J. Kleeman and C. A. Jakober, *J. Air Waste Manage. Assoc.*, 2007, **57**, 1429–1438.

45 M. A. Robert, S. VanBergen, M. J. Kleeman and C. A. Jakober, *J. Air Waste Manage. Assoc.*, 2007, **57**, 1414–1428.

Faraday Discussions

PAPER

Evaluating the performance of low cost chemical sensors for air pollution research

Alastair C. Lewis,*[ab] James D. Lee,[ab] Peter M. Edwards,[a] Marvin D. Shaw,[a] Mat J. Evans,[ab] Sarah J. Moller,[ab] Katie R. Smith,[a] Jack W. Buckley,[a] Matthew Ellis,[a] Stefan R. Gillot[c] and Andrew White[d]

Received 30th November 2015, Accepted 15th December 2015
DOI: 10.1039/c5fd00201j

Low cost pollution sensors have been widely publicized, in principle offering increased information on the distribution of air pollution and a democratization of air quality measurements to amateur users. We report a laboratory study of commonly-used electrochemical sensors and quantify a number of cross-interferences with other atmospheric chemicals, some of which become significant at typical suburban air pollution concentrations. We highlight that artefact signals from co-sampled pollutants such as CO_2 can be greater than the electrochemical sensor signal generated by the measurand. We subsequently tested in ambient air, over a period of three weeks, twenty identical commercial sensor packages alongside standard measurements and report on the degree of agreement between references and sensors. We then explore potential experimental approaches to improve sensor performance, enhancing outputs from qualitative to quantitative, focusing on low cost VOC photoionization sensors. Careful signal handling, for example, was seen to improve limits of detection by one order of magnitude. The quantity, magnitude and complexity of analytical interferences that must be characterised to convert a signal into a quantitative observation, with known uncertainties, make standard individual parameter regression inappropriate. We show that one potential solution to this problem is the application of supervised machine learning approaches such as boosted regression trees and Gaussian processes emulation.

Introduction

Long-term atmospheric monitoring of classical pollutants such as NO_x, O_3, SO_2, CO, Volatile Organic Compounds (VOCs) and Particulate Matter (PM) has typically been conducted to support air quality compliance and national frameworks for

[a]*Wolfson Atmospheric Chemistry Laboratories, University of York, Heslington, York, YO10 5DD, UK. E-mail: ally.lewis@york.ac.uk*
[b]*National Centre for Atmospheric Science, University of York, Heslington, York, YO10 5DD, UK*
[c]*Environment Department, University of York, Heslington, York, YO10 5DD, UK*
[d]*Department of Electronics, University of York, Heslington, York, YO10 5DD, UK*

pollution control. Such data has also made vital contributions that have supported research in more general atmospheric chemistry, for emissions assessment and in many epidemiological studies of health impacts.[1] Historically the users of air pollution instrumentation, and the resulting data, have been technical specialists in the field, supported by established protocols for gas and particle metrology. The most common observational strategy for urban atmospheric chemistry has been to make high accuracy air pollution measurements at small numbers of notionally representative urban/suburban locations, with modeling used to support interpolation to unmonitored locations.[2] The constituency and user-base for surface air pollution measurements has been largely static for perhaps fifty years, and changes in the methodologies and technologies employed for measurement have been incremental, most notably the evolution in methods used for regulatory monitoring of particulate matter. The situation for technologies that measure air pollution from space is rather different, but is outside of this discussion.

A fundamental change to the surface monitoring status quo, enabled by low cost air pollution sensors, has been mooted for more than a decade[3–5] and gained much media attention in recent years [in the last year, for example: The Times, 2014; *New York Times*, 2015; *The Guardian*, 2015].[6–8] Such a technology shift has enormous attractions for research users; dense or portable networks of measurement would give unprecedented insight into human exposure,[9] into CFD and high-resolution model performance, into emissions and much more. Possibly more significant than the research value is that such technologies could result in the democratization of air quality observations to the general public,[10] making local air pollution measurement as common, or more so, than amateur meteorological stations.

Whilst there is no single agreed definition of what constitutes a low cost sensor (the phrase naturally being rather subjective), in this paper we are referring to devices that purport to make autonomous observations of multiple pollutant parameters at a capital cost in the range of 100–10 000 USD per observing location. This contrasts with traditional multi-pollutant analytical capability, such that would meet prescribed US or European requirements for air quality compliance assessment, that have costs of the order >100 000 USD per observing location.

If one considers low cost air pollution sensors as just another type of atmospheric instrumentation, their development and user adoption has differed fundamentally from most other new analytical technologies. The traditional pathway for new analytical capability is for instruments to emerge from either technology-intensive private companies or research laboratories, and be used first by a small number of high-skill early adopters. The atmospheric early adopters typically test, refine and often improve the early versions of instruments, and publish peer-reviewed papers that give the early adopter scientific advantage, by the very fact they have measured something in a novel way. If successful, instruments then propagate into more widespread use, the extent of which depends on the parameter being measured. For classical air pollutants, instruments may ultimately meet requirements for legal compliance or regulatory emission measurements.

Sensors, however, have followed a somewhat different pathway. Many of the basic sensing technologies used in current commercial devices were created for other applications, often combustion or occupational health measurement, and have been co-opted into ambient measurements at a later date. There have been a range of developers including university labs,[11,12] private companies and even devices created *via* crowd-sourced funding[13] [*e.g.* AirQuality Egg]. There is a small body of literature that address issues around interferences, notably water vapour,[14] data usability[15] and field calibration,[16–19] and the conceptual advantages for developing countries.[20] We have struggled however to find examples in the peer-reviewed literature where a research user has unlocked new atmospheric understanding through the deployment of an air pollution sensor network, and this seems unusual when compared with historical technological precedent. Whilst academic usage has been modest, public uptake of air pollution sensors has been growing; for example in London, Breathe Heathrow[21] and Change London[22] are private initiatives that propose the use of large numbers of air pollution sensors, the latter suggesting that up to 10 000 may be installed. A consequence is that whilst the concepts and potential for such devices are well-established in the community – from researchers through to legislators – the peer reviewed literature on quantitatively successful deployments is limited.

Ultimately, low cost sensors are attempting something exceptionally challenging. To make a trace gas measurement to a usable degree of accuracy and precision, and with stability over time, is very hard. However since exposure to air pollution is a legally defined quantity in most countries it seems surprising that there are relatively few detailed published measures of performance, particularly for analytical technologies that already have the general public as a user.

In this paper we (i) report a detailed laboratory analytical study of common sensor sensitivities and a comprehensive analysis of cross interferences to other atmospheric constituents including other air pollutants and also variable species such as CO_2, H_2 and H_2O; (ii) report on initial performance of a commercial low cost air pollution sensing system when compared alongside more traditional measurement technologies; (iii) describe the development of a scientific concept that may exploit total VOC sensors and establish analytical performances and interferences. Our intention is to provide a snapshot of performance using current state of the art; we do not speculate on future advances, or attempt to make a valuation of non-research aspects of existing low cost sensor devices, for example in prompting public engagement with the topic area or related policy issues.

Results and discussion

Electrochemical sensor interferences

Air pollution sensors are typically not as analytically specific as current air pollution reference methods (*e.g.* UV, chemiluminescence, gas chromatography, mass spectrometry *etc.*), and there is potential for sensor measurements to suffer from interferences. Changes in ambient water vapour and temperature have long been known to affect sensor performance,[23] but there is also potential interference due to exposure and response to other co-pollutants.[24,25] A small number of previous studies have quantified some of these interferences and then applied corrections to ambient observations.[26] The problem is potentially exacerbated by high concentration co-pollutants, for example CO_2, that are in excess of the measurand often by many orders of magnitude. In an urban setting many gaseous pollutants are correlated in their temporal behaviour to some degree, and interferences may be difficult to distinguish from examination of ambient data alone.

The performance of five commercially available electrochemical sensors designed for the measurement of CO (CO-B4, Alphasense Ltd), O_3 (OX-B421, Alphasense Ltd), NO (NO-B4, Alphasense Ltd), NO_2 (NO_2-B4, Alphasense Ltd) and SO_2 (SO_2-B4, Alphasenase Ltd) in ambient air were evaluated in the laboratory. [Datasheets are available from http://www.alphasense.com/]. Our aim was to establish the selectivity of these sensors to their target compounds, and quantitatively characterise chemical interference to other pollutants. We then evaluated the scale of impacts of co-pollutants through an inter-comparison exercise alongside reference measurements of the same pollutants in ambient air.

All sensors were housed within a single 3D printed PLA flow cell (Makerbot) with calibration or ambient air introduced to the sensor heads simultaneously under controlled conditions. The data acquisition rate was 1 Hz (average to 5 min intervals) using a Labjack U6 (Labjack) measurement and automation device and Labview software (Labview 2012). All gas lines used were 1/4″ PTFE with stainless steel fittings (Swagelok). Since previous studies have indicated that pressure, humidity, temperature and flowrate can affect sensor response, we recorded these using an LM35 temperature sensor (Texas Instruments), MPX4200A absolute pressure sensor (Freescale Ltd) and an HIH-4000-001 humidity probe (Honeywell). During the analysis periods, in line gas temperature and pressure were consistent, 20.2 ± 0.7 °C and 1.0003 ± 0.0009 bar respectively, minimising their effects upon the sensors.

Chemical interference of co-pollutants

The experimental set-up used a pure air generator (PAG003, Eco-physics) to create the balance gas. Zero air from the generator during these experiments was measured and contained <10 ppt, <5 ppt, <10 ppt, <10 ppt and <50 ppt of NO, NO_2, O_3, SO_2 and CO respectively. Before work commenced each sensor's working electrode voltage (WE) and auxiliary electrode voltage (AE) offset were determined using the zero gas. Zero air was then humidity controlled using deionised water with a dew point generator (DG-3, Michell Instruments). Finally known gas phase mixtures of NO, NO_2, SO_2, CO_2, H_2 and CO were blended into air from calibration standards in N_2 (BOC) using a gas dilution device (Multi-gas calibrator, S6100, Monitor Europe). O_3 concentrations were generated *in situ* using a gas dilution device and mercury UV lamp, the output concentrations from which were confirmed using a separate online monitor (Thermo, model 49i). Humidity dependent WE and AE sensitivities toward the listed compounds were evaluated using five concentrations and linear regression analysis for each performed at 4 different humidities (Table 1). Mole fraction test ranges used were 0–200 ppb for CO, 0–350 ppb for O_3, 1–160 ppb for NO, 0–160 ppb for NO_2, 0–400 ppm for CO_2 and 0–40 ppb for SO_2, with each measurement point performed at 15, 30, 45 and 60% RH.

No observable change in AE voltage was observed during the experiments, and only the effect upon the WE voltage is shown here. We summarise the mV interference induced on a sensor for each co-pollutant in Table 1. There is a complex set of interconnections between co-pollutant and measurand. In some cases the *absolute* interferent signal induced by a co-pollutant is very small, *e.g.*

Table 1 Working electrode responses (in mV ppb^{-1} of co-pollutant) induced by the presentation of co-pollutants in zero air across five electrochemical sensors, with a four point humidity correction calibration applied *via* working electrode offset correction (mV/%RH)

	Compound							
Sensor	CO	SO_2	NO	O_3	NO_2	CO_2	H_2	%RH
CO-B4	0.378	−0.013	0.000	0.0200	0.032	0.000	−0.032	0.201
OX-B421	0.000	−0.016	−0.110	0.439	0.44	9.5×10^{-5}		0.560
SO_2-B4	0.013	0.210	0.023	−0.014	−0.32	9.8×10^{-6}		0.000
NO-B4	0	0.007	0.558	−0.011	−0.590	1.8×10^{-5}		−0.303
NO_2-B4	0	0.004	−0.008	0	0.148	2.3×10^{-5}		0.000

CO_2 impact on the SO_2 or NO_x sensors. However the relative amounts of certain co-pollutants to the measurand is such that the overall response of a given sensor could be dominated by the co-pollutant response.

The scale of interference in ambient air

We next performed a set of ambient measurements using the five sensors, plus separate reference measurements of the interferent co-pollutants shown in Table 1. We use the ambient measurements to provide us with a representative range of co-pollutant values for which possible sensor interferences may be calculated.

The reference instruments used in this study, and for the later Section 2, were as follows: for O_3, the reference instrument was a Thermo Environmental Instruments (TEI) 49C UV absorption analyser which is a United States Environmental Protection Agency (USEPA) equivalent method. Calibration of the instrument was carried out using a TEI Primary Ozone standard, which itself is certified yearly by the UK National Physical Laboratory (NPL). The instrument provided minute averaged data. NO_x was measured using a custom built, high sensitivity instrument (Air Quality Design Inc). The instrument uses two channels, with NO measured using the well known chemiluminescence technique. NO_2 was quantified using the second channel, with NO_2 being converted to NO using a blue light LED converter centred at 395 nm. The NO_2 mixing ratio is derived from the difference between total NO_x and NO mixing ratios. This method provides an accurate and largely interference free method of NO_2 detection, in contrast to commercially available NO_x analysers which typically use heated molybdenum catalysts to convert NO_2 to NO, which are known to be subject to significant interference from other reactive nitrogen species.[27] The instrument is calibrated *via* the addition of 5 sccm of a known NO concentration (from a 5 ppm NO in nitrogen standard cylinder (BOC – tied to the NPL scale)) to the ambient sample, which is scrubbed of NO_x during the calibration procedure. The conversion efficiency of the LED converter is measured in each calibration using gas phase titration of the NO to NO_2 on addition of O_3. A more detailed description of a similar system can be found in Lee *et al.*[28] The instrument produces data at a 1 second average. The O_3, NO and NO_2 reference data were subsequently averaged to 15 minutes for comparison with the sensors. Carbon dioxide concentrations were determined every 5 minutes using a Dual Column SRI 8610C GC. The dual channel GC has both a flame ionisation detector (FID) and electron capture detector (ECD); CO_2 measurements were made on the FID channel. Ambient air was flowed through the sample loop at a rate of 100 ml min^{-1} to flush and fill the sample loop. The sample was then injected onto the GC column (3 m Haysep D packed column) held at a temperature of 60 °C for the duration of the GC run. Hydrogen was used as the carrier gas. After exiting the GC column the effluent was passed through a methanizer packed with a nickel catalyst and held at 320 °C to convert the CO_2 into methane for detection by the FID. Ambient air was introduced to the flow cell from the main sample inlet using a stainless steel diaphragm metal bellows pump (Senior Aerospace, MB302) flow regulated to 1 slpm using a needle valve (Swagelok). Sensor and reference measurement data were averaged to 5 min intervals and evaluated over an 18 day period (7/8/2015–25/8/2015).

Average mixing ratios in ambient air determined from the reference measurements were 23.6 ± 12.3, 1.3 ± 7.2, 5.1 ± 0.2, 0.2 ± 0.1, 106 ± 24, 676 ± 161 ppb and 389 ± 24 ppm for O_3, NO, NO_2, SO_2, CO, H_2 and CO_2 respectively. Ambient humidity was 59.1% ± 12.1. Using the ambient measurements over a 18 day period we then calculate the interference induced on each sensor. We take the mean mixing ratio of each co-pollutant and then express the effect of this in inducing an artefact sensor signal. We show the scale of this interference signal as a percentage of the measurand, which was determined independently using a reference method (Table 2).

What is significant in our test study is the relative scale of impact that co-pollutants have on the induced sensor signal when the analyte pollutant is at typical European suburban values. In our study the ambient mean NO_2 mixing ratio is around 5 ppb which generates a mV sensor signal that is approximately the same as the artefact mV signal induced by the presence of ambient CO_2. SO_2 is a more extreme example: here our ambient measurements are very low, typically 0.2 ppb, such that the interference signals from the more abundant co-pollutants CO, O_3, NO_2 and CO_2 swamp any direct analytical signal from SO_2 itself. The impacts of co-pollutants in our test location are severe for certain sensors although we would stress that the scale of impact depends very much on the ratio of measurand to co-pollutant.

We show in Fig. 1 the relative responses that are induced in a NO_2 sensor arising from co-exposure to a range of different CO_2 and NO_2 mole fractions. At polluted roadside locations, for example central London, kerbside NO_2 can be as

Table 2 First row: observed mean ambient pollution mixing ratio and one sigma range over 18 days. Subsequent rows show the impact of the signal induced by a co-pollutant expressed as a percentage of the mean ambient mixing ratio of the measurand

Sensor	CO	SO_2	NO	O_3	NO_2	CO_2	H_2
Observed ppb	106 ± 24	0.2 ± 0.1	1.3 ± 7.2	23.6 ± 12.3	5.1 ± 0.2	389 ± 24 (ppm)	676 ± 161
CO-B4	—	−0.01	0.00	1.40	0.40	0.00	−53.98
SO_2-B4	4270.57	—	1.25	−9967.40	5194.90	12 063.45	
NO-B4	0	0.14	—	−34.12	−415.71	−985.32	
OX-B421	0.00	0.00	20.60	—	22.49	365.79	
NO_2-B4	0	0	−20.61	0	—	118.94	

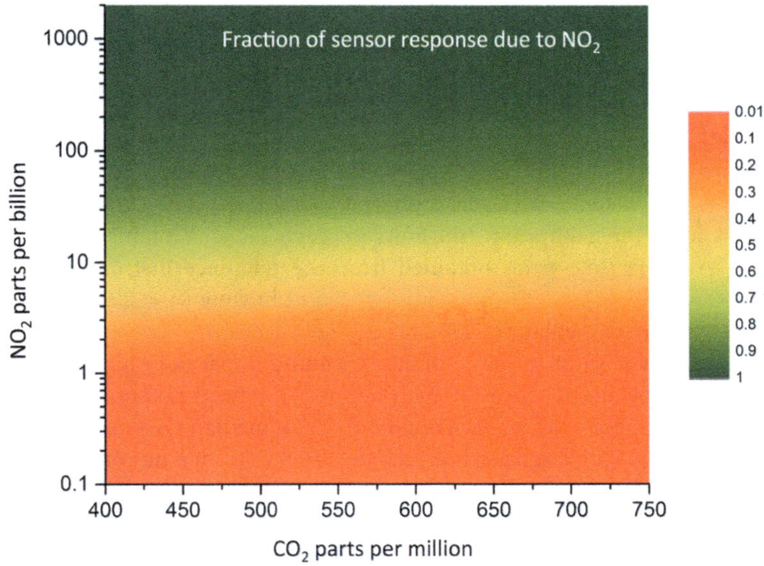

Fig. 1 Fraction of electrochemical sensor response due to [NO_2] for a given co-exposure to CO_2 over the CO_2 range 400 to 750 ppm.

high as 500 ppb, with corresponding CO_2 of around 750 ppm. With this pollutant mixture the sensor response would be overwhelmingly (>98%) due to the signal induced by NO_2 gas with only a small contribution from CO_2. The fractional contribution to this particular sensor signal then changes markedly as one moves from a polluted roadside scenario to suburban or background values. At typical suburban values of 450 ppm CO_2 and 10 ppb NO_2, the signal induced from the sensor is apportioned approximately 50 : 50 between the gases. We conclude that the influence of co-pollutants on sensor response requires very careful evaluation since sensors within a heterogenous network are likely to suffer from substantially differing degrees of analyte interference depending on the ratio of co-pollutant to measurand.

Evaluating a multi-sensor package

There are few published studies that compare emerging commercial air pollution sensor packages against existing reference methods in the field. We aim here to establish the utility of data provided by on-the-market devices, the unit to unit precision of measurement, and the comparability of this against reference instruments to establish accuracy. We co-located 20 identical sensor units with stated capabilities to measure urban concentrations of O_3, NO, NO_2, total VOC, and PM. For this paper we anonymise the commercial supplier of the devices, since we do not wish the reader to infer any specific advantage or disadvantage of these particular devices relative to any other current commercial products. We note that some of the core internal component sensing technologies in the system we report on are also found in other commercially available packages. The sensor units were co-located on a flat roofed building (53°56′52.08″N, 1° 2′46.17″W) in York, UK approximately 12 m above ground level and with unobstructed airflow.

The sensor inlets were around 30 m from a moderately trafficked single carriageway road. Sensors were placed next to a trace gas sample inlet manifold used as a supply of air to the standard reference instruments measuring O_3, NO and NO_2 housed in laboratories below.

Ozone. The calibration slope factors for the O_3 sensors were modified from initial factory settings following comparison against the reference photometric data and these values were applied on 11th Oct 2015. For the remaining three weeks of study the calibration constants for each sensor remained unchanged. Fig. 2 shows the time-series obtained from the reference instrument and the *median* O_3 sensor observation at any given point in time (n = 20 sensors). The agreement between median sensor and reference is clearly very good. Beyond the median value, a R^2 value in excess of 0.9 was found for virtually all sensors when examined one-by-one. There is a small negative intercept potentially due to an interference at higher NO levels. We note that the median O_3 sensor value does not derive from a single sensor, but is contributed to from a number of different units over the period.

We next examined the simultaneous performance of all sensors relative to the reference O_3 value and show this in Fig. 3. The highest and lowest sensor values (red and blue lines) deviate significantly from the sensor median value and reference values, although the agreement of trends remains good through the time period, where R^2 is around 0.9 ± 0.06 (1 sigma) for all individual sensors. To avoid negatively skewing any comparison with a small number of poor performing outlying sensors, we show on the figure a grey shaded region which covers those sensors whose data falls within the 25 to 75th percentile group of sensors. The agreement of this reduced dataset with the reference is also very good and the internal spread of sensor values would be compatible with meaningful measurement given expected atmospheric variability. Taking those sensors in the interquartile range, more than 70% of observations fall within 5 ppb and 95% of observations fall within 10 ppb of the reference values.

Our conclusion is that given an *in situ* initial normalization of calibration slope factors against a reference instrument, then the resulting O_3 sensor observations

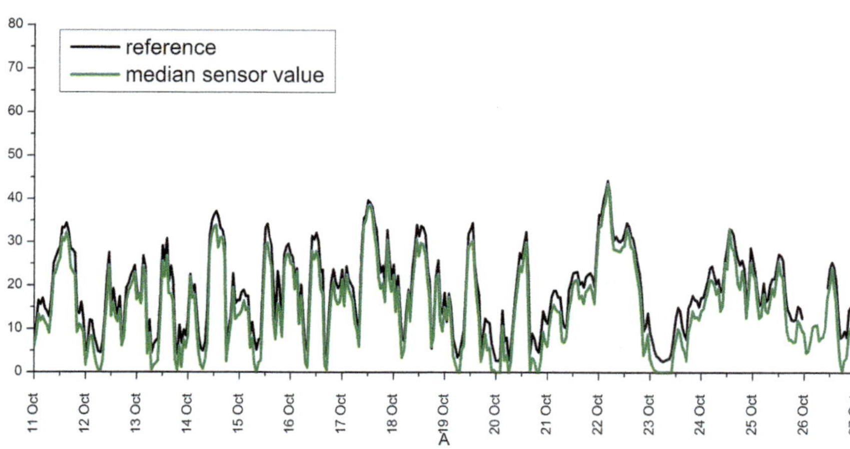

Fig. 2 Time series comparison in the field of photometric ozone (black line) and median O_3 sensor (green line) (n = 20), R^2 0.904, intercept − 4.6 ppb.

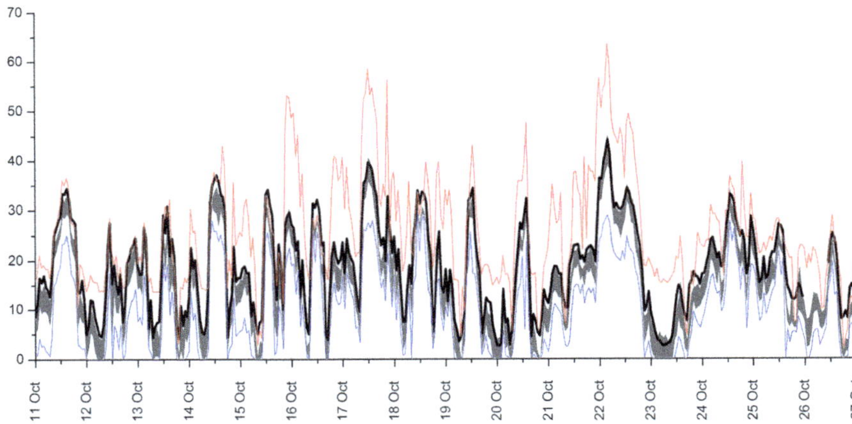

Fig. 3 A time-series comparison of reference photometric O_3 instrument (black line), highest O_3 sensor (red line), and lowest O_3 sensor (blue line). Grey shaded area shows those sensors lying in the 25th to 75th percentile range.

would be significant, *e.g.* in combination have precision that is better than typical daily atmospheric variability over our test period of three weeks, although numerical strategies to handle outlying sensors require development. An evaluation of longer-term stability, and issues such as frequency of field calibration would require further experiments. The analysis raises a wider implementation issue, suggesting that 'managed' deployments of sensors, where local calibration is coordinated and referenced to a single source, may be considerably more precise than that achieved from unstructured deployments of sensors without reference normalisation.

Nitrogen oxide. Our next analysis evaluates performance of the NO sensors against reference chemiluminescence observations and we apply a similar methodology that compares reference against minimum, maximum and those sensors in the 25th to 75th percentile range (Fig. 4). In general terms the NO sensors show good temporal agreement against reference measurement, but with poorer linear regression statistics than the O_3 sensors. We note that NO is often at low concentrations in the background York atmosphere and this challenges the detection limit of the sensors in many periods. NO sensor observations compared against chemiluminescence gave R^2 values of 0.73 ± 0.21, although with an overall sensor bias underestimating NO by a factor of 2.3 ± 1.1. It is possible that this underreporting bias for $[NO]_{sensor}$ may be correctable, although the inter-sensor spread would remain high.

Nitrogen dioxide. A comparison of reference NO_2 against sensor observation showed poor statistical agreement, suggesting that the sensor was not measuring this chemical parameter exclusively. This contrasts with O_3 and NO from Sections 2.1 and 2.2 where there was good evidence for an analytically selective measurement, even if calibration and sensor reproducibility were less than ideal. An analysis of correlation coefficients of each sensor against the reference gave a median ($R^2 = 0.25 \pm 0.13$). The sensors typically reported high concentrations, over-reporting against the reference by a factor of 3.2 ± 1.7. Setting to one side the poor agreement in trends between reference and sensor, this substantial over-reporting implied considerable exceedance of air quality standards, where

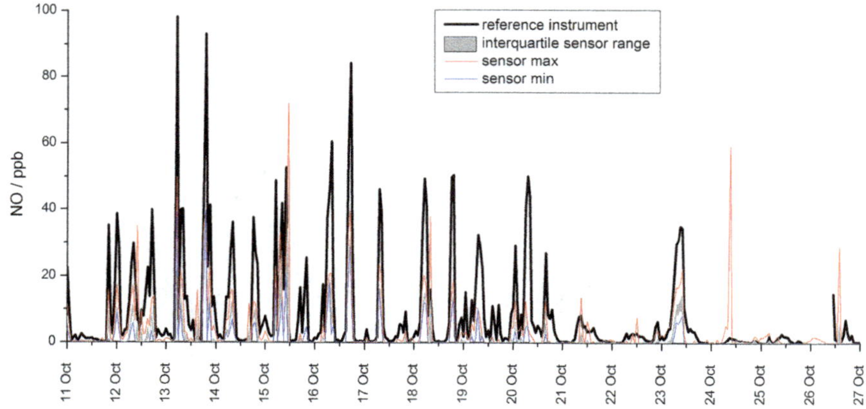

Fig. 4 A time-series comparison of reference chemiluminescence instrument (black line), highest NO sensor (red line), and lowest NO sensor (blue line). Grey shaded area shows those NO sensors lying in the 25th to 75th percentile range.

reference measurements showed values well inside regulatory limits. Whilst the direct comparison of NO_2 sensor against reference was poor, the 25th–75th percentile range of sensors was relatively narrow. This implies that the sensors are broadly reproducible in their response, even if they are not solely measuring the reported parameter of NO_2. Further work is required to determine the exact response characteristics of this particular NO_2 sensor, potentially leading to an operationally defined air quality measure, similar perhaps to a 'total' VOC observation (Fig. 5).

Particulate matter. The final two sensors evaluated (PM and VOC) do not have as precise a reference value to compare against. For NO, NO_2 and O_3 there is a single binary reference gas mixture that ties observations to amount of substance. For particulate matter a range of reporting metrics are available, *e.g.* $PM_{2.5}$ and PM_{10}, and these in turn are somewhat operationally defined in terms

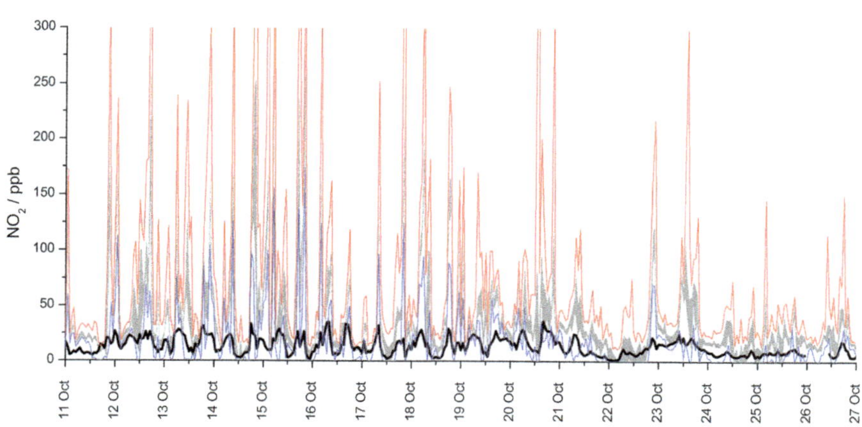

Fig. 5 A time-series comparison of reference chemiluminescence NO_2 instrument (black line), highest NO_2 sensor (red line), and lowest NO_2 sensor (blue line). Grey shaded area shows those NO_2 sensors lying in the 25th to 75th percentile range.

of individual instrument responses. For PM, we compared the 20 sensors co-located against one another to determine inter-sensor variability, and then compared two sensors against an air quality reference station in York to establish accuracy against a current standard air quality method for PM reporting.

There is a generally good statistical agreement between the individual sensor PM measurement ($R^2 = 0.8 \pm 0.11$), but with a significant spread in reported concentrations, even taking the reduced range of central sensor values. For PM the 25th to 75th percentile sensors (grey range in Fig. 6) span a concentration range ~20 μg m^{-3}. This measurement uncertainty is significant in the context of current regulatory limits (*e.g.* European Union Directive values: annual mean for $PM_{2.5}$ 25 μg m^{-3}, and for PM_{10} 40 μg m^{-3}). A comparison of sensor particulate matter measurements against a standard reference observations was made at a roadside location in York (Fishergate, TEOM-FDMS, following EN12341:1998). This co-located two sensor units alongside a City of York air quality monitoring site, affiliated to the Defra automatic urban networks for air quality. The data was taken from http://uk-air.defra.gov.uk. Fig. 7 shows this comparison, with modest agreement between sensor and reference PM_{10} ($R^2 = 0.35$, intercept ~10 μg m^{-3}). A positive bias was observed in the sensor data.

Volatile organic compounds (VOCs). The final sensor we evaluated was a measurement of Volatile Organic Compounds (VOCs) using a metal oxide sensing approach. This 'total VOC' definition is common in the sensor reporting of VOCs since metal oxide devices or photoionisation sensors do not speciate individual VOCs in the way that chromatographic or mass spectrometric methods do. The conceptual definition of 'total' is straightforward, but the conversion of a sensor signal to a quantitative value is difficult since sensors typically respond differently to individual organic compounds. Since there is no reference measurement for total VOC, in this section we evaluate only the inter-sensor variability in response, and then in Section 3 discuss in further detail compound specific responses and the potential application of total VOC sensors in the field (Fig. 8).

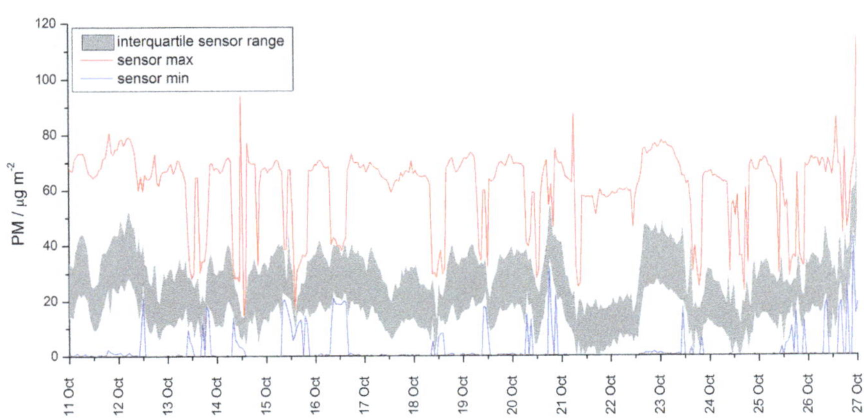

Fig. 6 A time-series comparison of highest PM sensor (red line) and lowest PM sensor (blue line). Grey shaded area shows those PM sensors lying in the 25th to 75th percentile range.

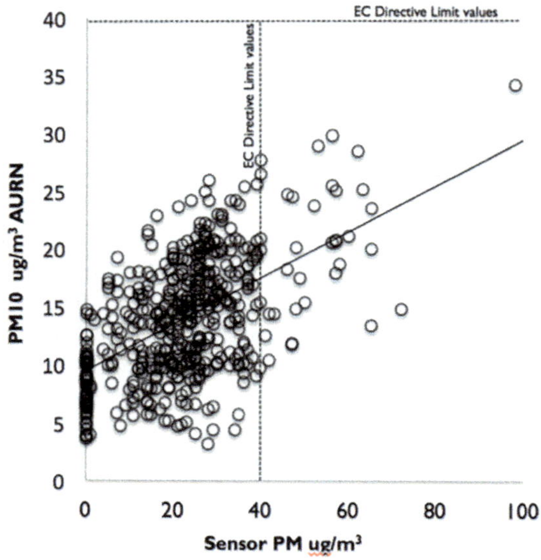

Fig. 7 Three week comparison of hourly sensor PM observations against a UK standard PM reference method at the Fishergate air quality station – data from http://uk-air.defra.gov.uk. EU Directive PM_{10} limit values are marked on the plot for information.

Whilst there is some considerable spread in the values reported between highest and lowest VOC co-located sensor in the study, the sensor data within the 25th to 75th percentile are relatively tightly constrained, with the sensor to sensor variability being modest compared to the atmospheric variability. Since the sensor reproducibility is encouraging, it demands some further thinking on how a relatively non-specific but reproducible observation could be used to address particular scientific problems, given that VOCs provide one of the principal controls over the functioning of the Earth's system and have a significant impact on air quality, climate change and the carbon budget.

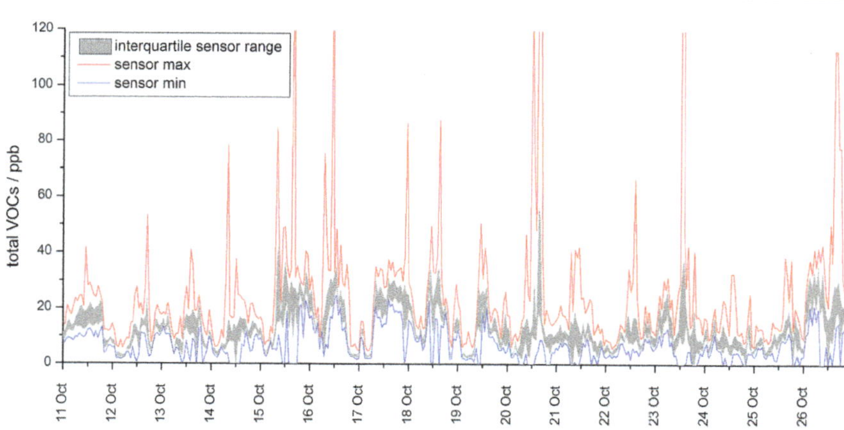

Fig. 8 A time-series comparison of highest total VOC sensor (red line) and lowest total VOC sensor (blue line). Grey shaded area shows those VOC sensors lying in the 25th to 75th percentile range.

New strategies for VOC sensors

Current methodologies for studying the impacts of VOCs on atmospheric chemistry and climate generally use process level or single point observations of concentrations or fluxes to provide constraints for regional and global chemistry models. The upscaling of emissions from process scale (typically that of cm to km scales) to that of the model grid scale (110 100 km^2 depending on model and application) are based on the very limited literature datasets since instrumental methods are highly complex. Of all the air pollution parameters considered in this paper, VOCs typically have the most limited observational datasets and by extension would benefit most from usable low cost sensors. In the UK for example there are in excess of 300 official online O_3 air quality measurements, 200 NO_x measurements, but only four online VOC measurement stations. This limited information on spatial and temporal variability makes it difficult to quantify uncertainties in model constraints or link emissions to process level controls.

The use of sensors for the detection of VOCs has the potential to deliver something different but complementary to the existing experimental approaches – that is, a direct measurement with a substantially degraded level of chemical detail compared to the single site observations, but with well-resolved time information and spatial resolution. Recent advances in sensors for the detection of VOCs, adopted from the hazard detection industry,[29] provide a particular opportunity to answer questions inaccessible to conventional research methodologies.

The non-selective nature of current VOC sensor technologies is not as limiting in terms of potential science applications as might first be presumed, as existing literature shows that in many environments VOC speciation is significantly less variable than the changes in total concentration. Our study from Section 2.5 shows that metal oxide sensor to sensor variability can be reasonable, when compared with atmospheric variability. Thus the use of a selective and sensitive VOC observation technique (*e.g.* GC) alongside a distributed network of sensors may enable a mapping of detailed speciated information onto an improved measure of overall temporal/spatial variability. For a VOC sensor to provide useful data on ambient variability, it must be able to detect small changes in VOC concentrations and requires limits of detection (LOD) in the 10–100 s of pptv range. Sensor response must also be stable over timescales significantly longer than that of the variability being observed (hours to days), and any interferences must be quantifiable and correctable.

Photo Ionisation Detectors (PIDs) are one available sensor technology for Σ[VOC] measurements. These small units ($\sim$2 cm diameter) use $\sim$100 mW and work *via* the ionisation of molecules whose ionisation energy is less than that of the photons from the UV lamp in the sensor (including most unsaturated VOCs). Current manufacturer quoted LODs of $\sim$5 ppb are however inadequate for ambient observations and improving this has not been a requirement for the majority of industrial PID applications. Reducing the LOD by approximately an order of magnitude, through thorough lab characterisation and signal optimisation, increases substantially the utility of the sensors for atmospheric chemistry applications.

Through detailed laboratory experiments we found the PID background signal to be highly sensitive to variables such as local electromagnetic interference, noise on the sensor supply voltage and ground, and noise introduced during the transmission of signal voltages to a suitable analogue to digital converter. Through a series of improvements to minimise the effects of these electronic interferences we were able to reduce the PID background signal standard deviation, for a measurement period of 60 hours, from approximately 10 mV to 3×10^{-2} mV. These improvements in electronics and signal handling enabled small signals from low VOC concentrations to be discernable from the background. Fig. 9a shows PID response to isoprene binary mixtures in N_2 over the range 0 to 2.5 ppb. We show the corresponding 3 sigma estimated LOD in Fig. 9b.

The sensor response shown in Fig. 9 suggests that order of magnitude improvements in sensor detection limits are possible with careful signal conditioning. However translating this quantitative performance to ambient operating conditions, compared with the highly optimised laboratory conditions used in Fig. 9, requires all possible interferences to be characterised over relevant ranges. Fig. 10 shows for example the non-linear PID background signal response to changing relative humidity (RH), which is of a comparable magnitude to that towards isoprene (Fig. 9). As well as the obvious complications of correcting for signals from multiple interferences, PID sensor sensitivity to VOCs has also been found to change with RH and will likely also be impacted by other interfering signals. Laboratory experiments have also shown PID response to be sensitive to operational parameters such as sensor supply voltage and temperature.

The results shown in Section 2 show the potential performance of sensor devices to illustrate trends and temporal behaviours – successfully so for O_3, NO, PM and VOC – but also the limitations in trying to obtain quantitative information – successful in our study only for O_3. The quantity, magnitude and complexity of analytical interferences that must be characterised in order to convert a VOC (or indeed any sensor) signal into a meaningful quantitative observation, with known uncertainties, makes standard individual parameter regression approaches

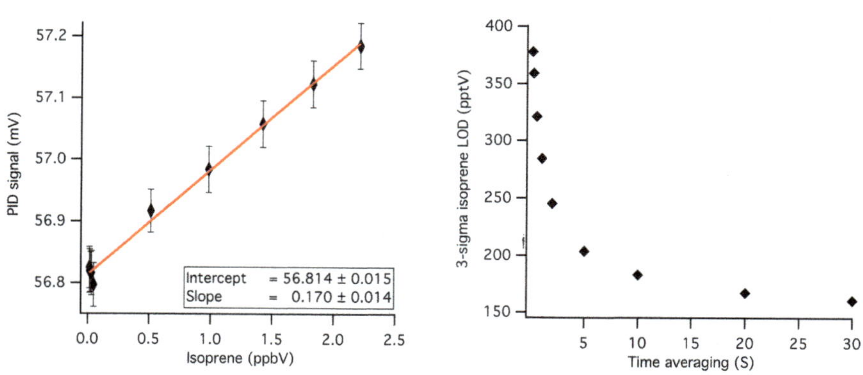

Fig. 9 (a) PID signal voltage against isoprene mixing ratio (in N_2 balance gas), mean PID signal response and calculated isoprene mixing ratio over a 1 minute sample time, with error bars representing ±1 standard deviation. Red line shows linear fit to data. (b) Estimated PID three-sigma LOD for a range of signal averaging times.

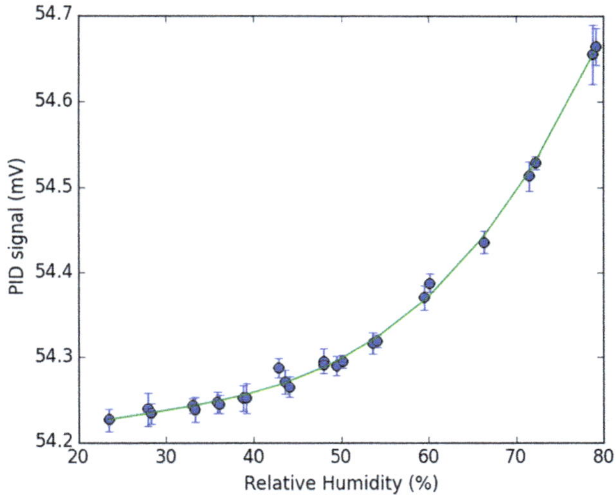

Fig. 10 Baseline PID response against relative humidity, with error bars showing ±1 standard deviation, highlighting potential artefact signal generated solely by a changing humidity. Green line is an exponential fit to the data.

extremely complicated and uncertain. One potential solution to this problem is the application of statistical methods combined with supervised machine learning approaches, such as boosted regression trees (BRT)[30,31] or Gaussian process emulation (GPE) or Kriging-based optimisation.[32] These algorithms have been shown to have significant predictive power once deterministic functions have been created using a training data-set, and have the advantage of being able to deal with complex nonlinearities and covariances.

One approach to characterising the multiple dependencies of a PID sensor signal could be to use these statistical methods to develop models of sensor response. For PIDs the four key variables that describe the majority of sensor output are [VOC], RH, temperature and supply voltage. Fig. 11 shows the raw sensor signal output (solid black line) compared with that calculated using BRT (red) and GPE (grey, plus shaded uncertainties). Both models were constructed using approximately 3 hours of training data and then used to predict a single sensor response based on a new set of observations of step-changing isoprene mixing ratio, relative humidity, temperature and sensor supply voltage. The model skill in reproducing the average sensor signal implies that the majority of the observed variability in this experiment can be explained by the variability in the 4 predictor variables. At ambient atmospheric concentrations [VOC] can in practice be only a minor contributor to the total signal response, and smaller than the variability introduced by the other three variables. Once confident that the model describes all factors controlling sensor response, these approaches can then be used to predict desired response variables. Fig. 12 shows the known isoprene mixing ratio (black line) compared with those calculated *via* BRT (red) and GPE (grey). Also shown is the calculated mixing ratio using individual parameter regression (blue), which is clearly poorer than the other two approaches. The short duration of the dataset meant that RH, temperature and supply voltage variability was small compared to isoprene, and thus all 3 methods

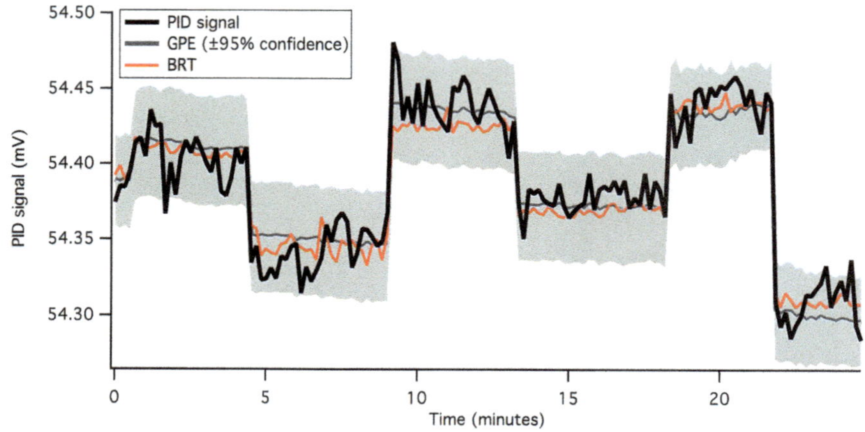

Fig. 11 Raw PID and signals corrected for humidity, temperature and supply voltage using boosted regression tree (red) and Gaussian process emulation (grey), for six different isoprene abundances in N_2 mixtures.

agree reasonably well and all have a mean deviation from the true isoprene value of <±1.5 ppb. However, as the quantity and dynamic range of interference parameters increases, the potential of these approaches to diagnose desired variables becomes much more significant.

Through the combination of these statistical models and the generation of comprehensive training data sets, using both laboratory and real-world experimental data, it is possible that meaningful information can be retrieved from complex sensor signals, providing that supporting observations of the major interferences are available. The success of these methods is, however, dependent on both the sensitivity and reproducibility of the sensor response, as well as the quality and scope of the training data used.

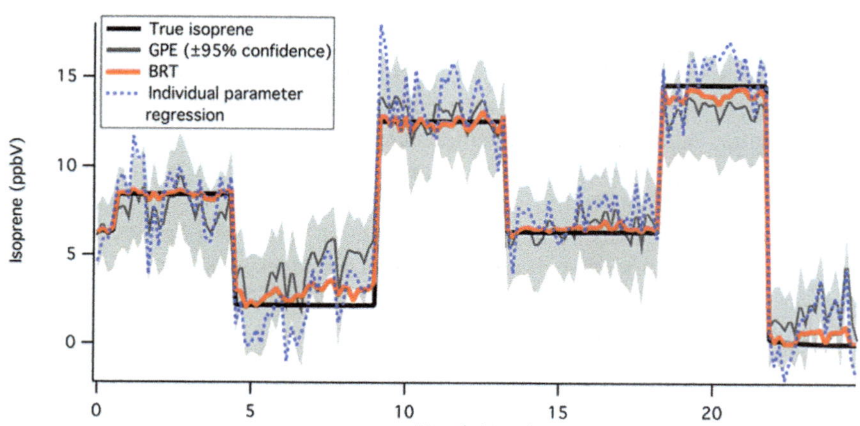

Fig. 12 Isoprene mixing ratio based on calculated calibration gas output (solid black line), PID signal corrected using individual regression parameters (blue dashed), boosted regression tree (red) and Gaussian process emulation (grey, with shaded uncertainties).

Conclusions

We conclude that for all air pollution sensor systems a full analytical evaluation of sensor performance is required using real-world pollutant abundances and in the presence of a full suite of realistic co-pollutants and interferences. Such evaluations are currently lacking in the literature. We highlight that in absolute terms the interference from stable trace gases (*e.g.* CO_2, H_2 *etc.*), expressed as mV ppb^{-1}, are generally very small, but that a high ratio of such co-pollutants to the measurand can render sensor observations prone to large artefact responses. A short-term evaluation of twenty commercial air pollution sensor packages provided some mixed results. For ozone a good overall agreement was found between reference measurements and sensors, once sensor calibration values had been linked to the *in situ* reference observation. Some challenges remain, however, in designing deployment and data strategies that can identify data from outlier sensors that deviate from the reference by more than the scale of expected atmospheric variability in urban environments. For other air pollution parameters, varying degrees of performance were found. For NO and PM, general trends in atmospheric pollution were recreated by the sensors, although with notable negative and positive biases compared to reference chemiluminescence or TEOM-FDMS. NO_2 sensor measurements from the commercial package did not track trends in reference NO_2, and this may be responding to some different air pollution metric that requires better definition. VOC sensors showed good unit to unit variability, but present a further challenge in defining firstly, what exactly they are responding to, and secondly, how this non-specific data may be used scientifically. We show that for PID-based VOC measurement, detection limits can be improved through electronic handling of signals, but that multiple interfering factors affect the overall VOC response. In some cases the sensor response to [VOC] is smaller than that rising from variability in humidity, temperature and supply voltage. We identify that when multiple signal corrections are required for a sensor then application of individual regression parameters is less successful than the statistical methods of boosted regression tree or Gaussian process emulation. Overall we conclude that the state of the art can at best be considered a 'work in progress', with promising performances for some chemicals and rather poor performances for others. Very extensive trace-level laboratory testing of multiparameter sensor responses is still required along with alternative numerical strategies than can correct for sub-optimal measurement. Only then will the sensor concept yield data that is reliable, fit for purpose in the public domain and potentially useful for research applications.

Acknowledgements

ACL, JDL, SJM, and MJE thank the National Centre for Atmospheric Science for support. ACL also acknowledges support for this work through NERC grant NE/F015240/1. PE acknowledges the financial support of a Marie Sklodowska-Curie fellowship. KS is currently in receipt of a NERC SPHERES DTP PhD studentship, ME and SG received support through NERC and EPSRC summer research studentship programmes. We grateful for funding provided by the University of York for some sensor devices tested in this study in a project led by Prof Alistair Boxall.

References

1 Review of the UK Air Quality Index, 2011, ISBN 978-0-85951-699-0.

2 http://uk-air.defra.gov.uk/reports/cat13/1309250915_130923_Review_of_air_quality_monitoring_station_classifications.pdf.

3 M. C. Carotta, G. Martinelli, L. Crema, C. Malagù, M. Merli, G. Ghiotti and E. Traversa, *Sens. Actuators, B*, 2001, **76**, 336–343.

4 W. Bourgeois, A. C. Romain, J. Nicolas and R. M. Stuetz, *J. Environ. Monit.*, 2003, **5**, 852–860.

5 W. Tsujita, A. Yoshino, H. Ishida and T. Moriizumi, *Sens. Actuators, B*, 2005, **110**(2), 304–311.

6 http://www.thetimes.co.uk/tto/environment/article4219453.ece.

7 http://www.nytimes.com/2015/04/16/business/experimenting-at-home-with-air-quality-monitors.html?_r=0.

8 http://www.theguardian.com/environment/2015/aug/10/lord-drayson-takes-on-uk-air-pollution-crisis-with-new-smart-sensor.

9 B. Predic, Y. Zhixian, J. Eberle, D. Stojanovic and K. Aberer, *IEEE International Conference on Pervasive Computing and Communications Workshops (PERCOM Workshops)*, 2013, pp. 303–305.

10 D. Hasenfratz, O. Saukh, S. Sturzenegger and L. Thiele, *2nd International Conference on Mobile Sensing*, 2012.

11 S. R. Utembe, G. M. Hansford, M. G. Sanderson, R. A. Freshwater, K. F. E. Pratt, D. E. Williams, R. A. Cox and R. L. Jones, *Sens. Actuators, B*, 2006, **114**, 507–512.

12 R. Piedrahita, Y. Xiang, N. Masson, J. Ortega, A. Collier, Y. Jiang, K. Li, R. P. Dick, Q. Lu, M. Hannigan and L. Shang, *Atmos. Meas. Tech.*, 2014, **7**, 3325–3336.

13 https://airqualityegg.wikispaces.com/AirQualityEgg.

14 J. H. Sohn, M. Atzeni, L. Zeller and G. Pioggia, *Sens. Actuators, B*, 2008, **131**, 230–235.

15 J. K. Hart and K. Martinez, *Earth-Sci. Rev.*, 2006, **78**, 177–191.

16 M. Kamionka, P. Breuil and C. Pijolat, *Sens. Actuators, B*, 2006, **118**, 323–327.

17 D. E. Williams, G. S. Henshaw, M. Bart, G. Laing, J. Wagner, S. Naisbitt and J. A. Salmond, *Meas. Sci. Technol.*, 2013, **24**(065803), 12.

18 D. Hasenfratz, O. Saukh and L. Thiele, *Wireless Sensor Networks*, 2012, **7158**, 228–244.

19 S. de Vito, P. D. Veneri, E. Esposito, M. Salvato, V. Bright, R. L. Jones and O. Popoola, *Proceedings of the 18th AISEM Annual Conference*, 2015.

20 J. Ikram, A. Tahir, H. Kazmi, Z. Khan, R. Javed and U. Masood, *Environmental Systems Research*, 2012, **1**, 1–8.

21 http://theodi.org/summer-showcase-breathe-heathrow-methodology.

22 http://www.changelondon.org/airpollution.

23 M. L. Hitchman, N. J. Cade, K. T. Gibbs and N. J. M. Hedley, *Analyst*, 1997, **122**, 1411–1418.

24 C. Austin, B. Roberge and N. Goyer, *J. Environ. Monit.*, 2006, **8**, 161–166.

25 C. Hamann, A. Hamnett and W. Vielstich, *Electrochemistry*, Wiley-VCH, 2nd edn, 2007, 978-3-527-31069-2.

26 M. I. Mead, O. A. M. Popoola, G. B. Stewart, P. Landshoff, M. Calleja, M. Hayes, J. J. Baldovi, M. W. McLeod, T. F. Hodgson, J. Dicks, A. Lewis, J. Cohen, R. Baron, J. R. Saffell and R. L. Jones, *Atmos. Environ.*, 2013, **70**, 186–203.

27 M. Steinbacher, C. Zellweger, B. Schwarzenbach, S. Bugmann, B. Buchmann, C. Ordóñez, A. S. H. Prevot and C. Hueglin, *J. Geophys. Res.*, 2007, **112**, D11307.
28 J. D. Lee, S. J. Moller, K. A. Read, A. C. Lewis, L. Mendes and L. J. Carpenter, *J. Geophys. Res.*, 2009, **114**, D21302.
29 G. Manes, G. Collodi, R. Fusco, L. Gelpi and A. Manes, *International Journal of Distributed Sensor Networks*, 2012, **820716**, 13.
30 D. C. Carslaw and P. J. Taylor, *Atmos. Environ.*, 2009, **43**, 3563–3570.
31 J. Elith, J. R. Leathwick and T. Hastie, *J. Anim. Ecol.*, 2008, **77**, 802–813.
32 O. Roustant, D. Ginsbourger and Y. Deville, *J. Stat. Software*, 2012, **51**, 1–55.

Faraday Discussions

PAPER

Atmospheric ethanol in London and the potential impacts of future fuel formulations†

Rachel E. Dunmore,[a] Lisa K. Whalley,[bc] Tomás Sherwen,[a] Mathew J. Evans,[ad] Dwayne E. Heard,[bc] James R. Hopkins,[ad] James D. Lee,[ad] Alastair C. Lewis,[ad] Richard T. Lidster,[a] Andrew R. Rickard[ad] and Jacqueline F. Hamilton*[a]

Received 23rd November 2015, Accepted 16th December 2015
DOI: 10.1039/c5fd00190k

There is growing global consumption of non-fossil fuels such as ethanol made from renewable biomass. Previous studies have shown that one of the main air quality disadvantages of using ethanol blended fuels is a significant increase in the production of acetaldehyde, an unregulated and toxic pollutant. Most studies on the impacts of ethanol blended gasoline have been carried out in the US and Brazil, with much less focus on the UK and Europe. We report time resolved measurements of ethanol in London during the winter and summer of 2012. In both seasons the mean mixing ratio of ethanol was around 5 ppb, with maximum values over 30 ppb, making ethanol currently the most abundant VOC in London air. We identify a road transport related source, with 'rush-hour' peaks observed. Ethanol is strongly correlated with other road transport-related emissions, such as small aromatics and light alkanes, and has no relationship to summer biogenic emissions. To determine the impact of road transport-related ethanol emission on secondary species (*i.e.* acetaldehyde and ozone), we use both a chemically detailed box model (incorporating the Master Chemical Mechanism, MCM) and a global and nested regional scale chemical transport model (GEOS-Chem), on various processing time scales. Using the MCM model, only 16% of the modelled acetaldehyde was formed from ethanol oxidation. However, the model significantly underpredicts the total levels of acetaldehyde, indicating a missing primary emission source, that appears to be traffic-related. Further support for a primary emission source comes from the regional scale model simulations, where the observed concentrations of ethanol and acetaldehyde can only be reconciled with the inclusion of large primary emissions. Although only constrained by one set of observations, the regional modelling

[a]*Wolfson Atmospheric Chemistry Laboratory, University of York, York, UK. E-mail: jacqui.hamilton@york.ac.uk; Tel: +44 (0)1904 324076*
[b]*School of Chemistry, University of Leeds, Leeds, UK*
[c]*National Centre for Atmospheric Science, University of Leeds, Leeds, UK*
[d]*National Centre for Atmospheric Science, University of York, York, UK*

† Electronic supplementary information (ESI) available. See DOI: 10.1039/c5fd00190k

suggests a European ethanol source similar in magnitude to that of ethane (~60 Gg per year) and greater than that of acetaldehyde (~10 Gg per year). The increased concentrations of ethanol and acetaldehyde from primary emissions impacts both radical and NO_x cycling over Europe, resulting in significant regional impacts on NO_y speciation and O_3 concentrations, with potential changes to human exposure to air pollutants.

1 Introduction

The global consumption of non-fossil fuels is increasing, particularly those made from renewable biomass such as ethanol. Gasoline in the UK currently contains 5% ethanol (E5) and this is expected to rise by 2020 to 10% in order to meet EU guidelines on the use of renewable fuel sources.[1,2] These guidelines were implemented to meet two of Europe's key energy policy challenges; first, to tackle climate change by replacing fossil fuel consumption with sources of energy that on balance emit less carbon, and secondly, to ensure there are adequate local supplies of energy to reduce reliance on imported fossil fuels.[1] In the UK, there is also the Renewable Transport Fuel Obligation (RTFO), in place since April 2008. This ensures that fuel suppliers use a minimum percentage of biofuel.[1]

The term biofuel is a generic definition to describe any solid, liquid or gaseous fuel source that is derived from biomass and some liquid fuels can be used as a replacement for gasoline and diesel fuels. For gasoline, this is usually ethanol, although butanol has also been considered and for diesel, this is fatty acid methyl esters (FAMEs). There are a variety of feedstocks that can be used to make biofuels, and commercially available fuels are categorised as first and second generation biofuels. The former are produced from biomass that is part of the food chain such as sugar and/or starch crops, *i.e.* maize, wheat or seeds such as oil seeds or palm oil for bioethanol. The latter are made from biomass that is not part of the food chain. These biofuels can be used in two forms for the transport sector, either in a pure form or, more usually, blended with gasoline and diesel.[1] Ethanol can be blended with gasoline in a variety of mixtures, the most common in Europe is E5, in North America E10 (10% ethanol) and Brazil E85 (85% ethanol) or pure hydrous ethanol. In fact, the US and Brazil combined represent more than 90% of the world's production and consumption of bioethanol.[3,4] Although, high strength blends are used around the globe, these are generally not used in the EU. The situation in Brazil is more complex due to the introduction of flexible fuelled vehicles (FFV). These vehicles can run on normal gasoline, pure ethanol and any blends of the two using only one fuel tank. The engines in FFVs can automatically adapt and make changes to the ignition timing based on the specific fuel blend using an ethanol/gasoline sensor.[5]

Blending gasoline with ethanol has many advantages; it increases the combustive potential of the fuel,[6] improves some of the engines running processes and can decrease the emission of certain pollutants to the air.[3] The emissions of carbon monoxide (CO), tailpipe particulate matter (PM) and hydrocarbons are generally reduced, however the emissions of nitrogen oxides (NO_x) are more variable, with some studies suggesting a reduction,[3] while others report increases.[7] Studies of low strength ethanol blended fuels suggest that there is no consistent change in the emissions of NO_x and that emissions of other regulated pollutants are decreased (CO −13%, total hydrocarbon −5%, benzene −12% and 1,3-butadiene −6%). However, there is a significant increase in the emission of acetaldehyde (+159%), an unregulated and toxic pollutant. This large scale emission of aldehydes during combustion is the main disadvantage of using alcohol blended fuels.[1,8] Despite the increase in use of ethanol, there have only been sporadic measurements around the globe, largely focused in Brazil and the US as they use large quantities of ethanol-blended fuels.[3]

The combustion of both pure ethanol and ethanol-blended gasoline emit significant concentrations of acetaldehyde, formaldehyde, vaporized ethanol and benzaldehyde.[9,10] There have been no large-scale studies on the toxicological effects of atmospheric concentrations of these pollutants. However, acetaldehyde and formaldehyde are suspected carcinogens due to their ability to form adducts with DNA.[11] These pollutants have also been associated with respiratory irritation, asthma aggravating effects[12] and bronchitis in both outdoor and indoor environments.[3] Benzaldehyde is also a known nerve-toxin that can cause brain damage when people are exposed to high concentrations.[13] Still, the main concern when using bioethanol blends remains the emission of acetaldehyde. By combining the degradation rates of acetaldehyde and its products, the exposure to ethanol combustion products can be estimated to be approximately 10 days.[3] This is a considerable time period where a large amount of the urban and rural population could be exposed, of specific concern to the increasing number of asthma and allergy sufferers present in urban areas. Depending on the local atmospheric conditions, acetaldehyde degradation can vary from 11 hours to up to 5 days.[14]

Observations of ethanol in the atmosphere are limited in number due to typically low atmospheric mixing ratios and associated measurement difficulties.[15] However, with the increased use of ethanol-blended fuels, it is expected that atmospheric concentrations will also increase. In fact, those urban areas that use ethanol-blended fuels have significantly enhanced atmospheric mixing ratios of ethanol, in some cases over an order of magnitude higher than those areas without the use of ethanol as a fuel.[15] Nguyen *et al.* (2001) compared the observed ethanol mixing ratios in Osaka, Japan (ethanol-blended fuels are not used) and Sao Paulo, Brazil (high use of ethanol-blended fuels).[16] They found that the observed mixing ratios of ethanol from Sao Paulo were enhanced by over a factor of 20.[16] Observations of atmospheric ethanol in those areas with high ethanol-blended fuel use were on average between 55–410 ppb.[16–20] However, the majority of these studies were conducted outside Europe, mainly in the United States or Brazil. There is a clear need for atmospheric measurements of ethanol in the EU and UK specifically.

The UK National Atmospheric Emissions Inventory (NAEI) currently does not include a road transport or evaporative gasoline source for ethanol,[21] rather its main source is spirit manufacture, (shown in ESI Fig. S1†). This is likely due to the fact that the volatile organic compound (VOC) source apportionment part of the emission inventory was last updated in 2002 and ethanol was first introduced in gasoline blends in 2003. An assumption is made that the VOC speciation profile of each emission source sector is unchanged each year, except for the specific VOCs benzene and 1,3-butadiene.[22] Acetaldehyde, in contrast has approximately 70% contribution from road transport to total UK emissions (ESI Fig. S2†). However, recent acetaldehyde flux measurements from London showed that the observations were approximately a factor of 6 higher than those estimated using the NAEI.[23] We believe that, not only will ethanol be underestimated given the current inventory estimates but that it is also likely that acetaldehyde will be.

The Master Chemical Mechanism degradation scheme for the reaction of ethanol with the hydroxyl (OH) radical in the presence of NO_x is shown in Fig. 1. Ethanol, acetaldehyde and formaldehyde are highlighted in the scheme in black, red and green boxes, respectively. This scheme shows that potentially 95% of ethanol emitted can directly form acetaldehyde through reactions with the OH radical. The degradation of ethanol also has multiple routes to form formaldehyde. In the calculations by Derwent and Jenkin (1990), ethanol shows a low photochemical ozone creation potential (POCP) value owing to its relatively slow reactivity with OH.[24] However, ethanol does show a high photochemical peroxyacetyl nitrate (PAN) creation potential (PPCP) value because it is oxidised directly to acetaldehyde, a major PAN precursor.

2 Experimental

As part of the Clean air for London (ClearfLo) project, two 5 week intensive operating periods (IOPs) were carried out at an urban background site, North Kensington, during the winter (January and February) and summer (July and

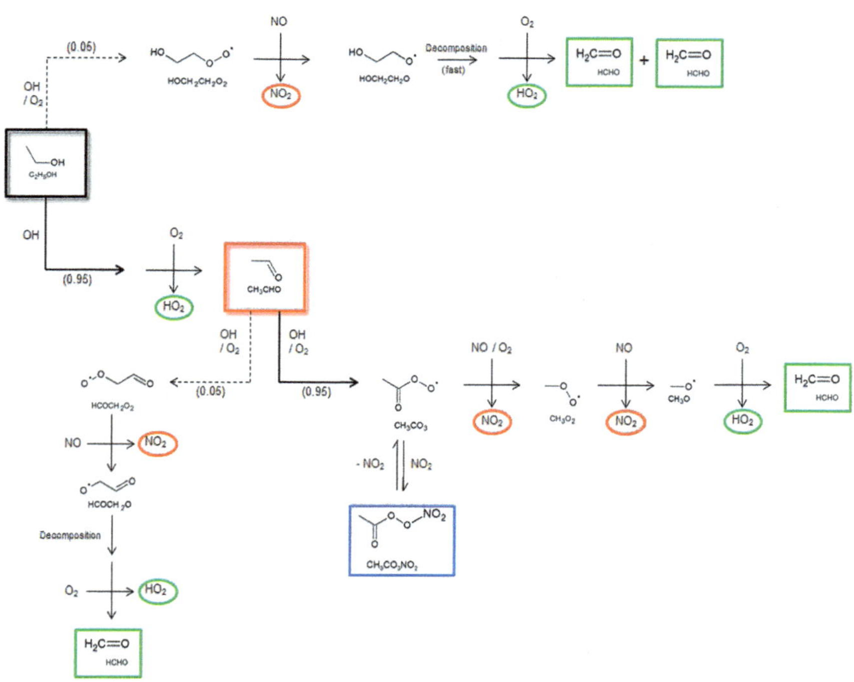

Fig. 1 Simplified schematic mechanism for the photo-oxidation of ethanol initiated by the OH radical taken from MCM.[25] Ethanol is boxed in black, with acetaldehyde in red, PAN (peroxyacetyl nitrate) in blue and formaldehyde in green. The NO to NO_2 conversions, highlighted by red circles, correspond to routes to photochemical ozone formation. HO_2 formation (highlighted in green circles) can also lead to NO to NO_2 conversions.

August) of 2012. For full details of the ClearfLo project and the IOP site refer to Bohnenstengel *et al.* (2014)[26] and Bigi and Harrison (2010)[27] respectively.

2.1 Gas chromatography measurements

A wide range of VOCs (C_1–C_{13}) were measured using two gas chromatography (GC) instruments, a dual channel GC flame ionisation detector (DC-GC-FID)[28] and a comprehensive two dimensional GC (GC × GC-FID). These measurements have been detailed extensively in Dunmore *et al.* (2015).[29] Briefly, the DC-GC-FID has two GC columns operating in parallel. After sampling and desorption, the flow is split equally onto; one aluminium oxide (Al_2O_3) Porous Layer Open Tubular (PLOT, 50 m, 0.53. mm id) column for the analysis of non-methane hydrocarbons (NMHCs); and a LOWOX column (10 m, 0.53 mm id) for the analysis of oxygenated VOCs (OVOCs). The GC × GC-FID system comprised of a Markes TT24-7 (Markes International, Llantrisant, UK) thermal desorption unit, a first dimension BPX-5 (SGE, Australia, 25 m × 0.15 mm, 0.4 μm df, 50 psi) column combined with a second dimension BP-20 (SGE, Australia, 5 m × 0.25 mm, 0.25 μm df, 23 psi) column and a total transfer flow valve modulator that incorporated a 6-port, 2-way diaphragm valve (Valco Instruments, Houston, TX, USA) where actuation was achieved using a solenoid valve, controlled by 'in house' written software. The modulation period was 5 seconds, with 4.7 second sample and 0.3 second injection times.

The calibration and quantification of VOC species measured using both GC instruments was detailed in the ESI of Dunmore *et al.* (2015).[29] The two main species of interest, ethanol and acetaldehyde were measured using the LOWOX columns on the DC-GC instrument. Both species were quantified using a relative response based on an initial quantification from a certified gas standard as detailed in Hopkins *et al.* (2011).[28]

2.2 Detailed chemical modelling using the master chemical mechanism

A zero-dimensional photochemical box model, based on MCMv3.2 was used to investigate the impact of current ethanol use on acetaldehyde concentrations. The MCM model was run in an analogous way as to that described in Lee *et al.* (2015).[30] Full details of the kinetic and photochemical data used as part of the model mechanism are available from the MCM website (http://mcm.leeds.ac.uk/MCM). A subset of the MCM was used in the model that treated the degradation of simultaneously measured trace VOC species, CH_4 and CO following oxidation initiated by OH, O_3 and NO_3, which included ~15 000 reactions and ~3800 species. The model was constrained to *in situ* measurements made at the North Kensington site of; NO, NO_2, O_3, HONO, CO, CH_4, 64 individually quantified VOC species from both GC systems, many more VOCs quantified in a grouped analysis (as detailed in Dunmore *et al.* (2015)[29]) PAN, HCHO, water vapour, temperature, pressure and photolysis rates. Model inputs were made every 15 minutes with the measured data either averaged or interpolated to 15 minute intervals if the species was measured more or less frequently, respectively. Only the summer IOP was modelled here, owing to the availability of extensive radical measurements.

2.3 Nested regional modelling using GEOS-Chem

To investigate the wider impacts of using ethanol blended gasoline, a regional version of the GEOS-Chem chemical transport model[31] (v10, http://www.geos-chem.org) was run at 0.5° × 0.666° resolution. Transport is driven by assimilated meteorology (GEOS-5) from NASA's Global Modelling and Assimilation Office (GMAO). Boundary conditions were provided by a global 4° × 5° resolution simulation run for the same period. The boundary conditions are provided to the regional model from a global run that has been 'spun up' for one year. The regional model is then 'spun up' for 3 months prior to the observation period to account for interannual meteorological variability. The model uses the Co-operative Programme for Monitoring and Evaluation of the Long-range Transmission of Air Pollutants in Europe (EMEP) emissions (http://www.emep.int) for NO_x, SO_x and VOCs. The base model has been extended to include a direct emission source of ethanol ('GEOS-Chem + EOH') as a simple multiplier on the ≥C3 alkenes emission rate and an additional scaling on acetaldehyde emissions ('GEOS-Chem + EOH + ACET'). The nested model is run from 10th December 2011 to 10th February 2012 for the winter campaign and 20th June to 20th August 2012 for the summer campaign, with the preceding month of output to the observational period discarded as 'spin up'.

3 Results

In terms of mixing ratio, ethanol was the most abundant VOC measured in London air during the ClearfLo campaigns. The winter and summer time series of atmospheric ethanol mixing ratios measured in London are shown in Fig. 2.

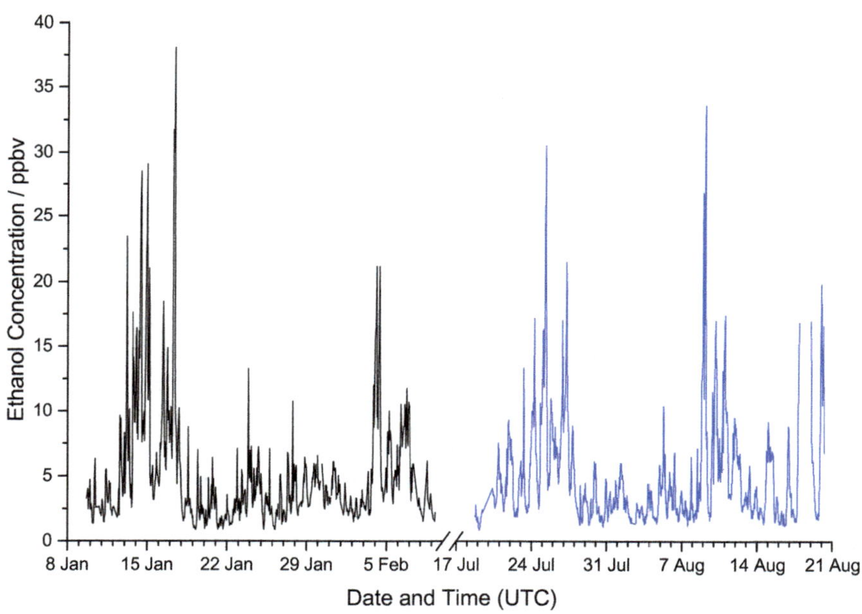

Fig. 2 Time series of ethanol in the winter (left panel, black) and summer (right panel, blue).

There were extended periods during both campaigns where very high levels of ethanol were observed (>10 ppb). During the winter and summer IOPs, the mean and maximum ethanol mixing ratios observed were: winter mean 5 ppb, winter maximum 38 ppb, summer mean 5 ppb and summer maximum 34 ppb.

The campaign average diurnal profiles of NO_x, CO, ethanol and acetaldehyde for winter and summer are shown in Fig. 3. These profiles suggest that all four species have a dominant traffic related source given the apparent 'rush-hour' peaks observed in winter (left hand plots) that peak at approximately 8/9 am and 5/6 pm, although part of this variability is also due to changing boundary layer depth (the mean boundary layer height as measured by LIDAR increases from 441 m (06:00–10:00) to 784 m (12:00–15:00) during the day).[32] During the summer (right hand plots) however, the average diurnals are largely influenced by meteorology (the mean boundary layer height increases from 1072 m (06:00–10:00) to 1757 m (12:00–15:00))[32] and photochemical processes. The boundary layer height observed was higher during the summer campaign,[32] leading to increased dilution of surface emissions. The ethanol winter diurnal profile (Fig. 3e) shows an increase at the end of the evening (21:00–23:00) which is not seen in the diurnal profiles of the other species. It is likely that there is a background source of ethanol, (*i.e.* from brewing and distilling processes), that is more prevalent during the later evening.

A comparison of coincident mixing ratios of ethanol and acetaldehyde is shown in Fig. 4. A strong linear correlation between ethanol and acetaldehyde is observed (R values = 0.89 in winter and 0.91 in summer). The data has been fitted using a least squares linear regression, with the gradient giving the ratio of the

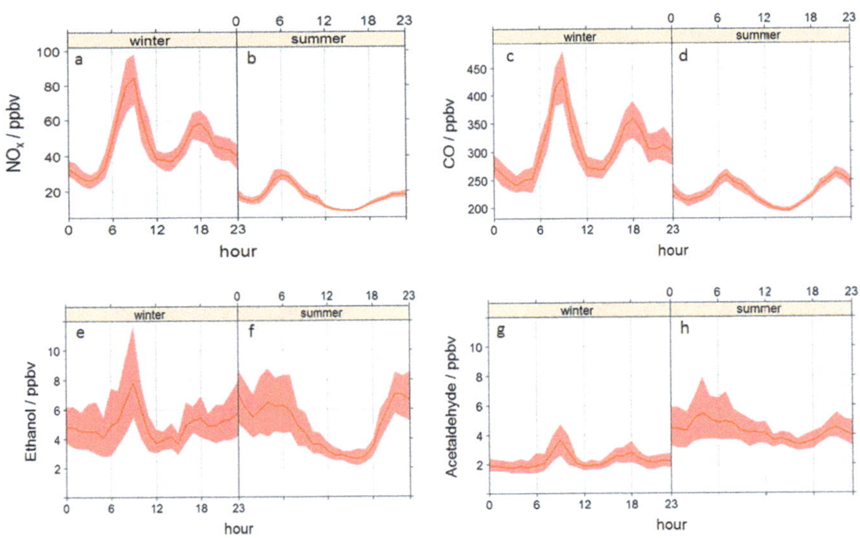

Fig. 3 Diurnal profiles of NO_x (a and b), CO (c and d), ethanol (e and f) and acetaldehyde (g and h) in winter (left-hand side of each plot) and summer (right-hand side of each plot). Each compound has the winter and summer plots on the same y-axis to aid direct comparison. This figure was constructed using the OpenAir project for R where the solid line represents the mean daily concentration and the shaded regions show the 95% confidence intervals surrounding the mean.[33–35]

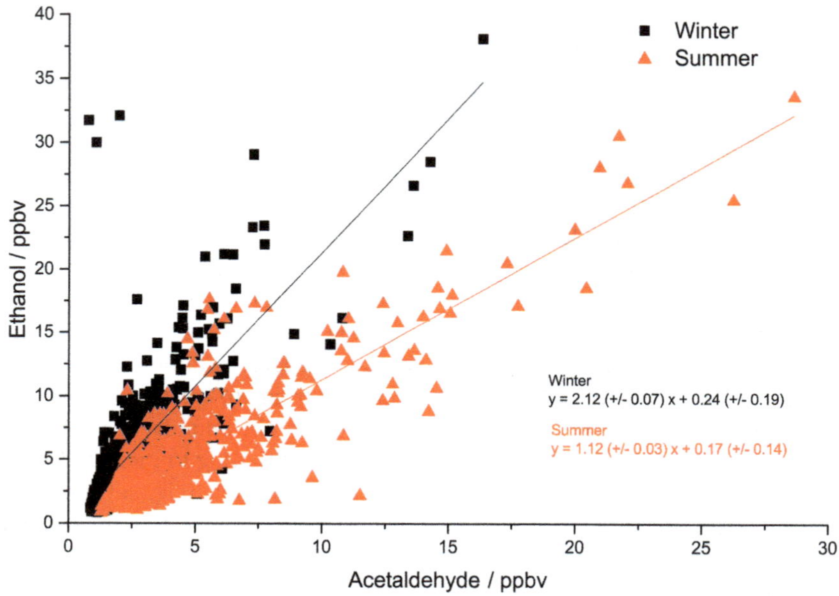

Fig. 4 Linear regression of ethanol and acetaldehyde in the winter (black squares) and summer (red triangles) campaigns.

observed concentrations (the equations of the linear regression are given on Fig. 4). The approximate ratio of ethanol : acetaldehyde changes from 2 : 1 in winter to 1 : 1 in summer. This may be due to the increased production of acetaldehyde in summer due to photochemical processing of ethanol and other VOCs or due to a change in the relative emission factors between the two seasons.

The linear regression correlation values of the observed ethanol mixing ratios with all VOC species measured during both seasons are given in ESI Tables S1 and S2.† During both campaigns, ethanol showed poor correlation with many of the other OVOC species, except acetaldehyde, likely due to the fact that ethanol has a direct anthropogenic source that is present in both seasons while the majority of the OVOCs, particularly in summer, are dominated by *in situ* photochemical production. The correlation values of selected species with ethanol are shown in Fig. 5, for winter (left column) and summer (right column). These species have first been ordered according to emission source, then by carbon number. In both seasons, ethanol has high correlations with gasoline related species, such as benzene, toluene, iso-octane (2,2,4-trimethyl pentane) and small carbon number aliphatic compounds, *n*-butane and 1,3-butadiene. Ethanol shows a stronger relationship in winter with petrol species that have a dominant combustion emission source (*n*-butane and 1,3-butadiene), whereas those with a more evaporation related emission have higher summer correlation values (benzene, toluene and iso-octane). The correlation between ethanol and CO, a combustion tracer, is stronger in winter than summer, but the summer value is still high (R of 0.90 and 0.83 in winter and summer respectively). Interestingly, the correlation between ethanol and isoprene is high in winter, due to a dominant traffic emission, but very low in summer when the biogenic emission source is dominant for isoprene.

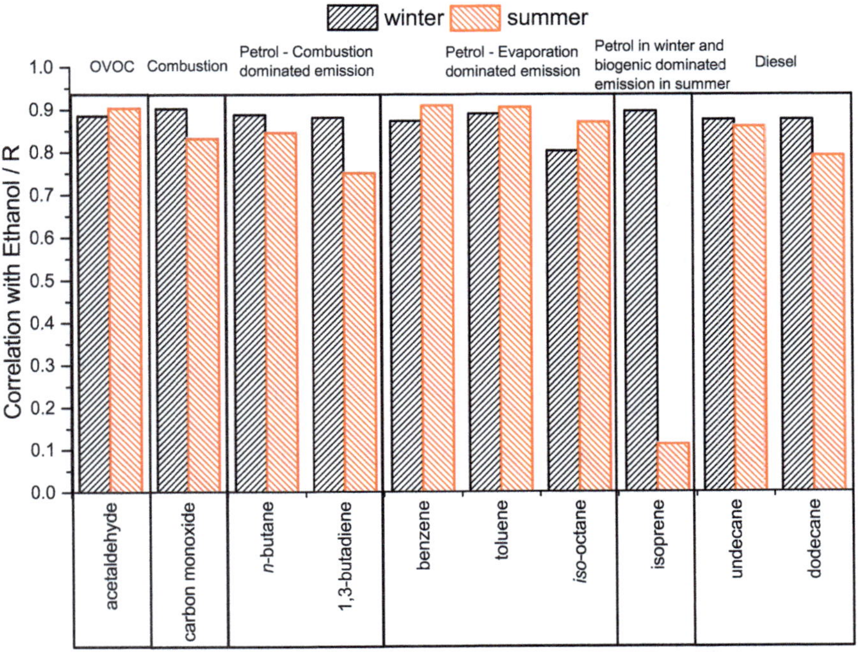

Fig. 5 Correlation of selected compounds with ethanol in winter (left column) and summer (right column), grouped by emission source and ordered by carbon number.

The most likely anthropogenic source of ethanol in London is related to gasoline use given the diurnal profiles (Fig. 3) and strong correlation with other gasoline related emissions (Fig. 5). The correlations between ethanol and benzene, toluene and iso-octane in summer are marginally higher than those in winter, which could be indicative of mutual evaporative emission from gasoline use and storage during periods of higher ambient temperatures (average summer temperature of 19.7 °C). In contrast, in winter (where temperatures are much lower, average winter temperature of 4.8 °C) a higher correlation is seen for compounds more likely to be from the combustion of gasoline, *i.e.* 1,3-butadiene and CO. This suggests that the emission of ethanol is a combination of both emission sources, with combustion of ethanol-blended fuels a larger source in winter and evaporation more important in summer.

3.1 Effect of ethanol content on acetaldehyde

Vasconcellos *et al.* (2005) studied the use of ethanol-blended fuels in Brazil. Generally, in urban areas there are higher observed mixing ratios of formaldehyde than acetaldehyde. However, in Brazil, where there is significant use of ethanol-blended fuels, the formaldehyde/acetaldehyde (f/a) ratio (ppb : ppb) observed was lower (in some cases less than 1), suggesting that large quantities of acetaldehyde are being directly emitted rather than formed in the atmosphere through photochemical processes.[36] Some studies use f/a ratios to suggest what the sources of these compounds are; a ratio of less than one is taken as direct emission of acetaldehyde while a ratio greater than one is representative of *in situ* photochemical formation.[36–38] Fig. 6 shows the f/a ratio from the winter (left, black) and summer (right, blue) ClearfLo campaigns in London, with the black line representing an f/a ratio of 1.

The measurements from London show a mix of the two different ratio conditions, with significant periods during the winter campaign where the ratio is <1. This suggests that the use of 5% ethanol-blended gasoline in the UK is directly emitting acetaldehyde into the atmosphere. During the winter campaign, the f/a ratio is below 1 during high-pressure systems, which resulted in a shallow boundary layer and stagnant air masses, allowing the accumulation of large concentrations of acetaldehyde. However, during the summer campaign, the ratio is much larger overall with very few points below 1. Interesting, during the night of the 8th August the f/a values are also <1. During this period, atmospheric concentrations at the North Kensington site were dominated by local emissions from London into a low boundary layer, with very low horizontal wind speeds and an unusual chemical regime, with low levels of oxidants (O_3, NO_3 and OH) leading to limited reactive chemistry. The higher ratio values in summer could suggest that the majority of acetaldehyde observed during this season is formed in the atmosphere through photochemical processing of a range of VOCs, however, this is contradicted by the lack of correlation between acetaldehyde and other OVOCs and the acetaldehyde diurnal profile observed. The higher ratios could also be due to the enhanced summer evaporative emission of ethanol from either gasoline use/storage and/or from distilling and brewing processes into the atmosphere, which then reacts with the OH radical to form acetaldehyde. It must be noted, however, that these ratios have been demonstrated to be most useful for determining the source of acetaldehyde in areas where there is a large use of high ethanol-blended gasoline, such as Brazil where the most used blend is E85. This is not the case in the UK, or London, where only E5 is used. Nevertheless, the fact

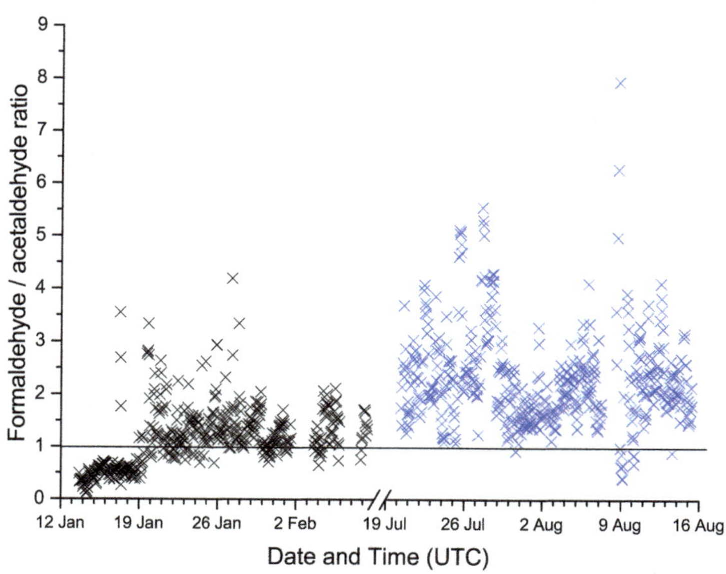

Fig. 6 Time series of the formaldehyde/acetaldehyde ratio from the winter (left, black) and summer (right, blue) campaigns, the black line indicates an f/a ratio of 1.

that f/a ratios below 1 are observed, indicates that there is a significant direct emission of acetaldehyde into London's atmosphere.

3.2 Impacts of ethanol blended fuel use on air quality

3.2.1 Local impacts. Detailed photochemical modelling incorporating the Master Chemical Mechanism (MCMv3.2) was used to determine the local impact of road transport-related ethanol emission on secondary species such as acetaldehyde. The model was constrained to the measured concentrations of all species, except acetaldehyde. Acetaldehyde formed photochemically from ethanol in the model simulations was tagged in the resulting data, so that it could be identified as being formed directly from ethanol.

Modelling results of the impacts of measured ethanol emissions on the concentrations of atmospheric acetaldehyde in London are shown in Fig. 7. On average, 6.5% of the acetaldehyde observed in London was a result of the photochemical oxidation of ethanol. Although the levels of acetaldehyde from the model degradation of ethanol (Fig. 7, black filled area) are relatively consistent and show a stable diurnal pattern throughout the campaign, there are periods where ethanol produces up to 18% of the observed acetaldehyde. The profile of the total modelled acetaldehyde (black and red filled areas) shows a diurnal variation typical of photochemical production and loss, with ethanol contributing

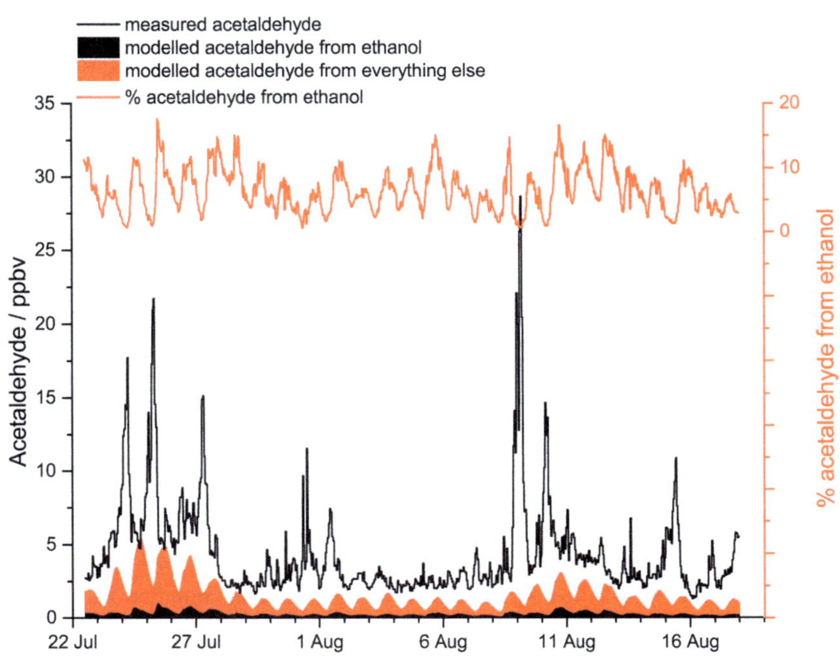

Fig. 7 Modelling results of the impacts of current levels of ethanol observed in London. The measured acetaldehyde during the summer campaign (black line), acetaldehyde formed in the model from the reaction of OH and ethanol (black filled area) and other photochemical acetaldehyde sources in the model (red filled area) are plotted on the left *y* axis. The percentage of the measured acetaldehyde that was directly formed from ethanol (red line) is plotted on the right *y* axis.

around 16% of the acetaldehyde produced in the model. However, it is clear that the detailed photochemical model simulation significantly under estimates the levels of acetaldehyde observed in London. This discrepancy is most likely due to a direct emission of acetaldehyde into the atmosphere from vehicle emissions.

A study by de Gouw *et al.* (2012) found that the ethanol emitted from the use of E10 in the US produced 20% of the acetaldehyde formed from other sources on the first day of photochemistry. They also found that on the second day this value increased to 90%, suggesting that the larger impact of ethanol emissions from fuel would be felt further away from the emission source.[4] It is worth noting that the US study only inputs 800 ppt of ethanol, while an average of 5 ppb of ethanol was observed during both ClearfLo campaigns implying that the impact of ethanol blended gasoline use in the UK could be more significant. Also worth noting is that while ethanol does not react particularly fast with the OH radical (kOH (298 K) $= 3.2 \times 10^{-12}$ cm^3 per molecule per s)[39] or produce significant quantities of O_3 (1.53 g O_3 per g VOC,[40] a relatively low maximum incremental reactivity (MIR) coefficient, a proxy for O_3 formation), the direct emission and photochemical production of large quantities of acetaldehyde (a compound which reacts rapidly with OH, kOH (298 K) $= 15 \times 10^{-12}$ cm^3 per molecule per s,[39] and has a high MIR coefficient, 6.54 g O_3 per g VOC)[40] indicates that the impacts of the use of ethanol fuels should not be judged solely on the reactions of ethanol.[18]

3.2.2 **Regional impacts.** The wider impacts of ethanol-blended gasoline use have been investigated using a regional version of the GEOS-Chem chemical transport model. Comparisons between model and measured CO, O_3, ethane, propane, $\geq C_3$ alkenes, and $\geq C_4$ alkanes are shown in ESI Fig. S3.† The calculated and measured concentrations of ethanol and acetaldehyde for the summer and winter campaigns are shown in Fig. 8. The base model simulation ('GEOS-Chem') does not consider ethanol, and hence is not plotted in Fig. 8A. Fig. 8 also shows a simulation that includes ethanol emissions ('GEOS-Chem + EOH', based on the emission of $\geq C_3$ alkenes as a proxy for vehicle emissions) tuned (multiplied by 50) to fit the observations. Appropriate ethanol degradation chemistry and deposition is included. Given the simplicity of the scaling, the model calculated ethanol concentrations appears to provide a reasonable estimate of the observations. From this simple scaling, the total annual European emissions of ethanol are equivalent to ~75 Gg per year which is ~134% of the ethane emission in the model.

The impact of the additional ethanol emission on acetaldehyde concentrations is shown in Fig. 8B. Consistent with the detailed chemical box modelling, a primary ethanol emission does increase concentrations of acetaldehyde but this increase is relatively small. A simulation was also run where the acetaldehyde emissions across Europe were increased by a factor of 40 ('GEOS-Chem + EOH + ALD2'). This effectively removes the bias between the acetaldehyde observations and the model during the winter but a significant bias remains during the summer. This result suggests that there is a more significant anthropogenic source of acetaldehyde than is currently considered but there may also be a secondary (potentially biogenic) source which is currently unrepresented in the model.

These changes (a primary ethanol source and an increased anthropogenic acetaldehyde source, 'GEOS-Chem + EOH + ALD2') increase the overall emissions of VOCs both for London but also across Europe. The average fractional increase in surface O_3 for Europe between 20th July and 20th August is shown in Fig. 9.

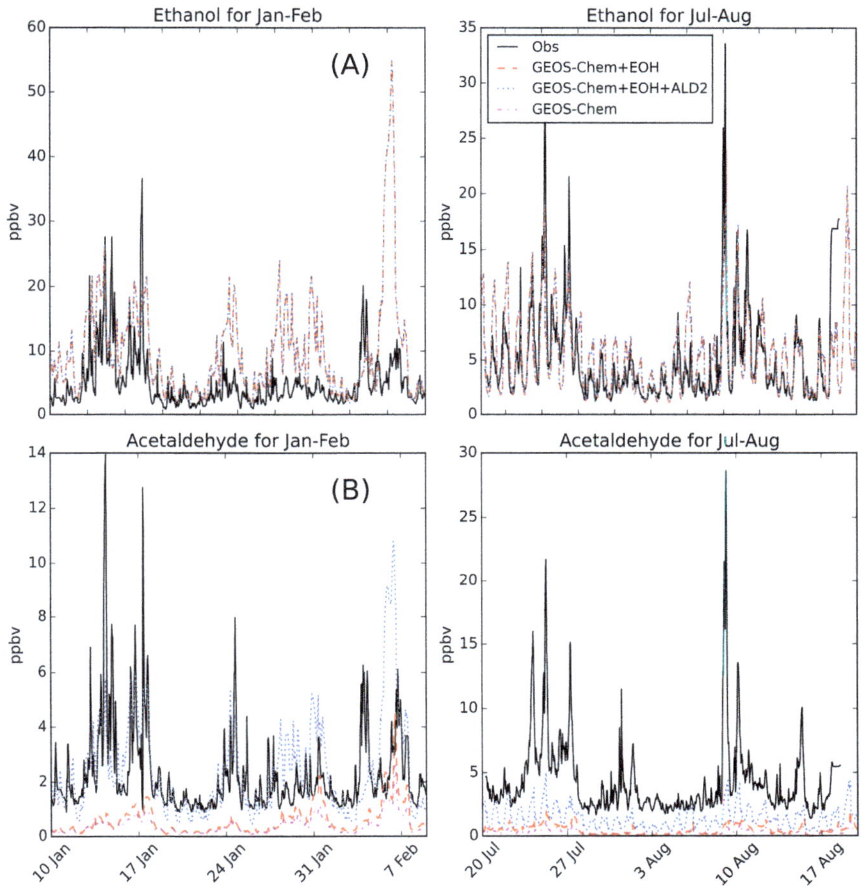

Fig. 8 Comparison between modelled values and observed for ethanol (A) and acetaldehyde (B) for winter (left) and winter (right) observation periods. Plots show the base model simulation without increased ethanol or aldehyde emissions ('GEOS-Chem') in magenta, with added ethanol emissions ('GEOS-Chem + EOH') in red, and with both additional ethanol and acetaldehyde emissions ('GEOS-Chem + EOH + ALD2') in blue.

Over and downwind of major emission regions (London, Benelux, Po Valley), large increases in acetaldehyde concentrations (>1000%) are simulated which is mainly due to the increase in emission of acetaldehyde. This leads to large increases in PAN concentrations (up to 140%) downwind of the emission regions (North Sea, Mediterranean). This increase in PAN leads to local reductions in NO_2 concentrations of up to 10% over polluted regions (Northern Europe, North Sea) but downwind from these regions the decomposition of the PAN leads to increased NO_2 concentrations (Mediterranean). The overall impact on O_3 concentration is complex. Increased OVOC emissions tend to increase O_3 production due to an enhanced availability of peroxy radicals. However the production of PAN leads to lower NO_2 concentrations close to source, reducing O_3 production, whilst increasing it downwind. Overall increases in surface O_3 are of the order of a few percent (up to 6.5%) over London and the Benelux countries. It is evident that the primary emission of ethanol and acetaldehyde can have a profound impact on the

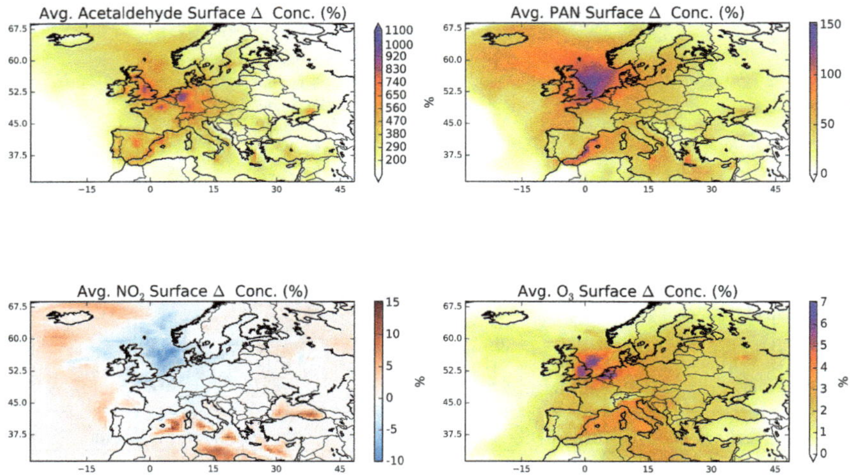

Fig. 9 Average fractional increases in surface acetaldehyde, PAN, NO_2 and O_3 concentrations with the inclusion of both a primary ethanol emissions and an increase in acetaldehyde emissions, ('GEOS-Chem + EOH + ALD2'). Shown averaged over the summer period.

concentration of OVOCs over Europe, which in turn can impact the concentration of both RO_x and NO_x, ultimately impacting the concentration of O_3.

4 Conclusion

We have shown that ethanol was the most abundant VOC in London's atmosphere during the winter and summer ClearfLo campaigns, and significant amounts of acetaldehyde were also observed. The diurnal profiles and correlation of these two species with other VOCs have shown that their main emission source is likely to be a combination of combustion and evaporative emissions from the use of ethanol blended gasoline. Detailed photochemical MCM simulations of the local scale impacts of measured ethanol has shown that, although ethanol does not seem to produce significant local levels of acetaldehyde, the model cannot accurately reproduce the observed profile of acetaldehyde, due to the presence of a direct emission into the atmosphere, with strong evidence that this is related to traffic emissions. It is clear that there are significant traffic related emissions of ethanol in urban areas and that these are not currently included in emission inventories.

In order to match the observed mixing ratios of ethanol and acetaldehyde in London, using GEOS-Chem model simulations (incorporating EMEP emissions), large additional sources were needed. This included the aforementioned unrepresented traffic related ethanol source and an increase in current acetaldehyde emissions by a factor of 40. This additional source would represent a European annual traffic-related emission of ethanol of around 75 Gg per year (*cf.* annual global estimates of anthropogenic emissions of ethanol of 2–10 Tg per year). This value assumes that similar scaling factors are required for all emissions inventories across Europe to match ambient concentrations. Further atmospheric observations are needed to accurately quantify the level of uncertainty in emission inventories.

Given the long lifetime of ethanol in the atmosphere, the full effects of its emission and the subsequent production of acetaldehyde may not be seen at the emission source location but rather downwind. The inclusion of previously unidentified traffic related ethanol and acetaldehyde sources in GEOS-Chem led to increases in PAN and a reduction in NO_2 downwind of emission sources. The subsequent impact on photochemical O_3 formation is complex, but small increases were predicted across Europe compared to the base case model. The missing acetaldehyde in current inventories is likely a result of (1) an underestimate of ethanol emissions and hence secondary photochemical production of acetaldehyde, (2) a direct emission source of acetaldehyde from ethanol combustion and (3) an underestimate of acetaldehyde emissions from gasoline combustion. Although increased use of bioethanol in European fuel may help reduce net carbon dioxide emissions we would suggest that there is the potential for an inadvertent impact on air quality, which should be monitored through the transition.

References

1 T. Murrells and Y. Li, *Road transport emissions from biofuel consumption in the UK*, Aea energy & environment, technical report, 2008.
2 S. López-Aparicio and C. Hak, *Sci. Total Environ.*, 2013, **452–453**, 40–49.
3 S. Manzetti and O. Andersen, *Fuel*, 2015, **140**, 293–301.
4 J. de Gouw, J. Gilman, A. Borbon, C. Warneke, W. Kuster, P. Goldan, J. Holloway, J. Peischl, T. Ryerson, D. Parrish, D. Gentner, A. Goldstein and R. Harley, *Geophys. Res. Lett.*, 2012, **39**, L15803.
5 H. Zhai, H. C. Frey, N. M. Rouphail, G. A. GonĂğalves and T. L. Farias, *J. Air Waste Manage. Assoc.*, 2009, **59**, 912–924.
6 W.-D. Hsieh, R.-H. Chen, T.-L. Wu and T.-H. Lin, *Atmos. Environ.*, 2002, **36**, 403–410.
7 S. H. Yoon and C. S. Lee, *Energy Fuels*, 2011, **25**, 3484–3492.
8 J. Gaffney and N. Marley, *Atmos. Environ.*, 2009, **43**, 23–36.
9 X. Pang, Y. Mu, J. Yuan and H. He, *Atmos. Environ.*, 2008, **42**, 1349–1358.
10 M. Costagliola, L. D. Simio, S. Iannaccone and M. Prati, *Appl. Energy*, 2013, **111**, 1162–1171.
11 P. J. Brooks and J. A. Theruvathu, *Alcohol*, 2005, **35**, 187–193.
12 S. Myou, M. Fujimura, K. Nishi, T. Ohka and T. Matsuda, *Am. Rev. Respir. Dis.*, 1993, **148**, 940–943.
13 W. Kluwe, C. Montgomery, H. Giles and J. Prejean, *Food Chem. Toxicol.*, 1983, **21**, 245–250.
14 J. Seinfeld and S. Pandis, *Atmospheric Chemistry and Physics: From air pollution to climate change*, J Wiley and Sons, INC, 2nd edn, 2006.
15 W. Kirstine and I. Galbally, *Crit. Rev. Environ. Sci. Technol.*, 2012, **42**, 1735–1779.
16 H.-H. Nguyen, N. Takenaka, H. Bandow, Y. Maeda, S. de Oliva, M. Botelho and T. Tavares, *Atmos. Environ.*, 2001, **35**, 3075–3083.
17 M. Colón, J. Pleil, T. Hartlage, M. Guardani and M. Martins, *Atmos. Environ.*, 2001, **35**, 4017–4031.

18 E. Grosjean, D. Grosjean, R. Gunawardena and R. Rasmussen, *Environ. Sci. Technol.*, 1998, **32**, 736–742.
19 P. Pereira, E. Santos, T. Ferreira and J. de Andrade, *Talanta*, 1999, **49**, 245–252.
20 M. Schilling, G. Voigt, T. Tavares and D. Klockow, *Fresenius' J. Anal. Chem.*, 1999, **364**, 100–105.
21 N. Passant, *Speciation of UK emissions of non-methane volatile organic compounds*, Aea technology report env-05452002 technical report, 2002.
22 A. Q. E. Group, *Ozone in the United Kingdom*, 2009.
23 A. Langford, E. Nemitz, E. House, G. Phillips, D. Famulari, B. Davison, J. Hopkins, A. Lewis and C. Hewitt, *Atmos. Chem. Phys.*, 2010, **10**, 627–645.
24 R. Derwent and M. Jenkin, *Atmos. Environ., Part A*, 1991, **25**, 1661–1678.
25 M. Jenkin, J. Young and A. Rickard, *Atmos. Chem. Phys.*, 2015, **15**, 11433–11459.
26 S. Bohnenstengel, S. Belcher, A. Aiken, J. Allan, G. Allen, A. Bacak, T. Bannan, J. Barlow, D. Beddows, W. Bloss, A. Booth, C. Chemel, O. Coceal, C. Di Marco, M. Dubey, K. Faloon, Z. Fleming, M. Furger, J. Gietl, R. Graves, D. Green, C. Grimmond, C. Halios, J. Hamilton, R. Harrison, M. Heal, D. Heard, C. Helfter, S. Herndon, R. Holmes, J. Hopkins, A. Jones, F. Kelly, S. Kotthaus, B. Langford, J. Lee, R. Leigh, A. Lewis, R. Lidster, F. Lopez-Hilfiker, J. McQuaid, C. Mohr, P. Monks, E. Nemitz, N. Ng, C. Percival, A. Prévôt, H. Ricketts, R. Sokhi, D. Stone, J. Thornton, A. Tremper, A. Valach, S. Visser, L. Whalley, L. Williams, L. Xu, D. Young and P. Zotter, *Bull. Am. Meteorol. Soc.*, 2014, **96**, 779–804.
27 A. Bigi and R. Harrison, *Atmos. Environ.*, 2010, **44**, 2004–2012.
28 J. Hopkins, C. Jones and A. Lewis, *J. Environ. Monit.*, 2011, **13**, 2268–2276.
29 R. Dunmore, J. Hopkins, R. Lidster, J. Lee, M. Evans, A. Rickard, A. Lewis and J. Hamilton, *Atmos. Chem. Phys.*, 2015, **15**, 9983–9996.
30 J. D. Lee, L. K. Whalley, D. E. Heard, D. Stone, R. E. Dunmore, J. F. Hamilton, D. E. Young, J. D. Allan, S. Laufs and J. Kleffmann, *Atmos. Chem. Phys. Discuss.*, 2015, **15**, 22097–22139.
31 T. Sherwen, M. J. Evans, L. J. Carpenter, S. J. Andrews, R. T. Lidster, B. Dix, T. K. Koenig, R. Volkamer, A. Saiz-Lopez, C. Prados-Roman, A. S. Mahajan and C. Ordóñez, *Atmos. Chem. Phys. Discuss.*, 2015, **15**, 20957–21023.
32 J. Barlow, C. Halios, S. Lane and C. Wood, *Environ. Fluid Mech.*, 2015, **15**, 373–398.
33 D. Carslaw and K. Ropkins, *Environ. Model. Software*, 2012, **27–28**, 52–61.
34 D. Carslaw and K. Ropkins, *R package version 0.6-2*, 2012.
35 R Development Core Team, *R: A Language and Environment for Statistical Computing*, R Foundation for Statistical Computing, Vienna, Austria, 2012.
36 P. Vasconcellos, L. Carvalho and C. Pool, *J. Braz. Chem. Soc.*, 2005, **16**, 1210–1216.
37 D. Grosjean, A. Miguel and T. Tavares, *Atmos. Environ., Part B*, 1990, **24**, 101–106.
38 J. de Andrade, H. Pinheiro and M. Andrade, *J. Braz. Chem. Soc.*, 1995, **6**, 287–290.
39 R. Atkinson, D. L. Baulch, R. A. Cox, J. N. Crowley, R. F. Hampson, R. G. Hynes, M. E. Jenkin, M. J. Rossi, J. Troe and I. Subcommittee, *Atmos. Chem. Phys.*, 2006, **6**, 3625–4055.
40 W. Carter, *SAPRC-07 Atmospheric Chemistry Mechanisms and VOC Reactivity Scales*, 2010, http://www.engr.ucr.edu/~carter/SAPRC/.

Faraday Discussions

PAPER

Biofuels, vehicle emissions, and urban air quality

Timothy J. Wallington,* James E. Anderson, Eric M. Kurtz and Paul J. Tennison

Received 1st December 2015, Accepted 7th January 2016
DOI: 10.1039/c5fd00205b

Increased biofuel content in automotive fuels impacts vehicle tailpipe emissions *via* two mechanisms: fuel chemistry and engine calibration. Fuel chemistry effects are generally well recognized, while engine calibration effects are not. It is important that investigations of the impact of biofuels on vehicle emissions consider the impact of engine calibration effects and are conducted using vehicles designed to operate using such fuels. We report the results of emission measurements from a Ford F-350 fueled with either fossil diesel or a biodiesel surrogate (butyl nonanoate) and demonstrate the critical influence of engine calibration on NO_x emissions. Using the production calibration the emissions of NO_x were higher with the biodiesel fuel. Using an adjusted calibration (maintaining equivalent exhaust oxygen concentration to that of the fossil diesel at the same conditions by adjusting injected fuel quantities) the emissions of NO_x were unchanged, or lower, with biodiesel fuel. For ethanol, a review of the literature data addressing the impact of ethanol blend levels (E0–E85) on emissions from gasoline light-duty vehicles in the U.S. is presented. The available data suggest that emissions of NO_x, non-methane hydrocarbons, particulate matter (PM), and mobile source air toxics (compounds known, or suspected, to cause serious health impacts) from modern gasoline and diesel vehicles are not adversely affected by increased biofuel content over the range for which the vehicles are designed to operate. Future increases in biofuel content when accomplished in concert with changes in engine design and calibration for new vehicles should not result in problematic increases in emissions impacting urban air quality and may in fact facilitate future required emissions reductions. A systems perspective (fuel and vehicle) is needed to fully understand, and optimize, the benefits of biofuels when blended into gasoline and diesel.

1. Introduction

Driven by a desire to improve energy security, address global climate change, and provide economic stimulus for rural communities, the use of biofuels in transportation has increased substantially over the past 10–20 years. Ethanol and

Research & Advanced Engineering, Ford Motor Company, Dearborn, Michigan 48121-2053, USA. E-mail: twalling@ford.com

biodiesel are the two most commercially important biofuels. Ethanol is produced commercially by the fermentation of sugars from corn, sugar cane, and cellulose. Biodiesel is produced from the trans-esterification of plant oils and animal fats. There are several routes to convert biomass into fuel for diesel engines and the nomenclature can be confusing. Fatty acid methyl esters (FAMEs) made from transesterification of plant oils or animal fats make up the majority of biomass-derived diesel fuel both in the U.S. and globally. The American Society of Testing Materials Standards (ASTM) defines biodiesel as "a fuel comprising mono-alkyl esters of long chain fatty acids derived from vegetable oils or animal fats that meets the requirements of ASTM D 6751".[1] Biodiesel is distinct from hydrocarbons produced from hydrotreating (and sometimes isomerizing) vegetable oils or animal fats which are referred to as "hydrotreated vegetable oil (HVO)" or "renewable diesel". There is interest in thermochemical conversion to produce biomass-to-liquid (BTL) fuels. One BTL option is gasification of biomass to give synthesis gas (mixture of CO and H_2) which can be converted into alkanes *via* the Fischer–Tropsch synthesis. The Fischer–Tropsch synthesis produces a mixture of straight-chain alkanes; it is well suited to the production of high-cetane diesel fuel. Another BTL option is the pyrolysis of biomass to produce bio-oil which could be blended with conventional oil and processed in a refinery. Renewable diesel and BTL are hydrocarbons and as such are effectively fungible with fossil diesel. Biodiesel is not fully fungible with fossil diesel as it has properties that differ substantially from fossil diesel.

The global production of biofuels increased by approximately a factor of four from 16 to 71 Mtonne oil equivalent between 2004 and 2014.[2] Although growing rapidly, the production of biofuels is still small when compared to the 4221 Mtonne of oil produced in 2014.[2] On a global basis ethanol accounts for the majority, approximately two thirds, of biofuel production, however there are important regional differences with biodiesel and HVO accounting for the majority, approximately 75%, of biofuel production in Europe. Ethanol is mainly used in blends with gasoline. Biodiesel is mainly used in blends with fossil diesel (commonly B2 [2% v biodiesel] or B20 [20% v biodiesel] in the U.S.). Ethanol use in fuel in the U.S. has increased approximately 4-fold over the last 10 years (2004–2014)[3] where it is now blended into most gasoline at a concentration of 10% v (E10). Ethanol is also available as E85, which is allowed to contain as much as 83% v and as little as 51% v ethanol.[4] In Brazil ethanol has been available as E18–E27 blends (gasohol) in recent years or as hydrous ethanol. Gasoline in the EU commonly contains 5% v ethanol, with E10 and E85 available in some locations. The U.S. Renewable Fuel Standard (RFS2) calls for an increase in use from approximately 13.4 billion gallons of ethanol and 1.4 billion gallons of biodiesel in 2014 to 36 billion gallons (136 billion liters) of renewable fuel by 2022, much of which would likely be ethanol. The European Union Renewable Energy Directive 2009/28/EC includes a target of 10% contribution by renewable sources such as biofuels in transportation fuel by 2020.

While the future trajectory of the global biofuel industry is hard to forecast, current regulations and concerns related to energy security, global climate change, and economic development for rural communities suggest long-term and increased use of biofuels. There are many important considerations for future fuel strategy, including fuel properties, refining sector implications, vehicle compatibility, refueling infrastructure compatibility, and transition timing.[5,6] An important consideration is the impact of biofuel content on tailpipe emissions of NO_x, volatile organic compounds (VOC), CO, and particulate matter (PM) and hence on urban air quality.[7,8] It is important to ensure that the current, and any future increased, biofuel use in automotive fuel does not hinder, and ideally supports, continuing progress in improving air quality.

There have been numerous studies on both engines and vehicles that have attempted to assess the influence of biodiesel on emissions. Almost universally, those studies have shown a reduction in PM with increasing biodiesel blend; however, there is significant variation in the effect of biodiesel on NO_x emissions. The most widely held belief is that there is a slight increase in NO_x with biodiesel.[9,10] However, several recent publications have suggested that the influence of biodiesel on NO_x is not a direct result of the fuel, but is due to the interaction of the properties of biodiesel with the engine calibration and control strategy.[11–13] To investigate this further we conducted vehicle testing to assess the change in emissions relative to diesel fuel when an oxygenated fuel is tested both with a production calibration and control strategy and with one that accounts and compensates for the properties of the oxygenated fuel.

Likewise, there have been numerous studies of the effect of ethanol fuel content on engine-out and vehicle tailpipe NO_x, non-methane hydrocarbons, PM, and mobile source air toxic emissions. Several of these studies include measurements from vehicles that were misfueled, such as using >E10 fuel in a vehicle designed and calibrated to operate on E0–E10. A transition to a future using more biofuel would include building the infrastructure to handle and use such fuels including vehicles designed to operate on the fuels. Studies of the effects of misfuelling are not relevant to addressing the question of how the future use of higher amounts of biofuels in vehicles designed for such fuels would impact emissions and hence air quality. To provide more clarity in future discussions of the effect of ethanol fuel on emissions and air quality we review and discuss the available data concerning the effect of ethanol blend content on vehicle emissions.

2. Experimental – vehicle testing with butyl nonanoate and fossil diesel fuels

Vehicle testing was performed with three different fuels: two fossil diesel fuels ("diesel A" and "diesel B") from a commercial supplier and butyl nonanoate (BN, $C_8H_{17}C(O)OC_4H_9$) synthesized by Michigan State University (MSU). Butyl nonanoate has been studied as a possible biodiesel fuel with desirable cold flow properties.[14] Diesel B is used for U.S. certification of vehicles for emissions and fuel economy. Diesel A is a diesel fuel with higher aromatic content. Properties of the three fuels are listed in Table 1.

A 2011 Ford F-350 with a 6.7 L diesel engine was tested on a chassis dynamometer using the three fuels over three different emissions test cycles: the cold-start Federal Test Procedure (FTP-75) representative of city driving, the highway fuel economy test (HWFET) representative of light highway driving, and the supplemental FTP test (US06) representative of aggressive driving including highway driving. Averages of emissions over 2 sets of tests are shown, with the exception of butyl nonanoate using the baseline calibration for which only 1 set of

Table 1 Properties of the fuels used in vehicle testing

	Diesel A	Diesel B	Butyl nonanoate
Net heating value [MJ kg^{-1}]	42.6	42.9	35.2
Density [g cm^{-3}]	0.856	0.846	0.864
Carbon [% wt]	0.871	0.870	0.740
Hydrogen [% wt]	0.129	0.130	0.124
Oxygen [% wt]	<0.005	<0.005	0.136
Aromatics [% v]	34	29	—
Saturates [% v]	63	68	—
Olefins [% v]	3	3	—
Kinematic viscosity, 40 °C [mm^2 s^{-1}]	2.39	2.43	2.10
Distillation, T_{10} [°C]	214	207	254
Distillation, T_{50} [°C]	253	256	254
Distillation, T_{90} [°C]	312	313	255
Cetane number	42	44	47

tests was conducted. FTP-75 data are bag-weighted averages for the three test phases. All tests were run at a vehicle test weight simulating a vehicle curb weight and payload totaling 9500 lbs (4300 kg). Standard emissions measurements of total hydrocarbon (THC), CO, and NO_x, were taken as well as PM measurements with a Dekati Mass Monitor (DMM). Measurements were taken at the engine-out location (also known as the catalyst feedgas location). For the two diesel fuels the production engine calibration was used. The production calibration was developed from emissions, performance, and fuel economy testing on multiple fuels, including diesel A and B. For the butyl nonanoate fuel, one set of tests was run utilizing the production calibration and two sets of tests were run with a modified calibration that compensated for differences in fuel net heating value (NHV), density, and oxygen content. The modified calibration adjusted the fuel quantities in each fuel pulse (for pilot, main, and post injections) to maintain an exhaust oxygen concentration equivalent to that with fossil diesel at the same operating conditions.[15]

3. Results – vehicle testing with butyl nonanoate and fossil diesel fuel

To better illustrate the emissions trends, the emissions data are normalized to those obtained using "Diesel A" using the production calibration. Fig. 1 shows the engine out, bag-weighted emissions of THC, CO, NO_x, and PM measured in the FTP-75 testing for the three fuels using the production calibration, and for butyl nonanoate using the adjusted calibration. As seen in Fig. 1, butyl nonanoate had lower THC emissions than diesel A with the production calibration. With the adjusted calibration, THC emissions for butyl nonanoate increased but were still lower than diesel A and similar to diesel B. Similar results are seen for CO, with lower CO emissions for butyl nonanoate with the production calibration, and CO emissions equivalent to those observed with diesel A with the adjusted calibration. The trend is very different for NO_x. With the production calibration, NO_x emissions for butyl nonanoate were higher than the fossil diesel fuels but with the

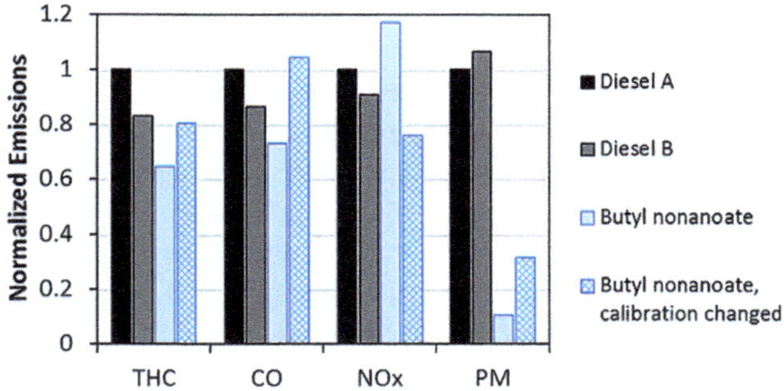

Fig. 1 Engine-out emissions of THC, CO, NO_x, and PM for the Federal Test Procedure (FTP-75 bag-weighted average) for the three fuels after engine calibration with diesel A. The right-most bar represents emissions for butyl nonanoate after adjusting the calibration to account for heating value differences as described in the text. All emissions (mg per mile) are normalized to the emissions for diesel A with the original calibration.

adjusted calibration were reduced compared to both fossil diesel fuels. The PM emissions benefits of butyl nonanoate are clearly evident in Fig. 1. PM emissions were approximately 90% lower than the fossil diesel fuels with the production calibration. The adjusted calibration results in an increase in PM emissions for butyl nonanoate, but they were still 70% lower than for the baseline diesel fuel.

Fig. 2 shows the results from emissions testing on the highway fuel economy test (HWFET). The overall trends in the HWFET data are similar to the FTP-75 data in Fig. 1, but with more substantial differences in the emissions changes. As can be seen in Fig. 2, butyl nonanoate with either calibration results in lower emissions of THC, CO, and PM relative to diesel A and B. Note that the engine responds differently to the initial engine and catalyst temperatures in the hot-start HWFET and cold-start FTP-75. When temperature sensors determine that the engine and catalyst are not starting in warmed-up condition, the calibration initiates engine operation in a catalyst warm-up mode which utilizes post injection and other actions to facilitate catalyst warm-up and thereby reduce cold-start

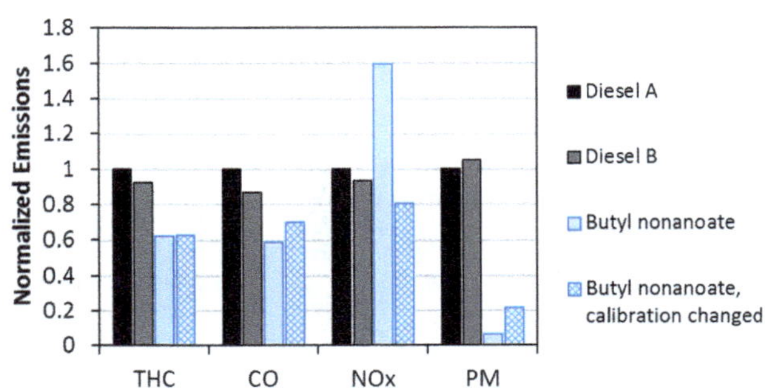

Fig. 2 Engine-out emissions of THC, CO, NO_x, and PM for the highway fuel economy test (HWFET) for the three fuels for the three fuels after engine calibration with diesel A. The right-most bar represents emissions for butyl nonanoate after adjusting the calibration to account for heating value differences as described in the text. All emissions (mg per mile) are normalized to the emissions for diesel A with the original calibration.

emissions. No such warm-up mode is employed for starting under warmed-up conditions. These operational differences have an impact on emissions for cold starting and hot starting. The NO_x data clearly show the dramatic effect the calibration changes can have on engine-out emissions. With no calibration changes NO_x emissions for butyl nonanoate are 60% higher than the baseline case, while with the calibration changes the NO_x emissions are 20% lower than baseline. This fuel property-based calibration adjustment leads to a 50% reduction in NO_x emissions and results in a substantial NO_x emissions reduction instead of an emissions increase relative to the baseline fuel. The PM emission trends for the HWFET cycle were similar to those for the FTP cycle; a >90% reduction in PM for butyl nonanoate with the baseline calibration and a slight increase (but still 80% lower than the baseline) for the tests with the adjusted calibration.

Results from the supplemental FTP test, US06, are shown in Fig. 3. The US06 is similar to the HWFET in that it involves warmed-up conditions for starting. Overall the results generally follow the same trends seen in both the FTP-75 testing and the HWFET testing. For both the US06 and HWFET cycles, THC and CO emissions for butyl nonanoate are lower than the fossil diesel fuels with both calibrations. Similarly, PM emissions are 90% and 80% lower with the production and adjusted calibrations, respectively. Emissions of NO_x with butyl nonanoate show an increase for the production calibration and a decrease back to the baseline fossil diesel level with the adjusted calibration.

In general, the emissions results for the biodiesel surrogate (butyl nonanoate) using the production engine calibration confirmed the typically reported findings of higher NO_x and lower PM relative to fossil diesel fuels,[10] though the NO_x emissions increase (20–60%) and PM emissions decrease (90–95%) were greater than typically reported, perhaps due to the lack of unsaturation in butyl nonanoate and its somewhat higher oxygen content (14% wt) compared to conventional biodiesel from plant oils and animal fats (typically ~11% wt oxygen). Emissions of THC and CO with butyl nonanoate were also reduced by 20–40% relative to the baseline diesel. However, after adjusting the calibration to compensate for the different fuel properties of butyl nonanoate, NO_x emissions

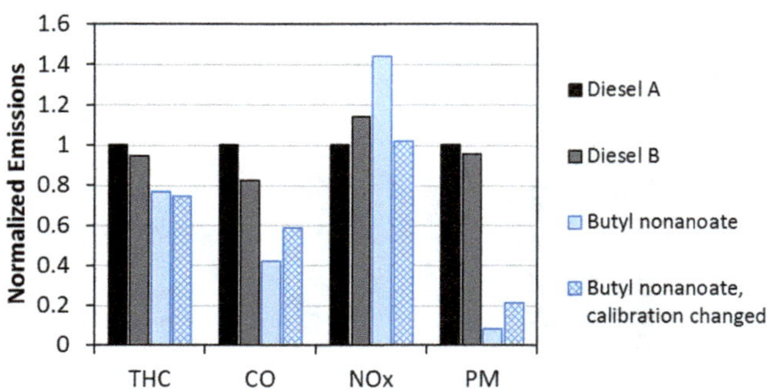

Fig. 3 Engine-out emissions of THC, CO, NO_x, and PM for the US06 driving cycle for the three fuels after engine calibration with diesel A. The right-most bar represents emissions for butyl nonanoate after adjusting the calibration to account for heating value differences as described in the text. All emissions (mg per mile) are normalized to the emissions for diesel A with the original calibration.

were significantly decreased relative to the baseline production calibration, causing NO_x emissions to change from being 20–60% higher than the fossil diesel baseline to 0–20% lower than the baseline on the three emissions test cycles. These calibration changes for butyl nonanoate also resulted in negligible to modest increases in THC and CO emissions (0–20% for THC and 10–30% for CO), but these emissions were still considerably less than or (in one case) similar to the fossil diesel baseline. Likewise, the 90–95% reduction in PM emissions for butyl nonanoate with the baseline calibration was somewhat diminished by the adjusted calibration, but were still 70–80% less than the baseline diesel fuel. These results highlight the importance of engine calibration effects on emissions for biodiesel or other biofuels which have properties that are different from fossil diesel fuel.

As described previously[11,13,16] NO_x emission increases reported for biodiesel are primarily related to higher peak temperatures during combustion. These higher temperatures are caused by an increase in the intake oxygen concentration for a given EGR rate with an oxygenated fuel, compounded by a shift in the calibration settings to lower EGR rate, higher fuel injection pressure, and higher boost pressure triggered by the pedal position required for increased fuel quantity to compensate for the energy density difference of the fuel. PM emissions reductions for biodiesel are primarily related to fuel oxygen content promoting higher oxygen content at the flame liftoff length as well as reduced content of aromatic hydrocarbons and other soot-prone chemical structures.

Therefore, increased biodiesel content for diesel engines can provide the well-accepted reductions in PM and other emissions, and need not result in problematic increases in NO_x emissions if, in a forward-looking fashion, vehicles have been appropriately designed and calibrated for these fuels. Optimal outcomes for emissions and efficiency for introduction of future fuels and vehicles requires a systems level perspective involving both fuel and vehicle considerations.

4. Literature review of effect of ethanol fuel content on vehicle emissions

Understanding the effects of ethanol fuel content on vehicle emissions reported in the literature is complicated by many factors, not least of which being the different ethanol blending strategies employed. When preparing ethanol blends there are two general approaches; splash-blending and match-blending. As the name implies, in splash-blending the base gasoline blendstock is fixed and the desired ethanol level is simply added (splashed) into the base gasoline. In splash-blending the ethanol-gasoline blend compositions are clearly defined, and effects on emissions are relatively straightforward to interpret. In contrast in match-blending the blendstock composition is modified for each ethanol-gasoline blend to match one or more fuel properties (*e.g.*, octane number, vapor pressure). For emission measurements using match-blending, the ethanol-gasoline blend compositions are often not clearly defined. The effects on emissions depend on which fuel properties are matched and what modifications are made, making emission trends difficult to interpret. As an example, Graham *et al.*[18] reported increases in emissions of 1,3-butadiene and benzene for E10 compared to E0 fuel. It is difficult to understand how increased ethanol fuel content would lead to increased 1,3-butadiene and benzene emissions. As noted by Graham *et al.*,[18] the observed emission trend probably reflected the higher benzene and aromatic content in the E10 fuel rather than its higher ethanol content.

Higher ethanol content in gasoline affects several fundamental fuel properties that can impact emissions, including increased oxygen content, decreased volumetric energy content typically measured as NHV, increased heat of vaporization (HoV), and other volatility changes.[18,19] These changes can have positive or negative effects that depend on engine design, hardware, and control strategy.[20] In addition to direct emissions impacts, higher ethanol content fuel can also provide more efficient combustion and overall engine operation under part-load conditions[21] and under knock-limited higher-load conditions.[22,23] Conventional gasoline vehicles in the U.S. are designed to operate on E0–E10 and have been certified for emissions compliance by testing with E0 fuel, but now transitioning to E10 as part of U.S. Tier 3 and California LEV-III emissions regulations. Flexible fuel vehicles (FFVs) are designed for operation on E0–E10 and E85 and have been certified for emissions compliance by testing with E0 and E85. While many studies have attempted to measure ethanol-related emissions changes, some studies are performed with vehicles not intended for such fuels, *e.g.*, using >E10 blends in vehicles designed to operate on E0–E10.[24–26] In Table 2 we exclude the results from such studies involving vehicle misfuelling. Finally, we note that there have been large advances in the efficiency of engine and emission control systems over the past few decades and we restrict our review to literature concerning 2000 model year (2000MY) vehicles or later. Given the different vehicle technologies involved it is convenient to separate the discussion of variation of ethanol fuel content over the ranges E0–E10 and E0–E85.

4.1 Effect of E0–E10 on vehicle emissions

The "EPAct/V2/E-89" study conducted by the U.S.E.P.A.[26] provides the most comprehensive dataset available to assess the effect of variation of ethanol blend over the range E0–E10 on vehicle emissions. The EPAct/V2/E-89 study used 15 different MY2008 vehicles with 17 different E0/E10 test fuels. Five fuel properties were varied in the study; ethanol volume, aromatic content, RVP, T_{50}, and T_{90}. T_{50} and T_{90} are the temperatures at which 50 and 90% of the fuel has evaporated for the standard ASTM D86 distillation test. A full chemical speciation was provided for each of the fuels. The chemical speciation allows calculation of the fuel particulate matter index[27] which is a predictive model of PM emissions based on the chemical composition of the fuel. Fig. 4 shows plots of vehicle NMOG and NO_x emissions *versus* fuel T_{50}, and Fig. 5 shows vehicle PM emissions *versus* fuel PM index,[27] for E0 and E10 fuels in the EPAct/V2/E-89 study. As can be seen from Fig. 4 and 5 there is a general trend of decreased NMOG and NO_x emissions with increasing T_{50}, and increased PM emissions with fuel PM index, but there is no discernable dependence of NMOG, NO_x, or PM emissions with ethanol content over the range E0 to E10. Within that data set, U.S.E.P.A.[28] showed an apparent PM increase for some vehicles with increasing ethanol content, however others showed no effect.

Recently, George *et al.*[29] reported VOC emissions from 1 conventional 2008MY and 2 flexible fuel 2008MY vehicles using summer and winter grade E0 and E10 for vehicle operating temperatures of −7 °C and 24 °C. George *et al.*[29] reported

Table 2 Literature data for E0–E85 ethanol fuel content impacts on VOC, CO, and NO_x vehicle emissions[a]

Study	Scope	Impact of increased ethanol on tailpipe emissions
Graham *et al.* (2008)[17]	One 2002MY, one 2004MY FFV, E0, E85, FTP cycle	NO_x and NMHC decreased by 45% and 48%, formaldehyde and acetaldehyde increased by 73% and 2540%. No discernable change in CO or NMOG emissions
Haskew and Liberty (2011)[32]	One 2006MY, six 2007MY FFVs, E6, E32, E59, E85, match blended (vapor pressure), FTP, US06, LA92 test cycles	Average NMHC, NMOG, CO, NO_x emissions did not exhibit an emissions trend with increasing ethanol content, with exception of decreasing NMHC and NMOG for US06 cycle
Karavalakis *et al.* (2012)[24]	One 2007MY FFV, E0, E10, E20, E50, E85, splash blended, FTP cycle	No statistically significant trends for CO or NO_x emissions; THC and NMHC increased for E85 but not for lower blends
Yassine and La Pan (2012)[33]	One 2006MY FFV, E0, E20, splash blended, FTP cycle	THC, NMOG, CO, NO_x decreased by 58%, 42%, 83%, and 60%, respectively, for E20 compared to E0
Yanowitz *et al.* (2013)[30]	Nine (2002–2011)MY FFVs, E10, E40, E76, LA92 cycle	Tested immediately after refueling with E40 having been previously adapted to either E10 or E76. Average decreases in NMOG, CO, and NO_x emissions of 5%, 10%, and 8% for E40 compared to E10 fuel
Karavalakis *et al.* (2014)[31]	One 2013MY, one 2014MY FFV, E10, E51, E83, match blended, FTP, UC cycles	NMHC, CO, and PM decreased, acetaldehyde increased, no trend for THC and NO_x with increased ethanol
Hubbard *et al.* (2014)[34]	One 2006MY FFV, E10, E20, E30, E40, E55, E80, splash blended, FTP cycles	Acetaldehyde, ethanol, and methane emissions increased, NO_x and NMHC decreased, no discernable effect on CO emissions. NMOG and THC emissions had minimum (30–35% lower than E0 or E85) for E20–E40
George *et al.* (2015)[29]	Two 2008MY FFVs, E0, E10, E85 (summer & winter), match blended (vapor pressure), LA92 cycle	NMHC decreased, acetaldehyde increased, no trend in VOC (NMOG) emissions, no trend in ozone-formation potential with increased ethanol
Suarez-Bertoa *et al.* (2015)[35]	One 2012MY FFV, E5, E10, E15, E85, splash blended, WLTC cycle	NO_x emissions decreased by 30–55%, CO, CH_4, formaldehyde, acetaldehyde, and ethanol emissions increased by 65%, 150%, 100%, 120%, and 350%, ozone-formation potential increased by factor of 2 for E85 *versus* E5–E15

Table 2 (*Contd.*)

Study	Scope	Impact of increased ethanol on tailpipe emissions

[a] FTP, federal test procedure; UC, unified cycle also known as the LA92 cycle; US06, high acceleration aggressive cycle also referred to as the supplemental FTP cycle; WLTP, worldwide harmonized light vehicles test procedure.

that there was no statistically significant difference in the sum of VOC emissions, the sum of the maximum incremental reactivity weighted ozone forming potentials of VOC emissions, or the sum of mobile source air toxic emissions between vehicles fueled with E0 or E10. The available data suggest that over the range E0 to E10 there is no discernable influence of ethanol fuel content on emissions of NMOG, NO_x, or PM from light-duty vehicles.

4.2 Effect of E10–E85 on vehicle emissions

The literature data upon which to assess the impact of ethanol concentration for blends intermediate between E10 and E85 on FFV emissions[24,29–35] are summarized in Tables 2 and 3. From inspection of Table 2 it is clear that there is a substantial range of results reported in the literature. Given that the bulk of organic compounds present in vehicle exhaust are unburned or partially burned fuel, it is to be expected that increasing the ethanol fuel content would lead to increased emissions of ethanol and its oxidation products acetaldehyde and formaldehyde, and decreased emissions of gasoline components and their oxidation products. Consistent with these expectations, there is general agreement in the literature data that increased ethanol fuel content leads to increased emissions of ethanol and acetaldehyde and decreased emissions of non-methane hydrocarbons (NMHC). There is also agreement that increased ethanol levels lead to comparable, or decreased, levels of NO_x and CO emissions. There is no consensus on the trend for total hydrocarbon emissions with one study reporting an increase,[24] one reporting a decrease,[33] and one reporting a minimum at about E40 (ref. 34) which probably reflects differences in conversion of acetaldehyde to methane on rhodium catalysts in the different vehicles.[34] There is also no

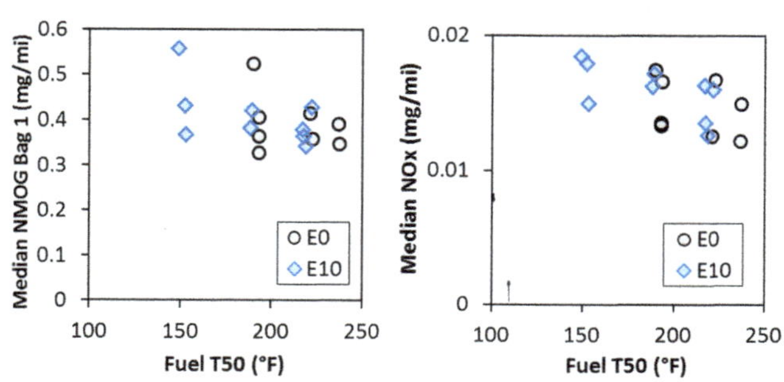

Fig. 4 NMOG and NO_x emissions from E10-capable vehicles on LA92 cycle when fueled with either E0 (black circles) or E10 (blue diamonds) as functions of fuel T_{50}.

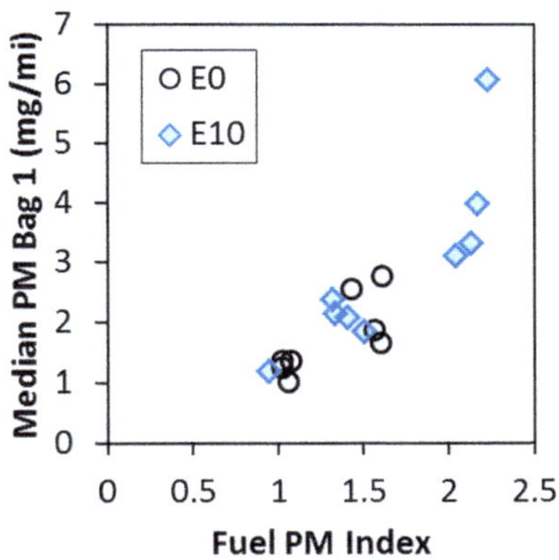

Fig. 5 PM emissions from E10-capable vehicles on LA92 cycle (bag 1) when fueled with either E0 (black circles) or E10 (blue diamonds) as functions of fuel T_{50}.

Table 3 Literature data for E0–E85 ethanol fuel content impacts on PM vehicle emissions

Authors	Scope	Tailpipe emissions
He *et al.* (2010)[37]	One 2009MY engine/emission system, E0, E10, E20, splash blended, 10 steady state engine operating modes	Compared to the baseline gasoline (E0), E10 does not significantly change PM emissions, while E20 and BU12 both reduce PM emissions under the conditions studied
Storey *et al.* (2010)[38]	One 2007MY vehicle, E0, E10, E20, splash blended, FTP, US06 cycles	No significant change in PM emissions from E0 to E10. From E0 to E20, the average mass emissions, declined 30% and 42% over the FTP and US06 cycles, respectively
Maricq *et al.* (2012)[39]	Light-duty truck, 3.5-L V6, E0, E10, E17, E32, E45 fuels, FTP test cycle	"As the ethanol level in gasoline increases from 0% to 20%, there is possibly a small (<20%) benefit in PM mass and particle number emissions, but this is within test variability. When the ethanol content increases to >30%, there is a statistically significant 30–45% reduction in PM mass and number emissions observed for both engine calibrations"
Karavalakis *et al.* (2014)[31]	One 2013MY, one 2014MY FFV, E10, E51, E83 fuels, FTP, UC cycles	"Particulate matter (PM) mass, number, and soot mass emissions showed strong reductions with increasing alcohol content in gasoline. Particulate emissions were found to be clearly influenced by certain fuel parameters including oxygen content, hydrogen content, and aromatics content"

consensus on the effect of ethanol fuel content on the ozone forming potential of VOC emissions calculated by weighting the VOC emissions by their maximum incremental reactivity factors.[36] One study reported no trend in ozone-forming potential[29] while another reported a factor of approximately 2 increased ozone-forming potential[35] with increased ethanol fuel content. Literature data for the effect of ethanol on PM emissions[31,37–39] are given in Table 3. As seen from Table 3, there is general agreement that for modest increases in ethanol content there is not a discernable effect on PM emissions but large increases in ethanol lead to decreased PM emissions.

The impact of increased ethanol fuel content on mobile source air toxic (MSAT) emissions has been reviewed recently by Stein *et al.*[20] Regulations in California require measurement of tailpipe emissions of the toxic compounds 1,3-butadiene, benzene, formaldehyde, and acetaldehyde and that the potency-weighted sum of these compounds is below a limit that is based on the corresponding value for the certification fuel. Tailpipe measurements show increased emissions of formaldehyde and acetaldehyde and decreased emissions of 1,3-butadiene and benzene as ethanol fuel content is increased. There was no discernable trend of the toxicity weighted sum of 1,3-butadiene, benzene, formaldehyde, and acetaldehyde with ethanol fuel content.[20] The available data indicate that emissions of NO_x, NMOG, PM, and MSATs from modern vehicles are not adversely affected by increased ethanol fuel content over the range for which the vehicles are designed to operate.

5. Vehicle emissions and urban air quality

To place into context the impacts of increased biofuel content on vehicle emissions and urban air quality it is useful to consider the historical trend of vehicle

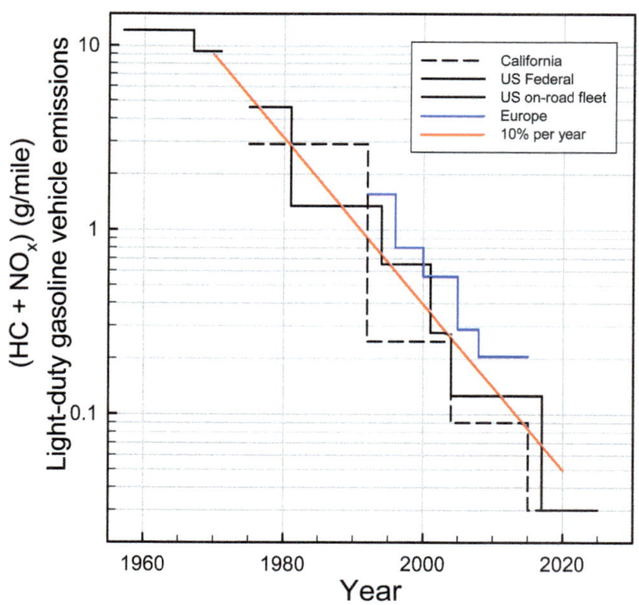

Fig. 6 Emission standards from gasoline light-duty vehicles. Data taken from Fegraus *et al.*[40] and TransportPolicy.net website (http://transportpolicy.net). See text for details.

emissions and urban air quality. To illustrate the trend for light-duty vehicles, Fig. 6 shows the combined hydrocarbon and NO_x emissions in units of g per mile. The data for 1957–1967 and 1967–1971 are emissions measured for vehicles representative of the on-road U.S. fleet.[40] The data for 1975–2025 are the California, Federal U.S., and EU regulatory requirements that vehicle manufacturers are required to meet for new vehicles.[41] The red line shows a 10% annual reduction to aid visual inspection of the data trend. As illustrated in Fig. 6, the per-vehicle per-mile emissions from light duty vehicles have been decreasing at a rate of approximately 10% per year, have decreased by orders of magnitude over the past 50 years, and will continue to decrease consistent with existing regulations. Reductions in emissions from the on-road fleet lag behind those shown for new vehicles in Fig. 6 reflecting the 10–20 years taken for turnover of the vehicle fleet.

As the result of actions taken to reduce emissions from vehicles and other sources there is a clear trend of improving air quality in U.S. cities.[42] From 1970 to 2014 the number of vehicle miles traveled in the U.S. increased by 172% but the aggregate emissions of CO, Pb, NO_x, VOCs, PM10, PM2.5 and SO_2 decreased by 69%.[42] Levels of O_3, PM, CO, NO_2, and SO_2 have decreased substantially in urban air in U.S. cities over the past several decades.[42] While remarkable progress has been made in improving air quality, more progress is needed. In 2014 more than 57 million people in the U.S. lived in areas which did not meet one or more of the National Ambient Air Quality Standards for ozone, PM, SO_2 or Pb.[42]

As discussed above the available data suggest that emissions of NO_x, non-methane hydrocarbons, PM, and mobile source air toxics from modern gasoline and diesel vehicles are not adversely affected by increased biofuel content over the range for which the vehicles are designed to operate. We highlight the importance of engine calibration as a critical factor determining vehicle emissions. Considerable engineering efforts go into the optimal design and calibration of engines and their aftertreatment systems to meet increasingly rigorous tailpipe emissions standards. Calibrations optimized for conventional fuels are not likely to be optimal for biofuels which typically have physical and chemical properties which differ significantly from fossil fuels. The fuel and the vehicle need to be thought of as one system. The available data suggest that increased biofuel content together with changes in engine design and calibration for new vehicles to accommodate increased biofuel content should not result in problematic increases in emissions impacting urban air quality and may in fact enable future emissions reductions in new vehicles designed for these fuels. As discussed above, vehicles need to be operated on fuels for which they were designed. Older vehicles in the on-road fleet need to be provided with appropriate fuels and there may be challenges to convert the existing infrastructure to accommodate substantially higher biofuel blends. A discussion of such challenges is beyond the scope of the present study. Further research is needed using a systems perspective to better understand, and optimize, the benefits of biofuels when blended into gasoline and diesel.

Disclaimer

This report was prepared as an account of work sponsored by an agency of the United States Government. Neither the United States Government nor any agency thereof, nor any of their employees, makes any warranty, express or implied, or assumes any legal liability or responsibility for the accuracy, completeness, or usefulness of any information, apparatus, product, or process disclosed, or represents that its use would not infringe privately owned rights. Reference herein to any specific commercial product, process, or service by trade name, trademark, manufacturer, or otherwise does not necessarily constitute or imply its endorsement, recommendation, or favoring by the United States Government or any agency thereof. The views and opinions of authors expressed herein do not necessarily state or reflect those of the United States Government or any agency thereof.

Acknowledgements

Butyl nonanoate was synthesized and provided by Prof. Dennis Miller and Dr Lars Peereboom of Michigan State University. This material is based upon work supported by the Department of Energy under Award Number DE-FC26-07NT43278.

References

1 ASTM D6751-15b, *Standard Specification for Biodiesel Fuel Blend Stock (B100) for Middle Distillate Fuels*, ASTM International, West Conshohocken, PA, 2015.
2 *BP Statistical Review of World Energy June 2015*, British Petroleum, 2015, http://www.bp.com.
3 U.S. Energy Information Administration, Monthly Energy Review March 2015, http://www.eia.gov/totalenergy/data/monthly.
4 ASTM International, *Standard Specification for Ethanol Fuel Blends for Flexible-Fuel Automotive Spark-Ignition Engines*, ASTM D, 2011, pp. 5798–11, Rev.
5 J. E. Anderson, R. E. Baker, T. J. Wallington and J. M. Ginder, Implications of the 2007 Energy Independence and Security Act for the US Light-Duty Vehicle Fleet, *SAE Tech. Pap. Ser.*, 2009, DOI: 10.4271/2009-01-2770, 2009-01-2770.
6 J. E. Anderson, D. M. DiCicco, J. M. Ginder, U. Kramer, T. G. Leone, H. E. Raney-Pablo and T. J. Wallington, High Octane Number Ethanol–Gasoline Blends: Quantifying the Potential Benefits in the United States, *Fuel*, 2012, **97**, 585–594.
7 U.S. Environmental Protection Agency, *Biofuels and the Environment: First Triennial Report to Congress. EPA/600/R-10/183F*, 2011.
8 J. S. Gaffney and N. A. Marley, The Impacts of combustion emissions on air quality and climate: from coal to biofuels and beyond, *Atmos. Environ.*, 2009, **43**, 23–36.
9 G. Knothe, C. A. Sharp and T. W. Ryan, Exhaust Emissions of biodiesel, petrodiesel, neat methyl esters, and alkanes in a new technology engine, *Energy Fuels*, 2006, **20**, 403–408.
10 A Comprehensive Analysis of Biodiesel Effects on Exhaust Emissions, EPA Draft Technical Report, EPA420-P-02-001, October 2002.
11 E. M. Kurtz, D. Kuhel, J. E. Anderson and S. A. Mueller, A comparison of combustion and emissions of diesel fuels and oxygenated fuels in a modern DI diesel engine, *SAE Int. J. Fuels Lubr.*, 2012, **5**, 1199–1215, DOI: 10.4271/2012-01-1695.

12 W. A. Eckerle, E. J. Lyford-Pike, D. W. Stanton, L. A. LaPointe, S. D. Whitacre and J. C. Wall, Effects of methyl ester biodiesel blends on NO_x emissions, *SAE Tech. Pap. Ser.*, 2008, 2008-01-0078.

13 D. B. Snyder, G. H. Adi, C. H. Hall and G. M. Shaver, Control-variable-based accommodation of biodiesel blends, *Int. J. Engine Res.*, 2011, **12**, 564–579.

14 A. L. Lown, L. Peereboom, S. A. Mueller, J. E. Anderson, D. J. Miller and C. T. Lira, Cold flow properties for blends of biofuels with diesel and jet fuels, *Fuel*, 2014, **117**, 544–551.

15 E. Kurtz, D. D. Brehob, C. W. Vigild, D. J. Dronzkowski, J. Pilbeam and J. E. Anderson, Compensation for oxygenated fuels in a diesel engine, *US Pat.* #8046153, issued on 10/25/2011.

16 W. Eckerle, E. Lyford-Pike, D. Stanton, L. LaPointe, S. D. Whitacre and J. C. Wall, Effects of methyl ester biodiesel blends on NO_x emissions, *SAE Int. J. Fuels Lubr.*, 2009, **1**, 102–118, DOI: 10.4271/2008-01-0078.

17 L. A. Graham, S. L. Belisle and C.-L. Baas, Emissions from light duty gasoline vehicles operating on low blend ethanol gasoline and E85, *Atmos. Environ.*, 2008, **42**, 4498–4516.

18 V. F. Andersen, J. E. Anderson, T. J. Wallington, S. A. Mueller and O. J. Nielsen, Distillation curves for alcohol–gasoline blends, *Energy Fuels*, 2010, **24**, 2683–2691.

19 V. F. Andersen, J. E. Anderson, T. J. Wallington, S. A. Mueller and O. J. Nielsen, Vapor pressures of alcohol–gasoline blends, *Energy Fuels*, 2010, **24**, 3647–3654.

20 R. A. Stein, J. E. Anderson and T. J. Wallington, An overview of the effects of ethanol–gasoline blends on SI engine performance, fuel efficiency, and emissions, *SAE International Journal of Engines*, 2013, **6**, 470–487, DOI: 10.4271/2013-01-1635.

21 H. Jung, M. H. Shelby, C. E. Newman and R. A. Stein, Effect of ethanol on part load thermal efficiency and CO_2 emissions of SI engines, *SAE International Journal of Engines*, 2013, **6**, 456–469, DOI: 10.4271/2013-01-1634.

22 R. A. Stein, D. Polovina, K. Roth, M. Foster, M. Lynskey, T. Whiting, J. E. Anderson, M. H. Shelby, T. G. Leone and S. VanderGriend, Effect of heat of vaporization, chemical octane, and sensitivity on knock limit for ethanol–gasoline blends, *SAE Int. J. Fuels Lubr.*, 2012, **5**, 823–843, DOI: 10.4271/2012-01-1277.

23 H. Jung, M. H. Shelby, T. G. Leone, J. E. Anderson and T. Collings, Fuel economy and CO_2 emissions of ethanol–gasoline blends in a turbocharged DI engine, *SAE International Journal of Engines*, 2013, **6**, 422–434, DOI: 10.4271/2013-01-1321.

24 G. Karavalakis, T. D. Durbin, M. Shrivastava, Z. Zheng and M. Villela, Impacts of ethanol fuel level on emissions of regulated and unregulated pollutants from a fleet of gasoline light-duty vehicles, *Fuel*, 2012, **93**, 549–558.

25 G. Karavalakis, D. Short, D. Vu, M. Villela, A. Asa-Awuku and T. D. Durbin, Evaluating the regulated emissions, air toxics, ultrafine particles, and black carbon from SI-PFI and SI-DI vehicles operated on different ethanol and iso-butanol blends, *Fuel*, 2014, **128**, 410–421.

26 U.S. EPA Office of Transportation and Air Quality, *Assessing the effect of five gasoline properties on exhaust emissions from light-duty vehicles certified to Tier 2 standards: Analysis of data from EPAct Phase 3 (EPAct/V2/E-89)*, April 2013, EPA-420-R-13-002.

27 K. Aikawa, T. Sakurai and J. Jetter, Development of a predictive model for gasoline vehicle particulate matter emissions, *SAE Int. J. Fuels Lubr.*, 2010, **3**, 610–622.
28 A. D. Butler, R. A. Sobotowski, G. J. Hoffman and P. Machiele, Influence of fuel PM Index and ethanol content on particulate emissions from light-duty gasoline vehicles, *SAE Tech. Pap. Ser.*, 2015, 2015-01-1072.
29 I. J. George, M. D. Hays, J. S. Herrington, W. Preston, R. Snow, J. Faircloth, B. J. George, T. Long and R. W. Baldauf, Effects of cold temperature and ethanol content on VOC emissions from light-duty gasoline vehicles, *Environ. Sci. Technol.*, 2015, **49**, 13069–13074.
30 J. Yanowitz, K. Knoll, J. Kemper, J. Luecke and R. L. McCormick, Impact of adaptation on flex-fuel vehicle emissions when fueled with E40, *Environ. Sci. Technol.*, 2013, **47**, 2990–2997, DOI: 10.1021/es304552b.
31 G. Karavalakis, D. Short, R. L. Russell, H. Jung, K. C. Johnson, A. Asa-Awuku and T. D. Durbin, Assessing the impacts of ethanol and isobutanol on gaseous and particulate emissions from flexible fuel vehicles, *Environ. Sci. Technol.*, 2014, **48**, 14016–14024.
32 H. M. Haskew and T. F. Liberty, Exhaust and evaporative emissions testing of flexible-fuel vehicles, *CRC Report E-80*, August 2011.
33 M. K. Yassine and M. La Pan, Impact of ethanol fuels on regulated tailpipe emissions, *SAE Tech. Pap. Ser.*, 2012, 2012-01-0872.
34 C. P. Hubbard, J. E. Anderson and T. J. Wallington, Ethanol and air quality: Influence of fuel ethanol content on emissions and fuel economy of flexible fuel vehicles, *Environ. Sci. Technol.*, 2014, **48**, 861–867.
35 R. Suarez-Bertoa, A. A. Zardini, H. Keuken and C. Astorga, Impact of ethanol containing gasoline blends on emissions from a flex-fuel vehicle tested over the Worldwide Harmonized Light duty Test Cycle (WLTC), *Fuel*, 2015, **143**, 173–182.
36 W. P. L. Carter, Development of ozone reactivity scales for volatile organic compounds, *J. Air Waste Manage. Assoc.*, 1994, **44**, 881–889.
37 X. He, J. C. Ireland, B. T. Zigler, M. A. Ratcliff, K. E. Knoll, T. L. Alleman and J. T. Tester, The impacts of mid-level biofuel content in gasoline on SIDI engine-out and tailpipe particulate matter emissions, *SAE Tech. Pap. Ser.*, 2010, DOI: 10.4271/2010-01-2125, 2010-01-2125.
38 J. M. Storey, T. Barone, K. Norman and S. Lewis, Ethanol blend effects on direct injection spark-ignition gasoline vehicle particulate matter emissions, *SAE Int. J. Fuels Lubr.*, 2010, **3**, 650–659, DOI: 10.4271/2010-01-2129.
39 M. M. Maricq, J. J. Szente and K. Jahr, The impact of ethanol fuel blends on PM emissions from a light-duty GDI vehicle, *Aerosol Sci. Technol.*, 2012, **46**, 576–583.
40 C. E. Fegraus, C. J. Domke and J. Marzen, Contribution of the vehicle population to atmospheric pollution, *Society of Automotive Engineers Technical Paper 730530*, 1973.
41 TransportPolicy, http://transportpolicy.net/, accessed November 2015.
42 U.S.E.P.A., Air Quality Trends, 2015, http://www.epa.gov/airtrends/, accessed November 2015.

Faraday Discussions

DISCUSSIONS

Chemical complexity of the urban atmosphere and its consequences: general discussion

Franz Geiger, Francis Pope, Rob MacKenzie, William Brune, Paul S. Monks, William Bloss, Gary Fuller, Nicolas Moussiopoulos, Matthew Hort, Alison Tomlin, Albert Presto, Dominik van Pinxteren, Athanasia Vlachou, Dwayne Heard, C. N. Hewitt, Urs Baltensperger, Alastair Lewis, Xavier Querol, Saewung Kim, Jacqueline Hamilton, Roberto Sommariva, Gordon McFiggans, Roy Harrison, Jose L. Jimenez, Eben Cross, John Wenger, Spyros Pandis, Astrid Kiendler-Scharr, Neil M. Donahue, Lisa Whalley, Brian McDonald, Simone Pieber, André Prévôt, Mohammed Salim Alam, Nivedita Krishna Kumar, Andreas Wahner, Andreas Skouloudis, Markus Kalberer, Timothy Wallington and Rachel Dunmore

DOI: 10.1039/C6FD90020H

Alastair Lewis opened discussion of the introductory lecture by Urs Baltensperger: Could you comment on functionality of the gas precursors needed to produce HOMs? The example you gave was for quite a fast reaction system – what about organic precursors that are less reactive with ozone (or other oxidants)?

Urs Baltensperger responded: The structure of the molecule needs to be such that after the formation of a peroxy radical multiple H shifts and O_2 additions can take place. This is not only possible for molecules like alpha-pinene but also *e.g.* for cyclohexene as shown by Rissanen *et al.*[1] In this way it is not so much about the reactivity but rather about the structure of the molecule. A molecule that is less reactive with ozone (or the OH radical) but does fulfill the above conditions can form HOMs as well.

1 M. P. Rissanen *et al.*, *J. Am. Chem. Soc.*, 2014, **136**, 15596–15606.

Nicolas Moussiopoulos said: I have a problem with the figures you and Roy Harrison showed regarding urban and local increments: species considered therein are in general products of chemical reactions involving compounds that may originate from local, urban or regional sources. How can such products be attributed to a specific scale? These figures can hardly be precise – do they represent compromises or simply averaged values? Or are these percentages based on additional assumptions?

Urs Baltensperger replied: This is indeed a simplified picture, which averages over many different species. For individual species the picture can look very different. A secondary species that is formed at a relatively low production rate such as sulphate would not show any urban or street site increment at all. On the other hand, species that are exclusively emitted by traffic, such as brake wear, will have an even higher urban increment than shown in Fig. 23 of Fuzzi *et al.*[1] The different urban and street site increments for different aerosol components are also nicely illustrated in Visser *et al.*[2]

1 S. Fuzzi *et al.*, *Atmos. Chem. Phys.*, 2015, **15**, 8217–8299.
2 S. Visser *et al.*, *Atmos. Chem. Phys.*, 2015, **15**, 11291–11309.

Franz Geiger commented: I'm excited about the results regarding the highly oxidized multifunctional organic compounds ("HOMs"). Have you explored the temperature dependence on SOA formation from HOMs in the absence of sulfate and/or ammonia?

Urs Baltensperger replied: I agree that this is an important question. However, we were only able to look at this temperature dependence in our last CLOUD campaign in fall 2015. The data are being evaluated, and no results are available yet.

Xavier Querol remarked: Thanks a lot for this outstanding introductory lecture on the chemistry of the urban atmosphere. My comment refers to your statement on the origin of the two types of nucleation mode bursts: with or without subsequent growth, referred to as secondary or primary in origin, respectively.

Several studies in urban areas show that nucleation episodes can be originated in urban areas, independently or not from the regional nucleation. A number of studies in the beginning of the last decade first pointed to the occurrence of new particle formation (NPF) episodes in urban environments.[1–4] These episodes occur very often at midday, coincidently with the lowest BC layers,[5] and in coastal areas also with sea-breeze flows transporting clean air masses to the cities, with noticeable levels of SO_2 from shipping emissions. Brines *et al.*[6] described these episodes using clustering tools for SMPS data sets from Barcelona, Brisbane, Los Angeles, Madrid and Rome. And they found that NPF episodes in these cities are mostly characterised by nucleation bursts without subsequent growth at midday. Dall'Osto *et al.*,[7] using simultaneous SMPS measurements in different sites inside and around Barcelona, Spain, also showed two types of NPF, one driven by regional nucleation episodes and another one starting in the city, independently of the regional one. Finally, Minguillón *et al.*[8] showed that (over 800–900 m.a.g.l.) during the traffic rush hour in Barcelona there were NPF episodes as the ground air ascends and is diluted. At midday both at ground level and over the city NPF may occur. On the other hand, in addition to the NPF, these nucleation mode bursts without growth have been also described in urban areas due to the influence of regional pollution plumes.[9]

Accordingly, in my opinion these nucleation mode bursts without growth might also be secondary in origin, especially in urban areas, although I agree that in regional background or remote sites maybe this is not the case.

1 K. S. Woo, D. R. Chen, D. Y. H. Pui and P. H. McMurry, Measurement of Atlanta aerosol size distributions: observations of ultrafine particle events, *Aerosol Sci. Tech.*, 2001, **34**, 75–87.

2 A. Alam, J. P. Shi and R. M. Harrison, Observations of new particle formation in urban air, *J. Geophys. Res.*, 2003, **108**, 4093, DOI: 10.1029/2001JD001417.
3 C. O. Stanier, A. Y. Khlystov and S. N. Pandis, Nucleation events during the Pittsburgh air quality study: description and relation to key meteorological, gas phase, and aerosol parameters, *Aerosol Sci. Tech.*, 2004, **38**, 253–264.
4 J. Pey, S. Rodríguez, X. Querol, A. Alastuey, T. Moreno, J. P. Putaud and R. Van Dingenen, Variations of urban aerosols in the western Mediterranean, *Atmospheric Environment*, 2008, **42**, 9052–9062.
5 C. Reche, X. Querol, A. Alastuey, M. Viana, J. Pey, T. Moreno, S. Rodríguez, Y. González, R. Fernández-Camacho, J. de la Rosa, M. Dall'Osto, A. S. H. Prévôt, C. Hueglin, R. M. Harrison and P. Quincey, New considerations for PM, Black Carbon and particle number concentration for air quality monitoring across different European cities, *Atmos. Chem. Phys.*, 2011, **11**, 6207–6227, DOI: 10.5194/acp-11-6207-2011.
6 M. Brines, M. Dall'Osto, D. C. S. Beddows, R. M. Harrison, F. Gómez-Moreno, L. Núñez, B. Artíñano, F. Costabile, G. P. Gobbi, F. Salimi, L. Morawska, C. Sioutas and X. Querol, Traffic and nucleation events as main sources of ultrafine particles in high-insolation developed world cities, *Atmos. Chem. Phys.*, 2015, **15**, 5929–5945.
7 M. Dall'Osto, X. Querol, A. Alastuey, C. O'Dowd, R. M. Harrison, J. Wenger and F. J. Gómez-Moreno, On the spatial distribution and evolution of ultrafine particles in Barcelona, *Atmos. Chem. Phys.*, 2013, **13**, 741–759, DOI: 10.5194/acp-13-741-2013.
8 M. C. Minguillón, M. Brines, N. Pérez, C. Reche, M. Pandolfi, A. S. Fonseca, F. Amato, A. Alastuey, A. Lyasota, B. Codina, H.-K. Lee, H.-R. Eun, K.-H. Ahn and X. Querol, New particle formation at ground level and in the vertical column over the Barcelona area. *Atmospheric Research*, 2015, **164–165**, 118–130.
9 H. C. Cheung, L. Morawska and Z. D. Ristovski, Observation of new particle formation in subtropical urban environment, *Atmos. Chem. Phys.*, 2011, **11**, 3823–3833, DOI: 10.5194/acp-11-3823-2011.

Urs Baltensperger answered: Thank you very much for these detailed comments. You refer to the example I showed from Baltensperger *et al.* (2002)[1] (Fig. 2). This figure presents nanoparticles formed by cooling of the hot exhaust gas in the morning; I call this primary particle emission because they are formed without oxidation of precursor molecules. Their growth occurs immediately and then stops, because after the cooling there is no further condensable material available for further growth. Therefore these events look like 'tomatoes' rather than 'bananas'. The formation of these particles is typically favoured by lower temperatures, as *e.g.* seen in Bukowiecki *et al.*[2] In addition, nanoparticles are seen to be formed in an event that starts around noon; this event has the usual 'banana' shape from new particle formation (NPF) where NPF occurs over a relatively large area, such that newly formed particles that show up at the measurement site at a later stage (and typically have been transported over some distance) exhibit a larger size, since they have been able to grow over a longer time. However, I fully agree with your statement that such secondary NPF events may also show up as 'tomatoes'. This is often associated with a spatial situation where there is a clear starting point of nucleation, such as a coastal area, like several of the examples that you give. Another good example is the NPF type described by O'Dowd *et al.*[3] which is related to low tides at the coast of Ireland. So, yes, if the size of the regional event does not allow for particles to arrive at the site that have had increasingly longer time to grow, bananas will not appear.

1 U. Baltensperger *et al.*, Urban and rural aerosol characterization of summer smog events during the PIPAPO field campaign in Milan, Italy, *J. Geophys. Res.*, 2002, **107**(D22), 8193, DOI: 10.1029/2001JD001292.
2 N. Bukowiecki *et al.*, Fine and ultrafine particles in the Zürich (Switzerland) area measured with a mobile laboratory. An assessment of the seasonal and regional variation throughout a year, *Atmos. Chem. Phys.*, 2003, **3**, 1477–1494.

3 C. D. O'Dowd, *et al.*, Coastal new particle formation: Environmental conditions and aerosol physicochemical characteristics during nucleation bursts, *J. Geophys. Res.*, 2002, **107**(D19), 8107, DOI: 10.1029/2000JD000206.

Gordon McFiggans asked: Concerning the ability to explain the aerosol formation in realistic mixtures, why should the sum of the yield of individual components match overall yield? Is this not counter-intuitive? Were the calculations carried out at the same condensation sink and same mass loading? Were you comparing final yield or the whole temporal trend? How do you reconcile your ability to close the yields to the strong dependence of yields on the experimental conditions (oxidants and precursor loading *etc.*), individual chambers, their wall behaviour, light behaviour and so on?

Urs Baltensperger replied: Indeed, all the parameters listed in your question (experimental conditions (oxidants and precursor loading *etc.*), individual chambers, their wall behaviour, light behaviour and so on) result in considerable uncertainties when calculating and comparing aerosol yields, and SOA formation experiments are far from being standardized. Therefore, we all hope that the submitted EU Infrastructure project EUROCHAMP 2020 will be funded such that we will have the possibility to develop such standard procedures for smog chamber experiments. We compared only final yields and not temporal evolutions. The yields were taken at the total aerosol mass, and not at the aerosol mass that would be present at the same mixing ratio of a single precursor only (which would be lower and would therefore result in a lower yield). We agree that the increased complexity of precursors does add additional uncertainty, however, we consider this minor in comparison to the uncertainties listed above. Despite of these uncertainties we consider it remarkable that we get into the range of 100% explained SOA mass when including the precursors that we were able to measure with the PTR-MS. Note that this was the case in five different experiments with quite diverse conditions, composition of precursors, *etc.* (Bruns *et al.*,[1]). This result is in clear contrast to Grieshop *et al.*,[2] who could only relate 20% of the observed SOA mass to traditional SOA precursors. Indeed, our explained mass was in the range of 20% as well when we only included these traditional precursors. This confirms our conclusion that these non-traditional precursors need to be taken into account in order to be able to predict the SOA formation.

1 E. A. Bruns *et al.*, Identification of significant precursor gases of secondary organic aerosols from residential wood combustion, *Sci. Rep.*, 2016, **6**, 27881, DOI: 10.1038/srep27881.
2 A. P. Grieshop *et al.*, Laboratory investigation of photochemical oxidation of organic aerosol from wood fires 1: measurement and simulation of organic aerosol evolution, *Atmos. Chem. Phys.*, 2009, **9**, 1263–1277.

William Brune opened discussion of the paper by Neil Donahue: In your talk you didn't get to the part in your paper where you were looking at the mixing of SOA and aged POA. You showed that there was some mixing. What's the influence of relative humidity on this mixing?

Neil Donahue answered: We performed these experiments at low relative humidity. We have explored the effect of relative humidity on mixing for SOA systems and are reporting that in a different publication.

Gordon McFiggans asked: Just one query for clarification: you state that the mixing behaviour is likely controlled by thermodynamic considerations – could it also be attributed to a kinetic limitation (from, for example, lower mass accommodation)?

Neil Donahue responded: When we have conducted experiments specifically designed to probe mass accommodation, we never see results consistent with values less than approximately 0.2. There is other evidence from condensation of, for example, extremely low volatility SOA, that for those constituents the mass accommodation coefficient is very close to 1.0. In my opinion it is far more likely that bulk diffusion limitations apply and that when we see signs of an apparent accommodation coefficient below 1 this is really an indication of diffusion limitations.

Francis Pope queried: Is there a good proxy for predicting miscibility between different aerosol components without having to resort to activity coefficients which are computationally heavy?

Neil Donahue answered: Yes. In our 2011 paper on the 2D Volatility Basis Set in Atmospheric Chemistry and Physics[1] we showed that one could formulate an activity coefficient based on the scant information we often have about organic aerosol composition (O : C and something about volatility). Word of mouth from colleagues employing much more sophisticated activity coefficient estimation methods (*e.g.* David Topping's UManSysProp) is that our simple estimation of the excess free energy works reasonably well. This is also supported by experimental studies showing that O : C is a good indicator of liquid–liquid phase separation in particles consisting of aqueous and organic mixtures from ETH Zürich[2] and the University of British Columbia.[3]

1 N. M. Donahue *et al.*, *Atmos. Chem. Phys.*, 2011, **11**, 3303–3318.
2 M. Song *et al.*, *Faraday Discuss.*, 2013, **165**, 289–316.
3 A. K. Bertram *et al.*, *Atmos. Chem. Phys.*, 2011, **11**, 10995–11006.

Dominik van Pinxteren opened discussion of the paper by Markus Kalberer: I was a bit surprised to read that the overlap of signals in the Birmingham background site samples to those in the tunnel was only 60%. Given that the tunnel is vented by urban background air, I would have expected most of the background compounds to be present in the tunnel as well. What do you think might have happened to the other 40% of background signals, not observed in the tunnel samples?

Markus Kalberer answered: The tunnel is not ventilated mechanically and the sampling site is near the end of the tunnel with respect to the direction of traffic. Since the surface area to volume and the turbulence within the tunnel is relatively high we would expect high deposition of aerosols entrained into the tunnel from the ambient atmosphere onto the tunnel surface. Hence we would expect that much of the PM from the urban background would deposit before it reached the measurement site.

In addition, as with all mass spectrometry analyses, it is important to realise that the most abundant components in a sample (and/or those which get ionised most efficiently) dominate the mass spectra. Thus, another possible reason why not all compounds measured at the background site were also detected in the tunnel samples is that some less abundant peaks in the ambient samples might not get ionised (and thus not measured) when analysing the tunnel samples due to the loss processes mentioned above.

André Prévôt said: You showed overlap of many ions with the ones found in the tunnel. One should mention that one cannot rule out that one would find a similar amount of overlap in ions when comparing with wood burning emissions. So is it not dangerous to infer from this analysis that traffic is a dominating source?

Markus Kalberer replied: As with all chemical analyses an unambiguous identification of compounds and their structure requires that several orthogonal analysis techniques are applied. The analysis presented here is only an initial step towards such a comprehensive characterisation.

Alastair Lewis addressed Mohammed Salim Alam: You showed a figure in your talk, not in the paper, of composition *vs.* lubricating oil composition. This had a number of components that appeared to be both more volatile and more oxidized than the original lubricating oil. Are they related to what Markus Kalberer was presenting, for example are these polycyclic aromatics? If they are, are they products from the degradation of the lubricating oil in the engine or are they secondary products from atmospheric oxidation?

Mohammed Salim Alam replied: These compounds are related to the diesel fuel and were also observed in the vapour phase adsorption tubes. These diesel fuel generated compounds were present in both phases indicating their active partitioning (Fig. 1). The presence of the two UCMs is a result of a more volatile diesel fuel component and a less volatile lubricating oil component. Some of these compounds were those that Markus spoke about, such as polycyclic aromatic hydrocarbons. However, these compounds are found in both the diesel

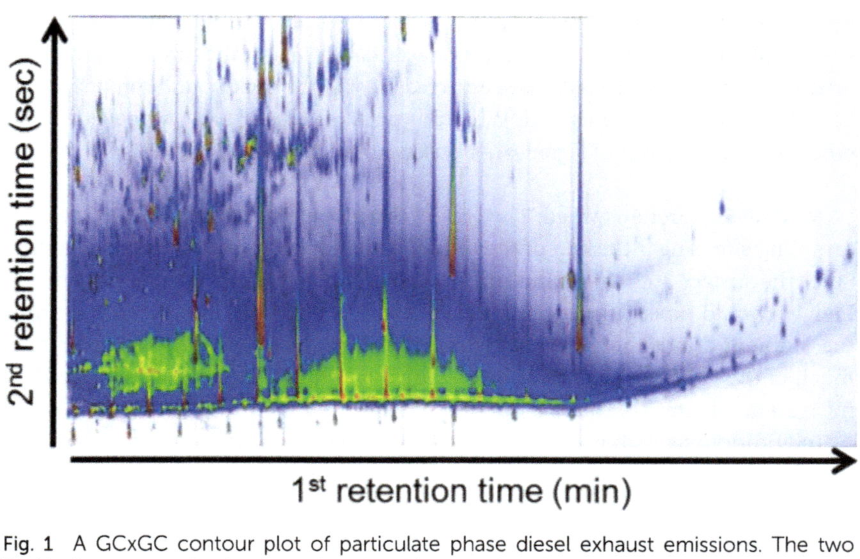

Fig. 1 A GCxGC contour plot of particulate phase diesel exhaust emissions. The two UCMs show a contribution from diesel fuel (more volatile) and lubricant oil (less volatile) to the total particulate emissions.

fuel and lubricating oil also. Some of the compounds observed may be oxidation products from the fuel formed in the engine, or possibly oxidised partially by the oxidation catalyst.

Urs Baltensperger remarked: A comment in support of Markus Kalberer's statement: I agree that with the partitioning concept where you allow molecules from particles to be evaporated then oxidized, not much similarity between the tunnel and the urban background site is to be expected. This is also confirmed by smog chamber experiments where you start with one single organic compound and get lots of different compounds within a short time.

Astrid Kiendler-Scharr addressed Markus Kalberer: You mentioned that nitrogen and sulphur containing compounds are much more prominent in the urban background than in the tunnel. Can you evaluate to what extent this is caused by either mixing or chemistry?

Markus Kalberer replied: Organic compounds containing nitrogen and sulfur (defined as "CHOS subgroup" and "CHONS subgroup" in this paper) are more abundant in the urban background aerosol samples than in the tunnel samples (Fig. 1 in our paper). We assume that the higher number especially of the S-containing compounds is largely due to atmospheric reactions leading to organosulfates and similar compounds, whereas these compounds are likely much less abundant in the tunnel samples which represent most directly emitted compounds.

Roy Harrison added: I would like to provide some clarification concerning the road tunnel where samples were collected. The road tunnel is 544 metres in length, rather poorly ventilated, and like all tunnels, it has a high surface area to volume ratio, hence favouring internal losses. This may well explain why we do not see a substantial number of the compounds observed outside of the tunnel. For example, retene, a PAH usually taken as a wood burning tracer is present in a much reduced concentration in the tunnel compared to outside air. On the other hand, for the traffic-generated PAH and nitro-PAH, concentrations are typically around ten-fold greater in the tunnel than in the outside air. This factor together with atmospheric chemistry causing chemical conversions in the external atmosphere is no doubt why there is a limited overlap of compounds between the tunnel and the outdoor site.

Xavier Querol commented: Thanks a lot for your excellent presentation and paper. You showed that S-organics are higher in the urban background than in the tunnel because of ageing, but in your paper you also stated the reverse for N-aromatics. How do you explain this? Is it because a major fraction of N-aromatics are primary? Is it because of a high interaction of high tunnel NO_x with high HCs? Usually we find in outdoor urban environments secondary N-organics in aged (and especially nocturnal by NO_3 reaction) air masses.

Markus Kalberer answered: We did observe almost four times more (poly-) aromatic peaks in the mass spectra with a CHON composition in the tunnel samples compared to the urban background samples. Our results support reports in the literature (see the paper for details) that a number of nitrogen-containing aromatics *e.g.* nitro-phenols and nitrocatechols are directly emitted by traffic sources. The reasons for the much higher number of N-aromatics in the tunnel could not be addressed in this study in detail. But likely reasons are the high NO_x levels in the tunnel as well as dilution and atmospheric reactions affecting the lifetime and concentration of these N-aromatics in the atmosphere likely leading to a decreased abundance at the background site.

Xavier Querol opened discussion of the paper by Mohammed Salim Alam: Thanks a lot for your excellent presentation and paper. You state in the paper that the proportion of the non-volatile fraction of PM increases towards the finest fraction. Why? Do you think offline sampling has produced artefacts driving to this? Is it because the dilution of emissions before sampling has already volatilised the SVOA of this fraction?

Mohammed Salim Alam replied: We hypothesize in the paper that this trend may be a consequence of the Kelvin effect which increases the equilibrium vapour pressure, and hence volatility of compounds, as particle surface curvature increases with diminishing particle size. Hence, the more volatile lower molecular weight compounds are destabilised, with an increased tendency to partition into the vapour phase as particle size decreases. Although this result is consistent with theoretical predictions and with thermodenuder experiments in the literature, and other studies have suggested that particles in the 5-100 nm diameter range consist mainly of HMW hydrocarbons associated with lubricating oil, it has been highlighted that the Kelvin effect may be unlikely to cause such a quantitative effect, except at very small particle sizes. It is very possible that there are sampling artefacts in the nano-MOUDI impactor, but these have not been studied in detail.

André Prévôt remarked: You were talking about partitioning – looking at Table 1 in your paper, I was wondering about the uncertainty of particulate phase partitioning. Some compounds would be expected only in the particulate phase or only in the gas phase, but you find them in both. Is this within your expected uncertainty? How do you explain this?

Mohammed Salim Alam responded: We focus on a range of SVOC between C12 and C33. These SVOC (particularly the low molecular weight (LMW) species) can readily partition into either phase. The higher molecular weight (HMW) compounds, particularly above C25, occur mostly in the particulate phase. For the n-alkanes we see reductions in concentrations on passage through the catalyst for LMW but not for HMW compounds, for which the data are more variable. As this test represents one speed and load, it would be interesting to repeat these measurements at different speeds and loads in order to more fully assess the variability of the HMW gas phase species behaviour. The unexpected results are for the HMW gas phase species where we observe increases in species after the DOC, for which there are a number of possible explanations. We have measured a vapour phase component of n-alkanes up to C34 in the atmosphere, so have evidence for partitioning throughout the range of volatilities studied.

Neil Donahue commented: I am surprised that the Kelvin effect works out quantitatively. The Kelvin term can be expressed in terms of a diameter for which the activity of a compound increases by a factor of 10, $K = 10^{(\mathrm{Dp/DK10})}$. Given likely surface tensions (maybe 30 mN m^{-1}) this is no more than 4.5 nm or so. Even if it is 5 nm, this means that for a 10 nm particle the Kelvin term is $10^{(.5)} = 3$, so for 20 nm it is $10^{(.25)} = 1.8$. While this may seem like a lot, let's consider larger particles with a large semi-volatile content (activity), say 0.5. At equilibrium the 20 nm particles will have the same activity, with a content of 0.25×2. This is obviously significant, as it means 25% less mass, but it is not that big a deal, especially when (if) the organics have a very wide volatility distribution.

Neil Donahue then asked: What are the implications of your findings regarding DOC efficiency for SOA formation from the emissions? The Robinson and Presto groups at CMU have shown that gasoline vehicles show higher SOA yields when the catalysts meet higher US emissions standards. The implication is that the catalysts are effective at removing total hydrocarbons but less efficient at removing the lower volatility organics that can produce SOA. Might this be true for diesel oxidation catalysts as well, especially if they are not functioning perfectly?

Mohammed Salim Alam answered: I believe that diesel emissions do have the potential to release organics that have a high SOA potential. In this study, we showed the release of many HMW compounds which were not effectively removed by the DOC. Our results also showed that the LMW species were reduced more efficiently than HMW compounds. Therefore, we envisage that these HMW, lower volatility organics, may lead to SOA production.

Roy Harrison remarked: The laboratory test rig is fitted with a total hydrocarbon analyser based on use of a flame ionisation detector which measures total vapour and showed an efficiency of around 98% during these experiments. There is clearly a much lower efficiency for high molecular weight compounds, largely associated with particles. The intercomparison mentioned by Mohammed Salim Alam relates to a comparison with hydrocarbons measured in ambient air.

Urs Baltensperger enquired: What was the temperature of your DOC? When you do idling, the temperature of your DOC will be lower than under real conditions on the street, and may not be sufficiently high for efficient oxidation.

Mohammed Salim Alam replied: I agree. The temperature of the exhaust gas at the inlet of the DOC for these experiments was approximately 360 °C and a reasonable total hydrocarbon conversion efficiency was observed. The engine operating speed and load was 1800 RPM and 5.9 bar BMEP, respectively. Thus, we would envisage that the temperature of the DOC was sufficiently high for efficient oxidation. Although the operating temperature of a DOC is the utmost influential operating parameter for controlling its conversion efficiency, the effects of other parameters such as exhaust gas composition, space velocity and OC/EC ratios should also be taken into account.

John Wenger addressed Mohammed Salim Alam and Markus Kalberer: There are some notable differences in the species identified in the ambient (Kalberer) and diesel exhaust (Alam) samples. The ambient aerosol contains a large number of oxygenated species, many of which also contain nitrogen and sulfur components, whereas the diesel samples mainly contain hydrocarbons. To what extent is this dependent on the type of analytical technique used and what are the implications for future strategies for analytical measurements of organic aerosol?

Markus Kalberer replied: The very different results obtained in the two studies are mostly due to the different analysis techniques used. All analytical techniques that are able to characterise complex organic samples on a compound-by-compound level have strongly varying sensitivities towards compounds with varying chemical-physical properties. This also affects quantification, and reliable and sound quantification of compounds in such samples can only be achieved when structurally identical or closely related standards are available. This becomes especially critical for compounds where the structure is not clearly known and thus any related quantitative results have to be dealt with very carefully.

The two studies by Tong *et al.* and Alam *et al.* use the only two techniques currently available to cope with the highly complex organic mixtures in organic aerosols on a molecular level: electrospray ionisation ultra-high resolution mass spectrometry (Tong *et al.*) is able to separate polar compounds and two-dimensional gas chromatography mass spectrometry (Alam *et al.*) separates non- and moderately polar compounds. Thus, the two techniques combined should allow us to obtain a uniquely comprehensive picture of the total organic aerosol composition on a molecular level and would be a highly interesting approach for future studies.

Mohammed Salim Alam added: Although the chromatography illustrates a large number of hydrocarbons, it does not mean that oxygenated species are not present. The colour plots of the 2D chromatogram illustrate the most dominant peaks and if not searched for, one can miss peaks that are present in smaller quantities. Some of the compounds that Markus spoke of are indeed present in the engine samples, particularly PAH and some sulphur containing compounds. Unfortunately there is no single method that can measure all species of interest, and it is the use of different analytical techniques that allows us to gather as much information as possible. As Markus mentioned, combining both their method and ours would provide unparalleled power in identifying different compounds. Furthermore, different extraction methodologies are usually utilised in order to target specific compound classes. The power of GCxGC is that we don't target a specific compound class, rather we try to determine what compound classes are present altogether. Ideally a fractionation approach should be taken enabling one to determine different classes of compounds more accurately. However, this is time consuming and cost ineffective.

John Wenger asked: The paper presented by Kalberer *et al.* reports a significant number of nitro-polycyclic aromatic hydrocarbons in organic aerosols collected in the tunnel. Did you observe these compounds in your diesel exhaust samples?

Mohammed Salim Alam responded: We did not observe many nitro-PAH (NPAH) compounds from the engine emissions, but did measure 1-nitropyrene, suggesting a diesel source for this compound. The lack of NPAH in these samples may suggest that these compounds are secondary in nature.

William Brune remarked: I'm really struck by the difference in the efficiency of the DOC system between PSI results presented by Urs Baltensperger and your results. Do you think that this information needs to be reconciled in some way before we move forward? What are the implications of the differences between the results? Second, a DOC system could be tuned up perfectly initially, but if you put it into a million cars, after a while it won't work well for some or many of them. How will this imperfect use of DOC systems affect the actual levels of particle mass mixing ratios?

Mohammed Salim Alam replied: There may be a number of reasons as to why the results of Urs Baltensperger are different to what we see in this study. As pointed out by Dr Baltensperger, the temperature of the DOC is an issue in itself, in that it does not convert efficiently unless operated at a specific temperature that is designed for that specific DOC. We are however confident that our DOC was operating above the "light off" temperature. It must also be stressed that no diesel particulate filter (DPF) was used in this study. The differences observed in both studies indicate that more tests need to be conducted in order to understand why these results are different. Furthermore, engine running conditions such as engine rotational speed, engine operating load, dilution *etc.*, all contribute to differing results. Therefore a direct comparison between two studies should be made with great care. It is agreed that after long use of a DOC it will no longer work as efficiently as its initial state and this will influence emissions. It is the introduction of the DPF that will greatly reduce levels of particle mass emissions, as these are designed to remove particulates. This work indicates that a DOC will not oxidise all hydrocarbons with the same efficiency, so the effect of an inefficient DOC would be different depending on what compounds are being emitted.

Neil Donahue said: We see lots of variability among different vehicles and even for one vehicle (or catalyst) at different times. Have you looked at more than one catalyst or varied the conditions?

Mohammed Salim Alam replied: We have observed the same. We have looked at two different engines and observed quite significant differences between them. When sampling from the engine before and after DOC we try to keep the engine running at the same conditions throughout and complete these experiments on the same day, to avoid the possibility of day-to-day variation of the engine. We are currently conducting experiments using a combined DPF/DOC to compare with the experiments completed with a DOC only. This involves sampling at different engine conditions.

Jacqueline Hamilton addressed Mohammed Salim Alam, Urs Baltensperger and André Prévôt: In much of the discussion around diesel emissions here, the assumption has been that the VOC emissions are low as a result of functioning emission controls such as the DOC or particle filter. The PSI studies were done on a warm engine traveling at high speed. However, in an urban environment, cars are often traveling at low speeds and engine conditions are not optimal. In a city like London, where residential areas are within the city centre, a significant amount of car journeys start and end within the city and so cold start emissions must also be taken into account. Recent results have shown that unburnt fuel emissions of large hydrocarbons are much greater during cold start and under idling or low load engine conditions.[1] Therefore, diesel cars do have the potential to release VOCs with a high SOA potential, which are not observed in highly controlled laboratory tests. In the recent ClearfLo project, large hydrocarbons were observed to constitute a significant fraction of the VOC loading in the atmosphere and had the potential to form significant amounts of SOA.[2,3] We must understand the emissions of these VOCs under sub-optimal and realistic driving conditions, in much the same way as for NO_x emissions.

1 E. Cross *et al.*, *Environ. Sci. Technol.*, 2015, **49**, 13483–13491.
2 R. E. Dunmore *et al.*, *Atmos. Chem. Phys.*, 2015, **15**, 9983–9996, DOI: 10.5194/acp-15-9983-2015.
3 R. Ots *et al.*, *Atmos. Chem. Phys. Discuss.*, 2016, DOI: 10.5194/acp-2015-920, in review.

Mohammed Salim Alam responded: I agree that high speed and high load may lead to a larger reduction of vapour and particulate phase compounds, due to a more efficient DOC/DPF. In our study we do not measure compounds < C12, so we cannot comment on VOCs smaller than those observed here, however, it is clearly seen that the vapour phase constituents are numerous, between C12 and C22. I agree that the potential to form significant amounts of SOA is possible from these measured large hydrocarbons, as outlined by the authors cited in your comment.

André Prévôt added: In my opinion it is important to distinguish new and old diesel cars. Old diesel cars without filters and oxidation catalysts may contribute large amounts to secondary organic aerosol formation while new cars with oxidation catalysts and particle filters do not seem to do that.

Urs Baltensperger said: I fully agree that the emissions at low temperatures are highly important. This however applies to both diesel and gasoline cars. Weilenmann *et al.*[1] reported VOC emission factors for diesel and gasoline cars at different temperatures, showing that the increase of the emission factors for VOCs with decreasing temperature was more pronounced for gasoline than for diesel cars. However, there is a clear need for papers that put this into the right perspective.

1 M. Weilenmann *et al.*, Cold-start emissions of modern passenger cars at different low ambient temperatures and their evolution over vehicle legislation categories, *Atmos. Environ.*, 2009, **43**, 2419–2429.

Urs Baltensperger commented: There is not necessarily a disagreement here. If the THC analyser showed 98% efficiency in the oxidation this means that only 2% are left for SOA formation. Even if you allow for a higher than usual yield of the remaining compounds the total SOA formation will be decreased by far more than 90% by the catalyst, despite a low oxidation efficiency of certain compounds. In this context it is important not only to look at the oxidation efficiency of an individual compound but also at the actual concentration of this compound.

André Prévôt remarked: What's important for diesel is the contribution of the whole fleet. We don't only have new cars on the roads which include particulate filter and an oxidation catalyst.

Roy Harrison commented: For clarification, a major difference between the work in Birmingham and that at PSI is that at Birmingham we have not yet worked with an engine fitted with a diesel particle filter. There is a small amount of literature which shows nucleated particles behind a diesel particle filter which certainly suggests that there are volatile hydrocarbons coming out of the system, although we do not know how much this contributes to SOA formation.

Matthew Hort said: Results from test beds where engines are run in perfect conditions indeed offer valuable data. However, in terms of real-world driving there is a lot of evidence that emissions are radically different. In producing a forecast or validating models against real world measurement data we need estimates of emissions that resemble those from real world driving. If we do not have those then we can't close the scientific challenge, we run the risk that different parts of the community are working towards different ends and we will not be able to understand the sources of error. Do you agree and/or think that test bed activities will develop into more real world type setups?

Mohammed Salim Alam replied: I agree that we need more real world measurement data and estimates of emissions that resemble those from real world driving. However, simulating this is an experimental challenge requiring funding and resources. Test beds on the other hand do offer very valuable data. This work corresponds to one test speed and load and does not take into account other speeds, loads, dilutions *etc.* Our other work includes many more scenarios which shed valuable light on emissions. This study provides comprehensive details of the composition of diesel emissions and also sheds light on what compounds we would expect to be present in the atmosphere. I also think that test bed activities will evolve to better reflect the real world and I think this is an important issue if we are to answer many of the remaining outstanding questions.

Alison Tomlin then remarked: Following on from the discussion about lab *vs.* real world I agree that capturing the impact of urban driving cycles is particularly important. One of the issues that crops up under urban conditions is the potential difficulty in reaching the temperatures required for trap regeneration. Therefore the condition of the trap could degenerate over time. Lab tests probably therefore need to use long time horizons in order to consider the influence of whether traps can generate. How were the particle filters in your work regenerated?

Mohammed Salim Alam answered: We have not completed any work with diesel particulate filters. This study has used a diesel oxidation catalyst and we will be performing tests with diesel particulate filters shortly. I agree that under urban driving conditions, it may be difficult to reach temperatures required to regenerate these DOC/DPF traps. However, there are some regeneration techniques that car manufacturers utilise to account for these shortcomings. In this study we monitor the temperature close to the DOC and we reach temperatures that are sufficient for efficient operation, as indicated by the total hydrocarbon conversion efficiency data.

Andreas Wahner asked: Could you test different fuels? Could biofuels have different emissions? We've seen emissions change drastically.

Mohammed Salim Alam responded: We can test different fuels including biofuels. Although this isn't in the remit of the current proposal, we have conducted some experiments previously and found quite significant differences between the emissions. When using biofuels the exhaust emissions have a huge amount of methyl esters present.

Eben Cross commented: Accurately measuring the emission index and composition of tailpipe I/SVOCs from a light-duty diesel vehicle is very challenging, primarily due to the propensity of I/SVOC species to be lost to surfaces prior to reaching the collector apparatus. In this context, it is important to obtain an empirically measured background concentration of I/SVOCs while sampling 'exhaust-free' air at the same temperature as the steady-state engine-on condition characterized in the experiment. This way, the I/SVOC contribution from the apparatus (in this case exhaust lines + DOC + sampling lines) can be assessed and then subtracted from the pre and post-DOC engine-on I/SVOC measurements. The authors should comment on how the lack of any background I/SVOC measurement may impact their results.

Mohammed Salim Alam answered: Particle losses within the sampling lines were minimised by using heated lines and conductive tubing. Hot exhaust gas, sampled from the exhaust pipe, was delivered to the dilution system *via* a heated line with a temperature maintained at 191 °C. This practice reduces the chance of condensation of volatile species onto the delivery and sampling lines from engine exhaust. From this point conductive tubing was used to minimise particle losses.

Furthermore, the scrubbed air used for the dilution system was free from moisture and particulates, by utilising both a high flow rate moisture trap and HEPA filters, respectively. This will reduce the chance of introducing any impurities in the clean air.

In addition, before sampling from the engine, all of the flow stream lines were purged with hot pure compressed air, to remove any impurities left in the lines. We performed some preliminary experiments which showed that the background noise was minimal.

Nivedita Krishna Kumar asked Neil Donahue: Did you do similar experiments with docosane? On flash vaporization docosane tends to have an ideal behavior. Additionally docosane forms needle shaped particles instead of spherical particles. Does this have an effect on the miscibility?

Neil Donahue answered: In our original paper describing this method[1] we presented an example where we combined docosane and deuterated docosane populations. At room temperature they remained totally distinct for hours, but heated to approximately 40 °C they mixed more or less instantly. Our interpretation was that the small lattice mismatch of the deuterated and non-deuterated molecules rendered them immiscible when they formed needle-like solid particles.

1 E. Robinson *et al.*, *J. Phys. Chem. A*, 2013, **117**, 13935–13945.

Brian McDonald addressed Markus Kalberer: I think it is really important to get real world measurements of vehicle fleets due to the variability of emission factors due to driving conditions, faulty catalytic converters, *etc.* There are advantages and disadvantages to different approaches used to measure tailpipe emission factors in roadway settings. An advantage of tunnel measurements is that these measurements are able to characterize the composition of volatile organic compound (VOC) emissions with high specificity for the average vehicle fleet. However, since individual plumes are not captured, these VOC profiles cannot be attributed to vehicles that may meet different emission standard categories. An advantage of remote sensing data is that they are able to capture individual plumes from motor vehicles, and information about the engine can be attained through vehicle registration data. However, remote sensing data are unable to speciate VOC emissions coming out of the tailpipe. In the future, is it possible to merge the advantages of tunnel and remote sensing measurements, *i.e.*, roadway measurements that capture high chemical specification for individual vehicles or vehicle classes, under real-world driving?

Markus Kalberer replied: We agree that all past and current tunnel studies are mainly providing average profiles of traffic emissions at a specific site and time. The challenge of emission measurements of individual vehicles in a roadway setting is that the analysis needs to performed on very short time scales and with remote sensing techniques. Most analytical techniques that are able to resolve complex compound mixtures on a molecular level are not remote sensing techniques, *i.e.* air samples (gas or aerosol) are pumped into an instrument (*e.g.* a mass spectrometer) for analysis. This would make vehicle-specific roadway measurements highly challenging and we are not aware of any technique currently available that would combine remote sensing with detailed molecular-level gas or aerosol composition analysis.

Dominik van Pinxteren asked Neil Donahue: I am wondering about the implications of the described mixing processes between aged POA and SOA for source apportionment studies. What type of source should such mixed particles actually be ascribed to?

Neil Donahue replied: Precisely. We should not confuse the source of a particle with the source of the mass on a particle. The vast majority of the mass on most particles more than a few hundred meters from their source is from condensation, and that condensing mass has myriad sources. This won't have a huge effect on the source apportionment of PM mass (other than the Raoult's Law effect) but yes, attribution of individual particles is fraught.

Jose-Luis Jimenez addressed Markus Kalberer: There are ion sources such as DART that can produce ions in real time at atmospheric pressure. Instruments such as the Orbitrap should be operable under field conditions with some care, even if that's not commonly done. Have you thought about sampling exhaust from individual cars on the road side, and spraying it with these ion sources using these new MS instruments (*i.e.* similar to what was done in a paper by Ban-Weiss *et al.*[1])? That would allow high-resolution characterization of car-by-car emissions.

1 G. A. Ban-Weiss *et al.*, *Environ. Sci. Technol.*, 2009, **43**, 1419–1424.

Markus Kalberer responded: In recent years a number of soft-ionisation online mass spectrometry techniques have been developed but ultra-high resolution mass spectrometers (UHR-MS) have to the best of my knowledge not been coupled with these online ionisation sources for field measurements, mainly because UHR-MS instruments are usually built for standard laboratory applications rather than field deployments.

However, organic aerosols from smog chamber and flow tube experiments have been measured with online UHR-MS techniques, one example is our recent paper.[1] Gallimore *et al.* (2013), dx.doi.org/10.1021/es305199h.

1 P. J. Gallimore and M. Kalberer, *Environ. Sci. Technol.*, 2013, **47**, 7324–7331.

Urs Baltensperger remarked: I would like to support Neil Donahue on his previous statement – yes, single particle instruments have lots of usefulness looking at processes *etc.* but we need to be careful in using them for source apportionment in terms of overall mass contributions by the individual sources.

Athanasia Vlachou asked: In the conclusions of the POA–squalane mixing experiment, you suggest that in an urban environment the POA particles stemming from different sources may mix within a few minutes, making the source lose its identity. Does this generalization stand for a source where the semi-volatile fraction is different than that coming from diesel? Could one reach the same conclusion in the case where biomass burning is the dominant source?

Neil Donahue replied: There are two sides to this. In my opinion semi-volatile organics (SVOCs) will mix into particles from almost any source relatively rapidly, if there are SVOCs about. So we would expect to observe mixing of vehicular POA into biomass burning POA fairly quickly. The volatility distribution of biomass burning POA seems to be much more variable, and so the speed with which a significant fraction of mass leaves biomass-burning primary particles is likely quite variable.

Gordon McFiggans remarked: Following the lively debate concerning the extrapolation and use of laboratory data for interpretation of real world experiments, there appear to be worrying difficulties. The UK National Atmospheric Emissions Inventory estimate for 2003 was that about 49% of CO and 13% of NO_x were from cold start. It would appear to be more complex for IVOCs, SOA precursors and PM in general from traffic emissions under non-ideal running conditions across a heterogeneous fleet. Could you envisage and describe a tractable strategy to inform interpretation of real world emissions by lab experiments that can account for the huge diversity of conditions? For example, are there more easily measurable emissions that can be mapped across engine conditions and measured in the real world, to which the important but more complex PM factors can be related?

Neil Donahue responded: In part this depends on how much the composition of the emissions varies or just the magnitude. Variable magnitude is easy. Variable composition is more bothersome, but we tend to source samples over integrated portions of a drive cycle when we (this is really Albert Presto) are looking at vehicle emissions. It is a huge issue for biomass burning, where the combustion conditions are completely uncontrolled.

Simone Pieber asked: Could you please comment on the current implementation of those processes in the volatility basis set model, and whether you think the model needs to be extended to take time-scales and mixing of phases dependent on their polarity as seen in your results further into account?

Neil Donahue replied: We are developing a dynamic version of the VBS model and have published some results, for example in Donahue *et al.* (2011)[1] and Trump and Donahue (2014)[2]. We have not yet combined the dynamic model with the 2D VBS, where we could also implement non-ideal mixing. That is ongoing work but it is conceptually straightforward.

1 N. M. Donahue *et al.*, *Geophys. Res. Let.*, 2011, **38**, L16801.
2 E. R. Trump and N. M. Donahue, *Atmos. Chem. Phys.*, 2014, **14**, 3691–3701.

Franz Geiger remarked: To comment on the question regarding connecting emissions inventories to laboratory measurements, I point out here that it will be important to pair science with econometrics. Consumer behavior is critical for understanding urban air pollution, and traffic is in the end a consumption of fuels. Specifically, we need to know when people drive their cars and how they drive them. At the moment we don't know and I think this will be a really important component to increasing the predictability of models, the utility of emissions inventories, and the design of sensor networks. That such an approach can work for a large metropolitan area was recently shown by my colleague Alberto Salvo and myself.[1]

1 A. Salvo and F. M. Geiger, Reduction in local ozone levels in urban São Paulo due to a shift from ethanol to gasoline use, *Nature Geoscience*, 2014, **7**, 450–458.

Gary Fuller opened discussion of the paper by Alastair Lewis: There is an urgent need for this type of work and for its publication, so well done. We run London's regulatory network and we get many requests from people wanting to measure their own air pollution. Small sensors could have great application in increasing awareness of air pollution exposure and for assessing exposure in health impact or epidemiological studies, but they have to give accurate results or people will get a false impression of their risk and we will waste a tremendous amount of effort on bad data. USEPA and CEN have put together frameworks on how small sensors could be evaluated with lab and field assessment. Additionally CEN already have methods for testing equivalence to reference methods (PM_{10} and $PM_{2.5}$ are the best known). Have you thought about presenting your results in the way that these test structures suggest? CEN has acceptance criteria for indicative measurements for instance. It would mean that your results could more easily be compared to other tests. Also, for the ozone sensors, you adjusted these to a reference before field testing, but you didn't adjust the others. Why? Is this really fair?

Alastair Lewis answered: I think that the future will have to have a more rigorous framework for sensor evaluation and as you identify there is lots of history that can be called on to test air pollution instruments and understand their equivalence. There is a wider question about how indicative measurements might be used. One can imagine these have much to contribute in terms of awareness raising and engaging the public. But there is also potential for them to be used beyond their capabilities, for example individuals might make health choices, for example asthma inhaler usage, based on qualitative readings. The ozone data were all normalised at the start of the period to the reference instrument. It is a good point regarding whether we have given the other sensors a fair fight, but there was some thinking behind our reasoning. For NO the sensors were really at their detection limits and we had problems getting all reading non-zero values simultaneously. For NO_2 sensors we knew from lab work that at ambient NO_2 in York this wasn't responding solely to NO_2 and so a normalisation would be artificial. For PM, there really didn't seem to be a good metric to normalise to since the TEOM and sensor don't pretend to measure the same quantities. For VOCs there is no standard for 'total VOC', although we could have tied all the sensor data together to a single value at the start of the experiment.

Albert Presto remarked: I would like to get your opinion on public perception of these low cost sensors. You started with describing how the technology path for these sensors is inverted. People here at the conference know that they don't work perfectly but others may have a false sense of trust in them and rely on them completely because of blind faith in technology. What's the response to this? Public trust in low cost sensors could lead to people having health panics, or protesting, *etc.* due to the results of these imperfect sensors. That could be a dangerous place to be.

Alastair Lewis replied: I think there is probably too much general trust that once a technology is launched to the public it must be fit for purpose. For most sensor technology examples the public can test themselves whether the sensor is working or not – for example if a compact GPS receiver in a smart phone was persistently wrong by 1 km, you'd take it back to the shop. Air pollution sensors are unusual in that they are measuring a parameter that the public simply have no way of verifying for themselves, so there is absolute trust. The counter-balance to this is information that can be provided by those that do have the capability to test these devices – but this may be only a handful of labs in any given country. So we do have a duty to conduct robust and fair tests and publish what we find, and provide some guidance on how such devices might be used and how they should not be used.

Andreas Skouloudis commented: This work reminded me of the situation as it used to be in the 5th Framework Programme during 2004 in the project ADA (IST-2000-28452). It's not true that research has not advanced in this area, and it is mostly because efforts have focused in areas other than electrochemical sensors. For example, nano-crystaline thick and thin film sensors are far more promising and has been successfully tested over difficult ambients for continuous monitoring. The calibration process is sufficiently reliable but the measuring interval is very important in carrying out comparisons with conventional measurements. Specific metal oxides could be used in arrays for targeting different chemicals. Unfortunately, these could not be considered as low-cost devices. Certainly, these are significantly lower cost than conventional instruments but their cost is currently in the range of four digits.

Paul Monks asked: Is there merit in going back to look at the metal oxide sensors. Clearly, there are questions about the reproducibility of electrochemical sensors, their temperature dependence *etc.* Metal oxide sensors don't suffer from the same set of problems. Should we be looking at metal oxide sensors and seeing how they have improved over the last 15 years? Second, critically, reproducibility seems a real issue - did you look at this in your study and how do you deal with this on a sole *in situ* sensor?

Alastair Lewis responded: A short answer is yes. Metal oxide sensors have many interferences as well, including very strong dependence on temperature, but they have the great advantage of being very cheap - a few pounds each. The issues around sensor-to-sensor reproducibility can potentially be overcome by deploying clusters of sensors together and then taking a median sensor value. Their low cost may open up alternative ways to gain better unit to unit comparability and this is something we are working on.

Franz Geiger commented: We are trying to do this in Chicago, which exhibits the highest occurrence frequency of asthma in children, with the goal to abate diesel emissions. In the US, diesel traffic is mainly from the heavy-duty transportation vehicle fleet. One issue we've seen with the low-cost sensors is that any legal action connected to data derived from them is difficult to pursue as the sensor data is not certified. Two questions: 1) how is this issue being addressed in the UK? And 2): have you tried to regress out air pollutant concentration data from a wide network of sensors, some of which already exist in the form of a robust air monitoring grid in London, against traffic patterns at the street-hour level?

Alastair Lewis answered: It's almost certain that any attempt to raise a legal challenge on the basis of existing sensor data is likely doomed to failure in the courts, since the devices do not meet the relevant standards, and the rules are very clear. But that is only half the story. It is very likely that responding to such claims, either through formal legal actions, or to individual members of the public will take up significant resources. We have operated for nearly 50 years where the provenance of air quality data was rarely, if ever, a point of contention. We might like more data, or disagree about its interpretation, but not the quality of the data itself. We are entering a new world where we will start to also have to debate the data, as well as the interpretation. I fear that cities and even national governments will lack the resources to be able to respond – this is a new thing for them to have to work with. Of course resources spent on answering questions about sensor data means resources then not being spent on tackling the problem itself. It could become an enormous distraction and one with real costs. We have yet to be confident in the reproducibility of any sensor data, so we haven't attempted to extract information from a deployed network. There are however some limited examples in the literature, but it is notable that there has yet to be any 'breakthrough' science enabled by new air quality sensor technologies. It is different to many other atmospheric technologies, where a new instrument is often at the heart of unlocking new scientific understanding. Of course in London there is a really excellent monitoring network of traditional instruments and this has been heavily exploited in the way you describe.

Markus Kalberer said: I fully support the idea that more characterization work is needed on these sensors, but I think we need to carefully think where and for what purpose these sensors would make sense – they are not really useful if people just put them in their gardens. They could be used for epidemiological studies. But this assumes that you are using 50 or 100 or more of these sensors, not just one or two.

Alison Tomlin commented: I think you are right that no one would expect that these sensors would be used for legislative purposes. But I'm concerned about local authorities, some are facing infraction fines and they're quite keen to improve public perception and understanding of the spatial variability of air pollution within their regions and to show that outside of hot spot areas, there is less cause for concern. They are likely to have an interest in such sensor networks. What can the community do to support these groups? Should they basically be warned off?

Alastair Lewis replied: There is enormous interest from local authorities in this sort of technology since it offers an attractive solution to some long-standing and very significant environmental problems. I'm aware of several authorities already some way down the path of implementing these sensors and also of agencies that are planning to use them to provide real-time data that will control traffic flows. These are very significant interventions that will have real costs and cause disruption, so the data really does need to be correct. There is clearly a problem that at a city level in that there isn't the scientific capability to evaluate performance. There is also a lot of history that air quality instruments are almost by default made compliant with relevant standards, so this aspect may not be questioned when new devices enter the market. It is a little strong to suggest we should warn them off the technology, but we should make it clear what the capabilities actually are and help design experiments that are within the performance capabilities of a given technology.

Eben Cross commented: The authors state that no variability was observed in the auxiliary electrode response of any of the tested B4- electrochemical sensors and therefore, only the voltage signals obtained from the working electrode were used in their laboratory calibration experiments. As the authors point out, robust laboratory and field-based calibration of low(er)-cost air quality (AQ) sensors such as the B4-series of electrochemical sensors used here, are lacking. Given this lack of rigorous testing (alongside the proliferation of AQ sensor applications outside of the research community) it would be informative to provide (in the supplementary information or otherwise) explicit data demonstrating the null auxiliary electrode response across the RH-experimental matrix explored here. This is particularly important given that the manufacturer recommendation for the use of such sensors is to calculate concentrations based on the differential voltage response (working-auxiliary) as a first order correction for interference effects of variable RH and temperature. Given that the laboratory experiments spanned RH values from 15 to 60%, alongside systematic changes in pollutant concentrations and gas mixtures, it would be informative to see both the working and auxiliary electrode response. The highest concentration of CO tested in the laboratory was 200 ppb while the O_3 concentrations reached 350 ppb. Given expected concentrations of CO and O_3 in ambient urban environments, the lab CO seems too low, and O_3 too high to capture ambient variability. Can the authors comment on this apparent disconnect between the lab and field concentrations?

Table 1 in the paper shows that the NO_2-B4 sensor exhibits no cross-sensitivity to O_3, whereas the OX-B421 shows nearly 100% cross-sensitivity to NO_2. Is it possible that the authors are in fact using a NO_2-B42F sensor instead of the NO_2-B4 model? The NO_2-B42F sensor includes a filter to scrub ozone upstream of the working electrode, so if such a filter existed in the sensor used here, that would explain the lack of any cross-sensitivity to O_3. Given the 100% cross sensitivity between O_3 and NO_2 present in the Alphasense B4 OX sensor, were the authors at all surprised that the O_3 data acquired with their 20-node field-deployment did not show strong signs of an NO_2 interference? A more detailed discussion of the field observations in the context of potential NO_2 and O_3 cross-sensitivities is warranted. With OEM low-cost AQ sensing hardware components the selectivity, sensitivity, and stability (age) of each individual sensing component needs to be carefully assessed. While the authors have nicely demonstrated some of the 'selectivity' issues that can arise in ambient monitoring applications, they do not reveal any details regarding when the sensors were originally purchased (out-of-package). This is an important detail to consider given the shelf-life of electrochemical sensors (~18–36 months) due to their finite electrolyte volume. By providing additional context on the age of the sensors evaluated in this study the authors could help improve the state-of-knowledge regarding sensor stability over time.

Alastair Lewis answered: On the first point we saw no sensitivities, but plan to publish in more detail an expanded study of this aspect of sensor performance. The lab testing tried to mimic cross-interference concentrations that might be found in the real world whilst also controlling precisely the measurand - this is a technical challenge in terms of blending stable mixtures and then measuring all the necessary parameters. We were limited therefore in the range we could achieve based on the primary standards available in our lab and the ability to blend and measure, but nonetheless think the range we've used captures some of the key interferences. We would welcome others conducting similar tests at ambient concentrations. The ozone sensor in the 20 unit trial was not a B4 OX sensor - the two studies are completely separate and use different sub-components, and perhaps we should have made this clearer. We can only agree regarding the need for additional information, such as age, on components with low cost AQ packages. There is virtually no information available on when individual sensors within individual units were commissioned and a widespread insistence from manufacturers that device lifetimes run to many years – something which seems difficult to support given the knowledge we have of the subcomponents and their mode of operation.

Urs Baltensperger remarked: I very much support your statement that we need to test the sensors with a complex mixture but under controlled conditions. Atmospheric simulation chambers are a good way to do this. A consortium including a number of UK groups as well as PSI have just submitted the EUROCHAMP 2020 Infrastructure proposal to the European Commission. If this project is funded there will be the possibility to invite people to visit such a chamber and perform specific experiments to test the sensors in a specifically prepared complex environment. But of course this only makes sense after the sensors have passed initial tests showing that they are reliable. For example I wouldn't be interested in testing one of those aerosol sensors where you showed that they are highly unreliable.

Alastair Lewis replied: Chambers potentially have a big role to play in sensor evaluation, and I would like to see this articulated in the research strategies of the major international chambers. The ability to control and measure such a wide range of co-pollutants and other parameters will be hugely valuable.

Roy Harrison said: The remarks from Markus Kalberer concerning the use of sensors in epidemiological studies prompt some comment. Two of the main currently topical issues with population-based epidemiology are exposure misclassification (wrong estimates of people's exposure) and correlations between pollutants. Exposure misclassification is important because it tends to bias the results towards to the null (*i.e.* it underestimates the effects of pollutant exposure). If low cost sensors give a better estimate of exposure than the methods currently used (typically fixed site monitors or land use regression models), it would reduce exposure misclassification and be beneficial. With regard to correlations between pollutants, if two pollutants are highly correlated, then it provides considerable difficulties in elucidating the separate impacts of each of the pollutants upon health. This is not an issue specific to the low cost sensors and would be unlikely to arise as different methods are used for quantification of NO_2 and particles, which are two of the most strongly correlated pollutants in the urban atmosphere.

Dwayne Heard addressed Alastair Lewis: You state that of all the air pollution parameters considered in your paper, that VOCs have the most limited observational datasets. Unlike O_3 and NO_x, there is no reference measurement with which to easily test the performance of the sensors for total VOC. Clearly such a reference would be very useful so I wondered if you could comment on how to provide a reference for such a measurement in the future? Ideally something with a good time resolution would be needed as total VOC variability is likely to be highly variable in time in an urban environment, but developing a reference might be quite challenging. Measuring a suite of VOCs using a GC system and adding up the concentrations is one approach, but the time resolution would be more limited. For some selected VOCs, high time resolution could be achieved using a PTR-MS (but this is unlikely to cover a full range of VOCs). Given that in your paper you state that the speciation of VOCs is probably not very different from one location to another in a given urban environment, then another approach would be to measure the OH reactivity at a particular location (good time resolution is possible), and divide this by an average rate constant for reaction of VOCs with OH to obtain the total VOC concentration. However, this may not be easy to do routinely.

Alastair Lewis responded: OH reactivity is a interesting concept for a sensor; it would provide bulk information, like a TotalVOC sensor does, but carry with it slightly more powerful information. Total reactivity is a much more informative property that total carbon amount. Total VOC sensors do however seem a good prospect in the relatively short term, but they are only going to give useful information if there is a co-located speciation, probably by GC. Whilst VOC speciation changes enormously from one location to another – think tropical forest to city centre – there are many experiments within those regimes where spatial and temporal changes are the big uncertainty not speciation. There is much that sensors could offer for research here.

Saewung Kim opened discussion of the paper by Tim Wallington: What I'm hearing from US is that in late 2008 and 2009 everyone was talking about biofuels, but now people have moved on to electric cars and there's less interest in biofuels. Could you comment on the future of biofuels?

Tim Wallington replied: Ethanol is the most widely used biofuel in the US. From 2000 to 2010 there was an approximately factor of 8 increase (1.7 to 12.9 billion gallons) in the consumption of ethanol (http://www.afdc.energy.gov/data/10323). From 2010 to 2014 there was a small increase (12.9 to 13.5 billion gallons) in the US consumption of ethanol. The future of biomass-based fuels depends on the availability of refueling infrastructure capable of handling higher blends of biofuel, the availability of vehicles capable of using higher blends of biofuel, fuel regulations and specifications allowing higher blends of biofuel, and economic feasibility wherein these biofuel blends are attractive to consumers.

Paul Monks opened discussion of the paper by Rachel Dunmore: Thank you for the paper, it is very interesting. In relationship to the data and modelling approach in Fig. 7, is the comparison valid? The MCM as a constrained box-model will calculate the instantaneous acetaldehyde produced on a 15 min basis. There seems to be clear evidence for transport of a longer-lived background in the anti-phase behaviour between the model and the data which indicates transport and changing BL height. Further, the data seems to sit on a background not going less than 2.5 ppbV. It strikes me that the modelling approach shouldn't work in this instance. Given these shortcomings, how can you attribute the missing portion to traffic emissions?

Rachel Dunmore responded: The acetaldehyde is attributed to traffic related emissions on the basis of the correlations of the measurements with other traffic related VOCs. The model is not being used to attribute this missing portion to traffic but to investigate the potential size of the ethanol source as a photochemical acetaldehyde precursor *versus* all other VOCs. The MCM model shows the expected acetaldehyde profile that is consistent with that from photochemical production and loss, something that is not seen in the observed acetaldehyde. This combination of evidence suggests that the major source of the observed acetaldehyde is anthropogenic emission, as opposed to secondary production. We therefore attribute this anthropogenic source of acetaldehyde to traffic related emissions given the clear rush-hour peaks observed in the associated diurnal profile and correlations with other traffic-related VOCs. It may be that the high background levels of acetaldehyde observed (*i.e.* not falling below 2.5 ppbv) can be attributed to long range transport and further regional modelling is needed to determine this.

André Prévôt asked: Is it not surprising that the background level of acetaldehyde never goes below 2 ppb? This seems rather high.

Rachel Dunmore answered: It is not surprising that the baseline level of acetaldehyde does not fall below 2ppb. It is quite a long lived species (lifetime with OH radical at 1×10^{-6} molecules cm^{-3} is 1 and a half days) and it is likely that the background levels are largely influenced by long range transport. Similar levels of acetaldehyde have also been reported, a daytime average of 1.5 ppb during the summer TORCH 2003 campaign which rises to 3 ppb during the heatwave episode.[1]

1 J. D. Lee *et al.*, *Atmospheric Environment*, 2006, DOI: 10.1016/j.atmosenv.2006.06.057.

Alison Tomlin opened a general discussion of the papers by Tim Wallington and Rachel Dunmore: Firstly, you mentioned ethanol as a component within engine exhaust emissions. Is that from unburnt fuel? Secondly, with regards to blending ratios, do you see a future for other alcohols, *e.g.* biobutanol which have been shown to have potential benefits in terms of making higher blending ratios possible. Would that impact on exhaust acetaldehyde emissions?

Tim Wallington answered: Yes, ethanol in the exhaust is unburnt fuel. We are open minded regarding future biofuels and indeed fuels in general. To address climate change we need low fossil CO_2 fuels. To address urban air pollution we need fuels which are compatible with the fuel/vehicle infrastructure and when used in compatible vehicles produce low emissions of air pollutants (NO_x, VOCs, PM, CO). Butanol has an advantage in that it has physical and chemical properties that are closer to those of gasoline. However, the availability of biobutanol is more limited than that of ethanol. Subject to the constraints above, we believe that the market will need to determine whether ethanol, butanol, or some other sustainably-produced low-CO_2 alternative fuel provides the most cost effective solution.

Rachel Dunmore responded: It is possible for ethanol to be emitted as part of engine exhaust through incomplete combustion within the engine system. Ethanol can also be emitted through evaporation from the vehicle, fuel stations and transport of fuel. It would be useful to fully investigate the possibility of using different fuel blends and compositions which may decrease the emission of certain pollutants.

Alastair Lewis added: It is impossible with the data we have to apportion the precise sources of ethanol from the vehicle fleet, but most likely it is a combination of unburnt fuel, fuel evaporation, fuel spillage and possibly other related emissions, for example screen-wash *etc.*

Rob MacKenzie asked: Are we sure that the MCM model study presented is the most appropriate test of the ethanol chemistry?

Re-setting the constraints of the longer-lived species every 15 min seems to me to be somewhat arbitrary, because it doesn't relate either to the reactivity lifetime of ethanol (ethanol has a lifetime of days against reaction by OH), or to the travel time of urban emissions reaching the measurement point (the associated Malkin *et al.* paper (DOI: 10.1039/C5FD00218D) suggests that this is approx 75 min, but there are probably ways of determining this point-by-point in the way that is done for biomass burning plumes).

The difficulty with using a 15-minute re-set of the model is that the chemistry is continually relaxing towards different steady-states and neither the 15-minute concentrations, nor the steady-state values, provide the answer to the question to which the modelling is addressed, which is what fraction of observed acetaldehyde is derived from ethanol (and other) emissions.

Rachel Dunmore responded: As part of the MCM model experiment, we were not investigating the long term or transportation effects of ethanol oxidation. The maximum acetaldehyde generated from ethanol (as a % of acetaldehyde observations) is the steady state concentration divided by the observed concentration, giving an average of 14%. After 15 min, the model is generating 6.5% of the acetaldehyde observed. Given the e-folding lifetime of acetaldehyde is ~6 h at noon and the estimated time since emission is ~1 h (from the $NO_x : NO_y$ ratio), somewhere between 6.5–14% of the acetaldehyde observed derives from ethanol oxidation. This range highlights (as discussed in the paper) that a large fraction of acetaldehyde observed is either transported in or directly emitted as transport emissions and thus are not included as secondary sources in the photochemical box model. We did implement a GEOS-Chem model experiment to determine what the impact of long range transport of emissions would be.

Spyros Pandis said: The measured acetaldehyde diurnal average profile suggests that most of it is not produced or emitted in London but rather that it is transported there from other areas.[1] The use of a box modeling framework on the other hand implicitly assumes that most of it is emitted by local sources or produced by chemical reactions over the city. Could the disagreement between the model and the measurements be due to the above mismatch and not to missing local sources or chemical reactions? Could the measured acetaldehyde still be a product of ethanol, but with the precursor emissions and chemistry taking place elsewhere?

1 M. Diamantopoulou *et al.*, *Atmos. Environ.*, 2016, DOI: 10.1016/j.atmosenv.2016.03.015.

Alastair Lewis replied: The urban centre concentrations of acetaldehyde are likely a combination of regional contributions with more localised direct emissions and small *in situ* production. The *in situ* box modelling was used to demonstrate that local secondary production would not come close to recreating the observed concentrations and that other direct sources greatly improve on closure. The relatively tight connection between other primary vehicular emissions (*e.g.* NO_2, benzene) and acetaldehyde temporal trends would suggest that a substantial component is, however, locally emitted, but almost certainly not all. With the data we have we can't make a good estimate of more regional contributions, but they will exist.

Alison Tomlin asked: Previous studies have shown significant acetaldehyde emissions from the combustion of ethanol blended fuels within engines. Did you account for such primary emissions in your model?

Alastair Lewis replied: No we didn't include primary emissions - that was partly the motivation for the modeling study. At present there are no primary emissions in most inventories from traffic. Our modelling calculated how much secondary acetaldehyde could be formed in the atmosphere, and it was well short of the observations, implying the primary source must be substantial.

Nick Hewitt addressed Tim Wallington: You say in your paper that a systems perspective that includes both vehicles and fuel is needed when considering the effects of biofuels on urban air quality. Do you think this systems thinking should also include the air quality impacts of growing crops for the fuel? Please could I refer you to our two recent papers that show that the large scale cultivation of poplar in Europe might lead to enhanced ground-level ozone concentrations.[1,2]

1 K. Ashworth, O. Wild and C. N. Hewitt, Impacts of biofuel cultivation on mortality and crop yields, *Nature Climate Change*, 2013, **3**, 492–496.
2 K. Ashworth, O. Wild, A. S. D. Ellers and C. N. Hewitt, The impact of biofuel poplar cultivation on ground-level ozone and premature human mortality depends on cultivar selection and planting location, *Environmental Science and Technology*, 2015, DOI: 10.1021/acs.est.5b00266.

Tim Wallington responded: Yes, a holistic systems perspective is required and this would include assessing impacts associated with fuel production. Our assessment focused on the question of whether emissions from vehicles are negatively impacted by the increased biofuel use.

Paul Monks asked: First, should we be disappointed that we can't get a benefit from using biofuels, rather than that it just doesn't have a negative impact? Do you think there is a hope of using biofuels in the fuel blend to reduce emissions?

Second, it is clear that biofuels in the fuel mix will lead to an increase in oxygenates and acetaldehyde as indicated in the AQEG report (https://www.gov.uk/government/publications/road-transport-biofuels-impact-on-uk-air-quality). These "oxygenates" have the potential to make more ozone – do you have the feel for the relative impact?

Tim Wallington replied: Our study focused on addressing the question of whether the current, and any future, increased use of biofuels would hinder progress made in reducing vehicle emissions and improving urban air quality. We find that increases in biofuel content when accomplished in concert with changes in engine design and calibration for new vehicles should not result in problematic increases in emissions impacting urban air quality and may in fact facilitate future required emissions reductions.

Thank you for bringing our attention to the AQEG report. The main VOCs in vehicle exhaust are unburnt or partially burnt fuel. Including more oxygenated biofuels (*e.g.* ethanol) in blends with gasoline will lead to increased emissions of oxygenates (*e.g.* ethanol and acetaldehyde) and decreased emissions of the gasoline fuel components and their partially burned products. The literature data (see Fig. 4 in our paper) indicate that there is no discernable effect of ethanol level over the range E0 to E10 on NMOG emissions. As indicated in Table 1 from our paper, George *et al.* (2015) measured emissions from E0, E15, and E15 vehicles and reported no trend in ozone-formation potential with increased ethanol, while Suarez-Bertoa *et al.* (2016) reported that ozone-formation potential increased by factor of 2 for E85 *versus* E5-E15 fueled vehicles.

William Bloss asked Rachel Dunmore: The rate constant for reaction of Cl atoms with ethanol is nearly two orders of magnitude faster than that for OH + ethanol. Over the past few years we have become aware of new sources of Cl to the atmosphere, particularly through heterogeneous $ClNO_2$ formation giving rise to an early morning photolytic burst of Cl atoms. Although the Cl abundance is low, it may be the dominant oxidant for ethanol early in the morning, giving rise to acetaldehyde production – with timing which may resemble morning traffic emissions. This chemistry is not included in the standard MCMv3.1 as used here – could this contribute to the acetaldehyde production in your study?

More generally, increased alcohol emissions will lead these species to sequester a larger fraction of Cl atoms, reducing the impact of Cl upon oxidation of other VOCs.

Rachel Dunmore responded: The impact of Cl atoms on the chemistry of the London atmosphere was investigated prior to the model parameters being fixed. This was found to not be as important as the OH radical in terms of VOC oxidation in the model.

Roberto Sommariva asked: The use of a constrained model for long-lived compounds is problematic, as has been noted by other participants. However, a study published in 2011[1] supports some of the results of this paper. The authors used an unconstrained model, based upon the MCM, initialized using emission ratios of VOC and oVOC measured in Boston and New York City. Interestingly, the fraction of acetaldehyde derived from ethanol photo-oxidation is quite similar (5–25%), although this number changes with the evolution of the urban plume. The New York data also suggest the presence of significant acetaldehyde background concentrations (in addition to primary local emissions and photochemical formation), which should be discussed in this paper as well.

On a related point, the higher concentration of ethanol in London, compared to Los Angeles and New York, requires a bit more explanation; if I am not mistaken the percentage of ethanol in gasoline in the US is in fact higher (10%) than in the UK (5%).

1 R. Sommariva, J. A. de Gouw, M. Trainer, E. Atlas, P. D. Goldan, W. C. Kuster, C. Warneke and F. C. Fehsenfeld, Emissions and photochemistry of oxygenated VOCs in urban plumes in the Northeastern United States, *Atmos. Chem. Phys.*, 2011, **11**, 7081–7096, DOI: 10.5194/acp-11-7081-2011.

Rachel Dunmore responded: If we were to de-convolute the acetaldehyde emissions sources, it may be that the high baseline levels are predominately from either secondary production or long range transport, whereas the spikes shown in the time series (with very high levels of acetaldehyde) are more related to traffic related emissions from ethanol blended fuel use. The higher concentration of ethanol in London is likely a result of the different locational conditions experienced in comparison to Los Angeles or New York. For example, London has a much higher population density (5354 inhabitants per km^2) than Los Angeles (963 inhabitants per km^2), and experiences more *in situ* vehicle cold and warm start emissions.[1] This may be a direct consequence of the high population density in London leading to more people starting and ending their journeys inside the city limits rather than the more radial commuting observed in many US cities.

1 "World Cities Culture Report 2015 Now Published", 2015, http://www.worldcitiescultureforum.com/

William Brune addressed Tim Wallington: To understand the effect of using biofuels on air pollution, we need to think about how the biofuels will be used in the real world where vehicles are not maintained as well as they should be. How do you think the increased use of biofuels will affect air pollution under those circumstances?

Tim Wallington answered: Regardless of fuel, regular vehicle maintenance is essential for as-designed vehicle function including long vehicle life, high fuel efficiency, and low emissions. Real world fuel efficiency and emissions vary depending on many factors such as road conditions, driving style, vehicle load, speed, traffic, and fuel quality. The available data show a long-term substantial decline in real world vehicle emissions of CO, VOCs, and NO_x (*e.g.*, for data in the US see Table 3 in the CRC E-106 report here: http://www.crcao.org/reports/recentstudies2015/E-106%20Chicago%202014/E-106%20Chicago%202014%20Final%20report%202015.06.18.pdf) that has not been interrupted by increased biofuel use (*e.g.*, the approximately four-fold increase in ethanol use in the US from 2004 to 2014).

William Brune asked Rachel Dunmore: Have you tried constraining your model to acetaldehyde and seeing what happens?

Rachel Dunmore replied: The main focus of the detailed MCM modelling in this paper was to look at the photochemical contribution of ethanol chemistry to the acetaldehyde budget, compared to other primary and secondary photochemical sources. However, we have also run the MCM model constrained to the observed acetaldehyde in order to look at the effects on OH reactivity as described in detail in Whalley *et al.*[1]

1 L. K. Whalley, D. Stone, B. Bandy *et al.*, *Atmos. Chem. Phys.*, 2016, **16**, 2109–2122.

Lisa Whalley added: In support of the observed acetaldehyde concentrations in London, we found that we needed to constrain a detailed model to these observations in order to reconcile the observed OH reactivity.[1]

1 L. K. Whalley, D. Stone, B. Bandy *et al.*, *Atmos. Chem. Phys.*, 2016, **16**, 2109–2122.

Astrid Kiendler-Scharr asked Tim Wallington and Rachel Dunmore: You point out that biofuels can be considered as sustainable. Yet the combustion of biofuels will emit CO_2 with respective climate effects and trace gases which affect air quality. Shouldn't there be a stronger effort and focus on developments towards alternatives such as cars powered by fuel cells or batteries?

Tim Wallington responded: Each alternative fuel and powertrain option has relative strengths and challenges and no clear winner has emerged. We believe a "portfolio approach" is needed in developing sustainable technologies and alternative fuel options. We are working to provide diversity in fueling options to meet customers' differing needs, while also improving vehicle energy efficiency and long-term sustainability. We aim to provide customers with a range of affordable, fuel-efficient vehicles, advanced powertrains and alternative-fueled vehicle options – including fuel-efficient gasoline engines, advanced diesel engines, hybrids, plug-in hybrids, all-electric vehicles and alternative-fuel vehicles.

Rachel Dunmore replied: Biofuels are considered to be a renewable energy source, particularly when compared to fossil fuels. It is likely that alternatives such as fuel cells and batteries would be a preferred energy source, in that it would remove a large local emission source from major cities and roads. However, this may be transferred to a larger emission from generating stations which can have impacts on regional scales. Before there is another large shift in energy sources, we need to further understand what the existing situation is and be able to predict how changes will affect the status quo.

Brian McDonald addressed Rachel Dunmore: Fig. 3 in your paper shows diurnal profiles of nitrogen oxides (NO_x), carbon monoxide (CO), ethanol, and acetaldehyde during wintertime and summertime periods. While the morning rush hour peak is clear for all four species in the wintertime, in the summertime the morning rush hour peak only appears distinct for NO_x and CO, and is less evident for ethanol and acetaldehyde. I am wondering if this is an indication that other non-vehicular sources of ethanol could be important, *e.g.*, cleaning products and toiletries, as well as windshield wiper fluid as previously mentioned.

Rachel Dunmore responded: In the summertime, we believe the profile of the VOCs to be largely influenced by meteorology, in particular from large fluctuations in the height of the boundary layer, and photochemical processes. A similar summertime diurnal profile is also shown in Dunmore *et al.*[1] for toluene, whose main emission source was determined to be from traffic related emissions. In this paper on ethanol, we also discuss the possible influence of increased summertime evaporation (rather than combustion) from the use of ethanol blended fuels. This is shown by the correlations in Fig. 5, where those other VOC species with a more evaporative dominated emission from petrol have a higher summer correlation with ethanol. However, it is likely that there is some contribution from other non-vehicular sources but these do not seem to be dominant.

1 R. Dunmore, J. Hopkins, R. Lidster *et al.*, *Atmos. Chem. Phys.*, 2015, **15**, 9983–9996.

Nick Hewitt remarked to Tim Wallington: You have helpfully reminded us of the great progress made in reducing emissions from motor vehicles by manufacturers. However, given the recent revelations uncovering the fraudulent emissions-related measures taken by at least one manufacturer, I wonder what steps you think vehicle manufacturers should now take to reassure both the public and the air quality scientific community that they are in fact taking all reasonable practicable steps to reduce tailpipe (and other) emissions from the on-road fleet.

Tim Wallington replied: As suggested in Fig. 6 of our paper, vehicle manufacturers working with fuel suppliers and regulatory agencies have made large reductions in vehicle emissions over the past few decades. Given the 10–20 years for the on-road fleet to turn over we can predict confidently that as new vehicles with modern emission treatment systems replace older vehicles we will see a continued reduction of emissions from the on-road vehicle fleet.

Dwayne Heard said: I would like to comment on the point made by William Bloss regarding oxidation by chlorine atoms in urban areas. During the summer ClearfLo campaign in Central London in 2012, there were measurements of nitryl chloride ($ClNO_2$) using a CIMS instrument.[1] The concentration of $ClNO_2$ builds up during the night and is then photolysed in the morning, and this generates a morning spike of chlorine atoms. The Master Chemical Mechanism (MCM) already contains chlorine atom chemistry for the oxidation of some alkanes, and in a recent study this was extended to include an inorganic mechanism for chlorine chemistry and oxidation of a wider range of VOCs by Cl atoms, including alkenes, alkynes (and also ethanol and acetylaldehyde).[1] The updated mechanism was then used within a box-model framework to calculate a diurnal variation of Cl atom concentrations to investigate the relative importance of Cl atom oxidation of various classes of VOCs compared with oxidation by O_3 and OH, both of which were measured directly. The impact of chlorine atom chemistry on ethanol and acetylaldehyde was minor. On average during the entire campaign the removal of VOCs by reaction with Cl atoms was minor in comparison with other oxidants, removing ~4%, 1%, and 7% of alkanes, alkenes, and alkynes, respectively. However, on the day when maximum levels of $ClNO_2$ were observed, and hence when maximum levels of Cl atoms were calculated (1×10^4 atom cm^{-3} at *ca.* 7.30 am local time), around 15%, 3% and 26% of alkane, alkene and alkyne species were removed by reaction with Cl atoms, respectively.[1] In modelling work currently in progress, the additional O_3 generated in London as a result of VOC oxidation by Cl atoms generating peroxy radicals is being investigated, but is relatively minor, accounting on average only for ~1 ppb or so of additional O_3 production per day.

1 T. J. Bannan *et al.*, The first UK measurements of nitryl chloride using a chemical ionization mass spectrometer in central London in the summer of 2012, and an investigation of the role of Cl atom oxidation, *J. Geophys. Res. Atmos.*, 2015, **120**, 5638–5657, DOI: 10.1002/2014JD022629.

Andreas Wahner opened another general discussion of the papers by Tim Wallington and Rachel Dunmore: Both studies present measurements of emissions from biofuels – do they agree?

Tim Wallington replied: Dunmore *et al.* report measurements of ethanol in London air in 2012. Mean mixing ratios of ethanol were approximately 5 ppb during the winter and summer of 2012. The ethanol concentration peak correlating with the morning rush hour in winter (07:00–10:00) in Fig. 3e suggests the importance of a road transport related source. The concentration peak at the end of the evening (21:00–23:00) in summer in Fig. 3f suggests that sources unrelated to road transport are also important. The work of Dunmore *et al.* shows that there are substantial concentrations of ethanol in London air. However, the relative contribution of road transport and other sources is unclear. In the absence of historical data it is not clear from the data presented by Dunmore *et al.* how the transition to ethanol containing gasoline blends impacted the ethanol concentrations in ambient air in London. In our study we considered the impact of increased biofuel content in automotive fuels on vehicle emissions and urban air quality. We presented results from laboratory measurements of emissions from a vehicle using either fossil diesel or biodiesel fuel and we reviewed the literature addressing the impact of ethanol blend levels (E0–E85) on emissions from gasoline vehicles. We concluded that the available data suggest that emissions of NO_x, non-methane hydrocarbons, particulate matter (PM), and mobile source air toxics (compounds known, or suspected, to cause serious health impacts) from modern gasoline and diesel vehicles are not adversely affected by increased biofuel content over the range for which the vehicles are designed to operate. It is difficult to directly compare the results from the two studies. Vehicle emissions include fuel compounds (hydrocarbon and ethanol) and products resulting from fuel oxidation. Studies evaluating effects of changes in fuel composition should address all vehicle emissions, including both those related to ethanol and hydrocarbons. Overall, the most we can say is that the results from Dunmore *et al.* are not inconsistent with our conclusion that (in aggregate) vehicle emissions are not adversely affected by increased biofuel content over the range for which the vehicles are designed to operate.

Rachel Dunmore added: With respect to the answer given by Wallington *et al.*, we agree that it is difficult to compare the two studies as they are quite different; one based on real-world observations of a variety of emission sources and the other investigating the effects of biofuel diesel use on vehicle emissions in the laboratory. Wallington *et al.* does suggest from the literature review of ethanol blended gasoline use that emissions of acetaldehyde are increased.

Faraday Discussions

PAPER

Ozone production chemistry in the presence of urban plumes†

W. H. Brune,*[a] B. C. Baier,[a] J. Thomas,[ab] X. Ren,[c] R. C. Cohen,[d] S. E. Pusede,[de] E. C. Browne,[df] A. H. Goldstein,[g] D. R. Gentner,[h] F. N. Keutsch,[i] J. A. Thornton,[j] S. Harrold,[jk] F. D. Lopez-Hilfiker[jl] and P. O. Wennberg[m]

Received 1st December 2015, Accepted 23rd December 2015
DOI: 10.1039/c5fd00204d

Ozone pollution affects human health, especially in urban areas on hot sunny days. Its basic photochemistry has been known for decades and yet it is still not possible to correctly predict the high ozone levels that are the greatest threat. The CalNex_SJV study in Bakersfield CA in May/June 2010 provided an opportunity to examine ozone photochemistry in an urban area surrounded by agriculture. The measurement suite included hydroxyl (OH), hydroperoxyl (HO_2), and OH reactivity, which are compared with the output of a photochemical box model. While the agreement is generally within combined uncertainties, measured HO_2 far exceeds modeled HO_2 in NO_x-rich plumes. OH production and loss do not balance as they should in the morning, and the ozone production calculated with measured HO_2 is a decade greater than that calculated with modeled HO_2 when NO levels are high. Calculated ozone production using measured HO_2 is twice that using modeled HO_2, but this difference in calculated ozone production has minimal impact on the assessment of NO_x-sensitivity or VOC-sensitivity for midday ozone production. Evidence from this study indicates that this important discrepancy is not due to the HO_2 measurement or to the sampling of transported

[a]Department of Meteorology, Pennsylvania State University, University Park, PA 16802, USA. E-mail: whb2@psu.edu
[b]Department of Earth and Planetary Sciences, Johns Hopkins University, Baltimore, MD 21218, USA
[c]NOAA Air Resources Laboratory, College Park, MD, 20740, USA
[d]Departments of Chemistry and Earth and Planetary Science, University of California, Berkeley, CA 94720, USA
[e]Department of Environmental Sciences, University of Virginia, Charlottesville, VA 22904, USA
[f]Department of Chemistry and Biochemistry, University of Colorado, Boulder, CO 80309, USA
[g]Department of Environmental Science, Policy, and Management, University of California, Berkeley, CA 94720, USA
[h]Department of Chemical and Environmental Engineering, Yale University, New Haven, CT 06511, USA
[i]Department of Chemistry and Chemical Biology, Harvard University, Cambridge, MA 02138, USA
[j]Department of Atmospheric Sciences, University of Washington, Seattle, WA 98195, USA
[k]Puget Sound Clean Air Agency, Seattle, WA 98101, USA
[l]Paul Scherrer Institut, Villigen, PSI, Switzerland
[m]The Linde Center for Global Environmental Science, Caltech, Pasadena, CA 91125, USA

† Electronic supplementary information (ESI) available. See DOI: 10.1039/c5fd00204d

plumes but instead to either emissions of unknown organic species that accompany the NO emissions or unknown photochemistry involving nitrogen oxides and hydrogen oxides, possibly the hypothesized reaction $OH + NO + O_2 \rightarrow HO_2 + NO_2$.

Introduction

Ground-level ozone (O_3) is a serious health hazard,[1,2] with no known safe limit, and is estimated to cause millions of deaths per year globally. In the United States, urban ozone levels have decreased dramatically in the past two to three decades due to the significant investments made in understanding the cause of ozone pollution and to air quality regulations.[3] These regulations have encouraged lower emissions of nitrogen dioxide (NO_2) and nitric oxide (NO) (collectively termed NO_x) and volatile organic compounds (VOCs), which are the raw ingredients needed to produce ozone. Although ozone reductions in some parts of the United States appear to have leveled off, ozone reduction in the United States is a success story – one that should be emulated globally. However, with limited resources, much of the world cannot afford the investment the United States has made. The solution is an optimized approach that can target the specific emissions to which ozone production is most sensitive.

An optimized approach to ozone pollution reduction requires a solid scientific understanding of the causes of ozone pollution. The basic chemistry has been known for decades.[4–6] Ozone production begins with the early morning production of the hydroxyl (OH) and hydroperoxy (HO_2) radicals in the presence of NO_x (NO_2 + NO) and VOC emissions. Hydroxyl reacts with the VOCs, causing a cascade of reactive organic compounds, including organic peroxy radicals (RO_2) and HO_2. The peroxy radicals react with NO to form peroxy radicals and NO_2. Nitrogen dioxide is decomposed by ultraviolet sunlight to form NO and atomic oxygen, which immediately reacts with molecular oxygen (O_2) to form ozone. This process is the dominant production pathway for tropospheric ozone.

However, NO also reacts with O_3 to make NO_2, NO and NO_2 come into a steady-state balance within tens of seconds, and NO_2 then reacts with OH to terminate NO_x and HO_x (OH + HO_2) cycling. By this theory, the instantaneous ozone production, $P(O_3)$, initially increases as NO increases, but then decreases with continued NO increase. The peak $P(O_3)$ is a sensitive function of HO_x production.[7–9] One simple test of the non-linear dependence of ozone production on NO_x is the so-called weekend-weekday effect, for which ozone is greatest on weekends when NO_x is less.[10]

Despite this well-accepted theory, there are discrepancies between measured and ozone calculated by regional air quality models. The models agree with the observations on average, but tend to be too high for values below 50 ppbv and too low for values approaching 80–90 ppbv and beyond. While models have had better agreement with observations for individual cities and times of the year, this discrepancy between modeled and measured ozone is found for several different models and urban areas.[11–13] The causes of the disagreements between measured and modeled ozone are usually attributed to errors in the emissions inventories for nitrogen oxides (NO_x) and volatile organic compounds (VOCs), but there is some evidence that the well-tested theory may need some modification.

The budget equation for ozone is

$$\frac{\partial[O_3]}{\partial t} = P_{chem} - L_{chem} + \frac{w_e \Delta O_3 - u_d[O_3]}{H} - \nabla(\boldsymbol{v}[O_3]), \quad (1.1)$$

where P_{chem} is the chemical O_3 production rate, L_{chem} is the loss rate, $w_e\Delta O_3/H$ is the ozone entrainment rate between the mixing layer and the free troposphere, $-u_d[O_3]/H$ is the ozone deposition rate, H is the mixing layer height, and $-\nabla(\boldsymbol{v}[O_3])$ is the ozone advection rate by the mean wind, $\boldsymbol{v}$. Net ozone production can be calculated from 1.2–1.4,

$$P(O_3) = P_{chem} - L_{chem} \quad (1.2)$$

$$P_{chem} = k_{NO+HO_2}[NO][HO_2] + \sum_i k_{NO+RO_{2i}}[NO][RO_{2i}] \quad (1.3)$$

$$L_{chem} = f_{H_2O}J_{O_3}[O_3] + k_{O_3+OH}[O_3][OH] + k_{O_3+HO_2}[O_3][HO_2] + k_{NO_2+OH}[NO_2][OH] + L(O_3 + \text{alkenes}) + L(O_3 + \text{halogens}) + P(RONO_2) \quad (1.4)$$

where k's are rate coefficients; NO is nitric oxide; HO_2 is the hydroperoxyl radical; RO_2 is the organic peroxyl radical; J_{O_3} is the photolysis frequency of O_3; f_{H_2O} is the fraction of the excited state $O(^1D)$ atoms from O_3 photolysis that react with H_2O; OH is the hydroxyl radical; and $RONO_2$ represents organic nitrates.

In several studies, the measured HO_2 often is less than modeled HO_2 at low NO, equals modeled HO_2 when NO is about 1 ppbv, and increasingly exceeds modeled HO_2 at increasingly higher NO abundances.[9,14–18] This measured-to-modeled difference is greatest when HO_x production is lowest, such as during morning rush hour, where the measured-to-modeled HO_2 ratio can be higher than ten. It is least when HO_x production is greatest, such as during the afternoon. This greater-than-expected HO_2 at higher NO is also inferred from peroxynitric acid (HO_2NO_2) measurements in Mexico City in 2006.[19] Therefore, ozone production (PO_3) calculated using the measured HO_2 can exceed $P(O_3)$ calculated using the modeled HO_2. In addition, a direct measurement of the ozone production rate shows that measured $P(O_3)$ is twice $P(O_3)$ calculated using modeled HO_2 and RO_2, although the timing of the $P(O_3)$ peak agrees better with the $P(O_3)$ calculated with modeled HO_2 than with $P(O_3)$ calculated with measured HO_2.[20,21] These observations are inconsistent with the current understanding of ozone production.

This discrepancy can be explained in several different ways. A first hypothesis is that the HO_2 measurement is being affected by the atmospheric NO levels. A second hypothesis is that calculation of $P(O_3)$ is being distorted by the averaging over plumes of NO_x-rich and HO_2-rich air so that the product of averaged measured $[HO_2]$ and [NO] is larger than the average of the products of the plume-scale NO and HO_2:

$$\overline{[HO_2]}\ \overline{[NO]} > \overline{[HO_2][NO]}. \quad (1.5)$$

A third hypothesis is that HO_x measurements made when NO_x is not in photostationary state (PSS) are being compared to models that calculate HO_x while assuming an NO_x photostationary state. Because NO_x is often emitted from combustion as NO, the HO_x loss due to OH + NO_2 would be less than in the model and measured HO_2 would appear to be greater than modeled. The fourth hypothesis suggests that unknown HO_x sources are co-emitted with the NO_x. The fifth hypothesis is that unknown chemistry is missing from the current understanding of the chemistry between nitrogen oxides and hydrogen oxides.

The multi-agency California Research at the Nexus of Air Quality and Climate Change-San Joaquin Valley (CalNex-SJV) study in Bakersfield CA during June 2010 meets the criteria needed to test these hypotheses. Bakersfield is situated in the southern San Joaquin Valley 180 km north-northwest of Los Angeles and experiences the second worst ozone pollution in California.[22] It is surrounded by agricultural land, but at the same time has active oil and gas fields, an oil refinery and significant traffic on highways that has a higher fraction of diesel powered vehicles than other urban areas in the US.[23] Its numerous point and distributed NO_x and VOC emission sources create a heterogeneous mix of plumes of different sizes and chemical composition. This combination of high VOC and NO_x emissions and the typically cloudless, hot weather causes high ozone abundances. Bakersfield CA typically has 90 exceedances of ozone air quality standards each year and, while the number has decreased during the past decade, it is still not in compliance with EPA air quality standards.[22]

The calculated ozone production has been examined for the San Joaquin Valley (SJV), including Bakersfield. In a study using midday (10:00–14:00 local time) measurements of temperature as a surrogate for OH reactivity from VOCs, NO_x, and frequency of O_3 exceedances, Pusede and Cohen showed that the calculated midday $P(O_3)$ is consistent with a transition from VOC-sensitive to NO_x-sensitive regimes for much of the SJV, starting first with higher maximum daytime temperatures (34–45 °C) followed by the beginning of a transition at more moderate temperatures (28–33 °C).[24] In a second study focused on the CalNex-SJV site, Pusede *et al.*[25] show that $P(O_3)$ calculated with an analog model is NO_x-sensitive for moderate-to-high temperatures, except for weekdays at moderate temperatures, for which it is VOC-sensitive. These conclusions appear to be inconsistent with the $P(O_3)$ calculated from the measurements of HO_2 and RO_2 as a function of NO described above. However, high NO and greater-than-expected $P(O_3)$ occur mainly in the morning before 10:00 PST and it is not clear how this difference will affect the averaged $P(O_3)$ sensitivity to NO_x and VOCs calculated for midday.

In this paper, we will test these hypotheses using primarily the HO_2 and NO abundances measured during CalNex-SJV. First we will compare the measured OH, HO_2, and OH reactivities to those calculated by a near-explicit box model that is constrained by all other simultaneous field measurements in Bakersfield. Measured and modeled OH production and loss rates and ozone production from HO_2 are also compared and discussed. Finally we add the evidence learned from this study to that from previous studies to assess the likelihood of the hypotheses for the greater-than-expected ozone production at greater NO abundances.

Methods

Measurement site

The CalNex 2010 field campaign in California consisted of aircraft and ship measurements, as well as two ground sites: one in Pasadena CA and a second in Bakersfield CA. This work focuses on the CalNex-SJV Bakersfield site (Fig. 1). A scaffolding tower 18 m high was erected in a field just to the east of a parking lot at the California Agricultural Experiment Station, which is located in south Bakersfield

(35.346153 N, 118.965519 W). The land to the south is fields and settling ponds. Light businesses were located to the north, along with highway 58, a highly trafficked corridor, about 760 m away. A school bus facility was also located 600 m to the west.

Measurements were made from 15 May to 28 June 2010. For that time of year, the weather in Bakersfield is typically hot and dry, with temperatures routinely exceeding 37 °C. However, 2010 was not a typical year. The temperature did not exceed 35 °C until the last three days, and during one week, the weather was cloudy with some light rain with temperatures ranging from 20 °C and to 25 °C. Winds were generally from the north at 4 ms^{-1}, although they were often calm and from the east in the hours around sunrise (Fig. 2).

HO_x measurements

OH and HO_2 were measured with Penn State's Ground-based Tropospheric Hydrogen-Oxides Sensor (GTHOS),[26] which uses laser-induced fluorescence (LIF). The hydroxyl radical (OH) is detected as the sampled air is pulled through a 1 mm inlet into low pressure (~6 hPa) and passes through the path of a laser tuned to the $Q_1(2)$ OH absorption line at 308 nm, and then fluoresces. This fluorescence is measured in a first detection axis by a gated microchannel plate detector positioned perpendicular to the sample flow and the laser beam. The sampled air flows through the first detection axis, and has NO injected into it to react with HO_2 to form OH, which is then detected by LIF in a second detection axis in order to detect HO_2.

A tunable dye laser pumped by a 532 nm Nd:YAG laser produces the 308 nm light used to detect OH. This laser wavelength is alternately tuned by an etalon to

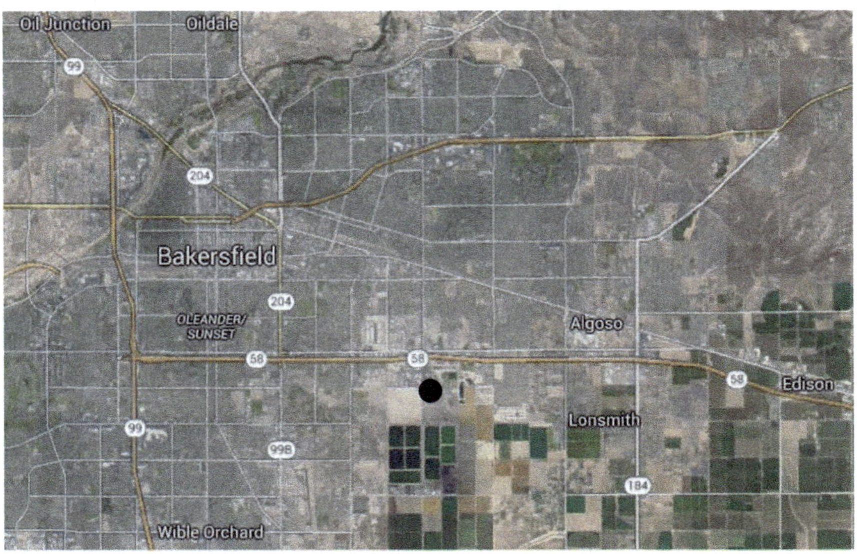

Fig. 1 Map of Bakersfield CA and location of the CalNex-SJV site (black dot) (Google, 2015). The site was at 35.346153° N and 118.965519° W. The heavily traveled four-lane highway 58 was located 760 m to the north of the site. The nearest local road, East Belle Terrace, was 50 m to the north and had light traffic during the study period.

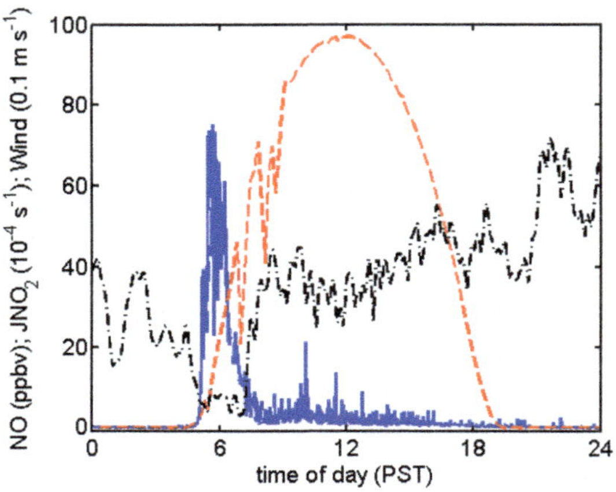

Fig. 2 Behavior of measured NO (blue solid line), $J(NO_2)$ (red dashed line), and wind speed (dot-dashed black line) for day-of-the-year (day) 160, 9 June. This behavior is typical for most days during the study period, although the NO bump on day 160 was one of the largest. The widths of the NO spikes range from seconds to hours.

a wavelength where OH absorbs and then fluoresces or to the background on alternate sides of the OH absorption line, all within a 30 second cycle. The difference between these two signals is proportional to OH. The proportionality constant is determined by laboratory and field calibrations.[26] This method of measuring OH is referred to as OH_{wave}. Wavelength modulation has been the most common method for measuring atmospheric OH by LIF.

A second OH measurement method involves injecting an OH reactant into the air to scavenge the OH before it is sampled through the instrument inlet. Reactant amounts are chosen to maximize the fraction of OH removed in the 10 ms between injection and entering the inlet and to simultaneously minimize the OH removed inside the instrument. Hexafluoropropylene (C_3F_6) was used as the OH scavenger. By turning C_3F_6 injection on and off, the OH signal is found by subtracting the signal when injection is on from the signal when injection is off. This method is called OH_{chem}. The difference between OH_{wave} and OH_{chem} is the OH from an interference, called OH_{int}. The OH interference has the spectral signature of OH, but all studies show that it is not laser-generated. To test the functionality of the OH_{chem} system, a UV lamp was affixed to the instrument near the inlet. The lamp, which photolyzed water vapor to make a large OH signal, was turned on for a few minutes three times a day to ensure that the C_3F_6 injection was scavenging OH properly. For this paper, the OH measurements were all made with the chemical removal method.

Two years after CalNex-SJV, it was reported that some alkene-based and aromatic-based organic peroxides were also detected along with HO_2 in most instruments that used NO to convert HO_2 to OH.[27] Methods have been found to minimize this interference, but these methods were not used in the CalNex-SJV study. This sum of HO_2 and a subset of RO_2 has been called HO_2*, which is then compared to modeled HO_2*.[28] In this paper, our approach is to bound the possible HO_2 values by subtracting all the modeled RO_2 from HO_2* and calling this value HO_2, which may be an overcorrection. This HO_2 is actually the lower bound of HO_2 and HO_2* is the upper bound. As will be seen, the main conclusions are not sensitive to the choice of HO_2 between these two bounds.

In addition to OH and HO_2 measurements, measurements were also made of OH reactivity, which is the inverse of the OH lifetime.[29,30] Approximately 150 LPM of ambient air is drawn into the instrument and flows through the aluminum flow tube (7.5 cm dia.). At the far end of the flow tube is a sampling inlet and an OH measurement system nearly identical to the one used in the main GTHOS system. Before the airflow reaches the sampling inlet, it flows past a movable source of OH called the wand. Inside the wand, 5 LPM of moist nitrogen flows past a mercury lamp, which photolyzes the water vapor to produce OH and HO_2 and then jets out the detection axis. As the wand is moved farther away from the sampling inlet, the OH has more time to react with trace gases in the ambient air flowing through the tube and the OH signal decreases exponentially. Moving 10 cm is equivalent to a decay time of 140 ms and the wand completes a cycle in 30 seconds. The OH reactivity is the slope of the change in the log of the OH signal divided by the reaction time.

The OH reactivity decay is affected by atmospheric NO because $HO_2 + NO \rightarrow OH + NO_2$ recycles OH that has decayed, thus causing curvature in the OH decay. A least-squares linear fit to this curved decay produces a decay slope that is less than the real slope and thus produces an OH reactivity value that is too low. The effects of atmospheric NO on the OH reactivity measurement are minimized by applying the correction algorithm described in Shirley *et al.*[31]

The suite of measurements at the CalNex-SJV site was extensive as documented in the CalNex overview paper.[32] It included meteorological parameters, inorganic species, volatile organic compounds (VOCs) and oxygenated VOCs, and many aerosol abundances and properties. Data used in this study were drawn primarily from measurements taken at or near the top of the measurement tower.

Photochemical box modeling

The simultaneous measurements of all available inorganic and organic species and meteorological parameters were used to calculate HO_x using the near-explicit Master Chemical Mechanism, Version 3.2 (MCMv3.2)[33] in a box model framework developed by G. Wolfe.[34] MCMv3.2 contains approximately 6700 unique chemical species and 17 000 reactions. Measured VOCs that are included in the model are treated explicitly; those that are not represented in the model are aggregated into appropriate MCMv3.2 species based on OH-reactivity or their molecular structure. It should be noted that isoprene chemistry in MCMv3.2 is replaced with explicit reactions detailed in Mao *et al.*[35] Chemical species were assumed to have a lifetime of one day to prevent buildup in the model. Data were then averaged into ten-minute time intervals for the modeling and the comparisons to measurements.

Photolysis frequencies were not measured during CalNex-SJV, so instead they were calculated using the NCAR Tropospheric Ultraviolet and Visible Radiation Model.[36] These calculations assume clear overhead skies with the overhead ozone column of 300 D.U. taken from satellite measurements; to correct for the effects of overhead cloud cover, photolysis frequencies were scaled by the ratio of the calculated JNO_2 to JNO_2 measured from a UV radiometer on the measurement tower. This same cloud correction factor was applied to the other calculated photolysis frequencies and has been shown to give accurate photolysis frequencies.[37]

Model results include days between 23 May and 28 June 2010 to allow for the greatest number of simultaneously measured chemical species to constrain the model. Data were taken from three periods when HO_x, OH reactivity, and other chemical species important for this analysis were being measured: day 145–150; day 156–164; and day 166–174. They were averaged or interpolated into 10 minute time intervals for the model runs. For analysis of HO_x and $P(O_3)$ as a function of NO, the 10 minute data were limited to hours between 7:00 and 17:00, a time period chosen to capture the portion of morning rush hour in which the photolysis and transport were well defined. All times are Pacific Standard Time even though the official time was Pacific Daylight Time during the field study.

CalNex-SJV measurement and model uncertainties

The absolute uncertainties for OH and HO_2 are approximately ±40% at the 2σ confidence level. However, the subtraction of the OH signals with and without C_3F_6 scavenging causes the OH limit-of-detection to be about $(2–3) \times 10^5$ cm^{-3}. In addition, with the RO_2 interference in the HO_2 measurement, there is an additional uncertainty that can be about a factor of two. In studies after CalNex, our measurement strategy was revised to minimize the RO_2 interference in the HO_2 measurement. The uncertainty for the OH reactivity instrument is estimated to be ±30% at 2σ confidence. The uncertainties in the other measurements are given in Table 7a of the overview paper.[32] The model uncertainty can be estimated from a global uncertainty and sensitivity analysis from a previous urban study using the RACM2 model.[38] These two studies have similar uncertainties in their input measurements, reaction rate coefficients, and products, so the uncertainties in the modeled OH and HO_2 for comparable urban areas should also be similar. Thus, the estimated 2σ uncertainty is approximately 40% for OH and HO_2. These uncertainties will be used to assess the significance of the comparisons between the measured and modeled OH, HO_2, and OH reactivity in this study.

Results

Comparison of measured and modeled OH and HO_2

Understanding ozone production requires an understanding of OH and HO_2. A first-order test of this chemistry is the comparison of the modeled and measured OH and HO_2 as a function of the time of day (Fig. 3).

The median measured and modeled OH display the same diel behavior but the measured OH peaked at $(7.3 \pm 2.5) \times 10^6$ cm^{-3} while the modeled OH peaked at $(10.4 \pm 4.0) \times 10^6$ cm^{-3}. This marginal statistical difference of ~30% persists during the midday, although at sunrise and sunset, the modeled and measured OH agree. At night, the median measured OH is 3×10^5 cm^{-3}, which is near the instrument detection limit and is consistent with the modeled nighttime OH of 3×10^4 cm^{-3}.

The OH interference signal peaks at midday at 2×10^6 cm^{-3}, is about 1×10^6 cm^{-3} at night, and is only about 25% of the OH determined by the chemical removal method. Interestingly, this OH interference signal is about the same as that observed in forests[35,39] and in other cities[17] and may have a common origin,

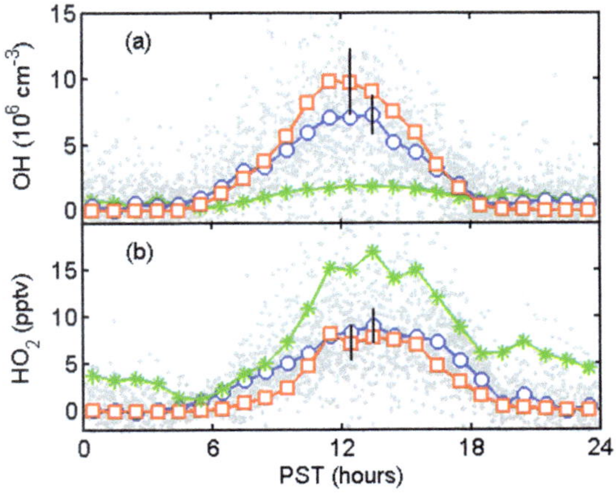

Fig. 3 Median diel variation of HO_x. (a) OH (10^6 cm^{-3}), measured (blue circles), modeled (red squares), and OH interference (green stars), with individual 10 min data for measured OH (gray dots); (b) HO_2 (pptv), measured (blue circles), modeled (red squares), and measured HO_2 with the RO_2 interference (green stars), with individual 10 min data for measured HO_2 (gray dots). Error bars are $\pm 1\sigma$ confidence.

but because the OH is about 10 times larger in cities than in forests, the interference's effect of the measured OH is much more significant in forests than in cities.

Measured HO_2 is on average about 7 pptv at midday and is within the combined uncertainties of the modeled HO_2. Measured and modeled HO_2 also agree at night, but in the morning, median measured HO_2 is more than twice median modeled HO_2. The HO_2 with RO_2 interference is about twice as large at midday and much larger at night. Thus measured and modeled HO_2 agree within uncertainties for much of the day, if true HO_2 is near the lower bound for measured HO_2.

When median measured and modeled OH and HO_2 are plotted against NO for the daytime, the modeled OH is similar to measured OH for NO up to about 10 ppbv (Fig. 4(a)). Measured HO_2 and HO_2*, on the other hand, begin deviating from modeled HO_2 when NO is 1 ppbv, and are ~10 times larger when NO is 10 ppbv (Fig. 4(b)). This result is consistent with the previous reports of higher-than-expected measured HO_2 for conditions where NO is above 1 ppbv. Note that this discrepancy does not depend on uncertainties in removing the RO_2 interference from the HO_2 signal because HO_2 and HO_2* behave the same way.

HO_2 increases with the increasing production rate of HO_x, $P(HO_x)$, for each value of NO (Fig. 4(c)). The 10 minute data are binned into three $P(HO_x)$ ranges: 1×10^5 cm^{-3} s^{-1} < $P(HO_x)$ < 5×10^6 cm^{-3} s^{-1}, 5×10^6 cm^{-3} s^{-1} < $P(HO_x)$ < 10^7 cm^{-3} s^{-1}, and $P(HO_x)$ > 10^7 cm^{-3} s^{-1}. In Bakersfield, these ranges correspond to low $P(HO_x)$ for early morning, late afternoon, or very cloudy days; medium $P(HO_x)$ for mid-morning, midafternoon, or cloudy days; and high $P(HO_x)$ for midday on sunny days. Note that both the measured and modeled HO_2 are greater for greater $P(HO_x)$, as would be expected, but modeled HO_2 is more sensitive to $P(HO_x)$ than measured HO_2 is. Furthermore, in general, the measured-to-modeled HO_2 ratio is least for the highest $P(HO_x)$ and greatest for the lowest $P(HO_x)$. Thus, regions with

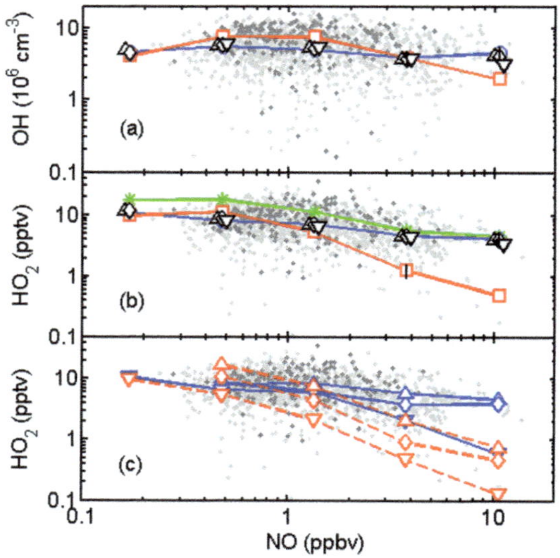

Fig. 4 Median behavior of HO_x as a function of NO. (a) OH (10^6 cm^{-3}), measured (blue circles), modeled (red squares), and measured averages of 20 seconds (upward pointing triangles), 1 minute (diamonds), and 1 hour (downward pointing triangles); (b) HO_2 (pptv), measured (blue circles), modeled (red squares), measured HO_2 with RO_2 interference (green stars), and measured averages of 20 seconds (upward pointing triangles), 1 minute (diamonds), and 1 hour (downward pointing triangles); (c) HO_2 (pptv), measured (solid blue lines) and modeled (dashed red lines), for $P(HO_x) > 10^7$ cm^{-3} s^{-1} (upward triangles), 5×10^6 cm^{-3} s^{-1} < $P(HO_x)$ < 10^7 cm^{-3} s^{-1} (diamonds), and 1×10^5 cm^{-3} s^{-1} < $P(HO_x)$ < 5×10^6 cm^{-3} s^{-1} (downward triangles). Gray dots are individual 10 min measured data; darker gray dots are data measured between 12:00 and 15:00. Data are filtered for daytime hours between 7:00 and 17:00.

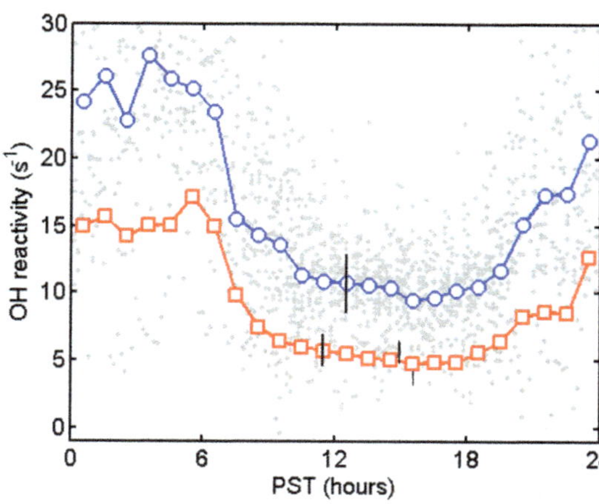

Fig. 5 Median diel variation of OH reactivity. Median measured (blue circles) and calculated from modeled chemical species (red squares), with individual 10 min data for measured OH reactivity (gray dots). A total of 339 out of 4392 10 minute data values are greater than 30 s^{-1} and are not shown. Error bars are 1s confidence.

higher $P(HO_x)$ likely have a measured-to-modeled HO_2 ratio that is closer to 1 even for higher NO values.

These differences between measured and modeled OH and HO_2 are within the widely varying range of previous studies.[40] Even this behavior as a function of $P(HO_x)$ is the same as in previous studies. The CalNex results are quite similar to what is observed in an April 2009 SHARP study in Houston TX.[17] An important difference between the Houston and Bakersfield studies is the higher daytime levels of NO in the Bakersfield study, levels that are closer to those observed in New York City in summer 2001.[15] This difference influences the $P(O_3)$ sensitivity to NO_x and VOCs.

Comparison of measured to modeled OH reactivity

The median measured OH reactivity was greatest at night when the atmospheric boundary layer height was lowest and was least during the late afternoon when the boundary layer height was greatest (Fig. 5). The median measured OH reactivity was as high at 26 s^{-1} at 4:00, dropped from 15 s^{-1} to 11 s^{-1} between 8:00 and 10:00, then slowly decreased to 9 s^{-1} at 17:00, after which it began a slow increase to nighttime values. Mean values are ~20% higher because of the few large spikes. From day-to-day, OH reactivity varied from 5 s^{-1} to more than 50 s^{-1} at night and from about 3 to 20 s^{-1} during the day. The lowest values were on the few cool rainy days in May. The spikes in OH reactivity generally correlate to NO spikes, suggesting that a large fraction of OH reactivity in Bakersfield is due to anthropogenic emissions and combustion,[41] but oxidation products, potentially from biogenic emissions from natural and agricultural sources, have large contributions to reactivity with high ambient temperatures.[25]

This measured OH reactivity is compared to the OH reactivity calculated from the measured chemical species and the modeled products of those measured chemical species. The ratio of model-calculated OH reactivity to measured OH reactivity is 0.59 at night and 0.53 during the day. In a cooler, rainy period at the beginning of the study (before day 150, 30 May), measured and calculated OH reactivity agreed to within 10%. The missing OH reactivity tends to be greater when the temperature and ozone are higher, but these correlations are weak, suggesting that other factors such as unmeasured species may be contributing to the measured OH reactivity.

If only measured OH reactants are included in the calculated OH reactivity total, the calculated OH reactivity is only slightly less (<10%) than the calculated OH reactivity that includes modeled chemical species. According to the model, 80% of the calculated OH reactivity is caused by 13 chemical species, with the most important being NO_2 (22%), carbon monoxide (11%), formaldehyde (6%), ethanol (8%), methanol (5%), and heptanal + nonanal (5%). The biogenic VOCs isoprene and limonene each contribute less than 2%, thus indicating the dominance of anthropogenic VOCs in the calculated OH reactivity during this study.[25] From this analysis, when NO is high, NO is a major OH sink, accounting for more than half the calculated OH reactivity in some plumes.

The OH reactivity is an indicator of the OH lifetime. For CalNex-SJV, the lifetime was 40 ms during the night and at sunrise and ~100 ms during the day. With winds of 0.5 m s^{-1} during the early morning and 4 m s^{-1} during the rest of the day, OH achieved steady state (>3 lifetimes) for air that had travelled less than

2 meters. However, this view is too simplistic; not only must the lifetime of OH be considered but also the lifetime of OH sources[42] and sudden changes in OH sources and sinks.

The effects on the OH value due to its sources and sinks can be seen by suddenly turning off the photolytic sources of OH and other radicals in a model. Using CalNex-SJV conditions and chemical species, a photochemical box model was run until it achieved steady state and then photolysis was terminated and the run continued for another 100 seconds to follow the decays of OH, HO_2, NO, and other short-lived chemical species (Fig. 6). Initially OH drops at a rate about equal to the OH reactivity until the OH loss rate matches the OH production rate from recycling, primarily HO_2 + NO in this case. The OH decay then roughly parallels the decay of its primary source – HO_2 times NO. The half-life of NO is 22 s, HO_2 110 s, and OH 13 s, which is two orders-of-magnitude longer than indicated by the OH reactivity. In this case, the HO_2 half-life is much longer than the NO half-life, but in the morning in CalNex-SJV, the half-life of NO is 60 s, HO_2 13 s, and OH 8 s. Thus, while OH is in steady state with its sources and sinks in less than a second, its abundance is tied to the half-life of its sources as they respond to changes in their sources or sinks.

Typically in model-to-measurement HO_x comparisons, the model is constrained to NO, NO_2, O_3, and all other measured chemical species except OH and HO_2. We have used this method in the analysis for this paper. The model calculates the steady-state values for OH and HO_2. However, HO_2 may not be in steady state because of surface deposition, upwind cloud shadowing, or other HO_x production or loss that occurred within the tens of seconds prior to the measurement. For our example, the typical afternoon wind was 4 m s^{-1}, so that in the HO_2 half-life of 110 s, the HO_2 in an air parcel could be affected during its travel over a distance of almost half a kilometer. Thus HO_2 could have values that are quite different from those calculated by a steady-state model constrained to the measurements made at the field site.

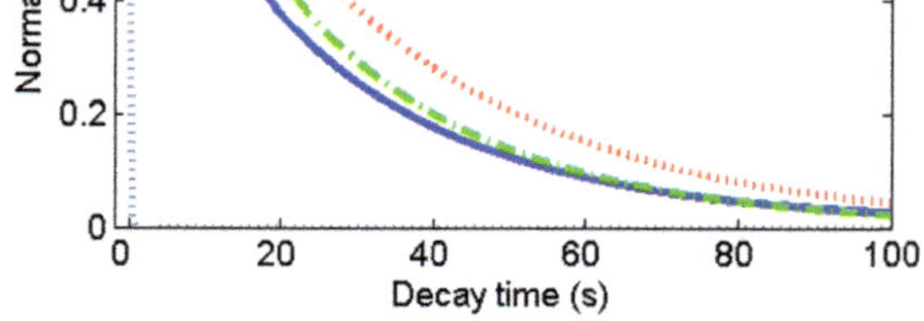

Fig. 6 Normalized modeled decays of OH (solid blue), HO_2 (dashed black), NO (dotted red), HO_2 times NO (dot-dashed green), and inverse of OH reactivity (leftmost dotted blue). This case is initialized with O_3 (60 ppbv), NO_2 (2.6 ppbv), NO (1.2 ppbv), and OH reactivity (10 s^{-1}).

OH production and loss

A critical consistency check of the measurements of OH, HO_2, and OH reactivity is the balance between production and loss for OH. The loss rate for OH is product of the measured OH and OH reactivity; the production rate is all the sum of all the OH sources, including recycling, mostly from HO_2 + NO, and primary production, mostly from O_3 photolysis followed by $O(^1D) + H_2O$ and HONO photolysis. Most of these quantities and reaction rate coefficients are measured, making this consistency check very close to being independent of the model. The mean measured and modeled OH production and loss rates peaked at (5–9) $\times 10^7$ cm^{-3} s^{-1}, (Fig. 7(a)). This OH production is lower by a factor of two compared to other US cities and by a factor of 6 compared to Mexico City in 2003.[43] As in previous studies, measured OH production and loss match in the afternoon but not in the morning. From sunrise to almost noon, measured OH production and loss are both two to three times the modeled production/loss and measured OH production is about twice measured OH loss. Because of the short OH lifetime, these two must always match.

To resolve this discrepancy, one possibility is that the OH reactivity is biased low when ambient NO is high, which can result from two factors. First, the high OH reactivity in the morning reduces even the initial OH signal by about a decade compared to midday values, so that the low signal-to-noise of the decay and the uncertainty in the subtracted OH background signal strongly influence the calculated OH decays. Laboratory experiments indicate that too little background signal tends to get subtracted when the OH signal is small, causing the calculated decay to be smaller than it should be. Second, the correction factor to account for the recycling of OH by the reaction of HO_2 and ambient NO in the OH reactivity instrument is dependent on NO and the OH reactivity, but it is typically 1.05 to

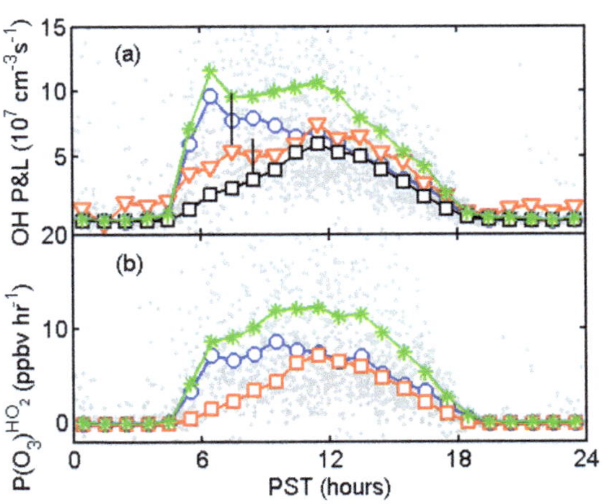

Fig. 7 Median diel variation of production and loss. (a) OH (10^7 cm^{-3} s^{-1}), measured OH production (blue circles) and loss (red triangles), measured OH production using HO_2* instead of HO_2 (green stars), modeled production and loss, which are equal (black squares), and individual 10 min production data using measured HO_2 (gray dots); (b) ozone production from HO_2 (pptv), $P(O_3)^{HO_2}$ calculated from measured HO_2 (blue circles), modeled HO_2 (red squares), and measured HO_2* (green stars), with individual 10 min data for $P(O_3)^{HO_2}$ calculated from measured HO_2 (gray dots).

1.1 for 1 ppbv of NO and increases to 6–10 for 100 ppbv of NO. The uncertainty in the correction factor is estimated to be ±40% (2σ confidence), based on repeated laboratory decays, although we think that this uncertainty can be cut in half. The combination of the tendency for insufficient background subtraction at low OH signals and uncertainty in the NO correction factor could be responsible for this discrepancy in the balance between OH production and loss. The other possibility is that the HO_2 measurement is in error, which is discussed in a following section.

Ozone production rate

The ozone production from HO_2, $P(O_3)^{HO_2}$, is only about half the total $P(O_3)$, the other portion coming from RO_2. We will confine our discussion only to calculated $P(O_3)^{HO_2}$ because HO_2 was measured and RO_2 was not. If HO_2 is greater than expected at higher NO values, then by eqn (1.2), $P(O_3)^{HO_2}$ should also continue to increase for NO greater than ~1 ppbv, in contrast to the modeled value, which peaks when NO is near 1 ppbv and then decreases. This difference in measured and modeled HO_2 translates directly into a difference in $P(O_3)$ values that are calculated from measured and modeled values of OH and HO_2 (Fig. 7(b)). $P(O_3)^{HO_2}$ calculated from measured HO_x is lower than $P(O_3)^{HO_2}$ calculated from modeled HO_x below ~1 ppbv and then becomes more than a decade larger for NO = 10 ppbv.

Most of this greater-than-expected $P(O_3)^{HO_2}$ occurs in the morning before 10:00. The cumulative median $P(O_3)^{HO_2}$ shown in Fig. 7(b) is 55 ppbv for the model and 97 ppbv for the measurement. This cumulative production can be compared with the mean diel peak O_3 of 58 ppbv (range ±20 ppbv). $P(O_3)$ is about twice $P(O_3)^{HO_2}$, so the cumulative $P(O_3)$ calculated from both modeled and measured HO_2 are more than twice the observed ozone. At any given location, neither the calculated instantaneous nor cumulative ozone production are necessarily related to the peak ozone, which comes from not only the local production but also from production throughout the planetary boundary layer and from transport (eqn (1.1)). The calculated $P(O_3)^{HO_2}$ suggests that the CalNex_SJV site is in an ozone source region that contributes to the ozone production in the Bakersfield plume as it moves south during the day.

How does this difference in measured and modeled $P(O_3)^{HO_2}$ affect the assessment of NO_x-sensitivity or VOC-sensitivity? In four urban areas, both measured and modeled $P(O_3)^{HO_2}$ were strongly VOC-sensitive during morning rush hour, but switched to weakly VOC-sensitive or NO_x-sensitive from mid-morning to mid-afternoon.[43] For CalNex-SJV, mean midday NO was typically 1.5 ppbv, so that measured HO_2 was less than twice modeled HO_2. As a result, the greater-than-expected $P(O_3)^{HO_2}$ that occurs primarily in the morning does not significantly change the $P(O_3)$ sensitivity to NO_x or VOCs during the 10:00 to 14:00 period used by Pusede *et al.*[25] in their analysis.

Possible cause of greater-than-expected HO_2 and $P(O_3)^{HO_2}$

What could be the cause of the greater-than-expected HO_2 and $P(O_3)^{HO_2}$ when NO exceeds 1 ppbv? We look at the five hypotheses presented in the Introduction.

A. HO_2 measurement error. Measured HO_2 could exceed modeled HO_2 if there were an error in the absolute HO_x calibration. However, the HO_2 calibration is tied to the OH calibration and measured OH is approximately equal to the modeled OH. In addition, measured and modeled HO_2 agree within uncertainties in the afternoon, especially if HO_2 is close to the lower limit of possible values. Second, HO_2 is detected by adding internally a few hundred ppmv of NO to convert HO_2 to OH, so the 10 ppbv of atmospheric NO is overwhelmed by the injected NO. Third, a small HO_2 signal offset is not responsible because the general differences between measured and modeled HO_2 in Fig. 4 persists even when an offset of 2 pptv is subtracted from the measured HO_2. Such an offset is not observed and is about a decade higher than the limit-of-detection. Fourth, the RO_2 interference due to alkene and aromatic peroxyl radicals is also unlikely for two reasons. The RO_2 is more rapidly converted to HO_2 as NO increases, but the HO_2 is recycled with OH, so the RO_2-to-HO_2 ratio decreases as NO increases (Fig. 4(b)). Next, a previous study showed these unexpectedly high HO_2 values at higher NO values in the upper troposphere where RO_2 abundances are small.[14] Further evidence is provided by other studies using other instruments and even another HO_2 measurement technique.[9,15,16,18] So far, no HO_2 measurement error has been found that would cause the greater-than-expected decrease in $P(O_3)^{HO_2}$ at high NO.

B. **Measurement averaging.** Does the average of measured HO_2 times the average of measured NO equal the average of HO_2 times NO? This question can be answered by using high-resolution HO_2 and NO measurements and averaging them together. The full width/half maximum (FWHM) of the NO plumes varied from a few seconds to well over an hour. However, for NO plumes with NO greater than 5 ppbv, 80% of time was in plumes that were sampled for more than 20 seconds. As a result, we will use 20 seconds as the minimum time resolution for comparing the products of the averages to the averages of the products.

The averages of HO_2 and NO were calculated for 20 seconds, 1 minute, 10 minutes, and 1 hour. These four different averages were plotted as a function of NO for HO_2 in Fig. 4 and for calculated $P(O_3)$ in Fig. 8. There are no significant differences among the four different averages. Thus, the greater-than-expected HO_2 and $P(O_3)$ at higher NO are not due to averaging over NO plume spikes.

In previous urban studies, point measurements have been compared to long-path absorption measurements to test the assumption that point measurements can adequately represent the integrated chemistry in urban plumes for measurement integration times of minutes or more. In a 1999 study in Nashville TN, long-path and point measurements generally agreed to within 15% or better.[44] In Mexico City in April 2003, point and long-path measurements of several VOCs were in good agreement for time scales of 5 minutes or more, although not for all VOCs.[45,46] While these averages are in space while our averages are in time, the conclusion is that, if there are no unique emission sources nearby, point measurements can represent the photochemistry of an urban region.

C. **Conditions when NO_x is not in photostationary state.** The theoretical steady-state curve for the HO_2 and $P(O_3)^{HO_2}$ *versus* NO assumes that HO_x production is constant and NO_x is in photostationary state (PSS) as NO changes. Even though NO emissions are typically out of PSS for at most tens of seconds during the day, they are continually coming from thousands of sources and these plumes are continually contributing to ozone production. The analysis presented here makes no assumptions about NO_x PSS. As a result, the measured and modeled curves in Fig. 4 and 8 are averages over different HO_x production regimes

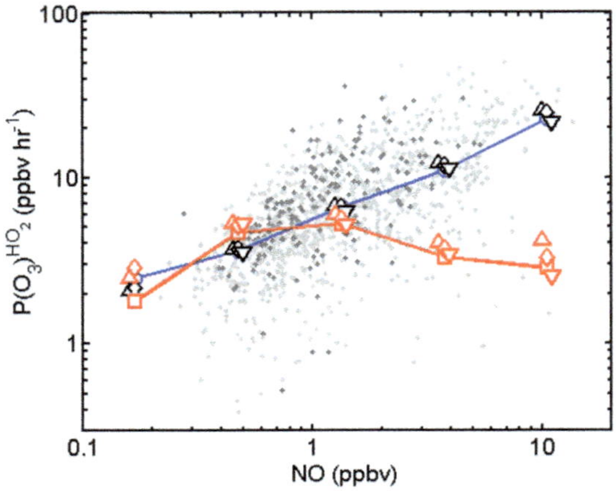

Fig. 8 Median behavior of the ozone production rate due to HO_2, $P(O_3)^{HO_2}$, as a function of NO. 10 minute averages of measured (blue circles) and modeled (red squares) $P(O_3)^{HO_2}$ are plotted along with 20 second (upward pointing triangles), 1 minute (diamonds), and 1 hour (downward pointing triangles) averages. Gray dots are individual 10 min measured data; darker gray dots are data measured between 12:00 and 15:00 when the wind was from the north, although the points with greatest NO generally occurred when the winds were light and out of the east. Data are filtered for daytime hours between 7:00 and 17:00.

and may include measurements during conditions when NO_x is not in PSS. However, while NO_x may be out of PSS if we constrain the model-to-measured NO, NO_2, and O_3, the model does calculate steady-state values for HO_2 and OH. So a valid comparison between measured and modeled HO_x requires that HO_x be in steady state but not NO_x.

We tested how close NO_x was to PSS using 1 minute averaged measurements. Comparisons were made between the model with NO, NO_2, and O_3 constrained (*i.e.*, measurements and no assumption of NO_x PSS) to calculations of NO_x PSS. These comparisons show that NO_x was generally within 10–20% of PSS from 8:00 to 14:00, suggesting that NO had decreased about $e^{-2} = 0.14$ from its initial emission. The typical NO lifetime is 20–60 seconds. When these lifetimes are multiplied by the wind speed and by 2, the resulting product is the approximate distance the air must travel from a NO source so that the NO has decreased to 0.14 of its initial value. This distance was in the range of 100–150 m, a distance great enough that several sources could contribute to keeping the NO_x slightly out of photostationary state. Because NO_x was close to steady state, HO_2 was likely to be in steady state in the morning when its lifetime was 13 s but not in the afternoon when its lifetime was 110 seconds. However, since the greater-than-expected HO_2 occurs primarily in the morning when HO_2 is in steady state, NO_x being out of PSS is not the cause of the greater-than-expected HO_2 and calculated $P(O_3)^{HO_2}$.

D. Unknown HO_2 sources accompanying NO emissions. Any unknown HO_x sources accompanying NO should show up in the OH reactivity measurement and in the comparison of the balance between OH production and loss (Fig. 7(a)). If the accompanying HO_x source is an HO_2 source, such as an aldehyde, then it would also likely be an OH loss and appear in the OH reactivity. However, the agreement between measured and calculated OH reactivity is better when NO is high than when NO is low; calculated OH reactivity accounts for ~50% of measured OH reactivity when NO was less than 1 ppbv but as much as 75% when NO is greater than 10 ppbv, just the opposite of expectations for an unknown HO_2 source that is also an OH loss. Furthermore, an unknown HO_2 source that acts also as an OH loss would cause the measured OH production to exceed the OH loss calculated from the known HO_2 sources, but just the opposite is seen. These arguments are based on OH reactivity measurements that have an uncertain NO correction, so an unknown HO_2 source cannot be entirely ruled out by these observations.

E. Missing chemical mechanism

For missing chemistry to cause the greater-than-expected HO_2 dependence on NO, it must have a few characteristics. First, in order for $P(O_3)$ to increase with NO as shown in Fig. 8, the HO_2 production must be approximately proportional to NO because the difference between measured and modeled $P(O_3)^{HO_2}$ increases about a decade for each decade of increase in NO (Fig. 8). Second, the HO_2 production rate from this chemistry would need to be approximately 5×10^7 cm^{-3} s^{-1} in order to balance the HO_2 loss due to the reaction of HO_2 with NO. This value comes from the difference in the OH production with measured and modeled HO_2 (Fig. 7(a)). We know of no known chemical mechanism that can satisfy these two constraints. Yet with the uncertainty associated with the products of reactions such as OH + NO_2, unknown HO_x-NO_x chemistry is a reasonable possibility for the greater-than-expected HO_2 and calculated $P(O_3)^{HO_2}$.

When we first observed this HO_2 discrepancy a decade ago, we hypothesized that it might be due to the reaction sequence

$$OH + NO \rightarrow HONO^* \quad (1.6)$$

$$HONO^* + N_2 \rightarrow HONO \quad (1.7)$$

$$HONO^* + O_2 \rightarrow HO_2 + NO_2 \quad (1.8)$$

The reaction to products HO_2 + NO is exothermic, $\Delta H = -94$ kJ mol^{-1}, and, to our knowledge, has never been tested. All studies of OH + NO have been done at either low pressure or in the absence of molecular oxygen, according to Sander *et al.* and references therein.[47] It is a difficult reaction to study because the HO_2 produced recycles immediately to OH by $HO_2 + NO \rightarrow OH + NO_2$. It is difficult to scavenge HO_2 in laboratory studies, so this recycling would appear as a slower rate coefficient for OH + NO + M, although laboratory studies of HONO formation have generally taken known recycling into account. This reaction can explain the observed HO_2 discrepancy only if the formation of HONO* is sped up in the presence of oxygen.

There is some evidence that this reaction could be occurring. Molecular energy calculations suggest that vibrationally excited HONO ($\nu_{OH} \geq 3$) can react with O_2 to form $HO_2 + NO_2$.[48] While this mechanism is a minor HO_2 formation pathway if HONO* is produced by the photo-excitation of HONO, it could be substantially greater if HONO* is produced by the OH + NO reaction. Other calculations suggest that the H atom can quickly hop from O atom to O atom in sub-nanosecond time scales, about five to ten times faster than the time between collisions with molecular oxygen and that the H atom could transfer to a colliding O_2 molecule.[49]

In another study of the OH + acetylene reaction, O_2 reacted with about 25% of the excited-state adducts before they could be collisionally relaxed.[50] These studies suggest that it may indeed be possible for this reaction sequence to occur.

The results of our laboratory studies on this possible reaction have been decidedly mixed. However, the reaction would need to proceed with an effective bimolecular reaction rate of $(3–15) \times 10^{-11}$ cm^3 $molecule^{-1}$ s^{-1} in order to explain the HO_2/OH ratio observed in this and previous ground-based studies made with GTHOS.

Conclusions

CalNex-SJV provides a good opportunity to compare measured and modeled oxidation chemistry in an environment with plumes of NO_x and their accompanying VOCs and OVOCs. The diel variation of the median measured and modeled OH and HO_2 were generally well represented by the box model constrained by other simultaneous measurements except for HO_2 during morning rush hour. As in many previous studies, the measured HO_2 and the ozone production rate calculated from measured HO_2 decreased much more slowly than the modeled HO_2 and calculated $P(O_3)^{HO_2}$ as a function of NO. The amount of missing OH reactivity was roughly 45% of the total measured OH reactivity. Also as seen in most previous studies, the OH production and loss rates balance in the afternoon and at night, but in the morning, the production greatly exceeds the loss. These discrepancies could arise from issues with transport or issues with chemistry.

The presence of frequent NO_x plumes that lasted from seconds to an hour suggests that transport timescales were comparable to chemical timescales for radicals like NO and HO_2. But even with strong winds (4 m s^{-1}) from a major highway and the urban core, NO_x was close to a photostationary state as it was sampled at the field site. So, despite the heterogeneity of the sampled air masses, the comparison between the measured and modeled OH and HO_2 was not significantly affected by these plumes in the morning, but may have been affected in the afternoon.

When the OH reactivity is between 5 and 25 s^{-1}, as it was in this study, OH comes into balance with a change in its sources and sinks in much less than a second. However, OH will change only as fast as its sources and sinks change. In this study, the main OH source was HO_2 + NO, so OH had a half-life of 13 seconds because $HO_2 \times$ NO had a half-life of 16 s. The results from this study show that point measurements can provide a valid test of urban oxidation chemistry if care is taken to ensure that HO_2 is in steady state. NO_x does not have to be in steady-state if NO, NO_2, and O_3 are constrained in the steady state model along with all other simultaneous measurements.

Greater-than-expected HO_2 at high NO results in higher ozone production calculated from measured HO_2, but it does not strongly influence the assessment of midday ozone production sensitivity to NO_x or VOCs. In the morning ozone production is VOC-limited whether modeled or measured HO_2 is used to calculate ozone production. At midday the difference between measured and modeled HO_2 is less, causing at most a small change in the assessment of NO_x-sensitivity or VOC-sensitivity for ozone production.

Unknown HO_x–NO_x chemistry emerged as one of the most likely causes for the greater-than-expected measured HO_2 seen in this and several other previous studies. The proposed reaction of $OH + NO + O_2 \rightarrow HO_2 + NO_2$ could resolve the HO_2 discrepancy; laboratory and modeling kinetic studies are needed to test this hypothesis. This conclusion re-emphasizes the need to re-investigate HO_x–NO_x photochemistry under atmospheric conditions.

Acknowledgements

We thank John Karlik, Rick Ramirez and the other scientists and staff at the University of California Kern County Extension Center for their hospitality and use of their facilities. Participation by WHB, BCB, JT was supported by grants from California Air Resources Board through the University of California, Berkeley (UCB subcontract 00007016) and from NSF (ATM-0706821).

Notes and references

1 M. L. Bell, A. McDermott, S. L. Zeger, J. M. Samet and F. Dominici, *JAMA, J. Am. Med. Assoc.*, 2004, **19**, 2372.
2 US EPA *2013*, *Final Report*, http://cfpub.epa.gov/ncea/isa/recordisplay.cfm?deid=247492#Download, accessed October 2015.
3 US EPA Air Trends, 2015, http://www3.epa.gov/airtrends/ozone.html, accessed November 2015.
4 A. J. Haagen-Smit and M. M. Fox, *J. Air Waste Manage. Assoc.*, 1954, **4**(3), 105–136.
5 National Research Council, *Rethinking the Ozone Problem in Urban and Regional Air Pollution*, National Academy Press, Washington, D.C., 1991.
6 J. Calvert, J. Orlando, W. Stockwell and T. Wallington, *The Mechanisms of Reactions Influencing Atmospheric Ozone*, Oxford University Press, New York, NY, 2015, ISBN: 9780190233020.
7 L. Kleinman, P. Daum, Y. Lee, L. Nunnermacker, S. Springston, J. Weinstein-Lloyd and J. Rudolph, *J. Geophys. Res.*, 2001, **107**, 4733.
8 J. Thornton, P. Woolridge, R. Cohen, M. Martinez, H. Harder, W. Brune, E. Williams, J. Roberts, F. Fehsenfeld, S. Hall, R. Shetter, B. Wert and A. Fried, *J. Geophys. Res.*, 2002, **107**, 4146.
9 M. Martinez, H. Harder, T. Kovacs, J. Simpas, J. Bassis, R. Lesher, W. Brune, G. Frost, E. Williams, C. Stroud, B. Jobson, J. Roberts, S. Hall, R. Shetter, B. Wert, A. Fried, B. Alicke, J. Stutz, V. Young, A. White and R. Zamora, *J. Geophys. Res.*, 2003, **108**, 4617.
10 I. Pollack, T. Ryerson, M. Trainer, D. Parrish, A. Andrews, E. Atlas, D. Blake, S. Brown, R. Commane, B. Daube and Coauthors, *J. Geophys. Res.*, 2012, **117**, D00V05.
11 D. Tong and D. Mauzerall, *Atmos. Environ.*, 2006, **40**, 3041.
12 K. Appel, A. B. Gilliland, G. Sarwar and R. C. Gilliam, *Atmos. Environ.*, 2007, **41**, 9603.
13 U. Im, R. Bianconi, E. Solazzo, I. Kioutsioukis, A. Badia, A. Balzarini, R. Baro, R. Bellasio, D. Brunner, C. Chemel and Coauthors, *Atmos. Environ.*, 2015, **115**, 404.
14 I. Faloona, D. Tan, W. Brune, L. Jaeglé, D. Jacob, Y. Kondo, M. Koike, R. Chatfield, R. Pueschel, G. Ferry, G. Sachse, S. Vay, B. Anderson, J. Hannon and H. Fuelberg, *J. Geophys. Res.*, 2000, **105**, 3771.

15 X. Ren, H. Harder, M. Martinez, R. Lesher, A. Oliger, J. Simpas, W. Brune, J. Schwab, K. Demerjian, Y. He, X. Zhou and H. Gao, *Atmos. Environ.*, 2003, **37**, 1352.

16 K. M. Emmerson, N. Carslaw, L. J. Carpenter, D. E. Heard, J. D. Lee and M. J. Pilling, *J. Atoms. Chem.*, 2005, **52**, 143–164.

17 X. Ren, D. van Duin, M. Cazorla, S. Chen, J. Mao, L. Zhang, W. Brune, J. Flynn, N. Grossberg, B. Lefer, B. Rappenglück, K. Wong, C. Tsai, J. Stutz, J. Dibb, B. Thomas Jobson, W. Luke and P. Kelley, *J. Geophys. Res.*, 2013, **118**, 5770.

18 Y. Kanaya, R. Cao, H. Akimoto, M. Fukuda, Y. Komazaki, Y. Yokouchi, M. Koike, H. Tanimoto, N. Takegawa and Y. Kondo, *J. Geophys. Res.*, 2007, **112**, D21312, DOI: 10.1029/2007jd008670.

19 K. M. Spencer, D. C. McCabe, J. D. Crounse, J. R. Olson, J. H. Crawford, A. J. Weinheimer, D. J. Knapp, D. D. Montzka, C. A. Cantrell, R. S. Hornbrook, R. L. Mauldin III and P. O. Wennberg, *Atmos. Chem. Phys.*, 2009, **9**, 3697–3707.

20 M. Cazorla and W. H. Brune, *Atmos. Meas. Tech.*, 2010, **3**, 545–555.

21 M. Cazorla, W. Brune, X. Ren and B. Lefer, *Atmos. Chem. Phys.*, 2012, **12**, 1203.

22 D. R. Gentner, G. Isaacman, D. R. Worton, A. W. H. Chan, T. R. Dallmann, L. Davis, S. Liu, D. A. Day, L. M. Russell, K. R. Wilson, R. Weber, A. Guha, R. A. Harley and A. H. Goldstein, *Proc. Natl. Acad. Sci. U. S. A.*, 2012, **109**, 18318–18323.

23 US EPA, 2015b, AirData, http://www3.epa.gov/airquality/airdata/, accessed November 2015.

24 S. E. Pusede and R. C. Cohen, *Atmos. Chem. Phys.*, 2012, **12**, 8323–8339.

25 S. E. Pusede, D. R. Gentner, P. J. Woolridge, E. C. Browne, A. W. Rollins, K.-E. Min, A. R. Russell, J. Thomas, L. Zhang, W. H. Brune and Coauthors, *Atmos. Chem. Phys.*, 2014, **14**, 3373–3395.

26 I. Faloona, D. Tan, R. Lesher, N. Hazen, C. Frame, J. Simpas, H. Harder, M. Martinez, P. Di Carlo, X. Ren and W. Brune, *J. Atmos. Chem.*, 2004, **47**, 139.

27 H. Fuchs, B. Bohn, A. Hofzumahaus, F. Holland, K. Lu, S. Nehr, F. Roher and A. Wahner, *Atmos. Meas. Tech.*, 2011, **4**, 1209.

28 S. M. Griffith, R. F. Hansen, S. Dusanter, P. S. Stevens, M. Alaghmand, S. B. Bertman, M. A. Carroll, M. Erickson, M. Galloway, N. Grossberg and Coauthors, *Atmos. Chem. Phys.*, 2013, **13**, 5403–5423, DOI: 10.5194/acp-13-5403-2013.

29 T. Kovacs and W. Brune, *J. Atmos. Chem.*, 2001, **39**, 105.

30 J. Mao, X. Ren, W. H. Brune, J. R. Olson, J. H. Crawford, A. Fried, L. G. Huey, R. C. Cohen, B. Heikes, H. B. Singh, D. R. Blake, G. W. Sachse, G. S. Diskin, S. R. Hall and R. E. Shetter, *Atmos. Chem. Phys.*, 2009, **9**, 163–173.

31 T. Shirley, W. H. Brune, X. Ren, J. Mao, R. Lesher, B. Cardenas, R. Volkamer, L. T. Molina, M. J. Molina, B. Lamb, E. Velasco, T. Jobson and M. Alexander, *Atmos. Chem. Phys.*, 2006, **6**, 2753–2765.

32 T. Ryerson, A. Andrews, W. Angevine, T. Bates, C. Brock, B. Cairns, R. Cohen, O. Cooper, J. de Gouw and F. Fehsenfeld, *J. Geophys. Res.*, 2013, **118**, 5830.

33 M. Jenkin, S. Saunders and M. Pilling, *Atmos. Environ.*, 1997, **31**, 81.

34 G. M. Wolfe and J. A. Thornton, *Atmos. Chem. Phys.*, 2011, **11**, 77–101.

35 J. Mao, X. Ren, L. Zhang, D. M. van Duin, R. C. Cohen, J.-H. Park, A. H. Goldstein, F. Paulot, M. R. Beaver, J. D. Crounse and Coauthors, *Atmos. Chem. Phys.*, 2012, **12**, 8009–8020.

36 S. Madronich and S. Flocke, *Environmental Photochemistry*, Springer Berlin Heidelberg, 1999.
37 R. E. Shetter, L. Cinquini, S. R. Hall and S. Madronich, *J. Geophys. Res.*, 2002, **107**, PEM 6-1–PEM 6-12, DOI: 10.1029/2001jd001320.
38 S. Chen and W. H. Brune, *Atmos. Environ.*, 2012, **55**, 288–296, DOI: 10.1016/j.atmosenv.2012.03.061.
39 K. Hens, A. Novelli, M. Martinez, J. Auld, J. R. Axinte, R. Bohn, B. H. Fischer, P. Keronen, D. Kubistin, A. C. Nölscher and Coauthors, *Atmos. Chem. Phys.*, 2014, **14**, 8723–8747, DOI: 10.5194/acp-14-8723-2014.
40 D. Stone, L. K. Whalley and D. E. Heard, *Chem. Soc. Rev.*, 2012, **41**, 6348–6404.
41 D. R. Gentner, T. B. Ford, A. Guha, J. Brioude, W. M. Angevine, J. A. de Gouw, C. Warneke, J. B. Gilman, T. B. Ryerson, J. Peischl, S. Meinardi, D. R. Blake, E. Atlas, W. A. Lonneman, T. E. Kleindienst, M. R. Beaver, J. M. St. Clair, P. O. Wennberg, T. C. VandenBoer, M. Z. Markovic, J. G. Murphy, R. A. Harley and A. H. Goldstein, *Atmos. Chem. Phys.*, 2014, **14**, 4955–4978, DOI: 10.5194/acp-14-4955-2014.
42 N. Bell, D. E. Heard, M. J. Pilling and A. S. Tomlin, *Atmos. Environ.*, 2003, **37**, 2193–2205.
43 J. Mao, X. Ren, S. Chen, W. H. Brune, Z. Chen, M. Martinez, H. Harder, B. Lefer, B. Rappengluck, J. Flynn and M. Leuchner, *Atmos. Environ.*, 2010, **44**, 4107.
44 J. A. Thornton, P. J. Wooldridge, R. C. Cohen, E. J. Williams, D. Hereid, F. C. Fehsenfeld, J. Stutz and B. Alicke, *J. Geophys. Res.*, 2003, **108**, 4496, DOI: 10.1029/2003jd003559.
45 E. Velasco, B. Lamb, H. Westberg, E. Allwine, G. Sosa, J. L. Arriaga-Colina, B. T. Jobson, M. L. Alexander, P. Prazeller, W. B. Knighton, T. M. Rogers, M. Grutter, S. C. Herndon, C. E. Kolb, M. Zavala, B. de Foy, R. Volkamer, L. T. Molina and M. J. Molina, *Atmos. Chem. Phys.*, 2007, **7**, 329–353.
46 B. T. Jobson, R. A. Volkamer, E. Velasco, G. Allwine, H. Westberg, B. K. Lamb, M. L. Alexander, C. M. Berkowitz and L. T. Molina, *Atmos. Chem. Phys.*, 2010, **10**, 1989–2005.
47 S. P. Sander, J. Abbatt, J. R. Barker, J. B. Burkholder, R. R. Friedl, D. M. Golden, R. E. Huie, C. E. Kolb, M. J. Kurylo, G. K. Moortgat, V. L. Orkin and P. H. Wine, JPL Publication, Pasadena, CA, 2012, pp. 10–6, http://www.jpldataeval.jpl.nasa.gov.
48 M. Staikova, D. J. Donaldson and J. S. Francisco, *J. Phys. Chem. A*, 2002, **106**, 3023–3028.
49 Y. Miller, G. M. Chaban, B. J. Finlayson-Pitts and R. B. Gerber, *J. Phys. Chem. A*, 2006, **110**, 5342–5354.
50 D. R. Glowacki, J. Lockhart, M. A. Blitz, S. J. Klippenstein, M. J. Pilling, S. H. Robertson and P. W. Seakins, *Science*, 2012, **337**, 1066, DOI: 10.1126/science.1224106.

Faraday Discussions

PAPER

On the interpretation of *in situ* HONO observations *via* photochemical steady state

Leigh R. Crilley,[a] Louisa Kramer,[a] Francis D. Pope,[a] Lisa K. Whalley,[bd] Danny R. Cryer,[b] Dwayne E. Heard,[bd] James D. Lee,[cd] Christopher Reed[c] and William J. Bloss*[a]

Received 15th December 2015, Accepted 22nd January 2016
DOI: 10.1039/c5fd00224a

A substantial body of recent literature has shown that boundary layer HONO levels are higher than can be explained by simple, established gas-phase chemistry, to an extent that implies that additional HONO sources represent a major, or the dominant, precursor to OH radicals in such environments. This conclusion may be reached by analysis of point observations of (for example) OH, NO and HONO, alongside photochemical parameters; however both NO and HONO have non-negligible atmospheric lifetimes, so these approaches may be problematic if substantial spatial heterogeneity exists. We report a new dataset of HONO, NO_x and HO_x observations recorded at an urban background location, which support the existence of additional HONO sources as determined elsewhere. We qualitatively evaluate the possible impacts of local heterogeneity using a series of idealised numerical model simulations, building upon the work of Lee *et al.* (*J. Geophys. Res.*, 2013, DOI: 10.1002/2013JD020341). The simulations illustrate the time required for photostationary state approaches to yield accurate results following substantial perturbations in the HO_x/NO_x/NO_y chemistry, and the scope for bias to an inferred HONO source from NO_x and VOC emissions in either a positive or negative sense, depending upon the air mass age following emission. To assess the extent to which these impacts may be present in actual measurements, we present exploratory spatially resolved measurements of HONO and NO_x abundance obtained using a mobile instrumented laboratory. Measurements of the spatial variability of HONO in urban, suburban and rural environments show pronounced changes in abundance are found in proximity to major roads within urban areas, indicating that photo-stationary steady state (PSS) analyses in such areas are likely to be problematic. The measurements also show areas of very homogeneous HONO and NO_x abundance

[a]School of Geography, Earth & Environmental Sciences, University of Birmingham, UK. E-mail: w.j.bloss@bham.ac.uk
[b]School of Chemistry, University of Leeds, UK
[c]Wolfson Atmospheric Chemistry Laboratory, Department of Chemistry, University of York, UK
[d]National Centre for Atmospheric Science, UK

in rural, and some suburban, regions, where the PSS approach is likely to be valid. Implications for future exploration of HONO production mechanisms are discussed.

1 Introduction

Chemical processes affect atmospheric composition, leading to the removal of primary species, including many air pollutants, and the formation of secondary products. The hydroxyl radical (OH) is the principal gas-phase tropospheric oxidant, responsible for initiating the degradation of most volatile organic compounds (VOCs), including the greenhouse gas methane (CH_4), and many nitrogen and sulphur species.[1] Products of OH-driven oxidation include pollutants such as ozone (O_3), and secondary inorganic and organic aerosol, which affect human health and radiation transfer, hence climate. Consequently, quantitative knowledge of OH abundance – and hence OH sources – underpins understanding of atmospheric chemical processing.[2] This paper concerns understanding the gas-phase sources and cycling of OH in the troposphere, and specifically the role of nitrous acid (HONO) in the atmospheric boundary layer.

Basic OH-driven VOC oxidation involves interconversion of OH radicals and organic- and hydro-peroxy radicals, RO_2 and HO_2, mediated by the abundance of nitrogen oxides (NO_x). Within this radical cycling, we can distinguish between primary sources of OH, and of HO_x (here, $HO_x = OH + HO_2 + RO_2$), and secondary sources of OH (which simply represent interconversion between OH, HO_2 and RO_2). In the free troposphere, the dominant primary OH (and HO_x) sources are the short-wavelength photolysis of ozone, followed by $O(^1D)$ reaction with water vapour; other primary sources of HO_x include photolysis of carbonyl species and alkene–ozone reactions, while the predominant secondary source of OH in most tropospheric environments is reaction of HO_2 radicals with nitric oxide (NO).

Nitrous acid (HONO) can be considered to be both a primary source of OH, and a HO_x (and NO_x) reservoir: HONO is formed in the gas phase, through the termolecular reaction between OH radicals and NO (1), and the principal fate of HONO in the sunlit atmosphere is photolysis (2) to reform OH and NO; HONO photolysis frequencies reach *ca.* $1.7 \times 10^{-3}\ s^{-1}$ in the mid-latitude boundary layer (value for SZA = 40°). Reaction with OH (3) represents a further sink of HONO, but in most situations is a minor channel given the broadly photolysis-driven abundance of OH.[3] HONO chemistry as described through reactions (1) and (2) has no net effect upon HO_x and NO_x (once steady state is achieved); however additional HONO formation through other processes (4) will lead to additional, primary, net OH production.[4]

$$OH + NO + M \rightarrow HONO + M \quad (1)$$

$$HONO + h\nu \rightarrow OH + NO \quad (2)$$

$$HONO + OH \rightarrow H_2O + NO_2 \quad (3)$$

$$[\text{Other sources}] \rightarrow HONO \quad (4)$$

HONO was first detected in the atmosphere by Differential Optical Absorption Spectroscopy, DOAS.[5] Subsequently, HONO has been measured using a range of direct and indirect approaches, including optical methods – open-path DOAS,[6] FTIR and various cavity approaches such as broadband cavity enhanced absorption spectroscopy,[7] wet chemical approaches,[8–10] and through applications of photo-fragmentation/laser induced fluorescence[11] and chemical ionisation mass spectrometry.[12] While technique equivalence remains a live issue,[13] the observations derived from modern instrumentation have highlighted that HONO is likely to play an important role in boundary layer (and potentially free troposphere) HO_x and NO_x chemistry.

The observed levels of HONO are such that it forms an important – sometimes dominant – component of the OH budget within the boundary layer (where air pollutants immediately impact upon health), *e.g.* accounting for 48–56% of OH production in New York City,[14,15] 33% of OH production in a deciduous forest near Jülich (Germany),[16] 35% of the OH source at a suburban site near Paris,[17] 80.4% of the OH source at a semi-rural site in Colorado[18] and 40% in central London.[19] We can identify two contrasting HONO formation mechanisms: within-atmosphere gas-phase HONO formation (with no net effect upon OH abundance) – through reaction (1), and other processes and heterogeneous formation and/or direct emissions (net HONO production, leading to increased OH – reaction (4)).

The existence of additional HONO sources, beyond reaction (1), was identified following the first observations of HONO by DOAS approaches[5,6] which showed HONO accumulation during nighttime and removal (by photolysis) during the day. Subsequent co-located observations of HONO, OH and NO have consistently found that additional HONO sources (*i.e.* reaction (4) in Fig. 1) are required to balance the *in situ* chemical HONO budget. The missing source strength is far from a minor correction: it has frequently exceeded the known (OH + NO) production by an order of magnitude. For example, in a European deciduous forest a missing source *ca.* 13-fold larger than OH + NO was identified;[16] at a rural site in southern China daytime HONO production exceeded known sources by a factor of 19.[20] Recent observations from the (highly homogeneous) SMEAR II Hyytiälä boreal forest site in Finland found that an additional HONO source, *ca.* 20–30 times larger than that predicted from OH + NO, was present.[21] In urban regions, with orders of magnitude higher NO_x levels, HONO formation *via* OH + NO is much faster – however substantial additional HONO sources are still evident: 50% of HONO production was unaccounted for at a suburban Paris site;[22] and 40% was unaccounted for in Santiago, Chile.[23]

A number of candidate mechanisms have been proposed to account for the missing HONO production: formation from dark NO_2–H_2O interactions occurs upon a range of surfaces[24,25] but is thought to be smaller than the missing (daytime) source (*e.g.* accounting for only 4% of the deficit in Santiago[23]). Laboratory studies have shown that photoenhanced HONO production can occur following NO_2 uptake to surfaces including soot,[26] aromatic species,[27] humic acids[28] and TiO_2.[29] Of these, photoenhanced NO_2-to-HONO conversion upon organic films has been shown to occur at a rate that can potentially account for the missing HONO source. Surface-enhanced nitrate photolysis may contribute to HONO formation upon forest canopies[30] although the yield is poorly constrained. Both ground and aerosol surfaces may contribute to heterogeneous HONO production, although aerosol surface areas are generally smaller than the geometric ground surface area within a typical boundary layer, and observations (vertical profiles of HONO and aerosol surface area) have consistently shown that the ground source dominates,[31–34] a conclusion supported by low HONO production observed in laboratory studies of NO_2–soot interactions.[35] Other insights into HONO formation include new gas-phase chemistry, for example the recently proposed reaction between water-complexed HO_2 radicals, $HO_2 \cdot H_2O$, and NO_2[36] and direct emissions from soil bacteria[37,38] and biocrusts.[39]

The identity of the additional HONO sources is still the subject of much attention, and has been variously discussed in the reviews of (for example) Lammel & Cape,[40] Kleffman,[4] Spataro & Ianniello[41] and George *et al.*[42] The focus of this discussion paper is not to identify the source(s) responsible for the net HONO formation frequently observed in the field, but rather to explore the strengths – and limitations – of one method used to identify the magnitude and suggest the identity of additional HONO sources, including in a number of the studies cited above – the application of photostationary steady state to the HONO–OH–NO system.

2 Photostationary steady state

The photostationary steady state (PSS) is a simple application of basic chemical kinetics, in which application of the assumption of chemical equilibrium permits derivation of the relative concentrations of more reactive species in terms of the kinetic and photochemical reaction parameters governing their interchange with less reactive components.[43] The classical example of PSS in atmospheric chemistry is the simple NO_x–O_3 or Leighton ratio:[44]

$$NO + O_3 \rightarrow NO_2 + O_2 \tag{5}$$

$$NO_2 + h\nu \rightarrow NO + O \tag{6}$$

$$[NO_2]/[NO] = k_5[O_3]/j_6 \tag{E1}$$

Deviations from the Leighton ratio arise from additional chemical processes (and have been used to infer, for example, RO_2 abundance and hence ozone production rates[45] and possible new chemistry[46]). Technically however, steady-state is never truly achieved – the concentrations are always relaxing towards equilibrium, on a timescale which may be characterised by the e-folding lifetime. In the case of the Leighton ratio PSS is a reasonable assumption for the sunlit atmosphere, where j_6 is of the order of (5–10) $\times$ 10^{-3} s^{-1} (SZA 60–20°), and $k_5[O_3]$ typically (5–50) $\times$ 10^{-3} s^{-1} (10–100 ppb O_3; 298 K) – and reactions (5) and (6) are the dominant controls on NO_x and O_3 abundance in most situations. The HONO PSS however displays more complex timescales. Neglecting additional sources (4):

$$[HONO] = k_1[OH][NO]/(j_2 + k_3[OH]) \tag{E2}$$

however, the abundance of OH and NO is not dominated by reactions (1)–(3): the principal fate for OH in the boundary layer is reaction with VOCs/NO_2, depending upon local conditions, and measured total OH reactivity ranges from 3–100 s^{-1}, with the lowest possible reactivity of the order of 1 s^{-1} (for remote background air where VOC reactivity is dominated by CH_4 and CO) – corresponding to an OH lifetime of 1 second or less. The lifetime of NO is dominated by the NO_x–O_3 PSS, and is typically of the order of 50 seconds (40 ppb O_3; 298 K), while the lifetime of

HONO is much longer: the photolysis frequency of $2.04 \times 10^{-3}\ s^{-1}$ at an SZA of 40° corresponds to an e-folding lifetime of 8.2 minutes – or more at higher SZA values. The photochemical lifetimes of the species within the HONO PSS thus range from <1 second to potentially tens of minutes.

Consequently, HONO levels observed at a given point in time reflect the integrated effect of reactions (1)–(3), and additional sources, over a period of at least several minutes of the air parcel history – a fetch over which OH and NO, with much shorter chemical lifetimes, may be highly variable depending upon local heterogeneity.

A number of studies have used point observations of HONO, OH and NO to infer the magnitude of additional HONO sources S:

$$S = \{[\text{HONO}]_{\text{obs}} \times (j_2 + k_3[\text{OH}])\} - k_1[\text{OH}][\text{NO}] \qquad \text{(E3)}$$

This approach is appropriate where the local environment is homogeneous and well mixed – for example remote ground based locations[16,21] or free tropospheric measurements[36] – but may be problematic for more heterogeneous boundary layer sites, where inappropriate application of eqn (E3) could lead to erroneous measures of additional HONO sources. A further frequently applied metric is the HONO/NO_2 (and HONO/NO_x) ratio – considered a better conserved quantity than HONO in isolation given the interchange between them – but again potentially subject to bias in heterogeneous environments given the different chemical lifetimes of HONO, NO and NO_2. It should be noted that eqn (E3) as presented here is the most simple representation of the dominant chemical terms in the HONO continuity equation; various studies have extended this to incorporate (for example) nitrophenol photolysis[47] and surface deposition of NO_2.[19]

Here, we briefly evaluate the "additional" HONO source S, and HONO/NO_2 and HONO/NO_x ratios, using a new set of field data from an urban background site. We then illustrate the errors in S and the HONO : NO_x ratios which may arise from non-PSS situations using a range of simple illustrative box-model simulations, building upon the insights of Lee *et al.*;[48] and present new on-road spatially resolved measurements of HONO and NO_x to investigate the likelihood of these effects being present in real-world, non-remote, continental boundary layer locations. The consequences for interpretation of *in situ* field data to infer missing atmospheric processes are discussed.

3 Field observations of HONO, OH and NO_x

Field measurements were performed on the campus of the University of York (UK) for a four-week period in early summer (June) 2014, as part of the NERC-funded "Missing OH Reactivity" campaign. The measurement location is considered an urban background site, with light passenger car/LGV traffic around University access roads to the East and West of the site, and bordering farmland to the South-East. The University campus is located 3 km from York city centre (to the North-West), and 1.75 km from the nearest major road(s) (the A64 to the South-East). Measurements were made *ca.* 3.5 m above ground level, with instrument inlets co-located horizontally (within 2 m) and vertically (within 0.5 m).

OH radicals were measured by laser-induced fluorescence (LIF), using the Fluorescence Assay by Gas Expansion (FAGE) technique; full details of the instrument and approach are given elsewhere.[2,49,50] OH was determined using the wavelength modulation methodology. NO and NO_2 were measured using chemiluminescence, with photolytic NO_2 conversion.[51] O_3 was measured using a conventional UV absorption monitor (TEI 49i). HONO was measured using a LOPAP (Long-Path Absorption Photometer) instrument.[8] Photolysis frequencies were determined using a calibrated spectral radiometer (Ocean Optics), alongside basic meteorological parameters.

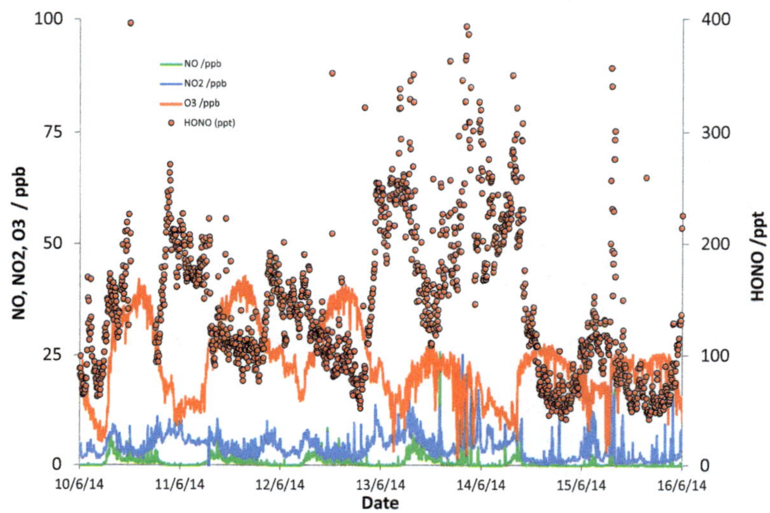

Fig. 1 (Subset of) measured NO, NO_2, O_3 and HONO data at the University of York campus.

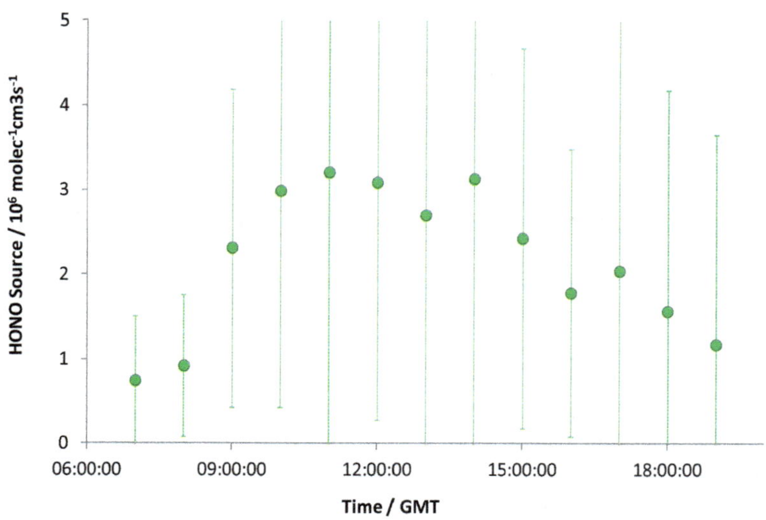

Fig. 2 Inferred additional HONO source, *S*, from York data – daytime data only. Errors are ±1 standard deviation and represent precision only, derived from all measurement values within each hourly bin.

Fig. 1 presents a representative sub-set of the measured timeseries for NO, NO_2, O_3 and HONO, to illustrate the typical conditions observed. NO_x levels ranged between 0.48 and 45 ppb [mean 6.3 ppb; median 4.8 ppb]; HONO levels ranged between 33 ppt and 1.15 ppb [mean 190 ppt; median 156 ppt], while the mean daytime maximum OH concentration was 2.2×10^6 molec cm^{-3}. Descriptive statistics here and subsequently relate to the 2526, 15 minute observations for which all of OH, NO, HONO and photolysis frequencies were satisfactorily measured (*i.e.* excluding missing data); this represents 48% of all measurement times.

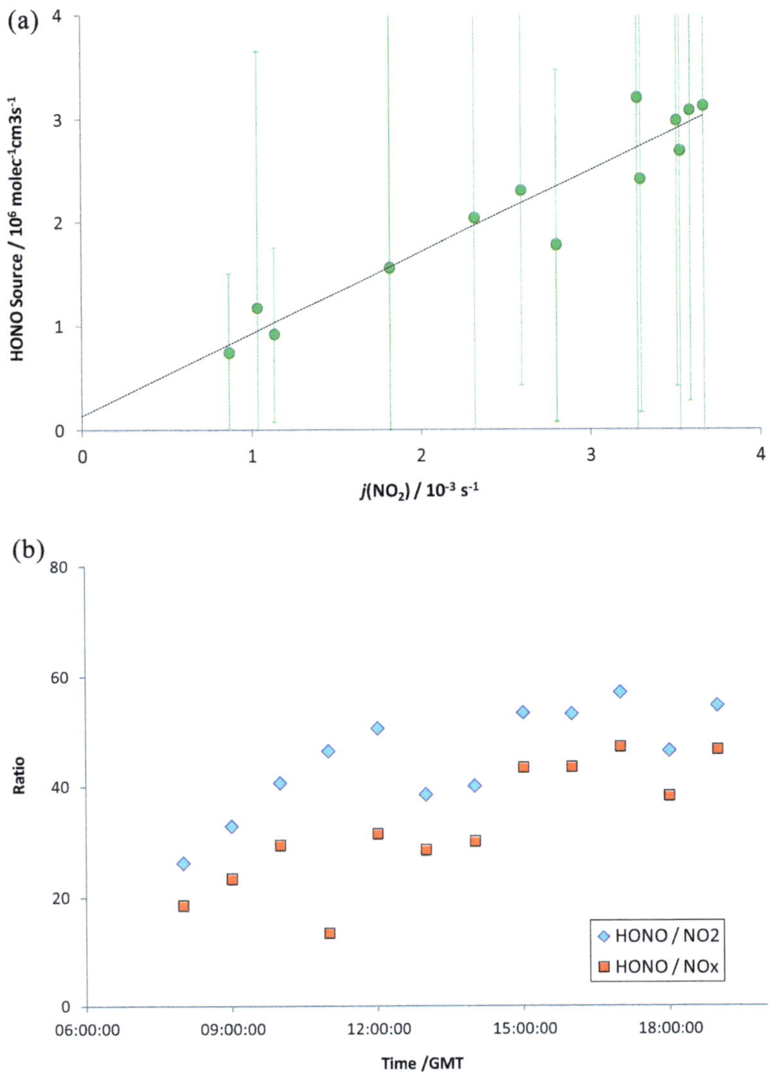

Fig. 3 (a) Variation in missing HONO source with NO_2 photolysis frequency, jNO_2. Error bars as for Fig. 2. (b) HONO/NO_2 and HONO/NO_x ratios (hourly mean values).

The observed HONO levels were higher than simple steady-state eqn (E2) would predict, by a factor of 1.8 (median). Eqn (E3) was applied to derive the inferred additional HONO source, under the approximation/assumption of the validity of PSS – as noted above, this is the most simple treatment of HONO chemistry, and we stress this neglects (for example) mixing, HONO deposition and other "known" HONO sources such as nitrophenol photolysis – but these have been shown to be minor terms in the urban background HONO budget.[19] The resulting "missing HONO source" is shown in Fig. 2, and follows a clear diurnal profile, peaking at *ca.* 3.2×10^6 molec cm^{-3} s^{-1} or approximately 1.3×10^{-4} ppb s^{-1}. This value is comparable to that reported in similar locations recently (*e.g.* Michoud *et al.*:[22] suburban Paris, $(0.7–1.9) \times 10^{-4}$ ppb s^{-1}; Lee *et al.*:[19] North Kensington, London, August: 2.5×10^{-4} ppb s^{-1}), although we note that a range of non-PSS sources (omission of which would increase the retrieved additional HONO production) were also considered in these studies.

The observed source is very strongly correlated with the NO_2 photolysis frequency (j_5) as shown in Fig. 3a (although many photolysis parameters will follow a similar pattern). The hourly mean diurnal measured HONO/NO_x and HONO/NO_2 ratios for daytime (here, defined as 07:00–20:00) are shown in Fig. 3b; the ratios averaged factors of 0.031 and 0.040 respectively. Interestingly, the mean ratios increase through the day, potentially indicating some degree of hysteresis. This is the opposite trend in behaviour to that expected for *e.g.* depletion of a surface HONO reservoir deposited during the night.[52]

Interpretation of the observationally-derived values of S and the HONO/NO_2 and HONO/NO_x ratios presented above depends upon the assumption that HONO, OH and NO, HONO and NO_2, or HONO and NO_x, are in photostationary steady state. To qualitatively assess this assumption, we have performed a series of simple model simulations to illustrate the evolution of OH, NO_x, HONO and hence S under a series of scenarios.

4 Illustrative model simulations

Model simulations were performed using a zero-dimensional box model, employing standard inorganic chemistry and oxidation of CO and CH_4 (only) as implemented in the Master Chemical Mechanism (MCM) v3.1.[53,54] The only HONO processes in this mechanism are reactions (1)–(3). The model was initiated with chemical conditions representative of a suburban background mid-latitude boundary layer airmass (CO = 200 ppb; CH_4 = 1850 ppb; O_3 = 40 ppb; NO_x = 5 ppb – no initial HONO), and allowed to spin-up for a 1 hour period prior to the perturbations outlined below. For simplicity, photolysis frequencies were held constant, and correspond to those calculated for Birmingham (UK) at midday on 1 July (SZA = 30°). Temperature, pressure and RH were set to 293 K, 1 bar and 70% respectively. No mixing or deposition processes were considered – solely photochemical reactions and emissions as outlined below. Many aspects of this simulation are therefore highly simplified in comparison to any real world environment – but serve to illustrate the impacts of the HONO–NO_x–OH system. In the discussion below, the following terminology is adopted:

$HONO_{MODEL}$	The model-calculated HONO concentration – as would be measured in the atmosphere, if only reactions (1)–(4) affected HONO abundance
$HONO_{PSS}$	The HONO concentration calculated using eqn (E2), from the modelled OH, NO and model j_{HONO} values. This is equivalent to the value which would be derived from measured OH, NO and j_{HONO} in the atmosphere, assuming PSS
Inferred source S	The "missing" HONO source calculated using eqn (E3), from the modelled OH, NO, HONO and model j_{HONO} values – as would be derived measured OH, NO, HONO and j_{HONO} in atmospheric field measurements, assuming PSS

Scenario 1

Scenario 1 represents an airmass arriving over a region of continuous HONO emissions – for example, a step change in soil emissions with different surface/vegetation type. A HONO source strength of 1.1×10^7 molec cm^{-3} s^{-1} was used, equivalent to the soil bacterial emission inferred for wheat fields of 258 ng N m^{-2} s^{-1} as reported by Oswald *et al.* (2013),[38] in a well-mixed boundary layer height of 1000 m; the HONO source was turned on at $t = 1$ hour and off again at $t = 2.5$ hours (in reality, the daytime lifetime of HONO is insufficient for even mixing throughout a deep boundary layer). Fig. 4 shows the resulting time series of $HONO_{MODEL}$, $HONO_{PSS}$ and the true (solid green line) and Inferred Source S, as a function of photolysis rates – for the base case, and increases/decreases in all photolysis frequencies of 50%.

Initially, HONO is zero within the model, and over the first 20 min of the simulation approaches steady state with OH and NO, as described by reactions

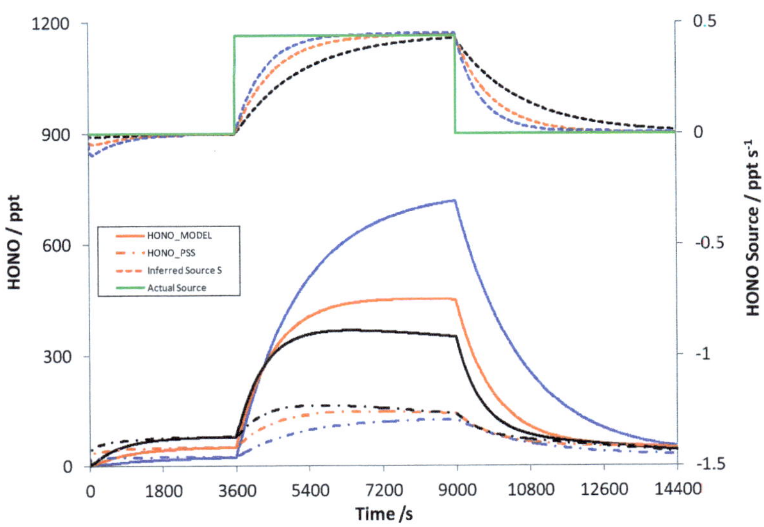

Fig. 4 Scenario 1 – simulation of the HONO abundance, and inferred HONO source strength, for an air parcel transiting a region of enhanced HONO production/emission. Red = base case photolysis frequencies; black/blue = increased/decreased (50%) photolysis frequency sensitivity studies. See text for details.

(1)–(3). Upon addition of the HONO emission (at $t = 1$ hour), $HONO_{MODEL}$ rises to reach *ca.* 450 ppt at the end of the source-on period (under base case photolysis rates – solid red line), while the steady-state derived HONO, which excludes the missing source, is substantially lower at *ca.* 140 ppt (dot-dash red line). The inferred HONO source (dashed red line) substantially underestimates the true source (solid green line) initially, approaching to within 90% of the true source after 34.5 minutes. The impact of increased/reduced photolysis frequencies is shown in red and blue respectively – variations of ±50% *versus* the base case. Decreased actinic flux increases the true HONO level, as would be expected, but also substantially increases the time for the inferred source strength to approach the true value, with 90% agreement only achieved after 61 minutes. Similarly the $HONO_{PSS}$, as derived from eqn (E2), increases with the increase in photolysis – as the combined increases in OH and NO (*i.e.* the rate of R1) exceed the increase in the HONO photolysis frequency j_2. The error in the inferred source S is substantial, persists for a significant time (of the order of tens of minutes) following a step-change in HONO emissions, and is greater under conditions of *lower* solar intensity. The model simulation here omits other potential HONO sinks, such as deposition. These will reduce the HONO lifetime, and hence shorten the time for PSS to be established; in this sense the results in Fig. 4 represent a worst-case scenario in terms of the time required to approach steady state.

Scenario 2

Scenario 2 represents an airmass receiving a transient injection of emissions (NO_x, and/or HONO, and/or VOCs) – for example a parcel of air making a perpendicular transect across a roadway. As for Scenario 1, the model was initialised and allowed to spin-up under typical continental suburban conditions; photolysis rates were held constant, and the evolution of modelled HO_x, NO_x and HONO was investigated with a one-off increase (implemented within the model as instantaneous step changes) of NO (8 ppb)/NO_2 (2 ppb)/HONO (0.1 ppb), representing an estimated present-day real-world mean UK fleet NO/NO_2 ratio with an f-NO_2 value of *ca.* 20%,[55] plus 1% HONO (at the upper end of the range recently reported for China[56] – no data are available for the UK).

Fig. 5a shows the modelled timeseries of OH, NO, NO_2, O_3 and HONO, for a 120 second period around the injection point. The initial increases in NO_x, their relaxation to PSS with O_3, and the response of OH (an initial increase from NO-driven CH_3O_2 and HO_2 titration, followed by reduction reflecting primarily the evolving additional NO_2 sink) are apparent. Fig. 5b shows the calculated $HONO_{MODEL}$ (red; with essentially a step-change driven by the emission), the calculated $HONO_{PSS}$ (black), and the resulting inferred HONO source strength (shown by the dashed blue line). At all points following the injection, the true HONO source strength is zero; however the non-equilibrium abundances lead to an initially negative inferred additional source, switching to positive after *ca.* 30 seconds, peaking at 2.5 minutes, and relaxing to a near-zero value on a timescale of tens of minutes (the e-folding lifetime of S here is 17 minutes). Fig. 5c shows the evolution of the model-derived HONO : NO_2 (blue dashes) and HONO : NO_x (solid red) ratios (multiplied by 1000, *i.e.* as ppt/ppb) – which may be compared with the emission ratios within the model simulation here of 0.05% and 1%

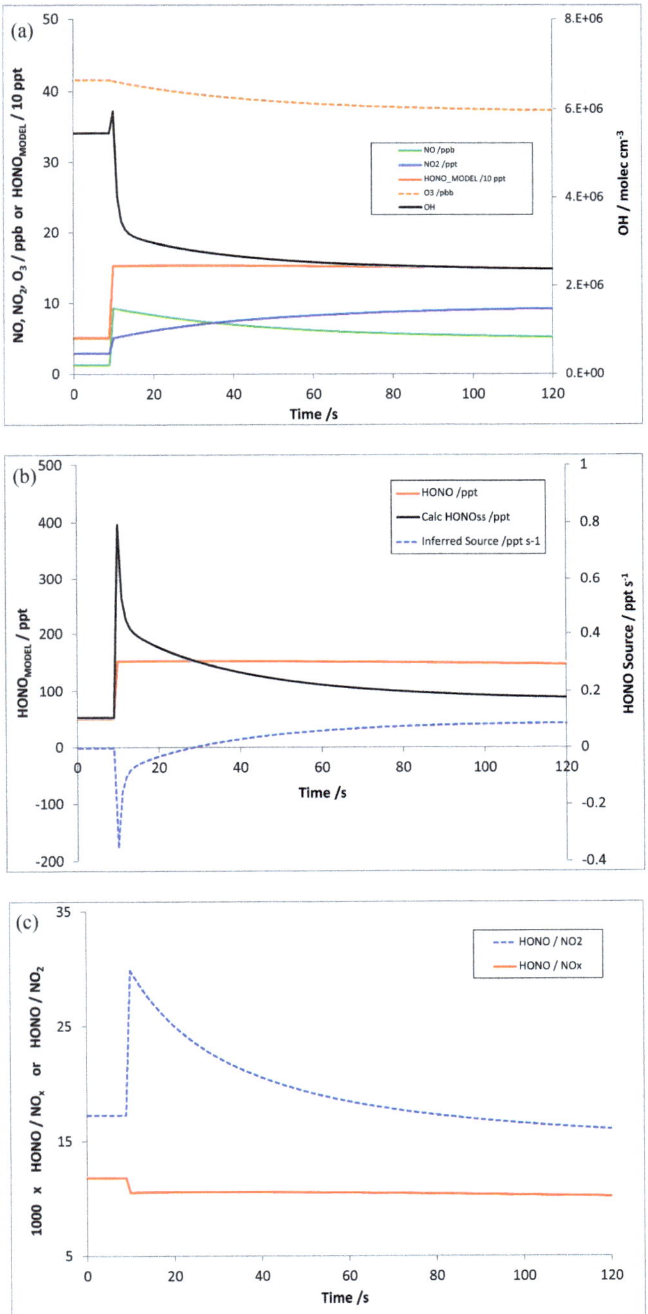

Fig. 5 Scenario 2: evolving composition chemical composition (panel a, top), inferred HONO source (panel b, mid) and HONO/NO_2 and HONO/NO_x ratios (panel c, bottom) for an air parcel transecting a point source (*e.g.* a roadway).

respectively (or 50 and 10, when scaled by 1000). The absolute values of these are of course directly dependent upon the pre-existing model NO_x as well as the emission ratio.

Scenario 3

Scenario 3 represents a repeat of Scenario 2, without the inclusion of the HONO source – *i.e.* just addition of NO and NO_2, at the levels outlined above. The resulting modelled HONO and inferred HONO source is shown in Fig. 6a – the equivalent of Fig. 5b. In this case, $HONO_{MODEL}$ is near-zero, rising slightly in response to the changing NO_x (and hence HO_x) abundance, but the calculated $HONO_{PSS}$, and hence the inferred HONO source, respond substantially to the NO_x addition – giving an inferred HONO source *S* which is negative at all times. The evolution of the corresponding model-derived HONO : NO_2 and HONO : NO_x ratios (multiplied by 1000) is shown in Fig. 6b – the true emission ratios in this case being zero, and the pre-injection values reflecting the background model steady state.

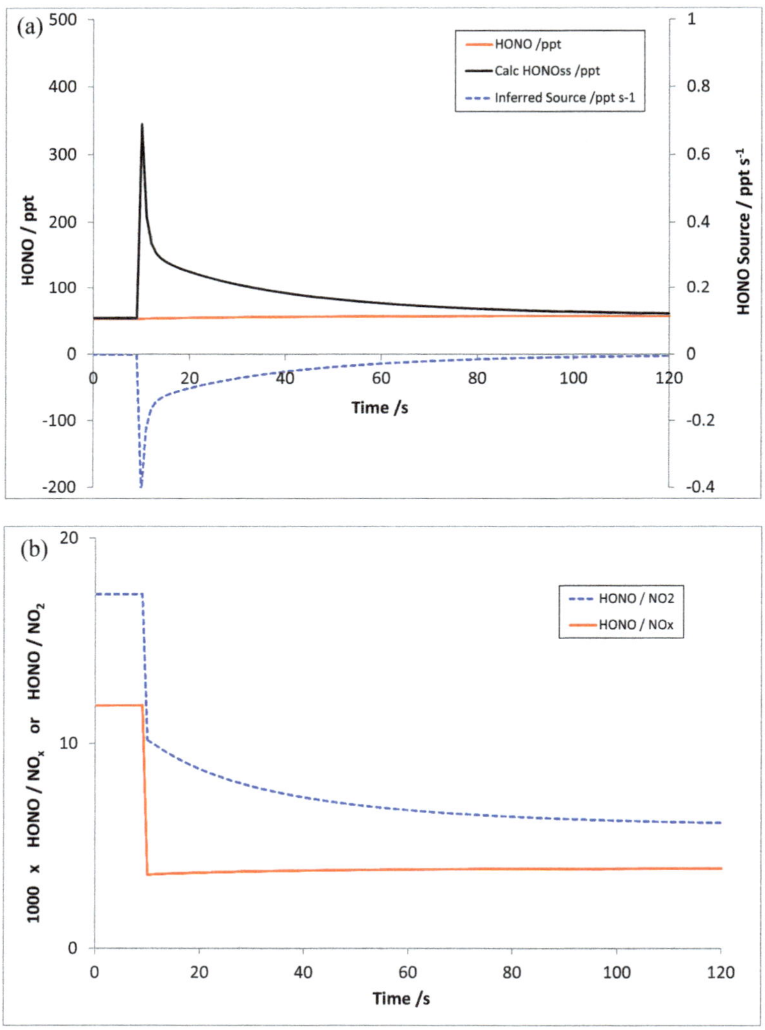

Fig. 6 Scenario 3 – addition of NO and NO_2 (only) to an air parcel. $HONO_{MODEL}$, $HONO_{PSS}$ and inferred HONO source strength (panel a, top), and calculated HONO/NO_2 and HONO/NO_x ratios (panel b, bottom).

Scenario 4

Scenario 4 explores the impact of (initially) OH-only perturbations, through addition of VOCs alone. The model was re-run, under the same initial conditions, with an "emission" or step-change in CH_4 and CO of 5550 and 600 ppb respectively – which represents a four-fold increase in OH reactivity (from parent VOCs – a slightly smaller increase overall considering degradation products and NO_y). Fig. 7a shows the resulting changes in modelled OH, and $HONO_{MODEL}$, while Fig. 7b shows the corresponding values of $HONO_{PSS}$ and the inferred HONO source strength S – the latter displaying a positive bias (from the true value of zero), although smaller than the magnitude of the initial signal in Scenario 3 (by

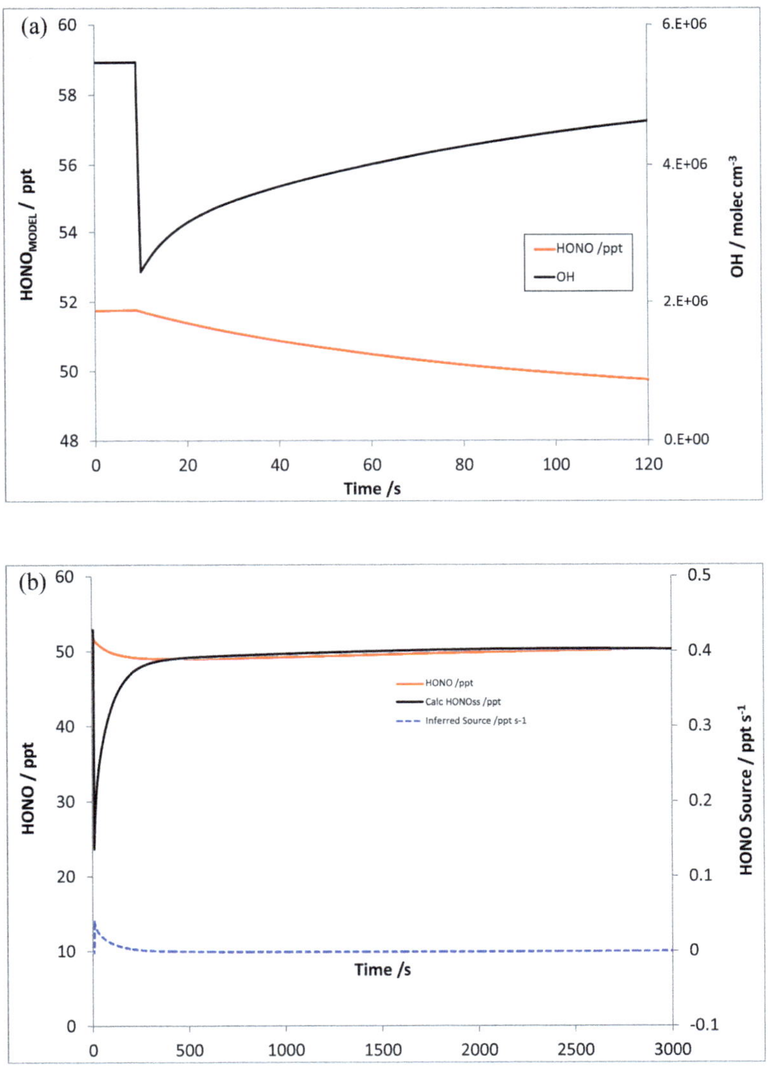

Fig. 7 Scenario 4 – increase in VOC reactivity only (no NO_x addition). Simulated $HONO_{MODEL}$ and OH timeseries (panel a, top) and $HONO_{PSS}$ and inferred HONO source strength (panel b, bottom).

a factor of 10), decaying rapidly to near-zero levels. The increase in VOC levels here is a modest change in the total OH reactivity, which at *ca.* 5.9 s^{-1} (CH_4 and CO) is substantially below that observed recently in London (*ca.* 20 s^{-1} (ref. 57 and 58)) and elsewhere – although the reduction in modelled OH following VOC addition (and hence the inferred source strength S) does not scale linearly with k'_{OH}.

The model scenarios therefore illustrate the substantial time required for the photostationary state approach to yield accurate results following substantial perturbations in the $HO_x/NO_y/NO_y$ chemistry. They also illustrate the scope for bias to the inferred HONO source from NO_x and VOC emissions, in the absence of any true HONO signal. The bias to the inferred HONO source strength may be either positive or negative, depending upon the airmass age following emission, for the more realistic mixed NO_x–HONO emission scenario.

The HONO/NO_2 ratio evolves substantially following emission in the road-transect Scenario (2), and will only reflect emissions when the background NO_x is negligible compared with the emission (a consequence of the ambient daytime NO_2/NO ratio favouring NO_2, but the vehicle-emitted NO_x being predominantly NO, to broadly equal but opposite extents – dependent of course upon O_3, j_{NO_2}, vehicle identity *etc.*). In contrast, ambient downwind HONO/NO_x ratios are a much better reflection of the emitted species (due to the NO/NO_2 relative abundance considerations), and evolve much more slowly (as they are independent of establishment of the NO_x–O_3 PSS) following emission.

The question which then arises is, how spatially variable is HONO (and the HONO/NO_x, HONO/NO_2 ratios *etc.*) in the real environment – in particular in non-remote environments where there is substantial landscape and built environment heterogeneity – are the challenges illustrated in Scenarios 1–4 evident in actual measurements? To explore this issue, we present exploratory spatially resolved measurements of HONO and NO_x abundance, performed using a mobile instrumented laboratory in Birmingham (UK) and the surrounding areas.

5 Real-world spatial variation of boundary layer HONO, NO and NO_2

The measurements reported here were performed around the University of Birmingham campus, Birmingham city centre and suburbs, and on a transect between Birmingham and Leicester, 50 km to the north east, along the M6/M69 motorway/freeway through an otherwise largely rural region. Instruments were mounted in a mobile laboratory, based upon a modified light goods van, with on-board power provided through batteries/inverters and a UPS system, data logging, GPS and camera facilities for chase vehicle emission "sniffing". HONO was measured using the LOPAP methodology as outlined above; for each measurement the system was thoroughly warmed up on mains power for several hours prior to on-road measurements. NO and NO_2 were measured using chemiluminescence, with a TE 42i-TL NO_x monitor – this unit is fitted with a molybdenum converter for NO_2 measurements; the approach suffers from known lack of specificity with respect to other nitrogen oxides in the NO_2 conversion unit, and the NO_2 data may consequently be more reflective of NO_y – although little time may have elapsed for NO_x oxidation. O_3 was measured using a conventional UV absorption monitor (TE 49i). All the monitors sampled from a roof-mounted inlet which protruded directly above and forwards of the front of the vehicle, with the exception of HONO which was sampled from a port on the vehicle side, with the sampling inlet unit located to protrude well beyond the vehicle boundary layer. The measurement time resolution (instrument response times) were 60 s for the NO_x instrument and 20 s for the O_3 instrument (with (shared) inlet residence time adding negligibly to this). HONO values were obtained every 30 seconds, but the instrument response time is *ca.* 4.5 minutes – the LOPAP approach is not optimised for high time resolution study, owing to the liquid residence/circulation time, and reagent diffusion. Data were corrected for all instrument lags.

Fig. 8a shows the observed HONO and NO_x time series for one deployment, which included the University of Birmingham campus (considered an urban background location), and a transect through Birmingham City Centre to the University of Leicester campus, and the return journey, performed on 23rd October 2015. There was a ~2 hour break in between the outward and return legs

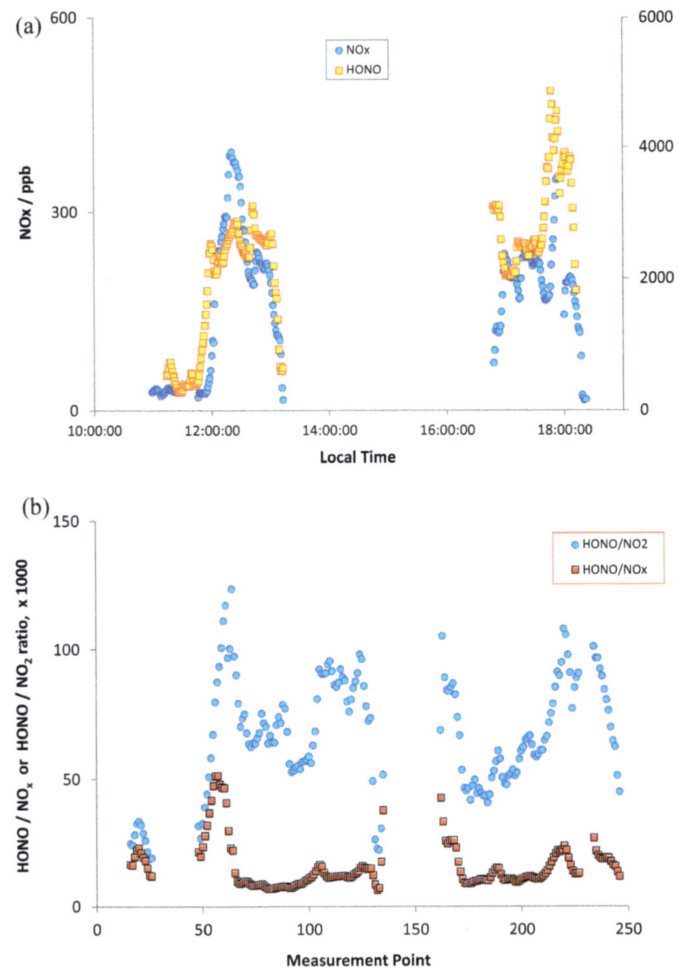

Fig. 8 Measured HONO and NO_x (panel a, upper) and $HONO/NO_2$ and $HONO/NO_x$ ratios (panel b, lower) during on-road measurements, transect from Birmingham to Leicester & return.

of the journey (battery change/recharge). A clear correlation between NO_x and HONO is apparent to the eye, although variable in scale (comparing *e.g.* 12:30 and 13:00). Fig. 8b shows the HONO/NO_2 and HONO/NO_x ratios, as a function of measurement time point (same dataset as Fig. 8a). The ratios are clearly highly variable, but with patterns reflecting similar values/trends between adjacent measurement points – showing the variation spatially during the journey. To illustrate this more clearly, the spatial trends in the HONO measurements, and HONO/NO_x ratio, are shown in Fig. 9a and b (for the whole journey) and Fig. 10a and b (for a zoom in of the region around Leicester).

The measured HONO and NO_2 data, and resulting HONO/NO_2 ratio, show substantial variation across the measurement sites. Elevated HONO is observed in the city centre, consistent with emissions from traffic, or from other urban activities (or from conversion of traffic-emitted NO_x, which displays a similar spatial variation). The "hot spots" around the city centres are more localised than might be expected – at the ground level windspeed of 14 km h^{-1}, the HONO lifetime of 17 min (value for 12:30 GMT 23 Oct Birmingham, as calculated by TUV – for clear skies, hence a lower limit to the lifetime) corresponds to a horizontal distance of around 4 km, and a greater spatial influence of the urban environments might have been expected. The relatively slow time response of the LOPAP analyser would be expected to exacerbate such a pattern in the retrieved values.

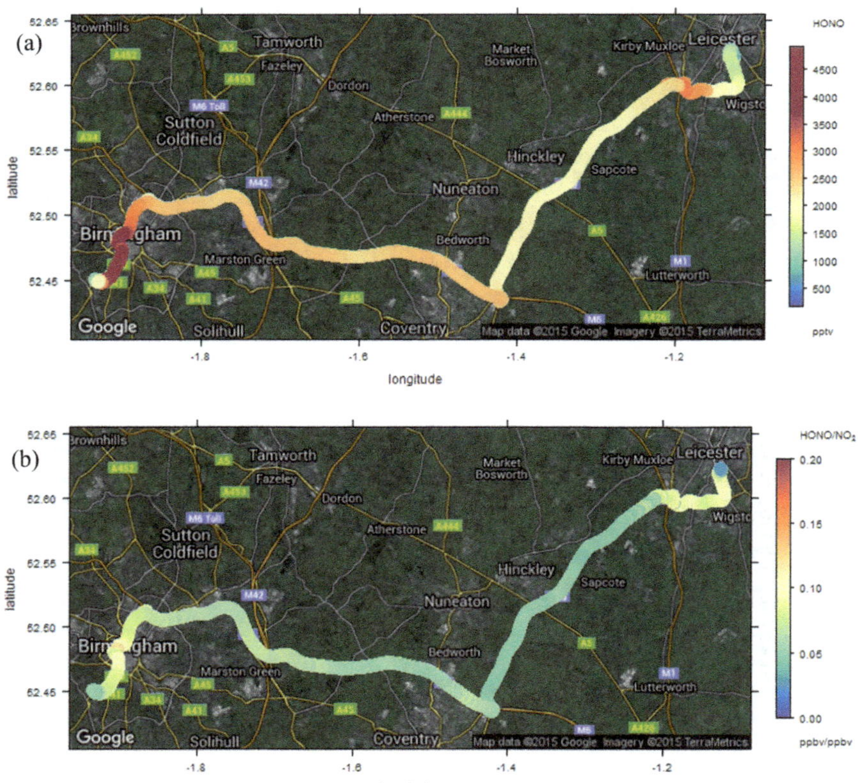

Fig. 9 Measured HONO (panel a, upper) and HONO/NO_2 ratio (panel b, lower) during Birmingham–Leicester transect.

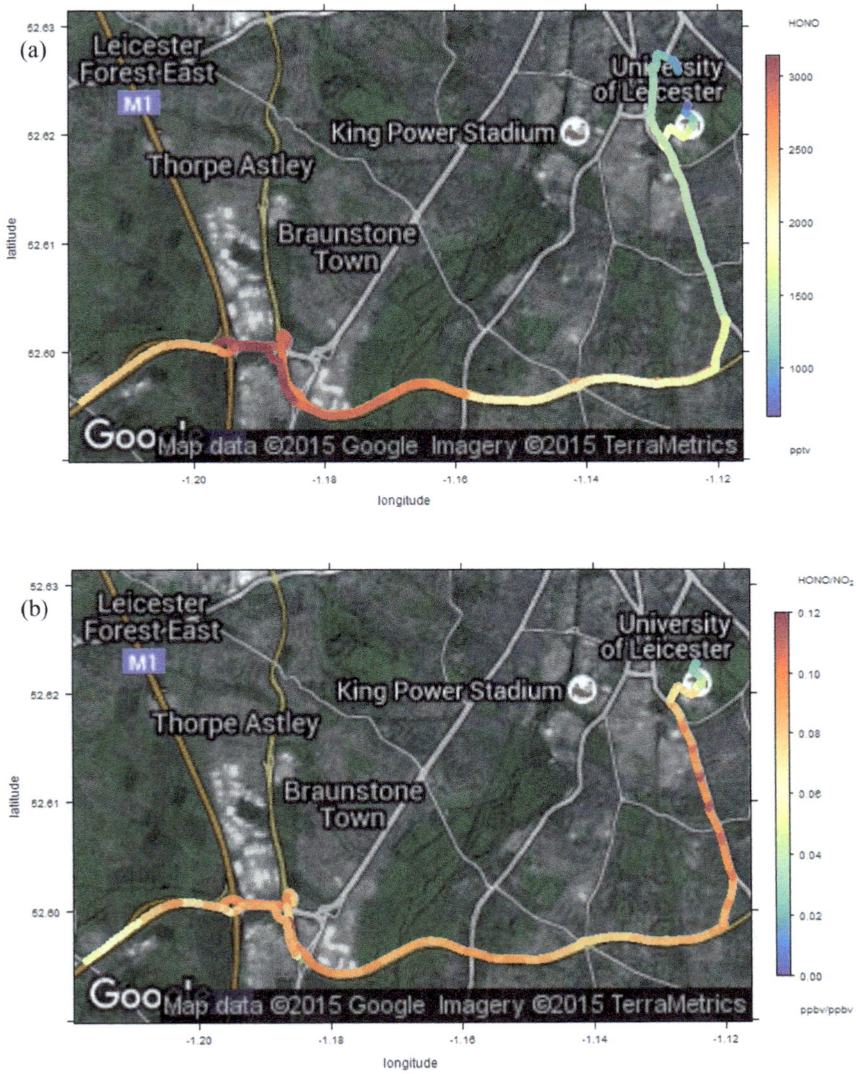

Fig. 10 Enlargement of Fig. 9 – measured HONO (panel a, upper) and HONO/NO_2 ratio (panel b, lower) around the Leicester end of the journey, including open highway, congested major urban artery and urban background sections, from bottom-left to top-right respectively.

Local topography may have contributed to this, *e.g.* "street canyon" type environments within the highway infrastructure concrete architecture.

HONO (and NO_x) levels are much more uniform on the open motorway/highway section of the journey; this is particularly clear from Fig. 11, which displays the mean ($\pm$1 standard deviation) HONO/NO_x ratios observed for each environment, spatially defined, during the measurements. The values for the two "motorway" periods are very similar, and with much smaller variability, while the urban centres – Birmingham, Leicester – have similar but much more variable values for the ratio. The decrease in variability apparent across the figure may also reflect the increasing HONO and NO_x lifetime, later in the day, as photolysis rates

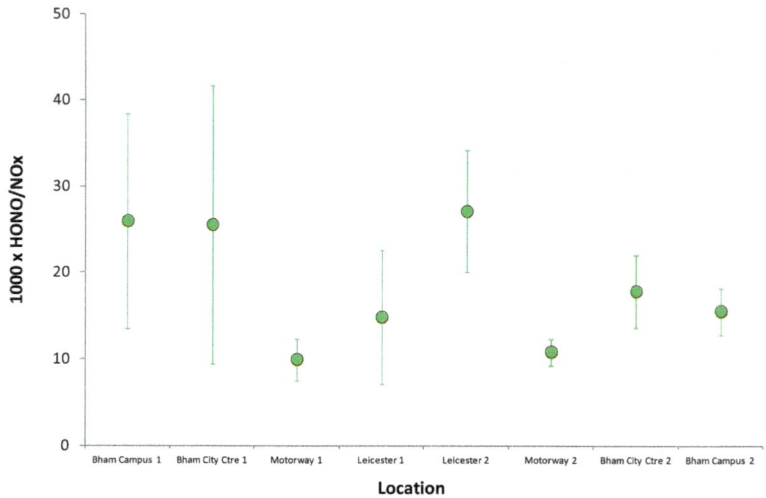

Fig. 11 Mean (±1 standard deviation) $HONO/NO_x$ ratios observed for each environment, spatially defined, during the on-road transect measurements.

declined. The $HONO/NO_2$ ratio is very highly variable (Fig. 8) – more so than the $HONO/NO_x$ ratio – reflecting the issues outlined above (impact of substantial evolution of the NO_x PSS following emission, reflecting the differing ratios of $NO : NO_2$ in exhaust and in background daytime air). For these reasons the $HONO/NO_2$ ratio is much less robust a metric of emissions, and much more sensitive to measurement location *vs.* post-emission airmass age, than the $HONO/NO_x$ ratio, for consideration of traffic sources. As all the environments probed here would be expected to be at least influenced, and in some cases dominated, by traffic emissions, this result may point to multiple sources contributing to the observed HONO levels. Variability in the $HONO–NO_2$ ratio with time (at the same location: weekday *vs.* weekend) was recently reported,[59] indicating the importance of non-direct-emission sources.

OH observations were not feasible for the mobile measurements, so analysis to determine the relationship between NO, OH and HONO and any additional HONO source is not possible; however the sharp spatial gradients in HONO in the urban centre (Fig. 10a) highlight the challenge of application of this approach to spatially heterogeneous regions, including near to major roadways (*e.g.* the M1–A42 junction region shown in the lower left of Fig. 10a) – the real-world equivalent of the problems demonstrated in Fig. 5b. In contrast, in environments which are homogeneous on spatial scales of a few km (*e.g.* the "motorway" segments of the transect in the centre of Fig. 9a), the approximation of the steady state approach is likely to hold. The situation in urban background regions, similar to those employed in a number of previous studies (including the data presented here) is more complex – within the Birmingham and Leicester environments, regions of consistent HONO and NO_x abundance are apparent, but also substantial deviations from these – in the vicinity of major road-ways, pointed to traffic sources of HONO (and/or, of NO_x which can be converted to HONO through the various mechanisms outlined above). Application of the PSS approach eqn (E3) in such locations may be problematic, because the substantial spatial heterogeneity indicates that local PSS is not achieved, and hence inferred additional source contributions may be inaccurate.

Alternative approaches are required to definitively constrain HONO sources, alongside analyses of point concentration measurements. These include flux measurements for surface sources, obtained using vertically resolved observations;[32–34] laboratory and tunnel investigations of specific processes,[42,60] and ensuring that where the PSS methods are applied to field data, the environments are sufficiently homogeneous[16,21] or decoupled[36] that local spatial variability concerns do not apply.

6 Conclusions

A substantial body of literature has shown that HONO levels are higher than can be explained by simple/established gas-phase chemistry (reactions (1)–(3)) supporting the existence of additional sources, and hence the role of HONO as an important net OH precursor in the boundary layer. Application of the PSS approach can quantify the missing source strength, but only where HONO, NO_x and OH are in photochemical steady state; the timescale for this to be established (tens of minutes) is non-negligible compared with spatial heterogeneity in many continental environments (although most remote environments will satisfy this condition). Non-chemical-equilibrium conditions can lead to such analyses inferring the presence of additional HONO sources where these do not exist, and can bias such analyses in both positive and negative directions. The HONO/NO_x ratio is a more robust tracer of emissions characteristics than the HONO/NO_2 ratio during daytime (although the latter is a key tool for constraint of NO_2–HONO surface conversion processes and ground fluxes away from local emission sources). Measurements of the spatial variability of HONO in urban, suburban and rural environments in the UK show that pronounced changes in abundance are found in proximity to major roads within urban areas – indicating that PSS analyses in such areas are likely to be problematic – but also areas of very homogeneous HONO and NO_x abundance in rural, and some suburban, regions, where the PSS approach may be valid.

Acknowledgements

This work was supported in part by the Natural Environment Research Council, grants NE/K012169/1 (Integrated Chemistry of Ozone in the Atmosphere), NE/M013545/1 (Sources of Nitrous Acid in the Atmospheric Boundary Layer) and NE/J008990/1 (Identification of missing organic reactivity in the urban troposphere). We thank Ajit Singh (UoB) for discussions regarding HONO variability. The Leeds FAGE instrumentation is supported by the NERC National Centre for Atmospheric Science.

References

1 H. Levy, *Science*, 1971, **173**, 141.
2 D. E. Heard and M. J. Pilling, *Chem. Rev.*, 2003, **103**, 5163.
3 F. Rohrer and H. Berresheim, *Nature*, 2006, **442**, 184.
4 J. Kleffmann, *ChemPhysChem*, 2007, **8**, 1137.

5 D. Perner and U. Platt, *Geophys. Res. Lett.*, 1979, **6**, 917.
6 J. Stutz, B. Alicke, R. Ackermann, A. Geyer, S. Wang, A. B. White, E. J. Williams, C. W. Spicer and J. D. Fast, *J. Geophys. Res.*, 2004, DOI: 10.1029/2003jd004209.
7 S. M. Ball and R. L. Jones, *Chem. Rev.*, 2003, **103**, 5239.
8 J. Heland, J. Kleffmann, R. Kurtenbach and P. Wiesen, *Environ. Sci. Technol.*, 2001, **35**, 3207.
9 G. Huang, X. Zhou, G. Deng, H. Qiao and K. Civerolo, *Atmos. Environ.*, 2002, **36**, 2225.
10 J. E. Dibb, M. Arsenault, M. C. Peterson and R. E. Honrath, *Atmos. Environ.*, 2002, **36**, 2501.
11 W. Liao, A. Hecobian, J. Mastromarino and D. Tan, *Atmos. Environ.*, 2006, **40**, 17.
12 T. Jurkat, C. Voigt, F. Arnold, H. Schlager, J. Kleffmann, H. Aufmhoff, D. Schäuble, M. Schaefer and U. Schumann, *Geophys. Res. Lett.*, 2011, DOI: 10.1029/2011gl046884.
13 M. Ródenas, A. Muñoz, F. Alacreu, T. Brauers, H.-P. Dorn, J. Kleffmann, and W. Bloss, *Assessment of HONO Measurements: The FIONA Campaign at EUPHORE, Disposal of Dangerous Chemicals in Urban Areas and Megacities, NATO Science for Peace & Security Series C: Environmental Security*, ed. I. Barnes and K. J. Rudzínski, 2012, ch. 4, pp. 45–58, DOI: 10.1007/978-94-007-5034-0_4.
14 X. Ren, H. Harder, M. Martinez, R. L. Lesher, A. Oliger, J. B. Simpas, W. H. Brune, J. J. Schwab, K. L. Demerjian, Y. He, X. Zhou and H. Gau, *Atmos. Environ.*, 2003, **37**, 3639.
15 X. Ren, W. H. Brune, J. Mao, M. J. Mitchell, R. L. Lesher, J. B. Simpas, A. R. Metcalf, J. J. Scwabb, C. Caib, Y. Lib, K. L. Demerjian, H. D. Felton, G. Boynton, A. Adams, J. Perry, Y. Hed, X. Zhou and J. Hou, *Atmos. Environ.*, 2006, **40**, S252.
16 J. Kleffman, T. Gavriloaiei, A. Hofzumahaus, F. Holland, R. Koppmann, L. Rupp, E. Schosser, M. Siese and A. Wahner, *Geophys. Res. Lett.*, 2005, **32**, DOI: 10.1029/2005gl022524.
17 V. Michoud, A. Kukui, M. Camredon, A. Colomb, A. Borbon, K. Miet, B. Aumont, M. Beekmann, R. Durand-Jolibois, S. Perrier, P. Zapf, G. Siour, W. Ait-Helal, N. Locoge, S. Sauvage, C. Afif, V. Gros, M. Furger, G. Ancellet and J. F. Doussin, *Atmos. Chem. Phys.*, 2012, **12**, 11951.
18 S. Kim, T. C. VandenBoer, C. J. Young, T. P. Riedel, J. A. Thornton, B. Swarthout, B. Sive, B. Lerner, J. B. Gilman, C. Warneke, J. M. Roberts, A. Guenther, N. L. Wagner, W. P. Dubé, E. Williams and S. S. Brown, *J. Geophys. Res.*, 2014, DOI: 10.1002/2013jd019784.
19 J. D. Lee, L. K. Whalley, D. E. Heard, D. Stone, R. E. Dunmore, J. F. Hamilton, D. E. Young, J. D. Allan, S. Laufs and J. Kleffmann, *Atmos. Chem. Phys.*, 2015, **15**, 22097.
20 X. Li, T. Brauers, R. Häseler, B. Bohn, H. Fuchs, A. Hofzumahaus, F. Holland, S. Lou, K. D. Lu, F. Rohrer, M. Hu, L. M. Zeng, Y. H. Zhang, R. M. Garland, H. Su, A. Nowak, A. Wiedensohler, N. Takegawa, M. Shao and A. Wahner, *Atmos. Chem. Phys.*, 2012, **12**, 1497.
21 R. Oswald, M. Ermel, K. Hens, A. Novelli, H. G. Ouwersloot, P. Paasonen, T. Petäjä, M. Sipilä, P. Keronen, J. Bäck, R. Königstedt, Z. Hosaynali Beygi,

H. Fischer, B. Bohn, D. Kubistin, H. Harder, M. Martinez, J. Williams, T. Hoffmann, I. Trebs and M. Sörgel, *Atmos. Chem. Phys.*, 2015, **15**, 799.
22 V. Michoud, A. Colomb, A. Borbon, K. Miet, M. Beekmann, M. Camredon, B. Aumont, S. Perrier, P. Zapf, G. Siour, W. Ait-Helal, C. Afif, A. Kukui, M. Furger, J. C. Dupont, M. Haeffelin and J. F. Doussin, *Atmos. Chem. Phys.*, 2014, **14**, 2805.
23 Y. F. Elshorbany, R. Kurtenbach, P. Wiesen, E. Lissi, M. Rubio, G. Villena, E. Gramsch, A. R. Rickard, M. J. Pilling and J. Kleffmann, *Atmos. Chem. Phys.*, 2009, **9**, 2257.
24 R. M. Harrison and A.-M. N. Kitto, *Atmos. Environ.*, 1994, **28**, 1089.
25 B. J. Finlayson-Pitts, L. M. Wingen, A. L. Sumner, D. Syomin and K. A. Ramazan, *Phys. Chem. Chem. Phys.*, 2003, **5**, 223.
26 M. Ammann, M. Kalberer, D. T. Jost, L. Tobler, E. Rossler, D. Piguet, H. W. Gäggeler and U. Baltensperger, *Nature*, 1998, **395**, 157.
27 C. George, R. S. Strekowski, J. Kleffmann, K. Stemmler and M. Ammann, *Faraday Discuss.*, 2005, **130**, 195.
28 K. Stemmler, M. Ammann, C. Donders, J. Kleffmann and C. George, *Nature*, 2006, **440**, 195.
29 J. M. Langridge, R. J. Gustafsson, P. T. Griffiths, R. A. Cox, R. M. Lambert and R. L. Jones, *Atmos. Environ.*, 2009, **43**, 5128.
30 X. Zhou, N. Zhang, M. TerAvest, D. Tang, J. Hou, S. Bertman, M. Alaghmand, P. B. Shepson, M. A. Carroll, S. Griffith, S. Dusanter and P. S. Stevens, *Nat. Geosci.*, 2011, **4**, 440.
31 C. J. Young, R. A. Washenfelder, J. M. Roberts, L. H. Mielke, H. D. Osthoff, C. Tsai, O. Pikelnaya, J. Stutz, P. R. Veres, A. K. Cochran, T. C. VendenBoer, J. Flynn, N. Grossberg, C. L. Haman, B. Lefer, H. Stark, M. Graus, J. de Gouw, J. B. Gilman, W. C. Kuster and S. S. Brown, *Environ. Sci. Technol.*, 2012, **46**, 10965.
32 B. Vogel, H. Vogel, J. Kleffmann and R. Kurtenbach, *Atmos. Environ.*, 2003, **37**, 2957.
33 N. Zhang, X. Zhou, P. B. Shepson, H. Gao, M. Alaghmand and B. Stirm, *Geophys. Res. Lett.*, 2009, **36**, DOI: 10.1029/2009gl038999.
34 T. C. VandenBoer, S. S. Brown, J. G. Murphy, W. C. Keene, J. C. Young, A. A. P. Pszenny, S. Kim, C. Warneke, J. A. de Gouw, J. R. Maben, N. L. Wagner, T. P. Riedel, J. A. Thornton, D. E. Wolfe, W. P. Dubé, F. Öztürk, C. A. Brock, N. Grossberg, B. Lefer, B. Lerner, A. M. Middlebrook and J. M. Roberts, *J. Geophys. Res.*, 2013, DOI: 10.1002/jgrd.50721.
35 J. Kleffmann and P. Wiesen, *Atmos. Chem. Phys.*, 2005, **5**, 77.
36 X. Li, F. Rohrer, A. Hofzumahaus, T. Brauers, R. Häseler, B. Bohn, S. Broch, H. Fuchs, S. Gomm, F. Holland, J. Jäger, J. Kaiser, F. N. Keutsch, I. Lohse, K. Lu, R. Tillmann, R. Wegener, G. M. Wolfe, T. F. Mentel, A. Kiendler-Scharr and A. Wahner, *Science*, 2014, **344**, 292.
37 H. Su, Y. Cheng, R. Oswald, T. Behrendt, I. Trebs, F. X. Meixner, M. O. Andreae, P. Cheng, Y. Zhang and U. Pöschl, *Science*, 2011, **333**, 1616.
38 R. Oswald, T. Behrendt, M. Ermel, D. Wu, H. Su, Y. Cheng, C. Breuninger, A. Moravek, E. Mougin, C. Delon, B. Loubet, A. Pommerening-Röser, M. Sorgel, U. Pöschl, T. Hoffmann, M. O. Andreae, F. X. Meixner and I. Trebs, *Science*, 2013, **341**, 1233.

39 B. Weber, D. Wu, A. Tamm, N. Ruckteschler, E. Rodríguez-Caballero, J. Steinkamp, H. Meusel, W. Elbert, T. Behrendt, M. Sörgel, Y. Cheng, P. J. Crutzen, H. Su and U. Pöschl, *Proc. Natl. Acad. Sci. U. S. A.*, 2015, **15**, 15384.
40 G. Lammel and J. N. Cape, *Chem. Soc. Rev.*, 2006, **25**, 361.
41 F. Spataro and A. Ianniello, *J. Air Waste Manage. Assoc.*, 2014, **64**, 1232.
42 C. George, M. Ammann, B. D'Anna, D. J. Donaldson and S. A. Nizkorodov, *Chem. Rev.*, 2015, **115**, 4218.
43 D. L. Chapman and L. K. Underhill, *J. Chem. Soc., Trans.*, 1913, **103**, 496.
44 P. A. Leighton, *Photochemistry of air pollution*, Academic Press, New York, 1961.
45 A. Volz-Thomas, H. W. Pätz, N. Houben, S. Konrad, D. Mihelcic, T. Klüpfel and D. Perner, *J. Geophys. Res.*, 2003, **108**, DOI: 10.1029/2001jd001255.
46 Z. Hosaynali Beygi, H. Fischer, H. D. Harder, M. Martinez, R. Sander, J. Williams, D. M. Brookes, P. S. Monks and J. Lelieveld, *Atmos. Chem. Phys.*, 2011, **11**, 8497.
47 I. Bejan, Y. Abd El Aal, I. Barnes, T. Benter, B. Bohn, P. Wiesen and J. Kleffmann, *Phys. Chem. Chem. Phys.*, 2006, **8**, 2028.
48 B. H. Lee, E. C. Wood, S. C. Herndon, B. L. Lefer, W. T. Luke, W. H. Brune, D. D. Nelson, M. S. Zahniser and J. W. Munger, *J. Geophys. Res.*, 2013, DOI: 10.1002/2013jd020341.
49 D. Stone, L. K. Whalley and D. E. Heard, *Chem. Soc. Rev.*, 2012, **41**, 6348.
50 L. K. Whalley, K. L. Furneaux, A. Goddard, J. D. Lee, A. Mahajan, H. Oetjen, K. A. Read, N. Kaaden, L. J. Carpenter, A. C. Lewis, J. M. C. Plane, E. S. Saltzman, A. Wiedensohler and D. E. Heard, *Atmos. Chem. Phys.*, 2010, **10**, 1555.
51 J. D. Lee, S. J. Moller, K. A. Read, A. C. Lewis, L. Mendes and L. J. Carpenter, *J. Geophys. Res.*, 2009, **114**, DOI: 10.1029/2009jd011878.
52 T. C. VandenBoer, C. J. Young, R. K. Talukdar, M. Z. Markovic, S. S. Brown, J. M. Roberts and J. G. Murphy, *Nat. Geosci.*, 2015, **8**, 55.
53 C. Bloss, V. Wagner, M. E. Jenkin, R. Volkamer, W. J. Bloss, J. D. Lee, D. E. Heard, K. Wirtz, M. Martin-Reviejo, G. Rea, J. C. Wenger and M. J. Pilling, *Atmos. Chem. Phys.*, 2005, **5**, 641.
54 S. M. Saunders, M. E. Jenkin, R. G. Derwent and M. J. Pilling, *Atmos. Chem. Phys.*, 2003, **3**, 161.
55 D. C. Carslaw and G. Rhys-Tyler, *Atmos. Environ.*, 2013, **81**, 339.
56 Z. Xu, T. Wang, J. Wu, L. Xue, J. Chan, Q. Zha, S. Zhou, P. K. K. Loouie and C. W. Y. Luk, *Atmos. Environ.*, 2015, **106**, 100.
57 R. E. Dunmore, J. R. Hopkins, R. T. Lidster, J. D. Lee, M. J. Evans, A. R. Rickard, A. C. Lewis and J. F. Hamilton, *Atmos. Chem. Phys. Discuss.*, 2015, **15**, 9983, DOI: 10.5194/acp-15-9983-2015.
58 L. K. Whalley, L. K. D. Stone, B. Bandy, R. Dunmore, J. F. Hamilton, J. Hopkins, J. D. Lee, A. C. Lewis and D. E. Heard, *Atmos. Chem. Phys. Discuss.*, 2016, **16**, 2747, DOI: 10.5194/acp-16-2747-2016.
59 S. E. Pusede, T. C. VandenBoer, J. G. Murphy, M. Z. Markovic, C. J. Young, P. Veres, J. M. Roberts, R. A. Washenfelder, S. S. Brown, X. Ren, C. Tsai, J. Stutz, W. H. Brune, E. C. Browne, P. Wooldridge, A. R. Graham, R. Weber, A. H. Goldstein, S. Dusanter, S. Griffith, P. S. Stevens, B. Lefer and R. C. Cohen, *Environ. Sci. Technol.*, 2015, DOI: 10.1021/acs.est.5b02511.
60 R. Kurtenbach, K. H. Becker, J. A. G. Gomes, J. Kleffmann, J. C. Lörzer, M. Spittler, P. Wiesen, R. Ackermann, A. Geyer and U. Platt, *Atmos. Environ.*, 2001, **35**, 3385.

Faraday Discussions

PAPER

Exploring the nitrous acid (HONO) formation mechanism in winter Beijing: direct emissions and heterogeneous production in urban and suburban areas†

Shengrui Tong,‡*[a] Siqi Hou,‡[a] Ying Zhang,[a] Biwu Chu,[b] Yongchun Liu,[b] Hong He,[b] Pusheng Zhao[c] and Maofa Ge*[a]

Received 13th November 2015, Accepted 11th December 2015
DOI: 10.1039/c5fd00163c

Continuous measurements of nitrous acid (HONO) were performed from December 12 to December 22, 2015 in both urban and suburban areas of Beijing to study the formation mechanism of HONO. The measurement campaign in both sites included a clean–haze–clean transformation process. HONO concentrations showed similar variations in the two sites, while they were always higher in the urban area. Moreover, correlations of HONO with NO_x, NO_2, NO, $PM_{2.5}$ and relative humidity (RH) were studied to explore possible HONO formation pathways, and the contributions of direct emissions, heterogeneous reactions, and homogeneous reactions were also calculated. This showed that HONO in urban and suburban areas underwent totally different formation procedures, which were affected by meteorological conditions, $PM_{2.5}$ concentrations, direct emissions, homogeneous reactions and heterogeneous reactions. $PM_{2.5}$ concentrations and RH would influence the NO_2 conversion efficiency. Heterogeneous reactions of NO_2 were more efficient in suburban areas and in clean periods while direct emissions and homogeneous reactions contributed more in urban areas and in polluted periods when the concentrations of NO_x and NO were at a high level.

Introduction

Nitrous acid (HONO) plays an essential role in photochemical cycles by providing hydroxyl radicals (OH), with contributions of 30–50% of OH during the daytime.[1–5] The OH radical, as one of the most important oxidants in the

[a]*State Key Laboratory for Structural Chemistry of Unstable and Stable Species, Beijing National Laboratory for Molecular Sciences (BNLMS), Institute of Chemistry, Chinese Academy of Sciences, Beijing, 100190, P. R. China. E-mail: tongsr@iccas.ac.cn; gemaofa@iccas.ac.cn*

[b]*Research Center for Eco-Environmental Sciences, Chinese Academy of Sciences, Beijing, 100085, P. R. China*

[c]*Institute of Urban Meteorology, China Meteorological Administration, Beijing, 100089, P. R. China*

† Electronic supplementary information (ESI) available. See DOI: 10.1039/c5fd00163c

‡ Shengrui Tong and Siqi Hou contributed equally to this work.

atmosphere, dominates the oxidation capacity in the troposphere by reacting with organic matter, leading to the formation of secondary pollutants and accelerating the formation of air pollution.[5–7]

The commonly accepted sources of HONO are direct emissions, homogeneous reactions and heterogeneous formations.[8–10] Studies by Kurtenbach[11] and Kirchstetter *et al.*[12] have recognized that HONO could be emitted by combustion processes and traffic emissions with quantities of about 0.3–0.8% of NO_x (NO + NO_2) obtained by tunnel studies. Homogeneous reaction between NO and OH can produce HONO, though this reaction was thought to be less important due to the low concentrations of OH radicals at night.[10] However, recent studies have strengthened the importance of this reaction source in polluted regions with high OH and NO concentrations.

Compared to the two sources mentioned above, the kinetics and mechanisms of HONO from heterogeneous reactions still remain unclear.[2,8,13] Studies have suggested HONO could be formed by NO_2 on wet surfaces,[14–16] and kinetics data showed the reaction is first order in NO_2 and dependent on surface water.[14,17,18] The hydrolysis of NO_2 could happen on various surfaces including ground, buildings and vegetation surfaces.[14] In addition, HONO could be formed by NO_2 reduction on the surface of soot and other aerosols, especially under sunlight.[13,16,19,20] However, the contribution of this reaction to the HONO budget is still in debate due to the rather small aerosol surfaces and the slow reaction rate.[9,21]

The effects of HONO have gained great attention due to the more and more frequent pollution events occurring in China recently.[8] Observations of HONO have been documented at many sites including Beijing,[22,23] Guangzhou[8,24,25] and Shanghai.[26] Results showed HONO concentrations were much higher in mega cities with heavy pollution than in remote places and clean areas. However, studies are still lacking, especially those focused on comparisons of HONO evolution between urban and suburban areas, and between haze and non-haze periods. The chemical reactions during haze episodes provide complicated interactions of HONO with other chemical species, which will be helpful to understand the formation and removal mechanisms of HONO.

In this study, a continuous measurement was performed in the urban and suburban areas of Beijing in December of 2014, when both urban and suburban areas suffered typical haze episodes and a following non-haze period. The evolution of HONO and other related gas pollutants like NO, NO_2, NO_x, O_3, SO_2 and CO was explored under changing meteorological conditions. Moreover, in order to investigate the impact factors of HONO, differences were compared between haze and non-haze periods of both urban and suburban sites. Finally, the influences of different HONO sources including direct emissions, homogeneous reactions and heterogeneous formation in both sites were discussed and formation mechanisms were analyzed in detail.

Experimental

Measurement instruments

HONO concentrations were measured by a home-made HONO analyzer, which has been used in several measurements previously.[27,28] The principle of the home-made HONO analyzer is introduced briefly: HONO was absorbed by a specific solution (0.06 M sulfanilamide in 1 M HCl)[5,29,30] in a glass device with a two-channel stripping coil. The absorption solution from two channels was pumped by a peristaltic pump to react with a dye solution [0.8 mM *N*-(1-naphthyl) ethylene diamine dihydrochloride][31] to form the azo dye and then detected by a mini-spectrometer using a diode array detector (Ocean Optics, SD2000). The true HONO concentrations could be obtained by subtracting the calibrated signal of the second coil from the first coil.[29,31] Comparison between the HONO analyzer and Long Path Absorption (LOPAP) yielded excellent agreements.

Gas phase pollutants of CO, SO_2, O_3, NO, NO_2 and NO_x were measured by commercial Thermo scientific analyzers: SO_2 analyzer (43i), O_3 analyzer (49i), NO_x analyzer (42i) and CO analyzer (48i) with detection limits of 1 ppbv, 1 ppbv, 1 ppbv and 0.05 ppmv, respectively.

Measurement site

The measurements were performed from 12 to 22 December in both urban and suburban areas in Beijing. The measurement sites were described in detail in a previous study.[28] The locations of the measurement sites are shown in Fig. 1. Here is a brief description.

Urban site (ICCAS): HONO concentrations were measured in No. 2 building of the Institute of Chemistry, Chinese Academy of Sciences (ICCAS, 116°19′21.58″E, 39°59′22.68″N). Concentrations of other gaseous species were measured at Baolian, 5.5 kilometers southwest of the ICCAS site. Both of the two measurement sites are located at the North Fourth Ring Road of HaiDian District with similar conditions, which are surrounded by dense population and heavy traffic. Meteorological parameters, including wind speed, wind direction, temperature, pressure, relative humidity and $PM_{2.5}$ concentrations were all obtained from Baolian station.

Suburban site: the HONO analyzer and other gas analyzers were set in the Lake Yanqi campus of the University of Chinese academy of sciences (UCAS, 40.4° N, 116.6° E). Concentrations of $PM_{2.5}$ and meteorological parameters were also

Fig. 1 The locations of the measurement sites.

obtained there. This site is a typical suburban representative, which is 50 km northeast of the city center. In addition, the environmental conditions at this site are more diverse due to being surrounding by a highway on the west side, a lake to the northeast, farm lands nearby and a railway station within 1.1 kilometers.

Results and discussion

Overview of the observation

Temporal variations of meteorological data and measured chemical species. $PM_{2.5}$ concentrations and the measured meteorological parameters of air temperature (*T*), relative humidity (RH), wind speed (WS) and direction (WD) are illustrated in Fig. 2. It can be seen from the plot that there were two increases of $PM_{2.5}$ concentration for both urban and suburban areas during the measurement campaign. Therefore, two groups were divided based on $PM_{2.5}$ concentrations (Table 1): haze period, when $PM_{2.5}$ concentrations were higher than 75 μg m^{-3}; and non-haze period, when $PM_{2.5}$ concentrations were lower than 75 μg m^{-3}. In detail, the haze period was from 8:00 of Dec 13 to 7:00 of Dec 15 and from 22:00 of Dec 16 to 8:00 of Dec 19. While the non-haze period was the remaining time. In addition, data with wind speed higher than 2.0 m s^{-1} were excluded from our discussion.

During the non-haze period, the average $PM_{2.5}$ concentrations of the urban and suburban sites were 29.0 and 23.1 μg m^{-3}, respectively. Such a low $PM_{2.5}$ concentration was due to the blowing wind with a mean speed of 3.1 and 1.5 m s^{-1}, respectively, which also caused a low temperature range in both observation sites. The average temperature during this period was −1.4 and −2.1 °C and the relative humidity was 24.9% and 25.7%, respectively. During the haze period, the

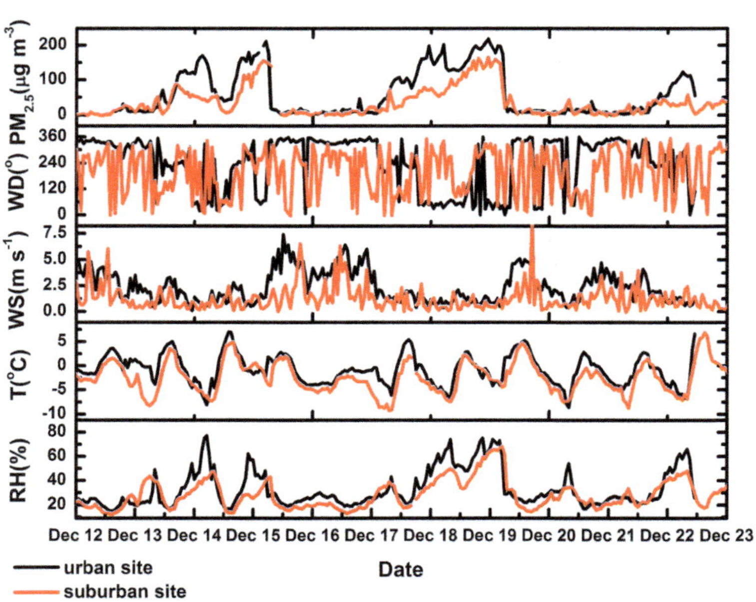

Fig. 2 Hourly-averaged air temperature (*T*), relative humidity (RH), wind speed (WS), wind direction (WD) and $PM_{2.5}$ concentrations during Dec 12 to Dec 22, 2014.

Table 1 Classification of meteorological conditions during Dec 12 to Dec 22, 2014

Location	Time period	Weather condition	$PM_{2.5}$ concentration ($\mu g\ m^{-3}$)	T (°C)	RH (%)	WS ($m\ s^{-1}$)	WD
Urban	Dec 14–15; Dec 18–19	Haze	144.0	−1.1	50.0	1.1	Northeast
	Dec 12–13; Dec 20–22	Non-haze	29.0	−1.34	24.9	3.1	North
Suburban	Dec 14–15; Dec 18–19	Haze	104.8	−0.5	40.8	0.9	North
	Dec 12–13; Dec 20–22	Non-haze	21.1	−2.1	25.7	1.5	No certain directions

mean $PM_{2.5}$ concentrations were enhanced to 144.0 and 104.8 $\mu g\ m^{-3}$ in urban and suburban sites, with RH increasing to 50.0 and 40.8%, respectively. In addition, the wind speed during this period went down to 1.1 and 0.8 $m\ s^{-1}$ on average. Variation tendencies of $PM_{2.5}$ concentrations and meteorological parameters in both haze and non-haze periods at the two observation sites were similar except the wind speeds and wind directions, which were mainly due to the geographical location differences.

Fig. 3 shows the variation of gas species measured during the observation campaign in the urban site and suburban site. Average concentrations of gas-phase species are shown in Fig. 4. The average values of HONO concentrations at the urban and suburban sites were comparable to our previous measurements at the same observation sites in October of 2014.[28] Moreover, they were also in accordance with measurements by Spataro *et al.*[22] (HONO concentration was 1.06 ppbv on average with NO_2 of 38.8 ppbv) and Hendrick *et al.*[23] (2 ppbv at urban sites and 1.25 ppbv at suburban sites in Beijing). It could be seen from Fig. 4 that during the haze period at the urban site, the concentrations of SO_2, CO, NO, NO_2, NO_x and $PM_{2.5}$ were enhanced 4.8, 4.2, 4.2, 4.0, 4.7 and 8.2 times, respectively, while concentrations of HONO were enhanced 3.8 times. For the suburban site, concentrations of SO_2, CO, NO, NO_2, NO_x and $PM_{2.5}$ were enhanced 1.4, 2.3, 3.5, 3.4, 3.6 and 5.0 times, respectively, while the enhancement of HONO concentrations was only 1.8 times. This indicates that some connections lie between these gas pollutants and HONO with different extents, which may be associated with various processes.

Diurnal variations of the measured chemical species. To further understand the connections between the gas-phase species and the characteristics of HONO variations in the two areas, diurnal variation of HONO concentrations and other gas-phase species were averaged by the weather conditions. Fig. 5 shows the daily variations of gas pollutants during the two periods in both sites. In the urban site, the concentration of HONO during the haze period showed a morning peak at around 7:00 (3.00 ppbv on average). Then it began to decrease due to the increase of irradiance. After reaching a minimum (0.39 ppbv on average) at around 15:00, it revived to 1.12 ppbv at 20:00 and kept increasing to 1.86 ppbv until 1:00. The concentrations of HONO then remained between 1.86–2.18 ppbv for the rest of the night. It should be noted that there was no evening peak of HONO concentrations during the evening traffic peak period. This may be covered by the cooperation of photolysis reduction and enhancement due to decreasing of

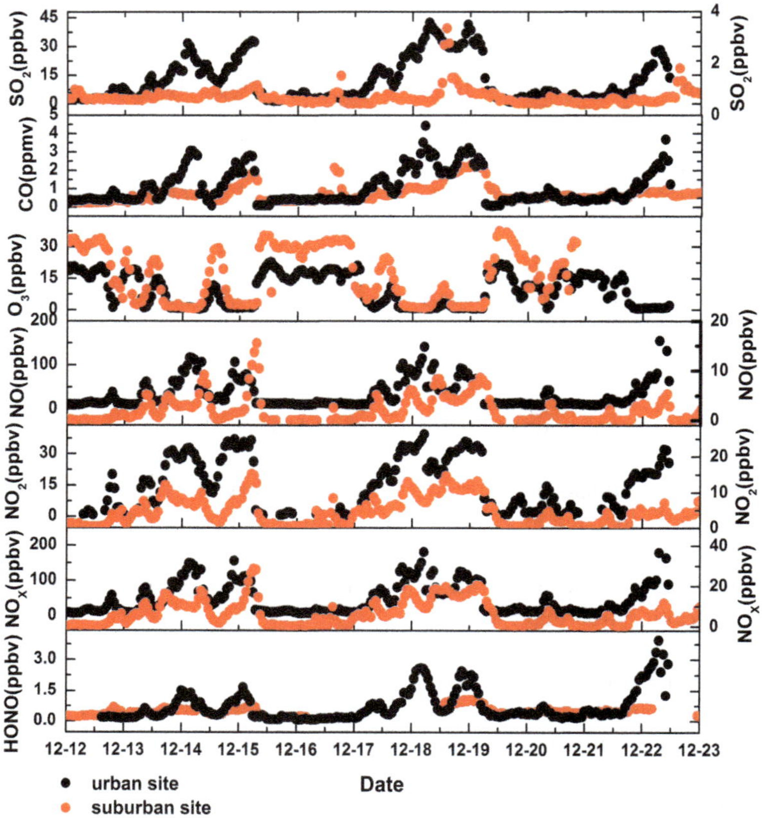

Fig. 3 Variation of gas species measured during the observation campaign at the urban site and suburban site (the black points show concentrations at the urban site, and red points show concentrations at the suburban site). The ordinate axes of SO_2, NO, NO_2 and NO_x in the suburban site were adjusted and are shown on the right side.

boundary layer height. While during the non-haze period, HONO concentrations decreased and remained at a low level (0.21–0.32 ppbv) during both day and night except for the morning traffic peak period, when concentrations of HONO could rise to 0.43 ppbv. Additionally, HONO concentrations did not show significant differences between daytime and nighttime. In the suburban site, HONO concentration varied in a similar manner to its variation in the urban area. During the haze period, it reduced to 0.40 ppbv around 13:00 and then rose to 0.72 ppbv at around 20:00. For the rest of the night, HONO concentrations remained between 0.72–0.75 ppbv until 6:00 in the morning. Unlike the variations during the haze period, HONO in the clean period showed two significant peaks in the traffic rush hour: one was of 0.52 ppbv at around 9:00 and the other was of 0.56 ppbv at 20:00. However, during the rest of the day, HONO concentrations remained at a low level with a nighttime range of 0.37–0.49 ppbv and a daytime range of 0.39–0.48 ppbv. After reaching the evening peak, HONO concentrations decreased and stayed at around 0.40 ppbv until the next morning. This value was comparable to the minimum value (0.39 ppbv) which appeared at 14:00. In general, the characteristics of HONO variations could be summarized as follows:

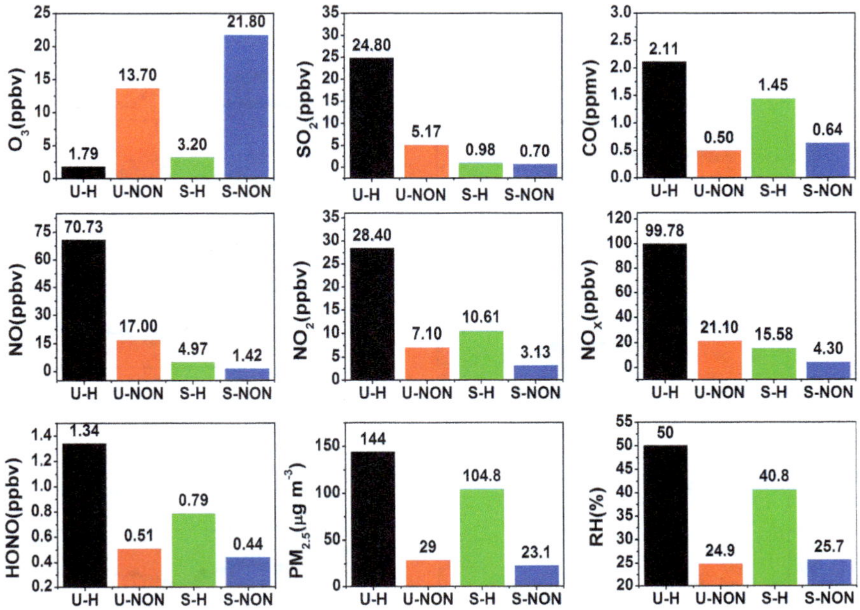

Fig. 4 Comparison of the average concentrations of gas-phase species as well as $PM_{2.5}$ and RH in urban and suburban sites during both haze and non-haze periods (U-H means urban with haze; U-NON means urban with non-haze; S-H means suburban with haze; S-NON means suburban with non-haze).

First, HONO concentrations in haze conditions were higher than those in clean conditions, no matter whether in urban areas or suburban areas. However, differences of HONO concentrations between haze and clean conditions were much bigger in urban sites than suburban ones. Second, unlike the variation tendencies of the suburban site, which showed an "M" shape with peaks at 9:30 and 17:00 in both conditions, HONO concentrations only showed one peak at morning rush hour. This suggested that HONO concentrations were influenced by complicated factors.

For variations of other species, significant differences existed between the urban and suburban sites as well as between the haze and non-haze periods. However, the results were consistent with our previous studies and other studies.[8,23,26,32] In the urban site, variations during the haze period were rather different from those during the non-haze period. For instance, the maximum value of O_3 concentrations was delayed to 15:00 from 13:00. Actually, concentrations of SO_2, CO, NO and NO_x in the non-haze period were so low that their diurnal variation characteristics were not obvious. However, they still showed morning peaks at around 8:00, and so did HONO. This indicated that the morning peak of HONO had some connection with SO_2, CO, NO and NO_x. For the situation at the suburban site, during the haze period concentrations of CO and NO_2 were higher in the night and kept decreasing with increasing sunlight intensity after dawn. Both of them reached a minimum at noon. NO and NO_x showed morning peaks around 8:00. Additionally, SO_2 concentrations increased significantly at 14:00 with low concentrations during the whole day. During the

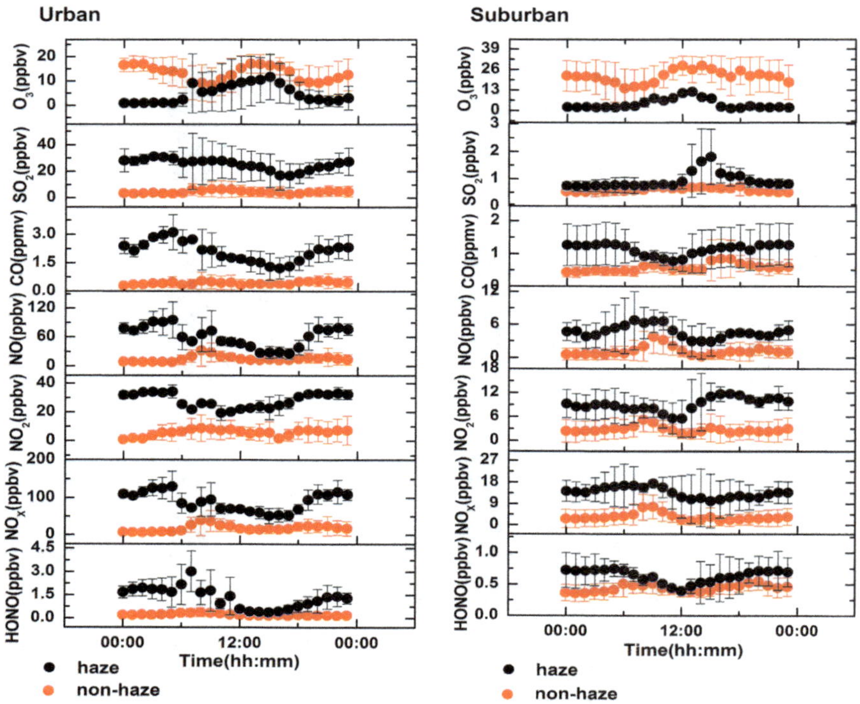

Fig. 5 Diurnal variations of HONO with O_3, SO_2, CO, NO, NO_2, and NO_x in urban and suburban sites, the black points show the haze period and red points show the clean period.

non-haze period, concentrations of gas pollutants remained low all day except during the morning peak. As mentioned above, it could be concluded that HONO concentrations were tightly connected with the gas-phase species. Therefore, more studies should be conducted to understand the factors which influence HONO.

Influence factors of HONO variations. To explore the influence factors of HONO, the impacts of meteorological conditions were studied first. HONO concentrations in the urban area were highly influenced by wind from the northeast. However, it has little influence in the suburban area (Fig. S1†). In the urban area, northwesterly wind blowing strongly and northeasterly wind with a low speed would be beneficial to HONO accumulation. Moreover, wind speed impacts on HONO concentrations without discrimination by enhancing the deposition.

Furthermore, correlations of HONO with gas pollutants are plotted in Fig. 6. The data were divided by locations and periods. As shown in Fig. 6(a), in the urban site, correlations of HONO with CO, NO and NO_x during the haze period were higher, at 0.503, 0.400 and 0.463, respectively. Correlations of HONO with O_3 were opposite, in that it was higher during the non-haze period. In addition, correlations of HONO with NO_2 were not obvious either during the haze period or the non-haze period. Although correlations of HONO with CO, NO and NO_x during the haze period were higher, the linear slopes between them were lower than

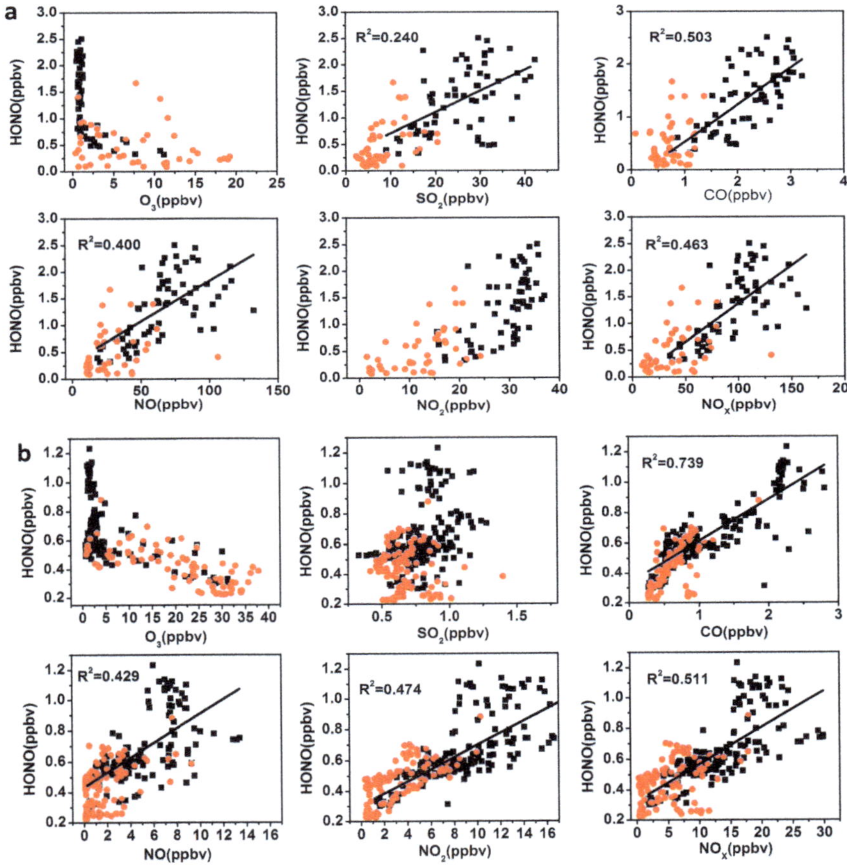

Fig. 6 (a) Correlation between HONO and O_3, SO_2, CO, NO, NO_2 and NO_x during the haze period (black) and clean period (red) in the urban site. (b) Correlation between HONO and O_3, SO_2, CO, NO, NO_2 and NO_x during the haze period (black) and clean period (red) in the suburban site.

those during the non-haze period. This indicated that the impacts of CO, NO_x and other gases may be more effective in the non-haze period, which strengthens the complication of HONO sources. In the suburban area (Fig. 6(b)), obvious correlations existed for HONO with CO, NO and NO_x during the haze period and HONO with O_3, CO and NO_2 during the non-haze period. No certain correlations existed for HONO with SO_2 in both periods. These facts suggested that HONO formation processes were different depending on locations and weather conditions, and the formation pathways in the suburban area were more complicated.

In general, CO and NO are mainly linked with combustion processes like vehicle emissions and the burning of fossil fuels and biomass, which could be seen as primary emission processes.[33–35] Meanwhile, NO_2 and O_3 are more likely come from secondary formations.[36] In the urban site, the correlations of CO with NO and NO_x during the whole observation were all higher than 0.8 (Fig. S2†), which means these three gases have a good chance to come from the same sources, such as traffic emissions.[34] Moreover, the distinction between environmental conditions at the observation sites would contribute to the correlation differences. Due to their distance from each other, the urban site is surrounded by dense traffic flows as well as human activities, while in the suburban area the train station, small factories, farmland and the Yanqi lake were all sources which made the environmental conditions complicated. As a result, though good correlations of HONO with CO, NO and NO_x appeared in both urban and suburban sites, the sources were different. In urban areas it may be caused by mobile sources[22] like vehicle emissions, while in suburban areas it may be caused by a combination of fossil fuel burning, biomass burning, vehicle emissions and other combustion and emission processes.

Direct emission. Apparently, HONO in the urban area was greatly affected by direct emissions. Moreover, due to the complicated emission sources and environmental conditions, HONO concentrations in the suburban area were much more associated with secondary formation processes.

To evaluate HONO amounts from direct emissions in the atmosphere, a factor of 0.008 was supported by Kurtenbach *et al.*[11] and is widely accepted in most observation studies. Therefore, 0.008 was used to calculate the HONO concentrations from direct emissions ($HONO_{emission}$) by the formula: $[HONO_{emission}] = [NO_x] \times 0.008$.[8] Where, $[HONO_{emission}]$ represents HONO concentrations from direct emissions, and $[NO_x]$ represents concentrations of NO_x.

The mean values of calculated $HONO_{emission}$ were 0.37 and 0.06 ppbv, respectively, which contributed 48.8% and 10.3% of HONO concentrations in the urban and suburban sites. The results were comparable with contributions of 49.7% when NO_x concentrations were 266 ppbv in urban Beijing[22] and 8.5% with NO_x concentrations of 11.9 ppbv obtained at the same observation sites in our previous work.[28] Moreover, the contributions of direct emissions in both sites were enhanced during the haze period. However, at the urban site, the contribution of direct emissions was higher at up to 40.4% with NO_x concentrations of 99.8 ppbv even during the non-haze period. This suggested direct emissions would be the major source of HONO in urban areas, and haze conditions would aggravate this contribution. On the other hand, there would be other important processes influencing HONO concentrations in suburban areas.

Heterogeneous conversion from NO_2

$HONO/NO_2$ ratio and NO_2 conversion rates. It is well-accepted that HONO could be formed on wet surfaces by the heterogeneous conversion of NO_2.[14–16] The $HONO/NO_2$ ratio is usually used to evaluate the heterogeneous conversion efficiency of NO_2 to HONO, assuming all of the HONO comes from NO_2 conversion.[8,37,38] However, this value would overestimate the conversion efficiency if the contributions from direct emissions and homogeneous production were large.[3,10,32,39] Hence, corrected HONO concentrations ($HONO_{corr}$), which had contributions from direct emissions subtracted, were used to study the heterogeneous reactions instead of the observed ones.

Fig. 7 shows the variations of $HONO_{corr}/NO_2$ during the haze and non-haze periods from the urban and suburban sites. Wind speeds were lower than 1.0 m s^{-1} during these selected periods. The low $HONO_{corr}$ concentrations showed mainly contributions of direct emissions. The variations between the two sites showed rather different tendencies. In the urban site, $HONO_{corr}/NO_2$ reached its

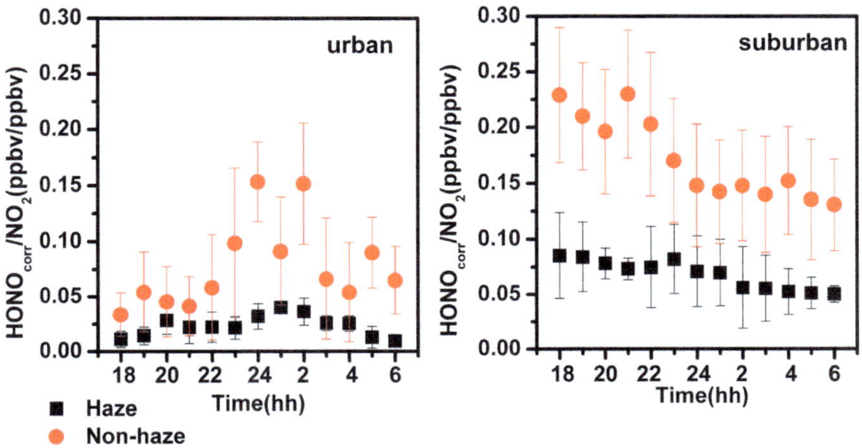

Fig. 7 Variations of mean $HONO_{corr}/NO_2$ ratios during nighttime in urban (left) and suburban (right) sites.

maximum at around midnight and then slightly decreased. The mean values of $HONO_{corr}/NO_2$ in the urban site were 0.027 in the haze period and 0.078 in the non-haze period. In the suburban site, the $HONO_{corr}/NO_2$ ratios during both periods kept decreasing after sunset and became stable in the second half of the night. The stable $HONO_{corr}/NO_2$ ratios indicated the conversion of NO_2 to HONO had reached a steady state,[8,18,32] which was 0.068 and 0.143 in the haze period and non-haze period of the suburban site, respectively. These results were in the range of 0.03 to 0.15 in many studies in China.[3,25,26,36] It could be seen that the $HONO_{corr}/NO_2$ ratios of urban sites were much lower in both the haze and non-haze periods. In addition, $HONO_{corr}/NO_2$ ratios in the non-haze period increased when compared to the haze period ones. The higher ratios of $HONO_{corr}/NO_2$ may connect with a higher conversion efficiency of NO_2 to HONO.

The HONO conversion frequency C_{HONO} was calculated by eqn (1)[3,32]

$$C_{HONO} = ([HONO_{corr}]_{t2} - [HONO_{corr}]_{t1})/(t2 - t1)[NO_2] \quad (1)$$

HONO concentrations were corrected by subtracting contributions of direct emissions.

As the nocturnal boundary layer was changing to be stable, the production and loss rates of HONO would finally reach a steady state, when the heterogeneous conversion of NO_2 would become dominant. Therefore, data from the second half of the night were used when NO_2 concentrations were stable and the conversion of NO_2 was steady. The average value of C_{HONO} was 0.0055 and 0.013 h^{-1} in the urban site during the haze and non-haze periods, respectively, and 0.008 and 0.016 h^{-1} in the suburban site during the haze and non-haze periods, respectively. C_{HONO} values obtained from the urban site showed typical urban values seen in Beijing,[22] Guangzhou,[25] Mainz and Milan[10] (0.004 to 0.012 h^{-1}), while the C_{HONO} values of the suburban site showed both farmland,[8,32] mountain and forest[38] characteristics which were in the range of 0.006 to 0.010 h^{-1} and 0.014 to 0.024 h^{-1}, respectively. This agrees with the discussions on $HONO_{corr}/NO_2$ ratios, which reflected higher conversion rates in suburban sites and during the non-haze period. The NO_2–HONO conversion rate was affected by several factors including the environment, particle concentrations and aerosol surfaces,[14,40] it is better to consider these parameters when exploring the heterogeneous conversion of NO_2.

Impact of $PM_{2.5}$ and RH. The heterogeneous conversion of NO_2 to HONO could occur on various types of surfaces as well as aerosols with the extinction of water, thus the impact of aerosol surfaces and surface water on the NO_2 conversion rates should be focused on. As already shown by several lab studies, the conversion was first order in NO_2 [40] and in functions of relative humidity[18] and surface area to volume ratio (S/V).[14]

Daytime data were ignored in order to avoid the influence of photolysis. HONO concentrations were corrected and denoted as $HONO_{corr}$, as well. Correlations of $HONO_{corr}$ and $HONO_{corr}/NO_2$ ratios with $PM_{2.5}$ and RH are illustrated in Fig. 8. This showed that the relationship between $HONO_{corr}/NO_2$ and $PM_{2.5}$ was obscure, especially when $PM_{2.5}$ concentrations were higher than 160 μg m^{-3}. However, the correlation coefficient of $HONO_{corr}$ with $PM_{2.5}$ (0.514) still showed some promotion effects. Heterogeneous conversion of NO_2 would occur on ground surfaces as well as aerosols, the higher surface density of ground was often supposed to play a main role in HONO formation over the aerosol surface. While, our measurement results showed that compared to the ground surface the high particulate matter concentrations during the haze period would promote the heterogeneous conversion of NO_2. Similarly, the correlations of $HONO_{corr}$ and $HONO_{corr}/NO_2$ with RH showed RH would promote the conversion of HONO, though

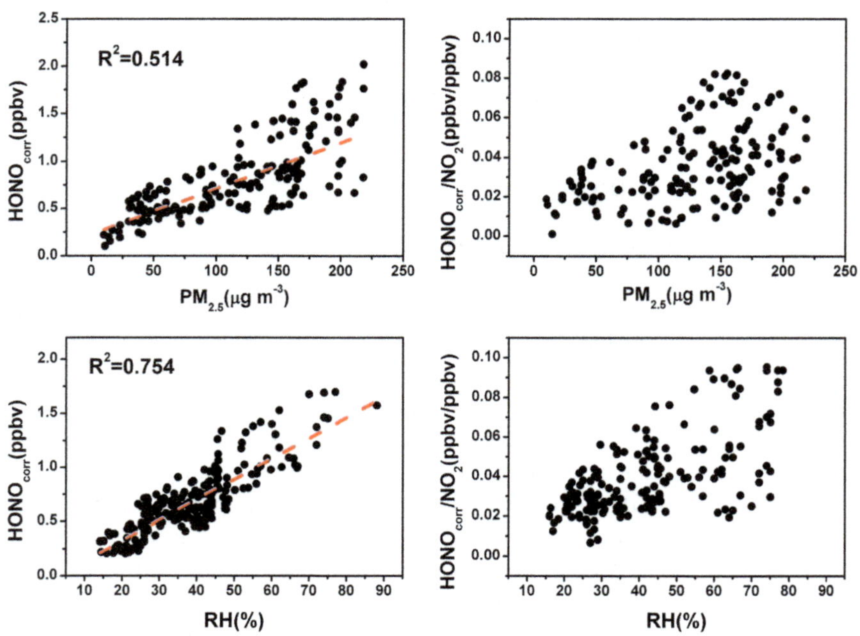

Fig. 8 Correlations of $HONO_{corr}$ and $HONO_{corr}/NO_2$ ratios with $PM_{2.5}$ (top) and correlations of $HONO_{corr}$ and $HONO_{corr}/NO_2$ ratios with RH (bottom).

$HONO_{corr}/NO_2$ ratios increased with growing RH and decreased when RH was higher than 60%. Additionally, the good correlations of $HONO/NO_2$ with particles[25,41,42] and RH[18,25] have been shown in many previous studies.

However, measurements by Kleffmann *et al.*[38] and Stutz *et al.*[18] showed weak aerosol impact on HONO formation, and this suggestion was supported by laboratory studies on soot and organic particles. As boundary layer height (BLH) decreased in the nighttime, this led to the increasing of the *S*/*V* ratio and concentrations of other gas-phase pollutants.[37,43] Meanwhile, RH would increase as well due to the decreasing of temperature. Therefore, the good correlations will be affected by these factors. And the impacts of aerosols and RH vary with environmental conditions and are still in debate. Given this measurement only provides a rough study on the impacts of particulate matter, gradient measurements of HONO with NO_2 and aerosol concentrations are highly recommended to prove the possible impacts of aerosols on HONO formation.[43]

The impacts of gas phase reaction between NO and OH. As discussed above, the known sources of HONO include direct emissions, homogeneous formation and heterogeneous conversions. Therefore, the impacts of reaction between NO and OH should not be neglected. The homogeneous production of nocturnal HONO was dominated by the reaction NO + OH → HONO ($k_{NO+OH} = 7.4 \times 10^{-12}$ cm^3 molecules^{-1} s^{-1}) and HONO + OH → NO_2 + H_2O ($k_{HONO+OH} = 6.0 \times 10^{-12}$ cm^3 molecules^{-1} s^{-1}), and the net HONO homogeneous production could be calculated by:[3,8,38]

$$P_{net} = k_{OH+NO}[OH][NO] - k_{OH+HONO}[OH][HONO] \tag{2}$$

where [ ] represents the concentrations of gases. For comparison, a nighttime OH concentration of 1×10^6 molecules per cm^3 was assumed to represent the measured ones in both urban and suburban areas as supported by previous studies[44–47] in Chinese urban areas during winter time.

In the urban site, the value of P_{net} was tightly connected with the NO concentration. When nocturnal NO was in the range of 52.7 to 141.4 ppbv in the haze period, the mean value of P_{net} could be 2.02 ppbv h^{-1} (Table 2). This showed the production of net NO + OH reaction would supply 59.50% of the $HONO_{corr}$ enhancement on average. While in the non-haze period, NO concentrations varied little with an average value of 12.33 ppbv, leading to a mean P_{net} of 0.32 ppbv h^{-1}. However, it still provided 43.92% HONO enhancement on average. In the case of the suburban area, the nocturnal NO concentration was too low to

Table 2 The nocturnal HONO budget during haze and non-haze periods in the urban and suburban sites

Location	Time period	P_{net}[a]	P_{heter}	$P_{heter}/[NO_2]$	C_{HONO}
Urban	Haze	2.02	0.070	0.002	0.005
Urban	Non-haze	0.32	0.013	0.010	0.013
Suburban	Haze	0.0051	0.058	0.005	0.008
Suburban	Non-haze	0.0023	0.025	0.019	0.016

[a] Units of P_{net}, P_{heter}, $P_{heter}/[NO_2]$ and C_{HONO} are ppbv h^{-1}, ppbv h^{-1}, h^{-1} and h^{-1}, respectively.

support HONO increase with a mean P_{net} value of 0.026 ppbv h^{-1}, which was an order of magnitude lower than the urban site. Even in the haze period, when NO concentrations increased from 1.2 ppbv to 6.4 ppbv, the P_{net} value was only 0.0051 ppbv h^{-1} on average. This indicated that the effects of homogeneous reactions did not work in the suburban area, and heterogeneous formation may be a critical supplement to nocturnal HONO concentrations, in light of the low contribution of direct emissions, which has been discussed above.

It is well accepted that the major HONO removal pathway is dry deposition during the night. HONO removal by dry deposition (L_{dep}) can be calculated by $V_d \times [HONO]/H$, where V_d represents dry deposition velocity of HONO and H is the mixing height. V_d was assumed to be 2.0 cm s^{-1} as used by Harrison *et al.*[47] Because the wind speed was low, horizontal and vertical transport could be neglected. The L_{dep} values were supposed to be 0.035 and 0.015 ppbv h^{-1} on average for the haze period and non-haze period, respectively. This means that at least 0.070, 0.013, 0.058 and 0.025 ppbv h^{-1} of heterogeneous production was needed after subtracting the contributions from direct emissions in the haze period in the urban area, non-haze period in the urban area, haze period in the suburban area and non-haze period in the suburban area, respectively. The results were in agreement with the C_{HONO} values of 0.005, 0.013, 0.008 and 0.016 h^{-1} calculated above.

Take the nights of Dec 18 and Dec 19 for example:

In the urban site, the integrated P_{net} values from 20:00 to 2:00 were 0.417 ppbv with OH concentrations of 1.0×10^6 molecules per cm^3, which was 69.14% of the measured increase of HONO after subtracting direct emissions (from 0.74 ppbv to 1.47 ppbv) for the night of Dec 18 (Fig. 9). During the night of Dec 19, the low NO concentrations could only provide a P_{net} of 0.292 ppbv h^{-1} with an integrating area of 0.048 ppbv, which was 43.67% of the HONO increase (0.21–0.32 ppbv). Thus, the heterogeneous productions of these two nights were at least 0.032 and

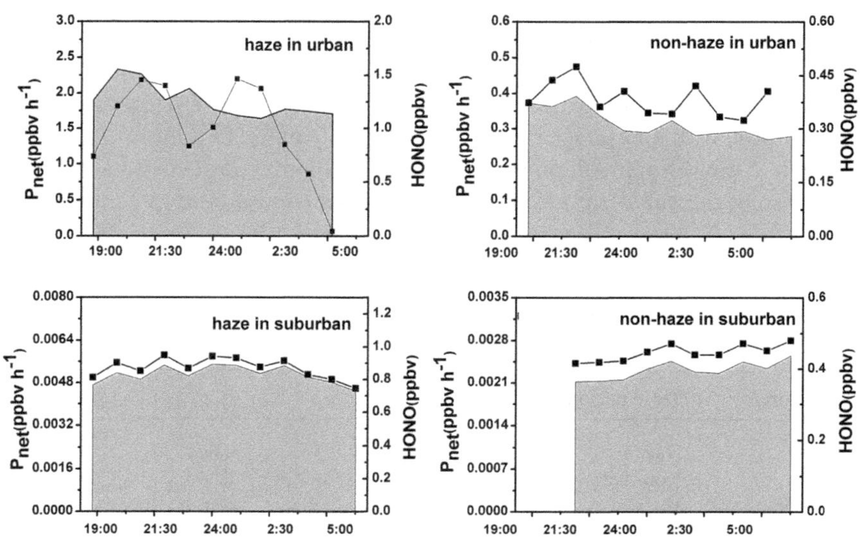

Fig. 9 Variations patterns of P_{net} and P_{heter} in the night of Dec 18 (haze period) and Dec 19 (non-haze) period in urban and suburban sites.

0.013 h^{-1}, respectively. In the same way, the calculated heterogeneous production P_{heter} was 0.058 and 0.025 ppbv h^{-1} for haze and non-haze periods in the suburban area. The heterogeneous production rates were almost 10 times higher than net homogeneous production. This suggested that the net production by homogeneous reactions could be dominant for the nocturnal increase of HONO in the urban site while production by heterogeneous reactions would be the major source of HONO in the suburban site.

In summary, the nocturnal HONO formation mechanisms in urban and suburban areas were different. The contributions of direct emissions were 48.8% and 10.3% in urban and suburban sites, respectively. Moreover, due to the rather high NO and NO_x concentrations in the urban area, the production of homogeneous reactions during both haze and non-haze periods were unneglectable with production rates of 2.02 and 0.32 ppbv h^{-1}, respectively. In the case of the suburban area, the concentrations of NO and NO_x were so low as to make the contributions of homogeneous production and direct emissions neglectable, which led to the heterogeneous reactions becoming the major source of HONO. Considering the removal of HONO by dry deposition, the heterogeneous production rates were calculated to be at least 0.070, 0.013, 0.058 and 0.025 ppbv h^{-1} for the haze period of the urban area, non-haze period of the urban area, haze period of the suburban area and non-haze period of the suburban area, respectively.

Conclusions

Measurements of HONO and other pollutants were performed in urban and suburban sites to explore the HONO formation mechanisms from December 12 to December 22 in Beijing. The measurements included haze and non-haze processes in both urban and suburban sites.

The hourly-averaged concentrations of HONO, CO, SO_2, O_3, NO, NO_2 and NO_x in the haze period were 0.86 ppbv, 1.54 ppmv, 12.0 ppbv, 9.4 ppbv, 33.1 ppbv, 16.5 ppbv, and 49.6 ppbv, respectively, much higher than concentrations in the suburban area (0.52 ppbv, 1.05 ppmv, 0.8 ppbv, 14.4 ppbv, 3.6 ppbv, 6.5 ppbv, and 11.1 ppbv). Moreover, HONO concentrations during the haze period were 2.63 times higher than in the non-haze period in the urban site, and 1.80 times higher during haze in the suburban site. Correlation studies showed HONO was highly connected with CO, NO_x, and SO_2, and the correlations varied with the locations and the environmental conditions. The results indicated that direct emission was the main factor affecting correlations of HONO with CO and NO_x in the urban site. Moreover, studies of $HONO_{emissions}$ (HONO from direct emissions) showed mean direct emissions contributions of 48.8% in the urban area, but were 10.3% in the suburban area, reflecting a less significant contribution.

Additionally, to evaluate the nocturnal heterogeneous conversion of NO_2, both $HONO_{corr}/NO_2$ ratios and C_{HONO} rates were calculated. The results agreed with each other and showed that higher $HONO_{corr}/NO_2$ ratios indicated a higher NO_2 conversion efficiency in the non-haze period in both areas. Moreover, the correlations of $HONO_{corr}/NO_2$ *vs.* RH and $HONO_{corr}/NO_2$ *vs.* $PM_{2.5}$ concentrations showed RH and particle concentrations may be important factors affecting the NO_2 conversion. $HONO_{corr}/NO_2$ ratios were proportional with RH and $PM_{2.5}$ concentrations when RH was below 60% and $PM_{2.5}$ concentrations were lower than 160 μg m^{-3}. However, $HONO_{corr}/NO_2$ ratios fell down when RH was higher than 60% and $PM_{2.5}$ concentrations were higher than 160 μg m^{-3}. Finally, the nocturnal productions from homogeneous reactions and heterogeneous reactions were compared. Due to the extremely high concentrations of NO in urban areas, the contributions of homogeneous reaction production would be significant in both haze and non-haze periods. While in the suburban area, the production from heterogeneous conversion was the major contributor, about 10 times higher than net homogeneous production in haze and non-haze periods.

Acknowledgements

This project was supported by the Strategic Priority Research Program (B) of the Chinese Academy of Sciences (Grant No. XDB05010400), and the National Natural Science Foundation of China (Contract No. 41475114, 21477134).

References

1 K. D. Lu, F. Rohrer, F. Holland, H. Fuchs, B. Bohn, T. Brauers, C. C. Chang, R. Häseler, M. Hu, K. Kita, Y. Kondo, X. Li, S. R. Lou, S. Nehr, M. Shao, L. M. Zeng, A. Wahner, Y. H. Zhang and A. Hofzumahaus, *Atmos. Chem. Phys.*, 2012, **12**, 1541–1569.

2 J. Kleffmann, *ChemPhysChem*, 2007, **8**, 1137–1144.

3 B. Alicke, U. Platt and J. Stutz, *J. Geophys. Res.: Atmos.*, 2002, **107**, DOI: 10.1029/2000JD000075.

4 K. Acker, D. Möller, W. Wieprecht, F. X. Meixner, B. Bohn, S. Gilge, C. Plass-Dülmer and H. Berresheim, *Geophys. Res. Lett.*, 2006, **33**, L02809.

5 B. Aumont, F. Chervier and S. Laval, *Atmos. Environ.*, 2003, **37**, 487–498.

6 B. Alicke, A. Geyer, A. Hofzumahaus, F. Holland, S. Konrad, H. W. Pätz, J. Schäfer, J. Stutz, A. Volz-Thomas and U. Platt, *J. Geophys. Res.: Atmos.*, 2003, **108**, 8247.

7 V. Michoud, A. Colomb, A. Borbon, K. Miet, M. Beekmann, M. Camredon, B. Aumont, S. Perrier, P. Zapf, G. Siour, W. Ait-Helal, C. Afif, A. Kukui, M. Furger, J. C. Dupont, M. Haeffelin and J. F. Doussin, *Atmos. Chem. Phys.*, 2014, **14**, 2805–2822.

8 X. Li, T. Brauers, R. Häseler, B. Bohn, H. Fuchs, A. Hofzumahaus, F. Holland, S. Lou, K. D. Lu, F. Rohrer, M. Hu, L. M. Zeng, Y. H. Zhang, R. M. Garland, H. Su, A. Nowak, A. Wiedensohler, N. Takegawa, M. Shao and A. Wahner, *Atmos. Chem. Phys.*, 2012, **12**, 1497–1513.

9 J. Kleffmann, T. Gavriloaiei, A. Hofzumahaus, F. Holland, R. Koppmann, L. Rupp, E. Schlosser, M. Siese and A. Wahner, *Geophys. Res. Lett.*, 2005, **32**, L05818.

10 G. Lammel and J. N. Cape, *Chem. Soc. Rev.*, 1996, **25**, 361–369.

11 R. Kurtenbach, K. H. Becker, J. A. G. Gomes, J. Kleffmann, J. C. Lörzer, M. Spittler, P. Wiesen, R. Ackermann, A. Geyer and U. Platt, *Atmos. Environ.*, 2001, **35**, 3385–3394.

12 T. W. Kirchstetter, R. A. Harley and D. Littlejohn, *Environ. Sci. Technol.*, 1996, **30**, 2843–2849.

13 C. George, R. S. Strekowski, J. Kleffmann, K. Stemmler and M. Ammann, *Faraday Discuss.*, 2005, **130**, 195–210.

14 B. J. Finlayson-Pitts, L. M. Wingen, A. L. Sumner, D. Syomin and K. A. Ramazan, *Phys. Chem. Chem. Phys.*, 2003, **5**, 223–242.

15 J. G. Calvert, G. Yarwood and A. M. Dunker, *Res. Chem. Intermed.*, 1994, **20**, 463–502.

16 M. Ammann, M. Kalberer, D. T. Jost, L. Tobler, E. Rössler, D. Piguet, H. W. Gaggeler and U. Baltensperger, *Nature*, 1998, **395**, 157–160.

17 V. H. Grassian, *Int. Rev. Phys. Chem.*, 2001, **20**, 467–548.

18 J. Stutz, B. Alicke, R. Ackermann, A. Geyer, S. H. Wang, A. B. White, E. J. Williams, C. W. Spicer and J. D. Fast, *J. Geophys. Res.: Atmos.*, 2004, **109**, D03307.

19 K. Stemmler, M. Ammann, C. Donders, J. Kleffmann and C. George, *Nature*, 2006, **440**, 195–198.

20 R. Broske, J. Kleffmann and P. Wiesen, *Atmos. Chem. Phys.*, 2003, **3**, 469–474.

21 M. Sörgel, E. Regelin, H. Bozem, J. M. Diesch, F. Drewnick, H. Fischer, H. Harder, A. Held, Z. Hosaynali-Beygi, M. Martinez and C. Zetzsch, *Atmos. Chem. Phys.*, 2011, **11**, 10433–10447.

22 F. Spataro, A. Ianniello, G. Esposito, I. Allegrini, T. Zhu and M. Hu, *Sci. Total Environ.*, 2013, **447**, 210–224.

23 F. Hendrick, J. F. Müller, K. Clémer, P. Wang, M. De Maziere, C. Fayt, C. Gielen, C. Hermans, J. Z. Ma, G. Pinardi, T. Stavrakou, T. Vlemmix and M. van Roozendael, *Atmos. Chem. Phys.*, 2014, **14**, 765–781.

24 M. Qin, P. H. Xie, W. Q. Liu, A. Li, K. Dou, W. Fang, H. G. Liu and W. J. Zhang, *J. Environ. Sci.*, 2006, **18**, 69–75.

25 M. Qin, P. Xie, H. Su, J. Gu, F. Peng, S. Li, L. Zeng, J. Liu, W. Liu and Y. Zhang, *Atmos. Environ.*, 2009, **43**, 5731–5742.

26 S. Wang, R. Zhou, H. Zhao, Z. Wang, L. Chen and B. Zhou, *Atmos. Environ.*, 2013, **77**, 718–724.

27 S. Hou, S. Tong, M. Ge and J. An, *Atmos. Environ.*, 2016, **124**, 199–206.

28 S. R. Tong, S. Q. Hou, Y. Zhang, B. W. Chu, Y. C. Liu, H. He, P. S. Zhao and M. F. Ge, *Sci. China: Chem.*, 2015, **58**, 1393–1402.

29 G. Huang, X. L. Zhou, G. H. Deng, H. C. Qiao and K. Civerolo, *Atmos. Environ.*, 2002, **36**, 2225–2235.

30 J. Kleffmann, J. C. Lörzer, P. Wiesen, C. Kern, S. Trick, R. Volkamer, M. Rodenas and K. Wirtz, *Atmos. Environ.*, 2006, **40**, 3640–3652.

31 J. Heland, J. Kleffmann, R. Kurtenbach and P. Wiesen, *Environ. Sci. Technol.*, 2001, **35**, 3207–3212.

32 H. Su, Y. F. Cheng, P. Cheng, Y. H. Zhang, S. Dong, L. M. Zeng, X. Wang, J. Slanina, M. Shao and A. Wiedensohler, *Atmos. Environ.*, 2008, **42**, 6219–6232.

33 M. Levy, R. Y. Zhang, J. Zheng, A. L. Zhang, W. Xu, M. Gomez-Hernandez, Y. Wang and E. Olaguer, *Atmos. Environ.*, 2014, **94**, 231–240.

34 J. Quan, X. Tie, Q. Zhang, Q. Liu, X. Li, Y. Gao and D. Zhao, *Atmos. Environ.*, 2014, **88**, 83–89.

35 Y. L. Sun, Q. Jiang, Z. F. Wang, P. Q. Fu, J. Li, T. Yang and Y. Yin, *J. Geophys. Res.: Atmos.*, 2014, **119**, 4380–4398.

36 K. D. Lu, Y. H. Zhang, H. Su, T. Brauers, C. C. Chou, A. Hofzumahaus, S. C. Liu, K. Kita, Y. Kondo, M. Shao, A. Wahner, J. L. Wang, X. S. Wang and T. Zhu, *J. Geophys. Res.: Atmos.*, 2010, **115**, 18.

37 Y. Yu, B. Galle, A. Panday, E. Hodson, R. Prinn and S. Wang, *Atmos. Chem. Phys.*, 2009, **9**, 6401–6415.

38 J. Kleffmann, R. Kurtenbach, J. Lörzer, P. Wiesen, N. Kalthoff, B. Vogel and H. Vogel, *Atmos. Environ.*, 2003, **37**, 2949–2955.
39 A. Febo, C. Perrino and M. Cortiello, *Atmos. Environ., Part A*, 1993, **27**, 1721–1728.
40 J. Kleffmann, K. H. Becker and P. Wiesen, *Atmos. Environ.*, 1998, **32**, 2721–2729.
41 J. L. An, W. Zhang and Y. Qu, *Atmos. Environ.*, 2009, **43**, 3454–3459.
42 S. S. Park, S. B. Hong, Y. G. Jung and J. H. Lee, *Atmos. Environ.*, 2004, **38**, 293–304.
43 N. Hao, B. Zhou, D. Chen and L.-m. Chen, *J. Environ. Sci.*, 2006, **18**, 910–915.
44 J. P. Pinto, J. Dibb, B. H. Lee, B. Rappenglück, E. C. Wood, M. Levy, R. Y. Zhang, B. Lefer, X. R. Ren, J. Stutz, C. Tsai, L. Ackermann, J. Golovko, S. C. Herndon, M. Oakes, Q. Y. Meng, J. W. Munger, M. Zahniser and J. Zheng, *J. Geophys. Res.: Atmos.*, 2014, **119**, 5583–5601.
45 K. D. Lu, A. Hofzumahaus, F. Holland, B. Bohn, T. Brauers, H. Fuchs, M. Hu, R. Häseler, K. Kita, Y. Kondo, X. Li, S. R. Lou, A. Oebel, M. Shao, L. M. Zeng, A. Wahner, T. Zhu, Y. H. Zhang and F. Rohrer, *Atmos. Chem. Phys.*, 2013, **13**, 1057–1080.
46 X. Ren, H. Gao, X. Zhou, J. D. Crounse, P. O. Wennberg, E. C. Browne, B. W. LaFranchi, R. C. Cohen, M. McKay, A. H. Goldstein and J. Mao, *Atmos. Chem. Phys.*, 2010, **10**, 6283–6294.
47 R. M. Harrison, J. D. Peak and G. M. Collins, *J. Geophys. Res.: Atmos.*, 1996, **101**, 14429–14439.

Faraday Discussions

PAPER

OH reactivity in urban and suburban regions in Seoul, South Korea – an East Asian megacity in a rapid transition

Saewung Kim,*[a] Dianne Sanchez,[a] Mark Wang,[a] Roger Seco,[a] Daun Jeong,[a] Stacey Hughes,[b] Barbara Barletta,[b] Donald R. Blake,[b] Jinsang Jung,[c] Deugsoo Kim,[d] Gangwoong Lee,[e] Meehye Lee,[f] Joonyoung Ahn,[g] Sang-Deok Lee,[g] Gangnam Cho,[g] Min-Young Sung,[g] Yong-Hwan Lee,[g] Dan Bi Kim,[g] Younha Kim,[h] Jung-Hun Woo,[h] Duseong Jo,[i] Rokjin Park,[i] Jeong-Hoo Park,[g] You-Deog Hong[g] and Ji-Hyung Hong[g]

Received 20th December 2015, Accepted 11th February 2016

DOI: 10.1039/c5fd00230c

South Korea has recently achieved developed country status with the second largest megacity in the world, the Seoul Metropolitan Area (SMA). This study provides insights into future changes in air quality for rapidly emerging megacities in the East Asian region. We present total OH reactivity observations in the SMA conducted at an urban Seoul site (May–June, 2015) and a suburban forest site (Sep, 2015). The total OH reactivity in an urban site during the daytime was observed at similar levels (~15 s^{-1}) to those previously reported from other East Asian megacity studies. Trace gas observations indicate that OH reactivity is largely accounted for by NO_X (~50%) followed by volatile organic compounds (VOCs) (~35%). Isoprene accounts for a substantial fraction of OH reactivity among the comprehensive VOC observational dataset (25–47%). In general, observed total OH reactivity can be accounted for by the observed trace gas dataset. However, observed total OH reactivity in the suburban forest area cannot be largely accounted for (~70%) by the trace gas measurements. The importance of biogenic VOC (BVOCs) emissions and oxidations used to evaluate the

[a]Department of Earth System Science, University of California, Irvine, Irvine CA 92697, USA. E-mail: saewungk@uci.edu; Tel: +1-949-824-4531

[b]Department of Chemistry, University of California, Irvine, Irvine CA 92697, USA

[c]The Division of Metrology for Quality of Life, Korea Research Institute of Standards and Science, Daejeon, South Korea 34113

[d]Department of Environmental Engineering, Kunsan National University, Kunsan, South Korea 573-701

[e]Department of Environmental Sciences, Hankuk University of Foreign Studies, Yongin, South Korea 449-791

[f]Department of Earth and Environmental Sciences, Korea University, Seoul, South Korea 02841

[g]Department of Climate and Air Quality, National Institute of Environmental Research, Incheon, South Korea 22689

[h]Division of Interdisciplinary Studies, Konkuk University, Seoul, South Korea 05025

[i]School of Earth and Environmental Sciences, Seoul National University, Seoul, South Korea 08826

impacts of East Asian megacity outflows for the regional air quality and climate contexts are highlighted in this study.

1. Introduction

One of the many consequences of the industrial revolution was urbanization. It was initiated in Europe and has spread out to the U.S. and other parts of the world such as Asia and South America.[1] Accelerating technological breakthroughs since the industrial revolution have brought different natures of environmental problems to different parts of the world.[2] For example, coal powered urban areas with limited emission control technologies in the late 1800s and early 1900s suffered public health issues caused by sulfur and soot pollution. A strong boundary layer inversion event that often happened in winter could trigger a massive smog event. In December of 1952, the Great Smog event in London caused ~100 000 casualties including 4000 premature deaths in only four days.[3,4]

A different type of air pollution was reported in Southern California in the early to mid-1900s. The successful implementation of factory smoke emission controls between 1905 and 1912 could not improve the air quality in the Los Angeles area as the size of the automobile fleet had been exponentially increasing.[4] In 1940, the research community realized that nitrogen oxides ($NO_X = NO + NO_2$) and volatile organic compounds (VOCs) emitted from tail pipes became precursors for photochemical reactions triggered by the intense solar radiation in Southern California to produce ozone and particles in the atmosphere.[5,6] Unique local meteorological and topographical characteristics such as a land-sea breeze pattern and a basin structure worsen the photochemical ozone and particle pollution issues in the Los Angeles basin.[4,7] This well known photochemical smog, also known as Los Angeles type smog, has became an issue in most cities in developing and developed countries as automobile use has increased in the mid to late 20th century.[4]

East Asian countries such as China, Japan, South Korea and Taiwan have emerged as important players in the global economy in the mid to late-20th century. Major economic indices show that this development is in its most mature stage in Japan and in a rapidly growing stage in China. The recent establishment of South Korea as a developed country has led to the implementation of aggressive air pollutant emission abatement policies in the past decade.[8] Geographically, South Korea is adjacent to China across the Yellow Sea. Strong continental outflow events often heavily impact the air quality on the Korean Peninsula.[9] The Seoul Metropolitan Area (SMA), the main urban region in South Korea, is home to 23 million inhabitants, approximately half of the total South Korean population. The majority of the land area (65%) in South Korea is covered by forests.[10] The roles of biogenic volatile organic compounds (BVOCs) from the surrounding forest in the SMA have been extensively discussed in the context of ozone and aerosol formation in both models and measurements. Observational results from Taehwa Research Forest (TRF), 30 km southeast from the Seoul city center indicate that isoprene is the dominant hydroxyl radical sink among the observed reactive gas species.[11–13] A regional model evaluation determined that 30–40% of ozone formation can be accounted for by isoprene photochemistry in the suburban region of the SMA. This suggests that anthropogenic and biogenic interactions need to be addressed when discussing regional air quality issues.[13] Ultra high-resolution regional modeling studies consistently show that the circulation pattern in the SMA brings isoprene and its oxidation products to the city center, leading to ozone enhancements.[14,15] Multiple reports have highlighted the critical roles of BVOCs in East Asian regional photochemistry. Bao and colleagues[16] presented CMAQ model simulation results indicating ozone enhancements of 20 to 26 ppb due to isoprene photochemistry in the Kinki region, Japan. In China, consistent reports emphasizing the importance of BVOCs over strong anthropogenic influences have been published from the Beijing-Tianjin area,[17] the Shanghai region,[18] and the Pearl River Delta region.[19]

Recently, at least two different analytical techniques have become available to directly constrain total OH reactivity (s^{-1}, reciprocal of OH lifetime) in the atmosphere.[20–22] It has long been suspected that our ability to constrain all the reactive trace gases in the atmosphere is limited, especially for VOCs as many species with different compositions and structural variations with various functional groups are emitted to the atmosphere. There is an estimated 10^4 to 10^5 VOCs in the atmosphere, but this is speculated to be a very conservative assessment, especially in the boundary layer.[23] Direct total OH reactivity observations provide quantitative constraints to this missing fraction, which is called missing OH reactivity. The majority of OH reactivity observations in environments with high BVOC concentrations have reported substantial missing OH reactivity (30–90%).[19,22,24–28] There have been two studies regarding total OH reactivity in East Asian megacity regions – Tokyo, Japan[29,30] and the Pearl River Delta, China.[19] Both observations commonly reported missing OH reactivity and speculated the main sources as unaccounted oxygenated VOCs (OVOCs).

We present total OH reactivity observations collected in the Seoul city center during late spring (May 20th to June 15th) and in the TRF in early fall (September 1st to September 15th) of 2015. Therefore, the observational dataset from the city center provides a good snapshot of photochemical environments with a heavy pollution influence. In contrast, the early fall observations were conducted at the TRF where BVOC photochemistry plays an important role in regional air quality in the aged pollution background. These measurements were conducted as part of the multi-year research initiative Megacity Air Pollution Study (MAPS)-Seoul which is designed to improve the level of understanding of emissions, atmospheric transformation, transport and removal processes over the Korean Peninsula. In 2016, an international collaboration aboard the NASA DC-8 airborne laboratory will take place as part of the KORUS-AQ (Korea-US Air Quality Study) field campaign (https://www.espo.nasa.gov/home/korus-aq/content/KORUS-AQ). This discussion provides an opportunity to critically review the current status of atmospheric reactive gas distributions in the urban and the suburban regions in the SMA. Moreover, as most East Asian megacities are characterized by dense population centers surrounded by rural forests, this case study of the SMA serves as a basis to predict future changes in air quality during the rapid development of East Asia that directly affect anthropogenic emission profiles.

2. Experimental

2.1. Total OH reactivity, auxiliary trace gas and meteorological parameters observations

Total OH reactivity was measured using a comparative reactivity method (CRM) – chemical ionization mass spectrometry (CIMS) which has been previously used to study urban,[22] clean,[25,31] and pristine[22] field environments. A glass flow reactor (~95 ml) was used which consists of two inlets introducing a mixture of ambient air and pyrrole (5 sccm in 5 ppm ± 5%, Air Liquid LLC) and humidified UHP N_2 (200 sccm) for OH generation *via* photolysis from a UV lamp (UVP LLC, Model 11SC-1). The flow rate in the reactor is maintained at 280 sccm by a MKS mass flow controller (M100B). A volume of 50 sccm is introduced into a custom built H_3O^+–CIMS system.[32] Total OH reactivity of ambient air is then experimentally determined by tracing the pyrrole signal in the reactor. The difference in the pyrrole signal with ambient air and ambient air scrubbed by a catalytic converter (Pt-wool kept at 350 °C) is caused by the competition between pyrrole and reactive gases in the ambient air for reactions with OH. Since OH and pyrrole concentrations are controlled by the experimental setup, the change in reactive gas content in the ambient air is the single variable driving the pyrrole signal change. Calibration was conducted in the field and laboratory using a synthetic air sample using a zero-air flow (1 slpm) with an addition of the NIST traceable standard gases (Airgas) in multiple flow rates to generate multi-point calibration curves. We tested with propane (C_3H_8), propene (C_3H_6), and isoprene (C_5H_8) standards for multi-point calibrations. The results indicate a good linearity ($R^2 = 0.998$) over a wide OH reactivity range from 5 s^{-1} to 100 s^{-1}. We estimate the observational uncertainty as 35% (3 sigma) and the lower limit of detection as 5 s^{-1} (3 sigma). We are currently preparing a separate manuscript with detailed characterizations of the CRM-CIMS instrumentation deployed.

Whole air samples[33] were collected for speciated VOC observations during the spring field campaign in Seoul. It was conducted for 12 days twice a day – in the morning (9:00 to 10:00 local time) and the afternoon (14:00 to 15:00 local time), for a total of 24 samples. Continuous VOC observations were also conducted using a proton transfer reaction-time of flight-mass spectrometer (PTR-ToF-MS, IONICON GmbH) system. The time series data for benzene, toluene, xylene, isoprene and acetone from PTR-ToF-MS indicate that canister sampling time periods are representative in the morning and afternoon as no significant spikes from local sources of these compounds are observed (a separate manuscript is under preparation). The analytical characteristics of other observational parameters including trace gases and meteorological parameters for both the city center and TRF are summarized in Table 1.

2.2. Emission and regional chemistry modeling

For emission analysis the National Institute of Environmental Research/Konkuk University-Comprehensive Regional Emissions inventory for Atmospheric Transport Experiments (NIER/KU-CREATE; CREATE)[34] was used. CREATE is a bottom-up emission inventory framework for Asia jointly developed by the National Institute of Environmental Research of South Korea and Konkuk University. This inventory includes anthropogenic, biogenic and biomass burning emission inventories from 2010, and an emission processing system to generate air quality model-ready gridded, temporally allocated, and chemically speciated emissions. The CREATE anthropogenic emissions were originally developed using the GAINS framework (http://www.gains.iiasa.ac.at/gains), allowing us to project base year (2010) emissions into the future using changing energy/control scenarios. We have used base year anthropogenic emissions for China, Korea, and

Table 1 Analytical characteristics for the trace gas and meteorological observations for the preMAPS campaign

Observables	Manufacturer and model number	Uncertainty	Lower limit of detection
CO	Thermo Scientific 48i TLE	10%	40 ppb
NO_X	Thermo Scientific 42i-TL with a photolysis converter	15%	50 ppt
SO_2	Thermo Scientific 43i-TLE	10%	50 ppt
Ozone	Thermo Scientific 49i	5%	<1 ppb
VOCs	Whole air sample with GC-MS Analysis (at UCI)[a]	1–5%	1–5 ppt
	PTR-ToF-MS[b]	12%	20 ppt
Meteorological parameters	LSI LASTEM meteorological sensors		

[a] More details are described in Colman *et al.*[51] [b] More details are described in Kim *et al.*[12]

Japan, from the CREATE inventory. The VOCs were chemically speciated using the SAPRC chemical mechanism (http://www.engr.ucr.edu/~carter/SAPRC/) and then categorized into alkane, alkene, aromatics and isoprene species. Total emissions were separated by country emissions of CO, NO_X, SO_2, alkane, alkene, aromatics, and isoprene.

Model simulations for the observation periods were conducted using the GEOS-Chem model version v9-02, which was driven by Goddard Earth Observing System-Forward Processing (GEOS-FP) assimilated meteorological data from the NASA Global Modeling and Assimilation Office (GMAO). The GEOS-FP meteorological data have a native horizontal resolution of $0.25° \times 0.3125°$ ($\sim 25 \times 25$ km^2) with 72 vertical pressure levels and 3 h temporal frequency (1 h for surface variables and mixed layer depths). In order to minimize the amount of memory required, we reduced the number of vertical levels to 47 by merging layers in the stratosphere. We use a nested version of GEOS-Chem with the native spatial resolution ($0.25° \times 0.3125°$) over Asia (70°E–140°E, 15°N–55°N). The lateral boundary conditions of the nested model are updated every 3 h from a global simulation at $2° \times 2.5°$ horizontal resolution. Detailed descriptions of gas and aerosol simulations can be found in the literature[35–39]

Biogenic VOC emissions are from the Model of Emissions of Gases and Aerosols from Nature (MEGAN) v2.1 inventory.[40] The emissions of these compounds are calculated on the basis of emission activity factors, canopy environment, leaf age, and soil moisture. Biogenic emissions of isoprene and monoterpenes over nested domains are 1.37 TgC per per month and 0.56 TgC per month, respectively. Anthropogenic emissions in Asia are obtained from the Intercontinental Chemical Transport Experiment Phase B (INTEX-B) inventory,[41] which was compiled on the basis of fuel use statistics for 2006 but scaled to 2015. INTEX-B includes four source sectors: power, industry, residential, and transportation. NO_X and anthropogenic VOC emissions were determined to be 0.62 TgN per month and 1.23 TgC per month, respectively. For biomass burning emissions, we use the Global Fire Emission Database version 3 (GFED v3) inventory for 2011.[42] The GFED inventory is a calculated biogeochemical model and satellite-derived estimate of area burned, fire activity, and plant productivity. Biomass burning emissions of NO_X and VOCs are 1.7 GgN per month and 5.3 GgC per month, respectively.

2.3. Observational sites

The spring observation was conducted in Korea Institute of Science and Technology (KIST, latitude 37° 36′ 10.4544″ and longitude 127° 2′ 46.0284″). The site is located 5 km northeast of the Seoul city center (Fig. 1). The fall observation is conducted in TRF, 30 km southeast of the Seoul city center, bordering the SMA, and is operated and maintained by the College of Agriculture and Life Sciences at Seoul National University. A trace gas flux tower facility is located in a conifer plantation (*Pinus koraiensis*) surrounded by a natural oak forest where several studies regarding atmospheric composition of the TRF have been conducted.[11–13] The site is under consistent influence from aged Seoul polluted air masses with episodic regional pollution outbreaks from trans-boundary continental pollution outflows. These elevated pollution plumes expedite the photochemical processes of freshly emitted BVOCs. Isoprene has been consistently assessed as the dominant OH sink at the site during the day.[11–13] At night, monoterpenes account for a considerable fraction of OH sinks.[11–13] Unaccounted HONO sources are also highlighted at the site, which may substantially enhance oxidation capacity.[11]

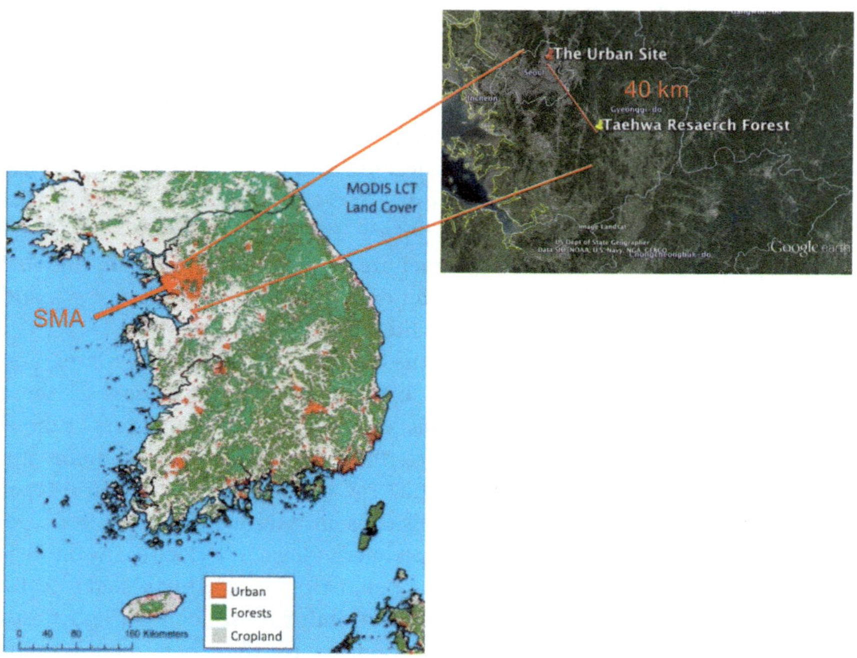

Fig. 1 MODIS (moderate resolution imaging spectroradiometer) land cover types, showing sharp distinction between urban and forested regions. A SMA map indicating the urban site and the suburban forest site (Taehwa Research Forest) is also presented as a subset (photo credit Google Earth).

Table 2 A summary of yearly emissions of anthropogenic trace gases from three East Asian countries in the unit of kmol per year

Countries	NO_X	CO	SO_2	Aromatics	Alkane	Alkene	Isoprene
South Korea	22 948 194.00	90 721 176.00	6 486 774.00	2 415 569.00	4 647 753.00	676 203.00	73 271.00
China	494 421 791.00	5 836 973 662.00	456 185 500.00	66 812 108.00	120 741 973.00	39 753 924.00	919 275.00
Japan	33 984 896.00	159 110 660.00	9 033 697.00	3 629 429.00	6 562 329.00	2 047 176.00	34 990.00

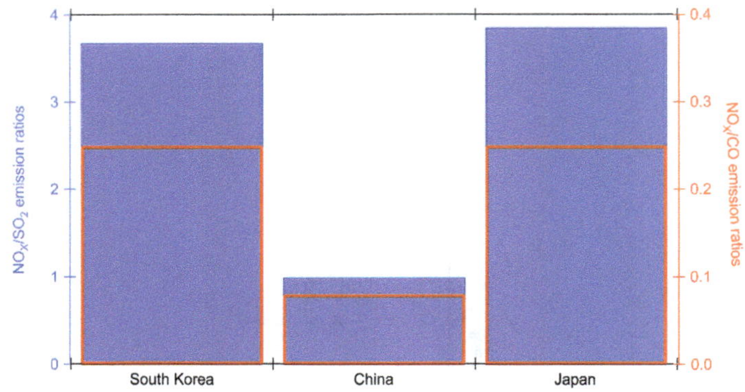

Fig. 2 Emission ratios of NO_X and SO_2 (in blue) and NO_X and CO (in red) assessed for three East Asian countries.

3. Results and discussion

3.1. Trace gas emission profiles

Table 2 summarizes major air pollutant emissions in China, Japan, and South Korea from the CREATE emission inventory. We argue that South Korea exhibits anthropogenic pollutant emissions as an emerging developed country. The ratios of NO_X and SO_2 (Fig. 2) clearly show the anthropogenic trace gas emission profiles in the East Asian countries in different stages of economic development. China shows the lowest ratio and Japan and Korea possess similar ratios. Stricter regulations in coal use become more effective in countries in the more developed stages of economic development such as South Korea and Japan. NO_X emissions from internal combustion engines tend to keep increasing due to increased

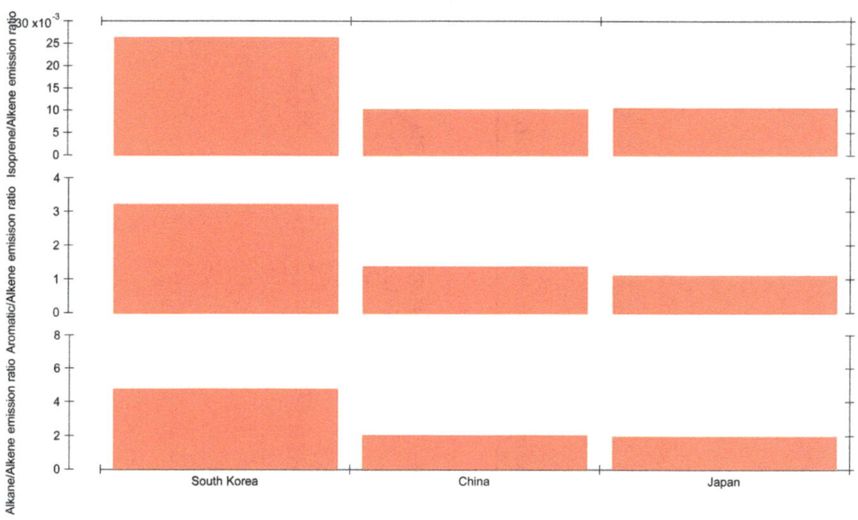

Fig. 3 Emission ratios of VOCs by chemical classes assessed for three East Asian countries.

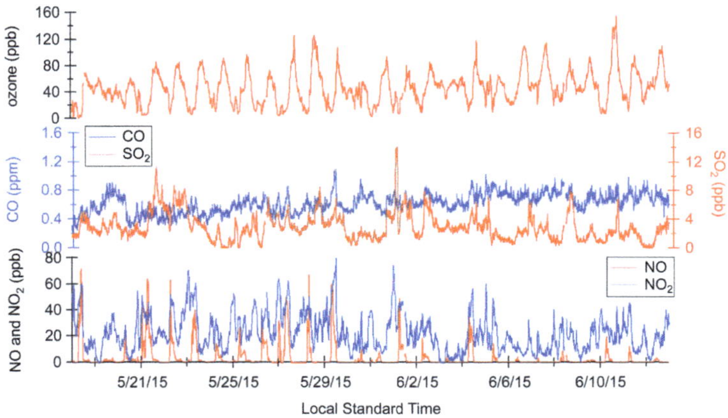

Fig. 4 Time series of observed trace gas variations during the spring campaign in the urban Seoul observational site.

vehicle usage until the economy reaches a mature stage as has been shown in China, Japan, and Korea.[43] In the U.S., several studies integrating satellite NO_X datasets, ground observational networks, and regional photochemical models to assess NO_X emission changes in early 2000 concluded that NO_X emissions have been substantially decreased.[44,45] We can speculate this similar trend may be observable in Japan and Korea in the foreseeable future. In Fig. 2, the NO_X/CO emission ratios are also shown. The highest emission ratio was assessed from Korean emissions. Japan also shows a substantially higher value than that from China. The differences in VOC emission profiles from China, Japan, and Korea can be also highlighted by the ratio of emissions in different chemical classes. Fig. 3 shows the emission ratios of alkane, aromatics and isoprene with alkene. The emission estimates in the CREATE system of alkane, aromatics and isoprene from Korea are all much higher than the alkene emissions.

3.2. Trace gas distributions and total OH reactivity in Seoul

The temporal trace gas variation during the MAPS-Seoul campaign in spring 2015 is shown in Fig. 4. This site has urban characteristics with heavy influences of fresh emissions from traffic. NO_X and CO were at typical levels reported from other air quality monitoring sites in Seoul,[8] which are substantially lower than other megacities such as Beijing,[46,47] Shanghai,[48] and Mexico City[49] (in the 1–2 ppm range) but similar with levels observed in Tokyo.[30] Ambient SO_2 in Seoul is affected by regional transport as the SMA does not have large SO_2 sources and sulfur content in diesel is heavily regulated.[50,51] Indeed, a coal-based Giga power plant cluster is located in the west coast of South Korea ~100 km from Seoul. The back trajectory analysis using the HYSPLIT model frame indicates that the Giga power plant source regions are associated with the enhanced SO_2 levels observed during the field observational period (a manuscript is in preparation). In general, SO_2 is also observed at a much lower level than the levels reported from mega-cities in China[52] (tens of ppb).

In Table 3, the speciated VOCs and their mixing ratio statistics over 12 days are summarized. As the emission inventory suggests, C2 to C4 alkane species were the

Table 3 A summary of observed VOCs with a UCI GC-MS system from canister samples for the 12 day period (24 samples overall) observed at an urban site during the spring campaign

Alkane	Min	Max	Median	Average (STD)	Alkene and alkyne	Min	Max	Median	Average (STD)
Ethane	1814	9620	3325	4143 (2486)	Ethene	245	2694	830	1176 (838)
Propane	532	155 573	2594	4317 (3770)	Ethyne	314	2441	888	1052 (599)
i-Butane	178	5959	1059	1710 (1518)	Propene	44	505	116	190 (150)
n-Butane	316	12 987	1804	3158 (3190)	1-Butene	9	101	23	29 (30)
i-Pentane	181	3431	688	1168 (989)	i-Butene	45	536	95	118 (103)
n-Pentane	126	2315	455	825 (704)	*trans*-2-Butene	7	53	16	20 (13)
n-Hexane	79	1293	329	478 (404)	*cis*-2-Butene	3	34	8	12 (9)
n-Heptane	22	317	98	132 (92)	1,3-Butadiene	5	55	19	23 (15)
n-Octane	12	186	53	68 (52)	1-Pentene	5	67	13	21 (18)
n-Nonane	22	265	72	94 (64)					
n-Decane	21	413	72	104 (89)					
2-Methylpentane	48	711	122	234 (210)					
3-Methylpentane	39	511	99	192 (169)					
2,2,4-Trimethylpentane	22	269	69	98 (75)					
Cyclopentane	17	230	48	85 (71)					
Cyclohexane	21	265	74	101 (76)					

Aromatics	Min	Max	Median	Average (STD)	OVOCs	Min	Max	Median	Average (STD)
Benzene	81	1444	277	324 (282)	Acetaldehyde	960	8927	1823	2351 (1643)
Toluene	798	8926	2035	3093 (2329)	Acetone	6466	21 760	12 041	12 379 (4108)
Ethylbenzene	190	1776	518	698 (474)	Butanal	70	1932	207	303 (127)
p-Xylene	69	834	231	317 (219)	Butanone	329	3096	871	1137 (767)
m-Xylene	87	1405	344	471 (352)	Methanol	3061	31 815	11 140	12 292 (7861)
o-Xylene	63	778	230	321 (211)	Ethanol	783	22 277	3863	6649 (6529)
1,3,5-Trimethylbenzene	3	66	11	18 (18)					
1,2,4-Trimethylbenzene	14	301	41	77 (86)					
1,2,3-Trimethylbenzene	5	105	14	26 (28)					

Table 3 (*Contd.*)

	BVOCs	Min	Max	Median	Average (STD)
	Isoprene	113	1863	625	690 (408)
	Alpha-pinene	7	175	56	76 (53)
	Beta-pinene	7	111	24	37 (29)

greatest fraction among the VOCs followed by aromatic compounds. Toluene (~3 ppb) was observed to be 10 times higher than benzene (~0.3 ppb), which is consistent with previous reports from South Korea.[11,13,53] However, in the Beijing-Tianjin area, VOC observations in the summer of 2009 indicate that the ratio of ambient toluene and benzene were observed close to unity.[17] A series of VOC observations in the Tokyo metropolitan area showed a substantially higher toluene level with respect to benzene (5 to 10 times),[29,30] which is consistent with the observational results from Seoul. As benzene and toluene ratios are often used to characterize photochemical age by taking advantage of substantial differences in their reaction constants with OH,[54] the observed near field toluene and benzene ratios would serve as an important constraint to evaluate photochemical aging of the pollution outflows. Acetone and methanol have been observed at high levels in the Beijing-Tianjin metropolitan area[17] and the Tokyo metropolitan area.[29,30] This is also a consistent observation in Seoul. The total sum of VOC mixing ratios is summarized in Fig. 5a. Large differences can be observed among the different chemical classes in the summed mixing ratios. Oxygenated VOCs (OVOCs) and alkane species composed most of the VOC pool.

Once we factor in different reactivity (rate constants[55]) toward OH for each chemical species in Table 3, the relative contributions from different chemical

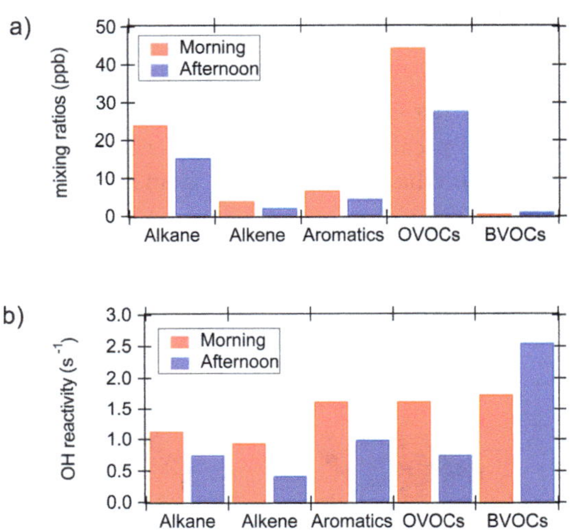

Fig. 5 Averaged (a) concentrations and (b) OH reactivity from different VOC classes during the Seoul urban observations. Morning and afternoon averages are shown in red and blue, respectively.

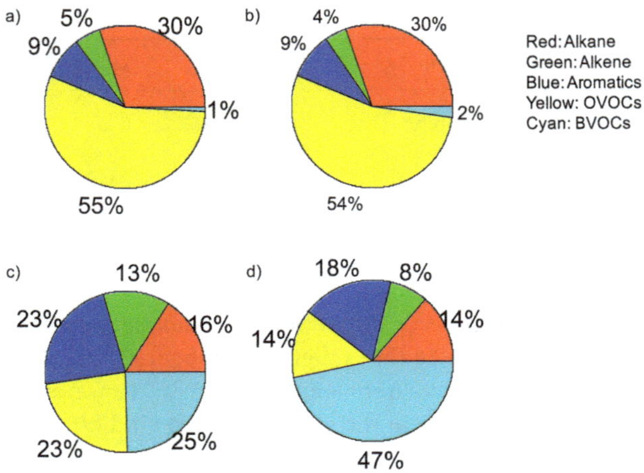

Fig. 6 Relative contributions to mixing ratios from different chemical classes (a) in the morning and (b) in the afternoon and relative contributions to calculated OH reactivity from different chemical classes (c) in the morning and (d) in the afternoon.

classes appear very different from the concentrations (Fig. 5b). Morning contributions to OH reactivity from aromatics, OVOCs and BVOCs primarily from isoprene are evaluated as comparable. BVOCs, again mostly isoprene, are assessed as the dominant OH reactivity contributor in the afternoon. The relative distribution of observed VOCs in the morning and the afternoon are presented in Fig. 6 in both the mixing ratio (Fig. 6a and b) and the OH reactivity scales (Fig. 6c and d). Such a high isoprene contribution (26% in the morning and 36% in the afternoon) to calculated OH reactivity from VOCs in the urban environment has not been commonly reported in the East Asian megacity region. In the suburban regions of the Pearl River Delta,[19] Beijing-Tianjin,[17] and Shanghai,[18] isoprene is an important contributor to OH reactivity. However, in urban environments such as Tianjin and Shanghai, China,[56] isoprene contributes less than ~5% of calculated OH reactivity, and 15% in urban Tokyo, Japan.[29,30] It should be noted that OH reactivity quantitatively represents the roles of chemical species in regional ozone and OVOC production that the concentration scale can often mislead.[57] The dominant contribution to calculated OH reactivity from the ambient observational dataset during the urban observations is from NO_X as shown in Fig. 7. The

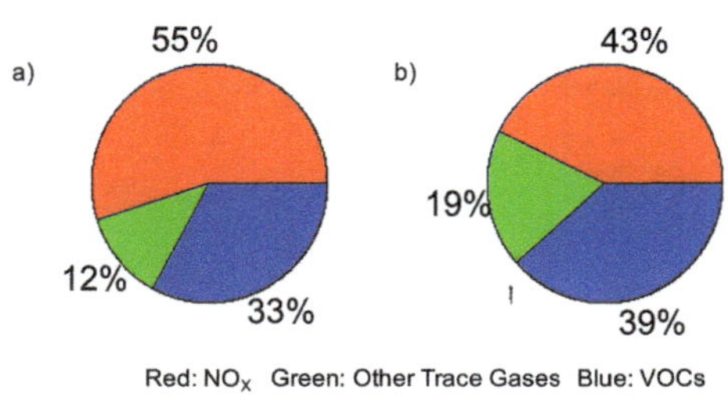

Fig. 7 Relative contributions to OH reactivity from NO_X, other trace gases, and VOCs (a) in the morning and (b) in the afternoon.

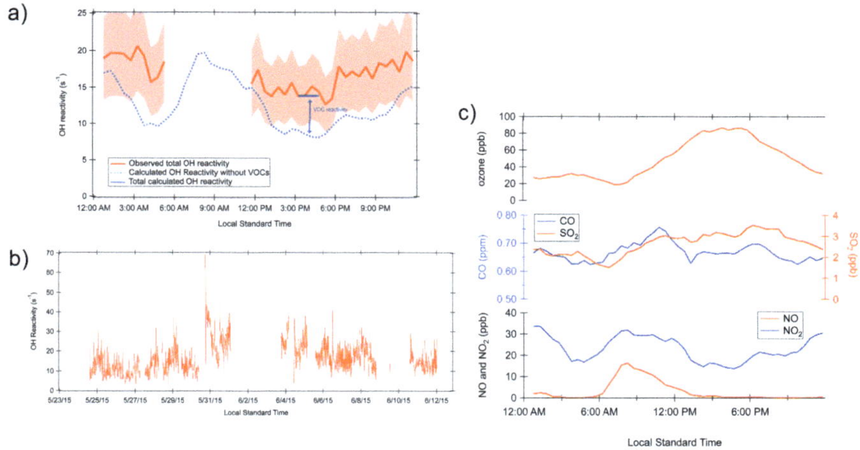

Fig. 8 (a) Averaged diurnal variations in observed OH reactivity, calculated OH reactivity from trace gases without VOCs and total calculated OH reactivity, (b) observed total OH reactivity, and (c) observed trace gas diurnal variation while OH reactivity observation was conducted during the urban observational campaign.

relative contributions from NO_X are significantly higher than those assessed from other East Asian megacities. For example, an observation in Tokyo indicates that NO_X contributes ~35% of calculated OH reactivity.

The averaged diurnal variations of observed and calculated total OH reactivity are shown in Fig. 8a. The average over the 16 days of observations is shown in Fig. 8b. As explained by Sinha and colleagues,[22] the CRM-OH reactivity method systematically underestimates total OH reactivity when NO is higher than 5 ppb

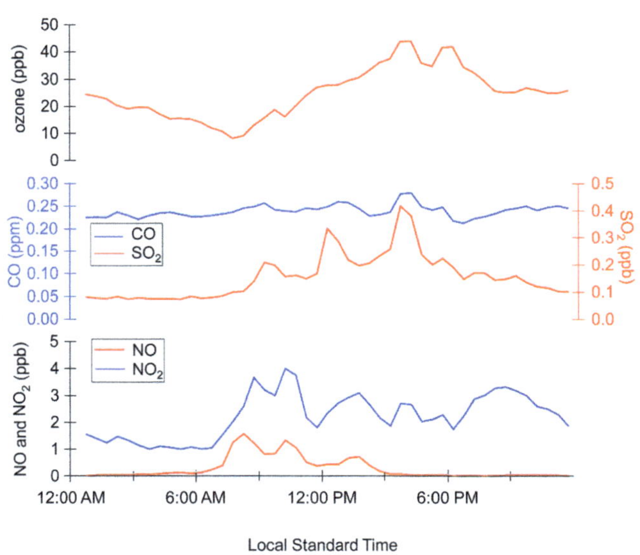

Fig. 9 Observed trace gas diurnal variation while OH reactivity observation was conducted at TRF.

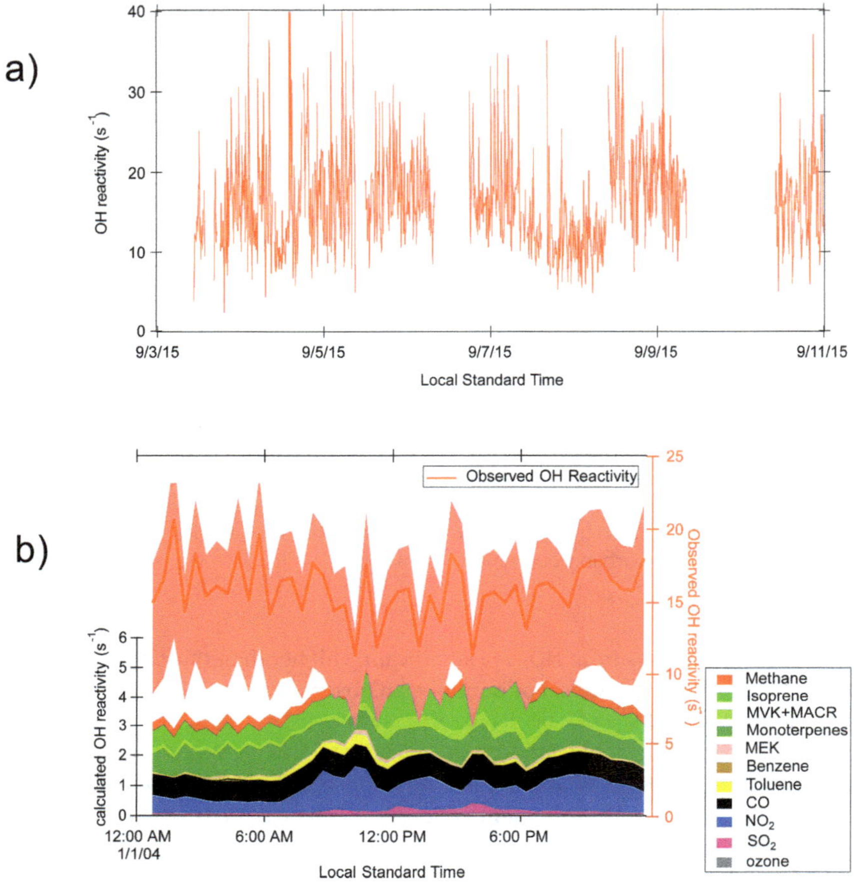

Fig. 10 (a) Observed total OH reactivity and (b) averaged diurnal variations of observed OH reactivity and calculated OH reactivity with a cumulative plot indicating contributions from different observed gas species.

due to fast OH recycling from HO_2. Therefore, total OH reactivity in the morning could not be reported when NO was greater than 5 ppb. Fig. 8c shows the averaged diurnal variations of trace gas observations while the OH reactivity observations were conducted. Despite missing morning data, it is clear that the diurnal variations show higher OH reactivity during the night than observed OH reactivity in the day. Similar results have been reported in the Pearl River Delta region and are most likely attributed to boundary layer evolution, while daytime averages in Seoul are similar to that of Tokyo[29,30], and the Pearl River Delta[19] is comparable. In contrast, the nighttime observed OH reactivity in the Pearl River Delta[19] region was reported to be much higher ($\sim$40 s^{-1})[19] than the observations in the SMA ($\sim$20 s^{-1}). Calculated OH reactivity from trace gases is shown as a dashed blue line in Fig. 8a. Once we add assessed OH reactivity from VOCs observed from the canister samples in the afternoon, the calculated OH reactivity generally accounts for the observed OH reactivity as shown by the blue solid line. This is consistent with the previous OH reactivity observations from urban and megacity environments that have indicated missing OH reactivity smaller than observational uncertainty.

3.3. Trace gas distributions and total OH reactivity in TRF

The September 3–9 trace gas analysis at TRF is presented in Fig. 9. It was consistently sunny during this time period. This rural site located 30 km from the Seoul city center shows a sharp decrease in trace gas concentrations, which is consistent with previous studies,[11–13] and implies fast photochemical processing of megacity pollution plumes. Ultra high-resolution regional model outcomes suggest that the city air pollution plume spreads out to TRF in a time scale of a few hours.[14,15] It should be noted that the sampling was conducted in the late summer/early fall time frame when the regional pollution level is usually much lower than that observed in the spring when the urban observation was conducted. The observed OH reactivity during this time frame is presented in Fig. 10a. Considering the substantially lower pollution levels in this suburban site, significant BVOC influences can be speculated to be the reason. Another noticeable difference from the urban observation is that no strong diurnal pattern could be identified. An identical diurnal trend in OH reactivity was reported from a boreal forest. Sinha and colleagues[25] justified that the observed diurnal variation is most likely caused by the combination of the diurnal boundary layer depth evolution and strong reactive BVOC emission during the daytime. As previously mentioned, air sampling at TRF was conducted in a conifer forest which may provide a plausible explanation of the observed diurnal variation seen in this study. Calculated OH reactivity is assessed based on the trace gas datasets and VOC dataset from PTR-Q(quadrupole)-MS. One of the limitations of PTR-Q-MS for atmospheric VOC quantification is that it cannot quantify most alkane and small alkene (<C_4) species. However, previous whole air sample observational results including a comprehensive suite of alkane and alkene species at the TRF have indicated that VOC species that can be quantified by PTR-QMS such as isoprene, toluene, and butanone (methyl ethyl ketone, MEK) account for the majority of VOC contribution towards OH reactivity among atmospheric VOCs.[11,13] Fig. 10b presents the averaged diurnal variation of calculated OH reactivity from the observed trace gas dataset when OH reactivity observation was conducted. The calculated OH reactivity accounts for 30–40% of observed total OH reactivity. Previous studies have reported 30–90% of missing OH reactivity in forested regions with higher missing OH reactivity (50–89%)[25,31] from boreal forests. The

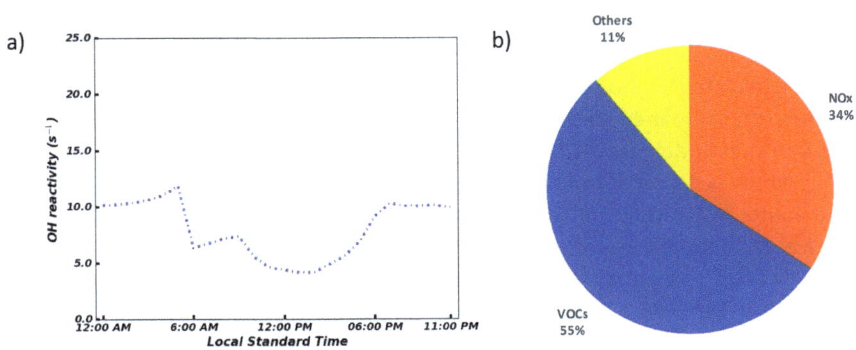

Fig. 11 (a) Simulated diurnal variation of total OH reactivity and (b) relative contributions to simulated OH reactivity from NO_X, other trace gases, and VOCs in Seoul, Korea.

most probable source for missing OH reactivity in forest environments is from either unaccounted BVOC emissions, especially reactive terpenoid species[25,27,31] or unaccounted oxidation products of known BVOCs, specifically isoprene.[19,24,58] In addition, as the roles of extremely low VOCs (ELVOCs) have been recently highlighted in forest canopies with high terpenoid emissions,[59,60] it is a reasonable speculation that ELVOCs which have not been quantified by conventional analytical techniques may contribute towards missing OH reactivity, although the magnitude of their contributions is difficult to assess. Previous observations have indicated that TRF is influenced by both isoprene from deciduous trees from surrounding ecosystems such as oak and mono- and sesquiterpenoids from the conifer tree plantation (200 m by 200 m) near the sampling tower.[11] Therefore, a reasonable assumption can be made that the combination of the different potential sources contributes to missing OH reactivity. Kim and colleagues[11] reported substantial emissions of reactive monoterpenes (α-myrcene) and sesquiterpenes (α and β-caryophyllene) at TRF that have not been quantified in previous studies of branch enclosure measurements. Additionally, the same study highlights potential contributions of unmeasured isoprene oxidation products to the missing OH reactivity (up to 50–70%). More advanced analytical techniques may need to be deployed at this site to qualitatively characterize the sources of missing OH reactivity.

3.4. Comparisons with predicted OH reactivity by a regional modeling system

Fig. 11 shows the diurnal variation of simulated total OH reactivity in the SMA averaged for the MAPS campaign period. Values are much lower than observed or calculated OH reactivity based on the observed species concentrations shown in Fig. 8. The observed high OH reactivity in the night-time and low OH reactivity in the day-time are captured by the model, reflecting the diurnal variation of PBL. However, the pronounced increase of the observed OH reactivity during the night is not captured by the model. The model spatial resolution of about 25 km may still not be fine enough to resolve concentrated pollution plumes from traffic on surrounding highways.

Fig. 11b shows the relative contributions of individual species to the simulated total OH reactivity. The simulated OH reactivity from NO_X is 2.76 s^{-1}, much lower than the observed OH reactivity from NO_X (7.2 s^{-1}), indicating that the model likely underestimates high NO_X concentrations from the traffic in the SMA. We find, however, that the simulated OH reactivity from VOCs is 4.41 s^{-1}, similar to the observed value, 5.3 s^{-1}. As a result, the simulated contribution of NO_X to the OH reactivity in the model is lower (34%) than that of the observation (43–55%), whereas the simulated contribution of VOCs to OH reactivity is high. Among the VOCs in the model, alkanes show the highest contribution to the total OH reactivity, which is different from the observation that isoprene is the most dominant contributor to OH reactivity. Currently, we are working on a separate manuscript to assess the model performances in simulating air quality in the SMA, specifically, detailed comparisons between the model outcomes and the observations. We realize that even a regional photochemical model frame with an enhanced spatial resolution cannot properly capture the photochemical characteristics of TRF, which is located on the border between urban and forest areas. However, the impacts of unconstrained OH reactivity in the suburban forest should be carefully examined in the perspective of their roles in regional ozone and aerosol formation, which may not be properly represented in the regional modeling system.

4. Summary

Air quality in megacities has been highlighted for its implications towards public health and short lived radiative forcers such as ozone and secondary aerosols.[61] We comprehensively discussed the observed trace gas and total OH reactivity datasets from the SMA. The SMA is the second largest metropolitan area in the world and presents developed pollutant emission profiles among the megacities in East Asia. This case study in the SMA provides an opportunity to test our current understanding of photochemistry in East Asian megacities. Particularly, how atmospheric composition will change in megacities and rapidly evolving energy usage and pollutant emission profiles through an ambitious government driven policy. The city center observation dataset indicates that NO_X accounts for 50% of OH reactivity among the observationally constrained reactive gas species. The NO_X contributions become much higher in the morning (55%) than those in the afternoon (43%). VOC contributions to OH reactivity are followed by NO_X with a greater contribution in the afternoon (39%) than in the morning (33%). In the concentration scales, alkane and OVOCs are the dominant VOC species. However, the importance of BVOCs, particularly isoprene, in urban photochemistry is highlighted by assessing relative contributions to OH reactivity. BVOCs mostly isoprene accounts for 25% of calculated OH reactivity among the observed VOCs in the morning and 47% in the afternoon. In contrast, the BVOC contributions are assessed to be 1% and 2% within the concentration scale, respectively. The observed OH reactivity in late May and early June of 2015 in the SMA city center indicates a similar range to other reported OH reactivity values (10–20 s^{-1}) from Tokyo, Japan[29,30] and Pearl River Delta, China.[19] However, nighttime OH reactivity was observed to be at a much higher level in Pearl River Delta, China. The overall calculated OH reactivity from observed trace gases and VOC species in this campaign account for the greatest fraction of the observed OH reactivity level in the afternoon. However, the observational results from TRF indicate a large missing OH reactivity, as calculated OH reactivity from observed trace gases and VOCs can only account for 24–32% of the total observed OH reactivity. The observed range of missing OH reactivity has been reported in other forest ecosystems where anthropogenic influences are minimal such as a boreal forest in Finland.[22,31] The pollution level in TRF is assessed as substantially lower than the center of Seoul, but the level of air pollutants such as CO and NO_X is still much higher in the TRF than that in the boreal forest. Therefore, how interactions between pollution and BVOCs in photochemistry would affect the formation of oxidation products not quantified by conventional analytical techniques (in our case PTR-Q-MS) should be carefully examined. In addition, potential uncharacterized BVOC emissions in the ecosystem should be carefully characterized. Overall, our presented observational results highlight a significant role of BVOCs in the local photochemical processes in the SMA. As similar land use characteristics – a densely populated urban area surrounded by a forested area – can be found in most megacities in East Asia, our analysis should be able to be applied to characterizing regional photochemical characteristics. Furthermore, several studies have interpreted the significant implications of East Asian air pollution into the global tropospheric composition influencing the global radiative budget.[62,63] Therefore, enhancing process level understanding of local photochemistry in the East Asian region will be essential in diagnosing and predicting global radiative forcing from short lived climate forcers.[64]

Acknowledgements

This research is supported by National Institute of Environmental Research (NIER) of South Korea.

References

1 UN, *World Urbanization Prospects*, United Nations, New York, New York US, 2014.
2 D. Y. Zhang, J. J. Liu and B. J. Li, *Sustainability*, 2014, **6**, 5322–5338.
3 D. L. Davis, M. L. Bell and T. Fletcher, *Environ. Health Perspect.*, 2002, **110**, A734–A735.
4 M. Z. Jacobson, *Air Pollution and Global Warming-History, Science, and Solutions*, Cambridge University Press, New York, New York U.S., 2012.
5 A. J. Haagen-Smit, *Ind. Eng. Chem.*, 1952, **44**, 1342–1346.
6 F. E. Blacet, *Ind. Eng. Chem.*, 1952, **44**, 1339–1342.
7 J. G. Edinger, *Environ. Sci. Technol.*, 1973, **7**, 247–252.
8 NIER, *Annual Report for Atmospheric Environment*, National Insititute of Environmetal Rsearch, 2010.
9 S. S. Yum, G. Roberts, J. H. Kim, K. Y. Song and D. Y. Kim, *J. Geophys. Res.: Atmos.*, 2007, **112**.
10 Y. J. Lim, A. Armendariz, Y. S. Son and J. C. Kim, *Atmos. Environ.*, 2011, **45**, 2202–2210.
11 S. Kim, S. Y. Kim, M. Lee, H. Shim, G. M. Wolfe, A. B. Guenther, A. He, Y. Hong and J. Han, *Atmos. Chem. Phys.*, 2015, **15**, 4357–4371.
12 S. Kim, M. Lee, S. Kim, S. Choi, S. Seok and S. Kim, *J. Atmos. Sci.*, 2013, **49**, 325–331.
13 S. Y. Kim, X. Y. Jiang, M. Lee, A. Turnipseed, A. Guenther, J. C. Kim, S. J. Lee and S. Kim, *Atmos. Environ.*, 2013, **70**, 447–453.
14 K. Y. Lee, K. H. Kwak, Y. H. Ryu, S. H. Lee and J. J. Baik, *Atmos. Environ.*, 2014, **96**, 209–219.
15 Y. H. Ryu, J. J. Baik, K. H. Kwak, S. Kim and N. Moon, *Atmos. Chem. Phys.*, 2013, **13**, 2177–2194.
16 H. Bao, K. L. Shrestha, A. Kondo, A. Kaga and Y. Inoue, *Atmos. Environ.*, 2010, **44**, 421–431.
17 L. Ran, C. S. Zhao, W. Y. Xu, X. Q. Lu, M. Han, W. L. Lin, P. Yan, X. B. Xu, Z. Z. Deng, N. Ma, P. F. Liu, J. Yu, W. D. Liang and L. L. Chen, *Atmos. Chem. Phys.*, 2011, **11**, 4657–4667.
18 F. Geng, X. Tie, A. Guenther, G. Li, J. Cao and P. Harley, *Atmos. Chem. Phys.*, 2011, **11**, 10449–10459.
19 S. Lou, F. Holland, F. Rohrer, K. Lu, B. Bohn, T. Brauers, C. C. Chang, H. Fuchs, R. Haseler, K. Kita, Y. Kondo, X. Li, M. Shao, L. Zeng, A. Wahner, Y. Zhang, W. Wang and A. Hofzumahaus, *Atmos. Chem. Phys.*, 2010, **10**, 11243–11260.
20 S. Kim, A. Guenther and E. Apel, *Environ. Sci.: Processes Impacts*, 2013, **15**, 1301.

21 T. A. Kovacs and W. H. Brune, *J. Atmos. Chem.*, 2001, **39**, 105–122.
22 V. Sinha, J. Williams, J. N. Crowley and J. Lelieveld, *Atmos. Chem. Phys.*, 2008, **8**, 2213–2227.
23 A. H. Goldstein and I. E. Galbally, *Environ. Sci. Technol.*, 2007, **41**, 1514–1521.
24 P. M. Edwards, M. J. Evans, K. L. Furneaux, J. Hopkins, T. Ingham, C. Jones, J. D. Lee, A. C. Lewis, S. J. Moller, D. Stone, L. K. Whalley and D. E. Heard, *Atmos. Chem. Phys.*, 2013, **13**, 9497–9514.
25 V. Sinha, J. Williams, J. Lelieveld, T. M. Ruuskanen, M. K. Kajos, J. Patokoski, H. Hellen, H. Hakola, D. Mogensen, M. Boy, J. Rinne and M. Kulmala, *Environ. Sci. Technol.*, 2010, **44**, 6614–6620.
26 R. Oswald, T. Behrendt, M. Ermel, D. Wu, H. Su, Y. Cheng, C. Breuninger, A. Moravek, E. Mougin, C. Delon, B. Loubet, A. Pommerening-Roser, M. Sorgel, U. Poschl, T. Hoffmann, M. O. Andreae, F. X. Meixner and I. Trebs, *Science*, 2013, **341**, 1233–1235.
27 P. Di Carlo, W. H. Brune, M. Martinez, H. Harder, R. Lesher, X. R. Ren, T. Thornberry, M. A. Carroll, V. Young, P. B. Shepson, D. Riemer, E. Apel and C. Campbell, *Science*, 2004, **304**, 722–725.
28 Y. Nakashima, S. Kato, J. Greenberg, P. Harley, T. Karl, A. Turnipseed, E. Apel, A. Guenther, J. Smith and Y. Kajii, *Atmos. Environ.*, 2014, **85**, 1–8.
29 Y. Sadanaga, A. Yoshino, S. Kato and Y. Kajii, *Environ. Sci. Technol.*, 2005, **39**, 8847–8852.
30 A. Yoshino, Y. Nakashima, K. Miyazaki, S. Kato, J. Suthawaree, N. Shimo, S. Matsunaga, S. Chatani, E. Apel, J. Greenberg, A. Guenther, H. Ueno, H. Sasaki, J. Hoshi, H. Yokota, K. Ishii and Y. Kajii, *Atmos. Environ.*, 2012, **49**, 51–59.
31 A. C. Nolscher, J. Williams, V. Sinha, T. Custer, W. Song, A. M. Johnson, R. Axinte, H. Bozem, H. Fischer, N. Pouvesle, G. Phillips, J. N. Crowley, P. Rantala, J. Rinne, M. Kulmala, D. Gonzales, J. Valverde-Canossa, A. Vogel, T. Hoffmann, H. G. Ouwersloot, J. V. G. de Arellano and J. Lelieveld, *Atmos. Chem. Phys.*, 2012, **12**, 8257–8270.
32 D. R. Hanson, J. Greenberg, B. E. Henry and E. Kosciuch, *Int. J. Mass Spectrom.*, 2003, **223**, 507–518.
33 J. J. Colman, A. L. Swanson, S. Meinardi, B. C. Sive, D. R. Blake and F. S. Rowland, *Anal. Chem.*, 2001, **73**, 3723–3731.
34 NIER, *Development of the Asia Emission Inventory in Support of Integrated Modeling of Climate and Air Quality*, National Institute of Environmental Research, Incheon, South Korea, 2013.
35 I. Bey, D. J. Jacob, R. M. Yantosca, J. A. Logan, B. D. Field, A. M. Fiore, Q. B. Li, H. G. Y. Liu, L. J. Mickley and M. G. Schultz, *J. Geophys. Res.: Atmos.*, 2001, **106**, 23073–23095.
36 R. J. Park, D. J. Jacob, M. Chin and R. V. Martin, *J. Geophys. Res.: Atmos.*, 2003, **108**.
37 R. J. Park, D. J. Jacob, B. D. Field, R. M. Yantosca and M. Chin, *J. Geophys. Res.: Atmos.*, 2004, **109**.
38 H. Liao, D. K. Henze, J. H. Seinfeld, S. L. Wu and L. J. Mickley, *J. Geophys. Res.: Atmos.*, 2007, **112**.
39 H. O. T. Pye, H. Liao, S. Wu, L. J. Mickley, D. J. Jacob, D. K. Henze and J. H. Seinfeld, *J. Geophys. Res.: Atmos.*, 2009, **114**.

40 A. B. Guenther, X. Jiang, C. L. Heald, T. Sakulyanontvittaya, T. Duhl, L. K. Emmons and X. Wang, *Geosci. Model Dev.*, 2012, **5**, 1471–1492.
41 Q. Zhang, D. G. Streets, G. R. Carmichael, K. B. He, H. Huo, A. Kannari, Z. Klimont, I. S. Park, S. Reddy, J. S. Fu, D. Chen, L. Duan, Y. Lei, L. T. Wang and Z. L. Yao, *Atmos. Chem. Phys.*, 2009, **9**, 5131–5153.
42 G. R. van der Werf, J. T. Randerson, L. Giglio, G. J. Collatz, M. Mu, P. S. Kasibhatla, D. C. Morton, R. S. DeFries, Y. Jin and T. T. van Leeuwen, *Atmos. Chem. Phys.*, 2010, **10**, 11707–11735.
43 T. Ohara, H. Akimoto, J. Kurokawa, N. Horii, K. Yamaji, X. Yan and T. Hayasaka, *Atmos. Chem. Phys.*, 2007, **7**, 4419–4444.
44 J. H. Woo, S. He, E. Tagaris, K. J. Liao, K. Manomaiphiboon, P. Amar and A. G. Russell, *J. Air Waste Manage. Assoc.*, 2008, **58**, 1483–1494.
45 S. W. Kim, A. Heckel, S. A. McKeen, G. J. Frost, E. Y. Hsie, M. K. Trainer, A. Richter, J. P. Burrows, S. E. Peckham and G. A. Grell, *Geophys. Res. Lett.*, 2006, **33**.
46 K. D. Lu, A. Hofzumahaus, F. Holland, B. Bohn, T. Brauers, H. Fuchs, M. Hu, R. Haseler, K. Kita, Y. Kondo, X. Li, S. R. Lou, A. Oebel, M. Shao, L. M. Zeng, A. Wahner, T. Zhu, Y. H. Zhang and F. Rohrer, *Atmos. Chem. Phys.*, 2013, **13**, 1057–1080.
47 M. Shao, S. H. Lu, Y. Liu, X. Xie, C. C. Chang, S. Huang and Z. M. Chen, *J. Geophys. Res.: Atmos.*, 2009, **114**.
48 X. Tie, F. Geng, A. Guenther, J. Cao, J. Greenberg, R. Zhang, E. Apel, G. Li, A. Weinheimer, J. Chen and C. Cai, *Atmos. Chem. Phys.*, 2013, **13**, 5655–5669.
49 X. Tie, G. Brasseur and Z. Ying, *Atmos. Chem. Phys.*, 2010, **10**, 8983–8995.
50 T. M. S. S. Comittee, *Megacity Air Pollution Studies-Seoul (MAPS-Seoul)*, National Institute of Environmental Research, 2015.
51 T. K.-A. S. Group, *NASA Contributions to KORUS-AQ: In International Cooperative Air Quality Field Study in Korea*, NASA, 2015.
52 W. Y. Xu, C. S. Zhao, L. Ran, W. L. Lin, P. Yan and X. B. Xu, *Atmos. Chem. Phys.*, 2014, **14**, 7757–7768.
53 K. H. Kim, D. X. Ho, C. G. Park, C. J. Ma, S. K. Pandey, S. C. Lee, H. J. Jeong and S. H. Lee, *Environ. Eng. Sci.*, 2012, **29**, 875–889.
54 C. Warneke, S. A. McKeen, J. A. de Gouw, P. D. Goldan, W. C. Kuster, J. S. Holloway, E. J. Williams, B. M. Lerner, D. D. Parrish, M. Trainer, F. C. Fehsenfeld, S. Kato, E. L. Atlas, A. Baker and D. R. Blake, *J. Geophys. Res.: Atmos.*, 2007, **112**.
55 R. Atkinson, D. L. Baulch, R. A. Cox, J. N. Crowley, R. F. Hampson, R. G. Hynes, M. E. Jenkin, M. J. Rossi and J. Troe, *Atmos. Chem. Phys.*, 2006, **6**, 3625–4055.
56 L. Ran, C. S. Zhao, W. Y. Xu, M. Han, X. Q. Lu, S. Q. Han, W. L. Lin, X. B. Xu, W. Gao, Q. Yu, F. H. Geng, N. Ma, Z. Z. Deng and J. Chen, *Atmos. Chem. Phys.*, 2012, **12**, 7531–7542.
57 W. L. Chameides, F. C. Fehsenfeld, M. O. Rodgers, C. Cadelino, J. Martinez, D. D. Parrish, W. Loneman, D. R. Lawson, R. Rasmusse, P. Zimmerman, J. Greenberg, P. Middleton and T. Wang, *J. Geophys. Res.*, 1992, **97**, 6037–6055.
58 S. Kim, A. Guenther, T. Karl and J. Greenberg, *Atmos. Chem. Phys.*, 2011, **11**, 8613–8623.
59 M. Ehn, J. A. Thornton, E. Kleist, M. Sipila, H. Junninen, I. Pullinen, M. Springer, F. Rubach, R. Tillmann, B. Lee, F. Lopez-Hilfiker, S. Andres, I. H. Acir, M. Rissanen, T. Jokinen, S. Schobesberger, J. Kangasluoma,

J. Kontkanen, T. Nieminen, T. Kurten, L. B. Nielsen, S. Jorgensen, H. G. Kjaergaard, M. Canagaratna, M. Dal Maso, T. Berndt, T. Petaja, A. Wahner, V. M. Kerminen, M. Kulmala, D. R. Worsnop, J. Wildt and T. F. Mentel, *Nature*, 2014, **506**, 476.

60 T. Jokinen, T. Berndt, R. Makkonen, V. M. Kerminen, H. Junninen, P. Paasonen, F. Stratmann, H. Herrmann, A. Guenther, D. R. Worsnop, M. Kulmala, M. Ehn and M. Sipila, *Proc. Natl. Acad. Sci. U. S. A.*, 2015, **112**, 7123–7128.

61 T. Zhu, M. Melamed, D. Parrish, M. Gauss, L. G. Klenner, M. G. Lawrence, A. Konare and C. Liousse, *WMO/IGAC Impacts of Megacities on Air Pollution and Climate*, Geneva, Switzerland, 2012.

62 Y. Wang, M. H. Wang, R. Y. Zhang, S. J. Ghan, Y. Lin, J. X. Hu, B. W. Pan, M. Levy, J. H. Jiang and M. J. Molina, *Proc. Natl. Acad. Sci. U. S. A.*, 2014, **111**, 6894–6899.

63 J. W. Fan, D. Rosenfeld, Y. Yang, C. Zhao, L. R. Leung and Z. Q. Li, *Geophys. Res. Lett.*, 2015, **42**, 6066–6075.

64 D. T. Shindell, G. Faluvegi, D. M. Koch, G. A. Schmidt, N. Unger and S. E. Bauer, *Science*, 2009, **326**, 716–718.

Faraday Discussions

DISCUSSIONS

Timescales of mixing and of chemistry: general discussion

Francis Pope, Paul S. Monks, Jörg Kleffmann, Alison Tomlin, Lisa Whalley, Dwayne Heard, Andrew Rickard, William Bloss, Roy Harrison, Sasha Madronich, Gordon McFiggans, Neil M. Donahue, Ugo Molteni, Andreas Wahner, William Brune, Xavier Querol, Siqi Hou, Shengrui Tong and Saewung Kim

DOI: 10.1039/C6FD90023B

Neil Donahue opened discussion of the paper by William Brune: How do you know the high-NO signal isn't simply a secondary correlation with the early-morning timing? Are there data from other campaigns that are consistent with your hypothesized mechanism?

William Brune replied: We have seen the greater-than-modeled HO_2 in many different environments. Urban areas generally have the highest NO levels for our ground-based measurements, and we have seen this effect in every city. However, the values for the ratio of measured-to-modeled HO_2 as a function of NO are different for different cities. For example, in New York City, the ratio at highest NO levels was more than a factor of 10–20, while in Mexico City at equally high NO levels it was only a factor of three. The difference could be related to the level of OH reactivity, which was much greater in Mexico City than in New York City. We have also seen this effect in the North Atlantic flight corridor, well into the free troposphere, during the SONEX campaign. In fact, this is where we saw this effect first and reported it in a paper by Faloona *et al.*[1] In addition, there is a fairly recent paper on peroxynitric acid (PNA) in which the PNA mixing ratio is not consistent with modeled HO_2.[2] As to the mechanism and other campaigns, when we include this mechanism in the model for four different urban areas, the measured and modeled HO_2 agree much better at high NO and the large difference disappears in all of the different urban areas.

1 I. Faloona, D. Tan, W. Brune, L. Jaegle, D. Jacob, Y. Kondo, M. Koike, R. Chatfield, R. Pueschel, G. Ferry, G. Sachse, S. Vay, B. Anderson, J. Hannon and H. Fuelberg, Observations of HO_x and its relationship with NO_x in the upper troposphere during SONEX, *J. Geophys. Res.*, 2000, **105**, 3771–3783.
2 K. M. Spencer, D. C. McCabe, J. D. Crounse, J. R. Olson, J. H. Crawford, A. J. Weinheimer, D. J. Knapp, D. D. Montzka, C. A. Cantrell, R. S. Hornbrook, R. L. Mauldin III and P. O. Wennberg, *Atmos. Chem. Phys.*, 2009, **9**, 3697–3707.

Jörg Kleffmann commented: In William Brune's paper modelled HO_2 levels were much lower than measured values only during the morning rush hour period under high NO_x conditions. Could the reason for this observation be constraining the model by underestimated NO_2 when using typical data from chemiluminescence instruments with photolytic converters? Information about the NO_x instrument used is unfortunately not given in the paper. If the model is constrained by a too low NO_2/NO ratio, the model would react with lower HO_2 concentrations (needed to convert NO into NO_2). In a recent paper by Villena *et al.*[1] negative interferences of "blue-light" instruments were shown to be important only under high pollution levels and would thus be a further explanation besides those given by the authors.

1 G. Villena, I. Bejan, R. Kurtenbach, P. Wiesen and J. Kleffmann, *Atmos. Meas. Tech.*, 2012, **5**, 149–159.

William Brune replied: For the CalNex study, the NO_2 was measured by laser induced fluorescence. As far as I know, LIF is not affected by this interference that was found in instruments that use "blue-light" converters and chemiluminescence.

Lisa Whalley asked: Do you think that a HO_2 artefact from RO_2 species may contribute to the higher than expected HO_2 at NO concentrations greater than 1 ppbv? The correction you have applied seems to be negligible at NO concentrations over 3 ppbv – is it possible that you are under-correcting for this RO_2 interference at high NO?

William Brune responded: This might be a possibility. However, according to the model, RO_2 decreases faster than HO_2 as NO increases. If RO_2 were causing this HO_2 discrepancy, you would expect RO_2 to be increasing, or at least be constant, relative to HO_2 as NO increases.

Lisa Whalley said: Could you comment on the model's ability to predict RO_2 in light of fact that the model under-estimates the observed OH reactivity? I ask this as we observed contrasting behaviour during the London ClearfLo project: we model a decrease in the $RO_2 : HO_2$ ratio (from 2 to 1) between 0–2 ppbv NO and then this ratio begins to increase with NO concentration, reaching a ratio of around 2 at NO concentrations over 10 ppbv. We also did not observe higher than expected HO_2 under the high NO conditions we encountered.

William Brune answered: This missing OH reactivity could cause our modeled RO_2 to be too low. If this is the case, then we would need to re-examine whether RO_2 could be contributing to the HO_2 discrepancy. However, the measured HO_2 exceeds modeled HO_2 by more than a factor of ten in several studies, not just our own, and this difference is much greater than the factor of 2 that your modeled RO_2/HO_2 would suggest. As to the London ClearfLo project, I have not looked at the ClearfLo HO_2 measurements in detail and so cannot really comment. However, at a recent HO_x measurement workshop in March 2015, it became clear that in some ways the measured-to-modeled HO_2 agreement was more variable than the measured-to-modeled OH agreement, suggesting that there might be more problems in the HO_2 measurements and/or models than there are for OH.

In addition, this HO_2 discrepancy has been seen by several different groups and by two different detection methods. Clearly this discrepancy needs to be resolved.

Lisa Whalley then asked: In recent campaigns where you have run with the RO_2 interference minimised, do you still see higher than expected HO_2 at NO concentrations over 1 ppbv?

William Brune replied: This is a good question. Our recent campaigns in which we have minimized the RO_2 have all been in forests where the NO levels were less than about a ppbv, so we don't have any information yet. We have an aircraft campaign coming up in which we will be flying some of the time at 1000 ft above a city, so we will have an answer soon.

Paul Monks commented: Looking at Fig. 8 in your paper, you have agreement at relatively low NO_x but a really sharp transition at 1.5 ppb. This is probably about the compensation point I'd guess, moving from one production regime to another. Have you mapped this out with NO_x control? This transition must be telling you what drives this discrepancy

William Brune answered: This transition between having measured HO_2 be much greater than modeled and having agreement does occur when the NO is transitioning between being much greater than 1.5 ppbv to much less. The ozone production is transitioning between VOC-sensitive to NO_x sensitive at this same time. I have to think about whether we can learn anything about the HO_2 discrepancy from this, but right now I would say that all these effects are related to the transition from high NO_x at morning rush hour to lower NO_x in the afternoon and thus we need to focus on the HO_2 discrepancy as a function of NO.

Ugo Molteni remarked: Yesterday Urs Baltensperger presented the mechanism of HOMs formation. This is not included yet in the model (*i.e.* the MCM). Could this pool of reactions like auto-termination, OH and HO_2 elimination affect the balance of HO_x?

William Brune answered: It is possible that the formation of high oxidized molecules (HOMs) could act as a sink for HO_x and thus probably reduce the mixing ratios of OH and HO_2. However, the formation of HOMs is competing with the reactions of isoprene and monoterpenes as sinks of OH, so the exact effect of HOM formation will be determined by this competition.

Dwayne Heard said: I have a comment and two questions. My comment: in the Clearflo campaign in Central London in 2012, the OH reactivity budget was essentially closed,[1] which is in contrast with Bakersfield, although the conditions of the two urban studies were quite different. My first question: did you try constraining the model to measured HO_2 to see the impact on modelled OH and ozone? My second question: you mentioned in your paper that you did some laboratory experiments to investigate the hypothesised reaction HONO* + O_2 but stated that the results were mixed. Could you elaborate on what you did and whether these experiments provided evidence for this reaction?

1 L. K. Whalley, D. Stone, B. Bandy, R. Dunmore, J. F. Hamilton, J. Hopkins, J. D. Lee, A. C. Lewis and D. E. Heard, *Atmos. Chem. Phys.*, 2016, **16**, 2109–2122.

William Brune replied: In cities, we have typically been able to close the OH reactivity budget to within 30%. For New York City, it was within 10%. The Bakersfield site has NO_x levels similar to those in New York City, but it also has substantial VOC emissions from the surrounding agricultural lands. These emissions are the likely contributors to the missing OH reactivity in Bakersfield. In answer to the first question, we did not run the model with constrained HO_2. However, we did look at the OH production and loss, and they balance for all times of the day except morning rush hour, for which OH production exceeds OH loss. We have consistently seen this difference, but we know that the OH production and loss must be in balance. Based on the results from a recent OH reactivity intercomparison study, we feel that the OH reactivity measurements that we have reported are too low when NO is above 5–10 ppbv and that our reported OH reactivity should be as much as twice as large when NO levels are high at morning rush hour. We are working to improve our analysis methods for OH reactivity measurements at these higher NO levels. So this balance between OH production and loss suggests that constraining the model to HO_2 will give us an OH value similar to our current case when HO_2 is not constrained. In answer to the second question, this reaction is difficult to study because the HO_2 recycles so quickly into OH. So we looked at the $[HO_2]/[OH]$ ratio as a function of NO and O_2. However, the laser was not working well when we did the studies, so the signals are low and the data are noisy. We started with a very simple system of UV light, water vapor, CO, and NO. It appears that there is an effect, but the data quality is too low to be sure. It would be good to take another look with better detection sensitivity.

Dwayne Heard said: To investigate the proposed mechanism shown in reactions (1.6)–(1.8) in your paper, could you perform some isotopically labelled experiments (*e.g.* using OD or ^{18}OH, or labelling on the nitrogen) and see if the isotopes follow through to the products?

William Brune answered: This seems to be a good idea. The recycling reaction of HO_2 + NO scrambles things, so we would need to do some modeling to see what would happen to the isotopes and whether it tells us anything.

Alison Tomlin opened discussion of the paper by Saewung Kim: Your paper talked about discrepancy in OH reactivity between model and measurements. How much do you take into account uncertainties in the model rate constants? Previous work *e.g.* by Sommariva *et al.*[1] has shown that if you account for such uncertainties this leads to quite a large envelope in predicted OH from such models. To what extent do you take this into account?

1 R. Sommariva, A.-L. Haggerstone, L. J. Carpenter, N. Carslaw, D. J. Creasey, D. E. Heard, J. D. Lee, A. C. Lewis, M. J. Pilling and J. Zádor, *Atmos. Chem. Phys.*, 2004, **4**, 839–856.

Saewung Kim replied: Uncertainty listed in recent rate constant compilations cannot account for the discrepancy we reported in the BVOC dominant environment.

Roy Harrison said: On the question of OH reactivity, recently published work from York University has indicated a substantial contribution from higher molecular weight VOCs that are not regularly measured. However, a paper to be given later in the meeting reports not seeing these compounds as being contributors. What do you see in your studies?

Saewung Kim replied: The whole air sampling VOC analysis results show that the concentrations of alkanes are exponentially decreasing as the number of carbons is increasing. However, we really need to comprehensively evaluate whether conventional sampling methods – such as inlet material and flow rates – are suitable to quantify those high-molecular weight compounds.

Andrew Rickard asked William Brune: In London during the ClearFlo ("Clean air for London") winter and summer field campaigns,[1] we observe that long chain, high molecular weight hydrocarbons (mainly alkanes) predominantly coming from diesel vehicle emissions[2] contribute significantly to the observed OH reactivity.[3] You mention in the paper that in Bakersfield the diesel emission are reasonably important. Did you constrain your detailed Master Chemical Mechanism box model simulations with respect to these species (*cf.* Whalley *et al.*[3])? How much do the model intermediates contribute to Fig. 5 of the paper (*i.e.* measured, calculated and modelled OH reactivities) and do the model intermediates show any diurnal variation?

1 S. I. Bohnenstengel *et al.*, *B. Am. Meteorol. Soc.*, 2015, **96**, 779–804.
2 R. E. Dunmore *et al.*, *Atmos. Chem. Phys.*, 2015, **15**, 9983–9996.
3 L. K. Whalley *et al.*, *Atmos. Chem. Phys.*, 2016, **16**, 2109–2122.

William Brune responded: We had extensive measurements in Bakersfield and we constrained the model to all of them except for OH and HO_2. I would have to look to see what the largest measured alkanes were. The modeled intermediates contributed about 10% to the calculated and modeled OH reactivity in Fig. 5 in our paper. I did not look at how much the modeled OH reactivity varied since it is such a small portion of the calculated OH reactivity.

Roy Harrison asked William Bloss: William Brune has hypothesised that a reaction of excited HONO can contribute to balancing the budget for HO_2. What are the implications of that reaction for the HONO budget?

William Bloss answered: We can "back-of-the-envelope" the implications for the HONO budget as follows, starting with the proposed mechanism from the paper by Brune *et al.* in this issue (DOI: 10.1039/C5FD00204D):

$OH + NO = HONO^*$

$HONO^* + N_2 = HONO + N_2$ or

$HONO^* + O_2 = HO_2 + NO_2$

The range of rate constants for the proposed OH + NO reaction is $(3–15) \times 10^{-11}$ molec^{-1} cm^3 s^{-1}. At typical urban conditions of $[OH] = 2 \times 10^6$ molec cm^{-3} and an NO level of 4 ppb (around the middle of the "high NO_x" regime where the ozone formation deviation from theory is observed) a HONO* production rate of $(2.4–12) \times 10^{-4}$ ppb s^{-1} is obtained. This may be compared with the "additional" non-PSS HONO formation rates reported of (1.3–2.5) × 10^{-4} ppb s^{-1} (for York – Crilley *et al.*'s paper in this issue (DOI: 10.1039/C5FD00224A), and London – Lee *et al.*,[1] respectively).

So, if even a modest fraction (10%) of HONO* were quenched to HONO this process could be of a comparable magnitude to the "missing" HONO sources reported, and we might expect >80% of the HONO* to end up as HONO on the basis of N_2/O_2 abundance. There may be further constraints on this from the existing laboratory data. However, ambient HONO is usually observed to be well correlated with NO_2 and $j(NO_2)$, rather than NO and $j(O^1D)$ (as would be expected for OH away from urban areas) suggesting that this reaction is not the dominant HONO source.

1 J. D. Lee, L. K. Whalley, D. E. Heard, D. Stone, R. E. Dunmore, J. F. Hamilton, D. E. Young, J. D. Allan, S. Laufs and J. Kleffmann, *Atmos. Chem. Phys.*, 2015, **15**, 22097.

Andreas Wahner asked Saewung Kim: I have the impression you compared measured k_{OH} to measured VOCs and didn't add *e.g.* oxygenated VOCs, except for two oxygenated VOCs, but not others. How large is the effect of other *e.g.* oxygenated VOCs, may this be estimated by model calculations? The diurnal k_{OH} plot in your paper hardly shows any diurnal variation – would you not expect high variation between night and day?

Saewung Kim replied: As mentioned, a previous modeling study indicated unobserved multi-generational oxidation products can account for up to 100% of the calculated OH reactivity from the observed trace gas dataset. A flat diurnal pattern was reported once in a boreal forest in Finland. The report justified the diurnal pattern as due to the temperature dependence of monoterpene emissions from conifers. The Taehwa forest is located in the middle of a pine plantation and is surrounded by an oak forest. Therefore, the isoprene contribution during the daytime and the monoterpene and other relatively long lived gas species contributions during the night may cause the flat diurnal observed OH reactivity pattern, while isoprene and other short lived reactive gases may maintain OH reactivity during the day time.

Dwayne Heard commented: You state in your paper that if the NO concentration is more than 5 ppb then you do not use the measurements of OH reactivity taken at those times. Could you state for what fraction of the time the NO concentration is greater than 5 ppb, as it looks from Fig. 4 in your paper that it could be a significant fraction of the time? I assume that these periods are not included in the average behaviour of OH reactivity that you discuss? Also, you mentioned in the paper that you are preparing a separate manuscript with detailed characterizations of the CRM-CIMS instrument. However, can you state here what magnitude of correction you are having to make for your OH reactivity measurements to account for any recycling from HO_2 to OH when NO levels are at or below 5 ppb?

Saewung Kim replied: Our laboratory calibration results with multiple standards of different VOC chemical classes such as alkane, alkene, and di-alkene indicate that for up to 5 ppbv of NO the calibration factor holds within the uncertainty (~ 20%) and shows bias from the calibration factor as recycling gets active. This is the justification for our NO correction.

Neil Donahue opened a general discussion of the papers by William Brune and Saewung Kim: Why does the biogenic environment seem to typically have a greater deficit in OH reactivity than the urban environment, even though it seems logical that the urban environment is more complex?

William Brune replied: For your 2001 New York City study, we were surprised when we compared the measured OH reactivity to the OH reactivity calculated with just the measurements of other chemical species and saw that they agreed to within 10% for exactly the reason you say – urban air chemistry should be more complex. I think that there are a few reasons for the seemingly greater missing OH reactivity in forests than in cities. First, there has historically been a much greater emphasis on urban pollution than on forest air chemistry, so the scientific understanding has been more mature for the most significant OH reactants and their products in urban areas than for those in forest environments. Second, the measurement capability for the urban VOCs is also more mature than the capability for the biogenic VOCs and their oxidation products. The net result is that, for urban areas, there is a better understanding of what needs to be measured and a better suite of measured significant OH reactants. There are exceptions for this agreement in urban areas, however, but this urban missing OH reactivity does not approach the very large amounts of missing OH reactivity that has been reported for some forests. I should also say that in an Alabama forest during the SOAS campaign in 2012, the measured and calculated OH reactivity, which included modeled intermediates, agreed to within 20%.

Saewung Kim added: Both the emissions and reactivity characteristics of BVOCs should be able to address this question. Although we have reasonably good understanding in general emissions such as isoprenoids, monoterpenoids, and sesquiterpenoids, speciation can be very different depending upon the tree species and even upon individual trees in identical tree species. In addition, in general, BVOCs tend to be more reactive and generate complex oxidation products that we do not necessarily constrain well with conventional field deployable instrumentation.

Paul Monks opened discussion of the paper by Siqi Hou and Shengrui Tong: You make a case around Fig. 8 in your paper for HONO coming from chemistry on the aerosol $PM_{2.5}$ surface *vs.* (land) surface processing. Does this hold that up given a relative surface area to volume argument? It seems to me that aerosol surface area to volume is quite small. I would like to know if the plots are causal or just coincidental?

Shengrui Tong communicated in reply: First, this article compared the particulate matter effects between haze and non-haze periods rather than effects between ground surfaces and the surface of particulate matter.

Second, the particulate matter effect on HONO surface formation may be important, besides the ground surface, given such high concentrations recently in Beijing, which is also suggested by Notholt *et al.*[1] and Reisinger.[2]

Third, particulate matter loading in ambient air corresponds to surface to volume ratios and they often show coinciding trends.[3,4] Thus, it could be an index to evaluate surface to volume ratios.

Fourth, as mentioned in the article our study is a preliminary work and provides clues on the impacts of heterogeneous conversion of NO_2. In fact, the correlation between the particulate matter and HONO may also be due to the decreasing of BCH or coming from similar sources. Further and more comprehensive work should be undertaken.

Fifth, besides the results in this article, our previous studies and measurements in Beijing from 2014[5,6] to now show very similar relationships between $PM_{2.5}$ mass concentration and HONO in different measurement sites and periods, which may be proof of a causal result.

1 J. Notholt, J. Hjorth and F. Raes, *Atmos. Environ., Part A*, 1992, **26**, 211–217.
2 A. R. Reisinger, *Atmos. Environ.*, 2000, **34**, 3865–3874.
3 S. S. Park, S. B. Hong, Y. G. Jung and J. H. Lee, *Atmos. Environ.*, 2004, **38**, 293–304.
4 M. Qin, P. Xie, H. Su, J. Gu, F. Peng, S. Li, L. Zeng, J. Liu, W. Liu and Y. Zhang, *Atmos. Environ.*, 2009, **43**, 5731–5742.
5 S. R. Tong, S. Q. Hou, Y. Zhang, B. W. Chu, Y. C. Liu, H. He, P. S. Zhao and M. F. Ge, *Sci. China: Chem.*, 2015, **58**, 1393–1402.
6 S. Hou, S. Tong, M. Ge and J. An, *Atmos. Environ.*, 2016, **124**, Part B, 199–206.

Jörg Kleffmann communicated: While in Fig. 7 in the paper the night-time behavior of the urban HONO/NO_2 ratio is very typical (continuously increasing during the night), its temporal trend is very untypical for the suburban conditions (decreasing HONO/NO_2 ratio). In addition, in the early evening (18:00) and for the haze conditions the HONO/NO_2 ratio is extremely high (over 20%). Typically the ratio reaches a minimum (a few percent at maximum) in the late afternoon (which has the highest boundary layer height and smallest S/V). Can you explain this very untypical data for the suburban measurements?

Siqi Hou communicated in reply: I wondered if the untypical variation of HONO/NO_2 ratios in suburban refers to the higher values in the late afternoon and keeping decreasing at night? To interpret this question, we checked the data measured in several periods and found this variation of HONO/NO_2 ratios was quite common. Actually, this phenomenon occurred mostly in non-haze periods. During these periods (Fig. 5 in the paper), concentrations of HONO and NO_2 were quite low. And NO_2 kept increasing slightly compared to the decreasing HONO. During the haze period, HONO concentrations rose to a maximum at around 20:00 and kept almost stable for the rest of the night. While in non-haze periods, after reaching the evening peak around 20:00, HONO concentrations decreased at first and then became stable during the second half of the night. This type of HONO variation led to a decreasing $HONO_{corr}/NO_2$ ratio in both periods after sunset which finally became stable in the second half of the night. In fact, the peaks during the evening time may be mainly attributed to the local emissions as mentioned in the article. Although we have corrected HONO concentrations by subtracting contributions from direct emissions, the low NO_x concentrations made the contribution of direct emissions less important. That implied there are other processes influencing the emission of HONO such as processes connected to CO, which showed evening peaks as well. This also reflects the influences of traffic type on the emissions. $HONO_{corr}/NO_2$ ratios finally became stable in the second half of the night. As the nocturnal boundary layer was changing to be stable, the production and loss rates of HONO would finally reach a steady state, when the heterogeneous conversion of NO_2 would become dominant. Therefore, data of the second half of the night were used when NO_2 concentrations were stable and the conversion of NO_2 was steady. The results were in the range of 0.03 to 0.15 in many studies in China, and could indicate the differences of NO_2 conversions between urban and suburban areas.

Jörg Kleffmann commented: In your paper, HONO source reactions were inferred from correlation studies of HONO (corrected for direct emissions and homogeneous sources) with different variables, for example NO_x, PM, humidity, CO, SO_2. This may highlight the general risk of over-interpretation of correlation studies caused by the variability of vertical mixing in the atmosphere. Here all near ground sources (emissions, heterogeneous ground sources, *etc.*) will correlate during night-time with the variation of the boundary layer. For example, in previous studies by Febo *et al.* excellent correlations of night-time HONO with Radon were observed, and there is definitely no chemical link between both trace gases. Thus, simple correlations of HONO with PM will not prove any heterogeneous production of HONO on particles. To answer the question of whether HONO is formed on particles or on ground surfaces gradient measurements of HONO, NO_x and PM are necessary. Here our first studies[1,2] clearly showed that particle reactions cannot explain night-time HONO levels in the atmosphere, in excellent agreement with all laboratory studies in which dark uptake kinetics on particle surfaces (soot, SOA, aqueous particles, dust...) were too slow to be important.

1 J. Kleffmann *et al.*, *Atmos. Environ.*, 2003, **37**, 2949–2955.
2 B. Vogel *et al.*, *Atmos. Environ.*, 2003, **37**, 2957-2966

Siqi Hou responded: As shown in Fig. 4 and Fig. 5 of the paper, the concentrations of HONO did not vary simultaneously with other gas phase species, RH and $PM_{2.5}$, actually, HONO concentrations undulate by a combination of these several factors. Thus, it is necessary to do correlation studies, which could reflect the same sources to some extent. The plots of night-time HONO against NO_2 were studied in many previous papers.[1–8] The co-variation and correlation between HONO and NO_2 could reflect the relationship between HONO and NO_2 and be helpful to study meteorological effects on the NO_2–HONO conversion. Moreover, recent studies in China showed good correlations between NO_2 and HONO, which indicated that heterogeneous conversion was important and could not be neglected. Additionally, the ratio of HONO/NO_2 was not increasing all night from our observation. Though the HONO/NO_2 ratio was increasing early in the night, it would be stable once the physics processes had balanced with the heterogeneous conversion,[9,10] which is mentioned in our article.

We agree that the boundary layer height (BLH) is an important element that could affect the RH, particle S/V ratios and HONO concentrations. The chemical processes between HONO and NO_2 could not be ignored.[9] The correlation study refers to many previous studies: there are many previous studies have proved good correlations of HONO/NO_2 with PM_{10} [4,11,12] and HONO/NO_2 with RH,[4,13] but it is still not sufficient to illustrate their impacts on the heterogeneous formation of HONO. And we illustrated this in our article. Additionally, for further studies, the gradient measurements of particles/HONO/NO_2 will help to understand the impact effects of particles and RH. We do appreciate this suggestion and will take note of it in future. The high concentrations of particulate matter in Beijing may show a distinct possibility for the surface formation of HONO, and our previous discussion did not make a quantitative consideration of the relationship between HONO/NO_2 and $PM_{2.5}$.

As we discussed in the article (DOI: 10.1039/C5FD00163C): "However, correlation coefficient of $HONO_{corr}$ with $PM_{2.5}$ (0.514) still showed promotion effects. Heterogeneous conversion of NO_2 would occur on ground surfaces as well as aerosols, the higher surface density of ground was often supposed to play a main role in HONO formation over the aerosol surface. While, our measurement results showed that compared to the ground surface the high particulate matter concentrations during the haze period would promote the heterogeneous conversion of NO_2. Similarly, the correlations of $HONO_{corr}$ and $HONO_{corr}/NO_2$ with RH showed RH would promote the conversion of HONO, though $HONO_{corr}/NO_2$ ratios increased with RH growing and decreasing when RH higher than 60%. Additionally, the good correlations of HONO/NO_2 with particles[25,40] and RH[18,25] has been shown in many previous studies. However, measurements by Kleffmann *et al.*[38] and Stutz *et al.*[18] showed weak aerosol impact on HONO formation, and this suggestion was supported by laboratory studies on soot and organic particles. As boundary layer height (BLH) decreased in the nighttime, this led to the increasing of the S/V ratio and concentrations of other gas-phase pollutants.[18,37] Meanwhile, RH would increase as well due to the decreasing of temperature. Therefore, the good relations will be affected by these factors. And the impacts of aerosols and RH vary with environmental conditions and are still in debate. Given this measurement only provides a rough study on the impacts of particulate matter, gradient measurements of HONO with NO_2 and aerosol concentrations are highly recommended to prove the possible impacts of aerosols on HONO formations.[18]"

1 B. Alicke, A. Geyer, A. Hofzumahaus, F. Holland, S. Konrad, H. W. Patz, J. Schafer, J. Stutz, A. Volz-Thomas and U. Platt, *J. Geophys. Res.-Atmos.*, 2003, **108**, 8247, DOI: 10.1029/2001JD000579.
2 N. Hao, B. Zhou, D. Chen and L. M. Chen, *J. Environ. Sci. -China.*, 2006, **18**, 910–915.
3 X. Li, T. Brauers, R. Häseler, B. Bohn, H. Fuchs, A. Hofzumahaus, F. Holland, S. Lou, K. D. Lu, F. Rohrer, M. Hu, L. M. Zeng, Y. H. Zhang, R. M. Garland, H. Su, A. Nowak, A. Wiedensohler, N. Takegawa, M. Shao and A. Wahner, *Atmos. Chem. Phys.*, 2012, **12**, 1497–1513
4 M. Qin, P. Xie, H. Su, J. Gu, F. Peng, S. Li, L. Zeng, J. Liu, W. Liu and Y. Zhang, *Atmos. Environ.*, 2009, **43**, 5731–5742.
5 J. Stutz, H.-J. Oh, S. I. Whitlow, C. Anderson, J. E. Dibb, J. H. Flynn, B. Rappenglueck and B. Lefer, *Atmos. Environ.*, 2010, **44**, 4090–4098.
6 H. Su, Y. F. Cheng, P. Cheng, Y. H. Zhang, S. Dong, L. M. Zeng, X. Wang, J. Slanina, M. Shao and A. Wiedensohler, *Atmos. Environ.*, 2008, **42**, 6219–6232.
7 T. Wu, Q. Z. Zha, W. D. Chen, Z. Xu, T. Wang and X. D. He, *Atmos. Environ.*, 2014, **95**, 544–551.
8 Y. Yu, B. Galle, A. Panday, E. Hodson, R. Prinn and S. Wang, *Atmos. Chem. Phys.*, 2009, **9**, 6401–6415.
9 K. Acker, D. Möller, R. Auel, W. Wieprecht and D. Kalass, *Atmos. Res.*, 2005, **74**, 507–524.
10 J. Stutz, B. Alicke and A. Neftel, *J. Geophys. Res.-Atmos.*, 2002, **107**, DOI: 10.1029/2001JD000390.
11 J. L. An, W. Zhang and Y. Qu, *Atmos. Environ.*, 2009, **43**, 3454–3459.
12 S. S. Park, S. B. Hong, Y. G. Jung and J. H. Lee, *Atmos. Environ.*, 2004, **38**, 293–304.
13 J. Stutz, B. Alicke, R. Ackermann, A. Geyer, S. H. Wang, A. B. White, E. J. Williams, C. W. Spicer and J. D. Fast, *J. Geophys. Res.-Atmos.*, 2004, **109**, DOI: 10.1029/2003JD004135.

Gordon McFiggans asked William Bloss: Linking to the previous two questions – in previous studies looking at interactive systems with interactions between reactive gases and surfaces, it's likely to be necessary to use a "moving column" or "stacked box trajectory" framework rather than a simple box model, owing to the strong vertical variability of heterogeneous sources and sink terms. Could you comment on how a box, without the vertical variability included in a column, could affect your interpretations and how robust the conclusions can be from a box model study? In particular, could you comment on how the spatial variability in the surface layer depth across, for example, variable street canyon profiles and intercepting variable wind directions can allow you to find dynamical dependencies from changes in sources and sinks that could lead to erroneous conclusions

William Bloss replied: HONO will vary in the vertical dimension to an extent dependent upon the sources (ground-based *vs.* homogeneous throughout a canyon, urban canopy or boundary layer) and mixing timescales. Considering that the HONO lifetime is of the order of 10 min (for photolysis at SZA = 40′) or longer, we might expect HONO to be relatively well mixed in a street canyon with a mixing timescale of a few minutes, but rather less so in the urban canopy layer or wider boundary layer. The conclusions drawn in the paper – that erroneous HONO source terms may be derived from PSS analysis of field observations made in the presence of locally spatially variable HONO (and NO_x, VOC) sources – should not be qualitatively affected by considering the wider domain, although the quantitative impacts will of course vary from the very idealised scenarios presented.

Dwayne Heard asked Siqi Hou: You do an analysis where you assume that the OH concentration at night is 1×10^6 molecule cm^{-3}. Some papers have shown that the night-time OH concentration is not that high in urban areas. Did you perform a sensitivity test to see the effect on your results of using a different value for the night time concentration of OH?

Lisa Whalley also remarked: I understand that OH observations were not made during this project. However, considering the supporting observations that were made (for example NO_x), do you think that it's correct to assume the same OH concentration in the urban and suburban studies? Please could you comment on the uncertainty that this assumption makes to your conclusion that homogeneous HONO formation is more important in urban areas than in the sub-urban area?

Siqi Hou replied: The assumption of same OH concentrations is a choice that has been considered due to the complex weather conditions in different sites during the measurements. The measurements included both haze and non-haze processes in urban and suburban areas of Beijing, the complexities in weather conditions and measurement sites make the assumption of OH concentrations difficult. Moreover, there are no previous measurements locally to be referred to, so that we have no sense about the general levels of OH concentrations. Therefore, to reduce the error as far as possible, we take a constant value to represent OH concentrations at night in both urban and suburban sites by considering previous studies by Huang *et al.*,[1] Ren *et al.*,[2] Li *et al.* [3] (values between 1×10^6 molecules cm^{-3} and 5×10^6 molecules cm^{-3} during the night-time in the summer period), Lu *et al.*[4,5] (pointing out the night-time OH concentrations in both urban and suburban areas were at a low level and changed little), and Spataro *et al.*[7] (using 1×10^6 molecules cm^{-3} as night-time OH concentrations in the winter in urban Beijing).

Moreover, the calculation results from the TUV model combined with the equation of observed OH *vs.* jO^1D (OH [cm^{-3}] = 4.0×10^{11} [s cm^{-3}] $\times$ jO^1D [s^{-1}])[5,6] showed an OH concentration of 9.3×10^5 molecules cm^{-3} during the sunrise period, which coincides well with the assumption.

Therefore, the use of 1×10^6 molecules cm^{-3} is a reasonable way to represent night-time OH concentrations in urban and suburban areas of Beijing.

For the uncertainties of the calculations of homogeneous production, there are no previous measurements locally to be referred to, and we have no sense about the general levels of OH concentration as mentioned above, so the discussion about the uncertainties caused by the assumption is meaningless.

1 G. Huang, X. L. Zhou, G. H. Deng, H. C. Qiao and K. Civerolo, *Atmos. Environ.*, 2002, **36**, 2225–2235.
2 X. Ren, H. Gao, X. Zhou, J. D. Crounse, P. O. Wennberg, E. C. Browne, B. W. LaFranchi, R. C. Cohen, M. McKay, A. H. Goldstein and J. Mao, *Atmos. Chem. Phys.*, 2010, **10**, 6283–6294.
3 X. Li, T. Brauers, R. Haeseler, B. Bohn, H. Fuchs, A. Hofzumahaus, F. Holland, S. Lou, K. D. Lu, F. Rohrer, M. Hu, L. M. Zeng, Y. H. Zhang, R. M. Garland, H. Su, A. Nowak, A. Wiedensohler, N. Takegawa, M. Shao and A. Wahner, *Atmos. Chem. Phys.*, 2012, **12**, 1497–1513.
4 K. D. Lu, Y. H. Zhang, H. Su, T. Brauers, C. C. Chou, A. Hofzumahaus, S. C. Liu, K. Kita, Y. Kondo, M. Shao, A. Wahner, J. L. Wang, X. S. Wang and T. Zhu, *J. Geophys. Res.-Atmos.*, 2010, **115**, 18.
5 K. D. Lu, F. Rohrer, F. Holland, H. Fuchs, B. Bohn, T. Brauers, C. C. Chang, R. Haeseler, M. Hu, K. Kita, Y. Kondo, X. Li, S. R. Lou, S. Nehr, M. Shao, L. M. Zeng, A. Wahner, Y. H. Zhang and A. Hofzumahaus, *Atmos. Chem. Phys.*, 2012, **12**, 1541–1569.
6 K. D. Lu, A. Hofzumahaus, F. Holland, B. Bohn, T. Brauers, H. Fuchs, M. Hu, R. Haseler, K. Kita, Y. Kondo, X. Li, S. R. Lou, A. Oebel, M. Shao, L. M. Zeng, A. Wahner, T. Zhu, Y. H. Zhang and F. Rohrer, *Atmos. Chem. Phys.*, 2013, **13**, 1057–1080.
7 F. Spataro, A. Ianniello, G. Esposito, I. Allegrini, T. Zhu and M. Hu, *Sci. Total Environ.*, 2013, **447**, 210–224.

Ugo Molteni said: I want to go back to Neil Donahue's comment and the difference between biogenic and anthropogenic scenarios. What we observed, and we have a poster on this topic, is that in HOMs formation, when the precursor is an aromatic compound, more than one OH free radical could be involved in the autoxidation process.

Jörg Kleffmann asked Siqi Hou: Whereas HONO and NO_x were measured at the same place at the suburban site, HONO and precursor measurements were separated by 5.5 km distance at the urban site. As shown in the paper by Crilley *et al.* (DOI: 10.1039/C5FD00224A) HONO levels can be highly variable throughout an urban environment. Could you comment on the risk of comparing apples with oranges when correlating non-co-located HONO and precursor data at the urban site?

Siqi Hou responded: Actually, it is not a question of apples and oranges. The variations of NO_2 concentrations in the two sites were compared to be synchronous and the accuracy of the measurements has been proved in our previous study. In detail, first, the concentrations and variations of NO_2 in the two sites were in good agreement. Comparison showed that NO_2 and NO_x concentrations at the two sites were at the same levels and had the same variation features (NO_2 Baolian = 0.97 × NO_2 ICCAS − 0.83; NO_x Baolian = 1.09 × NOX ICCAS − 8.35). This may because of the similar surrounding conditions at ICCAS and Baolian station with parallel sources of traffic and human activities around both as they are typical representatives of the urban area of Beijing.

Second, comparisons in Baolian station and IAP station (separated by 8.2 km distance) showed that the concentrations and variations of NO_2 in these two sites coincided well (NO_2 IAP = 0.92 × NO_2 Baolian + 9.254; NO_x IAP = 0.92 × NO_x Baolian + 18.415). This also reflects the similar NO_2 variations in IAP and ICCAS stations (separated by 5.2 km distance). In other words, NO_2 concentrations in Baolian station represent typical NO_2 concentrations of urban areas, and could be representative of NO_2 concentrations in ICCAS. Third, other previous studies and data from the environment ministry also showed little differences in NO_2 concentrations in the north of urban Beijing. For example, observation data from Spataro *et al.*[1] at Beijing University and Zhu *et al.*[2] at IAP from 17th August to 21st August showed similar concentrations and variations. Chen *et al.*[3] also point out little difference in the spatial distribution of NO_2 and NO_x concentrations. Fourth, the purpose of our article is to explore the HONO production and removal mechanisms by comparing the differences between urban and suburban areas in Beijing. Based on this purpose, we chose Huairou station to represent a suburban area and ICCAS and Baolian stations to represent an urban area. Since the concentrations of NO, NO_2 and NO_x vary a lot between urban and suburban areas, the differences of NO, NO_2 and NO_x concentrations between Baolian station and ICCAS stations are much smaller and could be ignored. Moreover, their differences have little influence on our exploration of the production and removal mechanisms of HONO. All in all, the differences of NO_2 and NO_x concentrations between Baolian station and ICCAS station have little influence on our exploration of HONO production and removal mechanisms between urban and suburban areas of Beijing, and data from Baolian station could be used to represent data in ICCAS station.

1 F. Spataro and A. Ianniello, *Journal Of the Air and Waste Management Association*, 2014, **64**, 1232–1250.
2 Y. W. Zhu, Q. Fu, P. H. Xie, W. Q. Liu, F. M. Peng, M. Qin, Y. H. Lin, F. Q. Si and K. Dou, *Spectroscopy and Spectral Analysis*, 2009, **29**, 1390–1393.
3 P. F. Chen, Q. Zhang, J. N. Quan, Y. Gao and M. Y. Huang, *Huanjing Kexue*, 2012, **33**, 4141–4150.

Roy Harrison asked: A critical value used in data interpretation in the Beijing study is the primary emission of HONO from road traffic which is taken as 0.8% of NO_x. How much do the HONO/NO_x experts feel that they can rely on this number? How much variability is associated with it and how is this number evolving as vehicle technology evolves? Most of the published data are now fairly old.

Siqi Hou replied: The ratio of 0.8% is first proposed by Kurtenbach *et al.*[1] in 2001 with tunnel measurements by considering the differences in combustion processes and fuel compositions. Meanwhile, tunnel measurements by Ackermann[2] (2000) showed a HONO/NO_x ratio of 0.65% in a European car fleet. In recent years 0.65% has been used in urban areas of China, including Beijing.[3,4] The choice of HONO/NO_x ratio is a confusing problem and it depends on the real traffic situation at the measurement site. However, there is another way to evaluate the influence of direct emissions referred to by Li *et al.*[5] Due to the conversion from NO_2 to HONO, the HONO/NO_x ratio would increase. And the lowest HONO/NO_x would be the upper limit for the emission factor when NO $\geq$ 20 ppb was used to catch the fresh emissions. For this measurement, the lowest HONO/NO_x ratio was higher than 0.8%. Thus, 0.8% is reasonable to evaluate the direct emission of HONO, which also reflects the final results discussed among the consortium.

1 R. Kurtenbach, K. H. Becker, J. A. G. Gomes, J. Kleffmann, J. C. Lörzer, M. Spittler, P. Wiesen, R. Ackermann, A. Geyer and U. Platt, *Atmos. Environ.*, 2001, **35**, 3385–3394.
2 R. Ackermann, Ph.D. thesis, Univ. Heidelberg, Heidelberg, Germany, 2000.
3 M. Qin, P. H. Xie, W. Q. Liu, A. Li, K. Dou, W. Fang, H. G. Liu and W. J. Zhang, *J. Environ. Sci. -China.*, 2006, **18**, 69–75.
4 J. L. An, W. Zhang and Y. Qu, *Atmos. Environ.*, 2009, **43**, 3454–3459.
5 X. Li, T. Brauers, R. Häseler, B. Bohn, H. Fuchs, A. Hofzumahaus, F. Holland, S. Lou, K. D. Lu, F. Rohrer, M. Hu, L. M. Zeng, Y. H. Zhang, R. M. Garland, H. Su, A. Nowak, A. Wiedensohler, N. Takegawa, M. Shao and A. Wahner, *Atmos. Chem. Phys.*, 2012, **12**, 1497–1513.

Neil Donahue asked: My understanding is that the ratio of cars to trucks in Beijing is a very strong function of time of day, with substantial truck traffic at night. What are the implications of this for your findings?

Siqi Hou responded: Stricter standards for diesel vehicles are implemented in Beijing, and the use of diesel trucks is forbidden in urban areas, which implies an effect of traffic types on the contribution of direct emissions. Moreover, research shows that vehicle exhaust emissions are diverse and affected by factors (including technical level, emission control devices and operation conditions) besides vehicle type.[1–3] And the characteristics of roads (altitude, temperature and humidity, road conditions, and traffic conditions) and the maintenance conditions (frequency, fuel type, levels) are also important. However, the traffic situations vary in different sites with local characteristics. As mentioned in our research, HONO, NO, NO_2, and NO_x concentrations in the suburban area often showed morning peaks around 9 : 00 and evening peaks around 20:00 in both haze and non-haze periods, which could be the result of direct emissions by local traffic.

1 L. Huang, P. K. Hopke, W. Zhao and M. Li, *Atmos. Pollut. Res.*, 2015, **6**, 1046–1054.
2 X. Shen, Z. Yao, H. Huo, K. He, Y. Zhang, H. Liu and Y. Ye, *Sci. Total Environ.*, 2014, **487**, 521–527.
3 H. Xue, S. Jiang and B. Liang, *Math. Probl. Eng.*, 2013, **2013**, 736285, 1-6.

Francis Pope asked William Bloss: If the percentage of HONO within vehicular exhaust has stayed reasonably constant over the last couple of decades, does this suggest that HONO output is not vehicle specific? Hence, does it suggest that HONO is related to something other than direct vehicle emissions?

William Bloss answered: The relationship between HONO abundance and vehicle emissions will be strongly dependent upon the environment and atmospheric conditions – contrasting, for example, an urban street canyon with the marine boundary layer. In the former environment we expect vehicle emissions dominate HONO formation; in the latter other processes will be responsible. The literature database on HONO emissions from vehicles is too sparse to confidently draw this conclusion. A limited number of measurements were reported in the 1980s, derived from vehicle emission technologies no longer representative of the modern fleet (these are concisely summarised in Kurtenbach *et al.*[1]).

Subsequently, Kirchstetter *et al.*[2] and Kurtenbach *et al.*[1] have reported mean fleet HONO/NO_x ratios of 0.3% and 0.8% respectively as whole fleet averages from (European) road tunnel measurements. More recently, Rappenglueck *et al.*[3] have reported HONO/NO_x fleet-averaged emission ratios of 1.7% derived from ambient measurements in Houston, Texas, while Xu *et al.*[4] found a range of 0.5–1.6% for traffic in Hong Kong.

Given the paucity of observations, and fleet composition and emissions technology differences with location and date, it is hard to draw conclusions over the existence of a trend in the values noted above. The limited data reported here (DOI: 10.1039/C5FD00224A) are suggestive of a higher HONO emission fraction relative to NO_x for the contemporary (2015) UK fleet – given the impact of even modest direct HONO emissions on urban air chemistry (*e.g.* Jenkin *et al.* [5]) further exploration is clearly required.

1 R. Kurtenbach, K. H. Becker, J. A. G. Gomes, J. Kleffmann, J. C. Lorzer, M. Spittler, P. Wiesen, R. Ackermann, A. Geyer and U. Platt, Investigations of emissions and heterogeneous formation of HONO in a road traffic tunnel, *Atmos. Environ.*, 2001, **35**, 3385.
2 T. Kirchstetter, R. Harley and D. Littlejohn, Measurement of nitrous acid in motor vehicle exhaust. *Environ. Sci. Technol.*, 1996, **30**, 2843.
3 B. Rappenglueck, G. Lubertino, S. Alvarez, J. Golovko, B. Czader, and L. Ackermann, Radical precursors and related species from traffic as observed and modeled at an urban highway 30 junction, *J. Air Waste Manage.*, 2013, **63**, 1270–1286.
4 Z. Xu, T. Wang, J. Wu, L. Xue, J. Chan, Q. Zha, S. Zhou, P. Louie and C. Luk, Nitrous acid (HONO) in a polluted subtropical atmosphere: Seasonal variability, direct vehicle emissions and heterogeneous production at ground surface, *Atmos. Environ.*, 2015, **106**, 100.
5 M. J. Jenkin, S. R. Utembe and R. G. Derwent, Modelling the impact of elevated primary NO_2 and HONO emissions on regional scale oxidant formation in the UK, *Atmos. Environ.*, 2008, **42**, 323.

Andreas Wahner communicated to Shengrui Tong: You show in Fig. 7 in your paper a steady decrease of the HONO/NO_2 mixing ratio after midnight and used data with constant NO_2 after midnight. How do you distinguish between changing loss processes and changing formation processes? How do you quantify conversion efficiency under these circumstances?

Shengrui Tong communicated in reply: As the nocturnal boundary layer was changing to be stable, the production and loss rates of HONO would reach a steady state.[1,2] In this steady state HONO would be formed by a first order conversion of NO_2 and the $HONO_{corr}/NO_2$ ratios would be constant.[3] Therefore, data of the second half of the night were used when $HONO_{corr}/NO_2$ ratios were changing less, which indicated that the conversion of NO_2 was steady. For calculating the formation and loss processes, we do it separately. Three main types of HONO formation pathways could be identified during the night by previous studies:[3–7] direct emission, homogeneous gas reactions and heterogeneous reactions. The contributions of the former two processes could be calculated directly using formulas (as shown in our manuscript, DOI: 10.1039/

C5FD00163C), thus the point is how to calculate the changing NO_2–HONO conversion process. Here we also calculated the NO_2–HONO conversion rate (C_{HONO}), which assumes HONO is linearly increasing with NO_2.[8,9] C_{HONO}, the combination of the two parameters, would be helpful to get detailed information on the NO_2–HONO conversion process.

1 J. Stutz, B. Alicke and A. Neftel, *J. Geophys. Res.-Atmos.*, 2002, **107**, DOI: 10.1029/2001JD000390.
2. K. Acker, D. Möller, R. Auel, W. Wieprecht and D. Kalass, *Atmos. Res.*, 2005, **74**, 507–524.
3 X. Li, T. Brauers, R. Häseler, B. Bohn, H. Fuchs, A. Hofzumahaus, F. Holland, S. Lou, K. D. Lu, F. Rohrer, M. Hu, L. M. Zeng, Y. H. Zhang, R. M. Garland, H. Su, A. Nowak, A. Wiedensohler, N. Takegawa, M. Shao and A. Wahner, *Atmos. Chem. Phys.*, 2012, **12**, 1497–1513.
4 R. Kurtenbach, K. H. Becker, J. A. G. Gomes, J. Kleffmann, J. C. Lörzer, M. Spittler, P. Wiesen, R. Ackermann, A. Geyer and U. Platt, *Atmos. Environ.*, 2001, **35**, 3385–3394.
5 T. W. Kirchstetter, R. A. Harley and D. Littlejohn, *Environ. Sci. Technol.*, 1996, **30**, 2843–2849.
6 J. Kleffmann, T. Gavriloaiei, A. Hofzumahaus, F. Holland, R. Koppmann, L. Rupp, E. Schlosser, M. Siese and A. Wahner, *Geophys. Res. Lett.*, 2005, **32**, L05818, DOI: 10.1029/2005GL022524.
7 G. Lammel and J. N. Cape, *Chem. Soc. Rev.*, 1996, **25**, 361–369.
8 B. Alicke, U. Platt and J. Stutz, *J. Geophys. Res.-Atmos.*, 2002, **107**, DOI: 10.1029/2000JD000075.
9 H. Su, Y. F. Cheng, P. Cheng, Y. H. Zhang, S. Dong, L. M. Zeng, X. Wang, J. Slanina, M. Shao and A. Wiedensohler, *Atmos. Environ.*, 2008, **42**, 6219–6232.

Francis Pope addressed William Brune and Saewung Kim: Does the influence of VOCs, which originate in areas external to the urban areas you are studying, suggest that future experiments should try and incorporate multiple areas both within and outside the urban area. This might allow for better understanding of the role of non-urban sources on urban pollution.

William Brune replied: This issue is important and has been a topic of studies for while. There is clearly an impact of the non-urban sources in some urban areas. However, I agree that more research is still needed to better quantify these impacts and to test regional models.

Saewung Kim answered: As we argued in the introduction section of our paper, in general the East Asian megacities have a dense population density and are surrounded by forest areas with little population. Therefore, our reports on the large missing reactivity in the suburban forest may represent an anecdotal snapshot of how poor our understanding is of BVOC photochemistry in heavy anthropogenic influences or BVOC emissions under air pollution stress. Follow up studies in these fronts are critical to address emerging regional and global air pollution issues.

Francis Pope asked Saewung Kim: Are there new analytical techniques that would enable more complete characterisation of the BVOCs present in your study?

Saewung Kim replied: There have been numerous technical breakthroughs in quantitative and qualitative analysis of atmospheric hydrocarbons and oxygenated volatile organic compounds. In this study, we deployed PTR-Q-MS for VOCs, the analysis serving as the basis for calculated OH reactivity estimations. The mass resolution for the analytical technique is limited to the unit mass. This prohibits estimating accurate reactivity from quantified VOC species as functional group information cannot be properly assessed at such a mass resolution. The recently developed high-resolution Time of Flight detector offers high mass resolution (up to 8000). Therefore, PTR-ToF-MS could allow us to identify compounds with different functional groups in the identical unit masses, which is critical for accurate assessments of OH reactivity from the observed VOC dataset. In addition, more information on VOC structural information has been recently investigated by advanced GCxGC or GC-ToF techniques. This is especially critical for the BVOC studies as many monoterpenoid ($C_{10}H_{16}$) and sesquiterpenoid ($C_{15}H_{24}$) species have a wide range of reactivity towards the oxidant depending upon their structures.

Dwayne Heard remarked: In central London, although the OH reactivity budget was pretty much closed during the summer ClearFlo project, the influence of biogenics and particularly their oxidation products (not measured but included in the modelled OH reactivity) was still important, and comprised a significant fraction of the OH reactivity.[1] I think that locations where you could look at the influence of biogenic emissions on the chemistry of urban areas, and the influence of anthropogenic emissions which travel into rural areas where there are significant emissions of biogenic species, would be very valuable.

1 L. K. Whalley, D. Stone, B. Bandy, R. Dunmore, J. F. Hamilton, J. Hopkins, J. D. Lee, A. C. Lewis and D. E. Heard, Atmospheric OH reactivity in central London: observations, model predictions and estimates of *in situ* ozone production, *Atmos. Chem. Phys.*, 2016, **16**, 2109–2122, DOI: 10.5194/acp-16-2109-2016.

Neil Donahue asked Sasha Madronich: Have you or Bernard Aumont ever looked at total OH reactivity in GECKO-A with some sort of Lagrangian model (including deposition and ventilation losses)?

Sasha Madronich replied: Total OH reactivity has been examined using the GECKO-A mechanism within Lagrangian box model simulations representative of

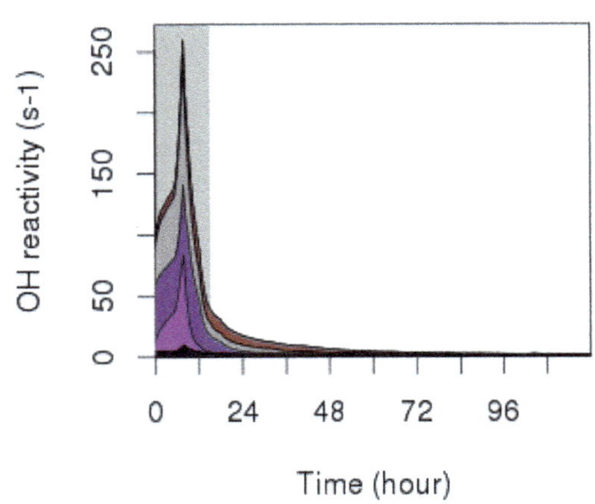

Fig. 1 OH reactivity for the Mexico City plume simulation. The grey area is for the Mexico City conditions. Black: CO, violet: NO, dark violet: NO2, grey: hydrocarbons, red: oxygenated organic compounds (Marie Camredon, private communication, 2016).

several environmental conditions. Fig. 1 shows the temporal evolution of the total OH reactivity in the simulated Mexico plume. Over Mexico City, the simulated total OH reactivity shows a morning peak of about 250 s^{-1} decreasing to about 50 s^{-1} in the mid afternoon. For this urban condition, more than 50 % of the total OH reactivity is due to the OH + NO_x reactions. The remaining reactivity comes from the reactions of OH with organic compounds, mostly hydrocarbons (i.e. primary) in the morning peak and both hydrocarbons and oxygenated (i.e. secondary) organic compounds during the afternoon. The OH reactivity decreases in the Mexico plume, reaching a value of 10 s^{-1} after 1 day of ageing. OH reactivity is then mostly due to organic compounds. It should be noted that the OH reactivity is expected to depend somewhat on the chemical mechanism being used, but not very sensitively. For typical intermediate compounds, the detailed kinetics are less important than the total material flow through that chemical pathway, and it is the product of rate coefficients and concentrations that determines OH reactivity.

Dwayne Heard asked William Brune: When using the model to calculate unmeasured oxygenated intermediates, can you give the value of the deposition rate that is used in the model, and whether the modelling results are sensitive to the value that is chosen?

William Brune replied: We chose a deposition rate of a day. However, we varied it from two hours to two days and did not see a change between the measured and modeled HO_x. So while many chemical species are sensitive to deposition rates, HO_x does not seem to be for this study.

Roy Harrison said: I would like to explore a different issue, which is based on the fact that most of the convincing atmospheric chemistry studies have been carried out on plumes that have become relatively homogeneous, so are not truly representative of the urban atmosphere. My question is what is the spatial variability of the OH radical in the urban atmosphere and how does this compare to its temporal variation? I assume that spatial variability is quite high. If we are going to model the urban atmosphere competently, what scale do we need to be working at to reflect this variation and its impact, or could we sensibly average out this variation? Has anyone studied this issue?

William Brune responded: We have wondered about the homogeneity of HO_x radicals in the boundary layer for a while. Fortunately, we have several aircraft studies in which we have flown at 300 m above ground level for long transects over cities. Our averaging time is equivalent to ~5 km, so these flight data do not give high spatial resolution. What we have seen is that OH and HO_2 are not highly variable, especially compared to NO_x, on the timescale and spatial scale of the measurement. There have been several studies that compare point measurements to long-path measurements of some VOCs and oxygenated VOCs or NO_y species in urban areas. In some cases, the point and long-path measurements are in good agreement, suggesting that the chemical environment is uniform enough despite the diversity of sources. So, for some aspects of the chemistry, it will be possible to average the measurements over the variations and still achieve a valid comparison of measured and modeled chemical species. Generally, I would think that OH, which is a highly buffered atmospheric chemical species (*i.e.*, its sources and sinks are often inter-related), can be modeled competently in an averaged sense. However, HO_2 may be harder to model competently because of its longer lifetime and thus greater potential for un-captured variability. Nonetheless, our study of ozone production as a function of averaging time (Fig. 8 in our paper) would suggest that even for non-linear ozone production chemistry, it can be possible to get valid comparisons of modeled and measured radicals in some urban environments.

Dwayne Heard asked: Measurements of VOCs with high time resolution are missing in urban areas, and would be useful to understand any fast changes in OH concentrations areas with heterogeneous sources and sinks. Typically, when a wide range of VOCs are measured with GC-FID systems, the averaging time can be as long as 30–60 min, and these values are often then interpolated to give 15 min data to use as input into models. This time resolution is usually fine for testing of chemical mechanisms *via* comparison of measured and modelled levels of radicals, but for calculation of ozone production (or other integrated quantities perhaps linked to personal exposure), the averaging periods used in the model may be important as the chemistry is highly non-linear, and using average values of VOCs and NO_x rather than the full range of short-term fluctuations may result in model error.

William Brune replied: It would be good to have higher time resolution VOC measurements in urban areas. Fortunately, the measurements of some VOCs and oxygenated VOCs can now be made on minute (or less) time scales. I agree that the averaging period used in the model can be quite important for calculating the ozone production rate. That is why we compared the calculated ozone production rates using 20-second HO_2 and NO measurements and the same measurements averaged into 1-minute, 10-minute, and 1-hour bins. As shown in Fig. 8 of our paper, it does make a difference in the calculated ozone production rate, but by less than about 30% for even the most extreme cases at higher NO. However, it would be easy to construct scenarios in which the difference could be much greater. So this issue of time scales for non-linear chemistry such as ozone production is very important.

William Bloss said: On the comparison of measured total OH reactivity, k_{OH}, with that calculated from individual measured VOCs and their degradation products *via* atmospheric chemical models: typically, it is not possible to measure more than a small fraction of the VOCs actually present. Comprehensive chemical mechanisms such as the MCM may be used to model the degradation of a small number of parent VOCs, to calculate the abundance of their degradation products and hence the contribution of these unmeasured species to the total OH reactivity. However, this approach may be biased due to the mechanism development protocols – for example, in the case of the MCM, for *e.g.* H-abstraction reactions, where a number of channels are possible for larger reactions. In this case the MCM Protocol[1] is to maintain up to four degradation channels, but for molecules with more degradation pathways, to either disregard minor channels (up to 10% of the total reaction) and scale the remaining channels proportionately to maintain the overall rate constant, or to gather together radical products possessing similarities in their subsequent chemistry. While this approach has no impact on the total OH reactivity of the species considered, it does alter the distribution of degradation products – and hence their total reactivity – as reaction at more substituted C atoms will proceed faster and be favoured, while other channels are neglected, producing slightly different daughter product reactivity.

However the impact on the overall reactivity is likely small, and the uncertainty also modest *cf.* the many other approximations involved, in most cases. It would be interesting to compare the MCM-predicted and GECKO-predicted reactivity in some test cases to quantify this effect.

1 M. E. Jenkin *et al.*, *Atmospheric Environment*, 1997, **31**, 81–104.

Saewung Kim replied: In our earlier publication at the Taehwa Research Forest, we illustrated that the unmeasured OVOCs mostly from isoprene oxidation could account for as much calculated OH reactivity as from all the observed trace gases. As suggested, constraining all the unobserved multi-generational oxidation products is definitely an important step towards closing the OH reactivity budget. However, including this study, most box model studies have a substantial uncertainty in dry deposition and aerosol uptakes, especially on liquid water surfaces, which also needs to be addressed.

Andrew Rickard addressed William Bloss and William Brune: With respect to Dr Bloss's comment on how the Master Chemical Mechanism (MCM; http://mcm.leeds.ac.uk/MCM) construction protocol rules[1] simplify the chemistry through disregarding any channels representing less than 5% of the total reaction, the remaining channels are scaled proportionately to maintain the same overall rate coefficient for the molecule of interest, thereby maintaining the overall OH reactivity for that compound. However, this simplification will have a relatively minor effect on the product distributions, and hence a minor effect on the overall OH reactivity of the system, with respect to other mechanistic uncertainties (i.e. the SAR estimations themselves) and other strategic chemical simplifications applied. The MCM is sometimes regarded as a highly detailed "semi-explicit" chemical mechanism as the construction protocols apply a number of necessary simplification rules in order to limit the chemical complexity to tens of thousands of reactions rather than the millions of reactions needed to describe the chemistry explicitly. These simplifications include: initiation criteria, channel probability, product degradation and parameterisation of peroxy radical permutation reactions ($RO_2 + RO_2/R'O_2$). The Generator for Explicit Chemistry and Kinetics of Organics in the Atmosphere (GECKO-A) is a unique tool for the automatic writing of explicit chemical schemes of organic species (based on similar mechanism construction rules as the MCM) and their partitioning between the gas and condensed phases.[2] (Auto)generation of explicit chemical mechanisms containing millions of reactions are useful for understanding the fundamental chemical mechanisms of individual compounds in the laboratory/chamber but reduction protocols need to be applied in order for them to be used in complex chemical environments. It would be interesting to evaluate the chemical evolution of organic compounds in chamber experiments using MCM and explicit GECKO-A simulations and compare the overall OH reactivity calculated under a range of chemical conditions. With regards to other chemical uncertainties with respect to understanding OH reactivity and hence photochemical ozone formation chemistry in urban environments, are we measuring all the major chemical contributors from important biogenic and anthropogenic sources and including them correctly in our models? (e.g. high molecular weight diesel related hydrocarbons[3]) and do we fully understand the chemical reactivity and ozone formation potentials of these important systems (e.g. aromatic photochemistry[4])?

1 M. E. Jenkin *et al.*, *Atmospheric Environment*, 1997, **31**, 81–104.
2 B. Aumont *et al.*, *Atmos. Chem. Phys.*, 2005, **5**, 2497.
3 R. E. Dunmore *et al.*, *Atmos. Chem. Phys.*, 2015, **15**, 9983–9996.
4 C. Bloss *et al.*, *Atmos. Chem. Phys.*, 2005, **5**, 623–639.

William Brune responded: It would be interesting to do chamber studies to examine the contribution of VOCs and their products, especially oxygenated products, to the OH reactivity. The comparisons between measured and calculated OH reactivity vary greatly from study to study. In some studies, the missing OH reactivity may be due to missing measurements of important VOCs, such as the high molecular weight diesel-related hydrocarbons. However, in other studies, such as our study in New York City where only a fairly standard suite of hydrocarbons was measured, there does not appear to be any significant missing OH reactivity, which would imply that the OH reactivity there is well understood. The greatest problem seems to be not in urban environments but instead in forests. Bakersfield is an interesting case because its OH reactivity has a significant urban component but also an equally significant biogenic component that has not been fully characterized. In this case, there are still questions about the chemical reactivity and ozone formation potentials of these chemical systems. This is why we developed the Measurement of Ozone Production Sensor (MOPS) so that we could measure the ozone production rate and then compare it to the calculated ozone production rate. Unfortunately, the MOPS did not work particularly well in the CalNex Bakersfield study. The MOPS measurement has technical challenges to resolve but is showing some promise for providing a constraint on calculated ozone production in the same way that OH reactivity measurements do on calculated OH reactivity.

Sasha Madronich asked William Brune: You showed in Fig. 8 in your paper that the production of ozone by the reaction $HO_2 + NO \rightarrow OH + NO_2$ would be greatly enhanced if HO_2 is regenerated rapidly from the reaction $OH + NO + O_2 \rightarrow HO_2 + NO_2$.

The concomitant regeneration of NO_2 from NO is also a net production of ozone. Could one surmise that the overall rate of ozone production is roughly twice the HO_2-driven values shown in the figure, *e.g.* approaching 50 ppb h^{-1}?

William Brune replied: The answer is "yes", the calculated ozone production rate will approach 50 ppbv h^{-1} for Bakersfield. We have developed an instrument, the Measurement of Ozone Production Sensor (MOPS), to measure the ozone production rate. We had one in Bakersfield but unfortunately it did not work well for that study. But in other urban areas on some high ozone days, MOPS measured ozone production rates that exceed 50 ppbv h^{-1}. These rates are not generally sustained for more than an hour or two, but they do contribute to high levels of ozone. If the measured ozone production rate is accumulated over the day, then our measurements suggest that the cumulative ozone produced exceeded the observed ozone by a factor of two or so. This difference would suggest that ozone is being made at the location of the MOPS measurement and then transported both vertically through the boundary layer and horizontally to other regions.

Francis Pope asked Saewung Kim: Can you comment on the possible effects of climate change (both increasing CO_2 and temperature) on BVOCs. What are the likely effects on urban pollution?

Saewung Kim replied: Although I am not a climate scientist, in my opinion, our community has done a better job in developing future CO_2 emission scenarios than predicting BVOC emission changes from different biotic and abiotic environmental changes that the future would bring on. First of all, it is well known that under the extreme weather conditions that may be caused by climate change, BVOC emission models such as MEGAN do not perform well. In addition, the socioeconomical changes in the future will be certainly reflected in land use changes such as in recreational and economic plantation practices. In the model frameworks, these uncertainties certainly can cause substantial error bars for predicting future air pollution. Last but not least, increasing temperature will certainly affect reaction rates in the tropospheric photochemical system, which is known for being highly non-linear. Therefore, these multi-faceted impacts from biotic and abiotic changes should be carefully evaluated with a comprehensive model framework that is able to constrain all the potential environmental changes for future air quality predictions.

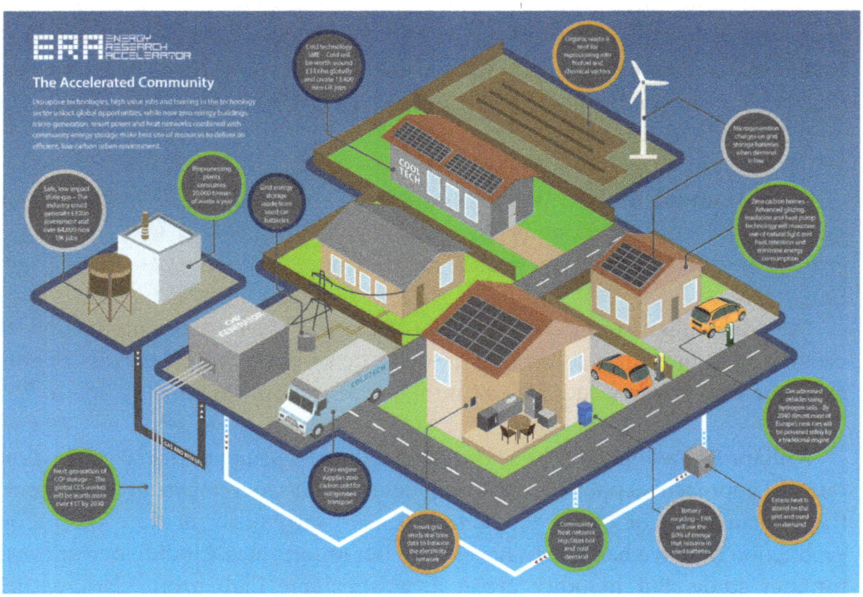

Fig. 2 Illustration of future smart grids and micro-generation, courtesy of the ERA consortium (www.era.ac.uk).

Paul Monks added: When looking at the impact of climate change on air quality it is clear as shown in a recent review[1] that the near term is dominated by emission reductions over the influence of climate change.

1 E. von Schneidemesser, P. S. Monks, J. D. Allan, L. Bruhwiler, P. Forster, D. Fowler, A. Lauer, W. T. Morgan, P. Paasonen, M. Righi, K. Sindelarova and M. A. Sutton, Chemistry and the Linkages between Air Quality and Climate Change, *Chem. Rev.*, 2015, **115**, 3856–3897, DOI: 10.1021/acs.chemrev.5b00089.

Nick Hewitt said: I don't think it's correct to make assumptions about climate change and biogenic VOC emissions in future - there are too many remaining uncertainties about the effects of water and nutrient availability and elevated carbon dioxide and other pollutant concentrations on BVOC emission rates. In addition native tree species are being replaced by high-yield crop species and exotic species and this change in species composition might hugely affect biogenic VOC emission rates.

Andreas Wahner remarked: In future, with improved exhaust treatment (Euro-6) and perhaps more and more electric cars, current emissions will most probably be reduced. But this will not solve the air quality problem in cities because, as we already know, other substances will be emitted (for example HNCO, NH_3 (some forms of EURO-6)), which will show different chemistry and impacts on air quality and health. Therefore, additional research and observational measurement is most likely needed.

Neil Donahue said: Going back to Astrid Kiendler Scharr's comment yesterday – if we really are going to seriously address climate change, the developed world needs to reduce CO_2 emissions by 5 or 10% per year indefinitely. One third of these emissions come from mobile sources right now – so I can't see a solution to this without having mostly electric cars in the very near future, with the electricity coming from non-fossil power sources. Unless of course we have biodiesels, but then we need too much land to produce the fuel.

Matthew Hort commented: Consolidation of emissions into specific locations (*e.g.* through a move to electric transport and continued large power station generation) does indeed present the risk of localised exceedances. However, it is easier to put abatement strategies in place and also the population density around such sites is likely to be far lower than that in cities (*i.e.* close to traffic) and so the actual impact could be less. The distribution of emissions and the impact of changes is highly complicated and this discussion goes to highlight how science and policy need to approach these challenges in a cross cutting and holistic way.

Xavier Querol said: Regarding the discussion on the issue of NO_x abatement and the electrification of the vehicle fleets. It was argued that this policy will move NO_x emissions from the urban areas to the place where power generation takes place.

This might not be the case. De-NO_x control systems in power plants, such as SCR systems, are implemented in power plants in countries such as the Netherlands. And these not only reduce NO_x but also Hg emissions. This will be increasingly implemented in Europe in the mid-term. Emission controls in both large- or mid-size combustion plants can be more easily implemented and monitored than those in private cars.

Paul Monks remarked: With respect to the question of air quality and future energy mixtures both in terms of displacement from roadside to "power station" from electric vehicles and the nature of future urban micro-generation. I think Fig. 2 (from ERA, http://www.era.ac.uk/) demonstrates a key challenge for air quality in terms of smart grids and micro-generation embedded in cities. There may not be a centralised future.

Faraday Discussions

PAPER

Urban particulate matter pollution: a tale of five cities

Spyros N. Pandis,*[ab] Ksakousti Skyllakou,[a] Kalliopi Florou,[a] Evangelia Kostenidou,[c] Christos Kaltsonoudis,[a] Erion Hasa[a] and Albert A. Presto[d]

Received 6th December 2015, Accepted 18th December 2015

DOI: 10.1039/c5fd00212e

Five case studies (Athens and Paris in Europe, Pittsburgh and Los Angeles in the United States, and Mexico City in Central America) are used to gain insights into the changing levels, sources, and role of atmospheric chemical processes in air quality in large urban areas as they develop technologically. Fine particulate matter is the focus of our analysis. In all cases reductions of emissions by industrial and transportation sources have resulted in significant improvements in air quality during the last few decades. However, these changes have resulted in the increasing importance of secondary particulate matter (PM) which dominates over primary in most cases. At the same time, long range transport of secondary PM from sources located hundreds of kilometres from the cities is becoming a bigger contributor to the urban PM levels in all seasons. "Non-traditional" sources including cooking, and residential and agricultural biomass burning contribute an increasing fraction of the now reduced fine PM levels. Atmospheric chemistry is found to change the chemical signatures of a number of these sources relatively fast both during the day and night, complicating the corresponding source apportionment.

1. Introduction

Large urban areas concentrate human activities and have been traditionally characterized by large air pollutant emission rates per surface area. Therefore a number of these cities have traditionally been air pollution hotspots. The air quality problems in these areas include high ozone levels, high particulate matter concentrations, low visibility, high concentrations of toxic compounds, *etc.* In this work we will concentrate on fine particulate matter (particles smaller than 2.5 μm, $PM_{2.5}$, and smaller than 1 μm, PM_1) because they represent probably the single most important air pollution threat to human health in urban areas.

[a]*Dept. of Chemical Engineering, University of Patras, Patras, Greece. E-mail: spyros@chemeng.upatras.gr*

[b]*Dept. of Chemical Engineering, Carnegie Mellon University, Pittsburgh, PA, USA*

[c]*Institute of Chemical Engineering Sciences (ICEHT/FORTH), Patras, Greece*

[d]*Center of Atmospheric Particle Studies, Carnegie Mellon University, Pittsburgh, PA, USA*

There have been excellent historical reviews of urban air pollution problems.[1,2] One of the main insights of these reviews is the evolution of air pollution in an urban area as it develops both in size but also from a technological point of view (Fig. 1). In this work we will focus on the change in the role of atmospheric chemistry during this evolution. We will draw data and insights from five urban areas that have had historically high pollution levels and have also been studied during the last decade with state-of-the-art scientific measurement and modeling tools. These cities are: Pittsburgh and Los Angeles in the U.S., Athens and Paris in Europe, and Mexico City in Central America.

2. The worst of times

All the urban areas in our case studies had at one point in time extremely high PM levels. The magnitude of these values remains uncertain because of either the lack of measurements at that time or the use of metrics (dust fall in Pittsburgh, total suspended particulates (TSP) in Los Angeles and Mexico City, black smoke in Europe) that are difficult to translate to $PM_{2.5}$ or PM_{10} values.

Pittsburgh was synonymous with "smoke" in the early 1900s. Estimated levels of TSP in 1912 were around 1000 μg m^{-3} according to Davidson.[3] The major source of pollution was coal burning in the steel mills of the city.

Los Angeles suffered from smog even at the beginning of the 20th century due to industrial emissions. The exact levels are difficult to estimate, but it has been reported that during one day in 1903 the residents mistook the smog for an eclipse of the sun. Serious control efforts, though, only started in the 1950s – targeting mainly industrial sources – and in the 1960s – targeting cars. During the 1973 California Aerosol Characterization Experiment (ACHEX) study TSP levels varied from a low of 100 μg m^{-3} in Pasadena to over 230 μg m^{-3} in Riverside with organics, sulfates, and nitrates all making significant contributions.[4]

Average black smoke levels in Athens in 1975 (ref. 5) were around 75 μg m^{-3} with certain stations in the center of the city reporting annual average values of above 200 μg m^{-3}. In the mid-1990s transportation was responsible for around 70% of the PM emissions, industry for 15%, and domestic heating for another 15%.[6]

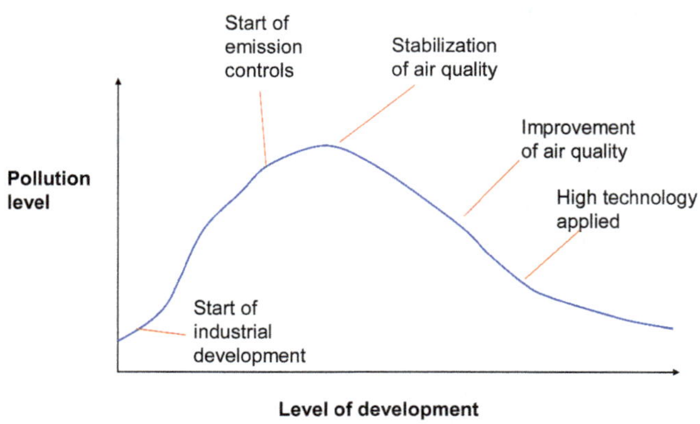

Fig. 1 Schematic of the pollution level in an urban area as a function of its level of development.

Average black smoke levels in Paris in the late 1960s were around 75 μg m^{-3} based on the earliest such measurements.[2] During 1987–92 (ref. 7) the average black smoke concentrations had been reduced to 32 μg m^{-3}. At that time the major PM_{10} emission sources were traffic and heavy industry.[8]

Annual average TSP levels in Mexico City in the late 1980s exceeded 300 μg m^{-3} and at that time it was considered the most polluted urban area in the world according to the United Nations.[9] The annual average PM_{10} value in 1997 was 80 μg m^{-3}. The major emission sources of PM_{10} according to a 2000 inventory were on-road motor vehicles (31%), other mobile sources (20%), stationary sources (27%), and vegetation and soils (17%). Mobile sources represented 76% of the $PM_{2.5}$ emissions.

3. The age of wisdom

Efforts to reduce the particulate matter concentration in Pittsburgh started in the 1910s with ordinances asking industry to improve its combustion processes but were in general ineffective.[3] It was not until the 1940s and the Donora air pollution disaster in October 1948 that serious progress was made. Annual average TSP levels were reduced to around 200 μg m^{-3} in the 1950s and approximately 100 μg m^{-3} in the 1960s. These improvements were made possible by strict control of local industrial emissions using ordinances by the county. Additional improvements have been the result of controls in regional industrial emissions (mainly coal-fired power plants) and national reductions of automobile emissions. The annual average $PM_{2.5}$ concentrations are now around 15 μg m^{-3}.

The efforts to improve air quality in Athens focused initially on transportation and included a retirement plan for old vehicles in 1991 (150 000 cars were retired), a national inspection programme, replacement of old buses, *etc.* These changes, together with the introduction of newer cars, resulted in significant reductions of the black smoke concentrations by the mid-90s.[10] PM_{10} emissions in Athens for 2003 were 21 ktons per yr and had a 13% contribution from transportation, 63% from industry, 18% from residential combustion, and the rest from smaller sources (waste disposal, off-road sources, *etc.*).[11] Major expansions of the subway system followed in early 2000 leading to average PM_{10} levels of around 55 μg m^{-3} in 2008 in the center of the city. The financial crisis was accompanied with a reduction of PM_{10} to around 30 μg m^{-3} in 2011. However, increased concentrations, especially in wintertime due to wood burning, have led to an increase of the annual average to 35 μg m^{-3} in 2012 and 40 μg m^{-3} in 2013.

Reductions in industrial and transportation emissions have also helped reduce the PM levels in our other case studies. The annual average concentration of $PM_{2.5}$ in Downtown Los Angeles in 2013 was 18 μg m^{-3}. The average Mexico City $PM_{2.5}$ levels have been reduced to approximately 25 μg m^{-3} in 2014. Annual average urban background $PM_{2.5}$ concentrations in Paris were approximately 17 μg m^{-3} during 2012–14 down from 21 μg m^{-3} in 2002–04.[12] The black smoke concentrations in Paris were reduced from 75 to around 20 μg m^{-3} from 1970 to 1995.

4. The age of foolishness

The mental model that has been traditionally used for urban air pollution is that of a practically closed system in which the local emissions by transportation and heavy industry lead to high particulate matter concentrations when the meteorological conditions are favorable. Most of the particulate matter is implicitly assumed to be primary. The reductions of PM concentrations that have taken place in the case studies discussed here have in general supported this model at least up to some point in time. However, its applicability as the levels of pollution are reduced may be questionable. Urban areas eventually move far enough to the right of the peak of the curve of Fig. 1 and as a result the above mental model still used by the public, policy makers, and even scientists starts having significant weaknesses or even collapses completely. There are a number of reasons for this collapse which are discussed below.

Primary *versus* secondary

Measurement campaigns in all case studies considered in this work show that most of the fine particulate matter (PM_1 or $PM_{2.5}$) is secondary. The secondary components include practically all the sulfates and ammonium nitrate, but also oxidized organic compounds. The major primary components in fine PM are elemental (or black) carbon, metal oxides and crustal material, and fresh organic particulate matter. The estimated primary and secondary components of fine PM for the five case studies are shown in Fig. 2. In all cases the secondary contribution exceeds 50% and is as much as 75% with the exception of recent winters in Athens. This is due to the high biomass burning (the corresponding emissions represent approximately 40% of the average fine PM) in Athens due to the financial crisis during that winter. It can actually be viewed as a movement towards the left in the level of development in Fig. 1.

These results do stress the problems created when one examines the particle emission inventories and the sources corresponding to them to reach broad conclusions about the causes of air quality problems in developed urban areas. There is little connection between the two because of the dominant role of atmospheric chemistry. One of the major implications of chemical production is the time delay between emission and production of the corresponding secondary

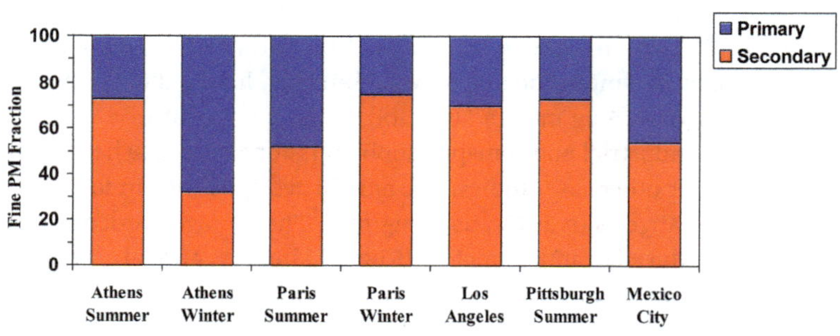

Fig. 2 Primary and secondary components of fine PM for the case studies. The Athens results are from the summer of 2012 and winter of 2013, the Paris ones are from the center of Paris during the MEGAPOLI campaigns,[13] the Los Angeles data correspond to measurements in Pasadena during CalNex,[14] the Pittsburgh results are from Wittig *et al.*[15] and the Mexico City results are from MILAGRO.[16] Sulfate, nitrate, ammonium, and oxygenated OA were assumed to be secondary with the rest of the fine PM components primary.

particulate matter. This implies that even if the reactions may take place over the urban area in question, the precursors may have originated elsewhere, contributing to the second misconception: the role of local emissions.

Long-range transport

The pioneering particulate matter studies in Los Angeles in the 1970s and 80s have helped shape our urban pollution mental model. However, Los Angeles is geographically a rather special urban area. The relatively clean Pacific Ocean is most of the time upwind of the city and as a result the background fine PM concentrations transported to Southern California are almost always only a few μg m^{-3}. These were negligible compared to the 50–100 μg m^{-3} over the most polluted parts of the basin thirty or forty years ago and therefore the assumption of a "closed" system was easily justifiable. However, all other urban areas discussed here have upwind (depending on wind direction) potentially polluted areas. The assumption of negligible background contributions was probably applicable at the historical peak pollution periods for each area, but can become more and more problematic as the urban area develops.

Fig. 3 illustrates the contributions of local, mid-range, and long-range sources to some of the major $PM_{2.5}$ components in the center of Paris. These source-area contributions have been calculated by combining the PSAT source apportionment algorithm[17] and the PMCAMx chemical transport model.[18] The concentrations resulting from emissions outside the Paris metropolitan area but within 500 km of the centre of Paris are considered contributions of mid-range transport. Those resulting from emissions more than 500 km away from Paris are considered long range transport.

The importance of the local sources is quite different depending on the PM chemical component. For primary components like elemental carbon the local sources dominate, contributing 64% during the summer and 57% during the winter. Most of the remaining elemental carbon (30% during the summer and 39% during the winter) is predicted to originate from sources within 500 km of the city which includes all of France, Belgium and the Netherlands, half of Germany, Switzerland, northern Italy and south U.K. The contributions were similar for fresh primary OA but a little lower (*e.g.*, 52% was local during the summer) due to the partial evaporation of these organic particles after emission. However, the situation was very different for the secondary components. Less than 10% of the secondary material was due to local emissions. For example, the sulfate due to local SO_2 emissions was 6% during the summer and 17% during the winter, while the local secondary OA was 7% and 2% respectively. More than half of the sulfate was due to long-range sources, while half of the SOA during the summer was due to medium range sources (Fig. 3).

The PMCAMx/PSAT prediction that on average 80% of the fine PM is due to emissions in other areas is consistent with the analysis of measurements of the MEGAPOLI campaigns inside and outside Paris by Beekman *et al.*[13] which suggested that 70% of the $PM_{2.5}$ is transported on an annual average basis.

Reduction in emissions of "traditional" sources

The success of national or international-scale control strategies during the last couple of decades in reducing pollution at the urban scale is often not well

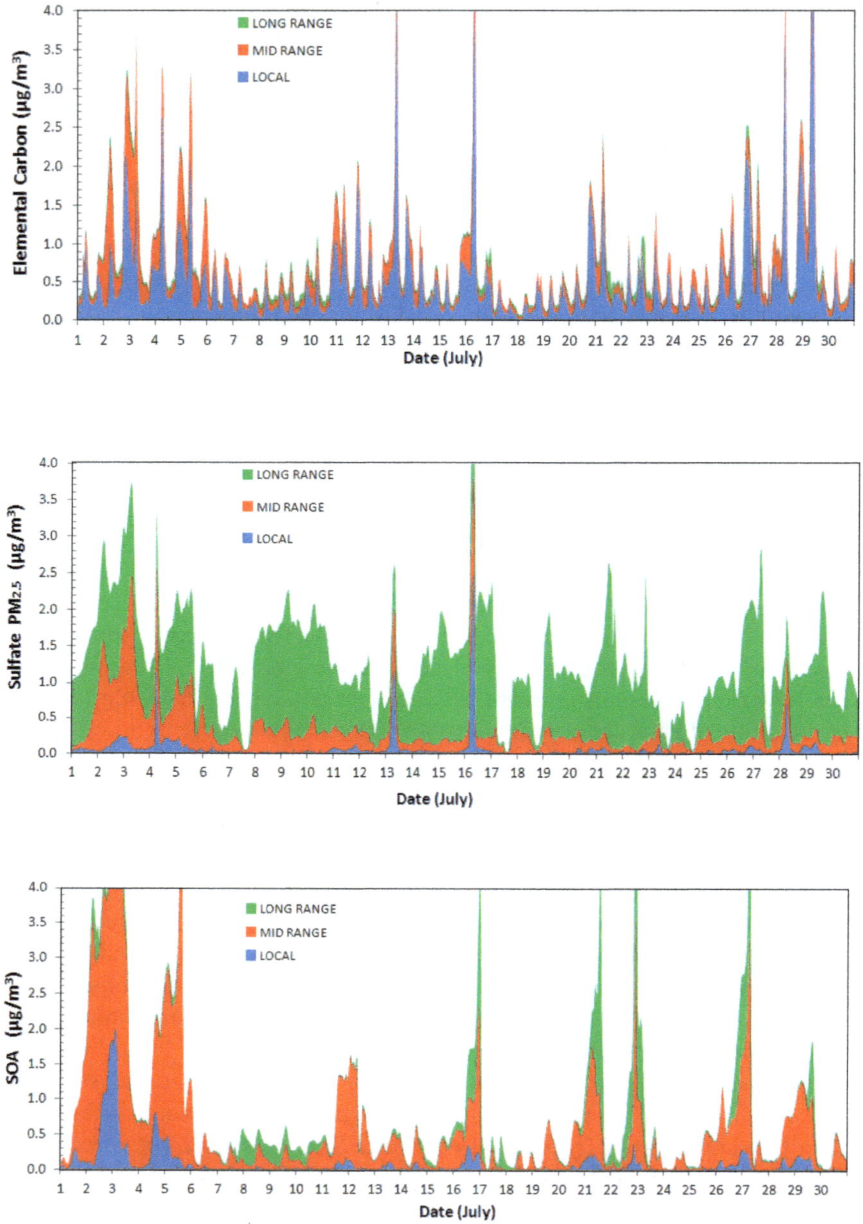

Fig. 3 The contributions from the different source areas (local, mid range and short range) to EC, sulfate and SOA concentrations in the center of Paris for the summer period using the approach of Skyllakou *et al.*[18]

appreciated. For example, there have been continuous reductions of emissions by cars per kilometre driven during the last twenty years in Europe. From the Euro 1 standards of 1992 to the Euro 5 standard in 2009 there has been a reduction of the allowed PM emissions per kilometre by a factor of approximately 30. Part of this reduction in emission factors has been offset by an increase of the activity levels (kilometres driven) resulting in an estimated reduction of $PM_{2.5}$ levels by cars in Europe of only approximately 30% from 1990 to 2010. The percentage reductions of emissions inside the urban centers often exceed those of the country on average. A rough estimate of the importance of these primary transportation and industrial emissions can be obtained by summing the hydrocarbon-like OA (HOA) of the AMS, the elemental carbon, and the metal oxide/crustal material contributions for the five case studies (Fig. 4). This is a rough approximation of the contributions of the "usual suspects" but is still instructive. The contributions of these sources are in all cases less than 40% of the fine PM and in some cases less than 20%.

Increase in importance of "non-traditional" sources

As the emissions of the traditional major sources have declined the emissions of other uncontrolled sources are becoming increasingly important as a fraction of the total. One of them appears to be emissions from cooking. These have a rather long history starting with Los Angeles and the pioneering work of the group of Glenn Cass in Caltech. Rogge *et al.*[19] and Schauer *et al.*[20] estimated that cooking (mainly barbecuing) was responsible for 20% of the organic aerosol in the 80s and early 90s in Los Angeles. More recently, Aerosol Mass Spectrometry (AMS) measurements have been used to quantify a cooking organic aerosol (COA) factor. Our measurements in Athens during the summer of 2012 and winter of 2013 suggest that this represents 15–20% of the total OA. The assignment of this factor to cooking has been mainly done based on its temporal pattern (peaks at traditional meal times). However, there is concern that contributions by other sources may be included in this factor. To test this hypothesis we performed similar AMS measurements in two Greek cities (Athens and Patras) both in summer and winter and performed the PMF analysis independently for each area and season (Fig. 5). The COA spectra found in Athens and Patras in the summer were the same (the theta angle between the corresponding vectors was less than 10°). The same was true for the winter COA spectra in the two cities. However the summer and winter spectra were different (the theta angle was equal to 25°). This is discussed in Section 6 below.

Relatively high COA concentrations have been found in the other urban case studies considered here. Cooking OA has been estimated to contribute 17% of the OA during winter and 13% during summer in the center of Paris.[21,22] The

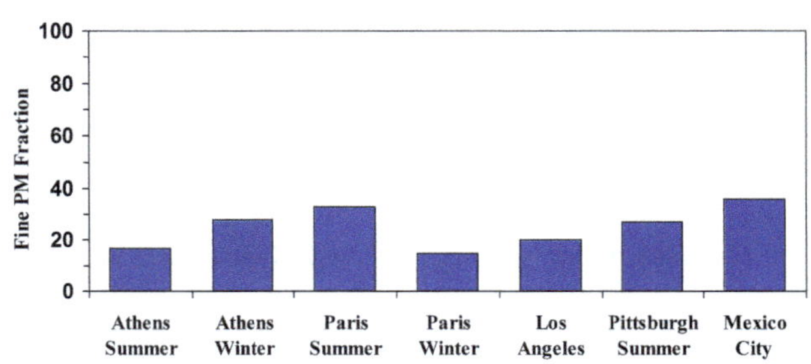

Fig. 4 Contributions of primary industrial and transportation contributions to fine PM levels in the five case studies.

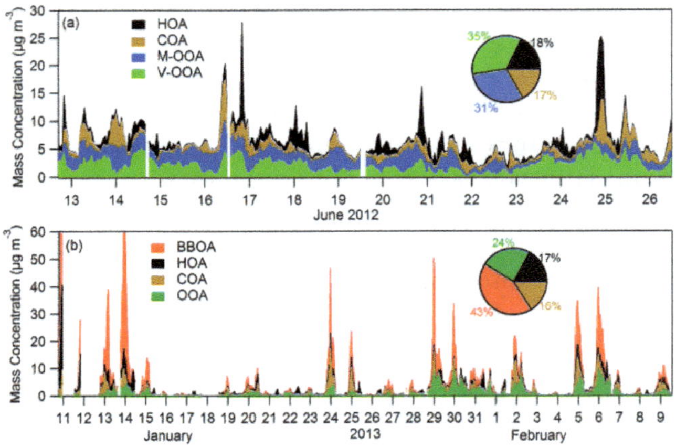

Fig. 5 Concentrations and composition of OA in Athens during the summer of 2012 and the winter of 2013. The various components shown are: hydrocarbon-like OA (HOA), cooking OA (COA), moderately oxidized OA (M-OOA), very oxidized OA (V-OOA) and biomass burning OA (BBOA).

corresponding contribution for Los Angeles (Pasadena) during CalNex was estimated to be 17% of the total OA.[23] Robinson *et al.*[24] estimated that cooking contributed on average 10% of the OA in Pittsburgh.

Biomass burning is another often underappreciated fine particulate matter source for urban areas. Residential heating during the winter can be a major source of particulate matter even in rather warm places. The example of Athens during the last few years is instructive. Due to the financial crisis the Greeks have increased their use of fireplaces and woodstoves for heating. The result is biomass burning OA concentrations that often exceed 50 μg m^{-3} during the night, reaching – when the meteorological conditions are right – levels as high as 150 μg m^{-3} (Fig. 5).

While the pollution in Athens can be viewed as an isolated case due to the financial situation, there is evidence that these biomass burning emissions are important for other areas in our case studies. For example, Crippa *et al.*[25] found that biomass burning for residential heating was responsible for around 33% of the OA in Paris during the wintertime MEGAPOLI study. This was due, according to the authors, not to emissions inside the city but to those in the surrounding areas. The contribution could be even higher if a fraction of the nighttime semivolatile oxygenated organic aerosol (representing 18% of the total) identified in the same study was chemically aged biomass burning OA. In Los Angeles, recent studies[26] estimated that biomass burning was responsible for 10% of the annual average OA during 2009–10 in both Central LA and Riverside. The wintertime contributions of biomass burning OA peaked during winter. For example, during December 2009 its contribution was approximately 35% of the total OA in Riverside. The biomass burning contribution to OA during wintertime in Pittsburgh was estimated to be around 10%.[27]

Open biomass burning (fires, agricultural burning, *etc.*) can be also a significant source of OA in urban centers that are downwind of such fire activity. Aiken *et al.*[28] concluded that biomass burning in areas just outside the Mexico City

Metropolitan Area contributed 15–23% of the OA in the center of the city during the MILAGRO campaign. Most of these emissions were not included in the Mexico City emissions inventory and were not related to the well known Yukatan fires. Papadakis *et al.*[29] estimated that agricultural burning contributes approximately 10–25% of the OA in Athens during late fall and winter periods during which burning of agricultural waste (*e.g.*, olive tree branches) is prevalent in southern Greece.

Despite the above information, one of the first reactions of policy makers in several of the case study cities like Paris and Athens is to ban traffic when the fine PM concentrations increase. For example, the emergency measures used in Athens during recent winters with the high biomass burning OA included prohibiting the use of diesel cars, non-catalytic gasoline cars, half of the taxis and heavy diesel trucks, a 30% reduction in industrial output, *etc.* While this reduction in emissions of traditional sources does help a little, it is clearly not the solution to the problem during periods when wood burning is contributing 80% of the fine PM levels in Athens (Fig. 3).

5. The best of times

During the last decade there has been dramatic progress in our ability to measure particulate and gas-phase pollutants automatically, *in situ*, and with high temporal resolution. Techniques like Aerosol Mass Spectroscopy (AMS) including applications to single particles, Proton Transfer Reaction Mass Spectrometry (PTR-MS), Chemical Ionization Mass Spectrometry (CIMS), *etc.* are prime examples of these capabilities. At the same time there has been rapid development of source-apportionment techniques that can use these large datasets (*e.g.*, positive matrix factorization, PMF) providing useful insights about source contributions but also source-receptor relationships.

Another major achievement is the development of techniques (small cost effective sensors, mobile monitoring approaches, *etc.*) allowing us to quantify the variability of pollution in urban areas in unprecedented spatial detail. As an example from Pittsburgh, Fig. 6 shows estimated average concentrations of BC inside and around the urban area, combining measurements and modeling. Measurements of BC were conducted at 70 locations in and around the city using a mobile laboratory during 2011–12. Using the approach of Tan *et al.*[30] statistical models were used to describe the spatial patterns of BC. The models consist of two layers: a plume layer to describe the contributions of high emitting vehicles using a near-roadway kernel, and an urban-background layer that predicts the spatial pattern of other sources using land use regression (LUR). This approach leverages unique information content of highly time resolved mobile monitoring data and provides insight into source contributions.

The near-roadway kernel apportions high emitting vehicle plumes along roadways using an essentially Gaussian fall-off curve with a characteristic length of 100 meters based on the work of Karner *et al.*[31] Thus high BC concentrations predicted near roadways fall off to background levels relatively quickly, within 100 m of the road edge. Descriptor variables in the base layer LUR include elevation, areas zoned for industrial use, and areas zoned for commercial use. When combined, the two-layer model represents the two primary sources of BC in Pittsburgh: traffic (as indicated by the plume layer and commercial areas in the

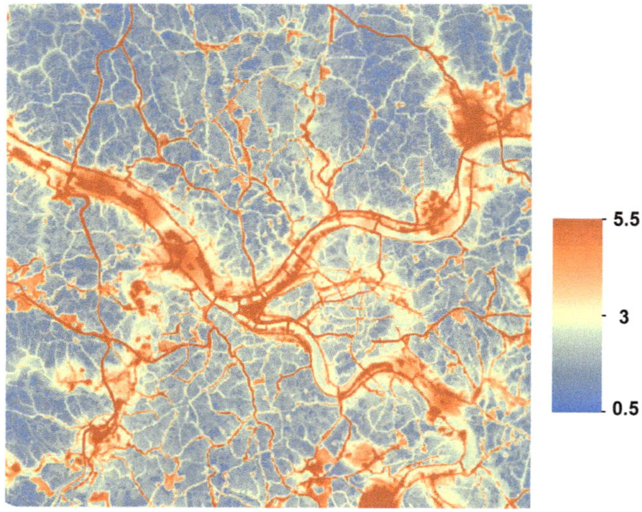

Fig. 6 Predicted ground-level average BC concentrations (μg m^{-3}) in the center of Pittsburgh during 2011–12, combining measurements and a land use regression model.

base layer) and industrial emissions. Concentrations are modulated by elevation, with consistently higher concentrations in the low-lying river valleys. Their peaks exceed 5 μg m^{-3} very close to busy major roads and drop to the background level of 0.5–1 μg m^{-3} relatively rapidly. The corresponding gradients for secondary fine PM components that represented most of the fine PM were minor in Pittsburgh.

The above approach can provide information about the average pollutant concentrations inside an urban area at scales of the order of ~100 m. It is possible to examine the exposure of residents at scales down to 1 m by combining the regular chemical transport models with computational fluid dynamics models simulating the dispersion and emission of pollutants. Fig. 7 presents such an example of simulated BC concentrations at an intersection of two roads in Athens with tall buildings. The concentrations vary from a peak of more than 10 μg m^{-3} to a little over the urban background value of 1 μg m^{-3}. The wind patterns inside the street canyon (the wind direction is assumed to be 60°) create stagnation zones with high pollution levels but also relatively clean areas.

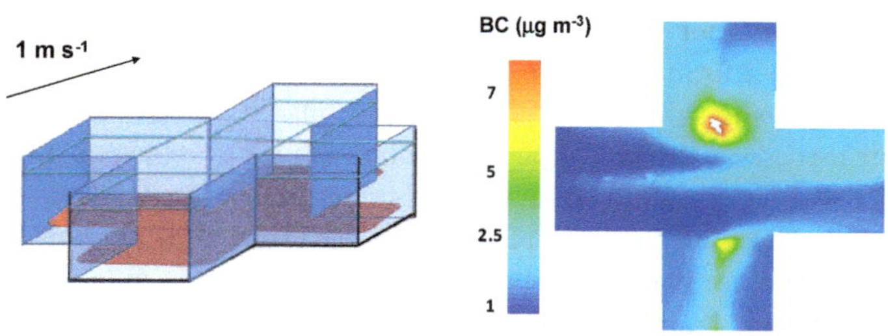

Fig. 7 Simulated BC concentration levels at an intersection of two street canyons in Athens with just one lane of traffic each. A wind speed of 1 m s^{-1} is assumed with a wind direction of 60°.

The available computational power, together with the development of numerical methods, make these simulations possible at the city level at least for inert pollutants. However, the simulation of urban atmospheric chemistry at these scales remains a major challenge.

6. The spring of hope and the winter of despair

While the tools available to the scientific community to address urban air quality problems have improved dramatically, their use suggests also their weaknesses and their need for improvement. These problems are especially challenging at the moderate to low pollution levels reached in most of our case study cities. They are also related to the increasing importance of chemical processes in determining the urban organic aerosol levels.

Chemical aging of primary OA

Most existing OA source apportionment techniques assume that the source fingerprints remain relatively constant at least during the timescales involved in urban air pollution studies. Recent work by our group, but also other efforts, provide strong evidence that this assumption may not be true for at least some sources, including cooking OA and biomass burning OA.

AMS measurements were performed in two major cities in Greece (Athens and Patras) during both summer and winter. PMF was used to quantify the OA sources in these campaigns. The PMF analysis suggested different COA AMS spectra in the two seasons (Fig. 8). The two corresponding vectors had an angle of 25°. The summer spectra were characterized by lower *m*/*z* 55, 57, 67 and 69 but higher *m*/*z* 44. Given that there is no significant change in the cooking habits of Greeks in summer and winter in Athens, our hypothesis was that this change is due to the chemical processing of the cooking OA in the atmosphere. To test this hypothesis we performed smog chamber experiments by filling the ICE-HT chamber with the emissions of souvlakia charbroiling. Souvlakia (pieces of pork on a stick) are the traditional Greek "fast food" and their cooking was hypothesized to be the major source of COA in Athens. The corresponding COA PMF mass spectra during winter in Athens were similar (a theta angle of 10°) to the laboratory spectrum of

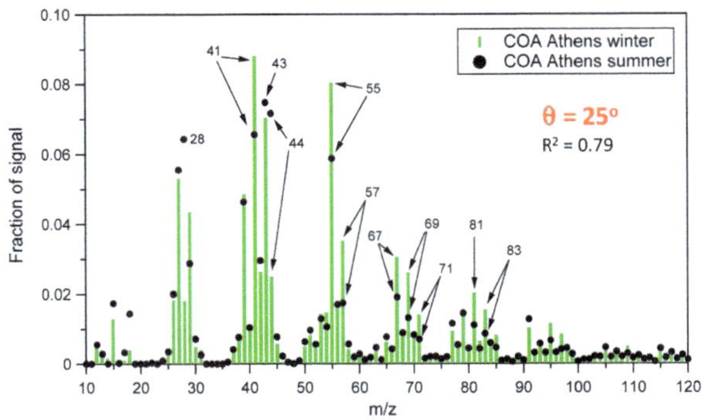

Fig. 8 Cooking OA spectra resulting from PMF analysis of AMS measurements in Athens during the summer of 2012 and winter of 2013.

fresh COA in the chamber. When the lights of the chamber were turned on, or when 40 ppb of ozone were added in the dark, there was a rapid change of the OA mass spectrum. After a few hours of reactions the AMS mass spectrum was similar (a theta angle of 9°) to that determined by the PMF analysis of the ambient measurements during the summer in Athens. This suggests that the chemical processing of the corresponding organic particulate matter is rapid and needs to be accounted for in source apportionment studies. The use of the same AMS spectrum for all periods for the same area may introduce significant errors in source apportionment.

In the same Athens field campaign, biomass burning OA also appeared to be processed in the atmosphere even during nighttime. The PMF analysis of OA in Athens during the winter suggested the existence of two biomass burning (BBOA) factors. One potential explanation is that these factors were related to different types of fuel burned or different stages of combustion. Chemical processing of the BBOA during nighttime is an alternative explanation. Once more we performed chamber experiments by filling the chamber with the smoke produced during the burning of oak and beech in a stove similar to the ones used in Greece for residential heating. The AMS spectrum of the OA was similar to the fresh BBOA spectrum determined by PMF in Athens (a theta angle of 13°). This similarity suggests that our experiments were a reasonable match for the average burning conditions in Athens during the campaign period. Then the OA was allowed to react with O_3 and NO_2 in the dark. The mass spectrum of the OA changed (a theta angle of 20° was created in just one hour), the O : C increased from 0.25 to 0.4, and there was also a reduction of the *m*/*z* values characteristic of biomass burning (*e.g.*, *m*/*z* 60). The new spectrum progressively looked like the aged BBOA spectrum determined in Athens (a theta angle of 15°). Based on these preliminary results, significant changes in the chemical characteristics of BBOA are possible even during periods with low photochemical activity and moderate to low oxidant levels.

Based on these results, we propose that when there is significant processing of OA from a source during a field study the AMS spectra of the primary OA may change significantly. Techniques like PMF may then try to describe this chemical evolution by producing two factors for the source: one corresponding to the fresh OA from the source and one to the most processed form of the OA from that source during the field campaign. Then PMF and similar source apportionment techniques can mathematically explain the chemical transformations by combining these two factors. To test this we performed PMF of the BBOA laboratory experiments. In all cases, PMF indeed explained all data with two factors corresponding to the initial and the final condition of the system.

Other sources of OA

Our understanding of ambient atmospheric organic aerosol has progressed dramatically during the last ten years. However, looking back at the Los Angeles case study with the longest history of chemical transport model applications, the results are rather surprising. Lurmann *et al.*[32] compared the predictions of the UAM-AERO chemical transport model against measurements in several Los Angeles sites during the 1987 Southern California Air Quality Study (SCAQS) study. The measured values of PM_{10} OA (4 h averages) were as high as 70 μg m^{-3} at the time and the daily average values were 15–20 μg m^{-3}. The model used assumed that primary OA was non-volatile and inert and used the one-product model of Pandis *et al.*[33] to describe secondary OA formation. The model performance was surprisingly good with a mean normalized bias of −16% and a mean normalized error of 32% for daily average PM_{10} OA concentrations. The model predicted that most of the OA was primary, something not unreasonable given the high primary emissions of the time. This good performance, despite the serious limitations of the model, could be at least partially due to the high primary OA emissions used and the dominance or potential dominance of primary OA over secondary sources. Of course, it could also be due to offsetting errors (*e.g.*, overestimation of the primary OA, which was assumed to be nonvolatile, and underestimation of the formation of secondary OA). Twenty years later the average OA levels measured in Pasadena during the CalNex study were only around 7 μg m^{-3}. Efforts to simulate the current lower OA levels in Los Angeles by a variety of modern chemical transport models has indicated significant problems. For example, CMAQ application to CalNex resulted in a fractional bias of −53% and a fractional error of 53%.[34] Hayes *et al.*[35] also presented model results for the same periods including a wide range of sensitivity studies. They concluded that IVOCs may be an important source of OA and also that a significant amount of background SOA appears to be transported to the Pasadena site. These results suggest that even for traditional sources like transportation, emitted intermediate volatility organic compounds may have become significant SOA precursors as the emissions of primary OA and VOCs have been reduced. This requires increased efforts in determining both their emissions and, more importantly, their atmospheric chemistry and aerosol formation potential.

7. Conclusions

The results from the five case studies summarized in the previous sections suggest that as all of them have been developing, they have passed through a maximum in particulate matter concentrations and their concentrations have been decreasing ever since. The importance of atmospheric chemistry during this period has been increasing. In all cases, most of the fine PM is now due to chemical reactions and is not just the result of primary particulate emissions any more. At the same time the relative importance of long range transport has been increasing. Our source apportionment results suggest that at least an area extending up to approximately 500 km from the city should be examined, together with the urban area itself, in efforts to further reduce fine particulate matter concentrations. The contribution of the "usual suspects" to fine PM, local transportation and heavy industry, has been decreasing due to the controls that have been successfully imposed on the corresponding emissions. This has resulted in the increase of the relative importance of other sources. The contributions of cooking, residential biomass burning, and nearby agricultural burning were highlighted by the various case studies analyzed in this study. Accounting for the complexities introduced by chemistry in urban locations is becoming increasingly important in further improvement of the air quality of large urban centers in the developed world.

Acknowledgements

This research was supported by the FP7 IDEAS project ATMOPACS and the US Environmental Protection STAR project RD-835405.

Notes and references

1 P. Brimblecombe, *The Big Smoke*, Routledge, 2011.
2 J. Fenger, *et al.*, *Urban Air Pollution: European Aspects*, Kluwer, 1998.
3 C. I. Davidson, *J. Air Pollut. Control Assoc.*, 1979, **29**, 1035–1041.
4 G. M. Hidy, *et al.*, *J. Air Pollut. Control Assoc.*, 1975, **25**, 1106–1114.
5 K. Katsouyanni, *et al.*, *Journal of Epidemiology and Community Health*, 1990, **44**, 321–324.
6 EEA, *Air pollution in Athens: Existing status and abatement practices*, European Environmental Agency, 2008, https://www.eea.europa.eu/publications/2599XXX/page018.html.
7 H. R. Anderson, *et al.*, *Eur. Respir. J.*, 1997, **10**, 1064–1071.
8 A. Jaecker-Voirol and P. Pelt, *Environ. Model. Software*, 2000, **15**, 575–581.
9 M. J. Molina and L. T. Molina, *J. Air Waste Manage. Assoc.*, 2004, **54**, 644–680.
10 P. D. Kalabokas, *et al.*, *Global NEST J.*, 1999, **1**, 157–167.
11 M. Kanakidou, *et al.*, *Atmos. Environ.*, 2011, **45**, 1223–1235.
12 Airparif, *Surveillance et Information sur la qualite' de l'air a Paris en*, Airparif, Paris, 2015, p. 2014.
13 M. Beekman, *et al.*, *Atmos. Chem. Phys.*, 2015, **15**, 9577–9591.
14 T. B. Byerson, *et al.*, *J. Geophys. Res.: Atmos.*, 2013, **118**, 5830–5866.
15 A. E. Wittig, *et al.*, *Atmos. Environ.*, 2004, **38**, 3107–3125.
16 A. C. Aiken, *et al.*, *Atmos. Chem. Phys.*, 2009, **9**, 6633–6653.
17 K. M. Wagstrom, *et al.*, *Atmos. Environ.*, 2008, **42**, 5650–5659.
18 K. Skyllakou, *et al.*, *Atmos. Chem. Phys.*, 2014, **14**, 2343–2352.
19 W. F. Rogge, *et al.*, *Environ. Sci. Technol.*, 1991, **25**, 1112–1125.
20 J. J. Schauer, *et al.*, *Atmos. Environ.*, 1996, **30**, 3837–3855.
21 M. Crippa, *et al.*, *Atmos. Chem. Phys.*, 2013, **13**, 961–981.
22 M. Crippa, *et al.*, *J. Geophys. Res.: Atmos.*, 2013, **118**, 1950–1963.
23 P. L. Hayes, *et al.*, *J. Geophys. Res.: Atmos.*, 2013, **118**, 9233–9257.
24 A. L. Robinson, *et al.*, *Environ. Sci. Technol.*, 2006, **40**, 7820–7827.
25 M. Crippa, *et al.*, *Atmos. Chem. Phys.*, 2013, **13**, 8411–8426.
26 J. Heo, *et al.*, *Atmos. Environ.*, 2013, **73**, 51–61.
27 A. L. Robinson, *et al.*, *Environ. Sci. Technol.*, 2006, **40**, 7811–7819.
28 A. C. Aiken, *et al.*, *Atmos. Chem. Phys.*, 2010, **10**, 5315–5341.
29 G. Z. Papadakis, *et al.*, *Atmos. Environ.*, 2015, **112**, 148–158.
30 Y. Tan, *et al.*, *Environ. Sci. Technol.*, 2016, submitted.
31 A. A. Karner, *et al.*, *Environ. Sci. Technol.*, 2010, **44**, 5334–5344.
32 F. W. Lurmann, *et al.*, *Atmos. Environ.*, 1997, **31**, 2695–2715.
33 S. N. Pandis, *et al.*, *Atmos. Environ.*, 1992, **26**, 2269–2282.
34 K. R. Baker, *et al.*, *Atmos. Chem. Phys.*, 2015, **15**, 5243–5258.
35 P. L. Hayes, *et al.*, *Atmos. Chem. Phys.*, 2015, **15**, 5773–5801.

Faraday Discussions

PAPER

Regional air quality in Leipzig, Germany: detailed source apportionment of size-resolved aerosol particles and comparison with the year 2000†

D. van Pinxteren,[a] K. W. Fomba,[a] G. Spindler,[a] K. Müller,[a] L. Poulain,[a] Y. Iinuma,[a] G. Löschau,[b] A. Hausmann[b] and H. Herrmann*[a]

Received 16th December 2015, Accepted 18th January 2016
DOI: 10.1039/c5fd00228a

A detailed source apportionment of size-resolved aerosol particles in the area of Leipzig, Germany, was performed. Sampling took place at four sites (traffic, traffic/residential, urban background, regional background) in parallel during summer 2013 and the winters 2013/14/15. Twenty-one samples were taken per season with a 5-stage Berner impactor and analysed for particulate mass, inorganic ions, organic and elemental carbon, water-soluble organic carbon, trace metals, and a wide range of organic species. The compositional data were used to estimate source contributions to particulate matter (PM) in quasi-ultrafine (up to 140 nm), accumulation mode, and coarse size ranges using Positive Matrix Factorisation (PMF) receptor modelling. Traffic (exhaust and general traffic emissions), coal combustion, biomass combustion, photochemistry, general secondary formation, cooking, fungal spores, urban dust, fresh sea/road salt, and aged sea salt were all found to contribute to different extents to observed PM concentrations. PMF derived estimates agreed reasonably with estimates from established macrotracer approaches. Quasi-ultrafine PM originated mainly from traffic (20–50%) and photochemistry (30–50%) in summer, while it was dominated by solid fuel (mainly biomass) combustion in winter (50–70%). Tentatively identified cooking aerosol contributed up to 36% on average at the residential site. For accumulation mode particles, two secondary sources typically contributed 40–90% to particle mass. In winter, biomass and coal combustion contributions were up to *ca.* 25% and 45%, respectively. Main sources of coarse particles were diverse and included nearly all PMF-resolved ones depending on season and air mass origin. For PM_{10}, traffic (typically 20–40% at kerbside sites), secondary formation (30–60%), biomass combustion (10–15% in winter), and coal combustion (30–40% in winter with eastern air mass inflow) were the main quantified sources. At the residential site, contributions

[a]*Leibniz Institute for Tropospheric Research (TROPOS), Atmospheric Chemistry Department (ACD), Permoserstr. 15, 04318 Leipzig, Germany. E-mail: herrmann@tropos.de*

[b]*Saxon State Office for Environment, Agriculture and Geology (LfULG), Pillnitzer Platz 3, 01326 Dresden, Germany*

† Electronic supplementary information (ESI) available. See DOI: 10.1039/c5fd00228a

from biomass combustion derived up to 60% from local emissions. Coal combustion as a significant source was only present during eastern air mass inflow and showed very similar concentrations at all sites, indicating the importance of trans-boundary air pollution transport in the study area. Overall, nearly half of the PM_{10} mass was attributed to urban sources by a simple subtractive approach with highest reduction potentials of up to 80% for local (urban) mitigation measures in ultrafine and coarse particles. Local increments of elemental carbon have decreased by about 50% as compared to the year 2000, corroborating results from a former study on the positive effects of a low emission zone, implemented in Leipzig in 2011.

Introduction

Air pollution has been of much concern to European societies for several decades. In order to reduce the adverse health effects of air pollutants, much action has been taken to abate emissions from transport and industry, including technological developments and tightened regulations on local, national and EU-wide levels. Despite the progress made, many cities in Europe and Germany, including Leipzig, are still challenged to comply with legal limit values, especially for pollutants such as NO_x and particulate matter (PM). While the main source of NO_x is high temperature combustion in motor engines, PM can have very diverse sources, including primary emissions and secondary formation from multiple precursor gases, which makes it more difficult to develop efficient mitigation strategies. These are required, however, not only to comply with present limit values, but also in view of the more strict guidelines of the World Health Organisation (WHO)[1] and the current knowledge from health-related PM studies that there is no threshold below which adverse health effects would not occur.[2] As a major prerequisite for sustainable air quality policies, detailed knowledge of the sources of PM and their impact at a given site is needed. Therefore, source apportionment studies using receptor modelling techniques have become increasingly popular in Europe[3,4] as they allow for quantitative estimations of source contributions, based on compositional data of PM obtained at receptor sites. Main source categories typically resolved include traffic (both exhaust and non-exhaust emissions), industry, resuspended crustal material, secondary inorganic aerosol, biomass burning, and sea/road salt.[3,4] Typically, these studies are performed on PM_{10} or $PM_{2.5}$ data at a single site, with only a few studies focussing on sources in different particle size ranges or at multiple sites (*e.g.* ref. 5 and references therein). As the health effects of PM can be very different depending on particle size and composition,[2,6] knowledge on the main sources of particles in different size ranges can be advantageous. For abatement strategies, knowledge on a more local or regional/transported character of a PM source can be important as well.

One of the most popular receptor models is Positive Matrix Factorisation,[7] a variant of multivariate factor analysis, describing a matrix of observations (constituent concentrations in ambient particle samples) by a linear combination of source contributions (weights) and source profiles (factors) implying non-negativity constraints. Even though these estimations of source contributions can carry substantial uncertainty[8] depending among many other things on how cleanly different sources can be resolved in the model, they provide an appropriate measure to quantitatively apportion PM to different sources or source categories. Generally, the ability of the PMF model to resolve different sources can increase with the availability of chemical marker species and the use of size information.[9]

In the present study, source apportionment of PM in 5 particle size ranges was performed, local contributions from within the city were estimated, and present-day data were compared with data from a similar campaign in 2000.[10] Implications of the results of this study for PM abatement strategies are briefly discussed in the Conclusions section.

Experimental

Particle sampling

Particle sampling was performed at 4 sites in parallel in the city of Leipzig, Germany (0.5 million inhabitants, *cf.* Fig. S1† for detailed locations): Leipzig-Mitte (LMI), Leipzig Eisenbahnstrasse (EIB), Leipzig TROPOS (TRO), and Melpitz (MEL). LMI is an official monitoring site in the air quality network of the Saxon State Office for Environment, Agriculture and Geology (LfULG). It is located next to a major crossroad in the centre of Leipzig with a traffic intensity of approx. 40 000–50 000 vehicles per day.[11] EIB is another traffic-influenced site in a residential area, approx. 2 km east of LMI. Traffic intensity is only about 25% of that at LMI,[11] but due to the position of the sampling inlets within a street canyon (approx. 18 m wide and high) and correspondingly less efficient dispersion of emissions, the observed impact of traffic emissions is high.[11] The TRO site is located at the TROPOS institute on a science campus in the urban background of Leipzig with a distance of 150 m to the next larger road. MEL is the TROPOS research site approx. 50 km north-east of Leipzig in the regional background (rural area).[12] Inlet heights were approx. 5, 7, 20, and 5 m above ground at LMI, EIB, TRO, and MEL, respectively.

Size-resolved sampling was done for 24 hours (00:00–24:00 CET) using a 5-stage Berner-type cascade impactor[13] with aerodynamic cut-off diameters at 0.05, 0.14, 0.42, 1.2, 3.5, and 10 μm and a flow rate of 75 L min^{-1}. Pre-heated (350 °C for at least 2 h) aluminium foils were used as impaction substrates, except for trace metals, which were determined from pieces of Nuclepore® filters (polycarbonate). These were placed on top of the aluminium foil and covered approx. 10% of total impactor stage area. Typically within 1–2 days after sampling, the loaded substrates were removed from the impactors and stored at −20 °C until further processing.

Sampling days were selected based on meteorological forecast data, aiming at precipitation-free days with air mass inflow from a western or eastern corridor (2 most typical inflow regimes in the Leipzig area[14]). Twenty-one sampling days were realised per season (summer 2013 and winters 2013/14 + 2014/15), which were categorized into western or eastern air mass origin based on a simple scheme[15] applying air mass back trajectories.

Chemical analyses

Mass concentrations of sampled PM were determined by differential weighing of the impaction substrates (after their respective conditioning at 20 °C and 55% RH for at least 48 hours) on a UMT-2 microbalance (Mettler-Toledo, Switzerland). Extraction of water-soluble compounds was done by shaking, ultrasonication, and shaking (10 min each) of fractions of the impaction substrates. After filtration through syringe filters (0.45 μm, Acrodisc 13, Pall, Dreieich, Germany), anions (chloride, nitrate, sulfate, oxalate) and cations (ammonium, sodium, potassium, magnesium, calcium) were analysed using ion chromatography (ICS3000, Dionex, Dreieich, Germany). Organic and elemental carbon (OC/EC) were determined applying a thermographic two-step method without charring correction. Water-soluble organic carbon (WSOC) was analysed with a TOC analyser (TOC-VCPH, Shimadzu, Kyoto, Japan) in the non-purgeable organic carbon (NPOC) mode after aqueous extraction and filtration of the samples. Several monosaccharidic compounds (sugars, anhydrosugars, sugar alcohols) were determined from the water extracts as well, using anion exchange chromatography with pulsed amperometric detection (HPAEC-PAD, ICS-3000, Dionex, Dreieich, Germany). A range of non-polar organic compounds, including *n*-alkanes, alkanones, polycyclic aromatic hydrocarbons (PAHs), oxy-PAHs and hopanes were determined directly from the impaction substrates by Curie point pyrolysis (for thermal desorption) gas chromatography mass spectrometry (CPP-GC/MS, 6890 N GC/5973 MSD, Agilent Technologies, Waldbronn, Germany). Finally, trace metals were analysed directly from the Nuclepore® filter pieces by total reflection X-ray fluorescence spectroscopy (TXRF, Picofox S2, Bruker Nano, Berlin, Germany). More details on the analytical methods can be found in related publications from our group for ions,[16] OC/EC,[17] WSOC,[18] monosaccharides,[19] non-polar organics,[20] and metals.[21]

Positive matrix factorisation

Positive matrix factorisation (PMF) was performed using the EPA PMF 5.0 software.[22] In order to increase the statistical robustness of the results, concentrations of PM constituents from all 4 sites were pooled into one dataset.[23] One sample from LMI (25/09/2013) was excluded due to extremely high concentrations of mass and some constituents, which were not observed in any of the other samples and likely caused by a short period of construction works close to the site. To increase chances of resolving sources which might be active in certain particle size ranges only, PMF was calculated separately for each impactor stage, leading to 5 final PMF solutions quantifying the PM sources in the respective particle size range. Species used for PMF included the weighed mass (as total variable), all determined anions and cations except potassium and calcium (included in metals), WSOC and water-insoluble carbon (WISC, calculated from total carbon by subtraction of WSOC and preferred over OC/EC due to non-correctable charring effects on the impactor substrates and corresponding misclassifications[17]), levoglucosan (anhydrosugar), arabitol (sugar alcohol), C22–C34 *n*-alkanes, a selection of PAHs and hopanes (with good data coverages) and 15 trace metals (K, Ca, Ti, Mn, Fe, Cu, Zn, As, Se, Ba, Pb, Ni, Cr, Sr, Sn). This range of species covers the main PM constituents as well as characteristic marker compounds for a variety of possible PM sources, which are indicated in Table S1.†

In the pre-processing of the concentration data, values below the instrumental limit of detection (LOD) were replaced by 0.5 times the LOD.[23] A few missing values (*e.g.* due to instrument failure) were replaced by their median within the corresponding site/season category (*e.g.* median Ca at MEL in summer). Their respective uncertainty (*cf.* below) was set to 4 times the median concentration, thereby eliminating their influence on the PMF solution.[22] All concentration data were corrected by mean field blank concentrations, obtained from unloaded aluminium substrates mounted into the impactors for at least 1 day before, within, and after the campaigns. Occasionally occurring small negative concentrations due to the field blank correction were kept after checking and assuring their corresponding uncertainty range included the zero value.[23]

Next to a concentration matrix, PMF requires a matrix of corresponding uncertainties, which are used for data weighting.[22] The uncertainty matrix was constructed from 3 components: (i) uncertainty of the LOD, defined as 5/6 of the LOD,[23] (ii) analytical uncertainty, obtained from relative standard deviations of signal intensities from repeated standard measurements, and (iii) uncertainty of the mean field blank concentration, defined as 3 times the standard deviation of the field blank. Total uncertainty was calculated from these components applying Gaussian error propagation.

The EPA PMF allows for a classification of species as "strong", "weak" or "bad" and correspondingly downweighting or excluding the respective species from the PMF calculation, which is based on their respective signal-to-noise ratios and other considerations of data quality.[22] A number of species were set as "weak" or "bad" in certain calculations (*i.e.* on some of the impactor stages), mainly due to poor data coverage with only few data above the LOD in corresponding particle size ranges. The classifications of all species for the 5 PMF solutions are given in Table S1† as well.

The number of factors in the final PMF solutions was identified based on several mathematical and physical criteria:[22,23] (i) decrease of the calculated *vs.* expected goodness-of-fit parameter Q/Q_{exp} with increasing number of factors, (ii) distribution of normalised residues within ± 3 for most elements, (iii) quality of correlation between observed and modelled total mass concentration, (iv) quality of correlation between observed and modelled species concentrations, (v) plausibility of chemical factor profiles (source profiles), and (vi) plausibility of patterns of factor concentrations during different seasons and at different sites. Considering these criteria, PMF solutions with 3 to 12 factors were calculated and evaluated for all 5 particle size ranges. The number of factors was considered too low mainly if total mass and/or characteristic marker concentrations were not well modelled (even in case the factor profiles would already allow for meaningful interpretations) and considered too high if physically/environmentally meaningful interpretations could not be found for all factors (even in case the correlation of observed *vs.* modelled mass and/or marker concentrations would be higher).

Data processing and back-trajectory calculations

Data processing, analysis, and plotting was done in R version 3.2.2,[24] including the packages plyr 1.8.3,[25] dplyr 0.4.3,[26] tidyr 0.3.1,[27] ggplot2 1.0.1,[28] and openair 1.6.4.[29] Calculation of 96 hour air mass back-trajectories was done using HYSPLIT 4.9 [30] with 1 degree GDAS input data and a starting height of 500 m above ground. For air mass classification trajectories for 10:00 and 18:00 CET were used,[14] while for statistical trajectory analysis hourly trajectories within the sampling intervals were calculated.[31]

Results and discussion

Chemical composition

A total of 42 sampling days were taken, which are listed together with meteorological data in Table S2.† Daily mean temperatures ranged between 9 and 25 °C at LMI in summer and −9 and 11 °C in winter. Both winters were comparably warm with most of the mean temperatures in the positive range. With respect to air mass origin, 13 out of the 21 summer days showed western and 6 eastern inflow. Two days could not be categorized and were ignored in the respective averaging. In winter these numbers were 11, 7, and 3 for West, East, and "uncategorized". In

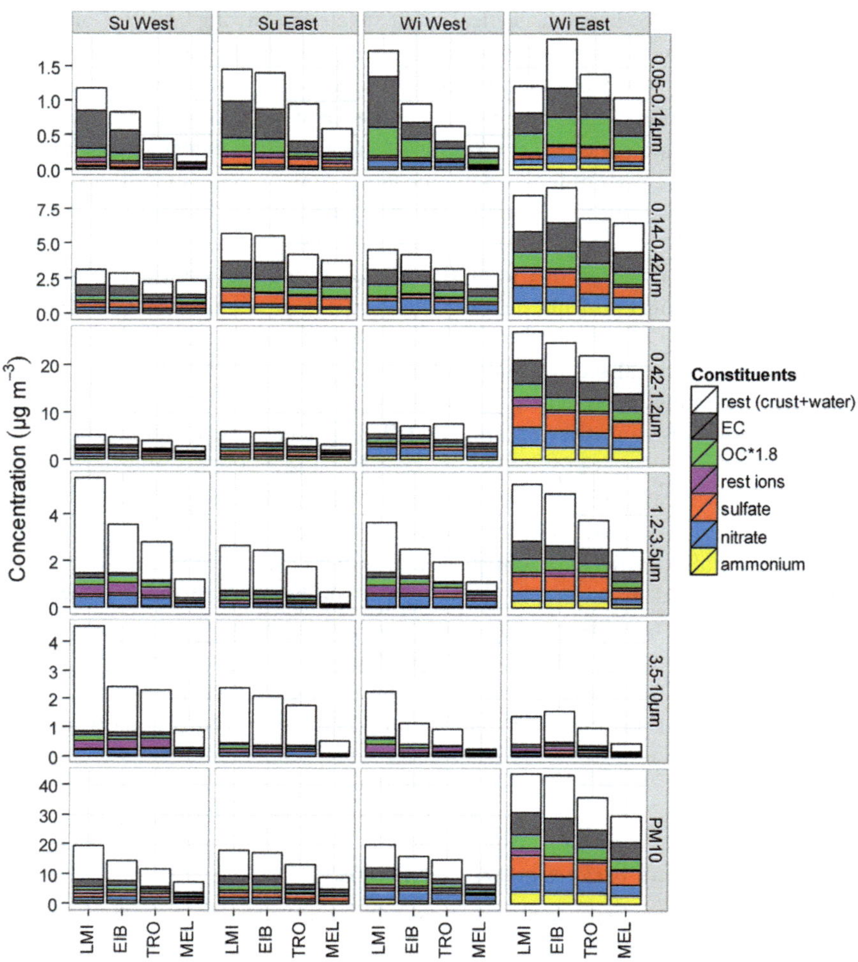

Fig. 1 Mean chemical composition of size-resolved particles (lines 1–5) and PM_{10} (line 6) during summer (Su) and winter (Wi) as well as Western and Eastern air mass inflow at the 4 sites.

Fig. 1 mean concentrations of mass and main constituents are shown for all impactor stages (particle size ranges) and PM_{10} as the sum of stage concentrations (note that such impactor PM_{10} concentrations are typically 10–20% smaller than filter concentrations due to different sampling artefacts with both devices[32]). Concentrations are divided into seasons and air mass sectors and then given for all sites in comparison. Within each category (season/sector/size range), a concentration gradient for particle mass consistent with the site characteristics was typically observed: LMI >≈ EIB > TRO > MEL, *i.e.* traffic sites > urban background > regional background. Within a season, particle mass concentrations were typically higher during eastern than during western inflow with the difference being especially pronounced in winter submicron particles. This trend has been commonly observed in the area of Leipzig and can be explained with a higher continental character of eastern air masses (western air masses typically spend considerable time above the Atlantic ocean) and higher PM pollution in Eastern European countries.[14,15,33] Between seasons, particle mass concentrations were significantly higher in winter than in summer only in the "East" category. In "Winter West", concentrations were often comparable to or only slightly higher than concentrations in summer. This has been observed in a similar way in long time series data at Melpitz.[14,15] The equally similar concentrations at the urban sites indicate a minor contribution from house heating emissions, at least for the relatively warm winters (all West days with daily mean temperatures >0 °C) within this study. For the largest particles, mass concentrations were higher in summer than in winter, which is likely related to higher resuspension of crustal material in summer (constituting a significant fraction of the given "undetermined rest" concentrations, *i.e.* weighed mass minus main constituents concentrations). Between the different particle size ranges, concentrations were highest in accumulation mode particles of impactor stages 3 (0.42–1.2 μm) and 2 (0.14–0.42 μm), followed by coarse (1.2–10 μm) and quasi-ultrafine particles (0.05–0.14 μm), consistent with typical mass size distributions.

In terms of composition, quasi-ultrafine particles were dominated by elemental and organic carbon, while both carbon and inorganic main ions (ammonium, sulfate, and nitrate) were important in the accumulation mode size range. In coarse mode particles, undetermined rest concentrations typically made up the major fraction, which can be expected to consist mainly of crust material and water (with some uncertainty due to the assumed conversion factor from OC to organic material of 1.8).

Generally, this brief overview reveals trends in particle mass concentration and composition which are consistent with expectations based on existing knowledge for seasonal, sectoral, size-dependent and station-dependent patterns.[34,35] Despite the relatively low number of sampling days, the data set thus seems to be appropriate for more detailed source apportionment and quantification of source contributions as done below.

Regional *vs.* local contributions

As a first step of source apportionment the so-called Lenschow approach[36] has been applied to particle mass concentrations, in order to reveal the contributions of the regional background (MEL concentration), the urban background (TRO minus MEL), and the local traffic (LMI minus TRO) to the observed concentrations at LMI. Calculations were done for all individual samples and then averaged in different categories as shown in Fig. 2. A clear size-dependent distribution of regional *vs.* traffic increments can be recognised, which is more pronounced during western than during eastern inflow. During western inflow, the regional background contributed approx. 20–30% to mass concentrations of quasi-ultrafine (0.05–0.14 μm) and coarse particles (1.2–10 μm) observed at this kerbside site. Local (traffic) emissions dominated the mass in the ultrafine size range with fractions of approx. 60%, while for coarse particles urban background and traffic increments were broadly similar in year-round averages (approx. 1/3 each) with higher traffic contributions in winter than in summer. During eastern inflow the traffic contributions in ultrafine particles were lower, in one case (ultrafine, winter, East) even negative on average (*i.e.* not resolvable by this simple approach). This is likely related to a dependency of traffic emissions on local wind direction at LMI, as will be further discussed below. At the EIB traffic site, traffic increments were more similar between "West" and "East" with mass fractions of

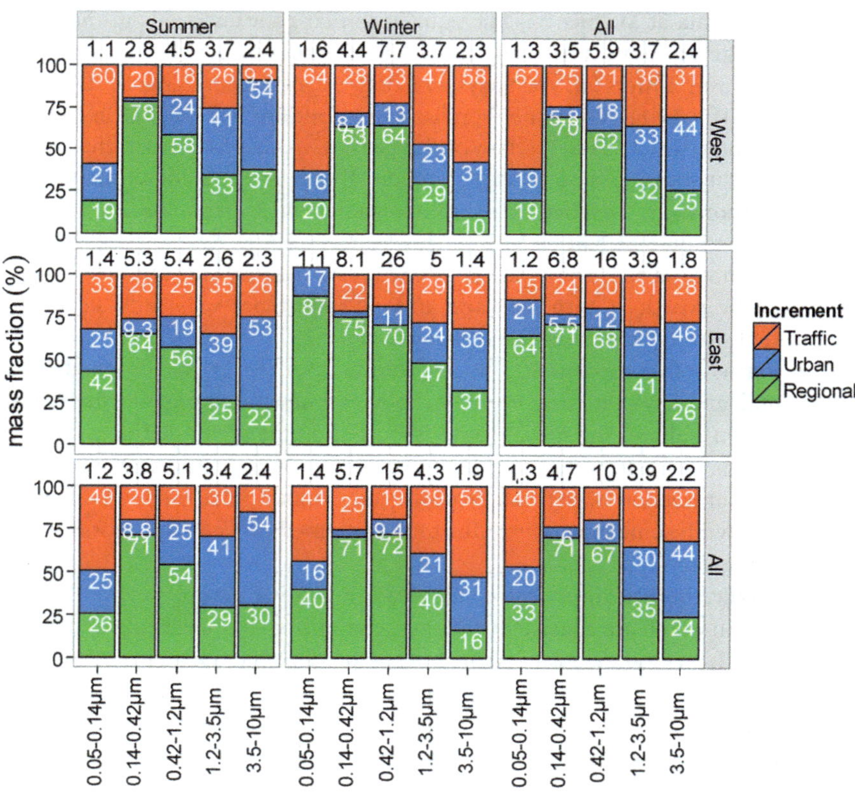

Fig. 2 Mean mass fractions of regional, urban and traffic increments from the Lenschow approach in 2 seasons and 2 main air mass inflow sectors. "All" refers to all data per season and/or sector (for sector including non-categorized samples). Numbers printed in black above the bars indicate mean particle mass concentrations at the LMI kerbside site, while numbers in white indicate the respective mass fractions (not printed if <5%). Note that in one case the mean traffic increment was negative, leading to a mass fraction sum slightly above 100%.

39 and 33% in ultrafine and typically around 25% in coarse particles as year-round ("all seasons") averages (data not shown). In accumulation mode particles (0.14–1.2 μm) the regional contribution was typically close to 60–70% at LMI regardless of season and air mass sector and thus dominated the particle mass concentrations in this size range. The average local traffic increments in the accumulation mode were only about 25%. The urban background increment showed mass fractions of typically about 10–20% for submicron and about 30–50% for supermicron particles.

These data imply that local measures to mitigate PM pollution (*e.g.* the clean air plans adopted in many cities, including Leipzig) have different potentials for reduction in different particle size ranges. The highest reductions of up to about 80% on average could theoretically be achieved in coarse and ultrafine particles, with the latter ones being highly respirable and therefore of special health concern, but the former ones being related to negative health impacts as well.[2] In terms of PM_{10} (Fig. S2†), which is the only legislative metric for particulate pollution at present, traffic and urban background increments still made up 26 and 18%, respectively, of the particle mass concentrations on total average (all seasons, all sectors) at LMI. This indicates that about half of the PM_{10} concentration at highly polluted sites in Leipzig could theoretically be addressed by local control measures. These values slightly decreased *via* 25 to 23% for traffic and *via* 15 to 12% for the urban background increment when considering $PM_{3.5}$ (sum of 4 stages as an approximation of $PM_{2.5}$) and $PM_{1.2}$ (sum of 3 stages as an approximation of PM_1) from the Berner impactor measurements, respectively (see Fig. S2† for more details). Thus, even for PM_1 there is a potential of on average about one third of locally controllable particle mass concentration at highly polluted sites in the city of Leipzig, as estimated from this study's data.

Source apportionment

Identified sources. A total of 12 different particle sources could be apportioned by PMF across the different particle size ranges. Many sources were active in some particle size ranges only (or could only be resolved by the PMF model in these size ranges). The number of factors within the particle size ranges differed between 7 and 9 (7, 7, 7, 9, 8 factors for impactor stages 1, 2, 3, 4, 5, respectively). The source identification was based on the chemical factor (source) profiles, given in Fig. 3 and the patterns of particle mass concentrations from the respective sources at the different sites in the two seasons ("time-series" of factor concentrations), shown in Fig. 4. Beyond these 2 summary figures, more detailed figures are given in the ESI† for each source, as indicated below. The chemical factor profiles indicated the main constituents (among the ones included in the analyses), *i.e.* species with high contributions to the factor mass (defined as the sum of all species concentrations), as well as characteristic marker compounds, *i.e.* species with high fractions of their total concentration (in a given particle size range) in the respective factor.

From Fig. 3 it can be seen that chemical profiles for a given source are generally quite similar in different particle size ranges. Deviations in mass contributions (bars) can derive either from real differences in the size-resolved chemical composition of particles from a source (or source category, as *e.g.* traffic, which includes several unresolved sources as discussed below) or from different

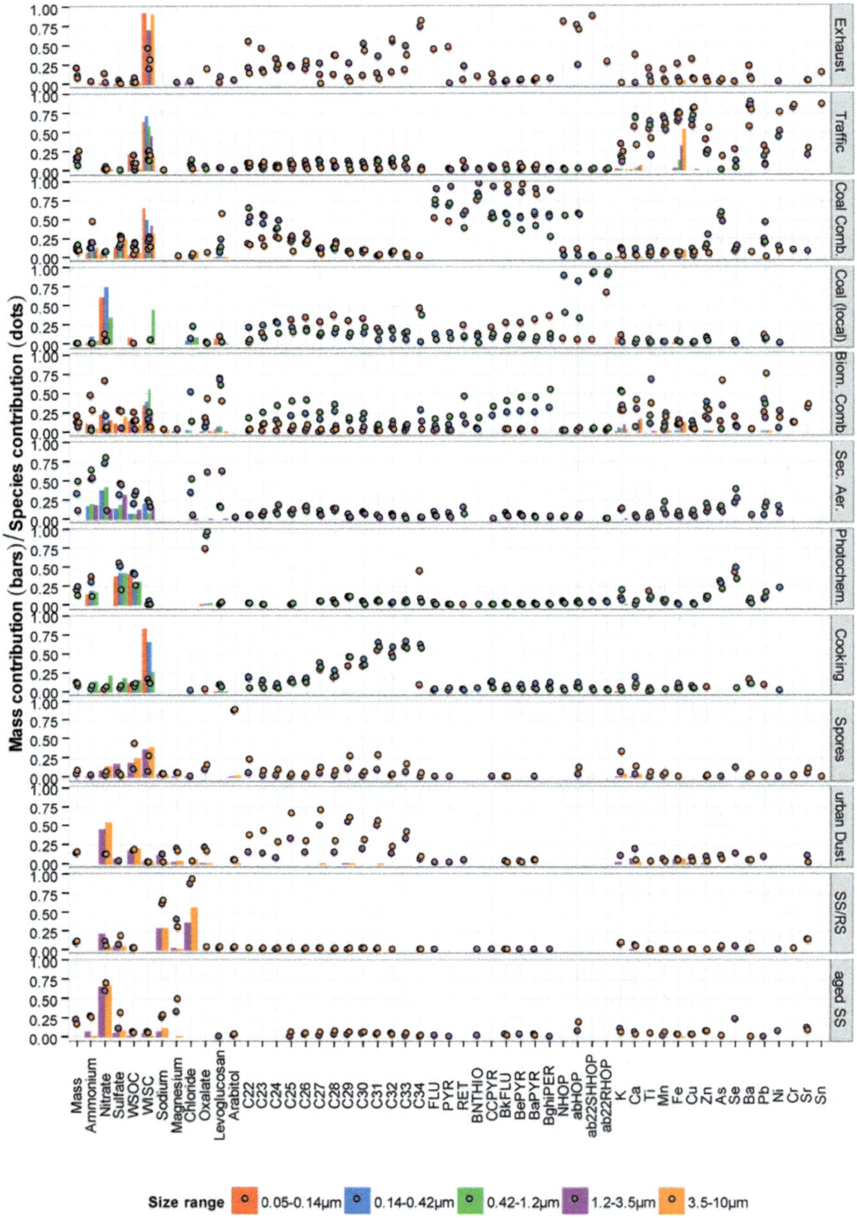

Fig. 3 Chemical profiles of PMF factors in 5 particle size ranges (colour-coded). Bars indicate contributions of species to total factor mass (sum of all species in factor equals 1), dots indicate contributions of species to total species concentration in particle size range (sum of all factors for given species equals 1).

qualities of the mathematical resolution of sources (differently "clean" profiles) in the particle size ranges. Deviations in species contributions (dots) can occur if species had different numbers of sources in different particle size ranges. From Fig. 4 it can be seen that source concentrations generally showed very similar

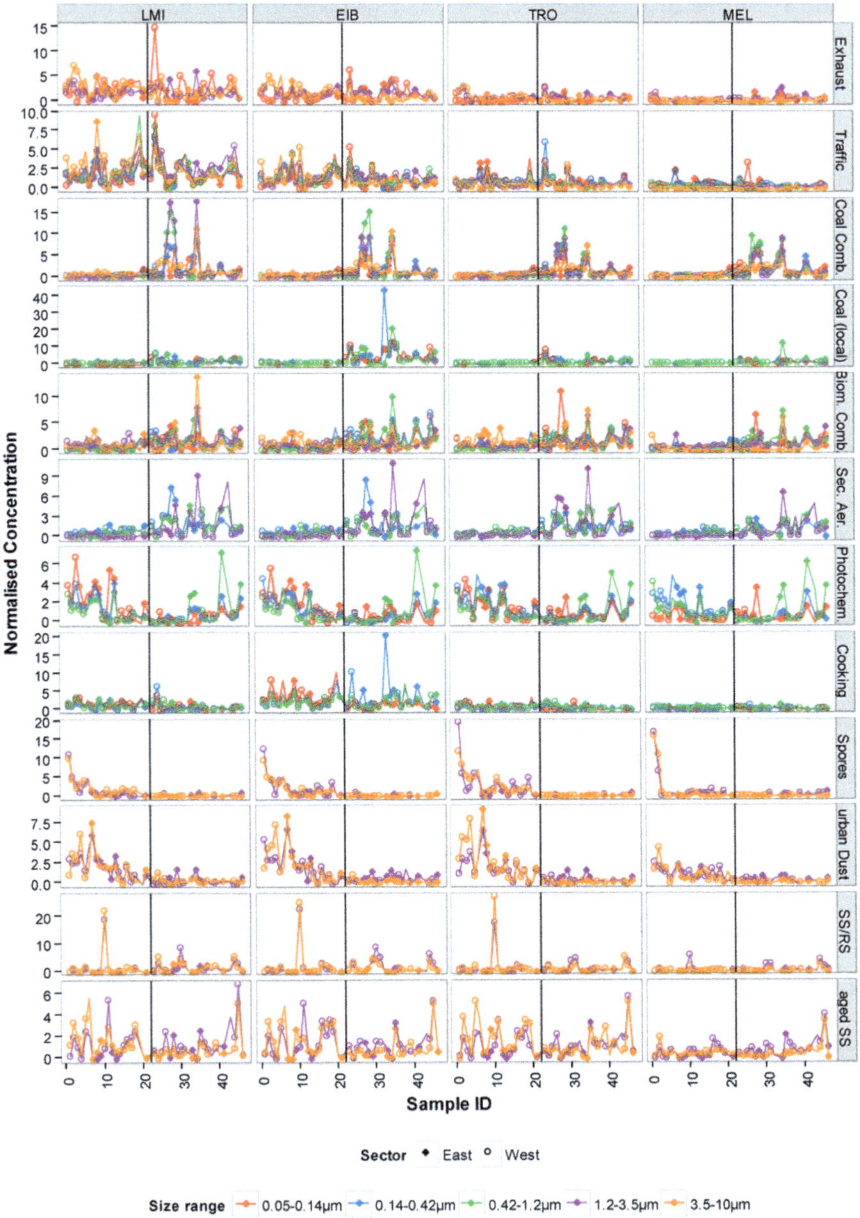

Fig. 4 Normalised particle mass concentrations (mean in factor equals 1) from sources resolved in 5 particle size ranges (colour-coded) by PMF. Samples are chronologically sorted at each site, black vertical lines at sample ID 22 indicate the beginning of winter samples. Full and empty dots indicate eastern and western air mass origin, respectively.

"time series" in different particle size ranges. Both these comparisons indicate consistency in the apportioned sources between the PMF solutions for the individual impactor stages.

Table 1 Particle sources identified with PMF together with the size ranges they were resolved in, their main constituents and characteristic marker compounds

Source	Size range	Main constituents	Marker compounds
Traffic exhaust	0.05–0.14 μm 1.2–3.5 μm 3.5–10 μm	WISC	Hopanes, <C25 *n*-alkanes, even *n*-alkanes
Traffic	0.05–0.14 μm 0.14–0.42 μm 0.42–1.2 μm 1.2–3.5 μm 3.5–10 μm	WISC, (Fe)	Cu, Ba
Coal combustion	0.05–0.14 μm 0.14–0.42 μm 0.42–1.2 μm 1.2–3.5 μm 3.5–10 μm	WISC, sulfate	PAHs, As
Coal combustion (local)	0.05–0.14 μm 0.14–0.42 μm 0.42–1.2 μm	WISC	Hopanes
Biomass combustion	0.05–0.14 μm 0.14–0.42 μm 0.42–1.2 μm 1.2–3.5 μm 3.5–10 μm	WISC, WSOC	Levoglucosan, K
Secondary (inorganic) aerosol	0.14–0.42 μm 0.42–1.2 μm 1.2–3.5 μm	Nitrate, ammonium, sulfate	WSOC, WISC, (oxalate)
Photochemistry	0.05–0.14 μm 0.14–0.42 μm 0.42–1.2 μm	Sulfate, WSOC	Oxalate
Cooking	0.05–0.14 μm 0.14–0.42 μm 0.42–1.2 μm	WISC	Odd *n*-alkanes
Fungal spores	1.2–3.5 μm 3.5–10 μm	WISC, WSOC	Arabitol
Urban dust	1.2–3.5 μm 3.5–10 μm	Nitrate, WSOC	Odd *n*-alkanes, Mg^{2+}, Ca, oxalate
Sea salt/road salt	1.2–3.5 μm 3.5–10 μm	Cl^-, Na^+	Mg^{2+}
Aged sea salt	1.2–3.5 μm 3.5–10 μm	Nitrate	Na^+, Mg^{2+}

In Table 1 all sources resolved by PMF in this study are summarised together with their size ranges, main constituents, and characteristic marker compounds. Their assignment based on factor profiles and concentration patterns will be discussed in the following.

The *traffic exhaust* factor was identified based on high mass contributions of WISC (*i.e.* EC + hydrophobic organics), making up >70% of the factor mass, as well as high species contributions of hopanes (>75% for ultrafine particles) with increasing species contributions either towards small (<C25) *n*-alkanes (for ultrafine particles) or towards larger, especially even chain length *n*-alkanes (coarse particles) as shown in Fig. 3 and S3.† The homohopane index, defined as

$\frac{\text{ab22SHHOP}}{\text{ab22SHHOP} + \text{ab22RHOP}}$ (see Table S1† for abbreviations) was 0.47 in ultrafine particles (calculation not possible in coarse modes due to concentrations below detection limits) and thus closer to typical values reported for mineral-oil-derived emissions (0.6) as compared to lignite combustion (0.1).[37] The carbon preference index (CPI), calculated from C22–C34 *n*-alkanes was 1.6 in ultrafine and 2.2 and 0.7 (stages 4 and 5) in coarse particles and thus also close to values observed from fossil fuel emissions (close to 1). Coarse soot particles are formed by deposition of smaller particles on cylinder and exhaust system surfaces and subsequent re-entrainment.[38] In accumulation mode particles, PMF was not able to resolve a direct traffic exhaust source, even though this is the particle size range where the mass maximum of diesel engine exhaust can be expected.[38] Direct emissions in this size range are thus mixed with non-exhaust emissions in the *general traffic* factor, resolved on all impactor stages. The chemical composition of this factor is comparably heterogeneous, with dominating mass contributions of carbonaceous material (mainly WISC, but also WSOC) in smaller particles and increasing contributions of metals, especially Fe and Ca, towards larger particles (Fig. 3 and S4†). Many other metals, including Ba, Cu, Mn, Ti, and Zn show high species contributions in this factor in many particle size ranges. The changing composition with particle size can be explained by varying contributions of direct and indirect traffic emissions to this factor, which the PMF model was not able to further resolve. Beyond direct exhaust emissions mixed in at least on impactor stages 2 and 3 (see above), brake wear, tyre wear, road wear, and resuspension of road dust are known sources of traffic-related particles with very diverse chemical compositions and different emission factors in different size ranges.[39,40] Brake wear can consist of up to 54% Fe, while Ba, Cu, and carbon can have significant mass fractions as well.[40] Such particles cover a wide size range, typically at larger sizes, but even in the ultrafine range their mass contribution can be 8–33%.[40] Tyre wear particles can also be found in all size ranges, with coarse particles originating from mechanical abrasion and fine and ultrafine being formed from volatilisation and recondensation.[40,41] They consist to a large extent of hydrophobic organic material and Zn has often been used as characteristic marker element.[40] The presence of Ca and Ti in the factor indicates a contribution from resuspended road dust. The temporal patterns of both traffic-related factor concentrations at the 4 sites (Fig. 4, S3 and S4†) with strongly decreasing values from the kerbside to the regional site and no apparent seasonal trend support the association of these factors with local traffic emissions. The day-to-day variations in traffic PM concentrations (note again the strongly unevenly spaced time series) were not related to weekday *vs.* weekend patterns, but rather originated from different road conditions (for resuspended particles) and different local wind patterns with more or less favoured transport of traffic emissions to the sampling inlets. The importance of such small-scale processes has already been reported for the EIB street canyon site,[42] and was observed in this study at LMI as well for both traffic particle and NO_x emissions (see polar plots in Fig. S5†).

Next to traffic, 3 more combustion particle sources were identified. A factor interpreted as emissions dominated by *coal combustion* was resolved in all 5 particle size ranges, showing high mass contributions of WISC, ammonium, sulfate, and WSOC (Fig. 3 and S6†). Species contributions were high for PAHs (typically <50%) and As (approx. 50%), which has often been used as a characteristic marker of coal combustion PM.[43,44] The CPIs of *n*-alkanes ranged from 1.0 to 1.2, while the homohopane index (calculated only on stage 3) was 0.24, both these indices thus supporting the assignment of the factor to a fossil fuel (CPI close to 1) and more specific coal combustion source (homohopane index of 0.1 for lignite combustion[37]). The high mass contributions of ammonium sulfate and WSOC indicate a substantial share of secondary material in the combustion plume, next to the primary emissions of WISC and metals. This is strongly related to the factor representing mainly transported emissions, as can be seen from the time series data in Fig. 4 and S6.† No strong concentration gradient was found between the sites and the source was only "active" during days with eastern air mass origin. In contrast, a second factor attributed to *local coal combustion* had highest factor concentrations at the EIB site, located in a residential area where non-refurbished flats with coal heating can still be found in a limited number. The identification of this factor is tentative, though, as its mass was mainly composed of nitrate instead of WISC in 2 out of 3 particle size ranges (Fig. 3 and S7†). Species contributions of hopanes were very high with homohopane indices of 0.02 for ultrafine and *ca.* 0.25 for fine particles (*i.e.* closer to coal combustion than diesel values, see above). The temporal patterns (Fig. 4 and S7†) reveal a winter-time source, which is why traffic could be excluded. Compared to the first (transported) coal combustion factor, source concentrations were very low for the local one (max. 0.5 $\mu g\ m^{-3}$ compared to max. 20 $\mu g\ m^{-3}$), indicating that this factor is of small importance for total PM apportionment.

A *biomass combustion* dominated source was resolved in all particle size ranges as well. In its chemical profile mass contributions of WISC were highest (typically *ca.* 25–50%), followed by WSOC, ammonium, nitrate, and sulfate (Fig. 3 and S8†). Mass contributions of biomass burning tracers potassium and levoglucosan were comparably high as well with values up to 10% of particle mass. Factor concentrations were much higher in winter, but still observable in summer (Fig. 4 and S8†), possibly related to either transported forest fire or local emissions from biomass or charcoal burning. A clear concentration trend between the sites could not be observed in either season, indicating the influence of mainly regional and/or transported biomass burning emissions to all the sampling sites.

Source areas of the two mainly regional solid fuel combustion sources were studied by applying the concentration-weighted trajectory method,[45] implemented in the "openair" R package.[29] Results are shown in Fig. S9† for particles in the size range 0.42–1.2 μm, which had highest factor concentrations and typically have longest lifetimes in the atmosphere. They broadly indicate South-East Europe as the main geographic source region of solid fuel combustion aerosol with other European areas having lower impacts on concentrations in the Leipzig area, especially for the coal combustion factor.

Beyond the contribution of secondary material in the discussed combustion factors, two separate sources related to secondary formation of particles were resolved in this study. A general *secondary aerosol* factor (resolved on impactor stages 2–4) consisted mainly of ammonium, nitrate, and sulfate (approx. 80% mass contribution in total, Fig. 3 and S10†) and WSOC and WSIC (approx. 20%). Its factor concentrations were highest in winter, but still significant in summer (Fig. 4 and S10†) and were very similar across all 4 sites (regional source). A *photochemistry* factor (resolved on impactor stages 1–3) showed high mass contributions of ammonium and sulfate (approx. 60% in total, Fig. 3 and S11†) and WSOC (approx. 30%) as well as high species contributions of oxalate (approx. 75–95% in ultrafine and fine particles). Factor concentrations (Fig. 4 and S11†) tended to be higher in summer and again showed no clear site-dependent trend. In contrast to the general secondary aerosol, the photochemistry factor thus seems to be more related to radiation-driven formation processes. The main geographic source regions (Fig. S9†) of both secondary aerosol and photochemistry factors somewhat resembled those of the solid fuel combustion factors. Together with partly high species contributions of combustion tracers (Se 20–50%, As in photochemistry 20–30%, levoglucosan in secondary aerosol 15–60%) this indicates that SO_2, NO_x, and VOC emissions from combustion sources in these areas might be major precursors of secondary particulate material observed in the Leipzig area. However, clear impacts from areas other than South-Europe can be observed for the secondary factors as well (Fig. S9†), making them even more of a larger-scale regional source.

A *cooking factor* was identified in ultrafine and fine particles with dominant mass contributions (Fig. 3 and S12†) of WISC (approx. 65–85% on stages 1 and 2, 25% on stage 3) and WSOC (approx. 15%), as well as major inorganic ions on impactor stage 3 (approx. 15–20% each). Longer-chain *n*-alkanes showed high species contributions with CPIs between 3.5 and 4.0 and thus in the biogenic region (>4). Factor concentrations (Fig. 4 and S12†) were highest at the traffic/residential site EIB, which has a number of fast food restaurants in close vicinity. Due to the lack of more characteristic chemical tracers for cooking emissions, the identification of this factor remained somewhat ambiguous.

A source representing *fungal spores* as a sub-fraction of primary biological material was also resolved for coarse particles. Mass contributions (Fig. 3 and S13†) were high for carbonaceous material (approx. 40% WISC and 20–25% WSOC), partly also for nitrate and/or sulfate. Arabitol as a unique tracer for fungal spores[46] showed very high species contributions (approx. 90%) and CPIs of *n*-alkanes were about 12 and thus clearly in the biogenic range (>4). Relevant factor concentrations were only observed in summer, consistent with the biological activity of this source.

An *urban dust* factor was resolved in coarse particle size ranges. It consisted mainly of nitrate (approx. 50%, Fig. 3 and S14†) and WSOC (approx. 20%), with comparably high mass contributions of Ca and Fe (approx. 3–7%). Species contributions of odd *n*-alkanes are partly very high and CPIs were 0.7 and 1.9 on impactor stages 4 and 5, respectively. The Ca/Fe ratio was between 0.8 and 1.4 and thus close to the value of 0.95 reported for the average composition of upper continental crust.[47] Factor concentrations (Fig. 4 and S14†) were highest in summer, similar at all 3 urban sites and lower at the regional MEL site. The high nitrate fraction and anthropogenic *n*-alkanes signature indicated a mixture of pure mineral dust with urban pollution, while the seasonal and site dependencies of concentrations pointed to an urban background source without direct association to traffic.

Finally, two sea salt related sources were identified on coarse particles. A *sea salt/road salt* factor had high mass contributions (Fig. 3 and S15†) of both sodium (approx. 30%) and chloride (approx. 40–60%) and Na^+/Cl^- ratios of 0.81 and 0.51 (stages 4 and 5, respectively), close to the value of 0.56 in sea water.[48] Elevated factor concentrations were frequently observed in winter (Fig. 4 and S15†), especially at the traffic sites (indicating an influence of road salt for de-icing), while a strong peak concentration was observed during one day in summer with presumably quick transport of marine air masses to the continental sites. An *aged sea salt* factor consisted mainly of nitrate (approx. 70% mass contribution) and sodium (approx. 10%) with no mass contribution of chloride (Fig. 3 and S16†), indicating efficient chloride depletion by exchange with nitrate during transport over the continent. The Na^+/Mg^{2+} ratios were 6.1 and 4.9 (stages 4 and 5, respectively) and thus also close to the sea water value of 8.3. The time series (Fig. 4 and S16†) indicated elevated concentrations predominantly during western air mass origin and the geographic source regions of both sea salt related sources were clearly above the Atlantic Ocean (Fig. S9†).

Comparison of source contributions between approaches. To assess the plausibility of the factor concentrations apportioned by PMF, PM_{10} values (sum of impactor stages) were compared to PM_{10} source contributions calculated from alternative established approaches, where possible. Data for comparison were calculated using the Lenschow approach for traffic-related sources (see above), and using the macrotracers levoglucosan for biomass combustion PM (factor 10.95),[49] arabitol for fungal spores (factor 27.5),[46] and the sum of ammonium, nitrate, and sulfate for secondary (inorganic) aerosol. The results are given in Table 2. Traffic contributions (sum of exhaust and general factor for PMF) typically agreed within 30% with the exception of "Winter East", where PMF apportioned only about 1/3 of the Lenschow mean contribution (and mass fractions were about half the Lenschow fractions). Contributions from biomass combustion agreed within 15% during winter, while relative deviations during summer were higher, but for rather small absolute concentrations. Secondary aerosol (sum of 2 PMF factors) showed contributions from PMF about 30–70% higher than the sum of 3 main secondary inorganic ions. Given that this PMF factor also included some contributions from secondary organic material as well as particle water, such differences seem plausible. Fungal spores showed somewhat higher contributions and mass fractions from PMF (factor 2–3), however at very low concentration levels with associated higher uncertainties.

Overall, this brief intercomparison of factor contributions showed reasonable agreement between the approaches, indicating an appropriate quality of the chosen PMF solutions and supporting their further use for quantitative source apportionment at the sites, as will be discussed in the next section.

Size-resolved source contributions. In Fig. 5 the relative mean contributions of PMF-resolved sources are given for all sites, size ranges, seasons, and air mass origins. Mean PMF-apportioned mass concentrations are indicated above the stacked bars. In addition, mean source concentrations are given in the ESI† as indicated below. *Ultrafine particles* in summer originated mainly from the 2 sources traffic (including exhaust and general traffic factors) and photochemistry with fractions of total ultrafine mass concentration of about 20–50% and 30–50%, respectively. While relative traffic contributions were often similar, absolute concentrations (Fig. S17†) strongly decreased towards the regional background site (*e.g.* total traffic contribution at LMI 0.51 *vs.* 0.04 $\mu g\ m^{-3}$ at MEL in "Summer West"). At EIB the (tentatively identified) cooking source contributed about 1/3 to ultrafine PM. Particles from solid fuel combustion were present in low mean concentrations in summer (max. about 0.1 $\mu g\ m^{-3}$), but strongly increased in winter (up to about 0.7 $\mu g\ m^{-3}$) typically explaining 50–70% of ultrafine particle mass (lower fractions only at traffic sites during "Winter West"). In contrast to

Table 2 Comparison of PM_{10} source contributions at LMI calculated from Lenschow and macrotracer approaches with PMF results. Median/mean concentrations and mass fractions (in bold) are given

Source		Lenschow/macrotracer conc. μg m^{-3} (mass fraction%)	PMF conc. μg m^{-3} (mass fraction%)
Traffic	Summer West	3.0/3.2 (**23/23**)	4.6/4.4 (**30/33**)[a]
	Summer East	5.3/4.7 (**30/26**)	3.9/4.7 (**26/27**)[a]
	Winter West	6.4/7.1 (**42/38**)	6.8/7 (**40/41**)[a]
	Winter East	5.7/8.5 (**15/18**)	2.2/2.7 (**6.9/8.2**)[a]
	All	5.2/5.7 (**25/27**)	5.0/5.2 (**28/29**)[a]
Biomass combustion	Summer West	0.16/0.18 (**1/1.2**)	0.29/0.27 (**1.8/2.1**)
	Summer East	0.47/0.57 (**2.9/3.5**)	0.71/0.86 (**5.3/5.6**)
	Winter West	1.1/1.4 (**6.2/6.7**)	0.96/1.3 (**6/6.9**)
	Winter East	4.2/3.7 (**9.7/9.1**)	3.6/3.6 (**9.1/11**)
	All	0.69/1.3 (**3.9/5**)	0.95/1.4 (**5/6.1**)
Secondary inorganic aerosol	Summer West	3.3/3.2 (**18/20**)[b]	4.2/4.2 (**32/30**)[c]
	Summer East	3.3/3.6 (**20/21**)[b]	5.7/6 (**43/41**)[c]
	Winter West	3.3/5.2 (**31/27**)[b]	5.1/6.7 (**33/33**)[c]
	Winter East	18/16 (**39/39**)[b]	17/17 (**46/49**)[c]
	All	3.7/6.6 (**24/27**)[b]	6.2/8.7 (**37/38**)[c]
Fungal spores	Summer West	0.1/0.18 (**0.7/1.2**)	0.26/0.48 (**2/3.3**)
	Summer East	0.06/0.07 (**0.34/0.51**)	0.1/0.11 (**0.75/0.85**)
	Winter West	0/0.009 (**0/0.04**)	0/0.008 (**0/0.04**)
	Winter East	0/0.03 (**0/0.07**)	0.002/0.029 (**0.01/0.11**)
	All	0.034/0.082 (**0.15/0.5**)	0.034/0.18 (**0.18/1.2**)

[a] Sum of 2 traffic-related factors "traffic exhaust" and "general traffic". [b] Sum of secondary ions ammonium, nitrate, and sulfate. [c] Sum of 2 secondary factors "secondary aerosol" and "photochemistry".

traffic, solid fuel combustion concentrations showed a much weaker (for biomass combustion) or no gradient at all (for coal combustion) between the sites in this particle size range (Fig. S17†). For ultrafine particle mass, traffic and – in part – cooking activities are thus the sources with highest potential for reduction by local mitigation measures, while solid fuel combustion in winter would require larger-scale efforts in the study area.

Main sources of *accumulation mode particles* in summer were of secondary nature. The sum of both secondary PMF sources typically contributed about 60–90% to PM in this size range –with higher shares from "photochemistry" on smaller particles (stage 2). Mean concentrations of the photochemistry source were very similar at all sites (about 1 μg m^{-3} during western and about 2 μg m^{-3} during eastern inflow on stage 2, Fig. S18†), while the "general secondary aerosol" source contributions tended to decrease towards the regional background, especially with eastern air mass origin. This likely indicates local secondary formation from traffic emissions, *e.g.* nitrate from NO_x, while the formation of secondary compounds in the "photochemistry" source (mainly sulfate and WSOC, see above) seemed to have a more regional character. Traffic contributed about 10–20% to the accumulation mode particle mass at the urban sites in summer (concentrations about 0.2–1 μg m^{-3}). On impactor stage 3, cooking aerosol constituted another main source of particle mass, with mean contributions up to about 50% at the residential EIB site and typically 20–30% at the other sites. Relative contributions of this source became lower in winter (while absolute concentrations often stayed similar, Fig. S19†) due to increased contributions from the general secondary aerosol (about 30–80%) and the solid fuel combustion sources (about 10–60%). Solid fuel combustion PM was dominated by biomass

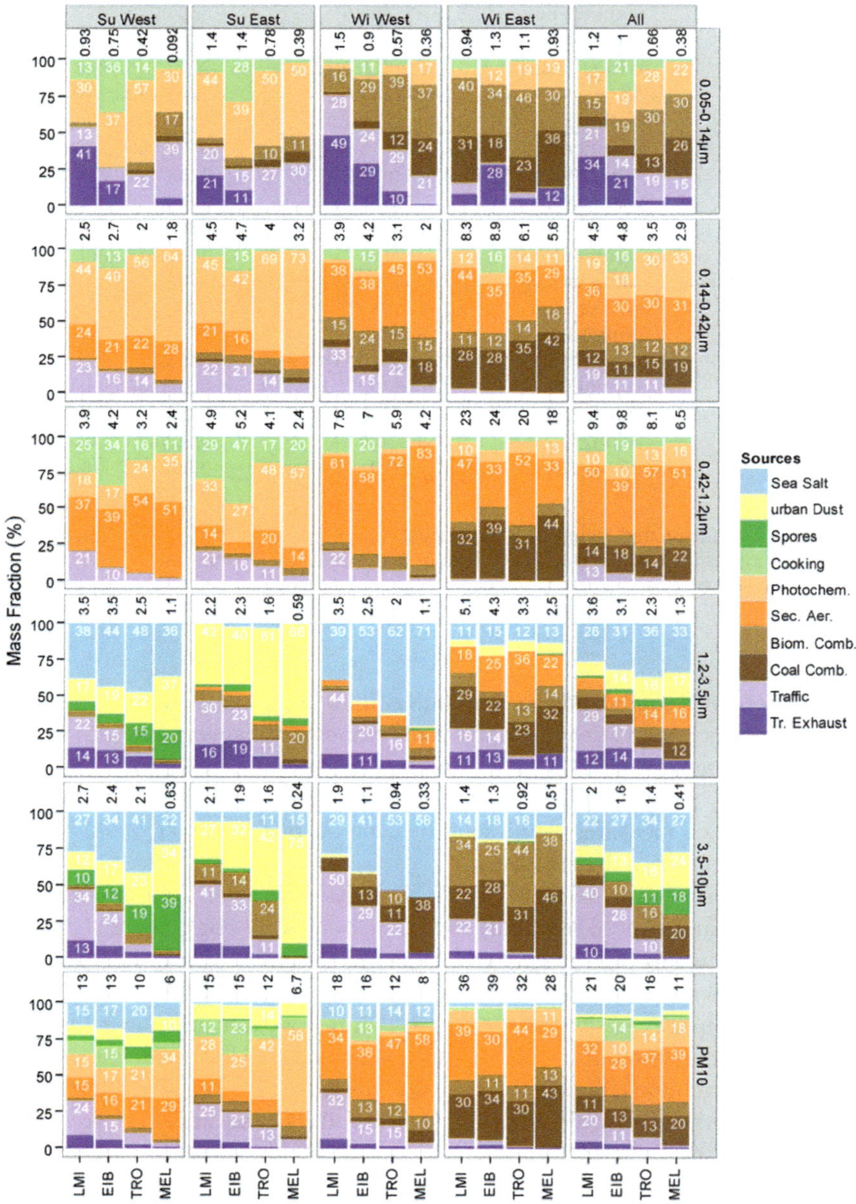

Fig. 5 Mean relative source contributions in different size ranges (including PM_{10}) at the 4 sites. Mean particle mass concentrations ($\mu g\ m^{-3}$) are printed in black on top of stacked bars. Numbers in white indicate the respective mass fractions (not shown if <10%). Note that 2 coal combustion and 2 sea salt related sources were combined for this plot.

combustion during western inflow with generally much lower concentrations and coal combustion during eastern inflow with significantly increased source contributions (*e.g.* coal combustion on impactor stage 3 ≤ 0.1 μg m^{-3} with western and about 6–9 μg m^{-3} with eastern air mass origin, Fig. S19†). Similar to the ultrafine size range, in accumulation mode particles no clear concentration trend between the sites was observed for solid fuel combustion concentrations.

For *coarse mode particles* the main sources are more diverse. In summer during western air mass origin sea salt (both fresh and aged) typically made up about 30–40% of PM at all sites, while it was hardly present during eastern inflow. In contrast, the urban dust source was about twice as active in "Summer East" (30–70% of coarse PM) than in "Summer West". Relative contributions of this source increased towards the regional MEL site, while concentrations were 1.5–4 times higher at the urban sites (Fig. S20 and S21†). Between the urban sites, however, no concentration gradient was observed, indicating no direct association of the source with traffic, but rather with a generally increased level of coarse polluted dust in the city. The urban dust source included secondary components in its chemical profile as well (nitrate, WSOC, oxalate, see above), which is likely the reason for the secondary PMF factors not contributing to summer-time coarse particles. Exhaust and general traffic sources had the highest contributions (both relative and absolute) at the traffic sites, explaining in total about 30–50% of coarse PM in summer (with concentrations around 1 μg m^{-3} in each of the coarse particle size ranges, Fig. S20 and S21†). Fungal spores constituted a non-negligible fraction especially during western air mass inflow with concentrations around 0.3 μg m^{-3}, explaining about 10–40% of stage masses depending on the site. During eastern inflow, the impact from biomass burning increased with mean concentrations of *ca.* 0.2–0.4 μg m^{-3} per impactor stage and mass fractions around 10–25% in the respective size ranges during summer. In winter, contributions from urban dust and fungal spores diminished, while the importance of biomass combustion for coarse PM increased, especially for eastern air mass origin. Coal combustion appeared as an additional important source in "Winter East" with concentrations around 1 and 0.3 μg m^{-3} on stages 4 and 5, respectively, explaining about 20–45% of respective particle masses at the different sites. A winter-time secondary aerosol source was resolved on impactor stage 4 only, with elevated contributions around 1 μg m^{-3} (mass fractions of 20–35%) during eastern air mass inflow. Traffic in winter showed broadly comparable mean coarse-mode concentrations and mass fraction as observed for the summer months.

In summary for PM_{10} mass concentrations, traffic, secondary, and solid fuel combustion particles were identified as the most important sources in Leipzig. Traffic was highest at the 2 kerbside sites, where it typically caused 3–7 μg m^{-3} (4.2 μg m^{-3} as total mean at both sites) or 20–40% of mean PM_{10} concentrations (Fig. 5 and also Fig. S23†). These figures are consistent with a recently reported average traffic contribution of 28% to PM_{10}, obtained from a large number of source apportionment studies at Western European urban sites[50] and they also lie well within the inter-quartile range (IQR) for traffic contributions of approx. 15–40% derived in a previous European review study.[4] Traffic shares decreased to an average of 9% at the urban background (about 1/3 of traffic sites values) and 3% at the regional background sites. Approximately half of the traffic-related particle mass was in the coarse size range, 35% in accumulation mode and 15% in ultrafine particles (Fig. S23†). Secondary particles added 4–17 µg m^{-3} to PM_{10} (*ca.* 30–60%) at all sites, with a total mean at the 3 urban sites of 8.3 µg m^{-3} (*ca.* 45%). Compared with other European sites, Leipzig thus lies at the upper end of the IQR of reported secondary inorganic aerosol contributions (*ca.* 15–40%).[4] On average, 95% of secondary material was found in accumulation mode particles and 80% of urban sites concentrations were from regional transport into the city (Fig. S23†). Biomass combustion particles were observed in both seasons with winter concentrations being a factor of 4–5 higher and explaining about 10–15% of mean PM_{10}. The IQR of biomass burning contributions at European sites was reported to be *ca.* 10–23%,[4] thus the impact of this source was lower than average during this study. About 2/3 of the biomass combustion contribution was from accumulation mode and 12% from ultrafine mode particles. At the residential EIB site, mean winter concentrations were about 1 µg m^{-3} higher than in the regional background, translating into about 60% local contribution during western air mass inflow (lower concentrations) and 20% during eastern inflow. Particle mass from coal combustion was abundant in the "Winter East" category only with very similar mean concentrations of 10–13 µg m^{-3} at all 4 sites (<1 µg m^{-3} in all other seasons/sectors), explaining about 30–40% of total PM_{10} mass concentrations in this category. This, together with increased regional concentrations of biomass combustion and secondary material, emphasizes the importance of transboundary pollution transport for regional air quality in the area of Leipzig.[14,15] In fact, 4 out of 5 campaign days with official PM_{10} mass concentrations at LMI exceeding the EU limit value of 50 µg m^{-3} fell into the eastern air mass origin category. An elevated impact of solid fuel combustion in eastern European regions was also derived in a comprehensive review of source apportionment studies, where "domestic fuel burning" was reported to contribute 7% on average to PM_{10} for western European countries, and 45% for Central and East Europe.[50]

Comparison with data from the year 2000. To assess changes in particulate air pollution within the past 15 years, a comparison with size-resolved particle composition data from a similar project performed at the sites LMI, TRO, and MEL in the year 2000 (ref. 10) is made. The former project focussed on $PM_{2.5}$, which is why data from only the first 4 impactor stages was available (up to 3.5 µm). Also, the number of samples per season was only 8 and thus much lower than in the 2013–15 campaign and eastern air mass inflow did not occur during the sampled days. In addition, discrepancies in the representativeness of campaign mean values for seasonal means were observed from daily quartz filter measurements in Melpitz (Fig. S24†): for winter, the 2000 campaign mean was significantly higher than the true mean and the 2013–15 campaign means were close to or slightly lower than the true means. In contrast for summer, the 2000 campaign mean was slightly higher, while the 2013 campaign mean was slightly lower than the respective true mean value. These deviations bias a direct comparison of concentrations in a way that observed differences are larger than observed from more representative data sets. Therefore, in Fig. 6, the size-resolved mean Lenschow increments for both projects instead of the mean concentrations are shown, which are "corrected" for poorly representative mean regional values. The assumption is that the 8 samples obtained per season in the former project are broadly representative for the local increments. As can be seen in Fig. 6, the regional concentrations (which are the observed MEL concentrations) have strongly decreased for mass and all constituents in basically all

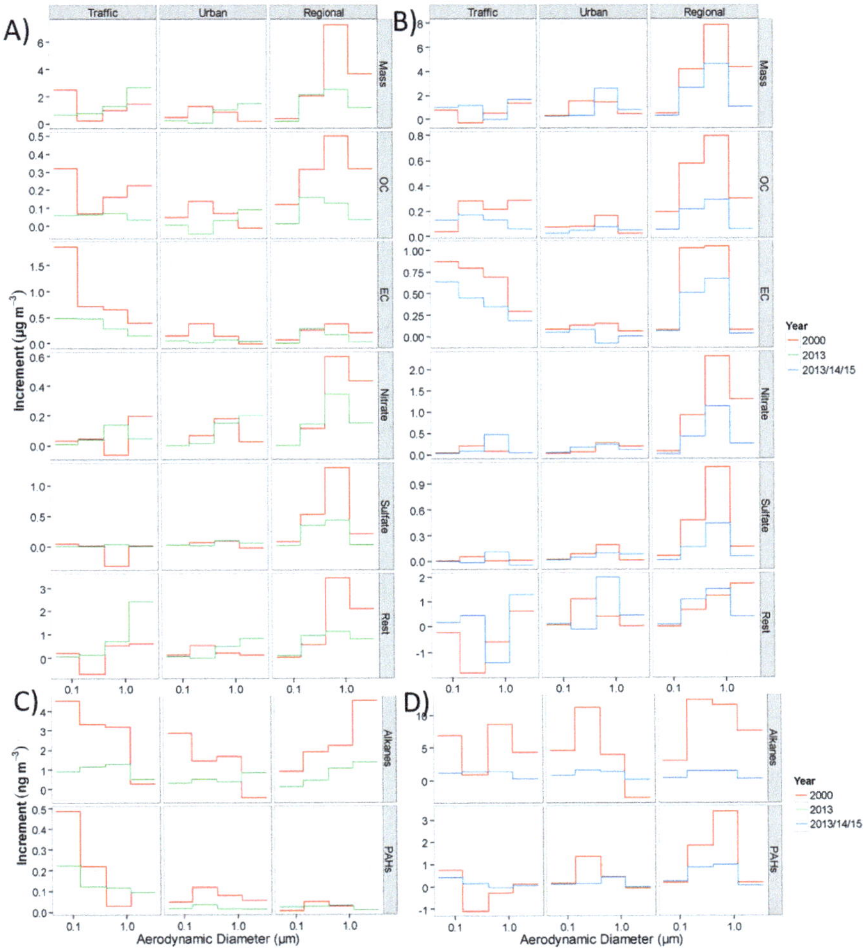

Fig. 6 Comparison of size-resolved means of regional background concentrations (MEL) and urban background and traffic increments (after Lenschow for sites TRO and LMI) between data from 2000 and the present project for main constituents (A and B) and sum of *n*-alkanes and PAHs (C and D) in summer (A and C) and winter (B and D) during western air mass inflow.

particle size ranges, but data comparability is limited here as discussed. However, the Lenschow increments have also decreased for many constituents and in most particle size ranges. This is most obvious for ultrafine EC, *n*-alkanes, and PAHs, but can be observed for these anthropogenic constituents in other size ranges as well. This is likely related to locally implemented mitigation measures, mainly the low emission zone which has existed since 2011 in the city of Leipzig. For EC in $PM_{3.5}$ particles the comparison shows a reduction of the local traffic increment at LMI from *ca.* 2–3 µg m^{-3} in 2000 to about half these values (*ca.* 1–1.5 µg m^{-3}) in 2013–2015 (Fig. S25†). For total mass (size-resolved and $PM_{3.5}$), the concentrations did not clearly decrease, because the undetermined rest (mainly crustal material) showed increased mean concentrations in the present project, either due to an increased traffic density (with on average cleaner cars) or due to different road conditions (dryness *etc.*) between the projects.

Overall, the limited comparability mainly due to a low number of samples, does not allow for firm conclusions about changes in particulate pollution. The analysis of size-resolved data at several sites in parallel, however, did indicate a reduced regional background level and especially a reduced local traffic increment for EC and other potentially harmful constituents.

Conclusions

Air quality with respect to PM has improved in the city of Leipzig over the years from 2000 to 2013–2015. Especially, the local traffic increments of soot and other potentially harmful constituents were lower in this study than in a similar one in the year 2000. This corroborates the results of a former study on the positive effects of a low emission zone, implemented in 2011 in the city of Leipzig.[11] However, further improvements would be beneficial and the discussed results can help to further develop efficient mitigation measures. On average in this study, nearly half the PM_{10} mass concentration originated from urban sources in contrast to regionally transported ones, with even higher values for ultrafine and coarse, and lower ones for accumulation mode particles. At the most polluted site, traffic typically contributed 30–40%, indicating a high potential for PM reductions. About half of traffic particle mass was in the coarse and 15% in the ultrafine size range. Technologies to reduce traffic emissions across the full particle size range would thus be favourable for substantial improvements of PM pollution at kerbside sites. Secondary material was another major source of PM at the urban sites (*ca.* 45% of PM_{10} on average). However, 80% of it originated from regional sources, indicating that larger-scale reductions of precursors (NO_x, SO_2, NH_3, VOCs) would be required to substantially reduce urban concentration values. Biomass combustion contributed about 10% to urban PM_{10} concentrations in winter with up to 60% of it originating from local emissions in a residential area. Given the comparably high winter temperatures in this study, its contribution can be expected to be much higher during cold winter periods and measures to minimise its impact on urban air quality seem reasonable. Finally, coal combustion was another major PM source when air masses in winter arrived from eastern regions to the study area, explaining 30–40% of PM_{10}. PM_{10} concentrations were generally highest in this scenario. Therefore, continued efforts to reduce emissions on an EU wide level will help to improve air quality not only locally in the emission regions, but also on a larger scale by reducing trans-boundary pollution transport. Further identified PM sources with less contribution to total particle concentrations were cooking, fungal spores, urban dust, and sea or road salt. Most of these sources are natural, originally. Due to interactions of naturally emitted particles with polluted air masses, however, their mass concentration can increase (*e.g.* by formation of particulate nitrate on dust or sea salt). Reductions of precursor gases could thus be expected to reduce mass contributions of such mixed natural/anthropogenic particles as well to a certain extent.

Acknowledgements

The authors gratefully acknowledge financial support from the Saxon State Office for Environment, Agriculture and Geology (LfULG) and technical support and cooperation with the operating company for environment and agriculture of the state of Saxony (BfUL). The authors also highly appreciate the technical support and the engagement in sampling and sample analysis by S. Fuchs, A. Dietze, R. Rabe, A. Grüner, S. Haferkorn, S. Liebig, and A. Rödger.

References

1 WHO, *WHO air quality guidelines for particulate matter, ozone, nitrogen dioxide and sulfur dioxide – global update 2005*, World Health Organisation, Geneva, Switzerland, 2006.
2 F. R. Cassee, M.-E. Héroux, M. E. Gerlofs-Nijland and F. J. Kelly, *Inhalation Toxicol.*, 2013, **25**, 802–812.
3 M. Viana, T. A. J. Kuhlbusch, X. Querol, A. Alastuey, R. M. Harrison, P. K. Hopke, W. Winiwarter, A. Vallius, S. Szidat, A. S. H. Prevot, C. Hueglin, H. Bloemen, P. Wahlin, R. Vecchi, A. I. Miranda, A. Kasper-Giebl, W. Maenhaut and R. Hitzenberger, *J. Aerosol Sci.*, 2008, **39**, 827–849.
4 C. A. Belis, F. Karagulian, B. R. Larsen and P. K. Hopke, *Atmos. Environ.*, 2013, **69**, 94–108.
5 S. Visser, J. G. Slowik, M. Furger, P. Zotter, N. Bukowiecki, F. Canonaco, U. Flechsig, K. Appel, D. C. Green, A. H. Tremper, D. E. Young, P. I. Williams, J. D. Allan, H. Coe, L. R. Williams, C. Mohr, L. Xu, N. L. Ng, E. Nemitz, J. F. Barlow, C. H. Halios, Z. L. Fleming, U. Baltensperger and A. S. H. Prevot, *Atmos. Chem. Phys.*, 2015, **15**, 11291–11309.
6 A. C. Rohr and R. E. Wyzga, *Atmos. Environ.*, 2012, **62**, 130–152.
7 P. Paatero and U. Tapper, *Environmetrics*, 1994, **5**, 111–126.
8 P. Paatero, S. Eberly, S. G. Brown and G. A. Norris, *Atmos. Meas. Tech.*, 2014, **7**, 781–797.
9 J. G. Watson, L. W. Antony Chen, J. C. Chow, P. Doraiswamy and D. H. Lowenthal, *J. Air Waste Manage. Assoc.*, 2008, **58**, 265–288.
10 H. Herrmann, E. Brüggemann, U. Franck, T. Gnauk, G. Löschau, K. Müller, A. Plewka and G. Spindler, *J. Atmos. Chem.*, 2006, **55**, 103–130.
11 F. Rasch, W. Birmili, K. Weinhold, S. Nordmann, A. Sonntag, G. Spindler, H. Herrmann, A. Wiedensohler and G. Löschau, *Gefahrstoffe - Reinhalt. Luft*, 2013, **73**, 483–489.
12 G. Spindler, K. Müller, E. Brüggemann, T. Gnauk and H. Herrmann, *Atmos. Environ.*, 2004, **38**, 5333–5347.
13 A. Berner and C. Lürzer, *J. Phys. Chem.*, 1980, **84**, 2079–2083.
14 G. Spindler, E. Brüggemann, T. Gnauk, A. Grüner, K. Müller and H. Herrmann, *Atmos. Environ.*, 2010, **44**, 164–173.
15 G. Spindler, A. Grüner, K. Müller, S. Schlimper and H. Herrmann, *J. Atmos. Chem.*, 2013, **70**, 165–195.
16 K. W. Fomba, K. Müller, D. van Pinxteren, L. Poulain, M. van Pinxteren and H. Herrmann, *Atmos. Chem. Phys.*, 2014, **14**, 8883–8904.
17 T. Gnauk, K. Müller, D. van Pinxteren, L. Y. He, Y. W. Niu, M. Hu and H. Herrmann, *Atmos. Environ.*, 2008, **42**, 6296–6309.
18 D. van Pinxteren, E. Brüggemann, T. Gnauk, Y. Iinuma, K. Müller, A. Nowak, P. Achtert, A. Wiedensohler and H. Herrmann, *J. Geophys. Res.: Atmos.*, 2009, 114.
19 Y. Iinuma, G. Engling, H. Puxbaum and H. Herrmann, *Atmos. Environ.*, 2009, **43**, 1367–1371.

20 C. Neusüβ, M. Pelzing, A. Plewka and H. Herrmann, *J. Geophys. Res.*, 2000, **105**, 4513–4527.
21 K. W. Fomba, K. Müller, D. van Pinxteren and H. Herrmann, *Atmos. Chem. Phys.*, 2013, **13**, 4801–4814.
22 EPA, *EPA Positive Matrix Factorization (PMF) 5.0-Fundamentals and User Guide*, US Environmental Protection Agency, Washington DC, 2014.
23 JRC, *European Guide on Air Pollution Source Apportionment with Receptor Models*, European Commission Joint Research Centre, Luxembourg, 2014.
24 R Core Team, *R: A language and environment for statistical computing*, http://www.R-project.org.
25 H. Wickham, *J. Stat. Software*, 2011, **40**, 1–29.
26 H. Wickham and R. Francois, *dplyr: A Grammar of Data Manipulation*, http://CRAN.R-project.org/package=dplyr.
27 H. Wickham, *tidyr: Easily Tidy Data with 'spread()' and 'gather()' Functions*, http://CRAN.R-project.org/package=tidyr.
28 H. Wickham, *ggplot2: Elegant graphics for data analysis*, Springer, New York, 2009.
29 D. C. Carslaw and K. Ropkins, *Environ. Model. Software*, 2012, **27–28**, 52–61.
30 R. R. Draxler, B. Stunder, G. Rolph, A. Stein and A. Taylor, *HYSPLIT4 User's Guide*, NOAA Air Resources Laboratory, Silver Spring, MD, USA, 2014.
31 D. van Pinxteren, E. Brüggemann, T. Gnauk, K. Müller, C. Thiel and H. Herrmann, *J. Atmos. Chem.*, 2010, **67**, 1–28.
32 R. Hitzenberger, A. Berner, Z. Galambos, W. Maenhaut, J. Cafmeyer, J. Schwarz, K. Müller, G. Spindler, W. Wieprecht, K. Acker, R. Hillamo and T. Makela, *Atmos. Environ.*, 2004, **38**, 6467–6476.
33 EEA, *Air quality in Europe — 2015 report*, report 05/2015, European Environment Agency (EEA), Luxembourg, 2015.
34 G. Spindler, T. Gnauk, A. Grüner, Y. Iinuma, K. Müller, S. Scheinhardt and H. Herrmann, *J. Atmos. Chem.*, 2012, **69**, 127–157.
35 J. P. Putaud, F. Raes, R. Van Dingenen, E. Brüggemann, M. C. Facchini, S. Decesari, S. Fuzzi, R. Gehrig, C. Huglin, P. Laj, G. Lorbeer, W. Maenhaut, N. Mihalopoulos, K. Mülller, X. Querol, S. Rodriguez, J. Schneider, G. Spindler, H. ten Brink, K. Torseth and A. Wiedensohler, *Atmos. Environ.*, 2004, **38**, 2579–2595.
36 P. Lenschow, H. J. Abraham, K. Kutzner, M. Lutz, J. D. Preuss and W. Reichenbacher, *Atmos. Environ.*, 2001, **35**, S23–S33.
37 J. Schnelle-Kreis, M. Sklorz, J. Orasche, M. Stolzel, A. Peters and R. Zimmermann, *Environ. Sci. Technol.*, 2007, **41**, 3821–3828.
38 D. B. Kittelson, *J. Aerosol Sci.*, 1998, **29**, 575–588.
39 P. Pant and R. M. Harrison, *Atmos. Environ.*, 2013, **77**, 78–97.
40 A. Thorpe and R. M. Harrison, *Sci. Total Environ.*, 2008, **400**, 270–282.
41 P. Kumar, L. Pirjola, M. Ketzel and R. M. Harrison, *Atmos. Environ.*, 2013, **67**, 252–277.
42 J. Voigtlander, T. Tuch, W. Birmili and A. Wiedensohler, *Atmos. Chem. Phys.*, 2006, **6**, 4275–4286.
43 R. M. Harrison, D. J. T. Smith and L. Luhana, *Environ. Sci. Technol.*, 1996, **30**, 825–832.
44 J. C. Chow and J. G. Watson, *Energy Fuels*, 2002, **16**, 222–260.

45 P. Seibert, H. Kromp-Kolb, U. Baltensperger, D. T. Jost, M. Schwikowski, A. Kasper and H. Puxbaum, in *Transport and transformation of pollutants in the troposphere*, ed. P. M. Borrell, P. Borrell, T. Cvitas and W. Seiler, Academic Publishing, The Hague, 1994, pp. 689–693.
46 H. Bauer, M. Claeys, R. Vermeylen, E. Schueller, G. Weinke, A. Berger and H. Puxbaum, *Atmos. Environ.*, 2008, **42**, 588–593.
47 K. H. Wedepohl, *Geochim. Cosmochim. Acta*, 1995, **59**, 1217–1232.
48 J. H. Seinfeld and S. N. Pandis, *Atmospheric chemistry and physics: from air pollution to climate change*, Wiley, New York, 2nd edn, 2006.
49 C. Schmidl, L. L. Marr, A. Caseiro, P. Kotianova, A. Berner, H. Bauer, A. Kasper-Giebl and H. Puxbaum, *Atmos. Environ.*, 2008, **42**, 126–141.
50 F. Karagulian, C. A. Belis, C. F. C. Dora, A. M. Prüss-Ustün, S. Bonjour, H. Adair-Rohani and M. Amann, *Atmos. Environ.*, 2015, **120**, 475–483.

Faraday Discussions

ROYAL SOCIETY OF CHEMISTRY

PAPER

The "Parade Blue": effects of short-term emission control on aerosol chemistry†

Haiyan Li,[a] Qiang Zhang,*[bc] Fengkui Duan,[a] Bo Zheng[a] and Kebin He[acd]

Received 11th January 2016, Accepted 2nd February 2016
DOI: 10.1039/c6fd00004e

The strict control on emissions implemented in Beijing, China, during the 2015 China Victory Day Parade (V-day Parade) to commemorate the 70th Anniversary of Victory in World War II, provided a good opportunity to investigate the relationship between emission sources and aerosol chemistry in a heavily polluted megacity. From August 11 to September 3, 2015, an Aerosol Chemical Speciation Monitor was deployed in urban Beijing, together with other collocated instruments, for the real-time measurement of submicron aerosol characteristics. The average PM_1 mass concentration was 11.3 ($\pm$6.7) μg m^{-3} during the V-day Parade, 63.5% lower than that before the V-day Parade. Differently to the relatively smaller decrease of organics (53%), secondary inorganic aerosols (sulfate, nitrate and ammonium) showed significant reductions of 65–78% during the V-day Parade. According to the positive matrix factorization results, primary organic aerosol (POA) from traffic and cooking emissions decreased by 41.5% during the parade, whereas secondary organic aerosol (SOA) presented a much greater reduction (59%). The net effectiveness of emission control measures was investigated further under comparable weather conditions before and during the parade. By excluding the effects of meteorological parameters, the total PM_1 mass was reduced by 52–57% because of the emission controls. Although the mass concentrations of aerosol species were reduced substantially, the PM_1 bulk composition was similar before and during the control period as a consequence of synergetic control of various precursors. The emission restrictions also suppressed the secondary formation processes of sulfate and nitrate, indicated by the substantially reduced SOR and NOR (molar ratios of sulfate or nitrate to the sums of the sulfate and SO_2 or nitrate and NO_2) during the event. The study also explored the influence of emission controls on the evolution of organic aerosol using the mass ratios of SOA/POA and oxygen-to-carbon ratios. The results

[a]*State Key Joint Laboratory of Environment Simulation and Pollution Control, School of Environment, Tsinghua University, Beijing 100084, China*

[b]*Ministry of Education Key Laboratory for Earth System Modeling, Center for Earth System Science, Tsinghua University, Beijing 100084, China. E-mail: qiangzhang@tsinghua.edu.cn*

[c]*Collaborative Innovation Center for Regional Environmental Quality, Beijing 100084, China*

[d]*State Environmental Protection Key Laboratory of Sources and Control of Air Pollution Complex, Tsinghua University, Beijing 100084, China*

† Electronic supplementary information (ESI) available. See DOI: 10.1039/c6fd00004e

showed that for northwesterly airflows, emission restrictions during the V-day Parade also reduced the oxidation degree of organic aerosol.

Introduction

Aerosol particles are of great importance not only because of their direct and indirect radiative forcing on climate but also because of their adverse effects on health. They can be emitted directly as particles from natural and anthropogenic sources, or formed in the atmosphere through gas-to-particle conversions.[1] According to Seinfeld and Pandis,[2] considerable amounts of aerosols are anthropogenic in origin, especially those with diameters <1 μm. Anthropogenic emissions of air pollutants in China are much greater than in European or North American countries, which intensifies China's severe air pollution.[3] Therefore, it is essential to understand how anthropogenic sources affect the aerosol load before implementing clean-air measures.

Beijing, the capital city of China, has suffered frequent haze episode events since the early 2000s.[4] In 2014, the city had a residential population of 21.52 million and the total number of motor vehicles exceeded 5.3 million.[5] Moreover, air quality in Beijing can be affected significantly by its surrounding regions, especially the heavily industrialized cities in the Hebei and Shanxi provinces.[6]

To improve the air quality for the 2015 Victory Day Parade, held on September 3, 2015, the Chinese government implemented a number of strict control measures to reduce anthropogenic emissions in Beijing and the surrounding areas (*e.g.* Tianjin, Hebei, Inner Mongolia, Shanxi, Shandong and Henan). The number of motor vehicles allowed on the roads of Beijing between August 20 and September 3, 2015 was reduced by half. In addition, more than 10 000 enterprises across all 7 municipalities and provinces were suspended or ordered to limit production during this period, and various construction activities were also forbidden. Consequently, the air quality in Beijing during the event was improved significantly and dubbed the "Parade Blue". The control measures implemented during the V-day Parade were much stricter than during the 2008 Olympic Games and 2014 Asia-Pacific Economic Cooperation (APEC) summit. Thus, this event served as a realistic "natural atmospheric chamber"[1] and provided a good opportunity to improve our understanding of the complex relationship between emission sources and aerosol chemistry to guide future control strategies.

Air pollution levels are affected not only by emissions but also by meteorological conditions. Various studies have indicated that air quality in Beijing depends strongly on weather conditions, especially the local wind speed and wind direction.[7,8] According to research focusing on air quality improvement during the 2008 Olympic Games and 2014 APEC summit, favorable meteorological conditions also played an important role in reducing PM levels in addition to emission reductions. Thus, it is not sufficient to simply compare aerosol concentrations prior to and during emission controls to evaluate the efficiency of reduction measures. A popular strategy to distinguish possible meteorological effects is to use comprehensive air quality models to simulate the aerosol formation processes under different emission and meteorological scenarios. Using emission inventories from June 2008 (without emission reductions) and August 2008 (with emission reductions), Wang *et al.*[9] simulated day-to-day $PM_{2.5}$

concentrations with the Community Multiscale Air Quality (CMAQ) model, and they found that more than 60% of the $PM_{2.5}$ reduction during the Olympic Games was due to emission controls. However, model predictions could be influenced strongly by the accuracy of the incorporated chemical mechanisms as well as the emission inventories, especially for $PM_{2.5}$, a complex mixture of chemical species.[10] As presented by Wang *et al.*,[9] the correlation coefficient between modeled and observed data was 0.7 for August 2008. With the application of remote sensing and satellite technology, an alternative approach is the development of statistical models based on pollution level and meteorological parameters. Cermak and Knutti[11] took aerosol optical depth (AOD) as a surrogate for pollution level and developed a neural network to relate log(AOD) to the wind vector, precipitation and relative humidity (RH). They predicted log(AOD), assuming no emission controls, using meteorological observations obtained during the Olympic Games and then compared them with the observed values. However, they also highlighted that the AOD used in their approach might not be representative of surface conditions. Liu *et al.*[12] developed a nonlinear statistical model based on both ground and satellite observations. They estimated ground-level $PM_{2.5}$ concentrations using satellite AOD and local meteorological parameters, and found that 70% of the temporal variability in $PM_{2.5}$ could be explained by the fitted model. Their model results indicated that emission controls during the Olympic Games resulted in an approximate 27–33% reduction in $PM_{2.5}$ concentrations. In addition to the unexplained variability in $PM_{2.5}$ concentrations, their small sample size might also have been a limitation of their approach considering such short-term political measures.

A more straightforward strategy is to regard the atmosphere as a natural laboratory and to estimate the influence of emission reductions using measured changes in air quality under comparable weather conditions. Cheng *et al.*[1] used source conditions, local meteorological parameters and regional transport characteristics as the selection criteria to distinguish the effects of weather conditions on air quality. Under similar local and regional meteorological factors, they estimated that traffic restrictions imposed during the 2006 Sino-African Summit in Beijing reduced the number concentrations of particles in the Aitken and accumulation modes by 20–60%. Several studies have applied a real-time aerodyne aerosol mass spectrometer (AMS) to explore the changes of aerosol chemistry due to emission restrictions during the 2008 Olympics and the 2014 APEC in Beijing.[13–16] These studies largely compared the aerosol loading during the emission control period with that of a pre-control period or the same period in previous years. However, meteorological conditions (*e.g.* wind and precipitation) can be significantly different between these periods. Thus the link between emission controls and aerosol chemistry is still not clearly understood.

In the present study, we conducted real-time measurements of submicron aerosol composition in urban Beijing, both before and during the V-day Parade (August 11–19 and August 20 to September 3, 2015, respectively), using an Aerodyne Aerosol Chemical Speciation Monitor (ACSM) and other collocated instruments. The changes of aerosol chemical species and organic aerosol (OA) sources prior to and during the control period were investigated in detail. In particular, data populations with similar local and regional meteorological conditions before and during the parade were selected. Under comparable weather conditions, the net effectiveness of emission restrictions on aerosol chemistry was elucidated by comparing the changes before and during the event.

Experimental methods

Sampling site

Field measurements were conducted on the campus of Tsinghua University (THU site: 40.0°N, 116.3°E), which is situated in the northwestern urban area of Beijing. Beijing is located at the northern tip of the North China Plain, with the Tianjin municipality on its eastern border and the Hebei province on its other three sides. Topographically, Beijing is characterized as a semi-basin region, bordered by the Yanshan Mountain to the northeast and the Taihang Mountain to the west. The basin-like topography plays an important role in the regional airflow of this area.[17] Sampling was performed approximately 10 m above ground level on the roof of a three-story building. Local emission sources within 1 km of the THU site included vehicles and fuel combustion for cooking. No major industrial sources are located in the locality.

Online instrumentation

The chemical composition of non-refractory PM_1 was measured *in situ* from August 11 to September 3, 2015, using an ACSM.[18] The ACSM uses the same particle lens and vaporizer as a research grade AMS; however, no size information is obtained because of the lack of a fast data acquisition and particle beam chopper system. Before sampling into the ACSM, a $PM_{2.5}$ cyclone (model URG-2000-30ED) was used to remove coarse particles (>2.5 μm), and a silica gel diffusion dryer was used to dry the aerosol particles. After passing through a 100 μm-diameter critical aperture, aerosol particles between 30 nm and 1 μm were focused into a narrow particle beam *via* an aerodynamic lens. Particles were then transmitted directly into the detection chamber, where non-refractory particulate materials were flash vaporized at oven temperature (~600 °C) and ionized by a 70 eV electron impact. In this study, the applied scan rate of the ACSM was 200 ms per amu from *m/z* 10 to 150. The time resolution of chemical species concentration was approximately 15 min, by alternating 14 cycles between the filter and sample modes. The ionization efficiency of the ACSM was calibrated at the beginning, in the middle and at the end of the field observations. A detailed description and calibration procedure of ACSM has been given in Ng *et al.*[18]

Because ACSM can only measure the chemical concentrations of non-refractory species (organic, sulfate, nitrate, ammonium and chloride), a multi-angle absorption photometer (MAAP; model 12, Thermo Electron Corporation) was deployed to obtain the concentration of refractory black carbon (BC). The MAAP[19,20] was operated at an incident light wavelength of 670 nm, with a PM_1 cyclone and a drying system incorporated in front of the sampling line. Simultaneously, PM_1 mass was measured based on a β-ray absorption method using a PM-714 Monitor (Kimoto Electric Co., Ltd., Japan).[21] Other collocated instruments included a suite of commercial gas analyzers from Thermo Scientific to measure the gaseous species (CO, O_3, NO, NO_2 and SO_2), and an automatic meteorological observation instrument (Milos520, VAISALA Inc., Finland) to monitor temperature, RH, precipitation, wind speed and wind direction. The height of the PBL (planetary boundary layer) was obtained from the ARL (Air Resources Laboratory) at the NOAA website (http://www.arl.noaa.gov/index.php).

ACSM data analysis

The ACSM data analysis was processed within Igor Pro (WaveMetrics, Inc., Oregon USA) using the standard ACSM data analysis software (v.1.5.3.0). To obtain the chemical concentration of the NR-PM_1 species, the default relative ionization efficiencies (RIEs) of 1.4 for organics, 1.08 for sulfate, 1.1 for nitrate, and 1.3 for chloride were used in this study. For ammonium, the RIE of 7.16 was derived from pure ammonium nitrate calibration. In most previous field studies, a constant collection efficiency (CE) value of 0.5 has been used to account for the incomplete detection of aerosol particles. However, Middlebrook *et al.*[22] proposed that the CE appeared significantly influenced by high aerosol acidity, high RH and high ammonium nitrate mass fractions (ANMF). In this study, RH plays a minor role in affecting the CE because a silica gel diffusion dryer was incorporated with the system to maintain RH < 40%. The aerosol particles were almost neutralized during the entire study, as indicated by the value of ${NH_4^+}_{measured}/{NH_4^+}_{predicted}$ (Fig. S1a†). However, occasionally, high ammonium nitrate was observed with ANMF higher than 0.4 (Fig. S1c†). Thus, variable CEs were applied based on the equation proposed by Middlebrook *et al.*[22] ($CE_{dry} = \max(0.45, 0.0833 + 0.9167 \times ANMF)$). As presented in Fig. S1b,† the mass concentration of NR-PM_1 plus BC using variable CEs correlated strongly with the total PM_1 mass measured by the PM-714 Monitor ($r^2 = 0.90$).

The ACSM mass spectra was processed using the PMF2.exe algorithm combined with the positive matrix factorization (PMF) evaluation tool (PET) to explore various sources of OA.[23] We considered only ions up to m/z 100 given the large interferences of the internal standard of naphthalene at m/z 127–129 and the low signal-to-noise ratio of larger ions. The PMF analysis was performed for 1 to 8 factors and for FPEAK values between −1 and +1 (step: 0.1). The results were evaluated further following the procedures detailed in Zhang *et al.*[24] A summary of the PMF diagnostic discussion and related plots is presented in Fig. S2–S5 in the ESI.† Finally, a 3-factor solution with FPEAK = 0 was chosen. The three factors were identified as hydrocarbon-like organic aerosol (HOA), cooking-related organic aerosol (COA), and oxygenated organic aerosol (OOA), based on the interpretation of the mass spectral signatures, the temporal and diurnal variation patterns and their correlations with external tracer compounds (see section 3.1.2 for details).

Results and discussion

General description

Submicron aerosol and meteorology. Fig. 1 depicts the time series of submicron aerosol composition and meteorological parameters for the entire study period. The concentrations of both PM_1 and the chemical species clearly decreased on August 20, when emission control measures were first implemented in Beijing. Relatively low ambient levels were maintained until September 3, during which far stricter emission reductions were imposed both in Beijing and its surrounding areas. Comparatively, the total PM_1 mass (= NR − PM_1 + BC) was

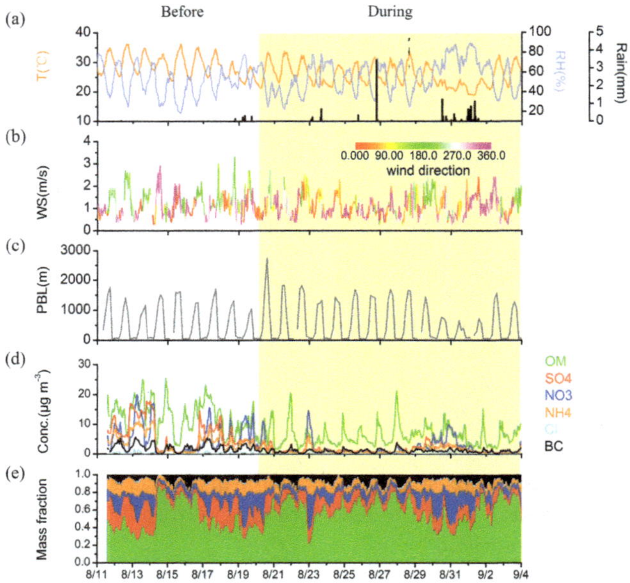

Fig. 1 Temporal variations of (a) T, RH and precipitation, (b) WS and WD, (c) PBL, (d) mass concentrations, and (e) mass fractions of chemical species in PM_1.

reduced by 63.5% during the parade period. The average PM_1 concentration (11.3 μg m^{-3}) was much smaller than that during the 2008 Olympics (61.3 μg m^{-3}) and 2014 APEC (41.6 μg m^{-3}) in Beijing, indicating a greater improvement of air quality during the V-day Parade because of the more stringent controls of emissions.[13,14] In fact, the good air quality during the parade (known as the "Parade Blue") could more or less mirror that observed over a 19 month period in northern central Oklahoma, USA (7.0 μg m^{-3}) and that reported in an urban background site in London, UK (9.9 μg m^{-3}).[25,26]

Compared with the sharp decline of inorganic species during the parade, changes of OA were relatively smaller. Correspondingly, the concentrations of sulfate, nitrate and ammonium (SNA) decreased by 65–78% during the emission control period, whereas organics showed a relatively smaller reduction of 53%. Different behaviors of inorganic and organic species also led to the variation of aerosol chemical composition before and during the parade. Although organics accounted for >50% of the total PM_1 mass during the parade, the contribution of SNA declined from 51.0% to 37.7%, possibly indicating their different responses to emission controls.[14]

In addition to the effect of emission controls, a detailed investigation of the meteorological variables and time series of aerosol species revealed that local wind direction might also play a dominant role in the variation of submicron aerosol characteristics. As indicated in Fig. 1, local meteorological parameters of temperature, RH, wind speed and the PBL height, showed no obvious changes during the entire study period, except for variations on August 31, which were influenced by accumulated precipitation; however, local airflow varied significantly. Although local weather conditions were dominated by southerly winds

(56%) before the parade, the frequency of southerly wind was only 18% during the parade. Previous studies have found that high PM concentrations in Beijing usually correspond to southerly wind, while low PM concentrations generally correspond to the opposite wind direction.[6,17] This was further verified in our study by the back trajectory (BT) analysis (Fig. S6†), which displayed that as the wind direction changed from southerly to northerly, the total PM_1 concentration decreased from 27.0 to 8.7 μg m^{-3}. These results suggested that favorable weather conditions helped considerably in the improvement of air quality during the parade. Moreover, we found that the variations of aerosol chemical composition were also closely related to changes of local wind direction. As shown in Fig. 1, the southerly airflow was always accompanied by a rapid increase of SNA, which was consistent with the fact that high concentration air masses from source locations south of the Beijing area had more abundant secondary aerosol species.[13] The BT clustering analysis also revealed that different PM_1 species showed different BT dependence. While SNA dominated the aerosol composition by 54.9% in southerly air masses, its contribution showed a gradual decrease as the wind direction changed from southerly to northwesterly and then to northeasterly. For the northeasterly BT group, SNA only accounted for 23.6% of the total PM_1 mass. Thus the influence of meteorological conditions, especially local wind directions, on submicron aerosol behavior is non-negligible in our study.

Investigation of OA sources. The PMF analysis of OA mass spectra measured during the entire study period allowed the identification of three organic components, HOA, COA, and OOA. These components were examined for their mass spectral signatures, and then for their correlation with other tracers, their diurnal variations and other characteristics.[23,27,28] Fig. 2 shows the mass spectra profiles of the three components and their time series during the campaign.

The HOA mass spectrum was characterized by the ion series of *m*/*z* 27, 41, and 55, belonging to cycloalkanes, and *m*/*z* 29, 43, and 57, belonging to alkanes.[29] The time series of HOA correlated moderately well with BC ($r^2 = 0.45$), consistent with

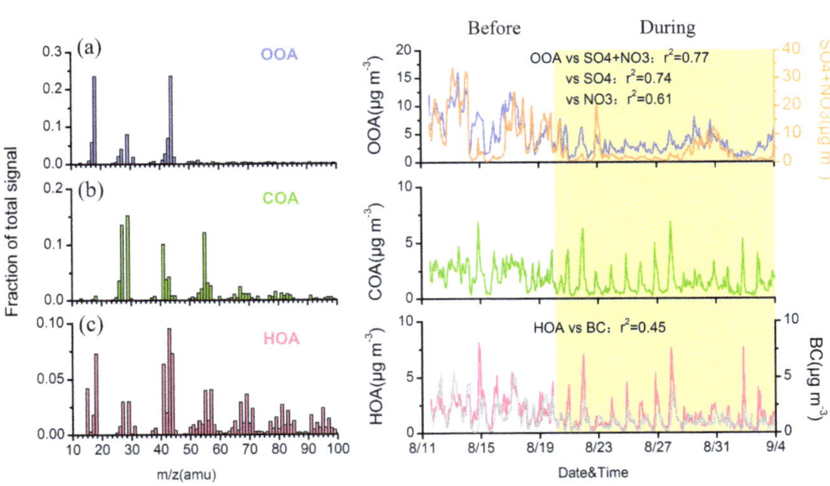

Fig. 2 Mass spectra and time series of (a) HOA, (b) COA and (c) OOA. Also shown are the time series of external tracers and their correlation coefficients with OA factors.

the results of previous studies that HOA was linked chiefly to primary combustion emissions, *e.g.*, gasoline and diesel exhausts.[30] Compared with the large diurnal peak at night, HOA only presented a tiny rush hour peak in the morning, which may be significantly decreased by the enhanced PBL height (Fig. 4). The same phenomenon has been observed and discussed by several studies in Beijing.[16,29,33] The results showed that HOA identified in our study might be associated mainly with traffic emissions, especially considering the minor contribution of coal combustion in Beijing during this season. However, relatively high fractions of ions at *m*/*z* 44 were observed in the HOA mass spectrum, indicating the presence of some oxygenated compounds in HOA. This could be attributed to the fact that a fraction of HOA was aged OA transported from regional sources, similar to the results conducted in Shanghai and Lanzhou, China.[13,28,31]

The most abundant ions of the COA mass spectrum (MS) were similar to those of HOA. However, consistent with the MS characteristics measured for primary Chinese cooking emissions, significantly larger peaks at *m*/*z* 41 and 55 were observed from the mass spectra of COA, indicating the large presence of unsaturated organic compounds (*e.g.* unsaturated fatty acids).[32] Another unique distinction between COA and HOA is their diurnal profiles. Unlike the morning and evening rush hour peaks of HOA, COA presented a small peak at noon and a large peak in the evening, corresponding to the lunch and dinner times of local residents (Fig. 3a). According to previous studies, the relative ion intensities of *m*/*z* 55 and 57 in the organic mass fragments could also be used to distinguish COA from HOA.[29,33,34] Compared with other POA components, COA was generally characterized by a higher ratio of *m*/*z* 55/57. The average ratio of *m*/*z* 55/57 of the

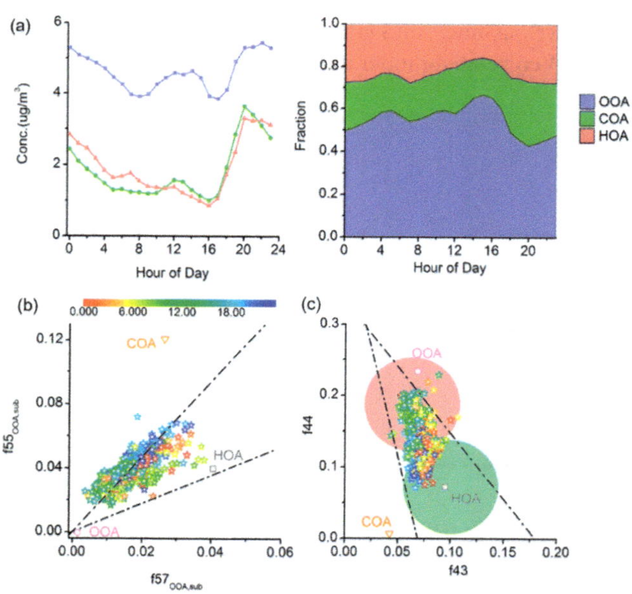

Fig. 3 (a) Diurnal variations of mass concentrations and mass fractions of HOA, COA and OOA, scatterplots of (b) $f55_{OOA,sub}$ *vs.* $f57_{OOA,sub}$, and (c) $f44$ *vs.* $f43$. The measured OA data points are colored according to time of the day. And the shaded ovals indicate regions of LV-OOA and SV-OOA reported in Ng *et al.*[36]

COA factor in this study was 2.4, close to the values observed previously in Beijing.[29,33] Mohr *et al.*[34] developed a V-shaped plot for distinguishing and quantifying HOA and COA, based on $f55_{OOA,sub}$ and $f57_{OOA,sub}$ (the mass fractions of *m*/*z* 55 and 57 after subtracting the interferences of OOA factors). The dashed lines in Fig. 3b corresponded to the edges of the COA and HOA factors from several urban AMS data sets.[34] The COA and HOA factors determined in this study were well separated in the scatter plot of Fig. 3b, with COA located on the left side and HOA on the right side of the V-shape region.[31] Some recently published work has attributed COA to some other sources such as agricultural practices and animal husbandry activities.[44] However, as the observation data is limited, we cannot identify if they are influential during this period.

OOA factor has shown to be a good surrogate for SOA.[35,36] Its mass spectrum was characterized by a prominent peak of *m*/*z* 44 (CO_2^+) (20.0% of the total OOA signal), which was mainly from the di- and poly-carboxylic acid functional groups. Two subtypes of OOA, SV-OOA and LV-OOA, have been identified in many ambient datasets, and LV-OOA is more oxidized than SV-OOA.[24] The OOA factor identified in this work tended to be highly aged according to the similarities between its spectral pattern and that of the highly oxidized LV-OOA. This was further evidenced by the better correlation of OOA with less volatile sulfate (r^2 = 0.74) than with semi-volatile nitrate (r^2 = 0.61). Ng *et al.*[36] developed a triangle plot of *f*44 (mass fraction of *m*/*z* 44 to total organic signal) *versus* *f*43 (defined similarly) to characterize the evolution of OOA components. In the well-defined triangular region, the less aged SV-OOA generally occupies the broader base while highly aged LV-OOA is concentrated in the narrowing top region of the triangle. The OOA factor determined here showed a higher *f*44 value but lower *f*43 value, close to the ambient LV-OOA (Fig. 3c). Data points colored according to time of day also revealed that photochemical processes would increase the oxidation degree of OA.

On average, HOA, COA and OOA accounted for 22.2%, 21.0% and 56.8% of the total OA during this campaign, respectively. Because of the strict emission controls implemented after August 20, 2015, the concentrations of all these components decreased significantly, with HOA decreasing from 2.3 to 1.4 μg m^{-3}, COA from 2.3 to 1.3 μg m^{-3}, and OOA from 7.0 to 2.9 μg m^{-3}, separately. The significant reduction of COA was mainly associated with the banning of open charcoal grills during the parade. And this control measure was also adopted during the 2014 APEC period. As presented by Zhang *et al.*, the concentration of COA decreased by 33.3% during APEC, from 7.8 to 5.2 μg m^{-3}.[16] The sum of HOA and COA could be regarded as POA, and the OOA corresponded to aged SOA. In general, POA and SOA were reduced by 41.5% and 59.0% respectively during the parade period.

Segregation of possible meteorological effects

The variation pattern of air pollution level reflects the combination of emissions, meteorological conditions, and secondary formation processes of PM.[1] Previous studies have indicated that air pollution variability in Beijing could be affected significantly by local weather conditions, *e.g.* temperature, RH, precipitation, wind speed and wind direction. Thus, it is insufficient to simply compare the pollution levels prior to and during the V-day Parade when evaluating the efficacy of emission reduction measures.

Fig. 1 depicts the time series of local meteorological parameters during the entire study period. In general, there were not obvious changes in the local temperature, RH, wind speed and PBL height before and during the parade, except for August 31, 2015, when accumulated precipitation strongly decreased the temperature and PBL height, and increased the RH. To better concentrate on the influence of emission control strategies on air quality, it is necessary to exclude possible meteorological effects by comparing pollution levels before and during the control period under similar weather conditions and similar regional transport characteristics.[37] To select comparable data populations, the following procedures were adopted. First, we removed the data points on 30–31 August and 1 September, 2015 to eliminate the interference of accumulated precipitation. While the other meteorological factors were very similar before and during the parade after that step, the local wind direction remained markedly different as indicated by the different colors in Fig. 1. We then used the BT analysis to select comparable conditions of wind direction and regional air mass transport. The 72 h BTs at 500 m above ground level were calculated every hour for the entire campaign and the trajectories were afterwards clustered using the HYSPLIT4 software according to their similarity in spatial distribution.[38] A three-cluster solution was adopted to better characterize the total spatial variance, as displayed in Fig. S6.† Although cluster 2 largely dominated the parade period, it showed no contribution before the parade. Thus, only cluster 1 and cluster 3 were chosen as two types of situations, both of which covered data points before and during the emission control period. Finally, other meteorological parameters prior to and during the parade, including temperature, RH, wind speed and PBL height, were compared in detail under cluster 1 and cluster 3 cases respectively. As shown in Fig. 4, weather conditions during the parade were very similar to those before the parade for both cluster 1 and cluster 3, and their data points were distributed uniformly throughout a 24 h day.

Therefore, according to the above selection procedures, data populations with similar weather conditions before and during the parade were formed for the

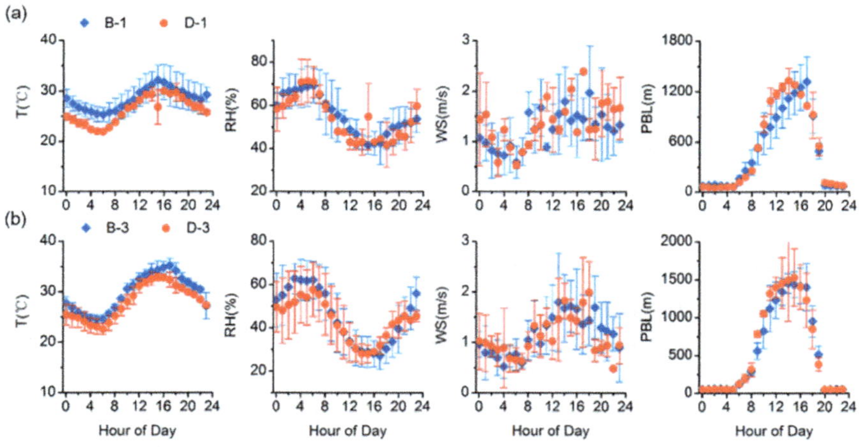

Fig. 4 Diurnal patterns of *T*, RH, WS and PBL height before and during the V-day Parade for (a) cluster 1 and (b) cluster 3.

cluster 1 and cluster 3 cases. Here, we defined B-1 and D-1 as periods before and during the V-day Parade respectively under cluster 1, and B-3 and D-3 were named similarly under cluster 3. The discussions that follow are based on these two cases.

Influence of emission restrictions excluding the effects of weather conditions

Mass concentrations and chemical composition. Under the comparable weather conditions of the cluster 1 and cluster 3 cases separately, the net effectiveness of emission control measures was estimated quantitatively for both PM_1 and its major chemical species (Fig. 5). Following the strict emission restrictions during the parade, the total PM_1 mass concentration was reduced by 57% and 52%, respectively, for cluster 1 and cluster 3. Compared with the non-control period, secondary inorganic aerosol (sulfate, nitrate and ammonium) also showed significant decreases of 42–67% because of the measures imposed during the parade period. The reduction of organics during the parade was 48% and 50% respectively under cluster 1 and cluster 3. Our results show that regional-scale control of precursor emissions is effective in improving air quality in northern China. However, it is worth noting that the magnitude of the restriction measures was different during different stages of the V-day Parade. As the event approached, the government continued to strengthen the control strategies to reduce the emissions of air pollutants. The days covered by D-1 and D-3 during the control period (20 August to 3 September, 2015) are listed in Table 1. As D-3 included days only in the early stage of the event, the reduction for the cluster 3 case might underestimate the effectiveness of control measures.

The significant reductions of secondary aerosol species were consistent with the heavily reduced emissions of their precursors. For example, under the cluster 1 case, the regional emissions of southern Hebei, where the airflow of cluster 1 came from, was reduced by 61–77% for SO_2, 48–57% for NO_x and 58–66% for VOCs during the parade, which was more proportional to reductions in their

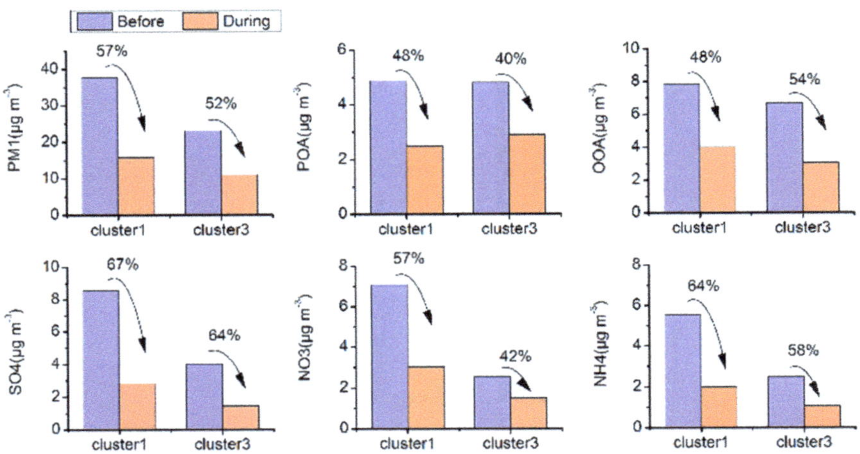

Fig. 5 Estimated influence of emission controls on mass concentrations of PM_1 and its major chemical species.

Table 1 Days covered by D-1 and D-3 during the V-day Parade

Periods	Days covered
D-1	Aug 22–23, 2015; Aug 28–29, 2015; Sep 3, 2015
D-3	Aug 20–22, 2015

secondary species. However, the primary pollutant (*e.g.* POA), mainly influenced by local emissions, was not observed to decline as much as secondary species. This is consistent with the fact that our monitoring site is located tens of kilometres away from the sources impacting cluster 1.

Fig. 6 displays the aerosol composition before and during the parade for cluster 1 and cluster 3. Under comparable weather conditions, the bulk PM_1 composition was similar between the control and non-control periods, with only a small increase of organics contribution and a slight decrease of sulfate contribution during the parade. The mass fractions of nitrate and ammonium remained almost stable. Although the mass concentrations of aerosol species were reduced substantially, these results highlighted that emission control during the parade in itself did not significantly affect the aerosol bulk composition, consistent with the conclusions of an APEC case study focusing on the aerosol characteristics above the 260 m urban canopy.[15] One possible explanation for this is the synergistic control of various precursors such as SO_2, NO_x and VOCs. According to the Beijing Municipal Environment Protection Bureau (Beijing EPB, http://www.bjepb.gov.cn/), emissions of SO_2, NO_x and VOCs decreased by 36.5%, 49.9% and 32.4% separately, due to emission strategies implemented during the event.

The reduced emissions of various precursors also suppressed the secondary formation process of aerosol. The SOR and NOR (molar ratios of sulfate or nitrate to the sum of sulfate and SO_2 or nitrate and NO_2) have been used as indicators of the secondary transformation of sulfate and nitrate.[21] In addition to the gas-phase

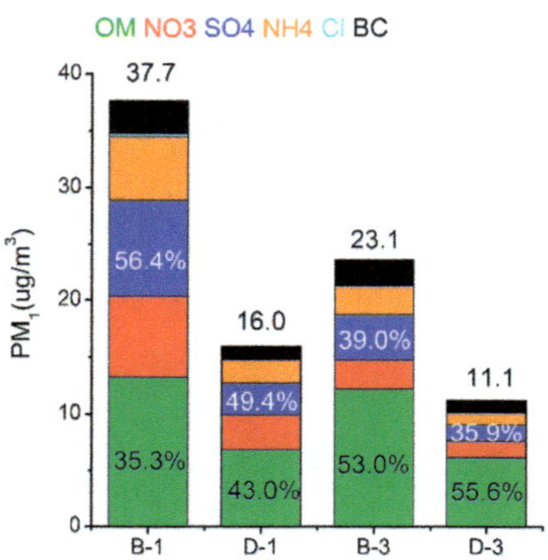

Fig. 6 Average composition of PM_1 before and during the V-day Parade for cluster 1 and cluster 3 respectively. Also shown are the average mass concentrations of PM_1, and the average ratios of SNA and organics.

oxidation by OH radicals, sulfate could also have been formed through heterogeneous reactions of SO_2 with dissolved H_2O_2.[2,39] The formation of nitrate is dominated by the gas-phase oxidation of NO_2 with OH radicals during daylight, and the heterogeneous reactions of nitrate radical (NO_3) during nighttime.[2] During the parade period, the SOR and NOR decreased substantially during both day and night, clearly demonstrating the abated secondary formation of sulfate and nitrate due to emission control measures (Fig. 7a and b). In addition, the mass ratio of NO_3^-/SO_4^{2-} could to some extent indicate the relative importance of mobile *versus* stationary sources.[40,41] A higher NO_3^-/SO_4^{2-} ratio likely reveals the predominance of mobile sources over stationary sources of pollutants. Fig. 7c shows the daytime and nighttime mass ratios of NO_3^-/SO_4^{2-} before and during the parade under similar meteorological conditions. Overall, NO_3^-/SO_4^{2-} fluctuated around 1, indicating the comparable importance of nitrate and sulfate to air pollution. Interestingly, we observed a significant decrease of the daytime NO_3^-/SO_4^{2-} during the emission control period, possibly due to the strict odd-even traffic control regulations. However, no such change occurred in the nighttime NO_3^-/SO_4^{2-}. The ratio of NO_3^-/SO_4^{2-} remained relatively stable or even slightly increased during the parade. This is consistent with the oxidation mechanisms of nitrate at night, when NO_3 radical production is not very sensitive to NO_2, varying between 10 ppb and 20 ppb with O_3 lower than 50 ppb.[45] For cluster 1, the mixing ratio of NO_2 decreased from 23 ppb to 13 ppb, and that of O_3 decreased from 53 ppb to 49 ppb, which lay in the above range. This resulted in little change to the conversion of NO_3 to nitrate *via* heterogeneous reactions. Consequently, the nighttime NO_3^-/SO_4^{2-} instead slightly increased during the control period under cluster 1.

Evolution of organic aerosol. Based on the PMF results of this study, OA was apportioned into HOA, COA and OOA. Here, we regarded the sum of HOA and COA as total POA, and OOA as aged SOA to investigate the influence of emission controls on OA characteristics. Under the similar weather conditions of cluster 1,

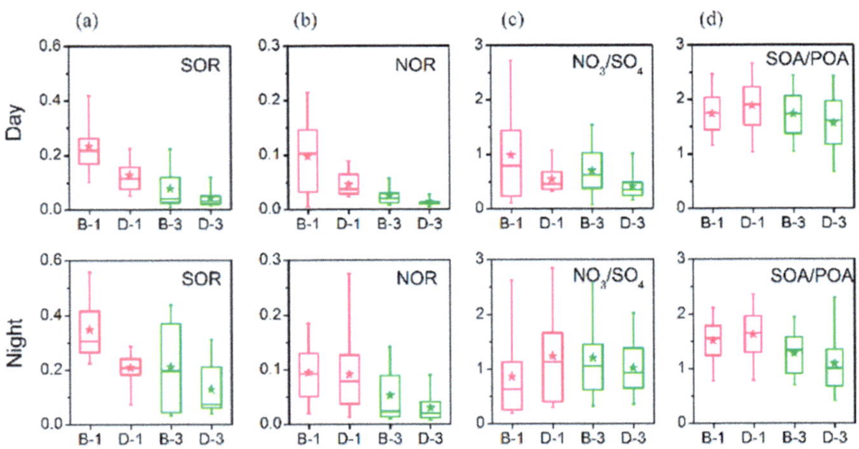

Fig. 7 Daytime and nighttime (a) SOR, (b) NOR, (c) NO_3^-/SO_4^{2-}, and (d) SOA/POA before and during the V-day Parade for cluster 1 and cluster 3 respectively. The mean (pentagram), median (horizontal line), 25th and 75th percentiles (lower and upper box), and 10th and 90th percentiles (lower and upper whiskers) are shown for each bin.

the mass concentrations of POA and SOA showed the same reduction of 48% during the control period (Fig. 5). However, for the cluster 3 case, SOA showed a higher reduction of 54% compared with the 40% reduction of total POA. We explored the relative importance of primary *versus* secondary sources of OA further by considering the mass ratio of SOA/POA. As indicated in Fig. 7d, the average SOA/POA ratios were >1 both before and during the parade, indicating the dominance of secondary sources of organics during the entire study period. However, under cluster 3, the mass ratio of SOA/POA decreased slightly during both daytime and nighttime for the parade period, reflecting the reduced contribution of secondary organic sources due to emission controls for north-westerly airflows over Beijing.

The oxidative properties of OA correlate closely with its density and water solubility, which could be characterized by the atomic oxygen-to-carbon (O : C) ratios.[42,43] According to Aiken *et al.*,[42] the O : C ratio of ambient OA can be estimated from *f*44 (the ratio of *m*/*z* 44 to total signal in the component mass spectrum) through a linear parameterization. Here, we calculated the O : C ratios following the least-squares fit of Aiken *et al.*[42] based on the *f*44 of the ACSM mass spectrum (Fig. 8). Consistently with the results of SOA/POA, the O : C ratios were similar both before and during the parade for cluster 1. However, for the north-westerly airflows of cluster 3, the O : C ratios showed a slight decrease during the control period. Similar results have also been observed by an APEC emission control study, where the average O : C level reduced from 0.43 to 0.36 during the control period.[14] This indicates that in addition to the large reduction of organic mass concentrations, emission control measures could also influence the oxidation degree of OA.

Conclusions

In this study, we undertook a detailed investigation of the impacts of emission controls implemented during the 2015 China Victory Day Parade in Beijing on submicron aerosol characteristics. We concluded that emission restrictions implemented during the V-day Parade, which focused mainly on vehicles and industry, were efficacious in reducing the mass concentrations of PM_1 and its chemical species. The average PM_1 mass concentration was 11.3 (±6.7) μg m^{-3}

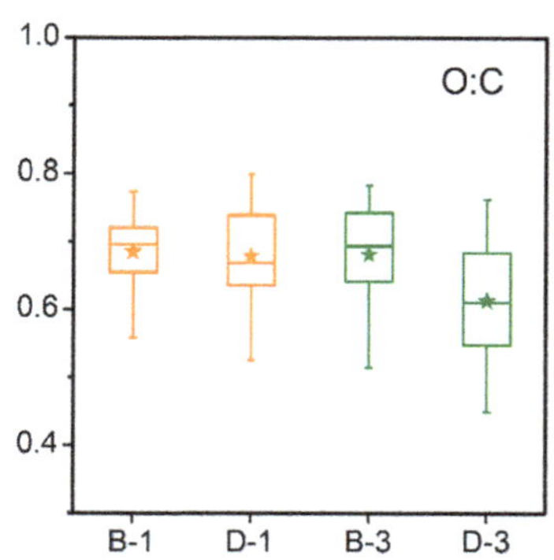

Fig. 8 O : C ratios of OA before and during the V-day Parade for cluster 1 and cluster 3.

during the parade, which was decreased by 63.5% compared with that before the parade. The aerosol composition also changed considerably. Whereas submicron aerosols were dominated by organics (41.3%), followed by sulfate (21.1%) and nitrate (16.4%) prior to the control period, the contribution of organics apparently increased to 53.2%, associated with a reduced contribution of SNA during the parade. A detailed characterization of meteorological parameters revealed that in addition to the emission restrictions, the influence of meteorological conditions on aerosol concentration and composition was non-negligible. Thus, the net effectiveness of emission controls was explored further by excluding the effects of weather conditions. Two cases, cluster 1 and cluster 3, resulting from the BT clustering analysis, were formed with similar weather conditions before and during the parade. Under comparable weather conditions, the total PM_1 mass concentration was reduced by 52–57% because of emission controls. However, the bulk submicron aerosol composition was relatively similar prior to and during the parade, possibly owing to the synergetic control of SO_2, NO_x and VOCs. The SOR and NOR decreased significantly during the control period both in daytime and nighttime, indicating the suppressed secondary formation of sulfate and nitrate because of emission restrictions. We also investigated the variation of OA composition using the SOA/POA mass ratios, which remained stable for cluster 1 and showed a slight decrease for cluster 3 during the parade. The oxidation degree presented corresponding changes to OA composition, with decreased O : C ratios for cluster 3 during the control period. This result indicated that emission controls implemented during the V-day Parade also reduced the oxidation degree of OA for airflows transported from the northwest of Beijing.

Acknowledgements

This work was supported by the National Science and Technology Support Program of China (2014BAC22B01), and by the National Natural Science Foundation of China (21107061, 21190054). The authors also thank the Energy Saving and Pollution control Association of East Asia (ESPA), for their help in the management of the field observation program.

References

1 Y. F. Cheng, J. Heintzenberg, B. Wehner, Z. J. Wu, H. Su, M. Hu and J. T. Mao, Traffic restrictions in Beijing during the Sino-African Summit 2006: aerosol size distribution and visibility compared to long-term *in situ* observations, *Atmos. Chem. Phys.*, 2008, **8**, 7583–7594.
2 J. H. Seinfeld and S. N. Pandis, *Atmospheric chemistry and physics: from air pollution to climate change*, John Wiley and Sons, Inc., Hoboken, New Jersey, 2nd edn, 2006.
3 H. Akimoto, Global air quality and pollution, *Science*, 2003, **302**, 1716–1719, DOI: 10.1126/science.1092666.
4 K. B. He, F. M. Yang, Y. L. Ma, Q. Zhang, X. H. Yao, C. K. Chan, S. Cadle, T. Chan and P. Mulawa, The characteristics of $PM_{2.5}$ in Beijing, China, *Atmos. Environ.*, 2001, **35**, 4959–4970, DOI: 10.1016/S1352-2310(01)00301-6.
5 State Statistical Bureau China, *Statistical Yearbook of China*, China Statistics Press, Beijing, 2015.

6 Y. Gao, X. Liu, C. Zhao and M. Zhang, Emission controls *versus* meteorological conditions in determining aerosol concentrations in Beijing during the 2008 Olympic Games, *Atmos. Chem. Phys.*, 2011, **11**, 12437–12451, DOI: 10.5194/acp-11-12437-2011.

7 Y. M. Zhang, X. Y. Zhang, J. Y. Sun, W. L. Lin, S. L. Gong, X. J. Shen and S. Yang, Characterization of new particle and secondary aerosol formation during summertime in Beijing, China, *Tellus, Ser. B*, 2011, **63**, 382–394, DOI: 10.1111/j.1600-0889.2011.00533.x.

8 Z. J. Wu, M. Hu, P. Lin, S. Liu, B. Wehner and A. Wiedensohler, Particle number size distribution in the urban atmosphere of Beijing, China, *Atmos. Environ.*, 2008, **42**, 7967–7980, DOI: 10.1016/j.atmosenv.2008.06.022.

9 S. X. Wang, M. Zhao, J. Xing, Y. Wu, Y. Zhou, Y. Lei, K. B. He, L. X. Fu and J. M. Hao, Quantifying the Air Pollutants Emission Reduction during the 2008 Olympic Games in Beijing, *Environ. Sci. Technol.*, 2010, **44**, 2490–2496, DOI: 10.1021/es9028167.

10 R. Zhang, J. Jing, J. Tao, S. C. Hsu, G. Wang, J. Cao, C. S. L. Lee, L. Zhu, Z. Chen, Y. Zhao and Z. Shen, Chemical characterization and source apportionment of $PM_{2.5}$ in Beijing: seasonal perspective (vol. 13, pg. 7053, 2013), *Atmos. Chem. Phys.*, 2014, **14**, 175, DOI: 10.5194/acp-14-175-2014.

11 J. Cermak and R. Knutti, Beijing Olympics as an aerosol field experiment, *Geophys. Res. Lett.*, 2009, **36**, L10806, DOI: 10.1029/2009gl038572.

12 Y. Liu, K. B. He, S. S. Li, Z. X. Wang, D. C. Christiani and P. Koutrakis, A statistical model to evaluate the effectiveness of $PM_{2.5}$ emissions control during the Beijing 2008 Olympic Games, *Environ. Int.*, 2012, **44**, 100–105, DOI: 10.1016/j.envint.2012.02.003.

13 X. F. Huang, L. Y. He, M. Hu, M. R. Canagaratna, Y. Sun, Q. Zhang, T. Zhu, L. Xue, L. W. Zeng, X. G. Liu, Y. H. Zhang, J. T. Jayne, N. L. Ng and D. R. Worsnop, Highly time-resolved chemical characterization of atmospheric submicron particles during 2008 Beijing Olympic Games using an aerodyne high-resolution aerosol mass spectrometer, *Atmos. Chem. Phys.*, 2010, **10**, 8933–8945, DOI: 10.5194/acp-10-8933-2010.

14 W. Q. Xu, Y. L. Sun, C. Chen, W. Du, T. T. Han, Q. Q. Wang, P. Q. Fu, Z. F. Wang, X. J. Zhao, L. B. Zhou, D. S. Ji, P. C. Wang and D. R. Worsnop, Aerosol composition, oxidation properties, and sources in Beijing:results from the 2014 Asia-Pacific Economic Cooperation summit study, *Atmos. Chem. Phys.*, 2015, **15**, 13681–13698, DOI: 10.5194/acp-15-13681-2015.

15 C. Chen, Y. L. Sun, W. Q. Xu, W. Du, L. B. Zhou, T. T. Han, Q. Q. Wang, P. Q. Fu, Z. F. Wang, Z. Q. Gao, Q. Zhang and D. R. Worsnop, Characteristics and sources of submicron aerosols above the urban canopy (260 m) in Beijing, China, during the 2014 APEC summit, *Atmos. Chem. Phys.*, 2015, **15**, 12879–12895, DOI: 10.5194/acp-15-12879-2015.

16 J. K. Zhang, L. L. Wang, Y. H. Wang and Y. S. Wang, Submicron aerosols during the Beijing Asia–Pacific Economic Cooperation conference in 2014, *Atmos. Environ.*, 2016, **124**, 224–231, DOI: 10.1016/j.atmosenv.2015.06.049.

17 T. Wang, W. Nie, J. Gao, L. K. Xue, X. M. Gao, X. F. Wang, J. Qiu, C. N. Poon, S. Meinardi, D. Blake, S. L. Wang, A. J. Ding, F. H. Chai, Q. Z. Zhang and W. X. Wang, Air quality during the 2008 Beijing Olympics: secondary pollutants and regional impact, *Atmos. Chem. Phys.*, 2010, **10**, 7603–7615, DOI: 10.5194/acp-10-7603-2010.

18 N. L. Ng, S. C. Herndon, A. Trimborn, M. R. Canagaratna, P. L. Croteau, T. B. Onasch, D. Sueper, D. R. Worsnop, Q. Zhang, Y. L. Sun and J. T. Jayne, An Aerosol Chemical Speciation Monitor (ACSM) for Routine Monitoring of the Composition and Mass Concentrations of Ambient Aerosol, *Aerosol Sci. Technol.*, 2011, **45**, 780–794, DOI: 10.1080/02786826.2011.560211, Pii 934555189.

19 A. Petzold and M. Schonlinner, Multi-angle absorption photometry – a new method for the measurement of aerosol light absorption and atmospheric black carbon, *J. Aerosol Sci.*, 2004, **35**, 421–441, DOI: 10.1016/j.jaerosci.2003.09.005.

20 A. Petzold, H. Schloesser, P. J. Sheridan, W. P. Arnott, J. A. Ogren and A. Virkkula, Evaluation of multiangle absorption photometry for measuring aerosol light absorption, *Aerosol Sci. Technol.*, 2005, **39**, 40–51, DOI: 10.1080/027868290901945.

21 G. J. Zheng, F. K. Duan, H. Su, Y. L. Ma, Y. Cheng, B. Zheng, Q. Zhang, T. Huang, T. Kimoto, D. Chang, U. Poschl, Y. F. Cheng and K. B. He, Exploring the severe winter haze in Beijing: the impact of synoptic weather, regional transport and heterogeneous reactions, *Atmos. Chem. Phys.*, 2015, **15**, 2969–2983, DOI: 10.5194/acp-15-2969-2015.

22 A. M. Middlebrook, R. Bahreini, J. L. Jimenez and M. R. Canagaratna, Evaluation of Composition-Dependent Collection Efficiencies for the Aerodyne Aerosol Mass Spectrometer using Field Data, *Aerosol Sci. Technol.*, 2012, **46**, 258–271, DOI: 10.1080/02786826.2011.620041.

23 I. M. Ulbrich, M. R. Canagaratna, Q. Zhang, D. R. Worsnop and J. L. Jimenez, Interpretation of organic components from positive matrix factorization of aerosol mass spectrometric data, *Atmos. Chem. Phys.*, 2009, **9**, 2891–2918.

24 Q. Zhang, J. L. Jimenez, M. R. Canagaratna, I. M. Ulbrich, N. L. Ng, D. R. Worsnop and Y. L. Sun, Understanding atmospheric organic aerosols *via* factor analysis of aerosol mass spectrometry: a review, *Anal. Bioanal. Chem.*, 2011, **401**, 3045–3067, DOI: 10.1007/s00216-011-5355-y.

25 D. E. Young, J. D. Allan, P. I. Williams, D. C. Green, M. J. Flynn, R. M. Harrison, J. Yin, M. W. Gallagher and H. Coe, Investigating the annual behaviour of submicron secondary inorganic and organic aerosols in London, *Atmos. Chem. Phys.*, 2015, **15**, 6351–6366, DOI: 10.5194/acp-15-6351-2015.

26 C. Parworth, J. Fast, F. Mei, T. Shippert, C. Sivaraman, A. Tilp, T. Watson and Q. Zhang, Long-term measurements of submicrometer aerosol chemistry at the Southern Great Plains (SGP) using an Aerosol Chemical Speciation Monitor (ACSM), *Atmos. Environ.*, 2015, **106**, 43–55, DOI: 10.1016/j.atmosenv.2015.01.060.

27 Q. Zhang, D. R. Worsnop, M. R. Canagaratna and J. L. Jimenez, Hydrocarbon-like and oxygenated organic aerosols in Pittsburgh: insights into sources and processes of organic aerosols, *Atmos. Chem. Phys.*, 2005, **5**, 3289–3311.

28 X. F. Huang, L. Y. He, L. Xue, T. L. Sun, L. W. Zeng, Z. H. Gong, M. Hu and T. Zhu, Highly time-resolved chemical characterization of atmospheric fine particles during 2010 Shanghai World Expo, *Atmos. Chem. Phys.*, 2012, **12**, 4897–4907, DOI: 10.5194/acp-12-4897-2012.

29 Y. L. Sun, Z. F. Wang, P. Q. Fu, T. Yang, Q. Jiang, H. B. Dong, J. Li and J. J. Jia, Aerosol composition, sources and processes during wintertime in Beijing,

China, *Atmos. Chem. Phys.*, 2013, **13**, 4577–4592, DOI: 10.5194/acp-13-4577-2013.

30 Q. Zhang, J. L. Jimenez, M. R. Canagaratna, J. D. Allan, H. Coe, I. Ulbrich, M. R. Alfarra, A. Takami, A. M. Middlebrook, Y. L. Sun, K. Dzepina, E. Dunlea, K. Docherty, P. F. DeCarlo, D. Salcedo, T. Onasch, J. T. Jayne, T. Miyoshi, A. Shimono, S. Hatakeyama, N. Takegawa, Y. Kondo, J. Schneider, F. Drewnick, S. Borrmann, S. Weimer, K. Demerjian, P. Williams, K. Bower, R. Bahreini, L. Cottrell, R. J. Griffin, J. Rautiainen, J. Y. Sun, Y. M. Zhang and D. R. Worsnop, Ubiquity and dominance of oxygenated species in organic aerosols in anthropogenically-influenced Northern Hemisphere midlatitudes, *Geophys. Res. Lett.*, 2007, **34**, L13801, DOI: 10.1029/2007gl029979.

31 J. Xu, Q. Zhang, M. Chen, X. Ge, J. Ren and D. Qin, Chemical composition, sources, and processes of urban aerosols during summertime in northwest China: insights from high-resolution aerosol mass spectrometry, *Atmos. Chem. Phys.*, 2014, **14**, 12593–12611, DOI: 10.5194/acp-14-12593-2014.

32 L. Y. He, Y. Lin, X. F. Huang, S. Guo, L. Xue, Q. Su, M. Hu, S. J. Luan and Y. H. Zhang, Characterization of high-resolution aerosol mass spectra of primary organic aerosol emissions from Chinese cooking and biomass burning, *Atmos. Chem. Phys.*, 2010, **10**, 11535–11543, DOI: 10.5194/acp-10-11535-2010.

33 Y. L. Sun, Q. Jiang, Z. F. Wang, P. Q. Fu, J. Li, T. Yang and Y. Yin, Investigation of the sources and evolution processes of severe haze pollution in Beijing in January 2013, *J. Geophys. Res.: Atmos.*, 2014, **119**, 4380–4398, DOI: 10.1002/2014jd021641.

34 C. Mohr, P. F. DeCarlo, M. F. Heringa, R. Chirico, J. G. Slowik, R. Richter, C. Reche, A. Alastuey, X. Querol, R. Seco, J. Penuelas, J. L. Jimenez, M. Crippa, R. Zimmermann, U. Baltensperger and A. S. H. Prevot, Identification and quantification of organic aerosol from cooking and other sources in Barcelona using aerosol mass spectrometer data, *Atmos. Chem. Phys.*, 2012, **12**, 1649–1665, DOI: 10.5194/acp-12-1649-2012.

35 J. L. Jimenez, M. R. Canagaratna, N. M. Donahue, A. S. H. Prevot, Q. Zhang, J. H. Kroll, P. F. DeCarlo, J. D. Allan, H. Coe, N. L. Ng, A. C. Aiken, K. S. Docherty, I. M. Ulbrich, A. P. Grieshop, A. L. Robinson, J. Duplissy, J. D. Smith, K. R. Wilson, V. A. Lanz, C. Hueglin, Y. L. Sun, J. Tian, A. Laaksonen, T. Raatikainen, J. Rautiainen, P. Vaattovaara, M. Ehn, M. Kulmala, J. M. Tomlinson, D. R. Collins, M. J. Cubison, E. J. Dunlea, J. A. Huffman, T. B. Onasch, M. R. Alfarra, P. I. Williams, K. Bower, Y. Kondo, J. Schneider, F. Drewnick, S. Borrmann, S. Weimer, K. Demerjian, D. Salcedo, L. Cottrell, R. Griffin, A. Takami, T. Miyoshi, S. Hatakeyama, A. Shimono, J. Y. Sun, Y. M. Zhang, K. Dzepina, J. R. Kimmel, D. Sueper, J. T. Jayne, S. C. Herndon, A. M. Trimborn, L. R. Williams, E. C. Wood, A. M. Middlebrook, C. E. Kolb, U. Baltensperger and D. R. Worsnop, Evolution of Organic Aerosols in the Atmosphere, *Science*, 2009, **326**, 1525–1529, DOI: 10.1126/science.1180353.

36 N. L. Ng, M. R. Canagaratna, Q. Zhang, J. L. Jimenez, J. Tian, I. M. Ulbrich, J. H. Kroll, K. S. Docherty, P. S. Chhabra, R. Bahreini, S. M. Murphy, J. H. Seinfeld, L. Hildebrandt, N. M. Donahue, P. F. DeCarlo, V. A. Lanz, A. S. H. Prevot, E. Dinar, Y. Rudich and D. R. Worsnop, Organic aerosol

components observed in Northern Hemispheric datasets from aerosol mass spectrometry, *Atmos. Chem. Phys.*, 2010, **10**, 4625–4641, DOI: 10.5194/acp-10-4625-2010.

37 Z. S. Wang, Y. T. Li, T. Chen, L. J. Li, B. X. Liu, D. W. Zhang, F. Sun, Q. Wei, L. Jiang and L. B. Pan, Changes in atmospheric composition during the 2014 APEC conference in Beijing, *J. Geophys. Res.: Atmos.*, 2015, **120**, 12695–12707, DOI: 10.1002/2015jd023652.

38 R. Draxler, B. Stunder and G. Rolph, *et al.*, *HYSPLIT4 user's guide, Version 4.9*, http://ready.arl.noaa.gov/HYSPLIT.php, 2009, last access: January 2010.

39 P. Saxena and C. Seigneur, On the Oxidation of SO_2 to Sulfate in Atmospheric Aerosols, *Atmos. Environ.*, 1987, **21**, 807–812, DOI: 10.1016/0004-6981(87)90077-1.

40 R. Arimoto, R. A. Duce, D. L. Savoie, J. M. Prospero, R. Talbot, J. D. Cullen, U. Tomza, N. F. Lewis and B. J. Jay, Relationships among aerosol constituents from Asia and the North Pacific during PEM-West A, *J. Geophys. Res.: Atmos.*, 1996, **101**, 2011–2023, DOI: 10.1029/95jd01071.

41 F. Yang, J. Tan, Q. Zhao, Z. Du, K. He, Y. Ma, F. Duan, G. Chen and Q. Zhao, Characteristics of $PM_{2.5}$ speciation in representative megacities and across China, *Atmos. Chem. Phys.*, 2011, **11**, 5207–5219, DOI: 10.5194/acp-11-5207-2011.

42 A. C. Aiken, P. F. Decarlo, J. H. Kroll, D. R. Worsnop, J. A. Huffman, K. S. Docherty, I. M. Ulbrich, C. Mohr, J. R. Kimmel, D. Sueper, Y. Sun, Q. Zhang, A. Trimborn, M. Northway, P. J. Ziemann, M. R. Canagaratna, T. B. Onasch, M. R. Alfarra, A. S. H. Prevot, J. Dommen, J. Duplissy, A. Metzger, U. Baltensperger and J. L. Jimenez, O/C and OM/OC ratios of primary, secondary, and ambient organic aerosols with high-resolution time-of-flight aerosol mass spectrometry, *Environ. Sci. Technol.*, 2008, **42**, 4478–4485, DOI: 10.1021/es703009q.

43 Y. Pang, B. J. Turpin and L. A. Gundel, On the importance of organic oxygen for understanding organic aerosol particles, *Aerosol Sci. Technol.*, 2006, **40**, 128–133, DOI: 10.1080/02786820500423790.

44 M. Dall'Osto, M. Paglione, S. Decesari, M. C. Facchini, C. O'Dowd, C. Plass-Duellmer and R. M. Harrison, On the Origin of AMS "Cooking Organic Aerosol" at a Rural Site, *Environ. Sci. Technol.*, 2015, **49**, 13964–13972, DOI: 10.1021/acs.est.5b02922.

45 S. E. Pusede, K. C. Duffey, A. A. Shusterman, A. Saleh, J. L. Laughner, P. J. Wooldridge, Q. Zhang, C. L. Parworth, H. Kim, S. L. Capps, L. C. Valin, C. D. Cappa, A. Fried, J. Walega, J. B. Nowak, R. M. Hoff, T. A. Berkoff, A. J. Beyersdorf, J. Olson, J. H. Crawford and R. C. Cohen, On the effectiveness of nitrogen oxide reductions as a control over ammonium nitrate aerosol, *Atmos. Chem. Phys.*, 2015, **15**, 27087–27136, DOI: 10.5194/acpd-15-27087-2015.

Faraday Discussions

PAPER

Secondary organic aerosol origin in an urban environment: influence of biogenic and fuel combustion precursors†

M. C. Minguillón,*[a] N. Pérez,[a] N. Marchand,[b] A. Bertrand,[b] B. Temime-Roussel,[b] K. Agrios,[c] S. Szidat,[c] B. van Drooge,[a] A. Sylvestre,[b] A. Alastuey,[a] C. Reche,[a] A. Ripoll,[a] E. Marco,[a] J. O. Grimalt[a] and X. Querol[a]

Received 18th November 2015, Accepted 8th December 2015
DOI: 10.1039/c5fd00182j

Source contributions of organic aerosol (OA) are still not fully understood, especially in terms of quantitative distinction between secondary OA formed from anthropogenic precursors *vs.* that formed from natural precursors. In order to investigate the OA origin, a field campaign was carried out in Barcelona in summer 2013, including two periods characterized by low and high traffic conditions. Volatile organic compound (VOC) concentrations were higher during the second period, especially aromatic hydrocarbons related to traffic emissions, which showed a marked daily cycle peaking during traffic rush hours, similarly to black carbon (BC) concentrations. Biogenic VOC (BVOC) concentrations showed only minor changes from the low to the high traffic period, and their intra-day variability was related to temperature and solar radiation cycles, although a decrease was observed for monoterpenes during the day. The organic carbon (OC) concentrations increased from the first to the second period, and the fraction of non-fossil OC as determined by ^{14}C analysis increased from 43% to 54% of the total OC. The combination of ^{14}C analysis and Aerosol Chemical Speciation Monitor (ACSM) OA source apportionment showed that the fossil OC was mainly secondary (>70%) except for the last sample, when the fossil secondary OC only represented 51% of the total fossil OC. The fraction of non-fossil secondary OC increased from 37% of total secondary OC for the first sample to 60% for the last sample. This enhanced formation of non-fossil secondary OA (SOA) could be attributed to the reaction of BVOC precursors with NO_x emitted from road traffic (or from its nocturnal derivative nitrate that enhances night-time semi-volatile oxygenated OA (SV-OOA)), since NO_2 concentrations increased from 19 to 42 μg m^{-3} from the first to the last sample.

[a]*Institute of Environmental Assessment and Water Research(IDAEA), CSIC, 08034 Barcelona, Spain. E-mail: mariacruz.minguillon@idaea.csic.es*

[b]*Aix Marseille Université, CNRS, LCE UMR 7376, 13331 Marseille, France*

[c]*Department of Chemistry and Biochemistry & Oeschger Centre for Climate Change Research, University of Bern, 3012 Bern, Switzerland*

† Electronic supplementary information (ESI) available: Sections 1 to 6, with 12 figures. See DOI: 10.1039/c5fd00182j

1 Introduction

Carbonaceous aerosols make up a large proportion of ambient PM_{10}, $PM_{2.5}$ and PM_1 (atmospheric particulate matter finer than 10, 2.5 and 1 μm, respectively). Worldwide tropospheric submicron PM may be from 20 to 90% organic aerosol (OA).[1] OA, also named organic matter (OM), may account for 15–26% of both PM_{10} and $PM_{2.5}$ in Europe, with the highest loads being recorded at urban and traffic sites.[2] For European regional background sites OM + EC (elemental carbon) ranged from 27% to 43% of $PM_{2.5}$.[3] OM concentration ranged from 2 to 4 μg m^{-3}, 3 to 7 μg m^{-3} and 3 to 7 μg m^{-3} in $PM_{2.5}$ for remote-rural, urban background, and traffic and industrial hotspots in Spain, respectively (approx. 25–30%, 20–40% and 30–40% of $PM_{2.5}$).[4]

In spite of these relatively high proportions in ambient PM, source contributions of OA are still not fully understood[5–7] due to the mixed source origins (anthropogenic or biogenic OA) and the atmospheric processes that cause the formation of OA. Moreover, OA may be primary (POA), emitted as solid/liquid PM from both anthropogenic or biogenic sources, or secondary (SOA), formed in the atmosphere from volatile organic compounds (VOCs), which can have an anthropogenic or biogenic origin (AVOCs or BVOCs).[7,8] Furthermore, in urban or industrial areas, the high concentrations of specific anthropogenic pollutants (*e.g.* NO_x, SO_2 and O_3) may enhance the formation of secondary organic aerosols from biogenic precursors (BSOA), giving rise to the so-called enhanced BSOA (eBSOA),[7] which, despite its biogenic origin, would be anthropogenically driven. The interaction between the BVOCs–O_3–NO_x–SOA systems is very complex.[9] Thus, several studies have reported higher yields of eBSOA formation from isoprene in the presence of relatively high concentrations of NO_x, although this SOA formation becomes lower for very high NO_x concentrations.[7,10] For the highly reactive sesquiterpenes the interaction with NO_x seems also to increase the SOA yields, probably due to the low volatility of the sesquiterpene nitrates.[7] Conversely, for monoterpenes increasing NO_x/VOC ratios have shown to reduce the BSOA formation from ozonolysis of α-pinene, and highly volatile N-containing organic species are preferentially formed under these conditions.[11] Other atmospheric species different to NO_x, OH radical and O_3 may have a large impact on BSOA formation. Thus, a higher BSOA yield from isoprene can occur under acidic conditions caused by the occurrence of sulphuric acid.[12] VOCs play a very relevant role in sulphuric acid related nucleation events and in the growth of newly formed particles.[13] In any case, one of the major BSOA formation pathways, especially under high NO_x and O_3 concentrations, seems to be related to the nocturnal oxidation of VOCs by the nitrate radical (a product of the NO_2 + O_3 interaction) reaction.[7] Thus, both organo-sulphates and organo-nitrates may constitute an important fraction of the SOA due to the prevalence of the above formation pathways.[14] Hence, it is expected that OA concentrations in urban environments will increase compared with those from rural areas due to (a) the enlarged anthropogenic POA and VOC emissions leading to anthropogenic SOA, but also to (b) eBSOA formation due to the interaction of BVOCs with other anthropogenic components.

The city of Barcelona, NE Spain, has one of the highest car densities in Europe, with diesel vehicles making up around 47% of the fleet.[15] In spite of this adverse fact, PM_x concentrations have decreased in the last decade,[16] going down from about 40 $\mu g\ m^{-3}$ to 22–30 $\mu g\ m^{-3}$ PM_{10}.

Urban background OA accounted for 20% of the PM_{10} and >30% of the $PM_{2.5}$ and PM_1 in 2003–2007.[17] The SOA contribution to OA, calculated with the EC tracer method,[18] was 36–40%. Hence, OA, including SOA, is a major component of PM_x in the city of Barcelona and, in general, in regional and urban environments in Europe.

Prior studies carried out at an urban background site during the DAURE campaign in March 2009 using a combination of HR-ToF-AMS and ^{14}C analysis of OC and EC reported a proportion of non-fossil OC of 60% of the total OC measured in PM_1, while 57% of the SOA was non-fossil.[19,20] This non-fossil SOA could be BSOA or eBSOA. Relatively strong correlations were found between dicarboxylic acids and other oxidation products from BVOCs, and the oxygenated OA (OOA) loading in the urban background of Barcelona.[21] Similar results concerning the high non-fossil OA contribution were obtained in Zurich,[22] Marseille,[23] and Birmingham,[24] among other locations.

In order to investigate the origin of OA in Barcelona, with special interest in determining the relative contributions from POA and SOA, and the anthropogenic and the biogenic fractions, a measurement campaign was devised in summer 2013, including the last weeks of August and the beginning of September, coinciding with a period of high insolation and high BVOC emissions. An important proportion of Barcelona's population takes holidays in August, thus traffic flow is substantially reduced compared with the rest of the year. Hence, the campaign comprised a low road traffic period and continued when the traffic resumed its usual flow. Through the combination of several analytical tools, such as offline ^{14}C analysis of OC, offline PM inorganic and organic chemical composition determination, online VOC measurements (using PTR-ToF-MS), online organic aerosol concentration and types/sources determination (using aerosol chemical speciation monitor (ACSM)), online black carbon (BC) measurements, and online NO, NO_2, O_3, CO, SO_2, PM_{10}, $PM_{2.5}$ and PM_1 concentration measurements, we investigated why and how road traffic emissions influence the concentrations and origin of OA in the city.

2 Experimental

An intensive monitoring campaign at the urban background of Barcelona was carried out from 16 August to 6 September 2013. Two periods are distinguished: a summer holiday period with low traffic intensity (16–25 August) named the low traffic (LT) period, and a typical urban traffic emissions period (28 August–6 September), named the high traffic (HT) period, both characterized by similar environmental conditions and high insolation.

Measurements were carried out at the urban background site Palau Reial (PR; 41°23′14″ N, 02°06′56″E, 78 m a.s.l., Fig. S1†), located 200 m from one of the main traffic avenues of the city (traffic density of 90 000 vehicles per working day). The main source of atmospheric PM_x is road traffic, although contributions from industry, regional secondary atmospheric pollutants, construction, and shipping are also relevant.[17,25] Atmospheric dynamics are driven by the breeze regime, with a NW wind component during the night and the development of breezes during the day turning progressively from SE to SW direction, with gradually increasing wind speeds reaching maximum levels around noon.[26]

2.1 On line measurements of VOCs and other gaseous pollutants

Measurements of VOCs were performed with a Proton-Transfer Reaction Time of Flight Mass Spectrometer (PTR-ToF-MS, Ionicon PTR TOF 8000, Ionicon Analytik, Innsbruck, Austria).[27,28] The instrument was operated in standard conditions, with a reaction chamber pressure fixed at 2.05 mbar, drift tube voltage at 500 V and drift tube temperature at 323 K corresponding to an electric field strength applied to the drift tube (E) to buffer gas density (N) ratio of 125Td (1Td $= 10^{-17}$ V cm^2). Mass spectra were recorded up to m/z 500 with a time resolution of 1 minute. Instrumental background signal was measured regularly by passing the sample flow through a catalytic converter that efficiently removes VOCs (platinum wool heated at 350 °C). VOC mixing ratios are expressed in parts per billion by volume (ppbv):[29]

$$[\mathrm{VOC\ H^+}] = [\mathrm{H_3O^+}]_0[\mathrm{VOC}]kt$$

where $[\mathrm{H_3O^+}]_0$ and $[\mathrm{VOC\ H^+}]$ are the transmission corrected densities of the primary ion H_3O^+ and of the protonated volatile organic compound ions respectively, in cps, k is the reaction rate constant between the VOC and H_3O^+ (cm^3 s^{-1}) and t is the residence time of the ions in the drift tube ($t = 100$ µs). A calibration gas standard (TO-14A Aromatic Mix, Restek Corporation, Bellefonte, USA) (100 ± 10 ppb in nitrogen) was used to assess the relative mass dependant transmission efficiency. The estimated accuracy is 30%. A specific proton transfer rate constant[30] was used for the quantification of each compound (Table 2). The inlet line was connected to a general sampling inlet through a splitter in counter-flow. Tentative assignment of ions to VOC was performed on the basis of the accurate mass determination of the fragment ions and on the known fragmentation patterns. Ion assignment was supported by offline GC-MS analysis.

Real-time concentrations of NO and NO_2 (Chemiluminescence, SIR S-5012), CO (Ecotech, GFC, EC 9830), O_3 (UV absorption, SIR S-5014), and SO_2 (Pulsed Fluorescence, Thermo Scientific Model 43C) were measured and supplied by the Department of Environment of the Autonomous Government of Catalonia.

2.2 Online aerosol measurements

Ultrafine particle number size distributions within the range 15–660 nm (N_{15-660}, 5 min time resolution) were measured with a Scanning Mobility Particle Sizer (SMPS), composed by a Differential Mobility Analyser (DMA, TSI 3081) coupled with a condensation particle counter (CPC, TSI 3772). Equivalent BC mass concentrations[31] were measured by a multi-angle absorption photometer (MAAP, model 5012, Thermo ESM Andersen Instruments, 1 min time resolution) with a PM_{10} inlet. Semi-hourly PM_{10}, $PM_{2.5}$, and PM_1 concentrations were measured by an optical particle counter (GRIMM 180) and corrected with simultaneous 24 hour gravimetric samples.

An Aerosol Chemical Speciation Monitor (Aerodyne Research Inc., ACSM)[32] was used to measure non-refractory submicron aerosol species (organic aerosol, nitrate, sulphate, ammonium and chloride) with a time resolution of 15 minutes.

Mass calibration of the ACSM was based on determining the instrument response factor (RF) using ammonium nitrate calibration aerosol.[32] Relative Ionization Efficiency (RIE) for ammonium was directly determined from the ammonium nitrate calibration (5.61). Sulphate RIE was determined by doing the aforementioned calibration exercise with ammonium sulphate monodisperse aerosol, and resulted in a value of 1.38. The default RIE for organic aerosol (1.4)[33] has been used. The data were acquired using Aerodyne Research software ACSM_DAQ_v1438 and processed with the ACSM data analysis software version 1.5.3.0 (Aerodyne Research Inc.) written in Igor Pro (WaveMetrics, Inc., Lake Oswego, OR, USA). A correction for the instrument performance limitations was applied based on the inlet pressure and N_2 signal. The aerosol mass concentrations were then corrected for particle collection efficiency (CE) following the Middlebrook approach.[34] Comparison with co-located measurements (real-time PM_1, and particle size distribution determined by an SMPS) allowed the identification of three periods with different correlation slopes. The ACSM data were then corrected according to these intercomparisons.

2.3 Offline aerosol analyses

$PM_{2.5}$ samples were collected in 8 h time periods (0:00 h–8:00 h, 8:00 h–16:00 h and 16:00 h–0:00 h UTC) on quartz filters using high volume samplers (30 m^3 h^{-1}).

A complete chemical characterization was performed.[35] Briefly, an acid digestion of half filter was carried out for each sample to yield solutions for the subsequent analysis of major and trace elements by Inductively Coupled Atomic Emission Spectroscopy (ICP-AES) and Mass Spectrometry (ICP-MS). Water leachates from a quarter of the filter were used to determine soluble ion concentrations by Ion Chromatography and NH_4^+ concentrations by a selective electrode. OC and EC concentrations were determined from 1.5 cm^2 filter sections by means of a thermal-optical carbon analyser (Sunset, EUSAAR_2 protocol).[36]

Molecular organic species were determined in the same filter samples after extraction by a mixture of methanol and dichloromethane (1 : 1 v/v), derivation of esters to TMS-esters and subsequently detection and quantification in a gas-chromatograph coupled to a mass-spectrometer (GC-MS) in full-scan mode.[21] Hence, several organic species in a wide range of polarity were analyzed, including levoglucosan, hopanes, and formation products of VOC oxidation, such as dicarboxylic acid, poly acids, and polyols.

Additional 24 h $PM_{2.5}$ samples were collected using high volume samplers and used for ^{14}C analysis on the OC fraction. The separation of the OC and collection as CO_2 was carried out according to the Swiss_4S method.[37] The ^{14}C isotopic ratio (fraction of modern carbon, fM) of each sample was determined using an Accelerator Mass Spectrometer MICADAS (MIni CArbon Dating System) at the Laboratory for the Analysis of Radiocarbon with AMS (LARA)[38] and corrected for the OC field blank.

2.4 Supporting measurements

Meteorological variables (atmospheric pressure, wind components, solar radiation, temperature and relative humidity) were recorded at a meteorological station located 300 m from our sampling site and provided by the Faculty of

Physics from the Universitat de Barcelona. Road traffic intensity data from the Diagonal Avenue were supplied by the Council of Barcelona (Servei de Mobilitat).

2.5 Source apportionment to the organic aerosol

The source apportionment to the organic aerosol was investigated by applying Positive Matrix Factorization (PMF)[39] using the Multilinear Engine (ME-2)[40] to the ACSM organic mass spectra, using the toolkit SoFi (Source Finder) version 4.8.[41] The dataset used comprised from 25 July to 19 September 2013. The ME-2 allows the user to introduce *a priori* information about sources *e.g.* using the *a*-value approach. Thus the user may 'constrain' a factor by introducing its source profile and determining the extent to which the output profile can differ from the one input to the model (the *a*-value). Only $m/z \leq 120$ were used given that those >120 have lower signal to noise and they account for a minor fraction of the total signal. Details on the different solutions are given in the results section.

The source apportionment of OC was also assessed based on the ^{14}C isotopic ratios (fM). The determination of the contribution from different sources to the total OC is done knowing their fM. The fM for fossil sources is zero. The fM of OC from non-fossil sources (fMOCnf) depends on the contribution from the different non-fossil sources, mainly biomass burning and biogenic sources, and it has to be defined in advance,[22] although subsequent iterations could be applied if necessary.[19] For the present study, the fMOCnf was calculated assuming a 10/90 share between biomass burning and biogenic source, in agreement with other studies[42] and based on the ACSM results from this study. The fMOC for biomass burning emissions is 1.063, corresponding to emissions from the burning of 25-year-old trees harvested in 2011 as determined with a tree-growth model,[43] and the fM for biogenic sources is 1.023, corresponding to the year 2013.[44] The resulting fMOCnf is 1.027. OC was then separated into fossil OC (OCf) and non-fossil OC (OCnf). OCf was attributed to POA and SOA from fossil fuel combustion. OCnf may have different origins, such as biogenic SOA (bio), biomass burning (bb), cooking, biofuel combustion, brake lining dust, natural rubber in tire dust, and others.[45]

3 Results and discussion

3.1 General conditions

Similar meteorological patterns were recorded during the LT period (means of 25 °C, 64% RH, and insolation of 239 W m^{-2}) and the HT period (23 °C, 59% RH, 230 W m^{-2}), which were separated by coincidence by a rainy inter-period (26 to 27 August, Monday and Tuesday), characterized by a clean atmosphere, low insolation and temperature and high humidity (Fig. S2†). However, in the HT period lower wind speed and higher pressure were recorded (1.7 *vs.* 2.1 m s^{-1}, and 1019 *vs.* 1016 hPa) as a consequence of an anticyclonic episode leading to atmospheric stagnation.

The HYSPLIT back-trajectory analysis evidenced that both periods were characterized by regional recirculation of air masses, with relatively short air mass transport distances, although in the LT period a North-western direction prevailed, whereas in the HT period this short transport had a European origin (Fig. S3 and S4†).

Mean traffic flow at the Diagonal Avenue increased +37% from the LT to the HT period. The rush traffic hours were more clearly identified during the HT period at 6 and 16–18 UTC, as well as the maximum traffic intensity/wind speed ratio at 6 UTC and 19 UTC (Fig. S5†).

The separation of the LT and HT periods was done based on these results. Hence, in the subsequent sections the influence of traffic emissions on PM composition and source contribution, with a special emphasis on OC, will be investigated.

3.2 PM_x, N and gases concentrations

PM_{10}, $PM_{2.5}$ and PM_1 concentrations increased in the HT *vs.* the LT period +18, +24 and +38%, respectively (Table 1). This increase is due to the more pronounced increase during traffic rush hours and the development of the high pressure episode, reaching the maximum PM_x on Tuesday 3 September (Fig. 1), one day before the maximum pressure.

Particle number concentrations increased +22% from the LT to the HT period for the N_{15-20} (the nucleation mode), and +41% and +49% for the N_{20-100} and $N_{100-500}$ (Table 1). N_{15-20} in the LT period showed intensive midday maxima, attributed to nucleation burst by prior studies in Barcelona,[17,46–48] whereas these midday nucleation burst maxima were lower in the HT period (Fig. 1).

CO, NO_2 and NO average concentrations increased from 333 μg m^{-3}, 16 μg m^{-3} and 3 μg m^{-3}, respectively, for the LT period to 495 μg m^{-3}, 29 μg m^{-3} and 5 μg m^{-3}

Table 1 Average of different parameters during the LT and the HT periods and difference in %

		LT	HT	Difference (%)
Meteorogical conditions and traffic				
T	(°C)	25	23	−6
RH	(%)	64	59	−8
SR	(W m^{-2})	239	230	−4
WS	(m s^{-1})	2.1	1.7	−20
P	(hPa)	1016	1019	<1
Traffic	(Vehicles per h)	1663	2271	37
Gases				
SO_2	(μg m^{-3})	2.8	3	7
NO	(μg m^{-3})	3	5	79
NO_2	(μg m^{-3})	16	29	81
O_3	(μg m^{-3})	80	85	6
CO	(μg m^{-3})	333	495	48
Aerosol				
PM_{10}	(μg m^{-3})	20	24	18
$PM_{2.5}$	(μg m^{-3})	13	16	24
PM_1	(μg m^{-3})	10	14	38
N_{15}	(# cm^{-3})	7741	9822	27
N_{15-20}	(# cm^{-3})	1227	1493	22
N_{20-100}	(# cm^{-3})	4610	6511	41
$N_{100-500}$	(# cm^{-3})	1180	1756	49

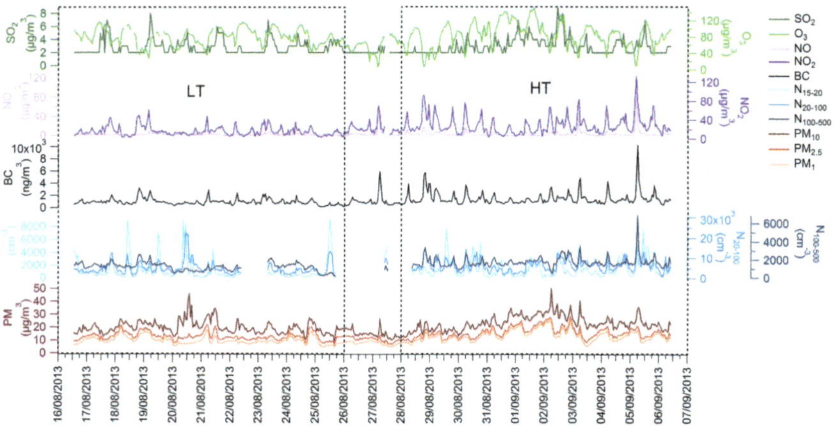

Fig. 1 Time series of gases and selected aerosol parameters along the campaign. The LT and HT periods are shown. Date format is dd/mm/yyyy.

for the HT period (+48, +81 and +79%, respectively, Table 1). O_3 concentrations were 80 μg m^{-3} (LT) and 85 μg m^{-3} (HT). SO_2 concentrations did not show clear variations between LT and HT periods (average around 3 μg m^{-3}), attributed to the main influence of shipping emissions, which are not dependent on the period, and resulting in a midday maximum coinciding with the maximum sea breeze intensity.

3.3 Volatile organic compounds

With the exception of formic acid, isoprene, methyl vinyl ketone (MVK) + MACR, and monoterpenes, all VOCs presented higher mixing ratios during the HT period. Aromatic hydrocarbons related to traffic emissions showed the highest differences between the LT and HT periods, increasing from +78% to +85% (C2, C3 and C4-alkylbenzenes), +63% (benzene), and +52% (toluene). VOCs were dominated by methanol (2.7 ppbv and 3.1 ppbv as averages for the LT and HT periods, respectively), acetone (2.2 ppbv and 2.7 ppbv), and acetic acid (0.9 ppbv and 1.1 ppbv). Isoprene average mixing ratios were 0.16 ppbv and 0.12 ppbv for the LT and HT periods, respectively, in the range measured during the winter DAURE campaign in Barcelona,[49] although mixing ratios of benzene and toluene were slightly lower than those registered in DAURE; and higher for acetone, acetic acid, or MVK (Table 2).

The traffic-related VOCs showed a marked daily cycle during the HT period, peaking during traffic rush hours and following the same daily pattern as BC (Fig. 2).

Daily variability of the biogenic VOCs is related to temperature and solar radiation cycles (Fig. 2). Monoterpenes showed a marked decrease during the day (Fig. 2), attributed to degradation by O_3.[50]

Other VOCs, such as methanol or acetone increased between +14% to +21% in the HT period (Table 2). These compounds did not show a clear diurnal pattern, although higher mixing ratios were recorded during the daytime (Fig. S6†). The

Table 2 Average of different VOC mixing ratios during the LT and the HT periods and difference in %. *k* rates according to Cappellin *et al.*[30]

m/*z*	Chemical formula (protonated ions)	Most probable VOCs	*k* rates (10^{-9} cm^3 s^{-1})	LT (ppbv)	HT (ppbv)	Difference (%)
33.033	$(CH_4O)H^+$	Methanol	2.22	2.7	3.1	14
47.012	$(CH_2O_2)H^+$	Formic acid	1.99	0.3	0.1	−46
59.049	$(C_3H_6O)H^+$	Acetone	3.32	2.2	2.7	21
61.028	$(C_2H_4O_2)H^+$	Acetic acid	2.25	0.9	1.1	19
69.069	$(C_5H_8)H^+$	Isoprene	1.96	0.16	0.12	−22
71.049	$(C_4H_6O)H^+$	MVK + MACR	3.39	0.03	0.03	−10
73.065	$(C_4H_8O)H^+$	MEK	3.28	0.06	0.08	34
79.054	$(C_6H_6)H^+$	Benzene	1.9	0.03	0.05	63
93.069	$(C_7H_8)H^+$	Toluene	2.08	0.13	0.20	52
107.085	$(C_8H_{10})H^+$	C2-Alkylbenzene	2.27	0.07	0.13	78
121.101	$(C_9H_{12})H^+$	C3-Alkylbenzene	2.45	0.02	0.04	84
135.117	$(C_{10}H_{14})H^+$	C4-Alkylbenzene	2.4	0.002	0.004	85
137.132	$(C_{10}H_{16})H^+$	Monoterpenes	2.37	0.017	0.014	−16

higher mixing ratios measured during the HT period are probably related to the higher oxidation of BVOCs and AVOCs.

3.4 PM chemical composition

The PM_1 relative composition was very similar during both the LT and HT periods (Fig. 3). It was dominated by OA (=OM) (59% and 57%, respectively), followed by sulphate (20%), ammonium (9% and 10%), nitrate (4% and 5%) and EC (8%). Table 3 shows the average concentration of the different PM components for the LT and HT periods, as well as the difference between both periods. To better assess the variation in the concentrations of the different PM components, a normalized concentration (relative to OA) was used, and the difference in these normalized concentrations is also shown in Table 3.

Average BC concentrations increased from 1.0 μg m^{-3} in the LT period to 1.6 μg m^{-3} in the HT period. BC concentrations were more driven by the traffic cycles than by the anticyclone development during the HT period (Fig. 2 and 3). Thus, maximum BC levels during traffic rush hours reached 4 μg m^{-3} on most days of the HT period (and up to 10 μg m^{-3}), whereas during the LT period these rarely exceeded 1 μg m^{-3}, and the daily pattern was less marked or even missing (Fig. 2 and 3).

OA concentrations increased +58% from the LT to the HT period. Nitrate and ammonium concentrations increased +99% and +72%, respectively, comparable to the BC concentrations (+62%). In terms of relative concentrations (with respect to OA), the increase was +26% and +9%. Part of this increase may be the result of the higher NO_x (+80%) and NH_3 exhaust traffic emissions in the HT period, resulting in the formation of ammonium nitrate. Nevertheless, and given that the increase for nitrate is higher than that for ammonium, part of the nitrate could potentially be attributed to the presence of organonitrates in the HT period. The sulphate concentrations increased +60%, similarly to the reference OA concentration. The higher sulphate concentration could be related

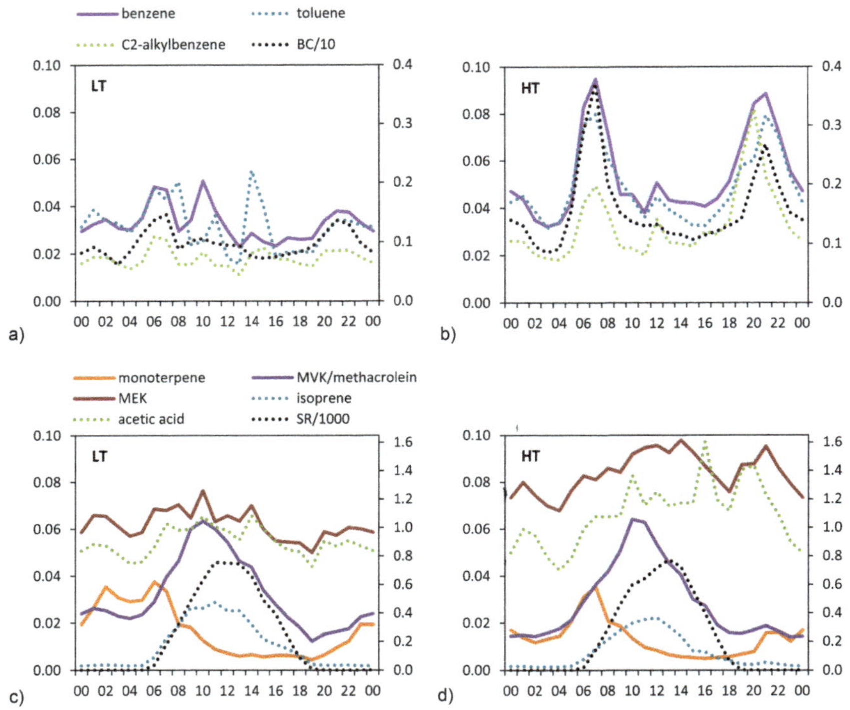

Fig. 2 Average daily cycle of the mixing ratios (ppbv) of selected VOCs (a and b: traffic related VOCs; c and d: biogenic VOCs) during the LT and HT periods. BC concentrations in $\mu g\ m^{-3}$ divided by 10 are also plotted. Solar radiation (SR) in $W\ m^{-2}$ divided by 1000 is also plotted. In each of the plots, the full lines correspond to the left axis and the dotted lines correspond to the right axis. The time is UTC.

to an increase in the sulphate formation at a regional scale (note that SO_2 concentrations only increased +7% in Barcelona, Table 1) or to the accumulation of pollutants over the Barcelona area. The intra-day variation of the different components varied from day to day, resulting in average daily patterns that were relatively flat for sulphate and ammonium (Fig. 4). Nitrate and OA concentrations partially followed the daily pattern observed for BC (Fig. 4), especially during the HT period.

Some mineral dust tracer concentrations, such as Al and Fe, increased in the HT period +60% and +18%, whereas Ca decreased by −27%, yielding a mineral dust increase of only +12% during the HT.

Sb and Sn concentrations, tracers of non-exhaust traffic emissions (mainly from brake pads),[51] increased in the HT period +78% and +79% (Fig. 5), similarly to BC concentrations, hence confirming their road traffic origin. Conversely concentrations of sea salt tracers such as Na decreased by −13%.

Also, Pb and Cd concentrations, industrial tracers, increased between +49% and +151% during the HT period (Fig. 5), probably due to an increase in industrial activity.

Table 3 Average concentrations of different PM chemical components during the LT and the HT periods, difference in %, and difference in % of the normalized concentrations with respect to OA

		LT	HT	Difference (%)	Difference of component/OA (%)
Chemical composition ($\mu g\ m^{-3}$)					
	BC	1.0	1.6	62	6
	Sulphate	2.4	3.8	60	1
	Nitrate	0.5	1.0	99	26
	Ammonium	1.1	1.8	72	9
	OA	6.8	10.8	58	—
OA types	HOA	0.6	1.1	79	14
	COA	1.2	1.9	62	3
	LV-OOA	2.5	3.5	40	−11
	SV-OOA	2.6	4.4	69	7
Organic compounds ($ng\ m^{-3}$)					
Diacids (urban SOA)	Succinic acid	16.8	21.3	27	−19
	Glutaric acid	2.9	3.7	29	−17
	Adipic acid	0.9	1.2	25	−20
	Pimelic acid	0.8	0.8	4	−34
	Azelaic acid	1.8	3.0	65	6
	Phthalic acid	3.9	5.2	32	−15
Pinene SOA	*cis*-Pinonic acid	12.1	42.6	251	125
	Malic acid	28.8	53.0	84	18
	3-Hydroxyglutaric acid	4.8	8.6	80	16
	MBTCA	4.2	10.0	135	51
Isoprene SOA	2-Methylglyceric acid	23.3	32.3	39	−11
	C5-Alkene triols	28.0	37.7	35	−14
	2-Methylthreitol	19.9	19.4	−2	−37
	2-Methylerythritol	67.5	57.6	−15	−45
Urban POA	Norhopane	0.5	0.9	85	19
	Hopane	0.5	0.7	45	−7
	Fluorathene	0.11	0.33	217	103
	Pyrene	0.12	0.42	249	124
	Benz[*a*]anthracene + chrysene	0.17	0.32	84	18

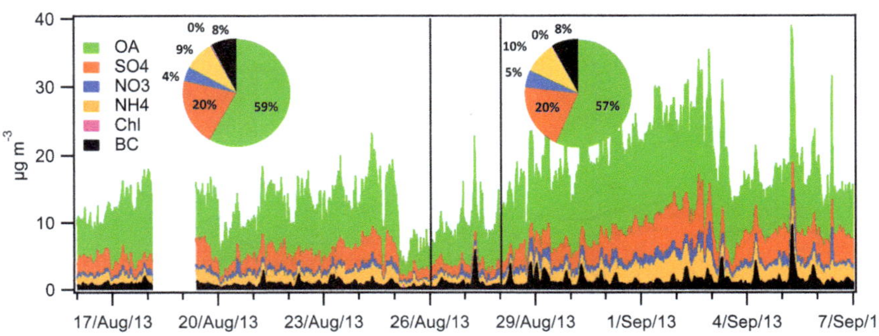

Fig. 3 Stacked OA, sulphate, nitrate, ammonium, chloride and BC concentrations in PM_1 (15 min time resolution) and average composition for the LT and HT periods.

3.5 Organic aerosol sources

Different source apportionment solutions were investigated. The 3-factor solution consisted of a hydrocarbon-like OA (HOA), a semi-volatile oxidized OA (SV-OOA) and a low-volatility oxidized OA (LV-OOA). The investigation of the residuals for this solution led to the identification of a diurnal pattern for the residuals of *m*/*z* 55. Hence, a cooking OA (COA) source was constrained based on a COA profile identified for Barcelona in March 2009.[52] The solutions for *a*-values from 0.025 to 0.2 in steps of 0.025 were investigated, and a final *a*-value of 0.1 was chosen based on the significance of the source profiles found and their correlation with literature profiles, the correlation of the factor contributions with external tracers, and the significance of the time variation (especially the daily patterns) (Section S3 of the ESI†). The contribution of biomass burning is expected to be low based on previous studies carried out in July 2009[19] and based on the low signal at *m*/*z* 60 registered in the present study, which is below the background threshold[53] for more than 95% of the data (Fig. S10†). The possible contribution of an isoprene epoxydiol derived OA factor (IEPOX-OA)[54] was investigated, but the solutions with a high number of factors resulted in a split of the HOA and/or the SV-OOA sources but a lack of a clear IEPOX-OA. More details on the solution are given in the ESI (Section S3†).

The final solution includes four organic sources: HOA, COA, SV-OOA, and LV-OOA (Fig. 6 and 7), which compares well with other mass spectral profiles from literature, and specifically with those identified for Barcelona in March 2009 (Fig. S9†).[52] The HOA is attributed to road traffic and its intra-day variation clearly follows the traffic rush hour times, although it is clearer in the HT period than in the LT period, similarly to the BC variation (Fig. 8). Its contribution was 0.6 and 1.1 μg m^{-3} for the LT and HT periods (9% and 10% of the total OA), respectively. The HOA/BC ratio ranged from 0.4 to 0.9, which is in agreement with ratios found close to road traffic emission sources (Fig. S11†).[55] The average daily pattern for COA shows an increase in the morning lasting until lunch time and an increase at dinner time (Fig. 8), although it may not be as well-defined as would be expected. This average daily pattern may indicate that this factor has some influence on sources other than COA, such as some road traffic influence, although it is

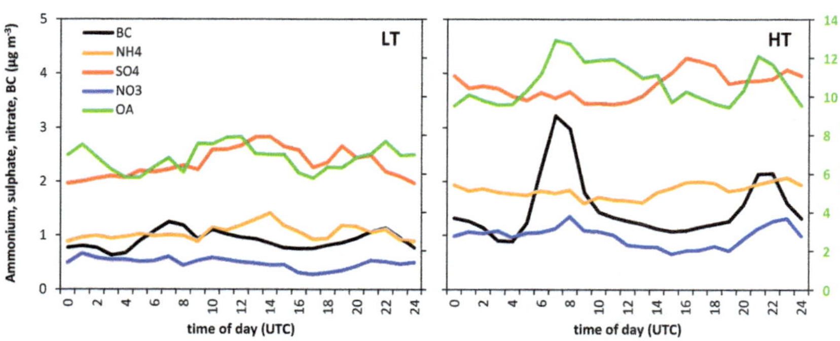

Fig. 4 Average daily pattern of OA (right axis), sulphate, nitrate, ammonium and BC concentrations (left axis) for the LT and HT periods. The time corresponds to the end of the hourly slot.

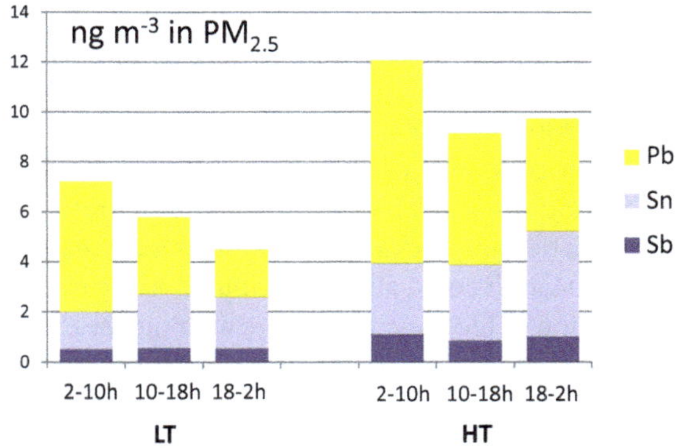

Fig. 5 Eight-hour averages of concentrations of selected trace elements in $PM_{2.5}$ during the LT and HT periods. Time is local time (UTC+2).

expected to be low, given that the mass spectral profile has been constrained according to the COA profile found specifically for Barcelona in a previous study. It contributed 1.2 and 1.9 $\mu g\ m^{-3}$ for the LT and HT periods, respectively, which implies an increase of +62%, similar to the bulk OA increase. The SV-OOA is attributed to recently formed SOA and its variation does not show a constant daily

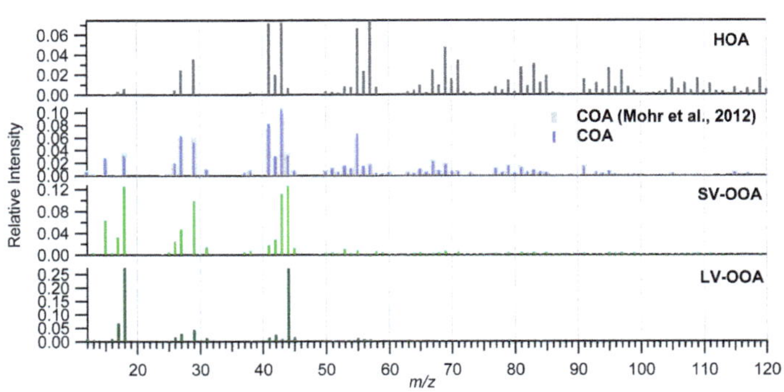

Fig. 6 Mass spectral profiles of the organic sources identified. In light grey is the profile used to constrain the COA factor.

pattern throughout the study period, hence resulting in an average daily pattern not representative of any of the days. It may have an influence of the prevailing wind condition, since the SV-OOA contribution under NW winds was 4.0 μg m^{-3}, while it was only 3.1 μg m^{-3} for other wind directions, whereas the difference in concentrations of other components depending on the wind direction is lower. The SV-OOA concentrations moderately correlate with nitrate concentrations ($R^2 = 0.43$). This correlation may be due to the similar gas-particle partitioning of both nitrate and SV-OOA, or it could also be attributed to the formation of organo-nitrates from the reaction of biogenic precursors with nitrate.[7,56,57] The fraction of SV-OOA with respect to total OA was similar for the LT and the HT periods (38% and 40%, respectively). The LV-OOA is attributed to aged SOA, hence reflecting the SOA contribution that can be in the atmosphere for a longer time. It is moderately correlated with sulphate concentrations ($R^2 = 0.41$). Although the absolute contribution of LV-OOA increased from the LT to the HT period, its relative contribution to total OA was similar (37% and 33%, respectively). The total OOA absolute increase (+2.8 μg m^{-3}, divided into +1.0 μg m^{-3} of LV-OOA and +1.8 μg m^{-3} of SV-OOA) is about 6 times that for HOA or BC.

3.6 Fossil *vs.* non-fossil OC combined with ME2-ACSM sources

A total of 5 samples were analysed for ^{14}C in OC (Fig. 7). An increase in the fraction of non-fossil OC from the beginning of the campaign towards the end, from 43% to 54% of the total OC, was recorded (Fig. S12†). The absolute concentrations of both the fossil OC and the non-fossil OC increased from the LT to the HT period.

Further information can be gained when combining ^{14}C and ACSM OA source apportionment results.[19,58] Relative contributions to OC determined by the ^{14}C and ME2-ACSM techniques are shown in Fig. 9. Fractional contributions are compared to avoid differences due to absolute concentration inaccuracies in either method. The OA source contributions from ME2-ACSM were converted into OC source contributions using the following OA-to-OC ratios: 1.21 for HOA, 1.42 for COA, 2.14 for LV-OOA and 1.61 for SV-OOA, as determined in previous studies for Barcelona.[52] The sources identified by ME2-ACSM were divided into fossil and non-fossil. HOC was included in the fossil sources and COC was included in the non-fossil sources. As LV-OOC and SV-OOC are thought to be

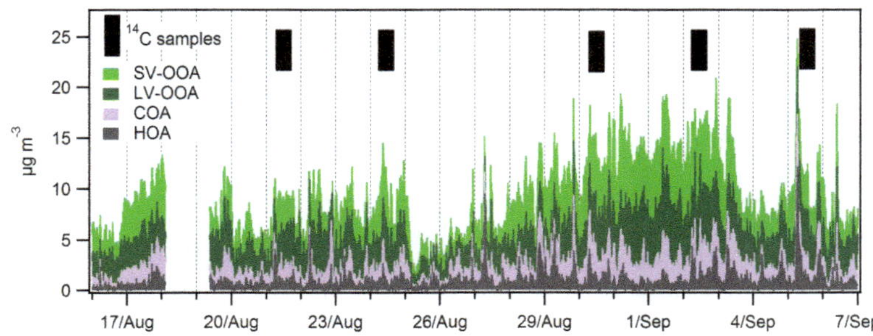

Fig. 7 Stacked HOA, COA, LV-OOA and SV-OOA. Selected days for ^{14}C analysis are indicated.

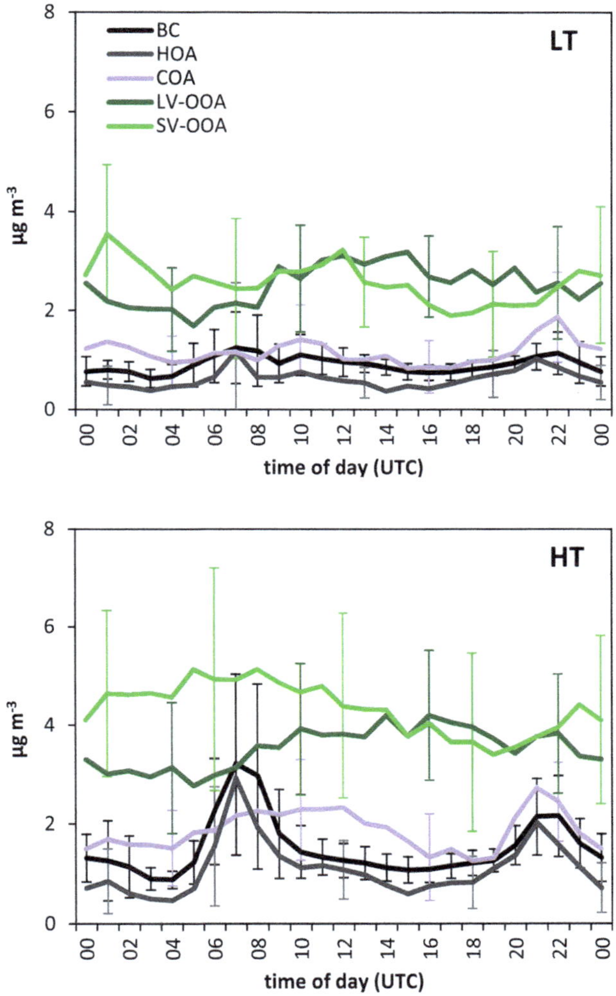

Fig. 8 Average daily pattern of HOA, COA, LV-OOA, SV-OOA and BC concentrations for the LT and HT periods. Error bars correspond to one standard deviation, only shown for some data points for clarity. The time corresponds to the end of the hourly slot.

a surrogate for total secondary OC and may originate from both fossil and non-fossil VOC precursors, these two factors were divided into fossil and non-fossil so that the resulting OCf/OCnf ratio from ME2-ACSM sources equals that from the ^{14}C method. The fossil OC was found to be mainly secondary (>70%) except for the last sample, where the fossil SOC only represented 51% of the total fossil OC (Fig. 10). This means that the amount of fossil SOC formed depended on the primary fossil OC for the first four samples, whereas it was influenced by something different for the last sample, in which the fossil SOC formation was lower with respect to the primary fossil OC. This last day resembles the results found in March 2009, when the fossil OC was found to be 53% secondary.[19] On the other hand, when analysing the SOC origin, it is observed that the fraction of non-fossil SOC increased from 37% for the first sample to 60% for the last

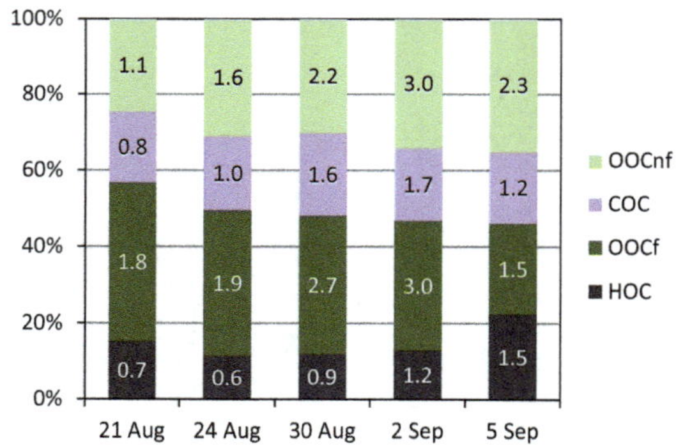

Fig. 9 Relative source apportionment to OC using the ^{14}C technique and ME2-ACSM. Numbers indicate the absolute concentrations in µg m^{-3}. HOC: hydrocarbon-like organic carbon; OOC: oxygenated organic carbon (secondary OC); COC: cooking organic carbon.

sample (Fig. 10). This enhanced formation of non-fossil SOC could be attributed to the reaction of BVOC precursors with NO_x emitted from road traffic (or from its nocturnal derivative NO_3 that enhances night-time SVOOA),[59] since NO_2 concentrations increased from 19 to 42 µg m^{-3} from the first to the last sample.

3.7 Organic compounds and relation with OA sources

The urban POA related organic compound concentrations increased from +45% to +249% in the HT period (Table 3). In relative terms, when considering their normalized concentrations with respect to total OA, they increased from +18% to +124%, except for the hopanes, attributed to their reactivity.[60] PAHs showed substantial correlations ($p < 0.05$) with typical traffic tracers, such as BC ($0.57 < R^2 < 0.81$) and NO_2 ($0.50 < R^2 < 0.71$), as well as the HOA contribution (the R^2 correlation coefficient of HOA and fluoranthene and pyrene was 0.68 and 0.79, respectively). Dicarboxylic acids, such as succinic acid, glutaric acid and azelaic acid, as well as poly- and aromatic acids, such as malic acid and phthalic acid are

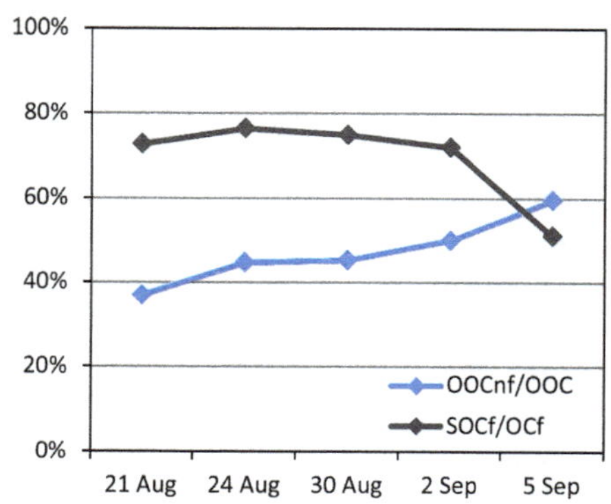

Fig. 10 Percentage of non-fossil secondary OC with respect to total secondary OC (represented by the oxygenated organic carbon (OOC)), and percentage of secondary fossil OC with respect to total fossil OC.

SOA-species in the absence of primary biomass burning emissions.[61,62] The very low levoglucosan concentrations (<1 ng m^{-3}) and the low signal at m/z 60 found by the ACSM suggest that these acids are mainly SOA. Their concentrations increased between +20% and +50% from LT to HT, although their normalized concentrations with respect to OA decreased from the LT to the HT period. The compounds related to SOA originated from isoprene showed a decrease of −11% to −45% in their relative concentrations. On the other hand, the compounds related to pinene SOA showed an increase in their normalized concentrations (+16% to +125%).

The precursors of the SOA species may be anthropogenic or biogenic. Thus, phthalic acid can be related to oxidation of naphthalene (*i.e.* anthropogenic fossil fuel combustion emissions), while azelaic acid can be related to ozonolysis of (particulate) oleic acid (*e.g.* an anthropogenic non-fossil carbon source from cooking). The increase in the normalized concentrations of some aged oxidation products of biogenic VOCs, such as 3-hydroxyglutaric acid and 3-methyl-1,2,3-butanetricarboxylic acid (MBTCA) in the case of α-pinene oxidation,[63,64] may reflect the enhanced production of BSOA at higher traffic intensities in the metropolitan area, as observed in other studies.[7,14,65] On the other hand, 2-methylglyceric acid, from oxidation of isoprene,[66,67] showed a decrease of −11% in its normalized concentration. While 2-methylglyceric acid is favoured in a high NO_x regime, 2-methylthreitol and 2-methylerythritol are preferentially formed under low NO_x conditions,[65] which explains the larger decrease in the HT period of normalized concentrations of 2-methylthreitol and 2-methylerythritol (−37% and −45%) compared to 2-methylglyceric acid (−11%).

Substantial correlations ($p < 0.05$) were observed among the different SOA species, as well as with the ME2-ACSM OA sources (Fig. 11), suggesting heterogeneous reactions and mixing of these compounds in the urban airshed. The long-chained dicarboxylic azelaic acid showed a strong correlation (R^2 = 0.61) with the COA component, suggesting that cooking activities could be involved, in agreement with a previous study in the urban area of Barcelona,[62] although the correlation between pimelic acid and COA was weaker ($R^2 = 0.21$). The short-chained dicarboxylic acids, *i.e.* succinic, glutaric and adipic acids, showed strong or relatively strong correlations with the LV-OOA component ($R^2 = 0.72$, 0.65 and 0.50, respectively), reflecting aged-SOA. These correlations were also observed for poly-acids, such as malic acid, 3-hydroxyglutaric acid, MBTCA and 2-methylglyceric acid ($R^2 = 0.49$, 0.41, 0.66 and 0.49, respectively). Nevertheless, in contrast to the former compounds these latter compounds also showed substantial correlations ($p < 0.05$) with the SV-OOA component ($0.43 < R^2 < 0.71$), supporting the idea of enhanced BSOA production at higher traffic intensities. The concentrations of the isoprene SOA products did not correlate to any other variable, including the ME2-ACSM OA sources. It is unlikely that these compounds are readily transformed to other compounds, due to their relatively high concentrations, although part may be transformed to organosulphates in urban atmospheres.[67,68] The decrease in their normalized concentrations from the LT to the HT period could be attributed to their formation in the region under relatively low NO_x conditions (LT period) and their transport to the urban area.[68]

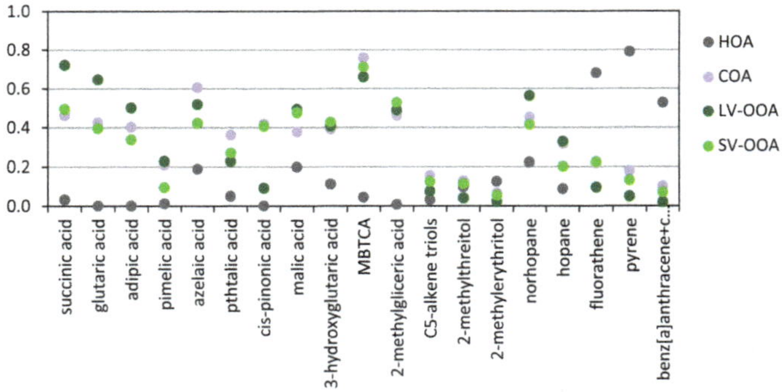

Fig. 11 Squared Pearson correlation coefficients (R^2) between organic tracer compounds and OA sources.

4 Final discussion and conclusions

The increase of traffic related organic species is in agreement with the increases observed for inorganic traffic indicators as well as the HOA. Moreover, an increase in OOA determined by ME2-ACSM from the LT to the HT period was simultaneously recorded for almost all analyzed SOA species, including those from BSOA. These findings may implicate an enhanced formation of BSOA in the urban airshed under higher traffic intensities (average NO_x concentrations were 20 μg m^{-3} for the LT period and 36 μg m^{-3} for the HT period). This is further suggested by the higher concentrations of non-fossil SOC found for the samples in the HT period. However, part of these SOA species and other possible oxidation products (including those from isoprene oxidation), may have been formed outside of the city and subsequently transported. This would mean that they were formed under conditions that were similar for both the LT and HT periods, in spite of the increase of traffic intensities in the HT period.

Maximum concentrations of C_4–C_6 dicarboxylic acid and phthalic acid were recorded during the daytime, similarly to the LV-OOA component. On the other hand, the products of pinene oxidation, *cis*-pinonic (a first generation product), 3-hydroxyglutaric acid and MBTCA (further oxidation products), and malic acid are sometimes registered in higher concentrations in the evening and night-time samples.

To further investigate how the different species, components, and parameters relate to each other, a factor analysis was applied to a combined dataset including: VOC-factors (obtained from the processing of the VOC dataset), ACSM data (inorganic compounds and OA sources), and meteorological and online air quality parameters. It resulted in the identification of 5 factors, which accounted for 72% of the variance:

- Factor 1: this reflects mainly the road traffic influence, with high factor loadings for the VOC traffic factor, N_{20-500}, BC, NO, NO_2, HOA and traffic/wind speed.

- Factor 2: this reflects mainly the biogenic VOC influence, with high factor loadings for the VOC biogenic factor, temperature, and the solar radiation.

- Factor 3: this reflects the components with the higher influence on the PM mass, like sulphate, nitrate, ammonium, COA, LV-OOA and SV-OOA.

- Factor 4: this reflects the fine particles, grouping the different particle number size ranges up to 500 nm.
- Factor 5: this reflects the more oxidized components, like LV-OOA and ozone.

Acknowledgements

This study was partially funded by the Spanish Ministry of Agriculture, Food and the Environment, the European LIFE+ AIRUSE project (LIFE11 ENV/ES/584), the European Union Seventh Framework Programme (FP7/2007-2013) through ACT-RIS (grant agreement no 262254), and the Generalitat de Catalunya (AGAUR 2014 SGR33 and DGQA). We thank Gary Salazar (University of Bern) for his support during the ^{14}C analysis. The publication fee was supported by the CSIC Open Access Publication Support Initiative through its Unit of Information Resources for Research (URICI).

References

1 Q. Zhang, J. L. Jimenez, M. R. Canagaratna, J. D. Allan, H. Coe, I. Ulbrich, M. R. Alfarra, A. Takami, A. M. Middlebrook, Y. L. Sun, K. Dzepina, E. Dunlea, K. Docherty, P. F. DeCarlo, D. Salcedo, T. Onasch, J. T. Jayne, T. Miyoshi, A. Shimono, S. Hatakeyama, N. Takegawa, Y. Kondo, J. Schneider, F. Drewnick, S. Borrmann, S. Weimer, K. Demerjian, P. Williams, K. Bower, R. Bahreini, L. Cottrell, R. J. Griffin, J. Rautiainen, J. Y. Sun, Y. M. Zhang and D. R. Worsnop, *Geophys. Res. Lett.*, 2007, **34**, L17811.

2 J. P. Putaud, R. van Dingenen, A. Alastuey, H. Bauer, W. Birmili, J. Cyrys, H. Flentje, S. Fuzzi, R. Gehrig, H. C. Hansson, R. M. Harrison, H. Herrmann, R. Hitzenberger, C. Hüglin, A. M. Jones, A. Kasper-Giebl, G. Kiss, A. Kousa, T. A. J. Kuhlbusch, G. Löschau, W. Maenhaut, A. Molnar, T. Moreno, J. Pekkanen, C. Perrino, M. Pitz, H. Puxbaum, X. Querol, S. Rodriguez, I. Salma, J. Schwarz, J. Smolik, J. Schneider, G. Spindler, H. ten Brink, J. Tursic, M. Viana, A. Wiedensohler and F. Raes, *Atmos. Environ.*, 2010, **44**, 1308–1320.

3 X. Querol, A. Alastuey, J. Pey, M. Cusack, N. Pérez, N. Mihalopoulos, C. Theodosi, E. Gerasopoulos, N. Kubilay and M. Koçak, *Atmos. Chem. Phys.*, 2009, **9**, 4575–4591.

4 X. Querol, A. Alastuey, M. Viana, T. Moreno, C. Reche, M. C. Minguillón, A. Ripoll, M. Pandolfi, F. Amato, A. Karanasiou, N. Pérez, J. Pey, M. Cusack, R. Vázquez, F. Plana, M. Dall'Osto, J. de La Rosa, A. Sánchez De La Campa, R. Fernández-Camacho, S. Rodríguez, C. Pio, L. Alados-Arboledas, G. Titos, B. Artíñano, P. Salvador, S. García Dos Santos and R. Fernández Patier, *Atmos. Chem. Phys.*, 2013, **13**, 6185–6206.

5 R. Volkamer, J. L. Jimenez, F. San Martini, K. Dzepina, Q. Zhang, D. Salcedo, L. T. Molina, D. R. Worsnop and M. J. Molina, *Geophys. Res. Lett.*, 2006, **33**.

6 J. L. Jimenez, M. R. Canagaratna, N. M. Donahue, A. S. H. Prevot, Q. Zhang, J. H. Kroll, P. F. DeCarlo, J. D. Allan, H. Coe, N. L. Ng, A. C. Aiken, K. S. Docherty, I. M. Ulbrich, A. P. Grieshop, A. L. Robinson, J. Duplissy, J. D. Smith, K. R. Wilson, V. A. Lanz, C. Hueglin, Y. L. Sun, J. Tian, A. Laaksonen, T. Raatikainen, J. Rautiainen, P. Vaattovaara, M. Ehn, M. Kulmala, J. M. Tomlinson, D. R. Collins, M. J. Cubison, E. J. Dunlea,

J. A. Huffman, T. B. Onasch, M. R. Alfarra, P. I. Williams, K. Bower, Y. Kondo, J. Schneider, F. Drewnick, S. Borrmann, S. Weimer, K. Demerjian, D. Salcedo, L. Cottrell, R. Griffin, A. Takami, T. Miyoshi, S. Hatakeyama, A. Shimono, J. Y. Sun, Y. M. Zhang, K. Dzepina, J. R. Kimmel, D. Sueper, J. T. Jayne, S. C. Herndon, A. M. Trimborn, L. R. Williams, E. C. Wood, A. M. Middlebrook, C. E. Kolb, U. Baltensperger and D. R. Worsnop, *Science*, 2009, **326**, 1525–1529.

7 C. R. Hoyle, M. Boy, N. M. Donahue, J. L. Fry, M. Glasius, A. Guenther, A. G. Hallar, K. Huff Hartz, M. D. Petters, T. Petäjä, T. Rosenoern and A. P. Sullivan, *Atmos. Chem. Phys.*, 2011, **11**, 321–343.

8 S. Fuzzi, M. O. Andreae, B. J. Huebert, M. Kulmala, T. C. Bond, M. Boy, S. J. Doherty, A. Guenther, M. Kanakidou, K. Kawamura, V. M. Kerminen, U. Lohmann, L. M. Russell and U. Pöschl, *Atmos. Chem. Phys.*, 2006, **6**, 2017–2038.

9 S. N. Pandis, S. E. Paulson, J. H. Seinfeld and R. C. Flagan, *Atmos. Environ., Part A*, 1991, **25**, 997–1008.

10 J. H. Kroll, N. L. Ng, S. M. Murphy, R. C. Flagan and J. H. Seinfeld, *Environ. Sci. Technol.*, 2006, **40**, 1869–1877.

11 A. A. Presto, K. E. Huff Hartz and N. M. Donahue, *Environ. Sci. Technol.*, 2005, **39**, 7046–7054.

12 F. Paulot, J. D. Crounse, H. G. Kjaergaard, A. Kürten, J. M. St. Clair, J. H. Seinfeld and P. O. Wennberg, *Science*, 2009, **325**, 730–733.

13 M. Kulmala and V.-M. Kerminen, *Atmos. Res.*, 2008, **90**, 132–150.

14 J. D. Surratt, Y. Gómez-González, A. W. H. Chan, R. Vermeylen, M. Shahgholi, T. E. Kleindienst, E. O. Edney, J. H. Offenberg, M. Lewandowski, M. Jaoui, W. Maenhaut, M. Claeys, R. C. Flagan and J. H. Seinfeld, *J. Phys. Chem. A*, 2008, **112**, 8345–8378.

15 DGT, 2013, https://sedeapl.dgt.gob.es/IEST2/menu.do?path=/vehiculos/parque/&file=inebase&type=pcaxis&L=0&js=1.

16 X. Querol, A. Alastuey, M. Pandolfi, C. Reche, N. Pérez, M. C. Minguillón, T. Moreno, M. Viana, M. Escudero, A. Orio, M. Pallarés and F. Reina, *Sci. Total Environ.*, 2014, **490**, 957–969.

17 N. Pérez, J. Pey, M. Cusack, C. Reche, X. Querol, A. Alastuey and M. Viana, *Aerosol Sci. Technol.*, 2010, **44**, 487–499.

18 I. Salma, X. Chi and W. Maenhaut, *Atmos. Environ.*, 2004, **38**, 27–36.

19 M. C. Minguillón, N. Perron, X. Querol, S. Szidat, S. M. Fahrni, A. Alastuey, J. L. Jimenez, C. Mohr, A. M. Ortega, D. A. Day, V. A. Lanz, L. Wacker, C. Reche, M. Cusack, F. Amato, G. Kiss, A. Hoffer, S. Decesari, F. Moretti, R. Hillamo, K. Teinilä, R. Seco, J. Peñuelas, A. Metzger, S. Schallhart, M. Müller, A. Hansel, J. F. Burkhart, U. Baltensperger and A. S. H. Prévôt, *Atmos. Chem. Phys.*, 2011, **11**, 12067–12084.

20 M. Pandolfi, X. Querol, A. Alastuey, J. L. Jimenez, O. Jorba, D. Day, A. Ortega, M. J. Cubison, A. Comerõn, M. Sicard, C. Mohr, A. S. H. Prévôt, M. C. Minguillõn, J. Pey, J. M. Baldasano, J. F. Burkhart, R. Seco, J. Peñuelas, B. L. van Drooge, B. Artiñano, C. Di Marco, E. Nemitz, S. Schallhart, A. Metzger, A. Hansel, J. Lorente, S. Ng, J. Jayne and S. Szidat, *J. Geophys. Res.: Atmos.*, 2014, **119**, 4978–5010.

21 B. L. van Drooge, M. Cusack, C. Reche, C. Mohr, A. Alastuey, X. Querol, A. S. H. Prevot, D. A. Day, J. L. Jimenez and J. O. Grimalt, *Atmos. Environ.*, 2012, **61**, 482–489.
22 S. Szidat, T. M. Jenk, H. A. Synal, M. Kalberer, L. Wacker, I. Hajdas, A. Kasper-Giebl and U. Baltensperger, *J. Geophys. Res.*, 2006, **11**, D07206.
23 I. El Haddad, B. D'Anna, B. Temime-Roussel, M. Nicolas, A. Boreave, O. Favez, D. Voisin, J. Sciare, C. George, J. L. Jaffrezo, H. Wortham and N. Marchand, *Atmos. Chem. Phys.*, 2013, **13**, 7875–7894.
24 M. R. Heal, P. Naysmith, G. T. Cook, S. Xu, T. R. Duran and R. M. Harrison, *Atmos. Environ.*, 2011, **45**, 2341–2348.
25 F. Amato, M. Pandolfi, A. Escrig, X. Querol, A. Alastuey, J. Pey, N. Perez and P. K. Hopke, *Atmos. Environ.*, 2009, **43**, 2770–2780.
26 C. Pérez, M. Sicard, O. Jorba, A. Comerón and J. M. Baldasano, *Atmos. Environ.*, 2004, **38**, 3983–4000.
27 A. Jordan, S. Haidacher, G. Hanel, E. Hartungen, L. Märk, H. Seehauser, R. Schottkowsky, P. Sulzer and T. D. Märk, *Int. J. Mass Spectrom.*, 2009, **286**, 122–128.
28 M. Graus, M. Müller and A. Hansel, *J. Am. Soc. Mass Spectrom.*, 2010, **21**, 1037–1044.
29 W. Lindinger, A. Hansel and A. Jordan, *Int. J. Mass Spectrom. Ion Processes*, 1998, **173**, 191–241.
30 L. Cappellin, T. Karl, M. Probst, O. Ismailova, P. M. Winkler, C. Soukoulis, E. Aprea, T. D. Märk, F. Gasperi and F. Biasioli, *Environ. Sci. Technol.*, 2012, **46**, 2283–2290.
31 A. Petzold, J. A. Ogren, M. Fiebig, P. Laj, S. M. Li, U. Baltensperger, T. Holzer-Popp, S. Kinne, G. Pappalardo, N. Sugimoto, C. Wehrli, A. Wiedensohler and X. Y. Zhang, *Atmos. Chem. Phys.*, 2013, **13**, 8365–8379.
32 N. L. Ng, S. C. Herndon, A. Trimborn, M. R. Canagaratna, P. L. Croteau, T. B. Onasch, D. Sueper, D. R. Worsnop, Q. Zhang, Y. L. Sun and J. T. Jayne, *Aerosol Sci. Technol.*, 2011, **45**, 770–784.
33 N. L. Ng, M. R. Canagaratna, J. L. Jimenez, Q. Zhang, I. M. Ulbrich and D. R. Worsnop, *Environ. Sci. Technol.*, 2011, **45**, 910–916.
34 A. M. Middlebrook, R. Bahreini, J. L. Jimenez and M. R. Canagaratna, *Aerosol Sci. Technol.*, 2012, **46**, 258–271.
35 X. Querol, A. Alastuey, S. Rodriguez, F. Plana, C. R. Ruiz, N. Cots, G. Massagué and O. Puig, *Atmos. Environ.*, 2001, **35**, 6407–6419.
36 F. Cavalli, M. Viana, K. E. Yttri, J. Genberg and J. P. Putaud, *Atmos. Meas. Tech.*, 2010, **3**, 79–89.
37 Y. L. Zhang, N. Perron, V. G. Ciobanu, P. Zotter, M. C. Minguillón, L. Wacker, A. S. H. Prévôt, U. Baltensperger and S. Szidat, *Atmos. Chem. Phys.*, 2012, **12**, 10841–10856.
38 S. Szidat, G. A. Salazar, E. Vogel, M. Battaglia, L. Wacker, H. A. Synal and A. Türler, *Radiocarbon*, 2014, **56**, 561–566.
39 P. Paatero and U. Tapper, *Environmetrics*, 1994, **5**, 111–126.
40 P. Paatero, *Journal of Computational and Graphical Statistics*, 1999, **8**, 854–888.
41 F. Canonaco, M. Crippa, J. G. Slowik, U. Baltensperger and A. S. H. Prévôt, *Atmos. Meas. Tech.*, 2013, **6**, 3649–3661.
42 P. Zotter, I. El-Haddad, Y. Zhang, P. L. Hayes, X. Zhang, Y. H. Lin, L. Wacker, J. Schnelle-Kreis, G. Abbaszade, R. Zimmermann, J. D. Surratt, R. Weber,

J. L. Jimenez, S. Szidat, U. Baltensperger and A. S. H. Prévôt, *J. Geophys. Res.: Atmos.*, 2014, **119**, 6818–6835.

43 J. Mohn, S. Szidat, J. Fellner, H. Rechberger, R. Quartier, B. Buchmann and L. Emmenegger, *Bioresour. Technol.*, 2008, **99**, 6471–6479.

44 I. Levin, T. Naegler, B. Kromer, M. Diehl, R. J. Francey, A. J. Gomez-Pelaez, L. P. Steele, D. Wagenbach, R. Weller and D. E. Worthy, *Tellus, Ser. B*, 2010, **62**, 26–46.

45 L. M. Hildemann, D. B. Klinedinst, G. A. Klouda, L. A. Currie and G. R. Cass, *Environ. Sci. Technol.*, 1994, **28**, 1565–1576.

46 J. Pey, S. Rodríguez, X. Querol, A. Alastuey, T. Moreno, J. P. Putaud and R. van Dingenen, *Atmos. Environ.*, 2008, **42**, 9052–9062.

47 M. Brines, M. Dall'Osto, D. C. S. Beddows, R. M. Harrison and X. Querol, *Atmos. Chem. Phys.*, 2014, **14**, 2973–2986.

48 M. C. Minguillón, M. Brines, N. Pérez, C. Reche, M. Pandolfi, A. S. Fonseca, F. Amato, A. Alastuey, A. Lyasota, B. Codina, H. K. Lee, H. R. Eun, K. H. Ahn and X. Querol, *Atmos. Res.*, 2015, **164–165**, 118–130.

49 R. Seco, J. Peñuelas, I. Filella, J. Llusia, S. Schallhart, A. Metzger, M. Müller and A. Hansel, *Atmos. Chem. Phys.*, 2013, **13**, 4291–4306.

50 M. Navazo, N. Durana, L. Alonso, M. C. Gómez, J. A. García, J. L. Ilardia, G. Gangoiti and J. Iza, *Environ. Monit. Assess.*, 2008, **136**, 53–68.

51 J. Sternbeck, A. Sjödin and K. Andréasson, *Atmos. Environ.*, 2002, **36**, 4735–4744.

52 C. Mohr, P. F. DeCarlo, M. F. Heringa, R. Chirico, J. G. Slowik, R. Richter, C. Reche, A. Alastuey, X. Querol, R. Seco, J. Peñuelas, J. L. Jiménez, M. Crippa, R. Zimmermann, U. Baltensperger and A. S. H. Prévôt, *Atmos. Chem. Phys.*, 2012, **12**, 1649–1665.

53 M. J. Cubison, A. M. Ortega, P. L. Hayes, D. K. Farmer, D. Day, M. J. Lechner, W. H. Brune, E. Apel, G. S. Diskin, J. A. Fisher, H. E. Fuelberg, A. Hecobian, D. J. Knapp, T. Mikoviny, D. Riemer, G. W. Sachse, W. Sessions, R. J. Weber, A. J. Weinheimer, A. Wisthaler and J. L. Jimenez, *Atmos. Chem. Phys.*, 2011, **11**, 12049–12064.

54 S. H. Budisulistiorini, M. R. Canagaratna, P. L. Croteau, W. J. Marth, K. Baumann, E. S. Edgerton, S. L. Shaw, E. M. Knipping, D. R. Worsnop, J. T. Jayne, A. Gold and J. D. Surratt, *Environ. Sci. Technol.*, 2013, **47**, 5686–5694.

55 H. L. DeWitt, S. Hellebust, B. Temime-Roussel, S. Ravier, L. Polo, V. Jacob, C. Buisson, A. Charron, M. André, A. Pasquier, J. L. Besombes, J. L. Jaffrezo, H. Wortham and N. Marchand, *Atmos. Chem. Phys.*, 2015, **15**, 4373–4387.

56 J. L. Fry, D. C. Draper, K. C. Barsanti, J. N. Smith, J. Ortega, P. M. Winkler, M. J. Lawler, S. S. Brown, P. M. Edwards, R. C. Cohen and L. Lee, *Environ. Sci. Technol.*, 2014, **48**, 11944–11953.

57 A. I. Darer, N. C. Cole-Filipiak, A. E. O'Connor and M. J. Elrod, *Environ. Sci. Technol.*, 2011, **45**, 1895–1902.

58 V. A. Lanz, M. R. Alfarra, U. Baltensperger, B. Buchmann, C. Hueglin, S. Szidat, M. N. Wehrli, L. Wacker, S. Weimer, A. Caseiro, H. Puxbaum and A. S. H. Prevot, *Environ. Sci. Technol.*, 2008, **42**, 214–220.

59 L. Xu, H. Guo, C. M. Boyd, M. Klein, A. Bougiatioti, K. M. Cerully, J. R. Hite, G. Isaacman-vanWertz, N. M. Kreisberg, C. Knote, K. Olson, A. Koss, A. H. Goldstein, S. V. Hering, J. de Gouw, K. Baumann, S. H. Lee, A. Nenes, R. J. Weber and N. L. Ng, *Proc. Natl. Acad. Sci. U. S. A.*, 2015, **112**, 37–42.

60 I. El Haddad, N. Marchand, H. Wortham, C. Piot, J. L. Besombes, J. Cozic, C. Chauvel, A. Armengaud, D. Robin and J. L. Jaffrezo, *Atmos. Chem. Phys.*, 2011, **11**, 2039–2058.
61 T. E. Kleindienst, M. Jaoui, M. Lewandowski, J. H. Offenberg and K. S. Docherty, *Atmos. Chem. Phys.*, 2012, **12**, 8711–8726.
62 M. Alier, B. L. van Drooge, M. Dall'Osto, X. Querol, J. O. Grimalt and R. Tauler, *Atmos. Chem. Phys.*, 2013, **13**, 10353–10371.
63 M. Claeys, R. Szmigielski, I. Kourtchev, P. van der Veken, R. Vermeylen, W. Maenhaut, M. Jaoui, T. E. Kleindienst, M. Lewandowski, J. H. Offenberg and E. O. Edney, *Environ. Sci. Technol.*, 2007, **41**, 1628–1634.
64 R. Szmigielski, J. D. Surratt, Y. Gómez-González, P. van der Veken, I. Kourtchev, R. Vermeylen, F. Blockhuys, M. Jaoui, T. E. Kleindienst, M. Lewandowski, J. H. Offenberg, E. O. Edney, J. H. Seinfeld, W. Maenhaut and M. Claeys, *Geophys. Res. Lett.*, 2007, **34**, L24811.
65 A. G. Carlton, C. Wiedinmyer and J. H. Kroll, *Atmos. Chem. Phys.*, 2009, **9**, 4987–5005.
66 M. Claeys, B. Graham, G. Vas, W. Wang, R. Vermeylen, V. Pashynska, J. Cafmeyer, P. Guyon, M. O. Andreae, P. Artaxo and W. Maenhaut, *Science*, 2004, **303**, 1173–1176.
67 J. D. Surratt, A. W. H. Chan, N. C. Eddingsaas, M. Chan, C. L. Loza, A. J. Kwan, S. P. Hersey, R. C. Flagan, P. O. Wennberg and J. H. Seinfeld, *Proc. Natl. Acad. Sci. U. S. A.*, 2010, **107**, 6640–6645.
68 Q. T. Nguyen, M. K. Christensen, F. Cozzi, A. Zare, A. M. K. Hansen, K. Kristensen, T. E. Tulinius, H. H. Madsen, J. H. Christensen, J. Brandt, A. Massling, J. K. Nøjgaard and M. Glasius, *Atmos. Chem. Phys.*, 2014, **14**, 8961–8981.

Faraday Discussions

PAPER

The relative importance of tailpipe and non-tailpipe emissions on the oxidative potential of ambient particles in Los Angeles, CA†

Farimah Shirmohammadi,[a] Sina Hasheminassab,[a] Dongbin Wang,[a] James J. Schauer,[b] Martin M. Shafer,[b] Ralph J. Delfino[c] and Constantinos Sioutas*[a]

Received 15th November 2015, Accepted 11th December 2015
DOI: 10.1039/c5fd00166h

This study examines the associations between the oxidative potential of ambient $PM_{2.5}$ and $PM_{0.18}$, measured by means of the dithiothreitol (DTT) assay, and their chemical constituents and modeled sources. Particulate matter (PM) samples were collected from 2012–2013 in Central Los Angeles (LA) and 2013–2014 in Anaheim, California, USA. Detailed chemical analyses of the PM samples, including carbonaceous species, inorganic elements and water-soluble ions, were conducted. Univariate analysis indicated a high correlation ($R > 0.60$) between the DTT activity and the concentrations of carbonaceous species at both sites. The strongest correlations were observed between DTT and organic tracers of primary vehicle tailpipe emissions including polycyclic aromatic hydrocarbons (PAHs) and hopanes as well as EC, with higher correlations for $PM_{0.18}$ *versus* $PM_{2.5}$ components. Moreover, metals and trace elements (*e.g.*, Ba, Cu, Fe, Mn, Pb and Sb) in both size ranges were also associated with DTT activity. Multiple linear regression (MLR) analysis was performed on DTT activity and PM sources identified by a Molecular Marker-Chemical Mass Balance (MM-CMB) model (*i.e.* major carbonaceous sources: vehicle tailpipe emissions, wood smoke, primary biogenic and secondary organic carbon) together with other typical sources of ambient PM (*i.e.* crustal material, vehicular abrasion, secondary ions and sea salt). Overall, our findings illustrate the relative importance of different traffic sources on the oxidative potential of ambient PM. Despite major reductions of tailpipe emissions, the lack of similar reductions (and possibly an increase) in non-tailpipe emissions makes them an important source of traffic-related PM in Los Angeles and their increasing role in the overall PM toxicity raises concerns for public health.

[a]*University of Southern California, Department of Civil and Environmental Engineering, Los Angeles, CA, USA. E-mail: sioutas@usc.edu*
[b]*University of Wisconsin-Madison, Environmental Chemistry and Technology Program, Madison, WI, USA*
[c]*University of California, Irvine, Department of Epidemiology, School of Medicine, Irvine, CA, USA*

† Electronic supplementary information (ESI) available. See DOI: 10.1039/c5fd00166h

1. Introduction

A large body of epidemiological and toxicological studies have investigated linkages between aerosol chemistry and health effects associated with exposure to particulate matter (PM).[1–5] In many of these studies, several adverse health outcomes have been attributed to the oxidative potential of ambient PM.[6–9] It has been hypothesized that these effects are caused by direct exposure to oxidants in PM, or the catalytic generation of reactive oxygen species (ROS) driven by certain PM components in human cells.[10–12] PM size also plays an important role in the severity of the health effects caused. Smaller particles have a higher pulmonary deposition fraction and higher surface area compared to larger particles, and therefore, they are considered major vectors for transport of toxic chemicals to the respiratory system.[13,14] Furthermore, studies have shown an increase in morbidity and mortality due to exposure to ambient fine particles ($PM_{2.5}$, particles with an aerodynamic diameter smaller than 2.5 μm).[7] Ultrafine particles (UFPs, traditionally defined as particles with an aerodynamic diameter smaller than approximately 0.1–0.2 μm), as a sub-fraction of $PM_{2.5}$, are strongly linked to systemic oxidative stress, inflammation, and platelet activation as well as cardiovascular effects in human and animal studies.[15–17]

Numerous studies have implemented biological[18,19] and chemical assays[20,21] to measure PM-induced toxicities with respect to PM oxidative potential. A widely used chemical assay is the dithiothreitol (DTT) assay in which DTT acts as a surrogate biological reducing agent,[22,23] with the rate of DTT consumption a proxy for the oxidative potential of a PM sample. Organic compounds such as quinones, which are believed to catalyze the electron transfer from DTT to oxygen, have been demonstrated to be associated with DTT activity.[23,24] In addition, certain metals such as Fe, Cu and Mn have been shown to be associated with DTT activity as well.[22,25–27] Furthermore, recent studies have shown that PM generated by specific sources such as vehicle tailpipe emissions, food cooking and secondary aerosol formation have different contributions to the overall oxidative potential of ambient PM and are strongly associated with health outcomes.[28–30] Therefore, how to target and apportion the toxicity of PM to specific sources remains a heated topic in current aerosol research.

As part of the Cardiovascular Health and Air Pollution Study (CHAPS) funded by the US National Institute of Health (NIH), ambient $PM_{2.5}$ and $PM_{0.18}$ (particles with an aerodynamic diameter smaller than 0.18 μm) were collected at two different locations of the Los Angeles (LA) Basin (*i.e.* Central LA, an urban area near Downtown Los Angeles and Anaheim, a residential area in Orange County, CA, USA). Chemical components along with the oxidative potential of the PM samples determined by DTT assay were quantified. In this paper, the temporal and spatial variations of DTT activity associated with PM in these two size ranges were investigated. In addition, major source contributions to ambient $PM_{2.5}$ and $PM_{0.18}$-bound organic carbon (OC) were determined using a novel hybrid Molecular Marker-Chemical Mass Balance (MM-CMB) model described in a companion paper by Shirmohammadi *et al.*[31] To provide more comprehensive insight into the dominant sources driving the DTT activity of $PM_{2.5}$ and $PM_{0.18}$, both primary and secondary CMB-derived sources contributing to ambient $PM_{2.5}$ and $PM_{0.18}$, along with some other sources not quantified by the CMB model (*e.g.*

vehicular abrasion, crustal material, *etc.*), were included in a multiple linear regression (MLR) analysis to identify possible predictors of the DTT activity. Vehicular abrasion has been estimated based on the source profile of brake wear particles reported by Schauer *et al.*,[32] and crustal material was estimated by summing up the oxides of Al, K, Fe, Ca, Mg, Ti and Si, as suggested by Hueglin *et al.* and Marcazzan *et al.*[33,34] Moreover, secondary ions were estimated as the sum of NO_3^-, NH_4^+ and SO_4^{2-}, while sea salt was also estimated using the concentration of soluble Na^+ and the sea salt fraction of typical sea water components such as Cl^-, Mg^{2+}, K^+, Ca^{2+} and SO_4^{2-} as suggested by Murphy *et al.*[35] The aforementioned source contribution estimations are discussed in more detail in the Experimental section. Furthermore, the DTT activity levels measured in this study were compared to several previous studies conducted at the same sampling site in Central LA. The analysis of historical trends in ambient PM toxicity enabled us to assess the relative importance of tailpipe and non-tailpipe emissions on DTT activity in this area.

2. Experimental

2.1. Sampling plan and sites

Time-integrated sampling was conducted every week from Monday to Friday, between July 2012 and February 2013 in Central LA, and from Sunday to Thursday, between July 2013 and February 2014 in Anaheim. PM sampling in Anaheim was discontinued in December 2013 and resumed in January 2014. Throughout this manuscript, "warmer months" refer to the July to September period, while "colder months" refer to October to February.

The Central LA site (referred to as an urban site) was located approximately 150 m to the east and downwind of a major freeway (I-110) at the Particle Instrumentation Unit (PIU) of the University of Southern California, about 3 km south of downtown Los Angeles. The other site in Anaheim (referred to as a suburban site), was situated in a residential area and about 500 m upwind of freeway I-5.

Ambient $PM_{2.5}$ and $PM_{0.18}$ were collected using two collocated Micro-Orifice Uniform Deposit Impactors (MOUDIs, Model 110 MSP Corporation, Minneapolis, Minnesota, USA), each operating at 30 L min^{-1} collecting particles in three stages: <0.18 μm (ultrafine), 0.18–2.5 μm (accumulation), and 2.5–10 μm (coarse). In this study we focused on ultrafine and fine (ultrafine + accumulation) PM only. For the purpose of chemical speciation, one MOUDI was loaded with Teflon filters (Teflon, 47 mm, pore size 2 μm, Pall Life Sciences, Ann Arbor, MI, USA) only, while the other one with aluminum-foil substrates in the coarse and accumulation stages and quartz microfiber filters (Whatman International Ltd, Maidstone, England) in the ultrafine stage.

2.2. Gravimetric and chemical analysis

Weekly samples were analyzed to quantify the mass concentrations of PM and its chemical constituents. The PM mass concentrations were determined by pre- and post-weighing of the Teflon filters, using a microbalance (Model MT5, Mettler Toledo Inc., Columbus, OH, USA; ±0.001 mg readability), after equilibration under a controlled temperature (22–24 °C) and relative humidity within the range of 40–50%. A 1.5 cm^2 punch of the aluminum and quartz filters was analyzed by the National Institute for Occupational Safety and Health (NIOSH) Thermal Optical Transmission (TOT) method in order to measure the elemental carbon (EC) and organic carbon (OC) content of the samples.[36] Furthermore, by means of gas chromatography mass spectrometry (GC-MS), organic constituents were quantified.[37] Total elemental composition of the samples was measured by high resolution inductively coupled plasma sector field mass spectrometry (SF-ICPMS) after microwave-aided solubilization of the PM in a mixture of acids (HNO_3, HF and HCl). Ion Chromatography (IC) was applied to measure the water soluble inorganic ions[38] as well. Metals, PAHs, hopanes and organic acids detected in this study are listed by their names in Table S1.†

2.3. Toxicological analysis

Reactive oxygen species (ROS) are produced in cells *via* reduction of oxygen to superoxides by biological reducing agents such as NADH (nicotinamide adenine dinucleotide) and NADPH (nicotinamide adenine dinucleotide phosphate).[20] This transformation catalytically occurs in the presence of either electron-transfer enzymes or redox-active chemical species such as organic chemicals (*e.g.* quinones) and metals.[39–41] The presence of excess ROS concentrations compared to the antioxidant capacity of the body to neutralize them leads to further oxidation of cellular components which can eventually lead to adverse health outcomes.[14,42] The dithiothreitol (DTT) assay was formulated to simulate the *in vivo* generation of ROS in which DTT acts as a surrogate of biological reducing agents (NADH and NDPH).[22,23] This is a commonly used cell-free approach to measure the oxidative potential of PM. The ability of a PM sample to catalyze the transfer of electrons from DTT to oxygen by generating superoxide radical anions is measured in this assay.[28,30] The concentration of redox-active species in a given sample determines the rate of depletion of DTT (in nmol min^{-1}) under a standardized set of conditions. The methodological details of this chemical assay can be found elsewhere.[20]

2.4 Statistical analysis

Univariate analysis was performed on weekly data sets in order to investigate the correlation of DTT activity with chemical species and sources. As the variables were log-normally distributed, Spearman rank correlations were obtained. These correlations also identified important species and sources that were included in the multiple linear regression (MLR) analysis. MLR was subsequently applied to investigate which species and sources still contributed significantly to the prediction of DTT activity (which was sufficiently normally distributed). Various combinations of species and sources, especially those with high correlation values with DTT activity in the univariate analysis, were tested with a stepwise selection procedure in the MLR models. Statistically significant ($P < 0.05$) species per sources, which were not co-linear with each other (if variance inflation factor < 2.5), were kept in the model to achieve the highest R^2 value and improve the prediction of the DTT activity.

2.5. Source apportionment of $PM_{2.5}$ and $PM_{0.18}$ mass concentrations

Sources of $PM_{2.5}$ and $PM_{0.18}$ OC have been quantified in a companion paper by Shirmohammadi *et al.*[31] using a hybrid CMB model in which the PMF-derived source profiles were combined with traditional source testing profiles. The model estimated the relative contributions from mobile sources (including gasoline, diesel, and smoking vehicles), wood smoke, primary biogenic sources (including emissions from vegetative detritus, food cooking, and re-suspended soil dust[43]), and anthropogenic secondary OC (SOC). The details of the CMB analysis can be found in the above companion paper.[31] OC source apportionment results from the CMB model were converted to $PM_{2.5}$ and $PM_{0.18}$ mass concentrations to evaluate the source contributions to total mass, using the OC-to-PM ratios for the wood smoke and mobile source profiles.[44,45] For primary biogenic and SOC sources, which were identified by the PMF model on organic species,[43] a PM/OC ratio of 2 was used for mass conversion of both sources. The un-apportioned OC (referred to as "other OC") was converted to "other OM" by multiplying by a factor of 1.6.[46] In addition to sources quantified by the CMB model, secondary ions (*i.e.* the sum of NO_3^-, NH_4^+ and SO_4^{2-}), crustal material, vehicular abrasion and sea salt contributions were also considered in $PM_{2.5}$ and $PM_{0.18}$ mass apportionment. Crustal material was calculated by summing the oxides of Al, K, Fe, Ca, Mg, Ti and Si based on the following equation:[33,34]

$$\text{Crustal material} = 1.89\text{Al} + 1.2\text{K} + 1.43\text{Fe} + 1.40\text{Ca} + 1.66\text{Mg} + 1.67\text{Ti} + 2.14\text{Si} \quad (1)$$

Si was not measured in this study, but was estimated as $3.14 \times \text{Al}$.[47]

Vehicular abrasions, especially particles emitted from brake wear, have been postulated to contribute significantly to the overall ambient $PM_{2.5}$ particle levels, especially in recent years and as tailpipe emissions have generally decreased.[48–50] Several studies have also reported the contribution of road dust and vehicular abrasion to ultrafine size fractions as well.[51–53] These studies have also acknowledged the importance of crustal elements such as Al, Ca, Fe and Ti in the ultrafine particle size range.[51–54]

Metals and elements such as V, Cr, Mn, Ni, Cu, Zn, As, Se, Sr, Ba and Pb have been identified as important tracers of vehicular abrasion emissions.[32,55,56] The contribution of tire wear to trace metal emissions from motor vehicles has been reported to be negligible compared with contributions of other sources (*e.g.*, brake wear) and it has been considered primarily a source of organic compounds.[32,57–59] In this study, vehicular abrasion was estimated based on a study by Schauer *et al.*[32] in which the sources of metals associated with motor vehicle traffic were determined by a CMB analysis. Intensive sampling was conducted to construct source profiles with a relatively large number of on-road vehicles in a tunnel test in Milwaukee along with brake wear and tire wear dust from Wisconsin vehicles. Schauer *et al.*[32] reported an averaged source profile for different compositions of brake pads. Taking Ba as our basis, with the assumption that all Ba emissions are from brake wear,[54] we estimated the brake wear contribution from the atmospheric concentrations of Ba and using the average mass ratio of this specie from the re-suspended brake composition reported by Schauer *et al.*[32] An average Ba mass fraction of 13.3 ± 0.14 mg g^{-1} PM of brake dust in $PM_{2.5}$ was applied for this conversion and was used for both $PM_{2.5}$ and $PM_{0.18}$ vehicular abrasion estimations. As noted above, Schauer *et al.*[32] indicated that tire wear may be a significant contributor to motor vehicle emissions of OC, but its contribution to metals emissions is negligible. Therefore, it should be

Table 1 Monthly-averaged mass concentrations (μg m^{-3}) of EC, OC, total metals and elements (*i.e.* sum of 50 species presented by name in Table S1), as well as secondary ions (*i.e.* sum of NO_3^-, NH_4^+ and SO_4^{2-}) in (a) Central LA and (b) Anaheim. Sampling was not conducted at Anaheim in December 2013

(a) Central LA

	EC		OC		Metals and elements		Secondary ions	
Sampling months	$PM_{2.5}$	$PM_{0.18}$	$PM_{2.5}$	$PM_{0.18}$	$PM_{2.5}$	$PM_{0.18}$	$PM_{2.5}$	$PM_{0.18}$
Jul 2012	0.40 ± 0.12	0.23 ± 0.07	1.86 ± 0.43	0.75 ± 0.21	1.35 ± 0.01	0.14 ± 0.00	4.14 ± 1.40	0.56 ± 0.23
Aug 2012	0.52 ± 0.18	0.31 ± 0.14	2.02 ± 0.42	0.86 ± 0.27	1.53 ± 0.01	0.19 ± 0.00	4.03 ± 1.01	0.38 ± 0.09
Sep 2012	0.59 ± 0.08	0.25 ± 0.03	2.51 ± 0.38	0.95 ± 0.17	1.33 ± 0.01	0.16 ± 0.00	3.58 ± 0.56	0.31 ± 0.06
Oct 2012	0.70 ± 0.30	0.37 ± 0.17	2.93 ± 0.93	1.31 ± 0.37	0.82 ± 0.01	0.25 ± 0.00	3.81 ± 2.60	0.30 ± 0.07
Nov 2012	0.45 ± 0.12	0.25 ± 0.06	3.00 ± 0.62	1.03 ± 0.36	1.00 ± 0.01	0.26 ± 0.00	6.96 ± 2.35	0.26 ± 0.08
Dec 2012	0.43 ± 0.10	0.26 ± 0.06	3.51 ± 1.23	1.22 ± 0.44	0.87 ± 0.01	0.39 ± 0.01	3.38 ± 1.38	0.15 ± 0.05
Jan 2013	0.51 ± 0.05	0.32 ± 0.02	3.97 ± 0.68	1.36 ± 0.11	0.55 ± 0.00	0.25 ± 0.00	1.89 ± 0.46	0.18 ± 0.03
Feb 2013	0.54 ± 0.32	0.30 ± 0.18	3.48 ± 1.86	1.21 ± 0.65	0.78 ± 0.01	0.20 ± 0.00	5.20 ± 1.69	0.25 ± 0.05

(b) Anaheim

	EC		OC		Metals and elements		Secondary ions	
Sampling months	$PM_{2.5}$	$PM_{0.18}$	$PM_{2.5}$	$PM_{0.18}$	$PM_{2.5}$	$PM_{0.18}$	$PM_{2.5}$	$PM_{0.18}$
Jul 2013	0.13 ± 0.03	0.11 ± 0.02	1.64 ± 0.35	1.01 ± 0.22	1.00 ± 0.01	0.17 ± 0.00	2.25 ± 0.48	0.40 ± 0.19
Aug 2013	0.18 ± 0.06	0.16 ± 0.05	2.43 ± 0.14	1.22 ± 0.14	0.95 ± 0.01	0.20 ± 0.00	3.17 ± 0.43	0.48 ± 0.11
Sep 2013	0.18 ± 0.03	0.15 ± 0.03	1.82 ± 0.15	1.12 ± 0.15	0.99 ± 0.01	0.17 ± 0.00	1.97 ± 0.34	0.44 ± 0.13
Oct 2013	0.31 ± 0.00	0.26 ± 0.00	2.95 ± 0.00	1.68 ± 0.00	1.12 ± 0.00	0.40 ± 0.00	2.29 ± 1.1	0.40 ± 0.11
Nov 2013	0.53 ± 0.19	0.42 ± 0.20	4.62 ± 1.71	1.85 ± 0.71	1.05 ± 0.00	0.43 ± 0.00	2.68 ± 0.67	0.32 ± 0.08
Dec 2013								
Jan 2014	0.46 ± 0.20	0.37 ± 0.17	3.74 ± 1.13	1.57 ± 0.49	0.94 ± 0.01	0.34 ± 0.00	4.46 ± 0.69	0.23 ± 0.07
Feb 2014	0.33 ± 0.18	0.28 ± 0.17	3.15 ± 0.99	1.54 ± 0.03	0.39 ± 0.00	0.16 ± 0.00	3.47 ± 0.22	0.31 ± 0.03

noted that vehicular abrasion source estimation in this study is mainly associated with contribution from brake wear.

Lastly, sea salt was also estimated as the sum of soluble Na^+ and the sea salt fraction of typical sea water components such as Cl^-, Mg^{2+}, K^+, Ca^{2+} and SO_4^{2-} (ref. 35) as follows:

$$\text{Sea salt} = [Na^+] + \text{ss}\,[Cl^-] + \text{ss}\,[Mg^{2+}] + \text{ss}\,[K^+] + \text{ss}\,[Ca^{2+}] + \text{ss}\,[SO_4^{2-}] \quad (2)$$

where ss $Cl^- = 1.8$, ss $Mg^{2+} = 0.12$, ss $K^+ = 0.036$, ss $Ca^{2+} = 0.038$ and ss $SO_4^{2-} = 0.252$.[35] The monthly-averaged source contributions (± standard deviation) are presented in Tables S2 and S3† for Central LA and Anaheim in both size ranges.

3. Results and discussion

3.1. $PM_{2.5}$ and $PM_{0.18}$ chemical composition

Table 1 presents the summary of monthly-averaged concentrations of EC, OC, and total elements, as well as secondary ions (as the sum of NO_3^-, NH_4^+ and SO_4^{2-}) for the two PM size ranges in Central LA and Anaheim. In general, $PM_{2.5}$ EC and OC constituted 4 and 24% of PM mass concentration in Central LA, while the contributions were 3 and 31% for EC and OC in Anaheim, respectively. Percent contribution of both EC and OC in the $PM_{0.18}$ size range increased to 13 and 48% in Central LA, respectively. In Anaheim, OC dominated the $PM_{0.18}$ composition with an average contribution of 55% while the contribution of EC to $PM_{0.18}$ was 10%. Inorganic elements constituted 8 and 10% of PM mass concentration in the $PM_{2.5}$

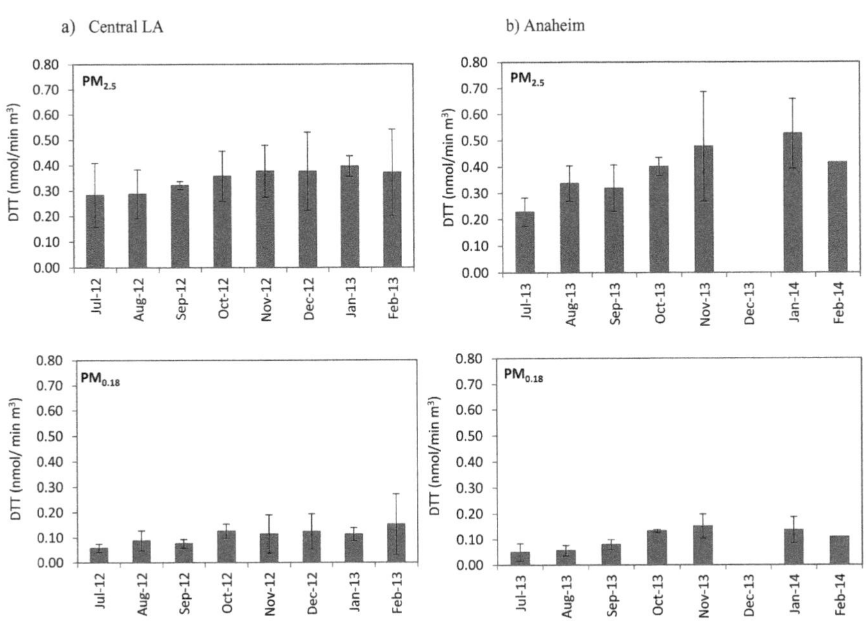

Fig. 1 Monthly-averaged volume-normalized dithiothreitol (DTT) activity (nmol min^{-1} m^{-3}) of ambient $PM_{2.5}$ and $PM_{0.18}$ in (a) Central LA and (b) Anaheim. February 2014 data corresponds to one sample. Sampling was not conducted in December 2013 at Anaheim. Error bars correspond to one standard deviation.

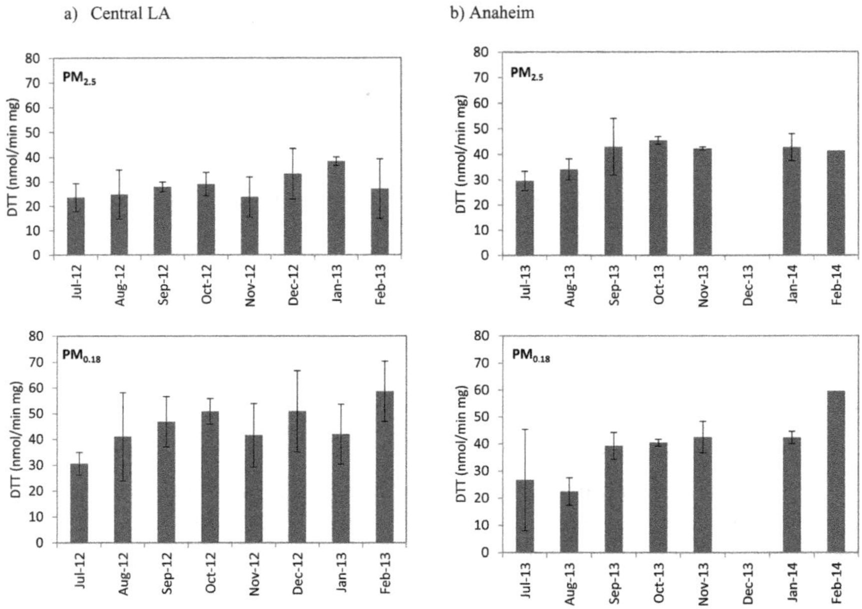

Fig. 2 Monthly-averaged mass-normalized dithiothreitol (DTT) activity (nmol min^{-1} mg^{-1}) of ambient $PM_{2.5}$ and $PM_{0.18}$ in (a) Central LA and (b) Anaheim. February 2014 data corresponds to one sample. Sampling was not conducted in December 2013 at Anaheim. Error bars correspond to one standard deviation.

and $PM_{0.18}$ size ranges, respectively, in Central LA. Similarly, the contribution of these species to PM mass was 10% for both size ranges in Anaheim. The contribution of secondary ions to $PM_{2.5}$ mass concentration was 33 and 31% in Central LA and Anaheim, respectively, while in the $PM_{0.18}$ size fraction they accounted for 13 and 14% in Central LA and Anaheim, respectively. More details on aerosol chemical composition and mass concentrations are discussed elsewhere.[31]

3.2. Spatial and temporal variations in DTT activity

The spatial and temporal variability of the DTT activity for both size ranges were investigated. The rate of DTT consumption normalized by volume of air sampled (expressed in units of nmol min^{-1} m^{-3}) is presented in Fig. 1, while the DTT consumption rate normalized by PM mass data (expressed in units of nmol min^{-1} mg^{-1}) are presented in Fig. 2. Our measurements showed that the volume-normalized DTT activity, as a metric for comparison of inhalation exposures, at both sites varied across a range of 0.05–0.15 nmol min^{-1} m^{-3} and 0.2–0.6 nmol min^{-1} m^{-3} for $PM_{0.18}$ and $PM_{2.5}$, respectively. Mass-normalized DTT activity, indicative of PM intrinsic toxicity, typically ranged from 20–60 nmol min^{-1} mg^{-1} and 20–45 nmol min^{-1} mg^{-1} for $PM_{0.18}$ and $PM_{2.5}$, respectively, at the two study sites. As evident from Fig. 1, within the same size range DTT activity is spatially uniform. Between both sampling sites, a seasonal variability with a 40 and 90% increase in volume-normalized DTT activity in colder months in comparison to warmer months was observed for $PM_{2.5}$ and $PM_{0.18}$, respectively. This increase was also seen in mass-based DTT activity with a 20 and 40% increase in colder months for $PM_{2.5}$ and $PM_{0.18}$, respectively. These trends are in general agreement with the recent study of Saffari *et al.*,[60] who investigated the seasonal variation of $PM_{0.25}$ in the LA Basin. as well as with the study by Verma *et al.*,[30] which also demonstrated higher levels of ambient $PM_{2.5}$ DTT activity in the southeastern USA during colder periods. The elevated levels of volume-normalized DTT activity during these months are mainly attributed to the higher atmospheric stability in addition to enhanced gas-to-particle partitioning of redox active semi-volatile organic compounds.[30,60]

3.3. Regression analysis and source apportionment of DTT

3.3.1. Correlations between DTT activity and PM chemical species. To investigate the association of DTT activity with $PM_{2.5}$ and $PM_{0.18}$ chemical

Table 2 Spearman's correlation coefficients (R) between the dithiothreitol (DTT) activity (nmol min^{-1} m^{-3}) and selected $PM_{2.5}$ and $PM_{0.18}$ components in Central LA and Anaheim. Bold numbers indicate values with $R > 0.50$ and $P < 0.05$. * indicates values with $R < 0.50$ and $P < 0.05$. PAHs, hopanes, *n*-alkanes and organic acids species are presented by their names in Table S1

	Central LA		Anaheim	
Species	$PM_{2.5}$ DTT	$PM_{0.18}$ DTT	$PM_{2.5}$ DTT	$PM_{0.18}$ DTT
EC	**0.54**	**0.67**	**0.72**	**0.86**
OC	**0.79**	**0.89**	**0.81**	**0.87**
PAHs	**0.63**	**0.69**	**0.60**	**0.78**
Hopanes	**0.62**	**0.68**	**0.84**	**0.81**
n-Alkanes	**0.74**	**0.57**	**0.57**	**0.85**
Organic acids	**0.66**	**0.69**	0.38	0.13
Levoglucosan	**0.56**	**0.64**	**0.72**	**0.58**
NO_3^-	0.32	0.35	**0.60**	0.40
SO_4^{2-}	−0.44	−0.11	−0.59	−0.73
NH_4^+	−0.019	−0.299	0.431	0.205
Al	0.21	**0.65**	**0.69**	**0.76**
P	**0.53**	**0.72**	**0.74**	**0.84**
S	−0.37	0.21	−0.49	0.15
K	**0.55**	**0.83**	**0.73**	**0.86**
Ca	0.36	**0.70**	**0.62**	**0.83**
Ti	0.48*	**0.64**	**0.66**	**0.75**
V	−0.20	0.06	−0.39	0.29
Cr	0.29	0.34	**0.64**	0.46*
Mn	0.46*	**0.70**	**0.84**	**0.82**
Fe	**0.54**	**0.75**	**0.81**	**0.83**
Co	0.26	0.45*	**0.59**	**0.69**
Ni	−0.11	0.04	0.34	0.39
Cu	**0.51**	**0.58**	**0.81**	**0.88**
Zn	**0.54**	**0.51**	**0.83**	**0.88**
Rb	**0.64**	**0.79**	**0.70**	**0.78**
Sr	**0.53**	**0.72**	0.48*	**0.62**
Pd	**0.57**	**0.78**	**0.72**	**0.77**
Cd	**0.61**	**0.78**	**0.66**	**0.81**
Sn	0.12	0.22	**0.76**	**0.82**
Sb	**0.60**	**0.75**	**0.66**	**0.84**
Ba	**0.58**	**0.80**	**0.79**	**0.85**
Pb	**0.58**	**0.64**	**0.68**	**0.78**

composition, univariate correlation analysis was performed between the weekly samples of volume-normalized DTT activity and chemical species concentrations at each sampling site separately. Univariate correlation also helped to identify important chemical species that should be considered as input species for the subsequent MLR analysis. Table 2 shows the Spearman's correlation coefficients between the air volume-normalized DTT activity and concentrations of chemical species including EC, OC, PAHs, hopanes, *n*-alkanes, organic acids, levoglucosan and selected elements for $PM_{2.5}$ and $PM_{0.18}$ in Central LA and Anaheim. Most notably, organic compounds showed overall high correlation values with DTT activity in both size ranges ($R > 0.60$), especially in Central LA. Similarly high correlations between the DTT activity and organic compounds such as OC, hopanes, and PAHs have also been reported in previous studies conducted in the LA Basin.[20,60–63] EC, which is a marker of combustion processes mainly from vehicle tailpipe emissions,[64] was also highly correlated with DTT activity at both sampling sites, especially in the $PM_{0.18}$ size range ($R = 0.76$ averaged over Central LA and Anaheim). Verma *et al.*[30] also demonstrated strong correlations between DTT activity and EC in southeastern United States. Overall, high correlations of these specific markers elucidate the underlying toxicity associated with the sources of these tracers. Levoglucosan, a tracer of biomass burning, showed relatively high correlation with DTT activity, with R values of 0.64 and 0.61 for $PM_{2.5}$ and $PM_{0.18}$, respectively, averaged over Central LA and Anaheim. This observation is also in line with the findings of Verma *et al.*[65,66] who also indicated biomass burning as an important source of reactive oxygen species and observed strong correlations between DTT activity and biomass burning characterized by a high concentration of levoglucosan. Some metals such as Ba, Cu, Fe, K, Mn, Pd and Pb were also significantly correlated with DTT activity in both size ranges in Central LA ($R > 0.60$) and Anaheim ($R > 0.70$). A number of previous studies have

Table 3 Output of multiple linear regression (MLR) analysis between the dithiothreitol (DTT) activity (nmol min^{-1} m^{-3}) and chemical species for $PM_{2.5}$ and $PM_{0.18}$ size ranges at Central LA and Anaheim

DTT activity	Species	Unstandardized coefficient	Units	Standard error	Partial R	P value	R^2
Central LA $PM_{2.5}$	(Constant)	0.12	nmol min^{-1} m^{-3}	0.057	—		0.50
	SOC tracers[a]	0.01	nmol min^{-1} ng^{-1}	0.004	0.432	0.022	
	Ba	0.012	nmol min^{-1} ng^{-1}	0.003	0.685	0.000	
Central LA $PM_{0.18}$	(Constant)	0.011	nmol min^{-1} m^{-3}	0.021	—		0.60
	EC	0.155	nmol min^{-1} ng^{-1}	0.076	0.37	0.053	
	Ba	0.014	nmol min^{-1} ng^{-1}	0.003	0.658	0.000	
Anaheim $PM_{2.5}$	(Constant)	0.121	nmol min^{-1} m^{-3}	0.051	—		0.83
	Organic acids	0.002	nmol min^{-1} ng^{-1}	0.001	0.490	0.033	
	Cu	0.023	nmol min^{-1} ng^{-1}	0.003	0.891	0.000	
Anaheim $PM_{0.18}$	(Constant)	0.032	nmol min^{-1} m^{-3}	0.009	—		0.84
	EC	0.136	nmol min^{-1} ng^{-1}	0.065	0.453	0.051	
	Pb	0.056	nmol min^{-1} ng^{-1}	0.022	0.533	0.019	

[a] Sum of phthalic acid, glutaric acid and succinic acid.[39]

also reported strong correlations of these metals with the DTT activity.[63,67–70] Metals such as Ba, Cu, Fe, Mn, Pb and Sr are primarily associated with vehicular abrasion and re-suspension of soil and road dust.[55,71,72]

MLR analysis was conducted between DTT activity and chemical species for both size ranges in Central LA and Anaheim in order to investigate the compounds driving the PM redox activity. The best fitted regression models for both size ranges and sampling sites are presented in Table 3. The optimum model for Central LA in the $PM_{2.5}$ size range included the sum of phthalic, glutaric and succinic acids as the tracers of SOC,[43] as well as Ba, a tracer of vehicular abrasion,[55] leading to an R^2 value of 0.50. EC, an important tracer of vehicular emissions,[64] and Ba for $PM_{0.18}$ were found to be the best fitted species in Central LA with an R^2 value of 0.60. In Anaheim, for the $PM_{2.5}$ size fraction, Cu and the sum of organic acids, indicative of vehicular abrasion[73,74] and SOC sources,[75,76] respectively, best predicted DTT activity ($R^2 = 0.86$). In the $PM_{0.18}$ size range at that site, EC and Pb (an indicator of vehicular abrasion[55,71]), are the main driving species of the DTT activity ($R^2 = 0.71$). It should be mentioned that other tracers of the same source were also examined in the MLR models, and they all led to similar results albeit with somewhat lower R^2 values, with either statistically significant ($P < 0.05$) or approaching significance ($P < 0.08$) contribution to DTT activity. For instance, in Central LA for the $PM_{2.5}$ size fraction, SOC tracers along with other tracers of vehicular abrasion such as Fe and Zn also had similar results with an R^2 of 0.38 and 0.31, respectively. Other combinations of the tracers of vehicle tailpipe emissions and vehicular abrasion in $PM_{0.18}$ at this site were also examined, such as Cu and EC ($R^2 = 0.37$) as well as PAHs and Mn ($R^2 = 0.59$). Similarly, in Anaheim other tracers of vehicular abrasion such as Zn and Ba with SOC tracers showed association with $PM_{2.5}$ DTT activity with an R^2 of 0.81 and 0.78, respectively. PAHs and Zn as other tracers of vehicle tailpipe emissions and vehicular abrasion, respectively, were associated with $PM_{0.18}$ DTT activity with an R^2 of 0.73.

3.3.2. **Correlations between DTT activity and sources.** Univariate analysis was performed to assess how individual sources (*i.e.*, the CMB-derived sources along with secondary ions, crustal material, vehicular abrasion and sea salt) correlate with the DTT activity (Table 4). Wood smoke showed positive significant

Table 4 Spearman's correlation coefficients (R) between the dithiothreitol (DTT) activity (nmol min^{-1} m^{-3}) and sources of $PM_{2.5}$ and $PM_{0.18}$ at Central LA and Anaheim. Bold numbers indicate values with $R > 0.50$ and $P < 0.10$ and * indicates values with $P < 0.05$

	Central LA		Anaheim	
Sources	$PM_{2.5}$ DTT	$PM_{0.18}$ DTT	$PM_{2.5}$ DTT	$PM_{0.18}$ DTT
Wood smoke	**0.56***	**0.61***	**0.55***	**0.57***
Primary biogenic	**0.68***	**0.76***	0.32	0.21
SOC	0.28	−0.14	**0.58***	−0.08
Secondary ions	−0.07	−0.05	0.49*	0.02
Crustal material	0.33	**0.72***	**0.73***	**0.81***
Vehicular abrasion	**0.58***	**0.80***	**0.79***	**0.85***
Sea salt	−0.46	−0.00	−0.32	0.42
Vehicle tailpipe emissions	**0.69***	**0.77***	**0.72***	**0.88***

correlations with DTT activity with R values of 0.56 and 0.61 in Central LA for $PM_{2.5}$ and $PM_{0.18}$, respectively, and similarly, in Anaheim with R values of 0.55 and 0.57 for $PM_{2.5}$ and $PM_{0.18}$, respectively. Vehicle tailpipe emissions also exhibited positive and significant correlations with DTT activity with R values of 0.69 and 0.72 for $PM_{2.5}$ in Central LA and Anaheim, respectively. For $PM_{0.18}$, stronger correlations between DTT activity and vehicle tailpipe emissions were observed in comparison to $PM_{2.5}$ in both Central LA ($R = 0.77$) and Anaheim ($R = 0.88$), underscoring the greater impact of vehicle tailpipe emissions on the DTT activity in this size range. Primary biogenic sources, which include emissions from vegetative detritus, food cooking and re-suspended soil dust[43] significantly correlated with DTT activity in both size ranges in Central LA with an average R value of 0.71. SOC also showed a positive association with the DTT activity in the $PM_{2.5}$ size range with a higher correlation in Anaheim ($R = 0.58$). Secondary ions showed low or negative correlations except for $PM_{2.5}$ in Anaheim. Crustal materials in $PM_{0.18}$ ($R = 0.58$) and vehicular abrasion emissions in both size ranges ($R = 0.73$) also showed statistically significant correlations in both size ranges in Central LA. In Anaheim, remarkably higher R values of 0.77 and 0.82 for crustal materials and vehicular abrasion, respectively, were observed for both size fractions on average. Sea salt, on the other hand, showed low or negative correlations with DTT activity in both size ranges at both sampling sites.

Furthermore, MLR analysis was also performed on DTT activity and the PM sources mentioned above to identify the best and statistically significant ($P < 0.05$)

Table 5 Output of multiple linear regression analysis between the dithiothreitol (DTT) activity (nmol min^{-1} m^{-3}) and sources of $PM_{2.5}$ and $PM_{0.18}$ size ranges at Central LA and Anaheim

DTT activity	Sources	Unstandardized coefficient	Units	Standard error	Partial R	P value	R^2
Central LA $PM_{2.5}$	(Constant)	−0.015	nmol min^{-1} m^{-3}	0.056	—		0.71
	Primary biogenic	0.081	nmol min^{-1} μg^{-1}	0.025	0.582	0.004	
	SOC	0.159	nmol min^{-1} μg^{-1}	0.036	0.697	0.000	
	Vehicular abrasion	0.069	nmol min^{-1} μg^{-1}	0.032	0.425	0.043	
Central LA $PM_{0.18}$	(Constant)	0.012	nmol min^{-1} m^{-3}	0.013	—		0.72
	Vehicle tailpipe emission	0.134	nmol min^{-1} μg^{-1}	0.037	0.630	0.002	
	Vehicular abrasion	0.093	nmol min^{-1} μg^{-1}	0.032	0.539	0.010	
Anaheim $PM_{2.5}$	(Constant)	0.175	nmol min^{-1} m^{-3}	0.031	—		0.81
	SOC	0.081	nmol min^{-1} μg^{-1}	0.034	0.487	0.029	
	Vehicular abrasion	0.181	nmol min^{-1} μg^{-1}	0.032	0.798	0.000	
Anaheim $PM_{0.18}$	(Constant)	0.046	nmol min^{-1} m^{-3}	0.008	—		0.75
	Vehicle tailpipe emission	0.068	nmol min^{-1} μg^{-1}	0.030	0.474	0.040	
	Vehicular abrasion	0.105	nmol min^{-1} μg^{-1}	0.029	0.658	0.002	

predictors of the DTT activity. The details of the output models are presented in Table 5. In Central LA, primary biogenic, SOC and vehicular abrasion sources contributed significantly to $PM_{2.5}$ DTT activity with an R^2 value of 0.71, while in the $PM_{0.18}$ size range, vehicle tailpipe emissions as well as vehicular abrasion were the main drivers of the DTT activity ($R^2 = 0.72$). Similarly, in Anaheim, SOC and vehicular abrasion in the $PM_{2.5}$ size range ($R^2 = 0.81$) and vehicle tailpipe emissions as well as vehicular abrasion in $PM_{0.18}$ ($R^2 = 0.75$) were identified to be the significant contributors of DTT activity.

3.4. Comparison with previous studies in Los Angeles

Table 6(a) and (b) presents a comparison of DTT activity measured in the current study with previous studies conducted in Central LA, all at the same sampling site, over the past decade for fine and ultrafine particles. Overall a small increase in the per PM mass-normalized DTT activity of ambient $PM_{2.5}$ can be seen from 2002 to 2012. The DTT activity normalized per m^3 of air volume has been more stable over time in Central LA, which may be attributed to an increase in intrinsic DTT activity, combined with the documented decrease in $PM_{2.5}$ mass concentration over time in Central LA, as discussed in following sections. Unlike $PM_{2.5}$, in the ultrafine size range, the cut point was not consistent among different studies. Nonetheless, both air volume and PM mass-based DTT activity have generally decreased over the past decade in Central LA. Verma *et al.*[63] reported

Table 6 Comparison of dithiothreitol (DTT) activity levels (±standard deviation) with previous studies conducted at Central Los Angeles for: (a) $PM_{2.5}$ and (b) $PM_{0.18}$ size ranges

(a) $PM_{2.5}$

Study	Size fraction	Sampling period	DTT activity (nmol min^{-1} μg^{-1})	DTT activity (nmol min^{-1} m^{-3})
Li *et al.* 2003 (ref. 89)	$PM_{2.5}$	Mar 2002	0.013	0.28
Hu *et al.* 2008 (ref. 87)	$PM_{2.5}$	Mar–May 2007	0.022	0.33
Verma *et al.* 2009 (ref. 70)	$PM_{2.5}$	Nov 2007	0.007 ± 0.003	0.35 ± 0.30
Current study	$PM_{2.5}$	2012–2013	0.028 ± 0.005	0.35 ± 0.04

(b) $PM_{0.18}$

Study	Size fraction	Sampling period	DTT activity (nmol min^{-1} μg^{-1})	DTT activity (nmol min^{-1} m^{-3})
Li *et al.* 2003 (ref. 89)	$PM_{0.15}$	Mar 2002	0.091	0.35
Verma *et al.* 2009 (ref. 63)	$PM_{0.18}$	Aug 2009	0.07 ± 0.03	0.33 ± 0.25
Saffari *et al.* 2014 (ref. 60)	$PM_{0.25}$	2008–2009	0.078 ± 0.007	0.82 ± 0.12
Current study	$PM_{0.18}$	2012–2013	0.045 ± 0.008	0.11 ± 0.03

$PM_{0.18}$ DTT activity of 0.07 ± 0.03 nmol min^{-1} μg^{-1} in June–August 2009, while three years later, this value decreased by approximately 66% to 0.024 ± 0.001 nmol min^{-1} μg^{-1} in July–August in Central LA. Even though Saffari *et al.*[60] reported $PM_{0.25}$ DTT activity levels, still a 42% decrease can be seen in comparison to $PM_{0.18}$ DTT levels measured at this study after four years.

Several studies have shown a major reduction in ambient $PM_{2.5}$ levels and its sources over the past decade in the Los Angeles Basin and demonstrated that stringent regulations on mobile sources, in particular from 2007, have resulted in major reductions in tailpipe emissions.[48,77–80] However, the small but consistent increase in $PM_{2.5}$ mass-based DTT activity level reveals that other factors besides tailpipe emissions may affect the $PM_{2.5}$ toxicity as well. The results discussed in the previous section revealed that vehicular abrasion, an important component of non-tailpipe emissions, together with secondary OC, were two major sources significantly contributing to $PM_{2.5}$ DTT activity, in addition to tailpipe emissions. Shirmohammadi *et al.*[31] recently showed that important tracers of tailpipe emissions in $PM_{2.5}$ and $PM_{0.18}$ such as PAH, hopane and sterane concentrations have decreased by 40–70% and 50–70% over the past decade in Central LA. These findings, along with the comparison of ultrafine PM redox activity with previous studies, corroborate the effectiveness of stringent regulations on vehicular tailpipe emissions in reducing the overall ambient ultrafine PM toxicity. However, tailpipe emissions are not the only contributor of PM toxicity in fine PM, as indicated and discussed above. Harrison *et al.*[81] showed that tracers of vehicular abrasion, which are generally more prevalent in the coarse ($PM_{2.5-10}$) size fraction, also partition into the upper size range of $PM_{2.5}$. Similar findings have been reported in other studies as well.[82,83] Recently, Shirmohammadi *et al.*[84] discussed the increase in metal and trace element concentrations in coarse PM over a 3 year period in Central LA, especially those emitted from vehicular abrasion and road dust. This was attributed to the increase in the vehicle's speed and number of trucks by 6 and 15%, respectively, which was also statistically significant (Mann–

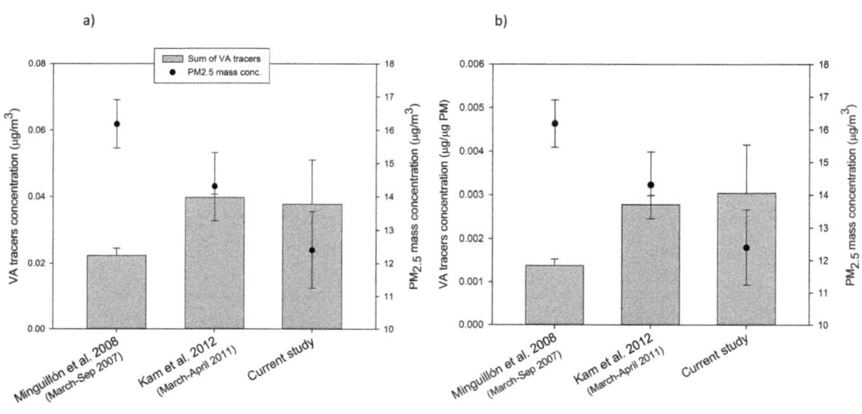

Fig. 3 Comparison of the sum of vehicular abrasion (VA) tracer concentrations (*i.e.* sum of Ba, Cu, Mn, Pb and Sr) (a) per volume of air ($\mu g\ m^{-3}$) and (b) per mass collected (µg per µg PM) in $PM_{2.5}$ size range with previous studies conducted at Central LA.[86,88] Error bars correspond to one standard deviation. The dates in the parentheses refer to the pertinent sampling dates.

Whitney rank sum test, $P < 0.001$) as the turbulence caused by the passing traffic highly contributes to a substantial amount of coarse particle emissions from re-suspension of soil and road dust,[85] in Central LA. Moreover, comparison with several previous studies conducted in Central LA also revealed a similar trend, with an increase in the contribution of vehicular abrasion tracers in $PM_{2.5}$ mass over time, unlike the $PM_{2.5}$ mass concentration decrease after implemented regulations on vehicle tailpipe emissions (Fig. 3a and b). For instance, as can be inferred from Fig. 3b, compared to a study conducted by Minguillón *et al.*[86] during a 13 week period between March–September 2007 (except June) in the same sampling site in Central LA, $PM_{2.5}$ mass concentration decreased by 24% in this study, while the contribution of the sum of VA tracers (*i.e.* Ba, Cu, Mn, Pb and Sr) to the $PM_{2.5}$ mass increased by a factor of nearly 2.2. In line with this comparison, the annual trends of individual VA tracers from 2008 to 2012, obtained from the Speciation Trends Network (STN) data at North Main street in downtown Los Angeles, showed that, although the per air volume concentrations of these species were relatively stable from 2008 to 2012 (Fig. S1(a)–(e)†), an increasing trend in the contribution of these tracers to $PM_{2.5}$ mass concentration is evident from their year-to-year per mass concentration levels (Fig. S2(a)–(e)†). The median value of the sum of per PM mass concentrations (μg per μg PM) of the VA tracers (*i.e.* Ba, Cu, Mn, Pb and Sr) increased statistically significantly by 11% from 2008 to 2012 (Mann–Whitney rank sum test, $P = 0.05$). This increase over these years can also be seen for individual species. For example, the per mass concentration of Ba (which was also used as the basis of our VA contribution estimation in the previous section) increased statistically significantly by 66% from 2008 to 2012 (Mann–Whitney rank sum test, $P < 0.01$). More information regarding the STN data is provided in the ESI.† Therefore, it can be hypothesized that the per PM mass increase in redox active metals and other trace elements from non-tailpipe emissions may counteract the reduction in vehicular tailpipe emissions in $PM_{2.5}$, leading to a slight increase in the overall DTT activity in this size fraction of PM with time. This is in contrast to the ultrafine PM, the toxicity of which seems to be largely dominated by tailpipe emissions, and appears to be decreasing with time, as shown in Table 6(b).

4. Conclusions

In this study, the oxidative potential of ambient $PM_{2.5}$ and $PM_{0.18}$, determined by the DTT assay, was linked to PM sources and chemical components in the LA Basin. Univariate analysis between chemical species and DTT activity suggested the potential influence of different sources and species on abiotic PM-induced toxicity measured by the DTT assay. MLR analysis performed on ambient PM sources in Central LA and Anaheim in both size ranges revealed the contribution from different sources to the oxidative potential of PM. Primary biogenic emissions, SOC and vehicular abrasion in the $PM_{2.5}$ size range, along with vehicular abrasion and vehicle tailpipe emissions in the $PM_{0.18}$ size range were found to be the major sources driving the DTT activity levels in Central LA. MLR yielded similar results for Anaheim, with higher contributions from vehicular abrasion and SOC in the $PM_{2.5}$ size range. Vehicular abrasion, followed by vehicle tailpipe emissions, were the two significant contributors to $PM_{0.18}$ DTT activity in Anaheim.

A comparison with previous studies conducted in Central LA revealed trends in DTT activity of ambient $PM_{2.5}$ and $PM_{0.18}$ in the past decade. Ambient $PM_{0.18}$, the toxicity of which was found to be mainly dominated by tailpipe emissions, showed a consistent decrease in DTT activity levels in the past decade, likely due to major reductions in tailpipe emissions in the LA Basin, as a result of stringent regulations on mobile sources. In the $PM_{2.5}$ size range, however, DTT activity levels showed overall a slight increase over the years, which is probably driven by the increase in the contribution of non-tailpipe emissions to ambient $PM_{2.5}$, counteracting reductions in the tailpipe emissions in this size fraction. Our findings illustrated the relative importance of different traffic sources on the oxidative potential of ambient PM. Despite rapid reduction of tailpipe emissions, the lack of similar reductions (and possibly an increase) in non-tailpipe emissions makes them an important source of traffic-related PM emissions in Los Angeles and a concern for public health.

Acknowledgements

The present project was supported by grant number ES12243 from the National Institute of Environmental Health Sciences, U.S. National Institutes of Health. The authors wish to thank the staff of the Wisconsin State Laboratory of Hygiene (WSLH) for their assistance with the chemical analysis. We also acknowledge the support of USC's Provost and Viterbi PhD fellowships.

References

1 A. Baccarelli, R. O. Wright, V. Bollati, L. Tarantini, A. A. Litonjua, H. H. Suh, A. Zanobetti, D. Sparrow, P. S. Vokonas and J. Schwartz, *Am. J. Respir. Crit. Care Med.*, 2009, **179**, 572–578.

2 R. D. Brook, R. L. Bard, R. T. Burnett, H. H. Shin, A. Vette, C. Croghan, M. Phillips, C. Rodes, J. Thornburg and R. Williams, *Occup. Environ. Med.*, 2010, DOI: 10.1136/oem.2009.053991.

3 R. T. Burnett, C. A. Pope III, M. Ezzati, C. Olives, S. S. Lim, S. Mehta, H. H. Shin, G. Singh, B. Hubbell, M. Brauer, H. R. Anderson, K. R. Smith, J. R. Balmes, N. G. Bruce, H. Kan, F. Laden, A. Prüss-Ustün, M. C. Turner, S. M. Gapstur, W. R. Diver and A. Cohen, *Environ. Health Perspect.*, 2014, **122**(4), 397.

4 W. J. Gauderman, H. Vora, R. McConnell, K. Berhane, F. Gilliland, D. Thomas, F. Lurmann, E. Avol, N. Kunzli, M. Jerrett and J. Peters, *Lancet*, 2007, **369**, 571–577.

5 R. Li, K. Navab, G. Hough, N. Daher, M. Zhang, D. Mittelstein, K. Lee, P. Pakbin, A. Saffari, M. Bhetraratana, D. Sulaiman, T. Beebe, L. Wu, N. Jen, E. Wine, C.-H. Tseng, J. A. Araujo, A. Fogelman, C. Sioutas, M. Navab and T. K. Hsiai, *Environ. Health Perspect.*, 2015, **123**, 34–41.

6 R. J. Delfino, N. Staimer, T. Tjoa, D. L. Gillen, J. J. Schauer and M. M. Shafer, *J. Exposure Sci. Environ. Epidemiol.*, 2013, **23**, 466–473.

7 C. A. Pope III and D. W. Dockery, *J. Air Waste Manage. Assoc.*, 2006, **56**, 709–742.

8 N. Li, T. Xia and A. E. Nel, *Free Radical Biol. Med.*, 2008, **44**, 1689–1699.

9 M. Strak, N. A. Janssen, K. J. Godri, I. Gosens, I. S. Mudway, F. R. Cassee, E. Lebret, F. J. Kelly, R. M. Harrison, B. Brunekreef, M. Steenhof and G. Hoek, *Environ. Health Perspect.*, 2012, **120**, 1183–1189.

10 L. Castro and B. A. Freeman, *Nutr. J.*, 2001, **17**(161), 163–165.
11 F. Tao, B. Gonzalez-Flecha and L. Kobzik, *Free Radical Biol. Med.*, 2003, **35**, 327–340.
12 K. Donaldson, V. Stone, P. J. A. Borm, L. A. Jimenez, P. S. Gilmour, R. P. F. Schins, A. M. Knaapen, I. Rahman, S. P. Faux, D. M. Brown and W. MacNee, *Free Radical Biol. Med.*, 2003, **34**, 1369–1382.
13 D. C. Chalupa, P. E. Morrow, G. Oberdörster, M. J. Utell and M. W. Frampton, *Environ. Health Perspect.*, 2004, **112**, 879–882.
14 R. J. Delfino, C. Sioutas and S. Malik, *Environ. Health Perspect.*, 2005, **113**, 934–946.
15 J. A. Araujo, B. Barajas, M. Kleinman, X. Wang, B. J. Bennett, K. W. Gong, M. Navab, J. Harkema, C. Sioutas, A. J. Lusis and A. E. Nel, *Circ. Res.*, 2008, **102**, 589–596.
16 R. J. Delfino, D. L. Gillen, T. Tjoa, N. Staimer, A. Polidori, M. Arhami, C. Sioutas and J. Longhurst, *Environ. Health Perspect.*, 2010, **119**, 196–202.
17 R. J. Delfino, N. Staimer, T. Tjoa, M. Arhami, A. Polidori, D. L. Gillen, M. T. Kleinman, J. J. Schauer and C. Sioutas, *Environ. Health Perspect.*, 2010, **118**, 756–762.
18 M. M. Shafer, D. A. Perkins, D. S. Antkiewicz, E. A. Stone, T. A. Quraishi and J. J. Schauer, *J. Environ. Monit.*, 2010, **12**, 704–715.
19 Y. Zhang, J. J. Schauer, M. M. Shafer, M. P. Hannigan and S. J. Dutton, *Environ. Sci. Technol.*, 2008, **42**, 7502–7509.
20 A. K. Cho, C. Sioutas, A. H. Miguel, Y. Kumagai, D. A. Schmitz, M. Singh, A. Eiguren-Fernandez and J. R. Froines, *Environ. Res.*, 2005, **99**, 40–47.
21 M. Y. Chung, R. A. Lazaro, D. Lim, J. Jackson, J. Lyon, D. Rendulic and A. S. Hasson, *Environ. Sci. Technol.*, 2006, **40**, 4880–4886.
22 Y. Kumagai, S. Koide, K. Taguchi, A. Endo, Y. Nakai, T. Yoshikawa and N. Shimojo, *Chem. Res. Toxicol.*, 2002, **15**, 483–489.
23 Y. Kumagai, T. Arimoto, M. Shinyashiki, N. Shimojo, Y. Nakai, T. Yoshikawa and M. Sagai, *Free Radical Biol. Med.*, 1997, **22**, 479–487.
24 V. Verma, Y. Wang, R. El-Afifi, T. Fang, J. Rowland, A. G. Russell and R. J. Weber, *Atmos. Environ.*, 2015, **120**, 351–359.
25 J. G. Charrier and C. Anastasio, *Atmos. Environ.*, 2011, **45**, 7555–7562.
26 Q. Li, A. Wyatt and R. M. Kamens, *Atmos. Environ.*, 2009, **43**, 1037–1042.
27 M. A. A. Schoonen, C. A. Cohn, E. Roemer, R. Laffers, S. R. Simon and T. O'Riordan, *Rev. Mineral. Geochem.*, 2006, **64**, 179–221.
28 J. G. Charrier, N. K. Richards-Henderson, K. J. Bein, A. S. McFall, A. S. Wexler and C. Anastasio, *Atmos. Chem. Phys.*, 2015, **15**, 2327–2340.
29 R. Lall, K. Ito and G. D. Thurston, *Environ. Health Perspect.*, 2010, **119**, 455–460.
30 V. Verma, T. Fang, H. Guo, L. King, J. T. Bates, R. E. Peltier, E. Edgerton, A. G. Russell and R. J. Weber, *Atmos. Chem. Phys.*, 2014, **14**, 12915–12930.
31 F. Shirmohammadi, S. Hasheminassab, A. Saffari, J. J. Schauer, R. J. Delfino and C. Sioutas, *Sci. Total Environ.*, 2015, **541**, 1083–1096.
32 J. J. Schauer, G. C. Lough, M. M. Shafer, W. F. Christensen, M. F. Arndt, J. T. DeMinter and J. S. Park, *Res. Rep.-Health Eff. Inst.*, 2006, 1–76, discussion 77–88.
33 C. Hueglin, R. Gehrig, U. Baltensperger, M. Gysel, C. Monn and H. Vonmont, *Atmos. Environ.*, 2005, **39**, 637–651.

34 G. M. Marcazzan, S. Vaccaro, G. Valli and R. Vecchi, *Atmos. Environ.*, 2001, **35**, 4639–4650.
35 D. M. Murphy, J. R. Anderson, P. K. Quinn, L. M. McInnes, F. J. Brechtel, S. M. Kreidenweis, A. M. Middlebrook, M. Pósfai, D. S. Thomson and P. R. Buseck, *Nature*, 1998, **392**, 62–65.
36 M. E. Birch and R. A. Cary, *Aerosol Sci. Technol.*, 1996, **25**, 221–241.
37 E. A. Stone, D. C. Snyder, R. J. Sheesley, A. P. Sullivan, R. J. Weber and J. J. Schauer, *Atmos. Chem. Phys.*, 2008, **8**, 1249–1259.
38 Y. Zhang, J. J. Schauer, M. M. Shafer, M. P. Hannigan and S. J. Dutton, *Environ. Sci. Technol.*, 2008, **42**, 7502–7509.
39 B. Dellinger, W. A. Pryor, R. Cueto, G. L. Squadrito, V. Hegde and W. A. Deutsch, *Chem. Res. Toxicol.*, 2001, **14**, 1371–1377.
40 P. J. O'Brien, *Chem.-Biol. Interact.*, 1991, **80**, 1–41.
41 G. L. Squadrito, R. Cueto, B. Dellinger and W. A. Pryor, *Free Radical Biol. Med.*, 2001, **31**, 1132–1138.
42 N. Li, M. Wang, L. A. Bramble, D. A. Schmitz, J. J. Schauer, C. Sioutas, J. R. Harkema and A. E. Nel, *Environ. Health Perspect.*, 2009, **117**, 1116–1123.
43 J. Heo, M. Dulger, M. R. Olson, J. E. McGinnis, B. R. Shelton, A. Matsunaga, C. Sioutas and J. J. Schauer, *Atmos. Environ.*, 2013, **73**, 51–61.
44 P. M. Fine, G. R. Cass and B. R. T. Simoneit, *Environ. Eng. Sci.*, 2004, **21**, 387–409.
45 G. C. Lough, C. G. Christensen, J. J. Schauer, J. Tortorelli, E. Mani, D. R. Lawson, N. N. Clark and P. A. Gabele, *J. Air Waste Manage. Assoc.*, 2007, **57**, 1190–1199.
46 B. J. Turpin and H.-J. Lim, *Aerosol Sci. Technol.*, 2001, **35**, 602–610.
47 B. Mason, *Principles of Geochemistry*, New York, Wiley, 3rd edn, 1966.
48 S. Hasheminassab, N. Daher, B. D. Ostro and C. Sioutas, *Environ. Pollut.*, 2014, **193**, 54–64.
49 R. F. Narváez, L. Hoepner, S. N. Chillrud, B. Yan, R. Garfinkel, R. Whyatt, D. Camann, F. P. Perera, P. L. Kinney and R. L. Miller, *Environ. Sci. Technol.*, 2008, **42**, 7330–7335.
50 P. Pant and R. M. Harrison, *Atmos. Environ.*, 2013, **77**, 78–97.
51 A. Dahl, A. Gharibi, E. Swietlicki, A. Gudmundsson, M. Bohgard, A. Ljungman, G. Blomqvist and M. Gustafsson, *Atmos. Environ.*, 2006, **40**, 1314–1323.
52 M. Gustafsson, G. Blomqvist, A. Gudmundsson, A. Dahl, E. Swietlicki, M. Bohgard, J. Lindbom and A. Ljungman, *Sci. Total Environ.*, 2008, **393**, 226–240.
53 A. Saffari, N. Daher, M. M. Shafer, J. J. Schauer and C. Sioutas, *Environ. Pollut.*, 2013, **181**, 14–23.
54 P. Sanderson, J. M. Delgado-Saborit and R. M. Harrison, *Atmos. Environ.*, 2014, **94**, 353–365.
55 A. Thorpe and R. M. Harrison, *Sci. Total Environ.*, 2008, **400**, 270–282.
56 P. Wåhlin, R. Berkowicz and F. Palmgren, *Atmos. Environ.*, 2006, **40**, 2151–2159.
57 L. M. Hildemann, G. R. Markowski and G. R. Cass, *Environ. Sci. Technol.*, 1991, **25**, 744–759.
58 H. Kumata, J. Yamada, K. Masuda, H. Takada, Y. Sato, T. Sakurai and K. Fujiwara, *Environ. Sci. Technol.*, 2002, **36**, 702–708.
59 C. M. Reddy and J. G. Quinn, *Environ. Sci. Technol.*, 1997, **31**, 2847–2853.

60 A. Saffari, N. Daher, M. M. Shafer, J. J. Schauer and C. Sioutas, *J. Environ. Sci. Health, Part A: Toxic/Hazard. Subst. Environ. Eng.*, 2014, **49**, 441–451.

61 K. L. Cheung, A. Polidori, L. Ntziachristos, T. Tzamkiozis, Z. Samaras, F. R. Cassee, M. Gerlofs and C. Sioutas, *Environ. Sci. Technol.*, 2009, **43**, 6334–6340.

62 M. D. Geller, L. Ntziachristos, A. Mamakos, Z. Samaras, D. A. Schmitz, J. R. Froines and C. Sioutas, *Atmos. Environ.*, 2006, **40**, 6988–7004.

63 V. Verma, Z. Ning, A. K. Cho, J. J. Schauer, M. M. Shafer and C. Sioutas, *Atmos. Environ.*, 2009, **43**, 6360–6368.

64 J. J. Schauer, *J. Exposure Sci. Environ. Epidemiol.*, 2003, **13**, 443–453.

65 V. Verma, R. J. J. Weber, T. Fang, L. Xu, N. L. Ng and A. G. Russell, *AGU Fall Meeting*, 2014, **53**, 3204.

66 V. Verma, T. Fang, L. Xu, R. E. Peltier, A. G. Russell, N. L. Ng and R. J. Weber, *Environ. Sci. Technol.*, 2015, **49**, 4646–4656.

67 J. G. Charrier and C. Anastasio, *Atmos. Chem. Phys. Discuss.*, 2012, **12**, 11317–11350.

68 T. Fang, H. Guo, V. Verma, R. E. Peltier and R. J. Weber, *Atmos. Chem. Phys. Discuss.*, 2015, **15**, 17189–17227.

69 L. Ntziachristos, J. R. Froines, A. K. Cho and C. Sioutas, *Part. Fibre Toxicol.*, 2007, **4**, 5.

70 V. Verma, A. Polidori, J. J. Schauer, M. M. Shafer, F. R. Cassee and C. Sioutas, *Environ. Sci. Technol.*, 2009, **43**, 954–960.

71 F. Amato, M. Pandolfi, T. Moreno, M. Furger, J. Pey, A. Alastuey, N. Bukowiecki, A. S. H. Prevot, U. Baltensperger and X. Querol, *Atmos. Environ.*, 2011, **45**, 6777–6787.

72 P. Pakbin, Z. Ning, M. M. Shafer, J. J. Schauer and C. Sioutas, *Aerosol Sci. Technol.*, 2011, **45**, 949–963.

73 A. Iijima, K. Sato, K. Yano, H. Tago, M. Kato, H. Kimura and N. Furuta, *Atmos. Environ.*, 2007, **41**, 4908–4919.

74 P. G. Sanders, N. Xu, T. M. Dalka and M. M. Maricq, *Environ. Sci. Technol.*, 2003, **37**, 4060–4069.

75 X. Ding, M. Zheng, L. Yu, X. Zhang, R. J. Weber, B. Yan, A. G. Russell, E. S. Edgerton and X. Wang, *Environ. Sci. Technol.*, 2008, **42**, 5171–5176.

76 R. Fisseha, J. Dommen, M. Sax, D. Paulsen, M. Kalberer, R. Maurer, F. Höfler, E. Weingartner and U. Baltensperger, *Anal. Chem.*, 2004, **76**, 6535–6540.

77 G. A. Bishop, B. G. Schuchmann and D. H. Stedman, *Environ. Sci. Technol.*, 2013, **47**, 9523–9529.

78 F. Lurmann, E. Avol and F. Gilliland, *J. Air Waste Manage. Assoc.*, 2015, **65**, 324–335.

79 B. C. McDonald, A. H. Goldstein and R. A. Harley, *Environ. Sci. Technol.*, 2015, **49**, 5178–5188.

80 C. V. Preble, T. R. Dallmann, N. M. Kreisberg, S. V. Hering, R. A. Harley and T. W. Kirchstetter, *Environ. Sci. Technol.*, 2015, **49**, 8864–8871.

81 R. M. Harrison, A. M. Jones, J. Gietl, J. Yin and D. C. Green, *Environ. Sci. Technol.*, 2012, **46**, 6523–6529.

82 M. D. Geller, P. M. Fine and C. Sioutas, *J. Air Waste Manage. Assoc.*, 2004, **54**, 1029–1039.

83 B. J. Majestic, J. J. Schauer, M. M. Shafer, P. M. Fine, M. Singh and C. Sioutas, *J. Environ. Eng. Sci.*, 2008, **7**, 289–298.

84 F. Shirmohammadi, S. Hasheminassab, D. Wang, A. Saffari, J. J. Schauer, M. M. Shafer, R. J. Delfino and C. Sioutas, *Environ. Sci.: Processes Impacts*, 2015, **17**, 2110–2121.
85 R. M. Harrison, J. Yin, D. Mark, J. Stedman, R. S. Appleby, J. Booker and S. Moorcroft, *Atmos. Environ.*, 2001, **35**, 3667–3679.
86 M. C. Minguillón, M. Arhami, J. J. Schauer and C. Sioutas, *Atmos. Environ.*, 2008, **42**, 7317–7328.
87 S. Hu, A. Polidori, M. Arhami, M. M. Shafer, J. J. Schauer, A. Cho and C. Sioutas, *Atmos. Chem. Phys.*, 2008, **8**, 6439–6451.
88 W. Kam, J. W. Liacos, J. J. Schauer, R. J. Delfino and C. Sioutas, *Atmos. Environ.*, 2012, **55**, 90–97.
89 N. Li, C. Sioutas, A. Cho, D. Schmitz, C. Misra, J. Sempf, M. Wang, T. Oberley, J. Froines and A. Nel, *Environ. Health Perspect.*, 2003, **111**, 455–460.

Faraday Discussions

PAPER

Oxidative potential of size-fractionated atmospheric aerosol in urban and rural sites across Europe†

Martin M. Shafer,*[ab] Jocelyn D. C. Hemming,[b] Dagmara S. Antkiewicz[b] and James J. Schauer[a]

Received 26th November 2015, Accepted 4th January 2016
DOI: 10.1039/c5fd00196j

In this study we applied several assays, an *in vitro* rat alveolar macrophage model, a chemical ROS probe (DTT, dithiothreitol), and cytokine induction (TNFα) to examine relationships between PM-induced generation of reactive oxygen species (ROS) and PM composition, using a unique set of size-resolved PM samples obtained from urban and rural environments across Europe. From April–July 2012, we collected PM from roadside canyon, roadside motorway, and background urban sites in each of six European cities and from three rural sites spanning the continent. A Hi-Vol sampler was used to collect PM in three size classes ($PM_{>7}$, PM_{7-3}, PM_3) and PM was characterized for total elements, and oxidative activity quantified in unfiltered and filtered PM extracts. We measured a remarkable uniformity in air concentrations of ROS and especially DTT activity across the continent. Only a 4-fold difference was documented for DTT across the urban sites and a similar variance was documented for ROS, implying that chemical drivers of oxidative activity are relatively similar between sites. The ROS and DTT specific activity was greater at urban background sites (and also rural sites) than at urban canyon locations. PM_3 dominated the size distribution of both ROS activity (86% of total) and DTT activity (76% of total), reflecting both the large contribution of PM_3 to total PM mass levels and importantly the higher specific oxidative activity of the PM_3 in comparison with the larger particles. The soluble fraction of total activity was very high for DTT (94%) as well as for ROS (64%) in the PM_3. However in the larger PM size fractions the contributions of the insoluble components became increasingly significant. The dominance of the insoluble PM drivers of activity was particularly evident in the TNFα data, where the insoluble contribution to cytokine production could be 100-fold greater than that from soluble components. ROS and DTT activity were strongly correlated in the PM_3 ($r = 0.93$), however oxidative activity was not correlated with any measured inorganic element in this size cut. In contrast, significant correlations of both

[a]*Environmental Chemistry and Technology Program, University of Wisconsin-Madison, 660 N. Park St., Madison, WI 53706, USA. E-mail: mmshafer@wisc.edu*
[b]*Wisconsin State Laboratory of Hygiene, 2601 Agriculture Drive, Madison, WI 53718, USA*

† Electronic supplementary information (ESI) available. See DOI: 10.1039/c5fd00196j

ROS and DTT oxidative activity with specific groups of chemical elements were documented in the larger PM size fractions.

Introduction

Adverse health effects of ambient air particulate matter (PM) include respiratory and cardiovascular diseases, as well as cancer, contributing to an estimated 3.2 million premature deaths annually worldwide.[1–3] The effect of near roadway air pollution on human health is especially significant in major cities, with respiratory and cardiovascular diseases being implicated in response to prolonged exposure to PM.[4–8] Current regulations aiming to limit roadway-related air pollution have focused on reducing tailpipe emissions, yet recent reports suggest that all major components of PM, including metals and trace elements can potentially contribute to toxicity.[9–11] However, tailpipe emissions from mobile sources are not the major contributor to those metals as their roadway emissions are largely associated with brake wear, tire wear, and re-suspension of road dust.[12] Important metals from brake wear include Fe, Cu, Mn, Sr and Ba, while the re-suspended road dust is an important source of Pb, V and Cr, as well as crustal elements.[13–15]

Oxidative stress due to excess reactive oxygen species (ROS) production is hypothesized to be one of the unifying mechanisms underlying the adverse health effects associated with exposure to ambient PM.[16] PM has been shown to generate ROS including H_2O_2, hydroxyl and superoxide radicals and to activate redox-sensitive signaling pathways including the mitogen-activated protein kinases (MAPKs), the nuclear factor kappa B (NFκB) cascade, and the antioxidant responsive element (ARE) systems.[17–19] The downstream effects of these redox-sensitive pathways include inflammatory responses such as cytokine production.[20]

Alveolar macrophages are a part of the innate immune system function and a first line of defense against particles and pathogens entering the lungs. They are equipped with a robust NADPH oxidase system (NOX_2) that generates ROS as means of defense against invading pathogens (process of "respiratory burst").[21,22] Macrophages have been shown to produce ROS in response to PM exposure in a dose-dependent fashion.[23,24] These cells are also particularly well suited to study the etiology of respiratory disease due to their instrumental role in mediating inflammatory response in the lungs. Multiple studies have demonstrated various macrophage cell lines to respond to PM exposure with a release of pro-inflammatory signals.[25–29] Recent studies reported that PM-derived water-soluble transition metals (*e.g.* Fe, Ni, Cu, Cr, Mn, Zn and V) correlate significantly with the oxidative potential of airborne PM across different urban areas and size ranges.[24,30] In contrast, little is known about the components of PM that are the main drivers of inflammatory signaling induction. In this study we apply several assays, an *in vitro* rat alveolar macrophage model, a chemical ROS probe (DTT), and inflammatory cytokine measure (TNFα) to examine relationships between PM-induced generation of reactive oxygen species (ROS) and PM composition, using a unique set of size-resolved PM samples obtained from urban and rural environments across Europe. We emphasize in this study geographic, site type, and particle-size contrasts in aerosol oxidative properties.

Experimental

Sample collection

PM samples were obtained from several sites in each of six geographically and climatically diverse major European cities (Amsterdam, Frankfurt, London, Milan, Stockholm and Thessaloniki) between April and July of 2012. Within each city samples were collected from sites representing roadside canyons, roadside motorways and background urban locations. We also placed samplers at several rural locations in Northern, Central and Southern Europe to characterize the regional background aerosol oxidative activity. The Northern site was located near the North Sea and thus had a marine influence as well as capturing background from Scandinavia; the Central site was located in rural central Germany and was positioned to capture continental background; and the Southern site was located on the Southern slope of the Alps in Northern Italy and primarily reflected aerosols from the Mediterranean region. Samples were collected using Tisch (TE-230) Hi-Vol Environmental Impactor Samplers configured for three size-cuts, and fitted with pre-cleaned mixed-cellulose ester (MCE) substrates. Samplers were operated continuously for 3–4 days at nominal flow rates of 1.2 m^3 min^{-1}, sampling 5000 m^3 of air, and collecting from 50–300 mg of PM onto the MCE filter media. Three particle size cuts were collected: a super coarse fraction including all total suspended particulate matter (TSP) with aerodynamic diameters greater than 7 microns ($PM_{<7}$), a coarse fraction containing particles with aerodynamic diameters between 3 and 7 microns (PM_{7-3}), and a fine fraction with particles less than 3 microns (PM_3).

Sample chemical characterization

The samples, along with an extensive set of field blanks, were analysed with a broad range of characterization tools including magnetic-sector ICPMS (SF-ICPMS) for 50 elements; ion chromatography (Dionex LC20 IC) for soluble ions (K^+, Na^+, NH_4^+, SO_4^{2-}, NO_3^-, Cl^-); and several toxicity assays. Gravimetric mass was determined using microbalances operated in a temperature and humidity regulated weighing room. Static was controlled by both electronic ionizers and polonium ionization sources. Filters were sectioned in a trace element clean laboratory under HEPA hoods using ceramic tools. Each section was weighed and dedicated to a specific analysis protocol. Total elements were determined by magnetic sector ICPMS (Thermo-Finnegan Element 2) after complete solubilization of the filter-collected PM using a microwave-aided mixed acid digestion. Quantitative recovery of over 50 elements was verified through inclusion of several certified reference materials in each analytical batch. Details of these methods are given in Okuda *et al.*, 2014;[31] Shafer *et al.*, 2012;[32] and Hu *et al.*, 2009.[33]

PM extract preparation

Sections of the MCE filter samples were extracted with high-purity Milli-Q (18 mΩ) water using an initial sonication period of 15 min followed by 16 h of continuous agitation at room temperature in the dark and another 15 min sonication, followed by 1 min of agitation by a vortex mixer. An aliquot of sample suspension (referred to here as "Unfiltered" samples) was subsequently portioned and distributed for various assays/analyses and then processed through 0.22 μm polypropylene (referred to here as "Filtered" samples) syringe filters. Filtration at 0.22 μm provides an operationally defined water soluble fraction, however the potential inclusion of colloidal material in the filtrate precludes defining this fraction as containing truly dissolved species. All samples were buffered with Salts Glucose Media (SGM) prior to use in the *in vitro* assays.

DTT

The DTT assay was used to measure the oxidative potential of particles and has frequently been applied to atmospheric PM and major PM sources. It is a purely chemical method, no cells are involved, and in the assay redox-active chemicals in the PM oxidize added DTT to its disulfide form and the linear rate of DTT loss is equated to the oxidative activity of the PM. In the process an electron transfer from these redox active species to molecular oxygen occurs, forming superoxide which can react to form other ROS species (*e.g.* hydrogen peroxide and hydroxyl radical).[34] This ROS cascade is similar in many respects to the oxidative burst initiated by superoxide formation in the mitochondria of cells. The technique is responsive to redox-active organic compounds (*e.g.* quinones) and to many metals; however the metals are significantly less efficient at oxidizing DTT compared to reactive quinones. The method used for this study was based on Cho *et al.* 2005.[35] Up to four dilutions of PM suspensions from water extracts were added to 1.2 mL cluster tubes with volume totaling 445 μL. Samples were allowed to equilibrate in a 37 °C water bath prior to initiating the reaction. The reaction was initiated by adding 55 μL of a mixture of 1 M KPO_4 buffer (pH 7.4) plus 10 mM DTT mixture, for a final concentration of 0.1 M and 0.1 mM, respectively. This incubation mixture (IM) was immediately vortexed and 25 μL per well of IM (in triplicate) was removed and added to a 96-well plate containing 25 μL 10% tri-chloroacetic acid to quench the reaction for the time = 0 time point. The cluster tubes containing the IM were returned to the water bath with subsequent 25 μL aliquots removed as described above at 15, 30, 45 and 60 minutes. After samples for all time points were collected, 100 μL Tris–HCl (0.4 M, pH 8.9) containing 20 mM of EDTA and 0.24 mM DTNB was added to each well. The plate was mixed for 20 seconds prior to absorbance reading at 412 nm (optical density of 2-nitro-5-thiobenzoic acid) and 650 nm (reference wavelength) on an M5e plate reader (Molecular Devices, Sunnydale, CA). The rate of DTT consumed was determined by converting OD readings to nmol DTT remaining. The slope of DTT consumed of type I water alone (untreated control) was subtracted from the slope of the PM-containing samples and corrected for the dilution used.

ROS production determination

The ROS activity of the PM extracts was measured using a previously described method employing an *in vitro* rat alveolar macrophage (NR8383) model and a wide spectrum ROS probe.[23] Cellular ROS encompass many chemical species, including the oxygen and hydroxyl radicals as well as other reactive forms of oxygen such as hydrogen peroxide and singlet O_2. Oxidative stress in cells results when ROS concentrations exceed the capacity of innate antioxidant systems. The NR8383 cell line manifests all of the typical characteristics of a normal primary macrophage cell and is highly responsive to particulate and microbial stimuli by phagocytosis and to soluble species. Importantly, these macrophages display oxidative burst and secrete a wide range of pro-inflammatory cytokines. In recent epidemiology studies, the assay has been shown to have a robust association with both airway and systemic inflammation biomarkers that lead to adverse health effects.[36,37] In summary, the membrane-permeable 2′,7′-dichlorodihydrofluorescein diacetate (DCFH-DA) probe enters the macrophage cell and is deacetylated by cytoplasmic esterases, which then allows ROS species to convert DCFH to its fluorescent form, DCF. The DCFH probe is sensitive to the broad spectrum of ROS species produced intra-cellularly. This enables the quantification of biologically mediated production of ROS in response to cellular stimulation (*e.g.* PM exposure). The rat alveolar macrophage NR8383 (CRL-2192™) cell line was obtained from ATCC (Manasses, VA, USA) and cultured according to the manufacturer's instructions: in Hams F12 medium (Sigma Aldrich, St. Louis, MO) supplemented with 15% fetal bovine serum (FBS; Hyclone, Fisher, USA) and sodium bicarbonate (1.176 g L^{-1}). Cells were maintained at 37 °C in a humidified incubator supplied with 5% CO_2. Non-adherent macrophage cells were harvested, centrifuged and resuspended in SGM to obtain a concentration of 1000 cells per μL. Subsequently 100 μL of this suspension was seeded into 96-well plates and cells were allowed to adhere for 2 h at the standard conditions of 37 °C and 5% CO_2. Following the cell adhesion period the buffer medium was aspirated and the SGM-buffered unfiltered and filtered PM extracts mixed with a small volume of DCFH-DA solution (45 μM final concentration) were added to the cells. This exposure was carried out for 2.5 h under standard culturing conditions, after which the fluorescence intensity of each well was determined at 488 nm excitation and 530 nm emission using an M5e microplate reader (Molecular Devices, CA, USA). Raw fluorescence data was blank-corrected and normalized to the zymosan (ZYM) positive controls (β-1,3 polysaccharide of D-glucose from Sigma Aldrich, MO, USA) to enable cross-study comparisons. Multiple dilutions of each sample were run, each in triplicate, to establish a linear portion of the dose response.

TNFα production determination

TNFα is a cytokine (cell signaling protein) involved in systemic inflammation, where its' primary role is regulation of immune cells. It is produced primarily by activated macrophages (like NR8383), though other cell types (*e.g.* neutrophils and endothelial cells) can also produce TNFα. The cytokine binds to two receptors (TNFR1 and TNFR2) found in many tissues and cells of the immune system and subsequently activates several major pathways (including NF-κB, MAPK/JNK) involved in cell survival, proliferation, inflammatory responses, and both anti- and pro-apoptotic factors. TNFα secreted by the NR8383 macrophages after 6 hours of exposure to the PM extracts was measured in the supernatant of the exposure solution by enzyme-linked immunosorbent assay (ELISA). Supernatants were pipetted from the cell exposure wells, centrifuged (1600*g* for 15 minutes at 4 °C) and the particle-free supernatants immediately frozen (−30 °C) until the ELISA was performed. The ELISA was carried-out with BD OptEIA Rat TNFα kits following the manufactures protocols. Absorbances were read in a Molecular Devices M5e plate reader at 450 nm and using 570 nm as correction/reference wavelength. Calibration was performed with authentic standards over the concentration range of 31 to 2000 pg TNFα per mL. Every standard, sample, and control sample was run in duplicate.

Statistical analysis

Basic statistical analyses, including Spearman correlation analysis were performed with the SigmaStat module integrated into SigmaPlot (version 13.0, Systat Software Inc. 2014).

Results

Aerosol PM concentrations

Aerosol PM concentrations (Fig. 1S, ESI†) ranged from <10 μg m^{-3} (Stockholm urban background) to over 75 μg m^{-3} in the urban roadside canyons of London and Thessaloniki. PM concentrations at the rural sites were relatively low (mean = 15.3 μg m^{-3}) and consistent (±1.5 μg m^{-3}). The average PM concentration across all cities and sites was 31.5 μg m^{-3} with urban background levels consistently in the 10–20 μg m^{-3} range. The large majority of the PM mass was present in the fine (PM_3) size cut (Fig. 1S, ESI†), with the coarse (PM_{7-3}) and super-coarse ($PM_{>7}$) size fractions contributing about equal to the remainder of the PM.

ROS and DTT activity – study site comparison

The size-resolved ROS activity of unfiltered and filtered PM extracts, expressed in terms of air concentrations is shown in Fig. 1 and 2. The ESI† presents this data in more detail (Fig. 2S and 3S, ESI†). The ROS activity of unfiltered extracts across all 17 sites ranged from 25 to 106 μg ZYM per m^3, averaging 69.6 ± 23.1, with a median value of 73.0. Filtered extract activity was substantially lower (range = 11 to 89 μg ZYM per m^3), averaging 42.6 ± 26.6, with a median value of 36.2. ROS activity was generally similar in all cities studied. The variation in ROS activity across all sites of 30% (unfiltered) and 60% (filtered) is relatively small, considering this reflects a wide range of anthropogenic influence, from heavily traveled urban roadway canyons to rural locations. Variation in ROS activity within site

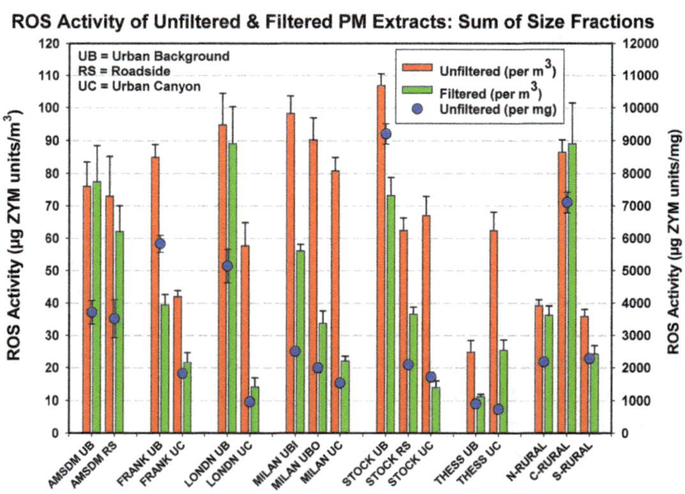

Fig. 1 ROS activity of PM extracts: sum of size fractions for unfiltered & filtered (per m^3) and for unfiltered (per mg).

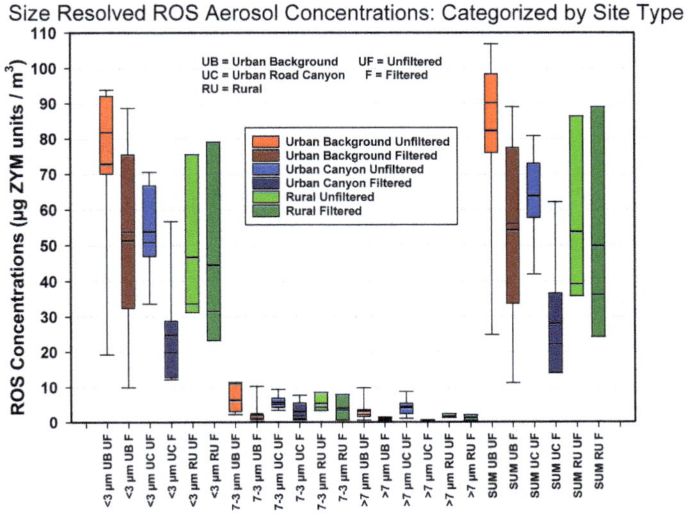

Fig. 2 Size resolved ROS aerosol concentrations: categorized by site type (per m^3). Median values are indicated by the narrower line within the box; mean values by the thicker line with the box. The 5th and 95th percentiles are indicated by the whiskers.

types (urban background, urban canyon/roadside, rural) was substantial (Fig. 2), however on average urban background sites exhibited the highest ROS concentrations (median = 90 µg ZYM per m^3), followed by urban canyon sites (64 µg ZYM per m^3) and rural sites (54 µg ZYM per m^3). In all cites except Thessaloniki, the measured ROS concentrations at the canyon sites was less than that at background sites – a trend that was particularly evident in the filtered extracts. The relatively high rural and urban background site concentrations (*e.g.* rural ROS activity is nearly twice as great as canyon site activity in filtered PM extracts) reflect primarily the higher specific ROS activity (see following section) of PM in rural and background sites.

The size-resolved ROS activity of unfiltered and filtered PM extracts, normalized to PM mass concentrations is shown in Fig. 1 and 3. The ESI† presents additional detail on the mass normalized ROS activity (Fig. 4S and 5S, ESI†). Normalizing the ROS activity to PM mass results in a metric that reflects the intrinsic oxidative activity of the PM (in contrast to the air volume normalization (above) that provides information on potential human exposure levels). The intrinsic activity of unfiltered extracts across all 17 sites ranged from 740 to 9200 µg ZYM per mg, averaging 3130 ± 2380, with a median value of 2175. The variation across sites in the mass normalized data (76%) is greater than the corresponding air volume data. Again, variation in ROS activity within site types was substantial (Fig. 3), however on average urban background sites exhibited the highest ROS intrinsic activity (median = 3700), followed by rural sites (2200) and then urban canyon sites (1750). The 2-fold lower activity in the canyon sites compared with background and rural sites is remarkable. In each city sampled the intrinsic activity of the background sites was greater than at canyon sites – in several cities by nearly 5-fold. PM specific activity was lowest in canyon sites of

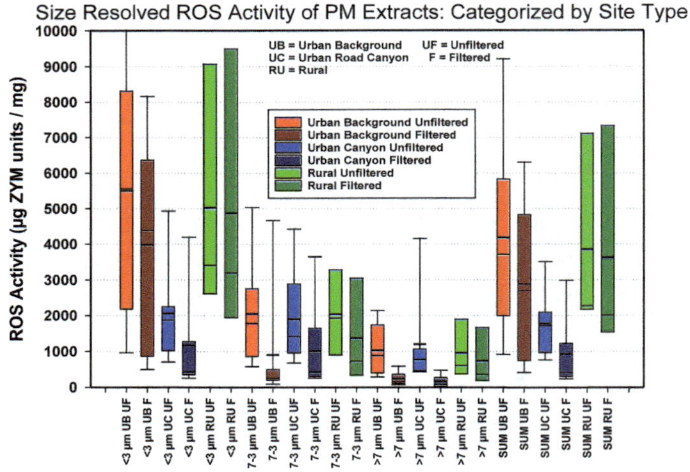

Fig. 3 Size resolved ROS aerosol concentrations: categorized by site type (per mg). Median values are indicated by the narrower line within the box; mean values by the thicker line with the box. The 5th and 95th percentiles are indicated by the whiskers.

London, Milan and Thessaloniki (those greatest density of traffic) and highest at urban background sites of Stockholm, Frankfurt and London. PM from the central Europe rural site also exhibited high specific ROS activity.

The size-resolved DTT activity of unfiltered and filtered PM extracts, expressed in terms of air concentrations is shown in Fig. 4. The ESI† presents this data in more detail (Fig. 6S and 7S, ESI†). The air concentrations of aerosol DTT activity of unfiltered extracts across all sites ranged from 0.14 to 0.62 nmol min^{-1} m^{-3}, averaging 0.287 ± 0.132, with a median value of 0.249. Filtered extract activity was only marginally lower (range = 0.11 to 0.51 nmol min^{-1} m^{-3}), averaging 0.254 ±

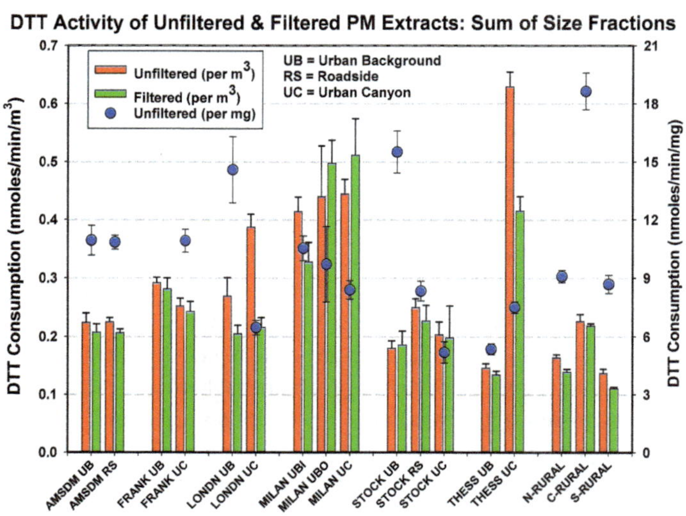

Fig. 4 DTT activity of PM extracts: sum of size fractions for unfiltered & filtered (per m^3) and for unfiltered (per mg).

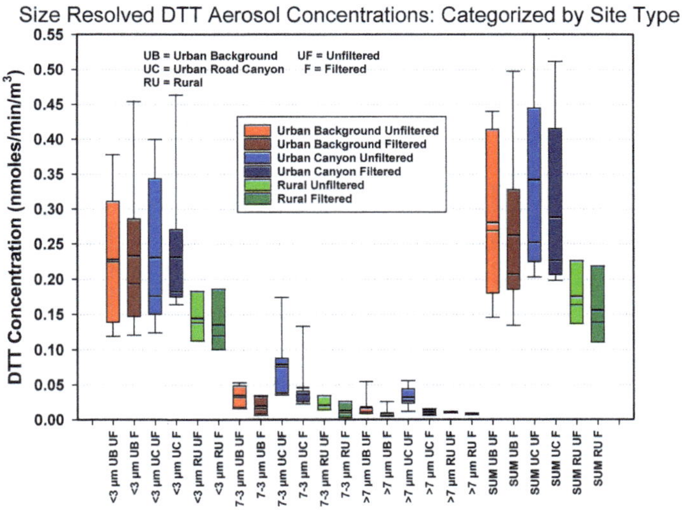

Fig. 5 Size resolved DTT aerosol concentrations: categorized by site type (per m^3). Median values are indicated by the narrower line within the box; mean values by the thicker line with the box. The 5th and 95th percentiles are indicated by the whiskers.

0.118, with a median value of 0.216 (just 13% lower than in unfiltered extracts). DTT activity was remarkably similar in most cities – only Milan and Thessaloniki exhibited activity significantly greater than the median. Even more so than in the ROS data, the variation across sites is exceptionally low. Variation in DTT activity within site types was evident (Fig. 5), but not as great as with ROS activity, and on average urban background sites exhibited the highest DTT concentrations (median = 0.27), followed by urban canyon sites (0.25) and rural sites (0.17). In all cites except Thessaloniki, the DTT concentrations at the canyon and background sites were nearly equivalent.

The size-resolved DTT activity of unfiltered and filtered PM extracts, normalized to PM mass concentrations is shown in Fig. 4 and 6. The ESI† presents additional detail on the mass normalized DTT activity (Fig. 8S and 9S, ESI†). The intrinsic activity of unfiltered extracts across all 17 sites ranged from 5.2 to 20.0 nmol min^{-1} mg^{-1}, averaging 10.6 ± 4.3, with a median value of 9.7. Unlike with the ROS data, where mass normalization increased the variation across sites, mass normalizing the DTT data actually reduced the variance (40% *versus* 46%) compared with the corresponding air volume data. Again, variation in DTT activity within site types was substantial (Fig. 6), however on average urban background sites exhibited the highest DTT intrinsic activity (median = 11.5), followed by rural sites (8.5) and then urban canyon sites (7.5). With the exception of Thessaloniki, in each city sampled the intrinsic activity of the background sites was equal to or greater than at canyon sites, but the contrast was not nearly as great as with ROS activity. In very similar findings to the ROS, PM specific activity was lowest in canyon sites of London, Milan and Thessaloniki and highest at urban background sites of Stockholm and London. PM from the central Europe rural site also exhibited high DTT activity. Overall, the pattern of DTT activity across study sites generally tracked that of ROS.

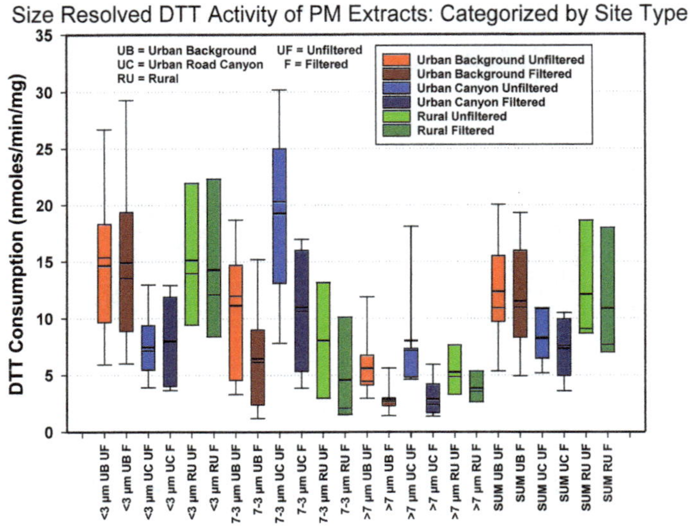

Fig. 6 Size resolved DTT aerosol concentrations: categorized by site type (per mg). Median values are indicated by the narrower line within the box; mean values by the thicker line with the box. The 5th and 95th percentiles are indicated by the whiskers.

TNFα production – site comparison

Expressed on an air volume normalized basis the TNFα production from the unfiltered PM_3 samples from Thessaloniki urban background (578 700 ng TNFα per m^3) and London urban canyon (525 300 ng TNFα per m^3) sites were over 5-fold greater than all other PM_3 sites (Table 1). In the filtered PM_3 samples, the highest TNFα induction was observed in the London sample (9300 ng TNFα per m^3), followed by the Milan urban background site (7000 ng TNFα per m^3). Within the rural site PM_3 samples, the TNFα production was lowest in the North rural site (unfiltered = 7400 ng TNFα per m^3, filtered = 989 ng TNFα per m^3), followed by the Central location and the highest values observed for the South rural site (both total and filtered fraction, normalized to mass, as well as to air volume).

The highest intrinsic (*i.e.* PM mass-normalized) TNFα-inducing activity of unfiltered PM_3 was observed at the Thessaloniki urban background site (28 900 pg TNFα per mg PM), followed by the London urban canyon site (10 900 pg TNFα per mg PM) and then the Frankfurt urban background site (8700 pg TNFα per mg PM). However, for the filtered PM_3 fraction the highest activity was observed at the Stockholm urban background site (591 pg TNFα per mg PM), followed by Frankfurt urban background (447 pg TNFα per mg PM) and Amsterdam urban background sites (Table 1). Interestingly, for the PM_{3-7} size fraction, the Thessaloniki PM_{3-7} site was the most active on both an air volume (33 200 ng TNFα per m^3) and PM mass normalized basis (9610 pg TNFα per mg PM).

Particle size-resolved analysis

As shown clearly in Fig. 7, the PM_3 dominates the air concentrations of ROS. Across all 17 sites, in unfiltered PM extracts, the PM_3 represents 86% of total activity, with the PM_{7-3} contributing on average 9% and $PM_{>7}$ just under 5%. The

Table 1 TNFα production in rat alveolar macrophages (AMs) exposed to PM from six major cities and three rural locations in Europe. Total and filtered PM_3 and PM_{3-7} extracts were incubated with AMs for 6 h. ELISA measurements of TNFα were normalized to PM mass and air volume (m^3). Data presented as mean ± standard deviation. "—" signifies missing data due to insufficient sample volumes

	pg TNFα per mg PM				ng TNFα per m^3 air			
Sampling location	PM_3 total	PM_3 filtered	PM_{3-7} total	PM_{3-7} filtered	PM_3 total	PM_3 filtered	PM_{3-7} total	PM_{3-7} filtered
AMSDM UB	3973	303	—	—	50 610	3863	—	—
±SD	74	92			944	1176		
FRANK UB	8671	447	—	—	73 127	3771	—	—
±SD	82	8			695	64		
LONDN UB	—	—	4542	425	—	—	11 605	1086
±SD			41	5			105	12
LONDN UC	10 922	193	—	—	525 269	9293	—	—
±SD	392	9			18 838	409		
MILAN UBI	2144	217	6058	440	68 937	6980	15 745	1142
±SD	133	7	328	15	4263	210	853	39
STOCK UB	3426	591	2038	155	25 998	4483	4457	340
±SD	53	14	20	5	401	107	43	11
THESS UB	28 881	225	9614	774	578 695	4509	33 198	2674
±SD	322	6	362	7	6460	129	1249	23
N-rural	754	100	—	—	7444	989	—	—
±SD	21	2			203	18		
C-rural	1405	153	198	155	11 701	1278	513	401
±SD	91	20	3	2	757	167	7	6
S-rural	3405	175	—	—	40 694	2090	—	—
±SD	428	25			5112	295		

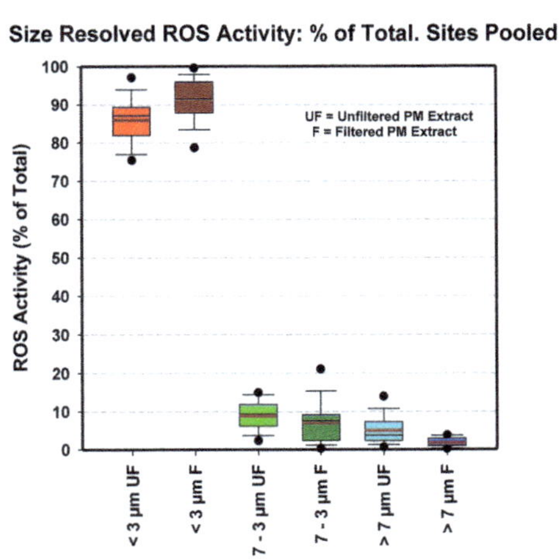

Fig. 7 Size resolved ROS activity: % of total activity. Sites pooled. Median values are indicated by the narrower line within the box; mean values by the thicker line with the box. The 5th and 95th percentiles are indicated by the whiskers and values outside the 5th and 95th percentiles are shown by dots.

corresponding percentages are 91%, 7% and 2% for the filtered extracts. The dominance of the PM_3 is evident in all site classes (Fig. 2) and the fraction of PM_3 is remarkable consistent; 87%, 84%, and 87% for urban background, urban canyon and rural sites respectively. The contribution of the super coarse ($PM_{>7}$) was greater than average at the Milan urban background (10%) and Thessaloniki sites (10% background and 14% canyon) (Fig. 2S, ESI†).

Similar to the ROS findings, air concentrations of DTT (Fig. 8) are also dominated by the PM_3. Across all 17 sites, in unfiltered PM extracts, the PM_3 represents 76% of total activity, with the PM_{7-3} contributing on average 17% and $PM_{>7}$, 7%. The corresponding percentages are 84%, 11% and 5% for the filtered extracts. The dominance of the PM_3 is particularly evident at the urban background and rural sites (Fig. 8 and 5) with contribution percentages of 81 and 83% respectively. A significantly lower PM_3 contribution was observed in the canyon sites (67%) with a corresponding large increase in the contribution of PM_{7-3} (23%), indicating a major coarse input from vehicle-sourced emissions (tire/brake wear and resuspended road dust). This coarse contribution was most evident at the canyon sites in London and Thessaloniki (the most vehicle influenced sites).

There is a clear trend of increasing ROS specific activity with smaller particle size (Fig. 9). Across all 17 sites, in unfiltered PM extracts, the median specific activity of the PM_3 was 2600 µg zymosan (ZYM) per mg, 1800 in the PM_{7-3} and just 750 in the $PM_{>7}$ (4-fold lower than in PM_3). Several high "outlier" values drive the arithmetic average higher than the median and these corresponding activities were 4000, 2000 and 1050 µg ZYM per mg. The particle-size dependence of ROS specific activity is evident in all three major site classifications (Fig. 3). The trend is seen most dramatically in the urban background sites (5600, 1800, 850 µg ZYM per mg in the PM_3, PM_{7-3}, and $PM_{>7}$ respectively); but also strongly evident in the rural sites (3400, 1950, 550) and canyon sites (1850, 1400, 750). The highest PM_3 specific activities were measured at urban background sites in Stockholm (12 400), Frankfurt (8300), London (6600) and Amsterdam (5600) (Fig. 4S†). ROS

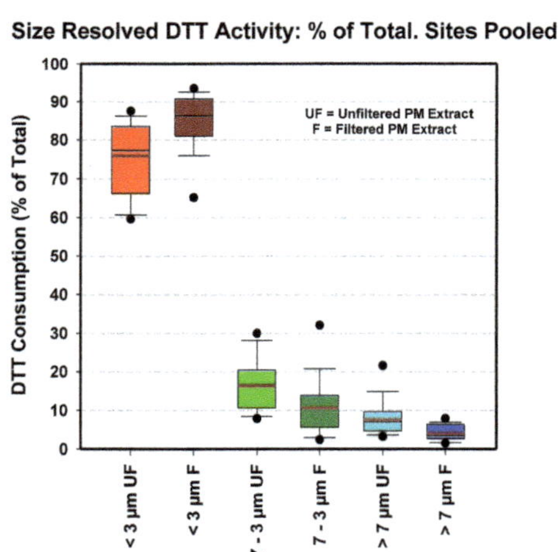

Fig. 8 Size resolved DTT activity: % of total activity. Sites pooled. Median values are indicated by the narrower line within the box; mean values by the thicker line with the box. The 5th and 95th percentiles are indicated by the whiskers and values outside the 5th and 95th percentiles are shown by dots.

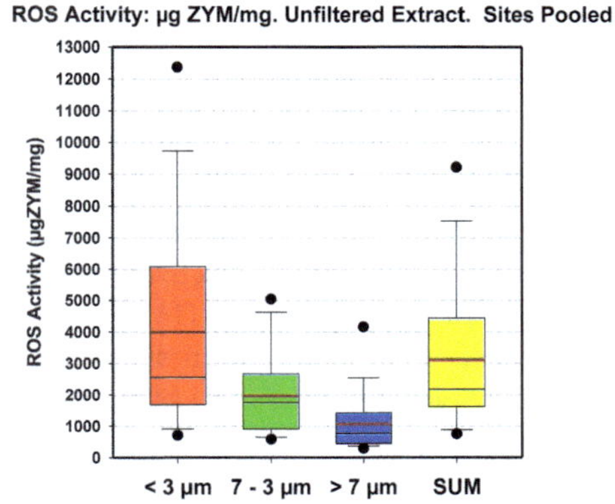

Fig. 9 ROS activity: µg ZYM per mg. Unfiltered extracts. Sites pooled. Median values are indicated by the narrower line within the box; mean values by the thicker line with the box. The 5th and 95th percentiles are indicated by the whiskers and values outside the 5th and 95th percentiles are shown by dots.

specific activity at most of the other urban sites was typically in the 1000–2000 µg ZYM per mg range. Rural sites exhibited ROS specific activities in the 1000–3000 range, though the PM_3 at the central Europe rural location had the 2nd highest measured activity (9050).

The trend of DTT specific activity with particle size (Fig. 10) is a bit more complex than that observed with ROS, resulting primarily from an input of DTT active materials into the PM_{7-3} size class in the urban canyon sites. Across all 17 sites, in unfiltered PM extracts, the median specific activity of the PM_3 was 9.5 nmol min^{-1} mg^{-1}, 13.5 in the PM_{7-3} and just under 5 in the $PM_{>7}$ (2-fold lower than in PM_3). As with the ROS data, several high "outlier" values drive the DTT arithmetic average higher than the median and these corresponding activities were 12, 13.5 and 7 nmol min^{-1} mg^{-1}. When these data are parsed into site types, the influence and unique trends in the urban canyon sites become very clear (Fig. 6). While the trends in specific activity as a function of particle size at the

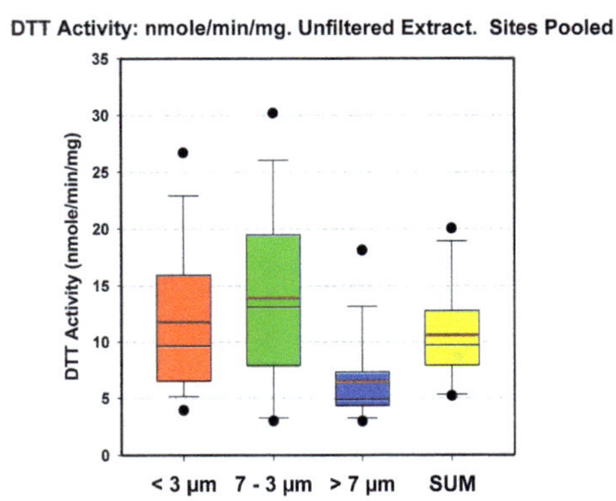

Fig. 10 DTT activity: nmol min^{-1} mg^{-1}. Unfiltered extracts. Sites pooled. Median values are indicated by the narrower line within the box; mean values by the thicker line with the box. The 5th and 95th percentiles are indicated by the whiskers and values outside the 5th and 95th percentiles are shown by dots.

urban background (15.5, 12.5, 4) and rural (14, 7.5, 5) (PM_3, PM_{7-3}, $PM_{>7}$ respectively) sites are very similar to that observed with the ROS data (consistent decrease with increasing particle size), the trend is dramatically different at the canyon sites with median PM_{7-3} activity of 20.5 compared with 7.5 in the PM_3 and 7 in the $PM_{>7}$. The canyon sites/cities contributing most significantly to this "elevated" PM_{7-3} specific activity were Stockholm (25–30), Frankfurt (22) and Thessaloniki (20.5). The highest PM_3 specific activities were measured at urban background sites in Frankfurt (27), Stockholm (18) and London (17) (Fig. 8S†). DTT specific activity at most of the other urban sites was typically in the 5–15 nmol min^{-1} mg^{-1} range. DTT activity at the rural sites was also typically in this 5–15 nmol min^{-1} mg^{-1} range, however just like with ROS the central Europe rural location exhibited a relatively high activity (22 nmol min^{-1} mg^{-1}).

TNFα. Overall, TNFα secretion was substantially higher in PM_3 than in PM_{3-7}, whether normalized to air volume or PM mass. The greatest size contrast was in the air volume normalized total fraction data, followed by air-volume normalized filtered fraction data and mass-normalized total PM (Table 1). Of note, the Milan UBI mass-normalized sample exhibited the opposite trend, with PM_{3-7} eliciting higher production of TNFα than PM_3 in both total and filtered fractions. However, this trend was not observed for the corresponding air-volume normalized data.

Unfiltered-filtered analysis

ROS. Averaged over all sites and particle size classes, the fraction of aerosol ROS activity that is water soluble was 60% (the median value, 57.5%, is similar, indicating a normal type distribution) Fig. 11. At several of the urban background sites, the soluble ROS percentage approached 100% (Amsterdam = 96%, London = 94%). Importantly, the soluble percentage was also consistently very high at the rural sites. Urban canyon sites exhibited the lowest water solubility, with London, Milan and Stockholm all in the range of 20–30%. When broken out by particle size class (across all 17 sites), there is a clear trend of decreasing median water solubility with increasing particle size (59.5% in PM_3, 30% in PM_{7-3}, and 25% in

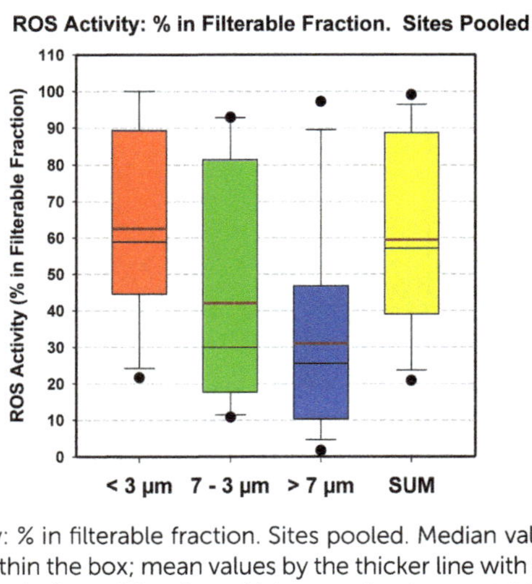

Fig. 11 ROS activity: % in filterable fraction. Sites pooled. Median values are indicated by the narrower line within the box; mean values by the thicker line with the box. The 5th and 95th percentiles are indicated by the whiskers and values outside the 5th and 95th percentiles are shown by dots.

$PM_{>7}$) (Fig. 11). Many of the $PM_{>7}$ samples exhibited exceptionally low water soluble ROS fractions of 10% or lower (Frankfurt and Stockholm urban background sites and London, Milan, Stockholm and Thessaloniki urban canyon sites). In contrast, the $PM_{>7}$ soluble percentage at the rural sites was very high (>90% in two locations).

DTT. Overall, water soluble metrics of the DTT data exhibited similar trends to the ROS data, though the absolute water soluble percentages were significantly higher. Averaged over all sites and particle size classes, the fraction of PM DTT activity that is water soluble was 83% (median = 84%) Fig. 12. At seven of the 14 sites, the soluble DTT percentage was greater than 90%. Water solubility was slightly higher in the urban background locations. As seen with the ROS data, the soluble percentage was also consistently very high at the rural sites. Urban canyon sites exhibited substantially lower water solubility, with London (55%) and Thessaloniki (65%), the lowest. When broken out by particle size class (across all 17 sites), there is, like with the ROS data, a clear trend of decreasing median water solubility with increasing particle size (93.5% in PM_3, 56% in PM_{7-3}, and 50% in $PM_{>7}$) (Fig. 12). With the exception of sites in London and canyon site in Thessaloniki, all PM_3 water soluble fractions were >90%. The lowest soluble DTT percentages were measured in the $PM_{>7}$ from urban canyon sites (London 38%, Milan 32%, Stockholm 27%, Thessaloniki 20%) (Fig. 11S, ESI†). In contrast, the $PM_{>7}$ soluble percentage at the rural sites was high (70–80% across all locations).

TNFα. Production of TNFα was especially sensitive to the presence of particles, with ratios of total to filtered values as high as 120 (Table 1). The filtered fraction was typically at least 10-fold lower in activity than the unfiltered samples in both the PM_3 and PM_{3-7} size classes. The one exception was at the Central rural site where the unfiltered and filtered activities of the PM_{7-3} were similar.

Association of oxidative activity with elemental composition of PM

In Table 2 we present a summary of the Spearman correlation coefficients between ROS and DTT activity and the elemental composition of the PM. From

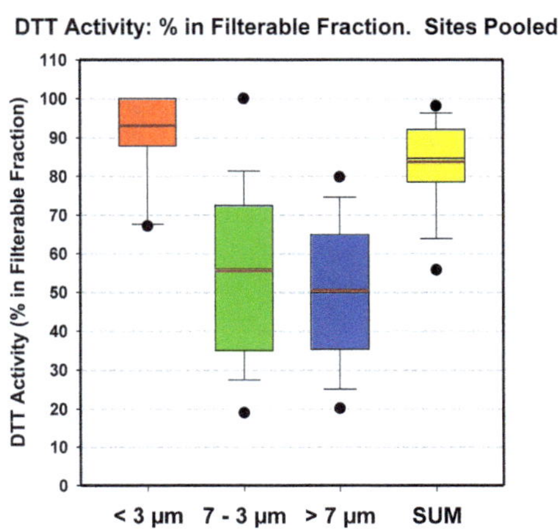

Fig. 12 DTT activity: % in filterable fraction. Sites pooled. Median values are indicated by the narrower line within the box; mean values by the thicker line with the box. The 5th and 95th percentiles are indicated by the whiskers and values outside the 5th and 95th percentiles are shown by dots.

the 50 element SF-ICPMS analysis of TOTAL elemental composition of the PM in the three particle size cuts (refer to Fig. 12S, ESI†), we selected the 28 elements with concentrations above 10 μg g^{-1} and generated the Spearman correlation coefficients on the size-resolved ROS and DTT data. Chemical elements are ranked in Table 2 by the median elemental concentration in the bulk (*i.e.* reconstructed) PM. The major PM bulk components, alkaline earths (Ca, Mg), alkali metals (Na, K), S (sulfate), and Al (aluminosilicate) are present at concentrations in the range of 1–7%. Iron (Fe) is by far the most abundant transition metal, with concentrations ranging from 2% in the PM_3 to over 5% in the coarser size fractions. Titanium (Ti), another aluminosilicate tracer is present at concentrations of 0.1–0.2%. The second most abundant 1st row transition metal is Cu, followed by Zn, and then Mn, with concentrations ranging from 800 to 2000 μg g^{-1}. Barium (Ba), a good tracer of brake wear is present at levels comparable to that of Zn and Mn. The 1st row transition elements, Cr, Ni, and V are found at

Table 2 Spearman correlation coefficients between total element content of PM and ROS and DTT activity in unfiltered extracts of size-resolved PM. Twenty-eight most abundant elements ranked by median concentrations in bulk PM. Bolded (red) = $r > 0.70$. Bolded italic (purple) = $r > 0.50$

Element	ROS			DTT		
	<3 μm	7-3 μm	>7 μm	<3 μm	7-3 μm	>7 μm
ROS	1.000	1.000	1.000	**0.929**	0.136	***0.691***
DTT	**.929**	0.136	***0.691***	1.000	1.000	1.000
Fe	-.221	-.082	.300	-.353	**.791**	***.545***
Ca	-.621	-.464	***.673***	-.535	.109	.391
Na	.400	.409	-.191	.338	.091	.064
S	***.571***	-.145	.091	***.585***	.036	.200
Al	-.118	***.564***	***.536***	-.256	.382	***.500***
K	.212	**.945**	.200	.050	.218	.309
Mg	-.015	***.609***	***.600***	-.044	.409	.491
P	.276	.445	.327	.259	-.018	.464
Ti	-.115	***.518***	***.545***	-.259	.427	***.627***
Zn	-.271	.264	.364	-.329	***.618***	***.564***
Cu	-.285	-.091	.173	-.391	**.764**	***.573***
Ba	-.188	-.064	.264	-.303	**.700**	***.536***
Mn	-.241	.055	**.727**	-.303	***.664***	***.618***
Cr	-.400	-.118	***.536***	-.474	***.545***	***.555***
Sn	-.156	.091	.255	-.268	**.791**	***.627***
Pb	-.056	-.264	***.636***	.135	***.536***	.464
Sr	-.088	.245	***.518***	-.129	***.573***	***.682***
Sb	-.244	-.127	.218	-.388	**.727**	***.555***
Ni	-.053	-.227	**.727**	-.062	.127	.291
B	-.350	-.082	.336	-.253	.182	.245
Mo	-.074	-.118	.136	-.279	***.682***	***.500***
V	.194	.164	***.655***	.103	.373	.427
Rb	.141	**.709**	.373	-.044	.382	.400
Ce	-.103	.336	.464	-.229	***.600***	***.564***
As	.388	-.009	**.709**	.382	***.591***	**.709**
La	.074	.218	.473	-.074	***.500***	***.609***
Li	-.015	***.609***	.373	-.124	.209	***.509***
Co	-.241	.173	***.636***	-.335	.491	***.564***

levels of 300, 100, 60 μg g^{-1}, respectively. Iron, Cu, Mn, Ni, and V are redox active in environmental PM. The semi-metals Sn and Sb, along with Pb fill out the elements present at levels >100 μg g^{-1} (200, 150, 200 μg g^{-1} respectively). With the exception of V and Zn ("soft" elements with greater mobility and water solubility), all the minor/trace elements mentioned above are strongly enriched in the coarser PM size fractions.

ROS and DTT activities were strongly (>0.9) associated in the PM_3, exhibiting a Spearman coefficient of $r = 0.93$ (Table 2 and Fig. 13S, ESI†). Association between ROS and DTT was also evident in the largest size cut ($PM_{>7}$, $r = 0.69$), but no correlation was found in the PM_{7-3} size cut (Table 2 and Fig. 14S, ESI†).

No chemical elements were associated with ROS and DTT activity at a level of correlation above $r = 0.7$ in the PM_3 size cut. In fact, S was the only element that exhibited a marginal correlation (>0.5) with ROS and DTT in the PM_3. In contrast, several elements exhibited good (>0.7) to marginal (>0.5) correlation with ROS and DTT in the larger PM size fractions. Potassium (K) ($r = 0.95$), a biomass burning tracer, and Rb ($r = 0.71$) were associated with ROS activity in the PM_{7-3} size cut. Magnesium ($r = 0.61$), Li ($r = 0.61$), Al ($r = 0.56$) and Ti ($r = 0.52$), all dust/aluminosilicate tracers exhibited marginal correlations with ROS. Interestingly, both K and Rb were not correlated with ROS in the super-coarse ($PM_{>7}$) fraction. Elements associated with ROS activity in the largest size cut ($PM_{>7}$) included most of the aforementioned crustal elements (Mg, Al, Ti) along with Ca ($r = 0.67$), but maybe more importantly a large suite of redox active transition metals exhibited good to marginal correlations (Mn: $r = 0.73$, Ni: $r = 0.73$, V: $r = 0.66$) with ROS. Arsenic (As) ($r = 0.71$) and Co ($r = 0.64$) also were associated with super-coarse ROS activity. A completely different suite of elements from that correlated with ROS was associated with DTT activity in the PM_{7-3} size cut. Iron ($r = 0.79$), Sn ($r = 0.79$), Cu ($r = 0.76$), Sb ($r = 0.73$), and Ba ($r = 0.70$) all non-tailpipe vehicle emission tracers, exhibited good correlations with DTT. Also showing marginal correlations to DTT in the PM_{7-3} were Mo ($r = 0.68$), Mn ($r = 0.66$), Zn (0.62), Ce ($r = 0.60$), As ($r = 0.59$), Cr ($r = 0.55$) and La ($r = 0.50$). Only one element, As ($r = 0.71$) exhibited a good correlation with DTT in the $PM_{>7}$ fraction, however a very large number of elements exhibited marginal correlation with DTT in this particle size cut. This included all of the trace elements correlated in the PM_{7-3} fraction, with the addition of several crustal tracers (Al, Ti, Sr, and Li).

Discussion

Air concentrations of aerosol ROS, DTT and TNFα activity

We measured a remarkable uniformity in air concentrations of ROS and especially DTT across the continent. Only a 4-fold difference was documented for DTT across the urban sites (0.14 to 0.62 nmol min^{-1} m^{-3}) and a similar variance was documented for ROS (25 to 106 μg ZYM per m^3). Rural site DTT activity clustered in a very narrow range, just 2-fold lower than at the urban sites. ROS activity at the rural sites exhibited a greater range, spanning a good portion of the range measured at the urban sites. This overall similarity reflects a relatively small variation in PM concentrations across the study sites and implies that chemical drivers of oxidative activity are not too dissimilar between sites. DTT levels of this order are also consistent globally, *e.g.* DTT concentrations in samples collected in Beijing by Liu *et al.*[5] ranged between 0.1 and 0.5 nmol min^{-1} m^{-3} – strikingly similar to the levels measured in the current study of European sites. The significant oxidative potential of the PM at urban background and rural sites suggests that the oxidative stress health implications of aerosol exposures may not necessarily be only an urban problem. In contrast to the ROS and DTT findings, the air concentrations of TNFα induction activity of the aerosol exhibited quite large contrasts between sites. Nearly an 80-fold range (7000 to 580 000 ng TNFα per m^3) was measured.

PM_3 dominated the size distribution of DTT activity (76% of total) with coarse and super-coarse size fractions contributing much smaller to total activity (17% and 7% respectively). Analogous trends were observed in the ROS data (86% of total activity in the PM_3), with even smaller average contributions from the larger size PM (9% and 7% in PM_{7-3} and $PM_{>7}$ respectively). This finding reflects both the dominance of PM_3 aerosol concentrations and higher specific activity of the PM_3 aerosols. At the urban canyon sites there is a significant contribution of oxidative activity from the PM $_{7-3}$ size fraction, likely sourced from non-tailpipe vehicle emissions. Clearly though, reduction of the risk from PM oxidative activity should focus on controlling PM_3 emissions.

Specific activity of aerosol ROS, DTT and TNFα

The aerosol at urban background locations exhibited greater specific activity than that present in urban canyon locations, though the contrast is maybe less than one might expect given differences in levels of potential drivers of oxidative activity. Averaged across all sites in the particle-size summed aerosol, a 2-fold contrast in ROS activity (3700 *versus* 1750 nmol min^{-1} mg^{-1}) was measured, and DTT activity at urban background sites averaged 35% higher than at urban canyon sites. Significantly, rural aerosol exhibited DTT and ROS specific activities comparable to that measured in urban background aerosol. These findings may belie expectations and clearly suggest that some of the major drivers of oxidative activity are regional in nature and distribution and not, for example, dominated by fresh emissions associated with vehicles. It is likely that aging of the aerosol is an important factor in generating oxidative potential. This hypothesis is supported by several recent publications that document that oxidation ("aging") processes in the atmosphere produce species with greater oxidative activity.[38,39] Platt *et al.* 2014 (ref. 38) demonstrated in dedicated chamber studies that oxidation processes produced a secondary organic aerosol from vehicle exhaust with dramatically greater ROS activity than primary emissions; and similarly Saffari *et al.* 2015 (ref. 39) documented that photochemical processing and secondary organic aerosol formation were the dominant drivers underlying the ROS activity in PM transported downwind to a location distant from the primary generation source.

In the fine PM fractions (*i.e.* $PM_{2.5}$ or PM_3) our European samples exhibited a range in PM mass normalized DTT activity from 4–27 nmol min^{-1} mg^{-1} PM, very similar, if slightly on the low side of what has been reported worldwide. For example: samples from Mexico city were between 5 and 40 nmol min^{-1} mg^{-1} PM;[40] samples from the Netherlands ranged from 40–214,[41] samples from the Los Angeles Basin ranged from 21–75 nmol min^{-1} mg^{-1} PM,[42] and samples from various sites in the Southeastern United States averaged about 35 nmol min^{-1} mg^{-1} PM.[43] A study conducted in Thessaloniki[44] measured DTT in PM_{10} at a heavy traffic site and results ranged from 18–60 nmol min^{-1} mg^{-1} PM. The samples we collected in Thessaloniki were lower, at 7.5 nmol min^{-1} mg^{-1} PM.

This finding of general consistency in DTT levels across large spatial and anthropogenic gradients (rural, urban, distance from industry and roadways) is significant and given the variability in extraction and assay techniques as well as samples collected worldwide under a wide range of environmental conditions, it is striking that DTT values are in the same general range. Research is needed to characterize further the chemical species in PM that the DTT assay is responsive to and whether these species are relevant drivers of adverse human health outcomes. The efficacy of an oxidative stress assay in this context depends in large part on its ability to capture the influence of a large spectrum of health relevant species, and if this consistency is linked to a small sub-set of species, then the utility of the assay may be reduced.

The contrast between sites in macrophage ROS activity was larger than that measured in the DTT assay. A greater than ten-fold difference was seen in the mass-normalized macrophage ROS for these European sites (using data from the reconstructed pooled PM sizes) compared with a four-fold difference for the DTT assay. Contrasts in ROS activity between sites are even larger within a specific PM size class. Saffari *et al.*[30] summarized macrophage ROS results from the literature and found large differences in mass-normalized ROS activity between sites. For example, the ROS activity from samples collected in Los Angeles and Thessaloniki were low (483 and 572 μg ZYM per mg PM), however, the ROS activity from samples collected in Milan and Beirut were 25–50 times higher. Two other measures of oxidative potential (electron spin resonance and ascorbate depletion) also show more contrast between sites than does DTT consumption.[45] While the macrophage ROS assay is a more biologically relevant model to address oxidative activity of aerosol PM, and several studies have shown direct predictive power in human health metrics,[36,46,47] further studies are needed to assess whether the greater sensitivity and selectively of the assay make it the preferred approach for assessment of potential human health impacts.

For urban background and rural locations there is a general increase in DTT specific activity with decreasing particle size. A 3-fold difference between PM_3 and $PM_{>7}$ was measured for DTT and an even greater difference (5-fold) was measured between these particle size classes for ROS. However, for the urban canyon sites PM_3 and PM_{7-3} are comparable (if not higher in PM_{7-3}, especially for DTT) for both DTT and ROS. Only in the $PM_{>7}$ is a small decrease observed. As noted above this likely reflects an input of non-tailpipe vehicle emissions into the coarse PM_{7-3} size fraction. The trends in specific activity of ROS and DTT as a function of particle size at the urban background and rural sites implicates chemical drivers that exhibit a preference for smaller aerosols, *e.g.* more volatile, combustion associated metals, black carbon, and certain organic species. Active surface area contrasts between the particle size classes may also play a role.

Our finding that DTT activity was greatest in the PM_3 fraction is consistent with numerous other studies in which fine PM fraction DTT activity was shown to be the highest. For example, de Vizcaya-Ruiz *et al.*[40] measured DTT activity in PM from three regions of Mexico City and found that DTT in the fine ($PM_{2.5}$) was about twice that found for the coarse $PM_{2.5-10}$, similar to the ratio documented in this study. PM of three size classes (0.18 μm, <2.5 μm, 2.5–10 μm) collected from diverse sites in the Netherlands were evaluated for DTT activity by Steenhof *et al.*[41]

The mass normalized DTT activity from the <2.5 μm size class was up to 3.7 times higher than the coarse size except for the stop and go traffic site where they were equal (a similar phenomenon was observed in our urban canyon sites (see above)). At two sites (stop and go traffic and an underground tunnel) the highest mass-normalized DTT activities were in the ultrafine <0.18 μm fraction. Studies in California by Cho *et al.* and Ntziachristos *et al.* found the ultrafine PM (<0.15 μm) generated the highest DTT activities, followed by fine (<2.5 μm) and coarse (2.5–10 μm) across all sites tested.[35,42] Charrier *et al.* further narrowed the ultrafine size class and determined that DTT activities were higher in 0.17–1 μm fraction than they were in the <0.17 size class for the samples they collected in the central valley of California.[48]

Similarly, our finding that cellular ROS activity was highest in the PM_3 fraction is corroborated in by other studies. In New York, coarse ($PM_{2.5-10}$) and fine ($PM_{<2.5}$) were collected from three rural and two urban sites.[49] ROS was significantly higher for the cells exposed to $PM_{<2.5}$ at the urban sites, but there was no significant difference between the coarse and fine fraction for the rural sites. Super coarse ($PM_{>10}$) samples were collected at one rural and one urban site. ROS response to the super coarse sample was lowest, although only significantly so for the urban site. In samples from Los Angeles, Wang *et al.*[50] collected coarse, $PM_{2.5}$, and ultrafine PM samples from an urban site. Alveolar macrophages (NR8383) exposed to $PM_{2.5}$ extracts exhibited higher ROS activity than macrophages exposed to either coarse or ultrafine PM on a per air volume basis.

Soluble *versus* insoluble activity of aerosol extracts

The soluble (filterable) activity was comparable to total (unfiltered) activity for DTT (94%) in the PM_3. The soluble ROS activity fraction was also relatively high (64%) in the PM_3, but a much greater range among sites in the soluble fraction was observed. These findings suggest that the drivers of activity in this size cut are readily soluble and that the insoluble component is either small and/or inactive. However in the larger size fractions the contributions of the insoluble components become increasingly more significant as PM size increases – this trend is particularly evident in the ROS data. The overall effect on oxidative activity may simply reflect a change in particle chemistry (to more insoluble components that are also active) but may also reflect a different mode of oxidative activity generation. Overall, in each of the size fractions, the ROS soluble activity fraction was 20–30% lower than with the DTT.

ROS–DTT interrelationships

As detailed in the results presentation, there are overall generally similar trends in the ROS and DTT data in regards to particle size distribution, with both exhibiting fairly strong inverse trends of oxidative activity with PM size. Trends in the water solubility of oxidative activity are also similar, though the absolute magnitude of the DTT activity solubility is substantially greater than with ROS activity. When parsed into site types, again many of the relationships are similar between ROS and DTT.

To provide a more quantitative picture of ROS–DTT relationships, the site pooled, particle-size resolved data (expressed both in terms of activity per air volume and activity per mg of PM) were regressed. There is a good correlation between intrinsic activity of ROS and DTT in the PM_3, both for unfiltered (r^2 = 0.75) and filtered extracts (Fig. 13S, ESI†). There is a significant intercept on the DTT axis, indicating that some chemical oxidative activity is present without any ROS activity. Correlations of intrinsic ROS and DTT in the coarser size fractions are poorer than in the PM_3, particularly in the PM_{7-3} (where DTT is especially active in the canyon sites), but $PM_{>7}$ correlations are fair, especially considering the low activity of much of the data (Fig. 14S, ESI†). Relationships between the air volume-normalized (*i.e.* per m^3) PM_3 ROS and DTT data are limited at best, with the unfiltered data exhibiting an r^2 of 0.46. The filtered data show little to no correlation. There are hints of underlying relationships in the coarser size fractions, however a few "outliers", particularly in the PM_{7-3} fraction, seriously degrade the correlations.

Drivers of ROS and DTT activity

In contrast to several studies,[30,51] which document associations between PM oxidative activity and inorganic elements (particularly those associated with traffic emissions and fuel/oil combustion) in fine PM, we observed only poor correlations between any of the 28 elements and ROS and DTT activity in the PM_3 size cut. Only S exhibited a marginal (r = 0.58) correlation (with both ROS and DTT) in this size fraction. ROS and DTT are strongly correlated (r = 0.93) however and it's probable that organic species are contributing to the observed oxidative activity. In contrast, we did document significant correlations between our oxidative activity metrics and several groups of elements in the larger PM size classes. In the PM_{7-3} size cut, K (r = 0.95), a biomass burning tracer, and several elements associated with dust/aluminosilicates Mg (r = 0.61), Li (r = 0.61), Al (r = 0.56) and Ti (r = 0.52) exhibited correlations with ROS. In the largest size cut ($PM_{>7}$) ROS activity was associated with a group of crustal elements (Al, Ca, Mg, Ti) and several redox active transition metals (Mn: r = 0.73, Ni: r = 0.73, V: r = 0.66). Anthropogenic V and Ni are typically sourced from fuel oil combustion. The profile of elements associated with DTT activity was entirely different from that associated with ROS activity. In the PM_{7-3} size cut, the dominant correlations with DTT were with a group of elements sourced from non-tailpipe vehicle emissions Fe (r = 0.79), Sn (r = 0.79), Cu (r = 0.76), Sb (r = 0.73), and Ba (r = 0.70). Arsenic (r = 0.71) along with all of the trace elements correlated in the PM_{7-3} fraction, and several crustal tracers (Al, Ti, Sr, and Li) were correlated with DTT in the $PM_{>7}$ size cut. Thus it would appear that in the larger particle size fractions ROS activity is likely associated with more soluble elements associated with biomass burning and fuel oil combustion and in the super coarse size fraction with crustal elements. In contrast DTT activity in the larger size PM is likely associated with less soluble vehicle/roadway sourced elements along with crustal elements.

Our TNFα data revealed that induction of this cytokine in our *in vitro* alveolar macrophage model was almost exclusively driven by the insoluble components of the PM extracts, which is in stark contrast to the macrophage ROS production data, where we show that a large majority of the ROS activity was driven by soluble species. The ratio of unfiltered to filtered extract induction (Table 1) exceeded 100 at several sites – dramatically higher than the maximum ratio of 2 observed in the ROS data. In a recent rat study by Snow *et al.* animals were treated with chemically characterized PM and showed significant increases in neutrophilic inflammation and TNFα expression due to the insoluble and total, but not soluble fraction.[26] Our data set revealed no significant correlations of TNFα levels with any of the elements measured, in agreement with reports in the literature.[52,53] Several hypotheses have been suggested to explain this TNFα production in the absence of correlation with the PM elemental composition, especially the high magnitude of expression in the unfiltered fraction of PM. One of the hypotheses suggests that the difference is due to the biogenic (particles like pollen) content of PM.[54] While some macrophage studies show a correlation between cytokine production and endotoxin levels,[52] others fail to establish that relationship and suggest other components of PM as the drivers, potentially associated with the surface of the core PM particles.[53] More research of coarse and fine PM toxicity, especially in relation to the PM's detailed chemical composition is needed to confirm any of these hypotheses.

Toxicological studies using human and animal exposure methods provide a more direct and potentially more robust approach to address aerosol driven health impacts, however the limitations of human/animal studies are many and it is clear that *in vitro* assays will play an increasing role in overall toxicity assessment and as tools in advancing our understanding of mechanisms of aerosol driven toxicity. Screening assays should be relatively rapid and inexpensive to implement and able to be standardized and implemented in laboratories in many countries. They should address the known major underlying pathophysiological mechanisms of PM-driven disease. The three assays applied in this study largely satisfy these criteria, though there are obviously substantive differences in approach and potential biological relevance as noted previously. These tools are focused primarily on measurement of oxidative stress potential (ROS and DTT) and inflammation (TNFα) (which is intimately tied to ROS) – clearly of relevance to aerosol driven human health concerns, and in that context are complementary. However they are not comprehensive in that other important pathways of aerosol toxicity (*e.g.* AhR activation and genotoxicty) are not addressed – that will require a larger suite of assays. A major thrust of ongoing research in the aerosol toxicity research community remains to advance our understanding of the specific chemical drivers, and complex interactions between those drivers, of pathway-specific aerosol toxicity. It is likely that pathway analysis using more specific biomarkers (*e.g.* expression of certain targeted genes) or reporter engineered cells will provide insight into these mechanisms. These "endpoints", along with ROS generation, will need to be coupled with physical and chemical manipulation/treatments of PM extracts designed to isolate the influence of specific classes of chemicals/elements, if not specific agents. A greater emphasis on characterization of redox pathways, *e.g.* through oxidation state specific elemental characterization or with electrochemical or fluorogenic probes, should aid in the identification of active chemical species. However application of the macrophage ROS and DTT assays to a much broader array of aerosol types from urban and rural areas across the globe – an application these assays are ideally adapted for – would also provide critical new information. Coupled with detailed chemical analysis this approach would not only reveal the overall oxidative potential of a wider range of aerosols, but also help direct more targeted laboratory studies into the role of specific chemical agents.

Acknowledgements

The authors offer many thanks to the chemists from the WSLH Trace Element Research group, especially Pam Skaar and Joel Overdier. We wish to also express our gratitude to many supportive colleagues and friends from the Environmental Toxicology and Flow Cytometry groups at the WSLH. We also appreciate the field sampling efforts of students and staff from the UW-Madison ET&C program and colleagues at each sampling location. We thank the International Platinum Association Health Science Research Group for funding assistance.

Notes and references

1 WHO, *The World Health Report 2002: Reducing Risks, Promoting Healthy Life*, WHO, 2002.
2 L. Curtis, W. Rea, P. Smith-Willis, E. Fenyves and Y. Pan, *Environ. Int.*, 2006, **32**, 815–830.
3 M. Kampa and E. Castanas, *Environ. Pollut.*, 2008, **151**, 362–367.
4 Ł. Adamkiewicz, A. J. Badyda, A. Gayer and D. Mucha, *Adv. Exp. Med. Biol.*, 2015, **834**, 15–20.
5 E. Liu, T. Yan, G. Birch and Y. Zhu, *Sci. Total Environ.*, 2014, **476–477**, 522–531.
6 S. Martin, E. Fernandez-Alanis, V. Delfosse, P. Evelson, J. S. Yakisich, P. H. Saldiva and D. R. Tasat, *Inhalation Toxicol.*, 2010, **22**, 1064–1071.
7 W. Yang and S. T. Omaye, *Mutat. Res., Genet. Toxicol. Environ. Mutagen.*, 2009, **674**, 45–54.
8 Y. F. Zhu, T. Kuhn, P. Mayo and W. C. Hinds, *Environ. Sci. Technol.*, 2006, **40**, 2531–2536.
9 V. Verma, T. Fang, H. Guo, L. King, J. T. Bates, R. E. Peltier, E. Edgerton, A. G. Russell and R. J. Weber, *Atmos. Chem. Phys.*, 2014, **14**, 12915–12930.
10 N. Daher, N. A. Saliba, A. L. Shihadeh, M. Jaafar, R. Baalbaki, M. M. Shafer, J. J. Schauer and C. Sioutas, *Sci. Total Environ.*, 2014, **470–471**, 417–426.
11 F. M. Arantes-Costa, F. D. T. Q. S. Lopes, A. C. Toledo, P. A. Magliarelli-Filho, H. T. Moriya, R. Carvalho-Oliveira, T. Mauad, P. H. N. Saldiva and M. A. Martins, *Toxicol. Pathol.*, 2008, **36**, 680–686.
12 J. J. Schauer, G. C. Lough, M. M. Shafer, W. F. Christensen, M. F. Arndt, J. T. DeMinter and J.-S. Park, *Res. Rep. - Health Eff. Inst.*, 2006, 1–76; discussion 77–88.
13 M. R. Riley, D. E. Boesewetter, A. M. Kim and F. P. Sirvent, *Toxicology*, 2003, **190**, 171–184.
14 J. Zhao, N. Lewinski and M. Riediker, *Aerosol Sci. Technol.*, 2015, **49**, 65–74.
15 J. J. Schauer, B. J. Majestic, R. J. Sheesley, M. M. Shafer, J. T. Deminter and M. Mieritz, *Res. Rep. - Health Eff. Inst.*, 2010, 3–75; discussion 77–89.
16 F. J. Kelly, *Occup. Environ. Med.*, 2003, **60**, 612–616.
17 S. K. Jalmi and A. K. Sinha, *Front. Plant Sci.*, 2015, **6**, 769.
18 R. Foncea, C. Carvajal, C. Almarza and F. Leighton, *Biol. Res.*, 2000, **33**, 89–96.
19 N. Li, J. Alam, M. I. Venkatesan, A. Eiguren-Fernandez, D. Schmitz, E. Di Stefano, N. Slaughter, E. Killeen, X. Wang, A. Huang, M. Wang, A. H. Miguel, A. Cho, C. Sioutas and A. E. Nel, *J. Immunol.*, 2004, **173**, 3467–3481.

20 L. Fialkow, Y. Wang and G. P. Downey, *Free Radical Biol. Med.*, 2007, **42**, 153–164.

21 K. E. Iles and H. J. Forman, *Immunol. Res.*, 2002, **26**, 95–105.

22 H. J. Forman and M. Torres, *IUBMB Life*, 2001, **51**, 365–371.

23 A. P. Landreman, M. M. Shafer, J. C. Hemming, M. P. Hannigan and J. J. Schauer, *Aerosol Sci. Technol.*, 2008, **42**, 946–957.

24 M. M. Shafer, D. A. Perkins, D. S. Antkiewicz, E. A. Stone, T. A. Quraishi and J. J. Schauer, *J. Environ. Monit.*, 2010, **12**, 704–715.

25 M. S. Happo, R. O. Salonen, A. I. Hälinen, P. I. Jalava, A. S. Pennanen, J. a. M. A. Dormans, M. E. Gerlofs-Nijland, F. R. Cassee, V.-M. Kosma, M. Sillanpää, R. Hillamo and M.-R. Hirvonen, *Inhalation Toxicol.*, 2010, **22**, 402–416.

26 S. J. Snow, A. de Vizcaya-Ruiz, A. Osornio-Vargas, R. F. Thomas, M. C. Schladweiler, J. McGee and U. P. Kodavanti, *J. Toxicol. Environ. Health, Part A*, 2014, **77**, 1164–1182.

27 T. Torvela, O. Uski, T. Karhunen, A. Lähde, P. Jalava, O. Sippula, J. Tissari, M.-R. Hirvonen and J. Jokiniemi, *Chem. Res. Toxicol.*, 2014, **27**, 1516–1527.

28 J. Brocato, H. Sun, M. Shamy, T. Kluz, M. A. Alghamdi, M. I. Khoder, L.-C. Chen and M. Costa, *J. Toxicol. Environ. Health, Part A*, 2014, **77**, 751–766.

29 Z. Sijan, D. S. Antkiewicz, J. Heo, N. Y. Kado, J. J. Schauer, C. Sioutas and M. M. Shafer, *Environ. Toxicol.*, 2015, **30**, 836–851.

30 A. Saffari, N. Daher, M. M. Shafer, J. J. Schauer and C. Sioutas, *Environ. Sci. Technol.*, 2014, **48**, 7576–7583.

31 T. Okuda, J. J. Schauer and M. M. Shafer, *Atmos. Environ.*, 2014, **97**, 552–555.

32 M. M. Shafer, B. M. Toner, J. T. Overdier, J. J. Schauer, S. C. Fakra, S. Hu, J. D. Herner and A. Ayala, *Environ. Sci. Technol.*, 2012, **46**, 189–195.

33 S. Hu, J. D. Herner, M. Shafer, W. Robertson, J. J. Schauer, H. Dwyer, J. Collins, T. Huai and A. Ayala, *Atmos. Environ.*, 2009, **43**, 2950–2959.

34 J. G. Charrier and C. Anastasio, *Atmos. Chem. Phys.*, 2012, **12**, 9321–9333.

35 A. K. Cho, C. Sioutas, A. H. Miguel, Y. Kumagai, D. A. Schmitz, M. Singh, A. Eiguren-Fernandez and J. R. Froines, *Environ. Res.*, 2005, **99**, 40–47.

36 R. J. Delfino, N. Staimer, T. Tjoa, M. Arhami, A. Polidori, D. L. Gillen, S. C. George, M. M. Shafer, J. J. Schauer and C. Sioutas, *Epidemiology*, 2010, **21**, 892–902.

37 S. Wittkopp, N. Staimer, T. Tjoa, T. Stinchcombe, N. Daher, J. J. Schauer, M. M. Shafer, C. Sioutas, D. L. Gillen and R. J. Delfino, *J. Exposure Sci. Environ. Epidemiol.*, 2016, **26**, 141–149.

38 S. M. Platt, I. E. Haddad, S. M. Pieber, R.-J. Huang, A. A. Zardini, M. Clairotte, R. Suarez-Bertoa, P. Barmet, L. Pfaffenberger, R. Wolf, J. G. Slowik, S. J. Fuller, M. Kalberer, R. Chirico, J. Dommen, C. Astorga, R. Zimmermann, N. Marchand, S. Hellebust, B. Temime-Roussel, U. Baltensperger and A. S. H. Prévôt, *Nat. Commun.*, 2014, **5**, 3749.

39 A. Saffari, S. Hasheminassab, D. Wang, M. M. Shafer, J. J. Schauer and C. Sioutas, *Atmos. Environ.*, 2015, **120**, 286–296.

40 A. de Vizcaya-Ruiz, M. E. Gutiérrez-Castillo, M. Uribe-Ramirez, M. E. Cebrián, V. Mugica-Alvarez, J. Sepúlveda, I. Rosas, E. Salinas, C. Garcia-Cuéllar, F. Martínez, E. Alfaro-Moreno, V. Torres-Flores, A. Osornio-Vargas, C. Sioutas, P. M. Fine, M. Singh, M. D. Geller, T. Kuhn, A. H. Miguel,

A. Eiguren-Fernandez, R. H. Schiestl, R. Reliene and J. Froines, *Atmos. Environ.*, 2006, **40**(suppl. 2), 583–592.
41 M. Steenhof, I. Gosens, M. Strak, K. J. Godri, G. Hoek, F. R. Cassee, I. S. Mudway, F. J. Kelly, R. M. Harrison, E. Lebret, B. Brunekreef, N. A. Janssen and R. H. Pieters, *Part. Fibre Toxicol.*, 2011, **8**, 26.
42 L. Ntziachristos, J. R. Froines, A. K. Cho and C. Sioutas, *Part. Fibre Toxicol.*, 2007, **4**, 5.
43 T. Fang, V. Verma, H. Guo, L. E. King, E. S. Edgerton and R. J. Weber, *Atmos. Meas. Tech.*, 2015, **8**, 471–482.
44 C. Samara, A. Kouras, K. Kaidoglou, E.-N. Emmanouil-Nikoloussi, C. Simou, M. Bousnaki and A. Kelessis, *Sci. Total Environ.*, 2015, **532**, 327–336.
45 N. A. H. Janssen, A. Yang, M. Strak, M. Steenhof, B. Hellack, M. E. Gerlofs-Nijland, T. Kuhlbusch, F. Kelly, R. Harrison, B. Brunekreef, G. Hoek and F. Cassee, *Sci. Total Environ.*, 2014, **472**, 572–581.
46 R. J. Delfino, C. Sioutas and S. Malik, *Environ. Health Perspect.*, 2005, **113**, 934–946.
47 R. J. Delfino, N. Staimer, T. Tjoa, D. L. Gillen, J. J. Schauer and M. M. Shafer, *J. Exposure Sci. Environ. Epidemiol.*, 2013, **23**, 466–473.
48 J. G. Charrier, N. K. Richards-Henderson, K. J. Bein, A. S. McFall, A. S. Wexler and C. Anastasio, *Atmos. Chem. Phys.*, 2015, **15**, 2327–2340.
49 J. Mirowsky, C. Hickey, L. Horton, M. Blaustein, K. Galdanes, R. E. Peltier, S. Chillrud, L. C. Chen, J. Ross, A. Nadas, M. Lippmann and T. Gordon, *Inhalation Toxicol.*, 2013, **25**, 747–757.
50 D. Wang, P. Pakbin, M. M. Shafer, D. Antkiewicz, J. J. Schauer and C. Sioutas, *Atmos. Environ.*, 2013, **77**, 301–310.
51 A. Saffari, N. Daher, M. M. Shafer, J. J. Schauer and C. Sioutas, *J. Environ. Sci. Health, Part A: Toxic/Hazard. Subst. Environ. Eng.*, 2014, **49**, 441–451.
52 S.-L. Huang, W.-L. Cheng, C.-T. Lee, H.-C. Huang and C.-C. Chan, *J. Toxicol. Environ. Health, Part A*, 2002, **65**, 1261–1272.
53 M. D. Ferguson, C. Migliaccio and T. Ward, *Inhalation Toxicol.*, 2013, **25**, 766–773.
54 I. Boldogh, A. Bacsi, B. K. Choudhury, N. Dharajiya, R. Alam, T. K. Hazra, S. Mitra, R. M. Goldblum and S. Sur, *J. Clin. Invest.*, 2005, **115**, 2169–2179.

Faraday Discussions

PAPER

Twenty years of ambient observations of nitrogen oxides and specified hydrocarbons in air masses dominated by traffic emissions in Germany

Christian Ehlers, Dieter Klemp, Franz Rohrer,* Djuro Mihelcic, Robert Wegener, Astrid Kiendler-Scharr and Andreas Wahner

Received 18th November 2015, Accepted 9th December 2015
DOI: 10.1039/c5fd00180c

The analysis of the individual composition of hydrocarbon (VOC) mixtures enables us to transform observed VOC-concentrations into their respective total VOC-reactivity *versus* OH radicals ($R_{VOC} = \Sigma(k_{OH+VOC_i} \times [VOC_i])$). This is particularly useful because local ozone production essentially depends on this single parameter rather than on the details of the underlying hydrocarbon mixture (Klemp *et al.*, *Schriften des Forschungszentrums Jülich, Energy & Environment*, 2012, **21**). The VOC composition also enables us to pin down the major emission source of hydrocarbons in urban areas to be petrol cars with temporarily reduced catalyst efficiency (the so-called cold-start situation) whereas the source of nitrogen oxides ($NO_x = NO + NO_2$) is expected to be nowadays dominated by diesel cars. The observations in the vicinity of main roads in German cities show a decrease in the ratio of OH reactivities of VOC and NO_2 (R_{VOC}/R_{NO_2}) by a factor of 7.5 over the time period 1994–2014. This is larger than the expected decrease of a factor of 2.9 taking estimated trends of VOC and NO_x traffic emissions in Germany (Umweltbundesamt Deutschland, *National Trend Tables for the German Atmospheric Emission Reporting*, 2015), during this time period. The observed reduction in the R_{VOC}/R_{NO_2} ratio leads to a drastic decrease in local ozone production driven by the reduction in hydrocarbons. The analysis reveals that the overall reduction of ozone production benefits from the low decrease of NO_x emissions from road traffic which is a consequence of the eventual absence of catalytic converters for nitrogen oxide removal in diesel cars up to now.

1 Introduction

Over the last 20 years (1994–2014), the compliance with ozone regulations of the European Union (EU) for the protection of human health has slowly improved in Germany. On the other hand, the average surface ozone concentration measured at the German Global Atmosphere Watch (GAW) station in Hohenpeissenberg has

Research Centre Jülich, Institute IEK-8:Troposphere, 52428 Jülich, Germany. E-mail: f.rohrer@fz-juelich.de

remained almost constant over the last 25 years (Fig. 1). To put this into a proper perspective, these observations in Germany can be compared to ozone data from the GAW station at the western coast of Europe in Mace Head, Ireland. Ozone is photochemically produced in the troposphere by the degradation of VOC in the presence of nitrogen oxides,[1] primarily over continents. In the troposphere, ozone is relatively long lived and has a life-time of about one month. As a consequence, its near-surface concentration shows only little variability over large areas. The predominantly westerly winds transport so-called background ozone from America to Europe's West Coast (see ref. 2). Observations at Mace Head give a very good picture of these ozone background concentrations which are then altered by photochemical production and surface deposition during the further transport of air parcels over France and Germany towards the East. At Hohenpeissenberg, the observed annually averaged surface ozone concentrations are larger then those at Mace Head during the 1990s, similar from 1999 to 2006, and smaller after 2007. The magnitude of this decrease at Hohenpeissenberg relative to Mace Head is on the order of 10–15% over 25 years. In fact, this decrease has not only been caused by changes in Germany since air parcels on their way from the Atlantic also pass France and other European countries. But nevertheless these observations show a general trend towards smaller ozone production in Western Europe.

The annually averaged concentrations themselves at Hohenpeissenberg are far lower than the target value for the protection of human health defined in the EU ozone regulations of 120 μg m^{-3} (for averaged episodes of 8 hours). A better insight into ozone reduction achievements in Germany is possible by looking at episodes of large ozone concentrations, popularly called ozone-alerts (see also Section 4.3). And here the improvements are obvious. Especially, the violations of the EU ozone alarm threshold (240 μg m^{-3}) vanished completely[4] (Fig. 2). These effects are likely to be caused largely by trends of pollutants emitted by road traffic in city districts. Here we show ambient observations of nitrogen oxides (NO_x) and hydrocarbons (VOC) in urban environments spanning this time period. To understand the observed trends of the ratio of VOC/NO_x in inner city districts, we

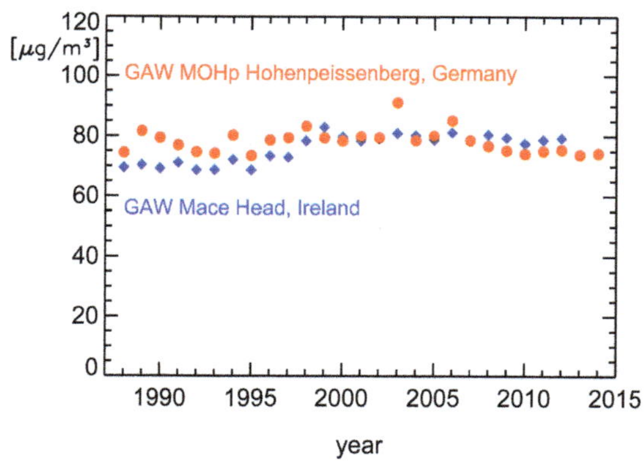

Fig. 1 Averaged annual surface ozone concentrations observed at the GAW stations Mace Head,[2] Ireland (blue diamonds), and Hohenpeissenberg, Germany (red circles). The data from Hohenpeissenberg for the time period until 2010 are taken from the literature.[3] The later data were kindly supplied by the German Meteorological Service (DWD) and are preliminary.

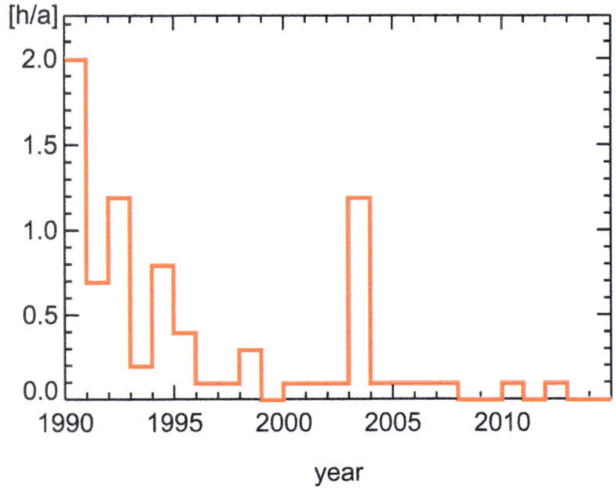

Fig. 2 Average number of hours per year for all survey stations in Germany exceeding the European ozone alarm value (240 μg m^{-3}, see ref. 4).

will first discuss the basic features of diesel and petrol motors, the effect of catalytic converters, and the trend of cars and fuel use in Germany.

1.1 Basic features of diesel and petrol engines

The major difference between diesel and petrol engines is the concentration of oxygen inside the cylinders when the fuel vapor is ignited. That depends on the fuel-to-air ratio λ which is determined by the engine management, nowadays handled by an on-board computer. If the amount of oxygen matches the requirement of stoichiometrical combustion, all of the fuel is burned and no oxygen is left over. This is the case in the engine of a petrol car. That specific fuel-to-oxygen ratio is called $\lambda = 1$. Its compliance is monitored at all times by a λ-sonde and adjusted by the on-board computer within very small limits. Diesel engines use a very large excess of oxygen at approximately $\lambda = 3$ which does not permit verification by a λ-sensor. Such an oxygen excess has important consequences for the emissions of pollutants by the car. Due to the larger amount of air mixed with the fuel, the temperature inside the cylinder of a diesel car is lower than that of a petrol car. That is responsible for the smaller NO_x emissions of a diesel compared to a petrol engine[5] because higher temperatures favor the production of nitrogen oxides.[5] In both types of engines NO_x is almost exclusively emitted in the form of NO. A higher amount of oxygen favors the oxidation of VOC so that their emission is also much smaller in diesel than in petrol cars. Fig. 3 illustrates these basic features of the two engine concepts. As a consequence, in the absence of catalytic converters, diesel cars emit much less VOC and NO_x than petrol cars. That was the case until 1985 (introduction of the G-Kat in Germany) when subsequently for both kinds of car catalytic converters were introduced to meet the requirements of a more stringent environmental policy.

1.2 Catalytic converters for diesel and petrol engines

For petrol cars, the so called 3-way-catalyst or G-Kat oxidizes CO and VOC with oxygen to CO_2 and reduces NO with CO to N_2 and CO_2. Since oxidation and reduction processes are taking place at the same time, this kind of catalytic

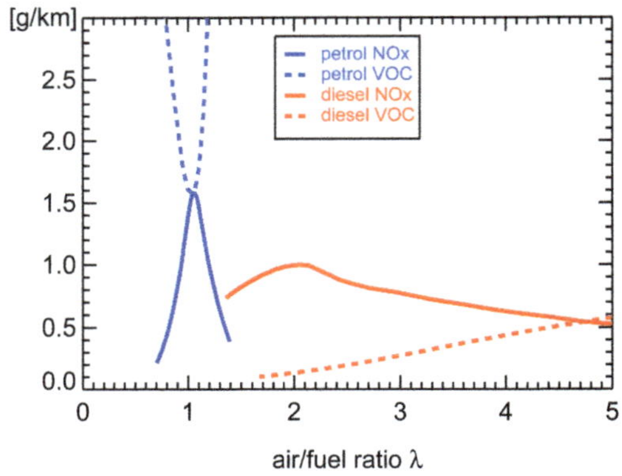

Fig. 3 NO_x and VOC emissions of diesel and petrol passenger cars without catalytic converters; replotted using published mixing ratio measurements (see ref. 6) multiplied with λ to be proportional to g km^{-1} and then scaled to the values of Table 1 at $\lambda = 1$ for petrol and $\lambda = 3$ for diesel passenger cars.

converter relies on the exact compliance of a stoichiometrical combustion at $\lambda = 1$ and therefore on the use of a λ-sonde. A little bit on either side of $\lambda = 1$ and either NO or CO-VOC are less efficiently converted by the catalyst. When this kind of converter is warm enough, it can reduce VOC and NO_x emissions by a factor of 100 (VOC) and 10 (NO_x) (see Table 1 and Fig. 4).

Since diesel engines operate at an oxygen excess ($\lambda = 3$), the so called diesel-oxidation-catalyst or DOC can only oxidize CO and VOC to CO_2, but cannot reduce NO_x at the same time. The reduction of CO and VOC by a warm DOC is also a factor of 100 (see Table 1 and Fig. 5). The higher oxygen amount in the exhaust of diesel cars is responsible for a shift of the NO_2/NO_x ratio by the DOC (depending on the catalytic material) towards larger NO_2 and smaller NO concentrations (Fig. 5). In comparison, petrol cars almost exclusively emit NO and only small amounts of NO_2.

Table 1 EU emission factors for passenger cars (g km^{-1})

	Petrol cars		Diesel cars	
Type of car	NO_x (as NO_2)	VOC (as HC)	NO_x (as NO_2)	VOC (as HC)
Without catalytic converter[8]	1.39–1.76	1.09–2.08	0.63–0.95	0.14–0.39
EURO-1 (1.1.1993[a])	0.97[b]		1.13[b]	
EURO-2 (1.1.1997[a])	0.5[b]		0.90[b]	
EURO-3 (1.1.2001[a])	0.15	0.20	0.50	0.56[b]
D4[d]	0.08	0.07		
(This paper[c], 2001–2002)	0.21	0.09	0.32	0.008
EURO-4 (1.1.2006[a])	0.08	0.10	0.25	0.30[b]
EURO-5 (1.1.2011[a])	0.06	0.10	0.18	0.23[b]
EURO-6 (1.9.2015[a])	0.06	0.10	0.08	0.17[b]

[a] Date of official implementation of the respective EURO norm.[9] [b] VOC + NO_x. [c] Observed values, see Fig. 4 and 5. [d] See ref. 10.

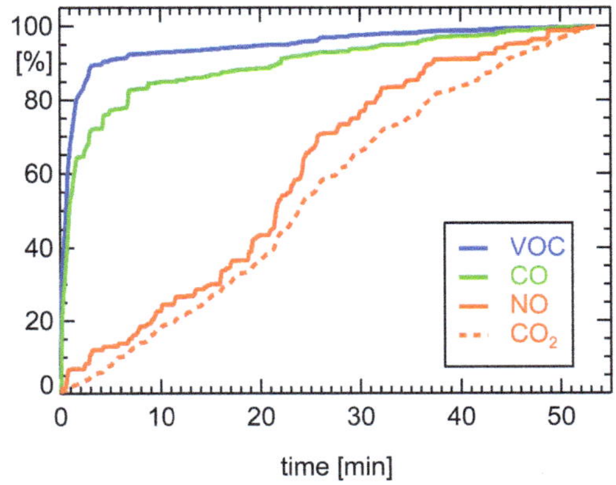

Fig. 4 Normalized cumulative emissions of CO (100% = 23.8 g corresponding to 0.80 g km^{-1}), CO_2 (262 g km^{-1}), NO (0.205 g km^{-1} as NO_2), and VOC (0.088 g km^{-1} as HC) derived during on-board measurements of a petrol car (see ref. 7) during a drive of 29.7 km length 11.07.2001 in Munich (passenger petrol car, year of manufacture: 2000, 85 kW, G-Kat, exhaust norm D4).

The efficiency of both kinds of converters, G-Kat and DOC, for the reduction of VOC depends on the temperature of their catalytic material. During the first part of a drive, when engine and converter are at ambient temperature, the converters have to be heated up by the exhaust gas to enable a significant catalytic effect. This warm-up period can last around 1–3 minutes (see Fig. 4 and 5). It is also termed "cold-start condition". At the beginning of this period, the car emits unconverted exhaust. After the warm-up, the full efficiency of the converters is reached, which offers a reduction of emitted VOC by a factor of 100. For a diesel car, the warming-up period is a little bit shorter than for a petrol car, since the excess of oxygen

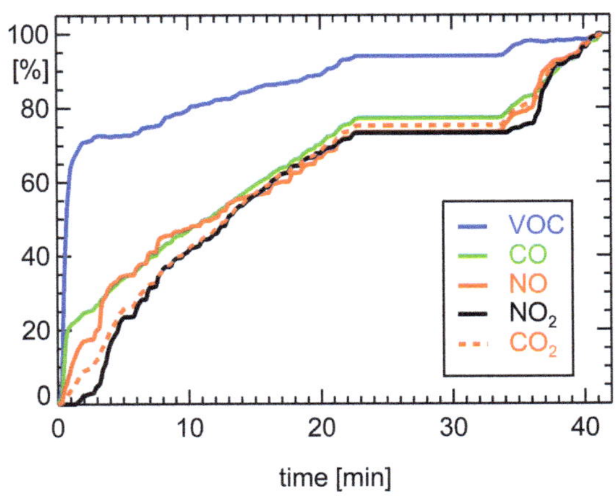

Fig. 5 Normalized cumulative emissions of CO (100% = 2.09 g corresponding to 0.116 g km^{-1}), CO_2 (180 g km^{-1}), NO (0.149 g km^{-1} as NO_2), NO_2 (0.173 g km^{-1}), and VOC (0.008 g km^{-1} as HC) derived during measurements of a diesel car (see ref. 7) during a drive of 18.0 km length 27.11.2002 executing a FTP-75-cycle (passenger diesel car, year of manufacture: 2002, 105 kW, DOC, exhaust norm EURO3).

favors the oxidation of VOC by a DOC. Taken together, the effective reduction of VOC for the full stretch of a 50 min drive which begins with a cold converter is approximately a factor of 10 for petrol and 30 for diesel passenger cars equipped with G-Kat or DOC, respectively. The situation is different when the converters are already warm at start-up. In this case, the converters are acting with full efficiency for the whole drive.

The efficiency of a G-Kat for NO_x reduction does not depend too much on the temperature of the converter (see Fig. 4). That seems to be against expectations. But one has to take into account the fact that during the warm-up period CO and VOC are not removed by the catalytic material so the reaction of NO with CO is favored by the large concentration of CO during that period of low temperature (see for example ref. 14). Later, when the catalytic material heats up, CO is removed but, on the other hand, the efficiency of the catalyst for the reduction of NO is growing. As a consequence, the removal of NO_x from the exhaust gas of a petrol car by a G-Kat seems not to have a significant warm-up period. But this overall reduction for NO_x is smaller than the potential reduction for VOC at optimal conditions. As a consequence, taking into account the warming-up period, the overall reduction efficiency of a G-Kat is a factor of 10 both for NO_x and for VOC for a petrol car. For a diesel car, the overall VOC reduction is a factor of 30. Since the initial VOC emission of the diesel engine before entering the DOC is ten times smaller than that of a petrol engine, a diesel passenger car emits approximately a factor of 30 less VOC than a petrol car during a drive inside a city district.

The NO_x emissions of petrol cars seem to be in accordance with EU regulations[11,12] (see Table 2). This is different for diesel cars. They show a significant inconsistency between allowed emission factors and real-world observations.[11,12] The real-world NO_x emissions of a diesel car today seem to be the same as 30 years ago (see Table 2). Together with the growing number of diesel cars with respect to the total number of passenger cars in Germany, the fraction of NO_x emissions of diesel cars relative to all passenger cars in Germany is estimated to be around 85% in the year 2015 (estimated from Table 3). Therefore, taking into account

Table 2 NO_x Real Drive Emissions (RDE) for passenger cars[11,12] (g km^{-1})[a]

Exhaust norm of car	Petrol cars (g km^{-1})[a]				Diesel cars (g km^{-1})[a]			
	CO_2[b]	RDE[c]	EU[d]	HBEFA[e]	CO_2[b]	RDE[c]	EU[d]	HBEFA[e]
None	228	2.38		1.22	186	1.04		0.97
EURO-1[f]	212	1.15		0.47	181	0.98		0.97
EURO-2[f]	204	0.74		0.33	171	0.94		1.02
EURO-3[f]	193	0.22	0.15	0.10	158	1.12	0.50	0.98
EURO-4[f]	178	0.08	0.08	0.10	152	0.96	0.25	0.85
EURO-5[f]	159	0.06	0.06	0.05	136	1.12	0.18	0.84
EURO-6[f]	143		0.06	0.05	122		0.08	0.84
EURO-6c[g]				0.05			0.08	0.14

[a] NO_x taken as NO_2. [b] UK emission factor estimates.[11] [c] Observed values in London by the RSD technique.[11] [d] EU NO_x emission factors for passenger cars. [e] HBEFA 3.2 (ref. 12) observed NO_x-emissions for CADC urban drive cycle. [f] EURO exhaust norm.[9] [g] EURO-6c expected value in HBEFA 3.2.[12]

Table 3 Ratio of the NO_x emissions for diesel/petrol passenger cars in Germany (estimated using real world observations, see ref. 11); the emissions of heavy-duty-cars are not included

Exhaust norm of car	Year	DPC/(DPC + PPC)[a] For number[b] [%]	DPC/(DPC + PPC)[a] For NO_x emissions[c] [%]
None	1980	5	2.2
EURO-1[d]	1993	13.5	10.3
EURO-2[d]	1997	13.8	14.9
EURO-3[d]	2001	13.9	41.5
EURO-4[d]	2006	21.9	72.4
EURO-5[d]	2011	26.6	83.2
EURO-6[d]	2015	31.2	

[a] DPC – diesel passenger cars, PPC – petrol passenger cars. [b] Inventory of passenger cars in Germany;[13] see Fig. 6. [c] Calculated from Table 2 (ref. 11) and from column 3, this table. [d] EURO exhaust norm.[9]

some heavy duty diesel vehicles, we can safely expect that the NO_x concentrations in German inner city districts in the last ten years were dominated by the emissions of diesel cars. In the year 1994, the situation was different. At that time, the majority of passenger petrol cars were not yet equipped with a G-Kat (see Fig. 8). Hence, the small number of diesel cars together with the basically lower NO_x emissions of diesel engines caused a low contribution of diesel cars to the NO_x concentrations in inner city districts in Germany.

1.3 Composition and emissions of the car fleet in Germany

The number of passenger cars in Germany rose from 32 million to 43 million cars in the time interval 1994–2014, if one accounts for some temporarily suspended/decommissioned cars in the numbers before 2008 (see Fig. 6). During the same time, the fraction of diesel cars increased[13] from 13.7% to 30.1%.

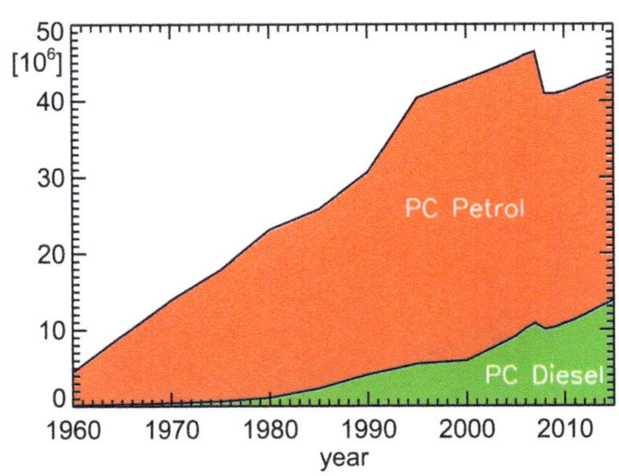

Fig. 6 Inventory of passenger cars (PC diesel and PC petrol) in Germany for the years 1960–2015 (see ref. 13); since January 1, 2008, only active vehicles are taken into account without counting temporary suspended or decommissioned cars.

If one wants to approach an estimate for NO_x and VOC emissions, one also has to take into account the average number of kilometers for each type of car (petrol or diesel car, passenger or heavy-duty car). This is not explicitly shown here, but reflected in the amount of fuel used by each class of cars (see Fig. 7). Another important boundary condition is the pollution classification of individual cars[15] (see Fig. 8) which also depicts the average lifespan of a technology change within the car fleet in Germany. Approximately 8 years after the implementation of a new pollution classification, half of the passenger car fleet still has the old classification.

If one takes suitable emission factors for NO_x and VOC for each pollution classification (see Table 1), one can estimate the amount of NO_x and VOC emitted by road traffic in Germany for each year (see Fig. 9, 10a and b). The shown inventories use emission factors called "HBEFA 3.1".[16] Astonishingly enough, these emission factors used to calculate the emission inventories are taken from observations at least since the year 2009 and are called "Real Drive Emissions"

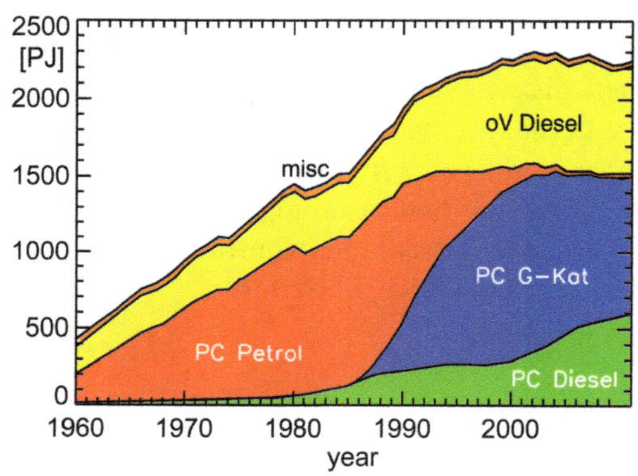

Fig. 7 Annual fuel consumption of road traffic in Germany for the years 1960–2011 (PC: passenger cars; oV: other vehicles including HGV, LCV, and bus (Heavy Goods Vehicles, Light Commercial Vehicles, buses); misc: miscellaneous petrol cars; natural and liquefied gas included in PC petrol; PJ: Peta Joule (=10^{15} Joule); replotted from TREMOD version 5.3 (see ref. 15)).

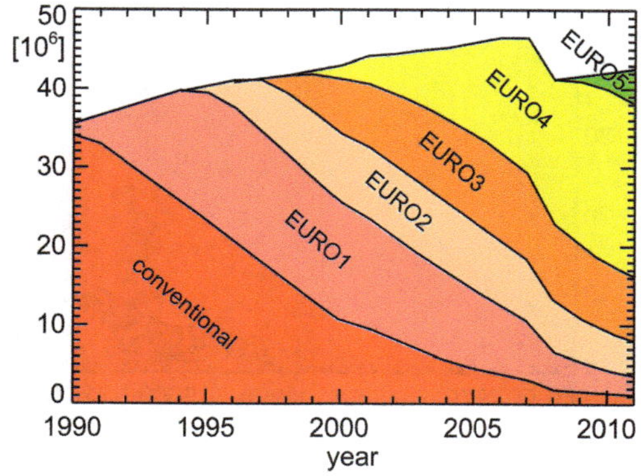

Fig. 8 Vehicle emission regulation categories for passenger cars in Germany for the years 1990–2011; replotted from TREMOD version 5.3 (see ref. 15 and Table 1).

(RDE). It appears that the current public discussion of the discrepancy between real world emissions and EU norm emissions (see for example Table 2) is already implemented into those emission estimates by using NO_x emission factors

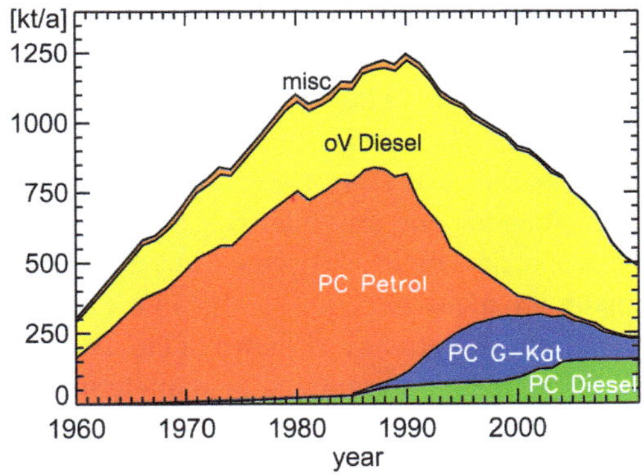

Fig. 9 NO_x emissions (as NO_2) of road traffic in Germany for the years 1960–2011 (PC: passenger cars; oV: other vehicles including HGV, LCV, and bus (Heavy Goods Vehicles, Light Commercial Vehicles, buses); misc: miscellaneous petrol cars and remaining types of road traffic); replotted from TREMOD version 5.3 (see ref. 15).

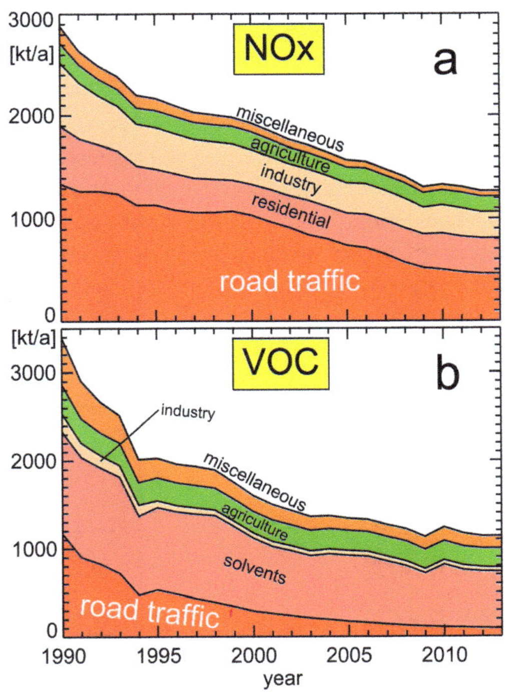

Fig. 10 NO_x (panel a) and VOC (panel b) emissions for different source categories (VOC without methane; replotted from a publication of the German Federal Environmental Agency UBA[18]); featured categories from bottom to top: (a) road traffic, residential, industry, agriculture, miscellaneous, and (b) road traffic, solvents, industry, agriculture, miscellaneous. The composition of the different categories is shown in Section 6.

around 0.8 g km^{-1} for all diesel passenger cars.[17] The intrinsic assumption in the following considerations is the expectation that the trend of concentrations in inner city districts will follow the trend of suitable emissions categories for the whole of Germany.

1.3.1 NO_x emissions in urban areas in Germany. For NO_x, one can compare the estimated emission inventories to observed concentrations, for example observations from inner city districts. If we focus on the time period 1995–2008, the estimated NO_x reduction by road traffic emissions in Fig. 10a is a factor of 1.9 in 13 years = 2.9 in 20 years. The observed trend of NO_x concentrations in inner city districts in Fig. 11 is a factor of 1.6 in 13 years = 2.4 in 20 years. That is close, but it might be coincidental. For example, the traffic in inner city districts contains a smaller proportion of heavy-goods-vehicles (HGV) and proceeds at a slower speed compared to highways and out-of-town traffic which all contribute to the total traffic emissions in Germany as shown in Fig. 10a. But the information at hand (TREMOD[15] and HBEFA[16]) gives enough parameters to try to make a better estimate for NO_x emissions in inner city districts ourselves.

If we take the published annual traveling distance of passenger cars (PC) in inner city districts in Germany of around 160×10^9 km a^{-1} (Fig. 12), the ratio of diesel to petrol passenger cars (Fig. 6), the emission regulation categories for passenger cars (Fig. 8), and the observed emission factors in London from Table 2, we can estimate the annual NO_x emissions of passenger cars in inner city districts of Germany (Fig. 13).

We then have to add the amount of NO_x emitted by heavy-goods-vehicles (HGV) and light-commercial-vehicles (LCV) by using the same kind of parameters for the traveling distance in inner city districts[15] (around 8×10^9 km a^{-1} for HGV and 17×10^9 km a^{-1} for LCV in 2010). All HGV and 93% of the LCV vehicles have a diesel engine.[15] The emission factor for these vehicles is estimated for a speed of 25 km h^{-1} to be 10 g km^{-1} for HGV and 1 g km^{-1} for diesel-LCV.[16] For simplicity we take 100% of LCV as diesel vehicles. The large emission factors for HGV at low speeds are explained in HBEFA 3.2:[12] "EURO V with SCR: the NO_x output in medium and high average engine loads (*e.g.* motorway and rural driving situations) is low. In contrast, NO_x emission factors are close to EURO III in urban

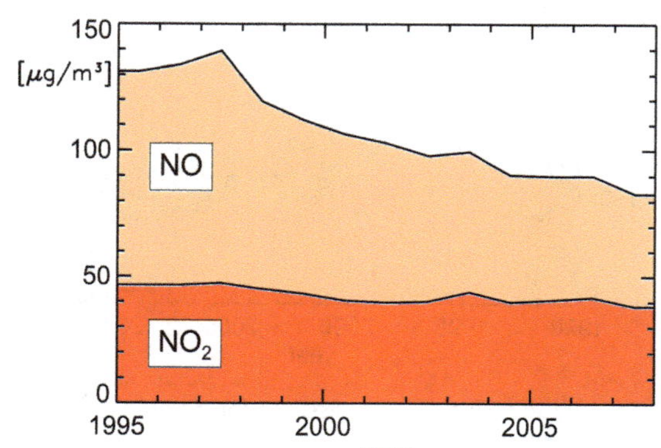

Fig. 11 Annually averaged NO_x concentrations (NO displayed as μg m^{-3} NO_2) close to traffic in inner city districts in Germany (replotted from UBA[19]).

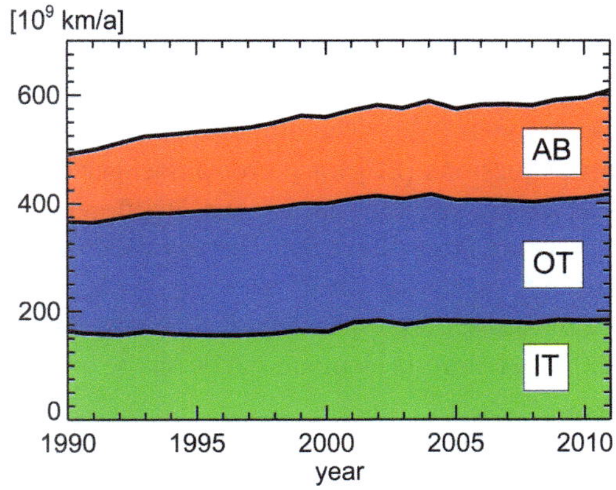

Fig. 12 Vehicle kilometers traveled by passenger cars in Germany; replotted from TREMOD version 5.3.[15] AB: Autobahn, OT: out-of-town, IT: in-town.

driving and in stop-and-go conditions due to low $DeNO_x$-performance of EURO V SCR after treatment in the low exhaust gas temperature range". Other categories for road traffic in urban areas (apart from "PC", "LCV", and "HGV") are not included because ref. 15 does not contain information about them. As a hint one can take ref. 20 which provides projected information for 2015. The vehicles within the categories "Bus", "Motorcycles", and "Vans" there have traveling distances in urban areas of 1.64, 5.31, and 2.85 $\times$ 10^9 km a^{-1}, respectively. The associated NO_2 emissions are 6.66, 0.77, and 2.34 kt a^{-1} which yield average emission factors of 4.1, 0.15, and 0.82 g km^{-1} NO_2. Taking these numbers, approximately 10 kt a^{-1} or 6% NO_2 would be missing in Fig. 13 around the year 2015.

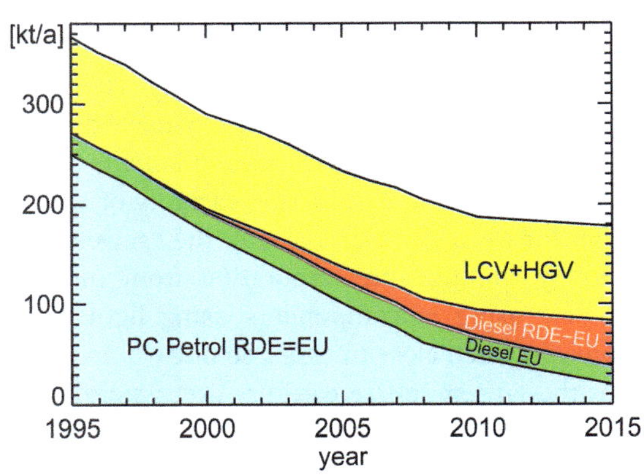

Fig. 13 Estimated NO_x traffic emissions in urban areas in Germany, this paper; for details see subsection 1.3.1. Featured categories from bottom to top are: "PC petrol RDE = EU": PC petrol car emissions which essentially show real-drive-emissions RDE in compliance with EU legislation, "PC diesel EU": intended PC diesel car emissions according to EU legislation, "PC diesel RDE-EU": surplus of PC diesel car real-drive-emissions RDE on top of the intended emissions according to EU legislation, "LCV + HGV": emissions of light-commercial-vehicles and of heavy-goods-vehicles. For other categories see text.

The final result in Fig. 13 shows the significant contribution of HGV and LCV diesel vehicles to the total amount of traffic NO_x emissions in urban areas in Germany, a point which already emerged from the published emission inventory for the whole of Germany in Fig. 9. The difference between real-world-emissions (RDE) and EU-norm-emissions (EU) for petrol passenger cars is very small and therefore cannot be seen in Fig. 13. That is entirely different for diesel passenger cars. As expected, there is a large discrepancy between real-world-emissions (RDE) and EU-norm-emissions (EU) for diesel passenger cars. Overall, the relative amount of NO_x emissions diesel passenger cars add up due to their non-compliance with EU regulations is estimated to be on the order of 25% for the year 2015 for Germany.

To test our simple approach we expanded the estimate for NO_x traffic emissions to the whole of Germany by adding portions for "out-of-town" and "Autobahn" of the different kinds of vehicles. For these situations we had to take a lower emission factor of 2.5 g km^{-1} for HGV allowing for the faster speed (see above). The total estimated NO_x traffic emission for the year 2012 is 467 kt a^{-1} from our simple approach. The published result from Fig. 10a is 463 kt a^{-1} in 2012, very close to our estimate. However, this good agreement should not obscure the simplicity of our approach. Another test of our estimate can be obtained from ref. 11, Table 5. This publication reports ratios of 26%, 25%, 26%, and 23% of NO_x RDE-emissions for the fleet of diesel PC, petrol PC, LCV, and HGV (including buses), respectively, encountered during campaigns in 2007–2010 in London, UK, quite comparable to Fig. 13. And the authors of this publication[11] explicitly state that their estimates "better reflect urban areas" than a national total.

The total decrease of NO_x emissions between the years 1995–2008 in Fig. 13 is a factor of 1.9 in 13 years = 2.9 in 20 years, or, directly using the years 1995–2015, it is a factor of 2.1 in 20 years. That is quite comparable to the observed trend of NO_x-concentrations in inner city districts in Germany (see Fig. 11). This analysis also confirms that the amount of diesel engine exhaust ($HGV + LCV_{diesel} + PC_{diesel}$) is expected to have a fraction of 90% with respect to NO_x traffic emissions in urban areas in Germany in the year 2015.

1.3.2 VOC emissions in urban areas in Germany. The situation is different for the VOC trend. First of all, we have no comprehensive data basis for observed VOC concentrations like for NO_x since VOC measurements are more demanding. Second, we have to deal with the dominating category of VOC emissions in the published inventory for Germany (see Fig. 10), the "solvents". These are VOCs which are emitted by spilling and evaporation from industrial sources, for example through degreasing of components using light petrol. In Germany, industrial zones are very often close to cities, so one would anticipate to find the VOC-fingerprint of the largest source category everywhere, thus in air samples from city districts too. As we show afterwards, significant concentrations are not found in ambient observations, neither in cities nor elsewhere (see also ref. 21). To give a more quantitative measure for this, the observed concentrations of solvents are less than 10% of the total VOC concentration in inner city districts. One would have to go very near to the industrial facilities to clearly detect the fingerprints of solvents.[21] This is astonishing, since the atmospheric lifetimes of the individual solvent species are on the order of days or weeks. That means, on the other hand, that their reactivity contribution will be very much less than 10%

which is the upper limit for their mass contribution. We therefore exclude the category "solvents" from the following considerations which aim at the air mass composition in inner city districts. If we only look at the trend of "road traffic" VOC-emissions, we see a reduction of a factor of 7.3 in 20 years. The other VOC emission categories do not seem appropriate ("agriculture") or are too small ("industry"). From these considerations, the emission inventories would suggest the trend of the VOC/NO_x mass ratio in inner city districts to be 7.3 in 2.5 years = 2.9 in 1 year for the time period 1994–2014.

2 Experimental

Measurements of VOCs and NO_x were conducted over a period of the last 20 years. Only the technical fundamentals will be described in the following. Individual conditions and optimization steps can be found in the listed references.

2.1 VOC detection

The gas chromatographic system for analyzing ambient air samples was developed in the late nineties[22] and has been continuously optimized and upgraded. In this context the current state of the system will be described.

The gas chromatographic system is based on a conventional gas chromatograph (Agilent 6890) which is equipped with a flame-ionization detector (FID) for the quantification, and with a mass-spectrometer (Agilent 5975C MSD) for the identification of the VOCs. To analyze concentrations at trace-gas level, a cryogenic pre-concentration system was developed, consisting of a sample loop (Silcosteel, 20 cm length, inner diameter 2 mm, packed with 0.25 mm glass beads) which was cooled down in the cold gas above liquid nitrogen (see Fig. 14). In the standard operation procedure, a volume of 800 mL was pre-concentrated in the sample loop at a flow of 80 mL min^{-1}. After thermal desorption at 120 °C, the sample was injected on a capillary column (type "DB-1", 120 m length, 0.32 mm inner diameter, 3 µm film thickness) at a temperature of −60 °C. The GC-oven was kept at −60 °C for 8 min, heated to 180 °C at a rate of 4 °C min^{-1}, then heated to 220 °C at a rate of 20 °C min^{-1}, and kept isothermal at 220 °C for 10 min. At the end of the capillary column, the carrier stream was split and fed towards the FID

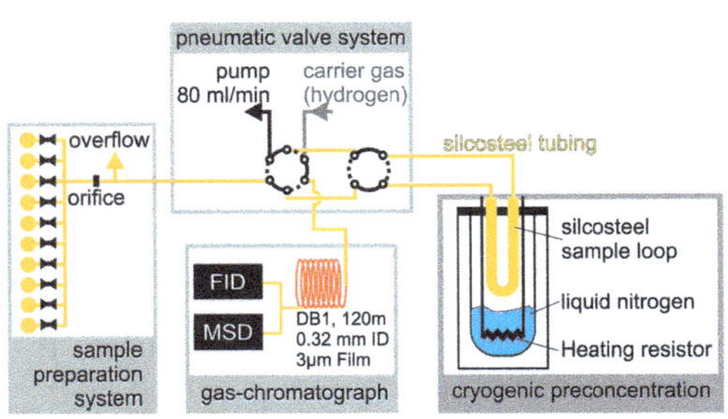

Fig. 14 Schematic overview of the gas chromatographic system optimized for VOC analysis ranging from trace level concentrations to exhaust concentrations consisting of a canister manifold, pneumatic valves, and a cryogenic preconcentration unit. The current setup allows the analysis of up to 10 canisters in unattended operation.

and MSD detectors. A typical FID chromatogram is shown in Fig. 15. The identification and quantification of the VOCs as well as routine checks for sensitivity by comparison to calibration reference standards (Apel-Riemer Environmental Inc., Denver, CO, USA) were performed in regular intervals. Overall, the precision of quantification was 5–6% and the limit of detection was 5–15 ppt (varying for the different VOC species[23]).

Using the previously described setup, online *in situ* measurements of VOCs were also performed aboard the mobile Laboratory-truck of the FZJ.[22,24–26] In other cases, when *in situ* measurements were not possible (*e.g.* measurements in a forest[27] to inspect biogenic emissions or measurements while driving on-road to inspect traffic emissions[23,28]), air samples were collected in evacuated Silcosteel canisters (SilcoCan, Restek).

If necessary, ozone was removed by a heated Silcosteel capillary prior to sampling into the cylinders. The canisters were then pressurized with synthetic air to reduce the relative humidity of the sample and hence possible water-

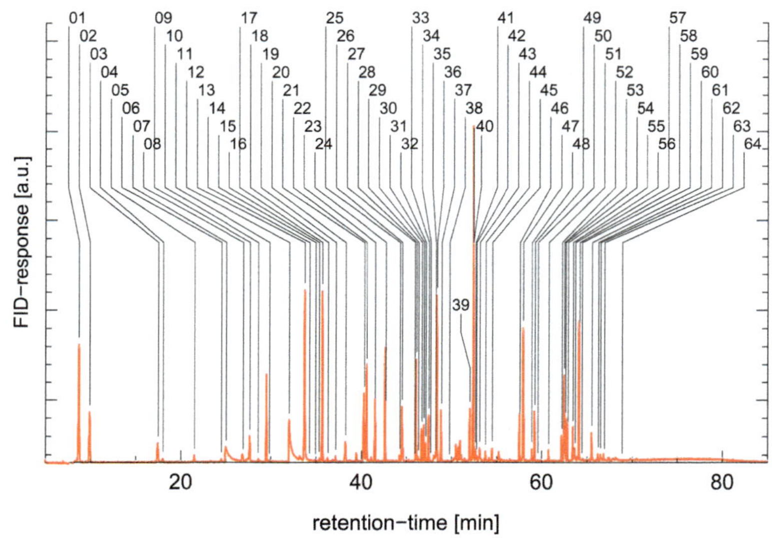

Fig. 15 Characteristic chromatogram from a cold petrol car. The peaks are labeled as follows: 01 – ethene; 02 – ethane; 03 – propene; 04 – propane; 05 – propyne; 06 – methylpropane; 07 – methanol; 08 – 1-butene/i-butene; 09 – *n*-butane; 10 – *trans*-2-butene; 11 – *cis*-2-butene; 12 – ethanol; 13 – 2-methylbutane; 14 – acetone; 15 – 1-pentene; 16 – 2-propanol; 17 – *n*-pentane; 18 – isoprene; 19 – *cis*-2-pentene; 20 – 2,2-dimethylbutane; 21 – cyclopentane/2,3-dimethylbutane; 22 – 2-methylpentane; 23 – 3-methylpentane; 24 – *n*-hexane; 25 – *trans*-1,3-hexadiene; 26 – methylcyclopentane/2,4-dimethylpentane; 27 – methylcyclopentene; 28 – benzene; 29 – 1-butanol; 30 – cyclohexane; 31 – 2-methylhexane; 32 – 2,3-dimethylpentane; 33 – 3-methylhexane; 34 – pentanal; 35 – 1-heptene; 36 – 2,2,4-trimethylpentane; 37 – heptane; 38 – 2,3-dimethyl-2-pentene; 39 – 2,3,4-trimethylpentane; 40 – toluene; 41 – 2-methylheptane; 42 – 4-methylheptane; 43 – 3-methylheptane; 44 – hexanal; 45 – *n*-octane; 46 – ethylbenzene; 47 – *m*/*p*-xylene; 48 – styrene; 49 – *o*-xylene; 50 – *n*-nonane; 51 – i-propylbenzene; 52 – *n*-propylbenzene; 53 – *m*-ethyltoluene; 54 – *p*-ethyltoluene; 55 – 1,3,5-trimethylbenzene; 56 – *o*-ethyltoluene; 57 – octanal; 58 – 1,2,4-trimethylbenzene/*t*-butylbenzene; 59 – *n*-decane; 60 – 1,2,3-trimethylbenzene; 61 – indane; 62 – 1,3-diethylbenzene; 63 – 1,4-diethylbenzene/butylbenzene; 64 – *n*-undecane.

interferences in the FID detector. The pressurized canisters were then connected to the canister manifold at the GC-system which allows the unattended analysis of up to 10 canisters at a time (see Fig. 14). Quantification was then performed using the same routines as for the *in situ* measurements.

The suitability of the Silcosteel canisters for VOC sampling and storage of humid samples for up to two days was extensively scrutinized and the stability of their content confirmed.[29] To analyze the VOC-composition in the exhaust of petrol- and diesel-vehicles, the gas was diluted with synthetic air prior to sampling in the canisters and then analyzed using the same gas-chromatographic system as for the ambient measurements.[7,28,30,31] If very high concentrations had to be expected, the sampling volume in the pre-concentration unit was limited to 80 mL to prevent overloading of the system.

The total hydrocarbon concentration in the exhaust of the different vehicles was analyzed with a commercially available total organic carbon detector (TOC; MICRO-FID 100, ABB Corp.) which is basically a flame-ionization-detector without a gas-chromatographic system in front of it.[7,28,30]

2.2 NO_x detection

NO_x-measurements in ambient air samples were based on the principle of chemiluminescence: NO reacts with ozone to an excited NO_2 molecule. Photons emitted during radiative disexcitation can then be detected by a photomultiplier and used for the quantification of the concentration of NO.[32,33] NO_2 was detected after conversion to NO using either a molybdenum converter[22,24] or a photolytic converter.[23,25,28]

To analyze the NO_x-concentrations in the exhaust of petrol- and diesel-vehicles, commercially available analyzers (LIMAS11, ABB corp.) based on UV-absorption were used.[7,28,31]

3 Results

3.1 Ambient observations

For the following considerations we will use NO_x and VOC observations not in the usual manner as concentrations or mixing ratios, but as reactivities. Reactivity denotes the reaction speed of a substance or of a group of substances with the OH radical, the principal agent in the photochemical production of ozone. The eminent significance of OH radicals within tropospheric chemistry is justified since during daytime more than 90% (ref. 1 and 34) of all VOC react with OH radicals.

Reactivity is expressed as reciprocal lifetime of OH in a certain reaction and has the unit s^{-1}. For example, VOCs are oxidized by OH to form more oxidized species. The final product of such an OH-oxidation pathway starting with VOC is, after several steps, CO_2. The important point here is that in each VOC-oxidation step ozone is produced as a by-product in the presence of nitrogen oxides, specifically in the presence of NO. The other nitrogen oxide species, NO_2, is also oxidized by reaction with OH to HNO_3 without producing ozone like VOCs. The reactivity of a certain NO_2 concentration, R_{NO_2}, can be calculated by multiplying the concentration value with the rate constant of the reaction of NO_2 with OH, k_{OH+NO_2}: $R_{NO_2} = k_{OH+NO_2} \times [NO_2]$. The same can be done for a specific VOC_i

substance: $R_{VOC_i} = k_{OH+VOC_i} \times [VOC_i]$. The sum of all individual VOC_i then gives the total VOC reactivity R_{VOC}. The individual rate constant parameterizations are taken from the literature[35,36] or from the chemical reaction mechanism MCM-3.1,[37] which is later-on also used to derive ozone production estimates for different air masses. For all reactivities shown in this paper we used a reference temperature of 298 K for input to the rate constant parameterizations. This has been done to ensure consistency with the ozone production approach discussed in Section 4.2. We note here, that the averaged ratio of the individual rate constants k_{OH+VOC_i}/k_{OH+NO_2} for a typical traffic VOC mix is temperature dependent with a slope of 10% per 10 K.

VOCs are typically measured using a gas-chromatographic system (GC). The single VOC_i are separated by such a GC-system and reach the detector at a specific time after injection of the gas sample, the so-called retention time. The retention time is only dependent on the physical properties of the individual VOC_i and on the GC system used. For a specific GC system, the retention times for the different substances are known by using reference air samples. The operator then transposes the raw signals of the different species, which establish the so-called chromatogram, into concentrations of the individual species and records their name and concentration in the form of a table. From this table one can then calculate the individual reactivity contributions of the individual VOC_i-species and add them up to obtain the total VOC reactivity, R_{VOC}. One can also transpose the original chromatogram showing raw signals into a chromatogram which shows peaks scaled to the reactivity of the different species. This is done by taking a Gaussian standard peak (which has an appropriate peak-width typical for the utilized GC-system), scale its maximum with the reactivity of a certain VOC_i and then add this synthetic Gaussian peak to a baseline data vector at the position of the retention time of that specific VOC_i. The appearance of such a reactivity-chromatogram is very different to the original chromatogram, because this technique emphasizes those VOC-species which are responsible for ozone production instead of showing a composition related to mass (FID raw signals are primarily related to the carbon atom content of the molecules). For that reason we will use these synthesized reactivity chromatograms (SRC) throughout this paper which still hold the look-and-feel of the original chromatograms and allow for a fast visual impression of the VOC-composition of an air sample. These SRCs will be used in a standardized form, that means they are divided by the largest occurring value in each particular chromatogram. The use of SRCs and the standardization procedure have the additional benefit of allowing comparison to different published VOC-patterns, if the species names and their concentrations are known from the publication (see Fig. 24).

3.1.1 VOC fingerprints. To show the specificity of GC patterns for individual sources we present reactivity chromatograms for a range of source situations and ambient observations. The look of such a chromatogram with its different peak positions and peak heights is called the "fingerprint". Since there might be hundreds of different VOC_i in a chromatogram, there are millions of different fingerprints possible. Fortunately, if one takes samples directly at the exhaust pipes of different cars, it is obvious that the engine type of a car generates a specific fingerprint which is very similar for cars of that engine type. There are slight differences from car-to-car, but the similarities between chromatograms of a certain engine type are striking. So those fingerprints, when known, can be used to determine the contribution of a certain type of car within a chromatogram of an ambient air sample. For example, measuring a large number of petrol cars, two kinds of fingerprints can be derived: the fingerprints for the cold-start and warm conditions of the car. These are clearly distinguishable from each other and from those of diesel cars as shown in Fig. 16–19.

It is important to note here that other sources for VOCs in ambient air masses have quite different but at the same time very specific fingerprints, for example the fingerprints of diesel fuel or of petrol fuel (both not shown here). The fingerprint from a forest site in Fig. 20 contains a single dominant peak caused by

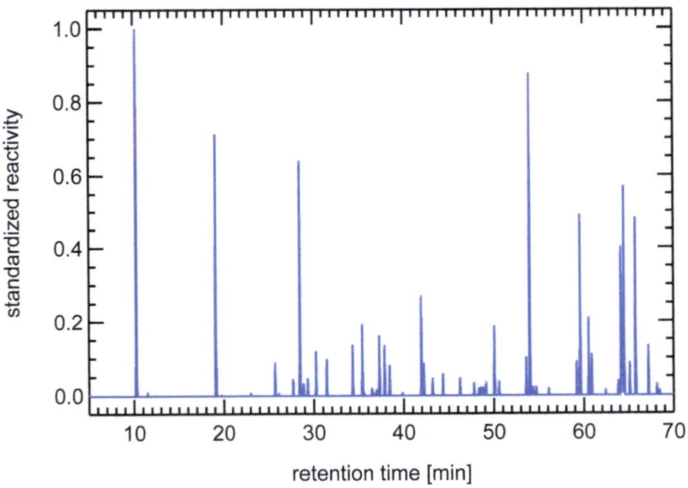

Fig. 16 Synthesized reactivity chromatogram of a petrol passenger car under cold-start conditions.

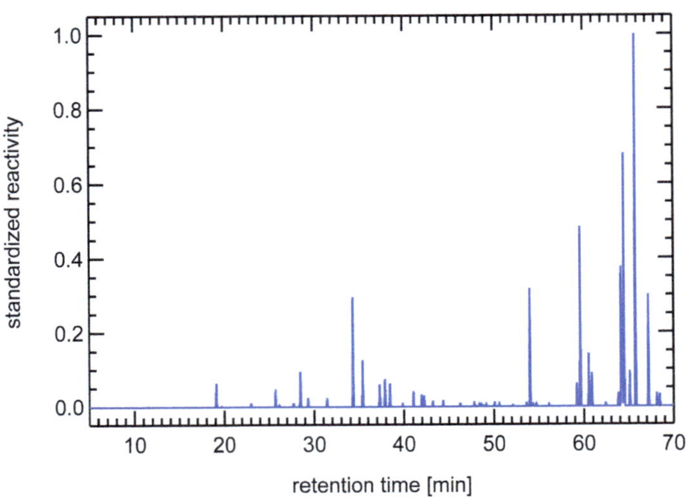

Fig. 17 Synthesized reactivity chromatogram of a petrol passenger car for warm conditions.

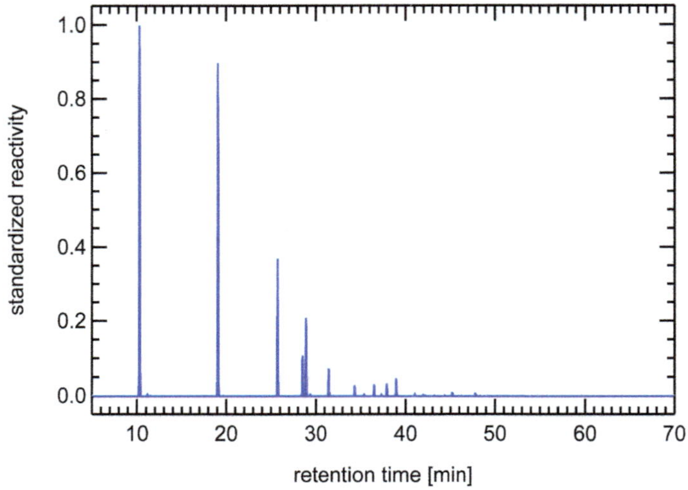

Fig. 18 Synthesized reactivity chromatogram of a diesel passenger car under cold-start conditions.

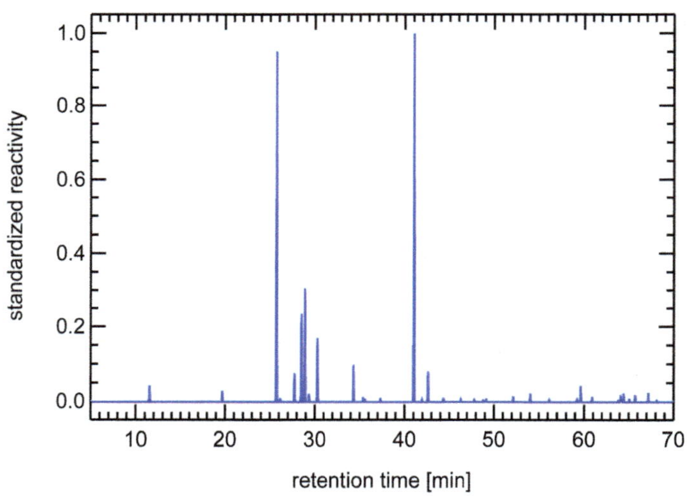

Fig. 19 Synthesized reactivity chromatogram of a diesel passenger car for warm conditions.

the large concentration and reactivity of isoprene. Isoprene is the most prominent VOC_i species from biogenic sources and is related to the emissions of broad-leafed trees like oaks and beeches. Other trees like conifers emit specific VOCs called mono-terpenes which are not shown here, but whose fingerprints are also clearly distinguishable from car emissions. The next large reactivity peak at a retention time of 65 min in Fig. 20 is such a monoterpene called sabinene which, as an exception, is also emitted by broad-leafed trees. Other fingerprints, for example from petrol cars or from solvents, are almost not visible, because airmasses originating in cities and traveling to remote locations are either

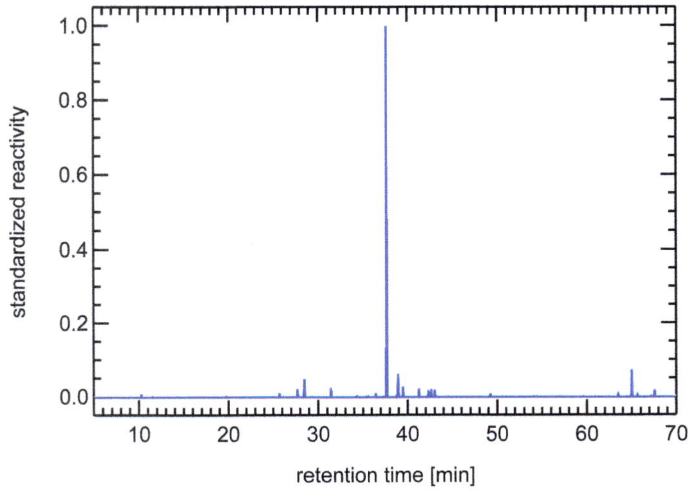

Fig. 20 Synthesized reactivity chromatogram of an air sample taken in summer 2003 in the forest near the Research Centre Jülich.

considerably diluted or their VOC content is degraded by reaction with OH radicals, a line-of-thought we will resume in Section 4.3.

3.1.2 VOC fingerprints found in air samples taken in inner city districts. With the information from the previous sections we can now try to match the source VOC fingerprints to those from ambient air samples, for example from inner city districts or roads. Just as an example, Fig. 21 shows the VOC fingerprint of an ambient air sample taken in 1997 in the inner city district of Munich. This fingerprint from Munich in 1997 can directly be compared to a fingerprint taken in 2014 in a street tunnel in Berlin (see Fig. 22) and to the exhaust of a petrol car

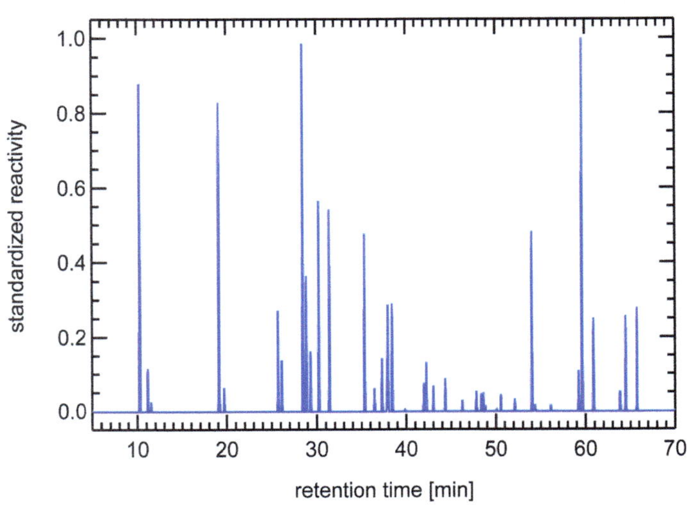

Fig. 21 Synthesized reactivity chromatogram of an air sample taken in 1997 in Munich, "Garmischer Straße".

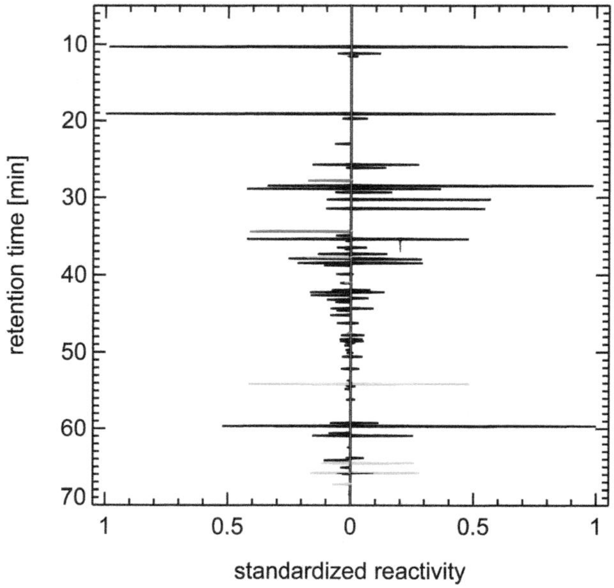

Fig. 22 Contrasting juxtaposition of synthesized reactivity chromatograms from air samples taken in 2014 in Berlin, "Tiergartentunnel" (left panel), and taken in 1997 in Munich, "Garmischer Straße" (right panel); red: methanol, ethanol, propanol; green: isoprene; blue: toluene, trimethylbenzenes.

under cold-start conditions (see Fig. 23). These juxtapositions show a striking concurrence between traffic dominated air masses in 1997 and 2014 and between air masses from a cold-start petrol car in 2015 and traffic emissions in 1997. The same similarity is found in the comparison of a single chromatogram of the exhaust of a petrol passenger car with the result of a literature study[38] (see Fig. 24). This proves the persistence of the fingerprint of petrol cars over the years regardless of car design, petrol fuel composition, or author. It also shows that the VOC fingerprint found in inner city districts is indeed related to the fingerprint of petrol cars under cold-start conditions, an observation which was already addressed in Section 1.2.

A close examination of the juxtapositions in Fig. 22–24 reveals one difference between the fingerprints of the 1990s and of today. That is the occurrence of methanol and ethanol in recent air samples of inner city districts and in petrol car exhaust. That feature is most likely caused by the addition of 5% or 10% alcohol to petrol fuel in Germany (fuel types called E5 and E10). But the overall effect of that change on the composition of the VOC reactivity in air masses dominated by traffic emissions seems to be negligible.

A recent publication[39] brought forward the argument that large hydrocarbons from diesel exhaust are underestimated by orders of magnitude because VOCs larger than C12 are not easily observed by current GC systems and therefore have been neglected. Including the reactivity contributed by these large VOCs, the authors of this publication[39] predict that diesel-related hydrocarbons contribute up to 50% of the ozone production potential in London during winter-time. We specifically designed a GC system to be able to observe VOCs in the region C8–C16

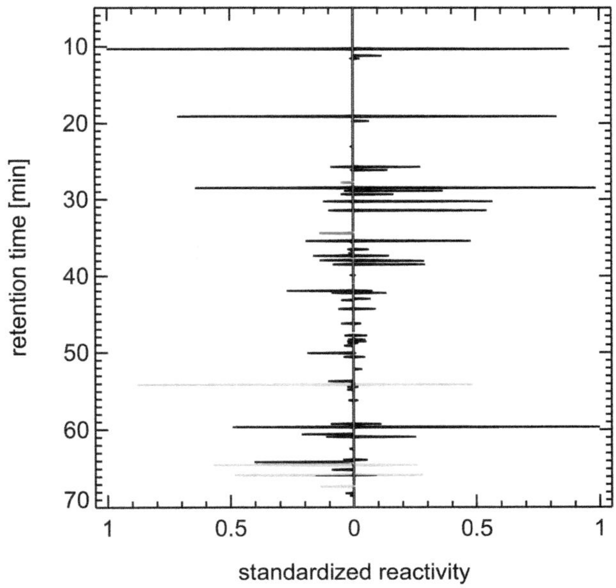

Fig. 23 Contrasting juxtaposition of synthesized reactivity chromatograms from air samples taken from the exhaust of a petrol car under cold-start conditions (left panel), and taken in 1997 in Munich, "Garmische Straße" (right panel); red: methanol, ethanol, propanol; green: isoprene; blue: toluene, trimethylbenzenes.

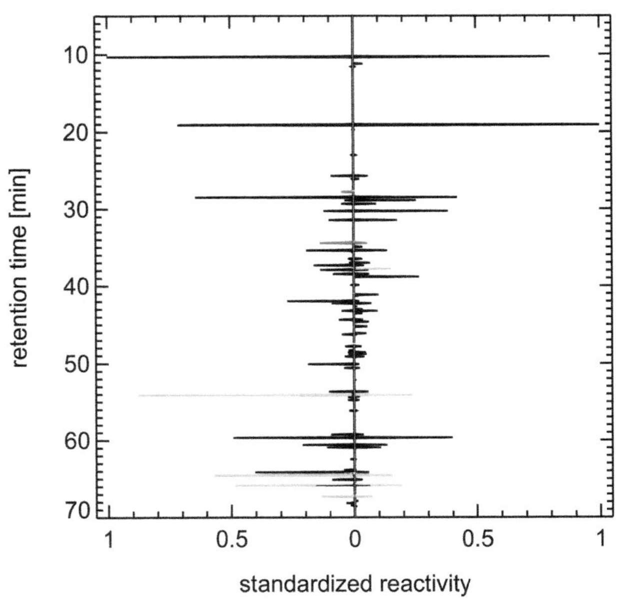

Fig. 24 Contrasting juxtaposition of synthesized reactivity chromatograms from a single air sample taken from the exhaust of a petrol passenger car under cold-start conditions (left panel), and from a literature study (evaluating 200 citations) for a typical petrol passenger car with GKat[38] (right panel); red: methanol, ethanol, propanol; green: isoprene; blue: toluene, trimethylbenzenes.

and found no evidence for such a statement ourselves. We took samples in a street tunnel in Bonn (Germany) and found only 0.3% of the total VOC mass above C11, the largest VOC included in our standard gas chromatographic analysis. Even if those hydrocarbons would have a larger OH rate constant than the typical VOC of car exhaust (about 1×10^{-11} cm^3 s^{-1}), their total contribution to the VOC reactivity in ambient samples dominated by traffic emissions is expected to be below 3%.

3.2 The ratio of VOC/NO_x in inner city districts of Germany

Here we show the result of ambient observations of the ratio of the reactivities of VOC/NO_2 from inner city districts of Wuppertal, Munich, Augsburg, Düsseldorf, Bad Homburg, and Essen over the years 1994 to 2015 (see Fig. 25). As mentioned before, the reactivities of VOC were determined by adding up the individual reactivities of all VOC_i compounds analyzed from air samples gathered by filling up appropriate Silcosteel canisters within the traffic of major roads of these cities. The reactivities of NO_2, R_{NO_2}, were derived for the same time periods from *in situ* observations of nitrogen oxide mixing ratios measured co-located to the canisters. To account for different ratios of NO and NO_2 in different traffic-, light-, and ozone-conditions, R_{NO_2} was calculated for an average NO_2 contribution of 75% of the sum of observed NO and NO_2. In that respect, the observed reactivity of NO_2 was calculated according to the function $R_{NO_2} = k_{OH+NO_2} \times (NO + NO_2) \times 0.75$. Following this approach, the ratio of R_{VOC}/R_{NO_2} in inner city districts of Germany was around 1.43 in the year 1994 and 0.19 in 2014.

This result can be compared to another long-term study[40] of traffic emissions from a spacious road tunnel, the Gubrist tunnel in Switzerland, but that comparison cannot be done straight forward. Since the traffic inside the Gubrist tunnel flows at an average speed of around 90 km h^{-1}, the ratio of R_{VOC}/R_{NO_2} is different compared to a city where the traffic flow is sometimes congested, the average speed is below 50 km h^{-1}, and the amount of cold-start petrol cars is larger. The fleet of passenger cars, diesel and petrol, and the contribution of heavy duty cars might not be the same as in Germany. A further difficulty is the type of data reported in the Gubrist tunnel study. That study reports emission factors for total VOC (TVOC) and NO_x and not reactivities. Due to all of these differences it is very difficult to derive and to compare absolute values of the R_{VOC} and R_{NO_2}

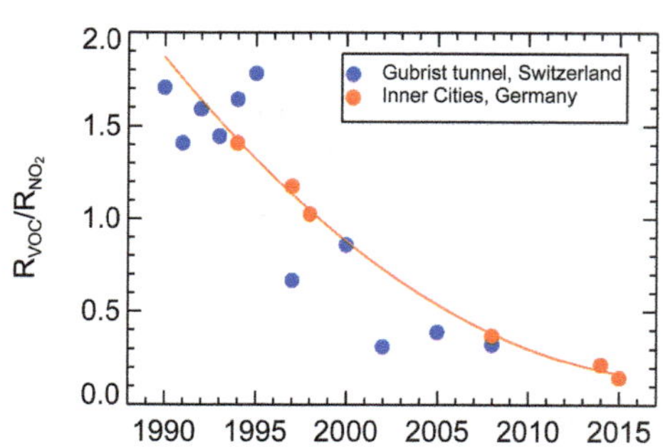

Fig. 25 Observed reactivity ratios of VOC and NO_2, R_{VOC}/R_{NO_2}, over the years 1990–2015 from inner city districts in Germany (this paper) and (scaled, see text) inside the Gubrist road tunnel[40] in Switzerland.

reactivities from the Gubrist tunnel. For our purpose it is enough to see the trend from this profound investigation offers an independent confirmation of our statement of a large decrease of the R_{VOC}/R_{NO_2} ratio in traffic emissions over the last two decades in Germany. To compare the trend of the Gubrist tunnel results to the trend in our own data set, one has to assume that the VOC composition of traffic emissions did not change over time, something we called a "persistent VOC fingerprint" before. With this assumption, one can simply scale the scattered mass ratios TVOC/NO_x from the Gubrist tunnel study of the time period 1990–1995 to match the best-fit line (the thin red line in Fig. 25) and compare the two trends. Basically, the trends for the Gubrist road tunnel in Switzerland and for inner city districts in Germany show the same behaviour, a decrease of a factor of 7.5 of the reactivity ratio R_{VOC}/R_{NO_2} within 20 years (derived for the thin red line in Fig. 25). That trend is much larger than what we expected from the VOC and NO_x emission inventories for Germany in Section 1.3.2, that is a factor of 2.9 in 20 years. For this comparison we again use the assumption that mass ratio and reactivity ratio for VOC/NO_x for traffic emissions have a simple and constant scaling factor. Since we have shown in Section 1.3.1, that the observed concentration trend in traffic dominated inner city districts of Germany follows the estimated trend for NO_x traffic emissions, this discrepancy has to come from the trend of traffic VOC emissions. There is no published observed average trend of VOC concentrations for inner city districts in Germany to compare to. Our observations of the R_{VOC}/R_{NO_2} ratios suggest a larger decrease of VOC traffic emissions than the published emission inventories. One reason might be the difficulty arising in the determination of the VOC emission trend from the "cold-start" burst of VOCs in the first minutes after starting a petrol car. This short transitory period during the drive emits on the order of 90% of all VOC emissions of a petrol car (see Section 1.2). It is very difficult to incorporate this kind of process into an inventory calculation. For NO_x, such a difficulty does not appear. Furthermore, it has to be expected that the car manufacturers have shortened that time period by optimizing the catalytic converter efficiency of the G-Kats over the years. Anyhow, such a large decrease of VOC relative to NO_x offsets the photochemical ozone production rate, as will be studied in Section 4.3.

For other countries, the VOC/NO_2 ratio and its trend might be different. It depends on the specific vehicle fleets for each country, *e.g.* if they are dominated more by petrol-engined vehicles rather than diesel or *vice versa*. For example, American vehicles have mostly used petrol engines for decades (3% (ref. 41) diesel contribution in 2012), whereas in France 78% (ref. 41) of vehicles use diesel fuel. And Switzerland and Germany are in-between with 22% (ref. 41) and 26% (ref. 13) of the passenger fleet using diesel fuel in 2010, respectively.

4 Discussion

The following section conceptually depicts the chemical changes an air parcel experiences when it is released inside a city, is loaded with traffic emissions inside a street tunnel or a main road, travels forward driven by the wind, passes minor streets and the city limit, going further into more rural environments, finally reaching a large forest or agriculturally dominated region. Here, we are not interested in details, just focusing on the concept. In reality, emissions might be different for different roads or cities, the wind might blow faster or slower, the light might change due to clouds and so on, there are a magnitude of possibilities which influence a specific location or point in time. We have to bear in mind that our concept wants to show only the generic features and does not have to be applicable for all possible conditions.

4.1 The photochemical production of ozone

In this paper, we want to explore, how the change of the VOC and the NO_x emissions from traffic over the last 20 years will affect the production of ozone, a secondary pollutant and a risk for human health. We have seen that for this time period the observed hazardous ozone concentrations decreased considerably in Germany and almost reached a state where they are no longer a risk for human health (Section 1). To substantiate the background for this observation, we have to understand how ozone is produced by photochemical processes, especially in smog episodes. Ozone is produced in the degradation of VOCs by OH radicals in the presence of nitrogen oxides. In this degradation process, peroxy radicals ROx are generated which react with nitrogen monoxide NO to form nitrogen dioxide NO_2. NO_2 is subsequently photolysed to give back NO and generate ozone. Besides this pathway, NO_2 can react with OH to form nitric acid. This latter process reduces the potential degradation of VOCs since the concentration of OH radicals is lessened. Nitrogen oxides have therefore a dual impact on ozone production. In the form of NO, they amplify the production rate, in the form of NO_2 they withdraw the major agent from the ozone production process, the OH radical. The production rate of ozone depends in a nonlinear way on the concentrations of nitrogen oxides, VOCs, ozone itself, light intensity, humidity, temperature, and other parameters. To understand these complex relationships we used the master chemical mechanism MCM 3.1.[37]

4.2 The box model approach

The master chemical mechanism MCM 3.1,[37] slightly extended with new information,[7] consists of 14 025 reactions and 4575 species. It is used here to calculate the concentration of OH radicals in the presence of certain amounts of VOC and NO_x for typical summertime conditions in Germany. The major ingredients of this set up were a temperature of 25 °C, a relative humidity of 75%, an ozone concentration of 60 ppb, and photolysis frequencies J_{O^1D} and J_{NO_2} of 2.9×10^{-5} s^{-1} and 8.4×10^{-3} s^{-1}, respectively. The ratio of NO_2/NO_x is on the order of 0.75 under these conditions. We started with NO_2 and VOC concentrations equivalent to R_{NO_2} and R_{VOC} reactivities each of 100 s^{-1}. The relative mix of the different VOCs was borrowed from the results of ambient observations in Munich 1997. The VOC fingerprint of these air masses, very similar to cold-start petrol car conditions, was already shown in Fig. 21. The fate of this NO_2 & VOC mixture was then calculated by the model for a time interval of 10 minutes to allow NO and NO_2 and the OH radicals to reach a quasi steady-state. After that time, the local ozone production rate P_{O_3} was calculated according to $P_{O_3} = \Sigma(k_{ROx_i+NO} \times [ROx_i] \times [NO])$, where the ROx_i denote the different peroxy radicals including HO_2. This ozone production rate and the related NO_2 and VOC reactivities were stored. The whole process started again with a slightly different NO_2 & VOC mixture until the desired surface area of NO_2 and VOC each spanning 0.1 to 100 s^{-1} was filled with enough supporting points to allow for a thorough representation of the internal structure within the ozone production process (see next the section).

4.3 The result of box model calculations

Fig. 26 shows the result of our approach to understand the relation between the reactivities R_{VOC} and R_{NO_2} with respect to photochemical ozone production. The important point here is that the use of the parameter "OH-reactivity" instead of concentrations simplifies the understanding of ozone production, since this integrative single parameter almost entirely determines the magnitude of the ozone production rate,[7] and not the complex peculiarities of the different degradation processes of the set of VOCs. This feature enables us to follow an air parcel within the same framework even if the VOC mixture is changing with time, for example by the addition of biogenic emissions. In Fig. 26, we start at reactivities for R_{NO_2} and R_{VOC} of 70 s^{-1} and 100 s^{-1}, respectively, to represent typical ambient concentrations in a street tunnel or main road in a German city and the observed reactivity ratio R_{VOC}/R_{NO_2} of 1.4 for the year 1994. On its way to leave the city, that air parcel picks up some more emissions when passing minor streets, but those are small contributions compared to the starting concentrations. The

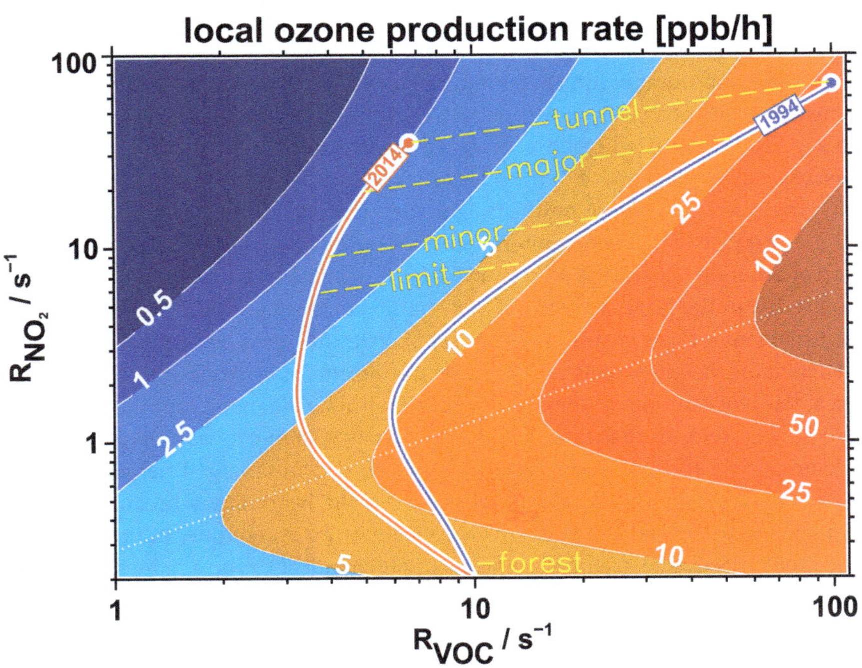

Fig. 26 Local ozone production rate (ppb h^{-1}) for typical summer time conditions in Germany (result of MCM-3.1-model calculations[7]) as a function of VOC and NO_2 expressed by their reactivity *versus* OH radicals (R_{VOC} and R_{NO_2}). The two trajectories denote typical air masses for 1994 (blue) and 2014 (red) traveling from inner city locations (street tunnel, major and minor streets, city limit) into regions dominated by biogenic emissions (forest). The dotted line separates the VOC (above the line) and NO_x (below the line) limited regimes of ozone production.

dominant process while travelling here is dilution, that means mixing-in air containing very much lower concentrations from the background. This will influence both the VOC and the NO_2 concentrations in an equal manner. So for the first part of the air trajectory shown for the year 1994 in Fig. 26, it is following a dilution path. As the reactivities approach background conditions (we have chosen $R_{VOC} = 3\ s^{-1}$ and $R_{NO_2} = 0.2\ s^{-1}$), the trajectory seems to turn around. The reason is that when the trajectory moves towards the forest in our example, we mix-in more and more biogenic VOC emissions. The end-point of the trajectory reaches background conditions for R_{NO_2} but approaches 10 s^{-1} for R_{VOC}, a value we would expect for a forest in Germany. Then we repeat that course of action to represent the year 2014, this time starting at 35 s^{-1} for R_{NO_2} and 6.5 s^{-1} for R_{VOC} to follow the observed trends from Section 3.2 (a factor of 2 for R_{NO_2} and a factor of 15 for R_{VOC}). The end-point of the 2014-trajectory is the same as that of the 1994-trajectory, since we assume that the background concentrations at the final point inside the forest did not change over time. But the trajectory itself to reach that end-point is very different. Due to the large decrease of the starting R_{VOC} reactivity, the trajectory seems to avoid the regions of large ozone production. The reason is that due to the small R_{VOC}/R_{NO_2} reactivity ratio, OH radicals primarily react with NO_2 (thereby producing HNO_3) instead of reacting with VOC (producing ozone in the latter case).

For sure, Fig. 26 only illustrates the concept of an air parcel starting in a city and moving to rural areas with conditions suitable to represent Germany. But the figure discloses the internal structure of photochemical ozone production. And that is the important point. No matter how a real-world trajectory would travel, as long as it starts at a location within the VOC limited regime (above the dotted line in Fig. 26) and has to reach a biogenic environment, the situation will be as discussed in the text above. If the wind has a fast speed or the forest is close to the city, the trajectory will be more straight than the one shown in Fig. 26. But the ozone production rate encountered and therefore the additional ozone the air parcel accumulates during its travel will be very much less nowadays than in the 1990s due to the large decrease of VOC traffic emissions, which is in the end an achievement of a political process. This VOC decrease would be especially beneficial if the air parcels would move around for a long time within the same area, like, for example, if they were trapped within a stagnant summertime high pressure system. In that case, the high pressure system can work like a reactor, accumulating and processing traffic emissions for an enlarged period of time. The normal background dilution process would be omitted because the air parcels would not be able to escape the "reactor" for some time. In other words, the "fuel" needed to feed a significant photochemical ozone production in the vicinity of cities, the VOCs released by traffic emissions, is nowadays almost gone.

There is one basic point concerning the validity of this simple box model approach which is not addressed until now. For the representation within a simple framework we have assumed a typical summertime NO_2/NO_x ratio of 75% everywhere. It is known that petrol car exhaust mainly[12] consists of NO ($NO_2/NO_x \approx 0.05$). The same is true for diesel cars from the 1990s. Nowadays, the ratio of NO_2/NO_x for diesel cars (above EURO-3) has been shifted[12] towards NO_2 ($NO_2/NO_x \approx 0.40$), for example to facilitate the oxidation of soot deposited by a diesel particulate filter (DPF). With the consequence that in the 1990s, NO from car exhaust in city districts did rapidly react with ambient ozone, diminishing ozone almost to zero. Since the NO emissions were very large in street tunnels or main roads, ozone was gone there and NO was left over. As a consequence, the NO_2/NO_x ratio in regions dominated by traffic emissions was low in the year 1994 (≈30%) and larger in the year 2014 (≈50%, see Fig. 11). The trajectory for the year 1994 in Fig. 26 would have started at a lower NO_2 reactivity and at a larger NO concentration as anticipated. This would have driven an even larger ozone production rate than illustrated in the figure. After a relatively short period, this virtual trajectory considering a large NO contribution would converge against the depicted trajectory and would follow it for the rest of the time. For the year 2014, the deviation between the virtual (representing more NO) and the depicted trajectory would have been very much less because the NO_2/NO_x ratio from diesel car exhaust would be near to the typical condition of 75% from the start. Overall, we would expect a small change if we use a more complex framework able to handle different NO_2/NO_x ratios for the calculation of ozone production rates. The result would even strengthen our conclusion of a strong reduction of photochemical ozone production over the last 20 years in Germany.

5 Conclusions

Traffic emissions of VOC have been significantly decreased in Germany over the last 20 years. Ambient observations suggest a reduction on the order of a factor of 15 in 20 years, which would mean 14.5% per year. That is a remarkable achievement. This decrease is most probably responsible for the overall improvement of the compliance with EU ozone regulations for human health in Germany.

Traffic emissions of NO_x did not show the same decline as the VOC emissions. In fact, they stayed almost the same over time. If the decrease of NO_x traffic emissions was parallel to the VOC decrease, the trajectory for the year 2014 in Fig. 26 would have been on top of the trajectory for 1994, just starting at a later stage down the 1994 trajectory. This would suggest a less favorable decrease of ozone production along the trajectory for that case. On the other hand, a faster reduction of NO_x traffic emissions would have avoided the frequently observed exceedance of the NO_2 limit for the improvement of human heath,[4] which is basically a problem of traffic emissions in inner city districts. If the decrease in NO_x traffic emissions would start now, both benefits can be achieved: the avoidance of large ozone production and that of large NO_2 concentrations. Fig. 13 suggests how this can be achieved. The emissions of heavy-duty-vehicles and buses (called HGV and LCV in that figure) are estimated to dominate NO_x traffic emissions in German cities nowadays. Their NO_x emissions are still large, although most of them are equipped with catalytic converters. The reason seems to be, that due to the relatively slow speed in city districts, the engine exhaust does not heat-up the converters enough to be efficient. If this heating-up could be managed by an additional power source at all vehicle speeds, NO_x traffic emissions in inner city districts would be considerably reduced. Of course one would have to ensure that the NO_x reduction techniques for heavy-duty-vehicles and buses do not introduce other pollutants with a similar potential for health-risks as NO_2. This has to be investigated yet. Maybe, in the end, other concepts like electric buses and fuel-cell-vehicles are a better choice.

6 Appendix

The emission categories for Fig. 10a and b contain the following subcategories:

6.1 Emission category "road traffic"

Passenger cars, light duty vehicles, heavy duty vehicles and buses, mopeds & motorcycles, gasoline evaporation, automobile tyre and brake wear, automobile road abrasion.

6.2 Emission category "residential"

Stationary, household and gardening (mobile), public electricity and heat production.

6.3 Emission category "industry"

Petroleum refining, manufacture of solid fuels and other energy industries, stationary combustion in manufacturing industries and construction – iron and steel & non-ferrous metals & chemicals, pulp, paper and print & food processing, beverages and tobacco & non-metallic minerals, mobile combustion in manufacturing industries and construction, stationary combustion in manufacturing industries and construction, international aviation LTO (civil), domestic aviation LTO (civil), cement production, lime production, glass production, quarrying and mining of minerals other than coal, construction and demolition, storage, handling and transport of mineral products, other mineral products, ammonia production, nitric acid production, adipic acid production, carbide production, titanium dioxide production, soda ash production, chemical industry, storage, handling and transport of chemical products, iron and steel production, ferroalloys production, aluminium production, magnesium production, lead production, zinc production, copper production, nickel production, other metal production, storage, handling and transport of metal products, pulp and paper industry, food and beverages industry, other industrial processes, wood processing, production of POPs, consumption of POPs and heavy metals, other production, consumption, storage, transportation or handling of bulk products, commercial/institutional: stationary, commercial/institutional: mobile.

6.4 Emission category "agriculture"

Manure management – dairy cattle & non-dairy cattle & sheep & swine & buffalo & goats & horses & mules and asses & laying hens & broilers & turkeys & other poultry & other animals, inorganic N-fertilizers (includes also urea application), animal manure applied to soils, sewage sludge applied to soils, other organic fertilizers applied to soils, urine and dung deposited by grazing animals, crop residues applied to soils, indirect emissions from managed soils, farm-level agricultural operations including storage, handling and transport of agricultural products, off-farm storage, handling and transport of bulk agricultural products, cultivated crops, use of pesticides, field burning of agricultural residues, agriculture other, agriculture/forestry/fishing: stationary, agriculture/forestry/fishing: off-road vehicles and other machinery, agriculture/forestry/fishing: national fishing.

6.5 Emission category "solvents"

Domestic solvent use including fungicides, road paving with asphalt, asphalt roofing, coating applications, degreasing, dry cleaning, chemical products, printing, other solvent use, other product use.

6.6 Emission category "miscellaneous"

This category is calculated from the category called "NATIONAL TOTAL″ minus the sum of the categories already used in the figures. It contains categories like "military", "fugitive", "waste" and "solvents" in case of NO_x and "residential" in case of VOC.

Acknowledgements

This work was supported by the "Federal Ministry of Education and Research" (BMBF) under grant numbers 07TFS45/3 and FKZ19B98118/0, by the "Federal Ministry for Economic Affairs and Energy" (BMWI) under grant number 03ET6036D, and by the town council of Bad Homburg *via* the project "Analyse der lokalen Luftqualität der Stadt Bad Homburg, 2014". The authors wish to thank Hermann-Josef Buers for his assistance in the set-up of the GC system and the analysis of its results. We also acknowledge the kind forwarding of the preliminary surface ozone data from the Global Atmosphere Watch station "Meteorological Observatory Hohenpeissenberg", German Meteorological Service, by Stefan Gilge.

References

1 P. J. Crutzen, *Annu. Rev. Earth Planet. Sci.*, 1979, **7**, 443–472.

2 R. G. Derwent, A. J. Manning, P. G. Simmonds, T. G. Spain and S. O'Doherty, *Atmos. Environ.*, 2013, **80**, 361–368.

3 O. R. Cooper, D. D. Parrish, J. Ziemke, N. V. Balashov, M. Cupeiro, I. E. Galbally, S. Gilge, L. Horowitz, N. R. Jensen, J.-F. Lamarque, V. Naik, S. J. Oltmans, J. Schwab, D. T. Shindell, A. M. Thompson, V. Thouret, Y. Wang and R. M. Zbinden, *Elementa*, 2014, **2**, 000029.

4 U. Dauert, S. Feigenspan, A. Minkos and M. Langner, *Luftqualität 2014, Vorläufige Auswertung*, Umweltbundesamt, Deutschland, Dessau-Roßlau, January 2015, pp. 11–12.

5 *Information über Abgase des Kraftfahrzeugverkehrs*, ed. P. Rabl, Bayrisches Landesamt für Umweltschutz, Augsburg, Germany, 2003.

6 R. Pischinger, M. Kell and T. Sams, *Thermodynamik der Verbrennungskraftmaschine: Der Fahrzeugantrieb*, Springer, Wien, Austia, New York, USA, 3rd edn, 2009.

7 D. Klemp, D. Mihelcic and B. Mittermaier, Messung und Bewertung von Verkehrsemissionen, Schriften des Forschungszentrums Jülich, *Energy & Environment*, Jülich, Germany, 1st edn, 2012, vol. 21.

8 D. Hassel, P. Jost, F.-J. Weber, F. Dursbeck, K.-S. Sonnborn and D. Plettau, *Abgas-Emissionsfaktoren von PKW in der Bundesrepublik Deutschland - Abgasemissionen von Fahrzeugen der Baujahre 1986 bis 1990 (Forschungsbericht*

10405152 und 10405509 im Auftrag des Umweltbundesamtes), Erich Schmid Verlag, Berlin, Germany, 1994.

9 *EC Commission Regulations No 91/441/EWG, 91/441/EWG, 94/12/EG, 98/69/EG, 98/69/EG, EG 715/2007 with EG 692/2008 and EG 459/2012 implementing and amending Regulations of the European Parliament and of the Council on type-approval of motor vehicles with respect to emissions from light passenger and commercial vehicles*, European Commission, Brussels, Belgium, 1991–2012.

10 *Gesetz zur stärkeren Berücksichtigung der Schadstoffemissionen bei der Besteuerung von Personenkraftwagen (Kraftfahrzeugänderungsgesetz 1997-KraftStÄndG 1997)*, Deutscher Bundestag, Bonn, Germany, 18 April 1997.

11 D. C. Carslaw, S. D. Beevers, J. E. Tate, E. J. Westmoreland and M. L. Williams, *Atmos. Environ.*, 2011, **45**, 7053–7063.

12 M. Rexeis, S. Hausberger, J. Kühlwein and R. Luz, *Update of emission factors for Euro 5 and Euro 6 vehicles for the HBEFA version 3.2. TUG report I-31/2013/Rex EM-I 2011/20/679*, TU Graz, Graz University of Technology, Graz, Austria, 2013.

13 *Fahrzeugzulassungen: Bestand an Kraftfahrzeugen nach Umwelt-Merkmalen; Statistische Mitteilungen des Kraftfahrt-Bundesamtes FZ 13*, Kraftfahrt-Bundesamt, Flensburg, Germany, 2015.

14 M. W. Lesley and L. D. Schmidt, *Surf. Sci.*, 1985, **155**, 215–240.

15 *Aktualisierung "Daten- und Rechenmodell: Energieverbrauch und Schadstoff-emissionen des motorisierten Verkehrs in Deutschland 1960-2030" (TREMOD, Version 5.3) für die Emissionsberichtserstattung 2013*, ed. W. Knörr, ifeu - Institut für Energie- und Umweltforschung Heidelberg GmbH, Heidelberg, Germany, 2012.

16 S. Hausberger, M. Rexeis, M. Zallinger and R. Luz, *Emission Factors from the Model PHEM for the HBEFA Version 3, Report Nr. I-20/2009 Haus-Em 33/08/679 from 07.12.2009*, TU Graz, Graz University of Technology, Graz, Austria, 2009.

17 ERMES: European Research Group on Mobile Emission Sources, *Information Paper"Diesel light duty vehicle NO_x emission factors", ERMES 2015-10-09*, INFRAS, Bern, Switzerland, 2015.

18 *National Trend Tables for the German Atmospheric Emission Reporting, ANNEX 1: National sector emissions: Main pollutants, particulate matter, heavy metals and persistent organic pollutants*, Excel spreadsheets DE_2015_Table_I_1990.xls - DE_2015_Table_I_2013.xls, ed. P. Gniffke, Umweltbundesamt, Deutschland, Dessau, 27 April 2015.

19 Umweltbundesamt, *Entwicklung der Luftqualität in Deutschland*, Umweltbundesamt Deutschland, Fachgebiet II 4.2, "Beurteilung der Luftqualität", Dessau-Roßlau, October 2009, pp. 5–7.

20 U. Kugler, J. Theloke and W. Jörß, *Verkehrsemissionsmodellierung - Modellvergleich und alternative Szenarien*, Umweltbundesamt, Dessau-Roßlau, Germany, 2013.

21 A. Niedojadlo, K. H. Becker, R. Kurtenbach and P. Wiesen, *Atmos. Environ.*, 2007, **41**, 7108–7126.

22 T. Schmitz, D. Klemp and D. Kley, *Messungen der Immissionskonzentrationen verschiedener Ozonvorläufersubstanzen in Ballungsgebieten und an Autobahnen - Charakterisierung der Emissionsverhältnisse des Straßenverkehrs unter verschiedenen Verkehrssituationen durch Messungen in Quellnähe*, Berichte des Forschungszentrums Jülich, Jül-3457, Jülich, Germany, 1997.

23 S. Urban, Charakterisierung der Quellverteilung von Feinstaub und Stickoxiden in ländlichem und städtischem Gebiet, Schriften des Forschungszentrums Jülich, *Energy & Environment*, Jülich, Germany, 2010, vol. 84.
24 T. Kern, N. Metz and D. Kley, *Untersuchungen von Verkehrsabgasemissionen insbesondere im Hinblick auf die Ozonbildung, Abschlussbericht*, Forschungszentrum Jülich, Jülich, Germany, 1997.
25 K. Mannschreck, *Experimentelle Bestimmung von städtischen Emissionen anhand von Konzentrationsmessungen im Lee einer Stadt*, Berichte des Forschungszentrums Jülich, Jül-3846, Jülich, Germany, 2001.
26 D. Klemp, K. Mannschreck, H. Pätz, M. Habram, P. Matuska and F. Slemr, *Atmos. Environ.*, 2002, **36**(1), 61–80.
27 B. Mittermaier and D. Klemp, EGU General Assembly 2004, Nice, France, 25 April 2004–30 April 2004.
28 C. Ehlers, Mobile Messungen - Messung und Bewertung von Verkehrsemissionen, Schriften des Forschungszentrums Jülich, *Energy & Environment*, Jülich, Germany, 2014, vol. 229.
29 M. Barker, M. Hengst, J. Schmid, H.-J. Buers, B. Mittermaier, D. Klemp and R. Koppmann, *Eur. Respir. J.*, 2006, **27**, 929–936.
30 T. Schmitz, D. Hassel and F.-J. Weber, *Atmos. Environ.*, 2000, **34**, 4639–4647.
31 B. Mittermaier and D. Klemp, *Gefahrstoffe, Reinhaltung der Luft*, 2004, vol. 64, pp. 487–494.
32 J. Drummond, A. Volz and D. Ehhalt, *J. Atmos. Chem.*, 1985, **2**, 287–306.
33 A. Volz and J. Drummond, *Fresenius' Z. Anal. Chem.*, 1984, **317**, 355–358.
34 D. H. Ehhalt, H.-P. Dorn and D. Poppe, *Proc. R. Soc. Edinburgh, Sect. B: Biol. Sci.*, 1991, **97**, 17–34.
35 R. Atkinson, S. M. Aschmann and J. Arey, *Atmos. Environ.*, 1990, **24**, 2647–2654.
36 A. K. Mollner, S. Valluvadasan, L. Feng, M. K. Sprague, M. Okumura, D. B. Milligan, W. J. Bloss, S. P. Sander, P. T. Martien, R. A. Harley, A. B. McCoy and W. P. L. Carter, *Science*, 2010, **330**, 646–649.
37 S. M. Saunders, M. E. Jenkin, R. G. Derwent and M. J. Pilling, *Atmos. Chem. Phys.*, 2003, **3**, 161–180.
38 T. Wiedmann, J. Kersten and K. Ballschmiter, *Art und Menge von stofflichen Emissionen aus dem Verkehrsbereich*, Akademie für Technikfolgenabschätzung in Baden-Württemberg, Stuttgart, Germany, 2000.
39 R. E. Dunmore, J. R. Hopkins, R. T. Lidster, J. D. Lee, M. J. Evans, A. R. Rickard, A. C. Lewis and J. F. Hamilton, *Atmos. Chem. Phys.*, 2015, **15**, 9983–9996.
40 U. Steinemann and F. Zumsteg, *Verkehrs- und Schadstoffmessungen 2008 im Gubristtunnel (Traffic and trace gas pollutant measurements 2008 in the Gubrist tunnel); report US 89-16-13*, OSTLUFT durch Amt für Abfall, Wasser, Energie und Luft des Kantons Zürich (AWEL), Zürich, Switzerland, updated 25 May 2010.
41 L. Ammoura, I. Xueref-Remy, V. Gros, A. Baudic, B. Bonsang, J.-E. Petit, O. Perrussel, N. Bonnaire, J. Sciare and F. Chevallier, *Atmos. Chem. Phys.*, 2014, **14**, 12871–12882.

Faraday Discussions

PAPER

Have vehicle emissions of primary NO_2 peaked?

David C. Carslaw,*[ab] Tim P. Murrells,[a] Jon Andersson[c] and Matthew Keenan[c]

Received 12th November 2015, Accepted 1st December 2015
DOI: 10.1039/c5fd00162e

Reducing ambient concentrations of nitrogen dioxide (NO_2) remains a key challenge across many European urban areas, particularly close to roads. This challenge mostly relates to the lack of reduction in emissions of oxides of nitrogen (NO_x) from diesel road vehicles relative to the reductions expected through increasingly stringent vehicle emissions legislation. However, a key component of near-road concentrations of NO_2 derives from directly emitted (primary) NO_2 from diesel vehicles. It is well-established that the proportion of NO_2 (*i.e.* the NO_2/NO_x ratio) in vehicle exhaust has increased over the past decade as a result of vehicle after-treatment technologies that oxidise carbon monoxide and hydrocarbons and generate NO_2 to aid the emissions control of diesel particulate. In this work we bring together an analysis of ambient NO_x and NO_2 measurements with comprehensive vehicle emission remote sensing data obtained in London to better understand recent trends in the NO_2/NO_x ratio from road vehicles. We show that there is evidence that NO_2 concentrations have decreased since around 2010 despite less evidence of a reduction in total NO_x. The decrease is shown to be driven by relatively large reductions in the amount of NO_2 directly emitted by vehicles; from around 25 vol% in 2010 to 15 vol% in 2014 in inner London, for example. The analysis of NO_x and NO_2 vehicle emission remote sensing data shows that these reductions have been mostly driven by reduced NO_2/NO_x emission ratios from heavy duty vehicles and buses rather than light duty vehicles. However, there is also evidence from the analysis of Euro 4 and 5 diesel passenger cars that as vehicles age the NO_2/NO_x ratio decreases. For example the NO_2/NO_x ratio decreased from 29.5 ± 2.0% in Euro 5 diesel cars up to one year old to 22.7 ± 2.5% for four-year old vehicles. At some roadside locations the reductions in primary NO_2 have had a large effect on reducing both the annual mean and number of hourly exceedances of the European Limit Values of NO_2.

[a]Ricardo Energy & Environment, Gemini Building, Fermi Avenue, Harwell, Oxon, OX11 0QR, UK. E-mail: david.carslaw@ricardo.com; Tel: +44 (0)1235 753 345
[b]Wolfson Atmospheric Chemistry Laboratories, Department of Chemistry, University of York, Heslington, York, YO10 5DD, UK
[c]Ricardo Ltd, Shoreham Technical Centre, Shoreham-by-Sea, West Sussex BN43 5FG, UK

1 Introduction

1.1 Background and aims

Despite several decades of vehicle emission control, atmospheric concentrations of nitrogen dioxide (NO_2) remain important across many urban areas of Europe. Ambient NO_2 concentrations in Europe are regulated by the Ambient Air Quality Directive (Directive 2008/50/EC), which sets an annual average Limit Value of 40 $\mu g\ m^{-3}$ and an hourly Limit Value of 200 $\mu g\ m^{-3}$ not to be exceeded more than 18 times a year, both to be attained by 2010 with the possibility of a time extension to 2015 in certain circumstances. The annual average Limit Value is the more stringent of the two, and a large fraction of EU Member States, including the UK, are not achieving it.[1] Vehicle emissions legislation has set increasingly stringent limits for the mass of emissions that can be emitted by both light and heavy duty vehicles.[2,3] The expectation has been therefore that considerable reductions in ambient concentrations of NO_x and NO_2 would have been achieved over the past decade or so.

Historically there has been little interest in the proportion of NO_x that is in the form of NO_2 in vehicle exhaust. In part this lack of interest is because NO is rapidly oxidised to NO_2 in the atmosphere through its reaction with O_3. Additionally, vehicle emissions legislation considers only total NO_x and does not speciate between NO and NO_2.[4] However, in 2003 at a major roadside monitoring site in central London concentrations of NO_2 were shown to have increased considerably compared with 2002 levels.[5] The increase in NO_2 concentration was particularly important for the hourly Limit Value where the number of hours exceeding 200 $\mu g\ m^{-3}$ increased from 2 in 2002 to 463 in 2003.

The reason for the increased NO_2 concentration was found to be largely related to the introduction of retrofit buses fitted with continuously regenerating particle traps (CRTs). These particle filters work by using a diesel oxidation catalyst (DOC) to oxidise NO to NO_2, using the excess NO_2 to further oxidise particle emissions. However, the CRTs fitted to London buses could not on their own explain all of the NO_2 observed.[5] Since that time many emission and atmospheric measurement studies have shown that the use of DOC and particle filters in both light and heavy duty vehicles have led to increased NO_2/NO_x ratios in vehicle exhaust.[6–9]

The increase in the proportion of NO_x emitted as NO_2 has had an important effect on roadside concentrations of NO_2 across Europe. However, UK projections of NO_x emissions from 2000 to 2015 and 2005 to 2015 suggested approximately 60 and 50% reduction in urban road vehicle emissions of NO_x, respectively, as a result of progressively tighter limits on vehicle NO_x emissions due to vehicle emissions legislation (so-called Euro limits).[10] Since that time it has become clear that emissions of NO_x have not decreased as expected.[11–14] The discrepancy between inventory-based NO_x emission estimates and actual emissions has also been revealed by Lee *et al.*[15] using an eddy covariance technique. These measurements were made in London and provide estimates of total surface NO_x emissions, and they also suggest that the largest discrepancy between emission inventory and the flux estimates were for wind sectors dominated by road vehicle emissions. The increase in NO_2/NO_x ratios in diesel vehicle exhaust combined with a lack of reduction in total NO_x emissions from road vehicles has led to a situation across Europe where many urban locations exceed EU limits for NO_2 concentration.

This paper considers the recent evidence concerning directly emitted NO_2 from road vehicles. We focus on London (UK) where there are a large number of ambient monitoring sites – and in particular sites located at roadside locations. These sites provide robust information on long term trends and also provide a way in which directly emitted NO_2 emissions can be quantified. Consideration is also given to the recent data from vehicle emission remote sensing measurements, also made in London, to help understand the observed trends in ambient concentrations and the changes in the NO_2/NO_x emissions ratio from vehicles. Because NO_2 plays an important role in modern emissions control technologies, we consider how these technologies have changed over time and how they have affected the emission of NO_2.

1.2 Changes in vehicle emissions control technology affecting NO_2 emissions

Over the past 15 years or so there have been many developments in vehicle emission after-treatment technology that have affected the emission of NO_2 – and in particular the NO_2/NO_x ratio. These developments are directly linked to observed changes in atmospheric composition. In gasoline and diesel vehicles most of the engine-out emission of NO_x is in the form of NO. However, as vehicle emission control technology has developed, NO_2 has become an important aspect of after-treatment technologies. NO_2 plays an important role in helping to burn soot (PM) from diesel vehicles and is also key to the efficient application of Selective Catalytic Reduction (SCR) technologies. The NO_x control in diesel vehicles is significantly more challenging than for gasoline vehicles.

Stoichiometrically operating gasoline vehicles use a three way catalyst (TWC) for emissions control. The role of the TWC is to oxidise hydrocarbons (HC) and carbon monoxide (CO) and simultaneously reduce NO_x. The reduction of NO_x to N_2 uses CO as the reductant. Gasoline vehicles operating under stoichiometric conditions produce NO engine-out and very little if any tailpipe NO_2, as there is no excess oxygen in the exhaust to oxidise NO to NO_2 over the TWC. Gasoline vehicles have used the TWC very effectively over the past 20 years for exhaust emissions control, as also shown by vehicle emission measurements.[11,16]

For Euro 3 diesel vehicles (introduced in 2000), DOCs were used to control HC and CO. The DOC can oxidise NO to NO_2 in excess oxygen over active precious metal catalysts containing a mixture of platinum (Pt) and palladium (Pd), which provide highly efficient oxidation reactions. This oxidation can lead to increased NO_2 at the tailpipe for DOC-only applications.

For Euro 5 diesel vehicles (introduced in 2009), the PM limits required the use of a Diesel Particulate Filter (DPF) to physically trap the PM. The stored PM needs to be oxidised in order not to completely block the filter, which could impact vehicle operation. There are two mechanisms for soot oxidation: high temperature oxidation at approximately 600 °C with oxygen and lower temperature oxidation (250–450 °C) using NO_2 as the oxidant. Hence, the NO_2 formed over the DOC can be used for soot control, also leading to an increase in the amount of NO_2 emitted.

The tighter limits for Euro 6 diesel emissions (introduced in 2014) has led to the development and introduction of after-treatment technologies to specifically reduce emissions of NO_x. Excess oxygen in lean operating diesel engines impacts NO_x reduction considerably, requiring the use of additional reductants and dedicated NO_x control catalysts. The catalysts for lean NO_x control are the Lean NO_x Trap (LNT) which requires reductants that are formed by the engine (HC/CO/H_2) and SCR, which requires the injection of urea into the vehicle exhaust to form ammonia, which is used as the reductant. LNT and SCR catalyst technologies remove both NO and NO_2 and hence reduce the tailpipe NO_2 emissions. For good low temperature NO_x control by SCR, NO_2 is required and hence has a positive benefit on NO_x control.

Fig. 1 provides an indication of how emissions of NO_2 have changed through pre-Euro to Euro 6, *i.e.* from pre-1993 to post-2014, for diesel cars. The most significant change in NO_2 emissions occurred when Euro 3 DOCs were introduced, resulting in a step-change increase in NO_2 emissions. Further increases in the emissions of NO_2 occurred when DPFs were introduced with late Euro 4 and Euro 5 vehicles. The reduction in Euro 6 NO_2 emissions is a result of the introduction of LNT and SCR technologies, which reduce absolute emissions of NO_x and not necessarily the ratio of NO_2/NO_x.

For heavy duty vehicles, exhaust NO_x control *via* SCR has been implemented since Euro IV in 2005 and has been continued to be used for Euro V in 2008, hence controlling both NO and NO_2 tailpipe emissions. Euro VI from 2014 has required the addition of a particulate filter for PM control. Here the PM can be controlled *via* direct oxidation with NO_2 and/or oxygen. Furthermore, local air quality requirements led to a significant amount of retrofit applications targeting soot control, for example buses – including buses in London. These applications rely on the direct oxidation of PM with NO_2 and do not use the high temperature reaction with oxygen. Hence the role of the retrofitted DOC is to produce NO_2 to control the PM, which led to an increase in the tailpipe NO_2 emissions compared to a non-retrofit application.

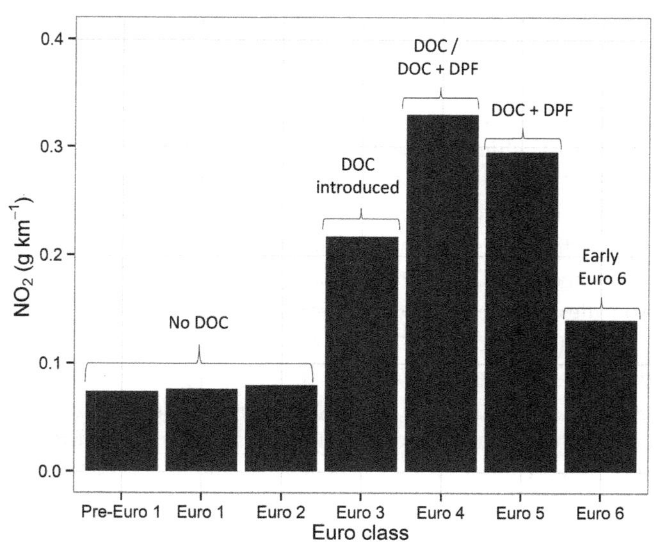

Fig. 1 Evolution of NO_2 emissions from diesel passenger cars based on emission factors used in the UK National Atmospheric Emissions Inventory.

There are other factors that will affect the NO_2/NO_x ratio in vehicle exhausts related to the catalysts used. For example, 'catalyst thrifting', where catalyst developers and Original Equipment Manufacturers (OEMs) reduce the amount of platinum group metal (PGM) used, will potentially affect the amount of NO_2 formed. Second, catalyst deactivation over time might also have an effect on the conversion of NO to NO_2 on the DOC. Thus, aged catalyst technologies may have substantially reduced oxidative capacity through thermal deactivation or poisoning. For these reasons, the exhaust from diesel vehicles when first introduced may have higher NO_2/NO_x ratios compared with vehicles introduced at a later time through catalyst thrifting. Similarly, the exhaust from older vehicles may also have lower NO_2/NO_x ratios than newly introduced vehicles. These effects are difficult to quantify in real vehicle fleets but are likely to play an important role in affecting the emission of NO_2.

2 Experimental

2.1 Ambient measurement sites

The analysis in the current work focuses on ambient monitoring sites in Greater London. London provides a useful case study for several reasons. First, it has a large number of ambient monitoring sites that form part of the Defra Automatic Urban and Rural Network (AURN) or the London Air Quality Network (LAQN). Second, as described in Section 2.3, there is a large data set of vehicle emission remote sensing data available from two campaigns. The two networks provide more than 70 ambient monitoring sites at urban background and roadside locations. In particular, there are a relatively large number of 'traffic-influenced' measurements sites that can be analysed for both trends and to estimate the NO_2/NO_x ratio from vehicle emissions. In this work we have used data from 35 long-term measurement roadside sites where there has been at least 10 years of measurements over the period 1996 to 2014.

2.2 Deriving NO_2/NO_x emission ratios

Techniques are available to estimate the NO_2/NO_x ratio from vehicle emissions through the analysis of ambient monitoring data.[17–19] Jenkin[18] showed that by considering the 'total oxidant' gradient, $(NO_2 + O_3)/NO_x$, estimates could be made of the primary (direct) NO_2/NO_x ratio from vehicle emissions. One of the limitations of the total oxidant approach for estimating the NO_2/NO_x ratio from vehicles is that roadside measurement of O_3 is also required. However, there are very few roadside measurement sites in London (and the UK) that measure O_3, making it difficult to build a comprehensive view of typical NO_2/NO_x ratios along different roads. To overcome this problem, Carslaw and Beevers[20] developed an approach that did not require roadside measurement of O_3.

Briefly, the increment in NO_2 concentration at a roadside monitoring site above an urban background site (at North Kensington) is partitioned into NO_2 that is chemically derived through the reaction between NO and O_3 and that which is emitted directly by road vehicles. The technique uses a constrained chemistry model that is based on a simple set of chemical reactions used to describe the time-dependent change in NO, NO_2 and O_3 concentrations as vehicle plumes are mixed with background air. By considering different values of the assumed NO_2/NO_x emissions ratio from road traffic and the time available for the $NO–O_3$ reaction to take place (τ), the best agreement between modelled and measured roadside concentrations of hourly NO_2 concentrations is sought. In practice, several hundred combinations of the NO_2/NO_x emissions ratio from road traffic sources and the time available, τ, for the $NO–O_3$ reaction to take place are considered before the best single combination, resulting in the minimum error, as determined by the residual sum of squares, between modelled and measured NO_2 concentrations, is identified.

The technique assumes that the increment in NO_2 concentration above a local background site is controlled by the availability of O_3 and directly emitted NO_2 only. The formation of NO_2 through other routes (*e.g.*, through reactions involving VOCs) is assumed to be negligible. To further minimise the potential influence of other photochemically-driven reactions, data are only considered from October through to March. Another source of uncertainty is the use of chemiluminescent NO_x analysers with molybdenum catalysts used in UK national networks. These instruments are affected by interference due to NO_y species as they will detect them as NO_2. However, at roadside locations – and for increments above local background concentrations – the influence of NO_y species is expected to be small. However, it is known that vehicle emissions are a direct source of nitrous acid (HONO), which would also be detected as NO_2 in these instruments. Measurements of HONO in vehicle exhausts does suggest that only low amounts are emitted. For example, Kirchstetter *et al.*[21] measured a HONO/NO_x ratio of $2.9 \pm 0.5 \times 10^{-3}$. Nevertheless, the direct emission of HONO is a potential interferent. The approach yields estimates of hourly NO_2 and O_3 at roadside sites, as well as an estimate of the road transport NO_2/NO_x emissions ratio for vehicles using the road adjacent to each roadside monitoring site. The analysis provides a monthly mean estimate of the NO_2/NO_x ratio for the vehicle emissions at the roadside site being analysed.

2.3 Vehicle emission remote sensing

Comprehensive vehicle emission measurements have been used from two vehicle emission campaigns carried out in London in 2012 and 2013. A comprehensive description of the instrument set-up and locations used for roadside sampling is given by Carslaw and Rhys-Tyler[16] and is not repeated here. The 2013 measurements were mostly focused on understanding the in-use emissions from Transport for London (TfL) Euro III buses retrofitted with a 'low NO_2' SCRT system described later in Section 3.3.[22] Nevertheless, the 2013 measurements also included a considerable number of measurements from the wider vehicle fleet. In total, there were approximately 105 000 valid measurements made of individual vehicles from the combined 2012 and 2013 campaigns.

A remote vehicle exhaust sensor developed at the University of Denver, named the Fuel Efficiency Automobile Test (FEAT), was used to collect all of the emission measurements over two six week campaigns (one each in June-July 2012/2013). The instrument consists of a light source and detector unit separated by a single lane of road. The detector is composed of four non-dispersive infrared (NDIR) detectors, including a reference channel (3.9 μm), CO (3.6 μm), CO_2 (4.3 μm), and hydrocarbons (HC, 3.3 μm), and two dispersive ultraviolet spectrometers. The first spectrometer measures NO, sulphur dioxide, and ammonia (NH_3) between 198 and 227 nm, while the second records NO_2 spectra between 430 and 450 nm. All of the detectors sample at 100 Hz and have been extensively described in the literature.[23–25] The measurements from the remote sensing detector provide fuel based emission factors (*e.g.* g NO_x per kg fuel) and a direct measure of the NO_2/NO_x ratio from individual vehicles.

3 Results and discussion

3.1 Trends in ambient concentrations of NO_x and NO_2

Trends have been calculated for ambient NO_x and NO_2 concentrations across 35 roadside monitoring sites in London that have at least 10 years of data capture. A summary of the trends from 1996 to 2014 is shown in Fig. 2. Over the full period from 1996 to 2014 concentrations of NO_x at roadside locations in London have decreased by $2.4 \pm 1.5\%$ per year. The corresponding change in NO_2 was $0.4 \pm 0.4\%$ per year. The trend analysis shows that on average across all roadside sites there has been little evidence of a decrease in concentrations of NO_2. The trends over the period 1996 to 2014 have however not been linear. For example, from around 2002 to 2014 the reduction in NO_x has been weakly downward at about $1.4 \pm 0.5\%$ per year. Similarly, for NO_2 Fig. 2(b) shows that there is evidence of a decrease in concentrations from around 2010. Clearly, the trends in NO_x and NO_2 at individual sites spans a range of behaviours but are not considered in detail here.

3.2 Trends in primary NO_2 emissions

The trend in estimated NO_2/NO_x emission ratios from ambient roadside data is shown in Fig. 3. The results are shown separately for inner and outer London, as defined in Fig. 3. When considering the data in this way it is clear that there are differences between these two regions of London. In the mid-1990s both inner and outer London had similar values for the NO_2/NO_x ratio of about 5 vol%. However, after this time, inner London is shown to have a higher NO_2/NO_x ratio than outer London. The patterns of change over the 19 years are revealing. From around 2003 inner London is shown to diverge from outer London and has

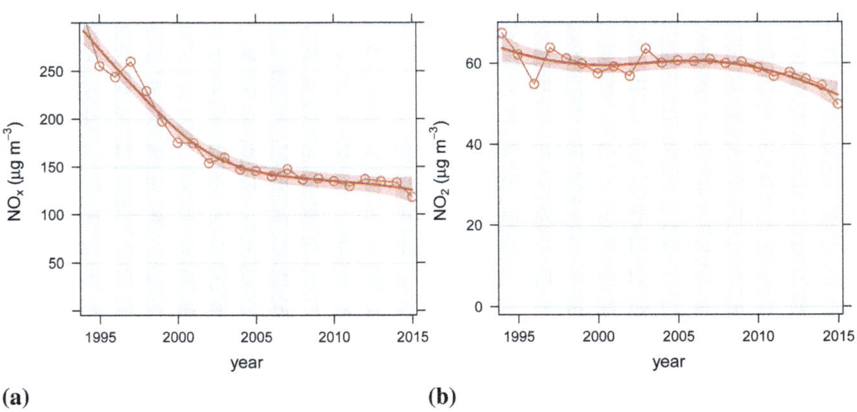

Fig. 2 (a) Trends in the mean concentration of NO_x across 35 roadside sites in Greater London with at least 10 years of data capture and (b) the same for NO_2. Trend estimates were made using the *openair* package.[26]

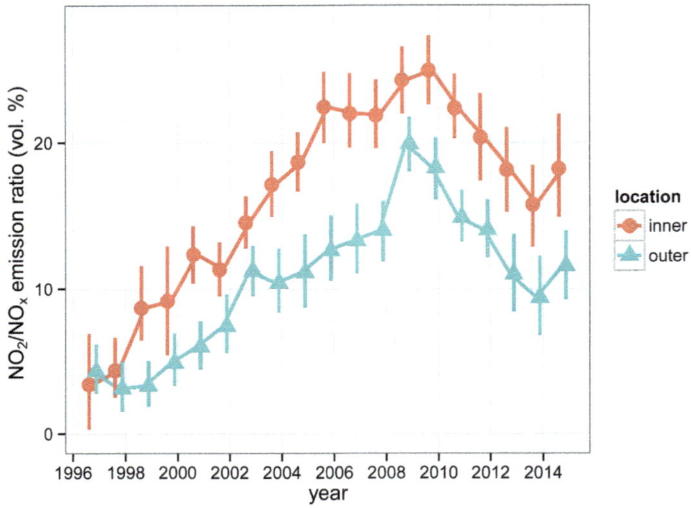

Fig. 3 Estimates of the NO_2/NO_x emission ratio calculated from ambient monitoring sites in London. The ratio has been calculated separately for inner and outer London. Inner London is assumed to include the following local authorities: Camden, City of London, Hackney, Hammersmith and Fulham, Islington, Kensington and Chelsea, Lambeth Southwark, Tower Hamlets and Wandsworth Westminster. The error bars represent 95% confidence in the mean estimates. Note that data for 2015 only include the first three months of the year.

a much higher NO_2/NO_x ratio than outer London. The likely reason for the divergence is related to the increased use of CRT after-treatment devices fitted to Transport for London (TfL) buses, which were retrofitted from 1999 onwards. The use of these filters increased from 60 to 90% over the period 2003 to 2005.

Also apparent in Fig. 3 for outer London sites is the increase in NO_2/NO_x ratio from 2008 to 2009, which is less apparent for the inner London data. It is difficult to reconcile this increase with known changes in Euro standards for vehicles. One explanation is that the increase is related to the introduction of the London Low Emission Zone (LEZ).[27,28] The LEZ came into operation on 4 February 2008 with a phased introduction targeting vehicles of more than 12 tonnes gross weight. From 4th February 2008 only Euro III compliant (for PM_{10}) vehicles were permitted into Greater London. From July 2008 more vehicles were restricted including HDVs from 3.5 to 12 tonnes, buses and coaches. These additional vehicle types were also required to be Euro III compliant. The TfL baseline monitoring report does show that the number of vehicles fitted with a particle filter through the 'Reduced Pollution Certificate' scheme increased when the LEZ was introduced in 2008.[29] It is likely therefore that the increased use of catalytic particle filters would have led to an overall increase in the NO_2/NO_x emission ratio from these vehicles. Specific information on the types of filter used are not available but given the proportionately greater number of these types of vehicle in outer London compared with inner London it would be expected that outer London roads would be more strongly affected. Neither Kelly *et al.*[28] nor Ellison *et al.*[27] specifically quantify the potentially important effect of increased particle filter use on primary NO_2 emissions.

Table 1 summarises the vehicle flows for the roads adjacent to the roadside monitoring sites based on Department for Transport manual count data from 2005 to 2013. On average the flow of vehicles is similar for inner and outer London – 34 274 and 36 285, respectively. The principal differences between the inner and outer London sites relate to the vehicle flows of buses and taxis. Buses are much more important in inner London, with flows being approximately double those in outer London. Similarly, taxis are also more important in inner London with flows three times those in outer London. Buses and taxis are all diesel vehicles and therefore inner London has a much greater influence due to diesel vehicles than outer London.

Table 2 shows the relative contribution to total emissions of NO_x along the same roads as considered in Fig. 2. These emission total estimates reveal some important differences by vehicle type between inner and outer London. In inner London, buses and taxis account for a large amount of the total NO_x emitted (31.8 and 7.1%, respectively) *i.e.* almost 40% of the total NO_x. By contrast, buses and taxis in outer London contribute a total of only 17% of the total NO_x. Consequently, the contributions of other vehicle types are also different – in particular there is a more significant contribution from passenger cars and articulated HDVs in outer London.

The changes in ambient concentrations of NO_x and NO_2 and primary NO_2 emissions have had important influences at specific monitoring sites. For example, at Marylebone Road in central London annual mean NO_x concentrations were very similar in 2008 (312 μg m^{-3}) and 2014 (316 μg m^{-3}). Despite the close similarity in concentrations of NO_x, concentrations of NO_2 differ markedly. The annual mean NO_2 concentration in 2008 was 115 μg m^{-3}, whereas in 2014 it was 79 μg m^{-3} *i.e.* over 30% less. However, the hourly Limit Value is much more sensitive to these changes. In 2008 there were 812 hours above 200 μg m^{-3} compared with only 60 exceedances in 2014. Such a dramatic difference in NO_2 concentrations, despite the similarity in NO_x concentration, is driven by directly emitted NO_2 emissions. In 2008 the NO_2/NO_x ratio was about 23% compared with about 13% in 2014 at the Marylebone Road site. The results for Marylebone Road illustrate how important primary NO_2 emissions can be in controlling ambient concentrations of NO_2 close to roads.

3.3 Changes in primary NO_2 emissions based on vehicle emission remote sensing

The preceeding analysis has shown that at London roadside monitoring sites the concentration of NO_x has changed little over the past decade. However, there is evidence that ambient NO_2 concentrations are decreasing at many sites. The

Table 1 Annual average daily flows of different vehicle types grouped by location in London. The data represent mean flows based on Department of Transport data from 2005 to 2013. Inner London is defined in Fig. 2. HDV are heavy duty vehicles and LGV light goods vehicles

Location	Car	Taxi	Bus	LGV	HDV Rigid	HDV Articulated	Motorcycle	All
Inner	21 201	2463	1603	4894	1146	133	2042	34 274
Outer	27 522	834	837	4650	1271	454	709	36 285

Table 2 Mean percentage contribution to total NO_x emissions split by inner and outer London. The emissions are based on NAEI estimates for the roadside sites considered in the trend analysis (see Section 2.1). Inner London is defined in Fig. 2

Location	Car	Taxi	Bus	LGV	HDV Rigid	HDV Articulated	Motorcycle	All
Inner	23.7	7.1	31.8	13.6	20.3	3.0	0.6	100.0
Outer	36.1	3.0	13.7	16.5	20.2	10.1	0.3	100.0

analysis of ambient measurements also shows that from around 2010 the mean NO_2/NO_x emissions ratio has shown clear decreases in both inner and outer London. It is important that the underlying causes of these decreases in the NO_2/NO_x ratio are understood.

Fig. 4 shows how the NO_2/NO_x ratio has changed through the Euro classes for the main types of vehicle. The ratios of petrol vehicles are low – and for Euro 4 to 6 unimportant because the absolute emissions of NO_x are very low. Diesel passenger car ratios of NO_2/NO_x were around 10 to 15% for pre-Euro to Euro 2 vehicles but increased to 16.1 ± 0.6% with the introduction of Euro 3 vehicles. The NO_2/NO_x ratio increased again with Euro 4 (27.6 ± 0.6%) and Euro 5 (25.5 ± 0.7%). LGVs have a similar pattern to diesel passenger cars. For both classes of HDV vehicle weights, the NO_2/NO_x ratio was around 10 to 20% for Euro II and Euro III vehicles (there were insufficient measurements of Euro I and pre-Euro

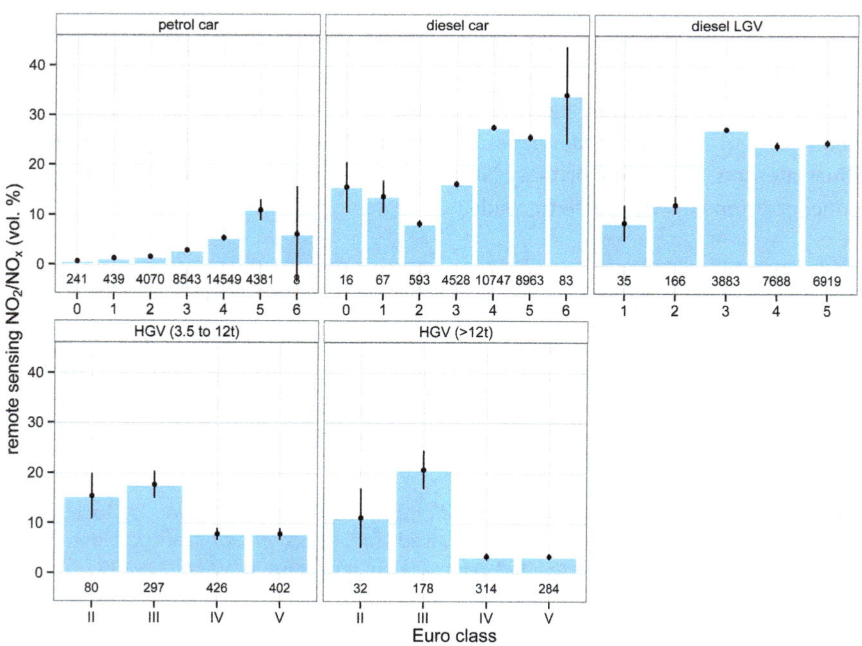

Fig. 4 Trend in NO_2/NO_x ratio by major vehicle and Euro class based on vehicle emission remote sensing data from the 2012 and 2103 London campaigns. The uncertainties show the 95% confidence interval in the mean. The numbers at the bottom of the plot show the sample size. Note that these ratios have not been weighted by vehicle number.

vehicles). However, the data clearly show that the NO_2/NO_x ratio decreased sharply for Euro IV and V vehicles in both weight categories. For vehicles 3.5 to 12 t, the NO_2/NO_x ratio for Euro IV/V vehicles is 7.8 ± 1.2%. However, for the heavier Euro IV/V vehicles >12 t the NO_2/NO_x ratio is only about 3.3 ± 0.7%.

The reasons for the sharp reductions in the NO_2/NO_x ratio from HDVs are unclear, although there are several factors that could affect the emission of NO_2. First, the requirement for a tailpipe NO_x sensor at Euro V for On-board Diagnostic (OBD) purposes may have reduced the number of vehicles with poor calibration. Second, as previously mentioned, there are reduced PGM levels through catalyst thrifting of the DOC, so as not to over-produce NO_2 for DOC + SCR vehicles and retrofitted DOC + DPF vehicles. Finally, the SCR efficiency improved, moving from Euro IV to Euro V where the SCR system is efficient at removing a 1 : 1 ratio of NO_2 : NO *via* the fast reaction ($NO_2 + NO + 2NH_3 \rightarrow 2N_2 + 3H_2O$). Even though there is little evidence of a reduction in NO_x from HDVs in London,[16] the reduction in the emission of NO_2 would likely have had an important effect on atmospheric concentrations of NO_2.

The data split by Euro class consist of a range of vehicle ages. For diesel cars for example (for which the sample sizes are large), Euro 5 vehicles span the years 2009 to 2013 (when the measurements were made). These data allow some consideration of the effect vehicle age has on the NO_2/NO_x ratio. Fig. 5 shows the results for Euro 4 and 5 diesel cars by vehicle age. It is clear that as the age of the vehicle increases the NO_2/NO_x ratio decreases. Four year old Euro 5 diesel cars have a NO_2/NO_x ratio of 22.7 ± 2.5%, whereas those up to one year old have a NO_2/NO_x ratio of 29.5 ± 2.0%. A similar behaviour is also seen for Euro 4 diesel cars, which are older overall than Euro 5 diesel cars.

These results show that older vehicles tend to have a lower NO_2/NO_x ratio than newer vehicles for the same Euro class. This is a potentially important finding suggesting that as vehicles age the amount of NO_2 emitted decreases. One reason for such a decrease is that the catalysts become less active over time, as indicated

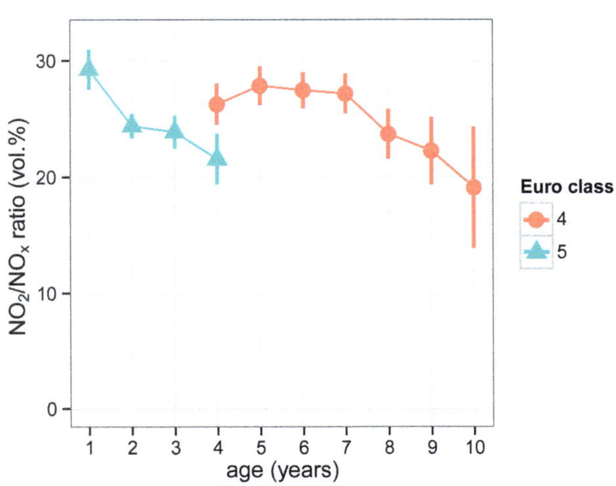

Fig. 5 Trend in NO_2/NO_x ratio for Euro 4 and 5 diesel cars by vehicle age based on a total sample of 25 721 vehicles from the 2012 and 2013 remote sensing data. The uncertainties show the 95% confidence interval in the mean.

in Fig. 5. Such degradation effects are not currently considered in emission factors for the NO_2/NO_x ratio but could nevertheless be important. Another influence could be 'catalyst thrifting' as discussed in Section 1.2. Four year old Euro 5 cars, for example, tend to have lower NO_2/NO_x ratios than four year old Euro 4 cars – which could indicate a change in the catalyst PGM composition between these two classes of vehicle.

Within the TfL bus fleet there have also been changes to the emissions control technology that affects the emissions of NO_x and NO_2. In particular, since 2012 approximately 2000 Euro III buses have been retrofitted to SCRT technology. The 'low-NO_2' SCRT technology used by TfL consists of a CRT (Continuously Regenerating Trap, to reduce particle emissions) and SCR system (to reduce NO_x emissions). The system consists of several closely coupled stages: a diesel oxidation catalyst (DOC), followed by a diesel particulate filter, urea/water injection, SCR catalysts and finally an ammonia slip catalyst. The performance of the TfL SCRT retrofit buses was measured by Carslaw *et al.*[22] where over 700 measurements were made using vehicle emission remote sensing. The SCRT on-road measurements showed that emissions of NO_x were reduced by 45% compared with similar vehicles using only a CRT. There was however a greater reduction in emissions of NO_2 of 61%.

By March 2014 there were 1479 TfL buses fitted with the SCRT system out of a total fleet of 8958 (Finn Coyle, personal communication). Furthermore, these bus conversions are not evenly spread across London but focused on certain bus routes. For this reason it would be expected that certain air pollution monitoring sites could be affected by the SCRT retrofits more than others. Such a change would also have the effect of potentially increasing the variation of NO_x and NO_2 emissions between monitoring sites and hence also atmospheric concentrations. Conversely, any changes to the majority of the rest of the vehicle fleet would be expected to be more consistent across the road network because the types of vehicle involved would tend to be uniformly mixed. The variation in the trends in NO_2 at individual roadside monitoring sites will likely be influenced by the variation in the bus fleet after-treatment technologies; although there is insufficient information available to verify whether this is the case.

3.4 Comparison with the COPERT emissions model

The comprehensive nature of the remote sensing measurements allows useful comparisons to be made with commonly used emissions models in Europe. The NO_2/NO_x ratio from the 2012 and 2013 remote sensing surveys can usefully be compared with those in the COPERT emission factor model and the emission factor advice considered in the EMEP/EEA air pollutant emission inventory guidebook.[30] COPERT is widely used throughout the world and forms the basis of vehicle emission factors used in most European national emissions inventories, including the UK. The comparison between the remote sensing-derived NO_2/NO_x ratios and COPERT is shown in Fig. 6, split by light and heavy duty vehicles. The comparison between the COPERT and remote sensing NO_2/NO_x ratios is a direct one and not dependent on the absolute, *e.g.* g km^{-1}, emission estimates of NO_x and NO_2.

The petrol vehicle NO_2/NO_x ratios for both data sources are low (typically below 5%). Furthermore, petrol vehicle NO_x emissions from post Euro 4 vehicles are very low and consequently the ratio of NO_2/NO_x will have little importance with respect to atmospheric concentrations of NO_x and NO_2.[16] For diesel cars pre-Euro 3 vehicles agree well, where the NO_2/NO_x ratio is around 15%. Where the estimates deviate is for Euro 3 to 5 vehicles. The remote sensing data suggests that Euro 3 vehicles have a NO_2/NO_x ratio of 16.1 ± 0.6%, whereas the ratio for COPERT is 27%. There are however larger discrepancies for Euro 4 and 5 vehicles, where the remote sensing data NO_2/NO_x ratio is 27.5 ± 0.6% and 25.5 ± 0.7% and COPERT is 46% in both cases. There is less certainty for diesel Euro 6 passenger cars because there were only 83 measurements from the remote sensing. Nevertheless the remote sensing NO_2/NO_x ratio of 34.0 ± 9.8% is similar to the COPERT value of 30%. The situation is also similar for diesel LGVs *i.e.* for Euro 3 to Euro 4 NO_2/NO_x ratios from remote sensing are considerably lower than those assumed in COPERT.

In general the HDV NO_2/NO_x ratios are less than those observed for light duty vehicles. However, the remote sensing data does show that the NO_2/NO_x ratios for larger Euro IV and V HDVs (>12 tonnes) are much lower than those reported by COPERT (about 3.3 ± 0.7% and 10 to 12%, respectively). As discussed in Section 3.3, there are several reasons why the remote sensing NO_2/NO_x ratios are low. The COPERT NO_2/NO_x ratios for HDVs also show very little variation, as shown in Fig. 6. It is known from previous work that HDV NO_2/NO_x ratios can be highly variable, depending on both the emissions technology used and the vehicle manufacturer.[16] The small variation in COPERT NO_2/NO_x ratios might reflect the relatively few HDVs that are tested as part of emission factor development. Additionally, the emission factor data for NO_2/NO_x at least, also reflect a mean of all driving conditions and not necessarily the urban conditions of the remote sensing data. The comparisons made in the current work therefore suggest that some refinement of the NO_2/NO_x ratios used in COPERT for HDVs is required.

Overall, the comparison between the remote sensing and the COPERT NO_2/NO_x ratios suggests that the ratios assumed by COPERT are higher than those observed from on-road measurements. As discussed in Section 3.3, there is

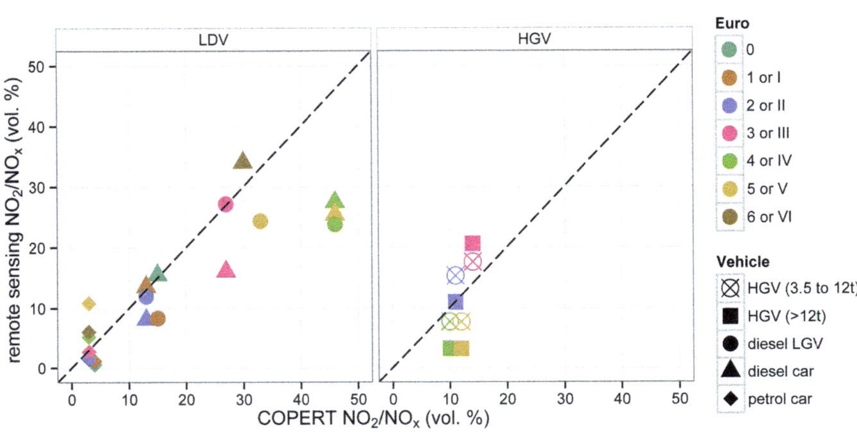

Fig. 6 Comparison of COPERT NO_2/NO_x ratios with those from vehicle emission remote sensing in London in 2012 and 2013. The data are shown by Euro class and major vehicle type and are split by light and heavy duty vehicles (LDV and HDV).

evidence that the NO_2/NO_x ratio decreases with vehicle age and such effects might explain some of the discrepancies observed. Similar to vehicle emission factors accounting for vehicle emissions deterioration – usually resulting in increased emissions with vehicle age – it is also important to specifically consider the effect of deterioration on the NO_2/NO_x ratio.

4 Conclusions

Exceedances of EU Limit Values for NO_2 remain important in many urban locations throughout Europe. Despite the introduction of increasingly stringent emissions legislation to reduce emissions of NO_x from vehicles, ambient concentrations of NO_x have not decreased as expected close to roads. Over the past 10 to 15 years directly emitted NO_2 from vehicles has become an important contributory factor affecting the near-road environment of NO_2 concentrations. The current work shows that in London, the NO_2/NO_x emissions ratio increased substantially from around 5 vol% in 1996 to around 20 to 25 vol% in 2010. Such a large increase has had an important effect on ambient NO_2 concentrations, especially close to roads. These increases have not been uniform along different roads due to the variation in vehicle fleet compositions. However, it is clear that the most significant increases in the NO_2/NO_x ratio have been in central and inner London where the proportion of diesel vehicles is high, and in particular, where there is a large number of buses fitted with CRTs. In outer London, the mean NO_2/NO_x ratio has been consistently lower than in inner London by about 5% (absolute).

This work shows for the first time that directly emitted NO_2 emissions have started to decline in London when expressed as a NO_2/NO_x ratio. The decreases have been substantial since around 2009/2010 *e.g.* reducing from a peak of around 25 vol% in 2010 to about 15 vol% at the end of 2014 in inner London. Furthermore, these decreases are similar in both inner and outer London. The recent decreases in the NO_2/NO_x emission ratio has led to a decrease in roadside NO_2 concentrations despite concentrations of NO_x remaining almost constant. The reasons for the decrease are likely related to considerable reductions in the NO_2/NO_x ratio for HDVs for Euro IV/V vehicles, confirmed by the comprehensive vehicle emission remote sensing emission measurements. Additionally, this work shows that for Euro 4 and 5 diesel passenger cars there is evidence that the NO_2/NO_x ratio decreases as vehicles age. Reductions in the NO_2/NO_x ratio are also seen in the TfL bus fleet. In particular, the increased use of 'low-NO_2' SCRT systems on TfL buses is also of note, as these systems have been shown under real-world conditions to reduce the NO_2/NO_x ratio by around 60% compared with that measured on buses before they were converted.

The analysis of the vehicle emission remote sensing data reveals that the NO_2/NO_x ratios assumed for many important classes of diesel vehicles used in European emission factors are likely too high. In particular, Euro 4 and 5 diesel passenger cars and vans have measured NO_2/NO_x ratios of around 25 to 27% on average compared with 46% assumed in the emission factors. Given the large numbers of these types of vehicle currently in service it is also likely that emission inventories also overestimate the amount of directly emitted NO_2 being emitted. This situation represents a reversal of that in the early 2000s where no consideration was given to NO_2 emissions and a default 5 to 10% value was assumed, which would have led to an underestimate of emissions of NO_2.

The future impact of Euro 6/VI technologies on the NO_2/NO_x ratio is currently unclear due to a lack of NO_x and NO_2 emission measurements from different vehicle technologies. However, if these vehicles deliver the expected substantial NO_x reduction, then the proportion of NO_2 in the exhausts of these vehicles will become less important because of the reduction in absolute emissions of NO_x and NO_2. For the existing diesel vehicle parc, this work shows that for diesel cars at least, there is evidence that the NO_2/NO_x ratio decreases as vehicles age. This decrease will have a potentially important effect on further reducing real-world NO_2/NO_x ratios from vehicles in the future, particularly if a similar behaviour is seen for other types of diesel vehicle.

Acknowledgements

We wish to thank Defra for the original funding of the vehicle emission remote sensing measurements.

References

1 EEA, Air quality in Europe – 2014 report. European Environment Agency, No 5/2014, ISSN 1977-8449, 2014.
2 EC, Regulation (EC) No 715/2007 of the European Parliament and the Council of 20 June 2007 on type approval of motor vehicles with respect to emissions from light passenger and commercial vehicles (Euro 5 and Euro 6), European Commission, Brussels, Belgium., 2007.
3 EC, Regulation (EC) No 595/2009 of the European Parliament and the Council of 18 June 2009 on type-approval of motor vehicles and engines with respect to emissions from heavy duty vehicles (Euro VI), European Commission, Brussels, Belgium., 2009.
4 D. C. Carslaw and S. D. Beevers, *Atmos. Environ.*, 2004, **38**, 1233–1234.
5 D. C. Carslaw, *Atmos. Environ.*, 2005, **39**, 4793–4802.
6 P. Anttila, J.-P. Tuovinen and J. V. Niemi, *Atmos. Environ.*, 2011, **45**, 986–992.
7 G. A. Bishop and D. H. Stedman, *Environ. Sci. Technol.*, 2015, **49**, 11234–11240.
8 D. E. Millstein and R. A. Harley, *Environ. Sci. Technol.*, 2010, **44**, 5042–5048.
9 R. Alvarez, M. Weilenmann and J.-Y. Favez, *Atmos. Environ.*, 2008, **42**, 4699–4707.
10 AQEG, Nitrogen Dioxide in the United Kingdom, Report prepared by the Air Quality Expert Group for the Department for Environment, Food and Rural Affairs; Scottish Executive; Welsh Assembly Government; and Department of the Environment in Northern Ireland, March 2004, technical report, 2004.
11 D. C. Carslaw, S. D. Beevers, J. E. Tate, E. Westmoreland and M. L. Williams, *Atmos. Environ.*, 2011, **45**, 7053–7063.
12 M. Weiss, P. Bonnel, R. Hummel, A. Provenza and U. Manfredi, *Environ. Sci. Technol.*, 2011, **45**, 8575–8581.
13 Y. Chen and J. Borken-Kleefeld, *Atmos. Environ.*, 2014, **88**, 157–164.
14 G. J. M. Velders, G. P. Geilenkirchen and R. de Lange, *Atmos. Environ.*, 2011, **45**, 3025–3033.

15 J. D. Lee, C. Helfter, R. M. Purvis, S. D. Beevers, D. C. Carslaw, A. C. Lewis, S. J. Moller, A. Tremper, A. Vaughan and E. G. Nemitz, *Environ. Sci. Technol.*, 2015, **49**, 1025–1034.
16 D. C. Carslaw and G. Rhys-Tyler, *Atmos. Environ.*, 2013, **81**, 339–347.
17 L. J. Clapp and M. E. Jenkin, *Atmos. Environ.*, 2001, **35**, 6391–6405.
18 M. E. Jenkin, *Atmos. Environ.*, 2004, **38**, 5131–5138.
19 D. C. Carslaw and S. D. Beevers, *Atmos. Environ.*, 2004, **38**, 3585–3594.
20 D. C. Carslaw and S. D. Beevers, *Atmos. Environ.*, 2005, **39**, 167–177.
21 T. W. Kirchstetter, R. A. Harley and D. Littlejohn, *Environ. Sci. Technol.*, 1996, **30**, 2843–2849.
22 D. C. Carslaw, M. Priestman, M. L. Williams, G. B. Stewart and S. D. Beevers, *Atmos. Environ.*, 2015, **105**, 70–77.
23 D. A. Burgard, G. A. Bishop, R. S. Stadtmuller, T. R. Dalton and D. H. Stedman, *Appl. Spectrosc.*, 2006, **60**, 135A–148A.
24 P. J. Popp, G. A. Bishop and D. H. Stedman, *J. Air Waste Manage. Assoc.*, 1999, **49**, 1463–1468.
25 D. Burgard, T. Dalton, G. Bishop, J. Starkey and D. Stedman, *Rev. Sci. Instrum.*, 2006, **77**, 14101.
26 D. C. Carslaw and K. Ropkins, *Environ. Modell. Software*, 2012, **27–28**, 52–61.
27 R. B. Ellison, S. P. Greaves and D. A. Hensher, *Transp. Res., Part D: Transp. Environ.*, 2013, **23**, 25–33.
28 F. Kelly, B. Armstrong, R. Atkinson, H. R. Anderson, B. Barratt, S. Beevers, D. Cook, D. Green, D. Derwent, I. Mudway and P. Wilkinson, *Res. Rep. Health Eff. Inst.*, 2011, 3–79.
29 TfL, London Low Emission Zone Impacts Monitoring Baseline Report, July 2008, Transport for London., 2008.
30 EEA, EMEP/EEA air pollutant emission inventory guidebook 2013, Technical guidance to prepare national emission inventories, European Environment Agency, EEA Technical report No 12/2013, ISSN 1725-2237, 2013.

Faraday Discussions

PAPER

Spatially resolved flux measurements of NO_x from London suggest significantly higher emissions than predicted by inventories

Adam R. Vaughan,[a] James D. Lee,*[b] Pawel K. Misztal,[c] Stefan Metzger,[de] Marvin D. Shaw,[b] Alastair C. Lewis,[b] Ruth M. Purvis,[b] David C. Carslaw,[bf] Allen H. Goldstein,[c] C. Nicholas Hewitt,[g] Brian Davison,[g] Sean D. Beevers[h] and Thomas G. Karl[i]

Received 16th November 2015, Accepted 30th November 2015
DOI: 10.1039/c5fd00170f

To date, direct validation of city-wide emissions inventories for air pollutants has been difficult or impossible. However, recent technological innovations now allow direct measurement of pollutant fluxes from cities, for comparison with emissions inventories, which are themselves commonly used for prediction of current and future air quality and to help guide abatement strategies. Fluxes of NO_x were measured using the eddy-covariance technique from an aircraft flying at low altitude over London. The highest fluxes were observed over central London, with lower fluxes measured in suburban areas. A footprint model was used to estimate the spatial area from which the measured emissions occurred. This allowed comparison of the flux measurements to the UK's National Atmospheric Emissions Inventory (NAEI) for NO_x, with scaling factors used to account for the actual time of day, day of week and month of year of the measurement. The comparison suggests significant underestimation of NO_x emissions in London by the NAEI, mainly due to its under-representation of real world road traffic emissions. A comparison was also carried out with an enhanced version of the inventory using real world driving emission factors and road measurement data taken from the London Atmospheric Emissions Inventory (LAEI). The measurement to inventory agreement was substantially improved using the enhanced version, showing the importance of fully accounting for road traffic, which is the dominant NO_x emission source in London. In

[a]*Department of Chemistry, University of York, York, UK*
[b]*National Centre for Atmospheric Science, University of York, York, UK. E-mail: james.lee@york.ac.uk*
[c]*University of California, Berkeley, USA*
[d]*National Ecological Observatory Network, Boulder, USA*
[e]*University of Colorado, Boulder, USA*
[f]*Ricardo Energy and Environment, Fermi Avenue, Harwell, Oxon, OX11 0QR, UK*
[g]*Lancaster Environment Centre, Lancaster University, Lancaster, UK*
[h]*Department of Analytical and Environmental Sciences, King's College London, UK*
[i]*Institute of Atmospheric and Cryospheric Sciences, University of Innsbruck, Innsbruck, Austria*

central London there was still an underestimation by the inventory of 30–40% compared with flux measurements, suggesting significant improvements are still required in the NO_x emissions inventory.

Introduction

Due to its adverse effects on human health,[1–4] nitrogen dioxide (NO_2) concentrations are regulated by the EU Air Quality Directive which sets limit values for hourly and annual mean ambient concentrations. The annual mean limit value of 40 μg m^{-3} is exceeded in many urban centres throughout the UK, including London. In addition to its direct health effects, NO_x (the sum of NO + NO_2) contributes to the formation of ozone and secondary particles through a series of photochemical reactions[5] and hence reductions in NO_x emissions are necessary to control the regional-scale ground level concentrations of ozone, which is itself a regulated pollutant under the Air Quality Directive.

Concentrations of carbon monoxide (CO) and hydrocarbons in urban centres in the UK have decreased by around an order of magnitude over the past 20 years, providing clear evidence of the effectiveness of both the legislative framework and the emission control technologies employed for these pollutants. However, whilst trends in ambient concentrations of NO_x and NO_2 in the UK generally showed a decrease in concentration from 1996 to 2002, this has been followed by a period of more stable concentrations from 2004 to 2014. This is not in line with the expected decrease suggested by the UK emission inventories (Fig. 1).[6]

It is known that ambient NO_2 does not respond linearly to reductions in the concentration of NO_x, in part due to changes in diesel emission control technology leading to increases in directly emitted NO_2,[7] and partly due to the complexities of atmospheric chemistry. This may partially explain why ambient concentrations of NO_x have not declined as rapidly as expected. Although it has been known for some time that on-road emissions of NO_x from diesel passenger cars are often higher than those measured during test cycles,[8] it has very recently emerged that this may be due to deliberate action by some manufacturers. The issues surrounding NO_x emission from diesel vehicles are exacerbated in Europe by the high proportion of diesel engine vehicles in the passenger car fleet.[9] Hence, not surprisingly, there remain considerable difficulties in reconciling predictions of changes in NO_x concentrations arrived at using emission inventories with actual measurements of ambient concentrations.

Air pollutant emission inventories provide input data for air pollution models, which in turn are used for predicting current and future air pollution and in developing strategies for improving air quality. One approach uses the so called 'bottom up' approach involving estimating emissions from different individual sources (*e.g.* emissions from a particular type of vehicle per km driven) and activity factors (*e.g.* number of vehicle km driven on a particular road) to produce annual emission estimates.[10] Therefore errors in the emissions from a large source sector (such as passenger cars) can lead to significant inaccuracies in the inventories, which then further propagate into forecasts of air pollutant levels. Evaluation of emission inventories can be carried out by comparing air quality model predictions (using inputs from the inventory) to observed concentrations. However this method does not provide a direct comparison with the emission rate as it requires knowledge of other parameters such as chemistry and meteorology, as well as the inherent uncertainty in the models themselves. In contrast, the eddy-covariance technique provides a direct measurement of an atmospheric pollutant flux from a particular 'footprint', providing a 'top down' approach for quantifying emissions.[11]

A recent study from a tall tower site in central London[12] directly compared the measured NO_x emission rate with estimates from UK emissions inventories.[10,13] The study found observed emissions of NO_x were on average 80% higher than standard inventory estimates for central London, suggesting the inventory was poorest at estimating NO_x where traffic is the dominant source. Agreement was found to be better when an inventory with more explicit treatment of traffic emissions was used, so showing the importance of correctly accounting for the traffic source in London. While this tower-based study well represents the vicinity of the measurement location, it cannot reveal spatial patterning across Greater London.

In the present work, we report measurements of NO_x fluxes taken from a low flying aircraft over London. Full details of the flight tracks, as well as measurement concentrations of NO_x and volatile organic carbons (VOCs) from the aircraft (including a comparison to various ground based measurements in London) can be found in Shaw *et al.* (2015).[14] Here, we will concentrate exclusively on the flux measurements. Spatially resolved data are compared to emissions inventories, providing a measure of how well the inventories reproduce measured emissions over a wide area of central and suburban London.

Experimental

Measurements

Research flights were conducted using the Natural Environment Research Council's Dornier 228 aircraft, based at the NERC Airborne Research and Survey Facility (ARSF) at Gloucester Airport. Each flight operated with a crew of two pilots, a mission scientist and up to three instrument scientists. Flight speed was maintained at 80 m s^{-1} over central London, with a flight altitude for flux measurements of ~360 m, ensuring measurements could be directly related to

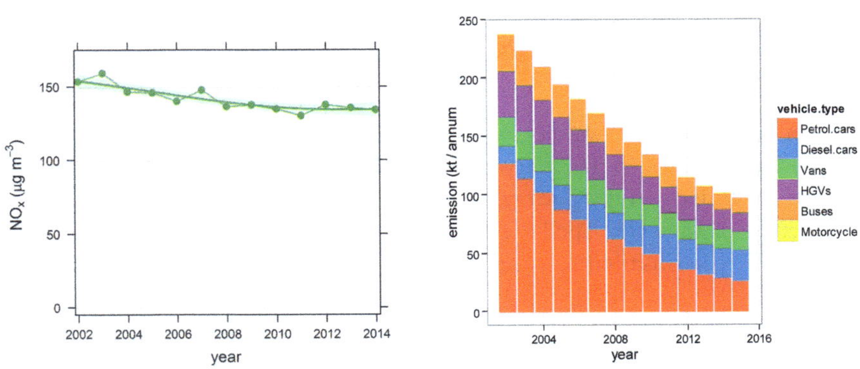

Fig. 1 (a) Trends in the mean concentration of NO_x across 35 roadside sites in Greater London with at least 10 years of data capture, (b) projected change in urban road transport emissions split by main vehicle type from the NAEI using 2002 as a base year.

the surface. The measurements ran over a two-week period during July 2013. In total 12 flights were completed across a set flight path, running from the south-western to north-eastern suburbs, of central London, crossing the River Thames in the vicinity of London Bridge and Tower Bridge. Each flight was designed to gain multiple replicate transects across the same flight path. The flight path was chosen to allow flux measurements to be made over a representative sample of land use types in London. Flight times were staggered over the two weeks to allow for study into the rush-hour periods, as well as weekday to weekend comparisons.

Measurement of NO_x (NO and NO_2) mixing ratios were made using a chemiluminescence instrument designed for 10 Hz measurements, *via* dual channel design to allow concurrent measurement of NO and NO_2.[15] The instrument quantifies NO *via* chemiluminescence upon its reaction with excess O_3, and NO_2indirectly using a photolytic converter to convert NO_2 to NO at 395 nm, followed by detection by chemiluminescence.[16] A more detailed description of the running, calibration and design of the instrument can be found in Shaw *et al.* (2015).[14] Mixing ratios used in this study are reported with respect to dry air + water vapour, due to constant stable humidification of the O_3 reactant used in the instrument. An accurate humidity value within the sample volume was not obtained. Humidification of the reaction O_3 is conducted to remove any changes in background signal caused by rapid water vapour changes in the ambient sample during the flights. Due to this the reported wet mixing ratios are lower compared to dry mixing ratios, and a potential underestimation of the NO_x flux was calculated to not exceed 4%.

During all flights, an Aircraft-Integrated Meteorological Measurement System (AIMMS 20) turbulence probe (Aventech Research Inc.) mounted underwing was used to obtain micrometeorological measurements at a 20 Hz acquisition rate.[17] The probe was calibrated for static and dynamic upwash.

Wavelet transformation

In this study, we build upon previous work described by Torrence and Compo (1998) and Thomas and Foken (2007) and use the continuous wavelet transform (CWT) methodology for quantifying eddy-covariance fluxes from an aircraft.[18,19] We conduct CWT *via* eqn (1), defining the transform of the discrete data sequence $x(n)$ *via* complex conjugate of the Morlet wavelet $\psi^*_{p,a,b}$, for N data points[20]

$$w_x(a,b) = \sum_{n=0}^{N} x(n)\psi^*_{p,a,b}(n) \tag{1}$$

ψ denotes the mother wavelet, with a and b acting as parameters to scale and localise the wavelet in frequency and time respectively, and p as the normalized factor. In order to fully portray our approach, it is useful to understand the properties of a wavelet. The base equation for a wavelet is coined by the mother wavelet after localisation has occurred. This is defined in eqn (2).[21]

$$\psi_{a,b}(t) = \frac{1}{\sqrt{a}}\psi\left(\frac{t-b}{a}\right) \tag{2}$$

Wavelets are localised both in the frequency domain as for Fast Fourier Transform (FFT) but also in the time domain.[22] This allows for a signal's properties to be explored in both domains. Another added strength of the method is the ability to tailor the wavelet's properties to suit the application. In this study we used the complex Morlet wavelet as expressed in eqn (3).[18]

$$\psi_0(\eta) = \pi^{-1/4} e^{i\omega_0\eta} e^{-\eta^2/2} \tag{3}$$

ω_0 represents the frequency and η the time parameter in non-dimensional state. A more in-depth description for wavelet principles and the Morlet wavelet can be found in Torrence and Compo.[18] The Morlet wavelet has been implemented in a number of previous studies specifically for analysing airborne measurements.[20,21,23–25]

Flux calculations

Flux is the measure of the rate of change of a quantity moving through an area per unit time, in this case, the amount of a chemical species (NO and/or NO_2) being uplifted at a point in space per unit time. In order to quantify this, we implement a flux calculation algorithm developed and previously used in airborne studies of isoprene fluxes.[25] Using CWT, the covariance between vertical wind and species of interest is evaluated. We first calculate the lag time difference between vertical wind speed (w') and analyte concentration (c'), where w' represents the difference of the instantaneous vertical wind measurement to the mean vertical wind and c' is the difference of the instantaneous concentration (mixing ratio) to the mean concentration value. This allows the two data sets to be aligned, giving normalised covariance.

The lag time between the two data sets was found to be in the range of 4–7 s. Due to the observed non-stationarity of the lag difference, each flight leg was analysed separately ensuring no additional bias. Before we calculated fluxes *via* CWT, de-spiking of the data was conducted. The CWT calculates the global cross-spectrum between analyte concentration and vertical wind speed. Fig. 2(b) depicts an example of the global cross-spectrum for a flight leg of NO_2 concentration data, with the y-axis giving the eddy contributions integrated over all frequency periods in seconds and the x-axis being the distance travelled along the flight leg. For each individual flight leg, we compared the co- and cumulative spectra across all frequencies for both the CWT and a standard Fast Fourier Transform (FFT). For all the flight legs, the majority of the flux contributions were found to be below 0.1 Hz. This inferred that high frequency spectra correction was not needed, with its contribution to the calculated flux being low. As a final data quality tool, we pad the edges of the wavelet with zeros to help remove any associated edge effects. After this padding is done we apply the cone of influence (COI). The COI is defined as the area outside of which edge effects in the wavelet cross-spectrum give rise to data of a lower quality than the rest.[18] For this reason we only consider data within the COI, and remove all other data. Due to the reactive chemistry between NO and NO_2 in the presence of O_3,[26] we calculated NO_x flux *via* separate flux calculations for NO and NO_2, with the combination of the two yielding total NO_x flux.

Error analysis

Error quantification for the instantaneous fluxes was conducted as previously described by Karl *et al.* (2009), with the total error for the calculated flux being due to both random (re) and systematic (se) errors.[23]

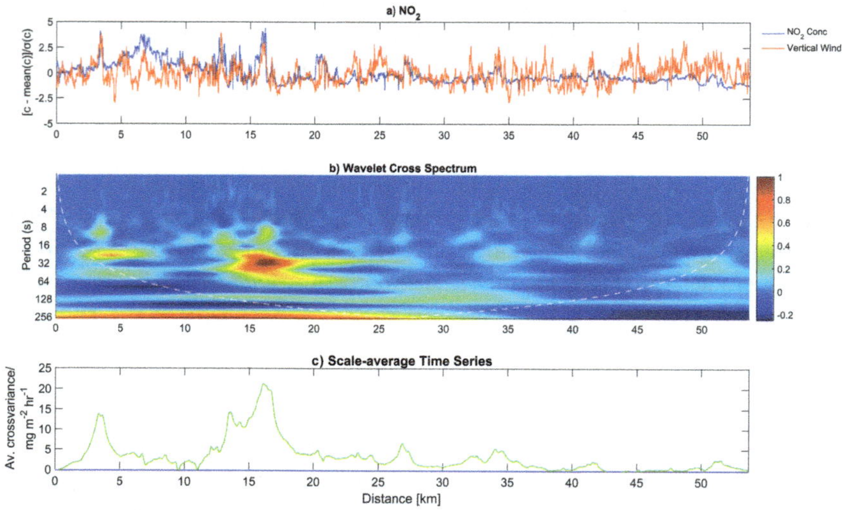

Fig. 2 (a) Variation of NO_2 mixing ratio from the mean and variance of vertical wind speed from the mean, (b) time resolved wavelet cross spectrum, (c) the average cross-covariance between NO_2 mixing ratio and vertical wind, for a typical NE to SW run across London.

$$\mathrm{re} \leq 1.75\left(\frac{z}{z_\mathrm{i}}\right)^{0.25}\left(\frac{z_\mathrm{i}}{L}\right)^{0.5} \tag{4}$$

$$\mathrm{se} \leq \frac{2.2z_\mathrm{i}\left(\frac{z}{z_\mathrm{i}}\right)^{0.50}}{L} \tag{5}$$

L denotes the flight leg length, z the flight altitude and z_i the height of the boundary layer. Boundary layer heights were obtained at the beginning and end of each flight *via* atmospheric soundings from Heathrow Airport, west of London. A more detailed study into the relative effects of errors for calculated flux measures can be found in Mann and Lenschow (1994).[27] We find calculated systematic sampling error for the entire flight leg to vary from 1–4%, with random sampling error being the major contributor to the overall error, varying from 15–25%. We also account for the 4% error associated with humidity, giving a total error for the flux estimates in the region of 20–35%. In order to provide meaningful comparison with emission inventories, the random sampling error was also calculated individually for each 1 km resolved flux measurement, and found to range from 60–250%. Errors associated with chemical and physical losses and storage of NO_x in the city canopy below flight level are discussed in detail below.

Footprint model

In order to carry out interpretation of the data and compare to emissions inventories, it is necessary to calculate a flux footprint for each measurement. For this we use a footprint model which quantifies the spatial area from which the emission originates from.[28–30] An in-depth review into footprint models and their continued development can be found in Leclerc and Foken (2014).[31] We use the footprint model described by Metzger *et al.* (2012),[32] due to its similar application.[21] This builds upon a cross-wind integrated model, which quantifies the flux contribution relative to the distance away from the measurement position, into the prevailing wind direction. This approach alone still leaves uncertainty due to the analysis being only in 1-dimensional space.[33] For this purpose Metzger *et al.* (2012)[32] coupled the model with a cross-wind function, allowing for non-perpendicular wind direction influences to be accounted for.

We parametrise the model using friction velocity, measurement height, standard deviation of the vertical wind and roughness length. Turbulence statistics for the footprint model are calculated at 1 km intervals from the wavelet cross-scalogram, with movement in both the x and y direction being 1 km for each new footprint. We use the Drew *et al.* (2013) study data to estimate expected roughness lengths for the London area.[33] For suburban area roughness lengths, values generally range from 0.4–0.6 m but up to 2.0 m for central London. The model evaluates the maximum influence distances in all directions, with the measurement point at its centre. From this it is possible to create a weighing matrix at the same temporal resolution as the inventories being quantified. The matrix when summed up gives a value of 1. This weighting matrix predicts, for every overflown 1000 m cell, the ground influence contributing to the observed emission flux. The matrix is set to the same coordinate system as the inventory. A separate weighting matrix is calculated for every measurement point along the flight track to allow for independent comparison.

For each point along the flight track, the emission inventory value for every cell within the footprint matrix is weighted accordingly and summed up to give a single emission estimate. All estimates from all source sectors are summed to give a total emission estimate every 1 km. Fig. 3 depicts an example of the area for which we can consider our measurements to be spatially representative. The footprint area ranges in distance from the flight track anywhere from 5 to 12 km into the prevailing westerly wind direction. Part of the footprint area includes part of the London Low Emissions Zone (LEZ) which was introduced in 2008 to help improve air quality in central London.

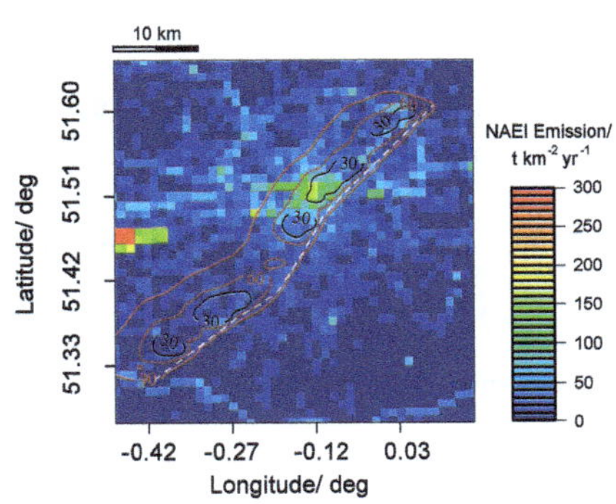

Fig. 3 The contributing footprint area to a typical flight track, dependent on the prevailing wind direction and altitude of flight. The footprint is overlaid onto the NAEI at 1 km^2 grid resolution, coloured to annual NO_x tonnage emission estimates.

Results

Flight descriptions

Due to problems with the AIMMS probe, here we only calculate NO_x fluxes from data collected on three flights (research flights 5, 7 and 12), from 3rd to 9th July 2013 (see Table 1). Each flight was planned to follow a set flight track over central London, containing multiple legs back and forth in north-easterly (NE) and south-westerly directions (SW), to allow for repeat measurements over the same area. Each flight started at the SW corner of the M25 ring road and proceeded across central London over the Tower Bridge area, to finally end at the NE corner of the M25. The leg was then reversed and repeated 7 or 8 times for each flight. Flight 5 was conducted during the morning rush hour from 08:00 until 10:00 am. Flights 7 and 12 were performed during the afternoon (12:00 to 16:00), thus providing some diurnal information of the emissions. Flights 5 and 7 had wind directions from the west, giving information about the emissions coming from central London. Flight 12 had wind directions prevailing from the NE giving emission information from more easterly areas of London. All the flights were performed during clear sunny days with the highest air temperatures observed (>23 °C) during flight 12. Further information of the flights including mixing ratios observed and comparisons to ground level observations can be found in Shaw *et al.* (2015).[14]

Spatially resolved flux observations

Calculated fluxes across all flights showed significant variability along the flight track. Fig. 4 shows all flight legs across London for flights 5, 7 and 12 (total of 17 legs). Each data point at 1 km resolution is coloured by measured NO_x flux. Consistently, the highest observed fluxes all coincide within the same spatial area, in central London. This area of London contains high traffic densities, a high density of large buildings and also the London Bridge railway station with a large number of diesel trains operating in the area. Measured NO_x fluxes over this area ranged from 30–90 mg m^{-2} h^{-1}, with the highest fluxes observed during flight 5. This corresponded to the morning rush hour period from 08:00 to 10:00 with high traffic densities in central London.[34] Fluxes measured on the other flights over central London during the afternoon showed consistently lower NO_x fluxes compared to the morning flights, in the range of 30–40 mg m^{-2} h^{-1}, with no clearly defined evening rush hour period obvious. For all flights, fluxes measured outside of central London (both in the SW and NE directions) were significantly lower, typically in the range 5–10 mg m^{-2} h^{-1}, corresponding to the lower traffic density in these parts of the city. Some spikes in NO_x fluxes, up to 20 mg m^{-2} h^{-1},

Table 1 Details of the flights where NO_x fluxes were calculated

Flight number	Number of flight legs	Day of week	Time of day	Flight altitude/m	Wind direction/deg	Air temperature/°C
5	9	Wed	Morning	365–380	250–280°	14.0–16.0
7	7	Thurs	Afternoon	340–380	220–260°	19.0–21.0
12	7	Tues	Afternoon	340–360	30–70°	22.0–23.5

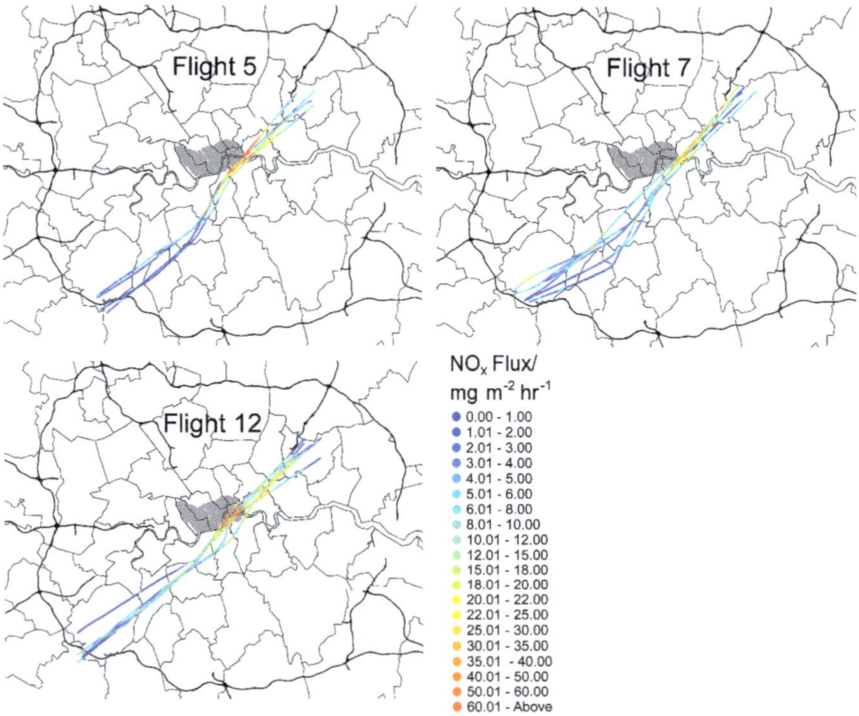

Fig. 4 Tracks of all the flight legs for flights 5 (3rd July 2013 morning), 7 (4th July 2013 afternoon) and 12 (9th July 2013 afternoon), mapped over Greater London, coloured to the measured NO_x flux. The grey area represents London's Low Emission's Zone (LEZ).

were observed on some legs, which seem to correspond with major roads (*e.g.* the M25 ring road).

Emission inventories

The UK's National Atmospheric Emissions Inventory (NAEI) is the standard tool for both reporting and understanding the emissions of a range of pollutants in the UK. For each pollutant the inventory is broken down into a series of source types which, when grouped and summed together, give an annual emission estimate at 1 km^2 resolution across the entire country. Emission source types include road transport, rail and aviation, domestic and industrial combustion, energy generation and other sources such as waste production.[35] In common with many other emission inventories in Europe, the NAEI is mainly based on the (Calculation of Emissions from Road Transport) COPERT 4 emission factor model for road vehicle emissions. The COPERT 4 methodology is part of the European Monitoring and Evaluation Programme/European Economic Area (EMEP/EEA) air pollutant emission inventory guidebook for the calculation of air pollutant emissions.[36]

To allow for real time comparison between our airborne flux measurements and the inventory estimates for the relevant flux footprint, each inventory source sector is assessed individually using the described footprint method in Section

2.5, giving an emission contribution. The emission inventories used here only provide annual emission estimates for each grid square from the wide range of source sectors – they do not directly provide temporally-resolved hour of the day or day of the year estimates. Therefore, we scale each source sector separately, using scaling factors[35] that take into account the temporal variation in emissions for any given month, day and hour. By scaling each source sector individually, we generate more realistic emission estimates for the specific time of day and day of the year of our flux measurements. Scaled estimates for each source sector are then summed up at each point along the flight track, to give 1 km resolution estimates for NO_x emission. Nevertheless, such temporal profiles represent the typical average profiles expected for different emission sources. Uncertainty will be introduced when comparing estimates obtained over short periods of time with the mean profiles used in atmospheric emission inventories.

Fig. 5 depicts the 1 km spatially resolved NO_x flux observations averaged for all three flights in blue and the NAEI emission estimates plotted in red for all flight legs, all against latitude of the sample position along the flight track. For all of the flights there is relatively good agreement between the NAEI estimates and the measurements in suburban outer London, but, there appears to be a significant underestimation in the inventory compared with the measurements, by a factor of 2–4, in central London (latitude 51.43–51.52). The data interpretation is complicated by the significant uncertainty (60–250%) in our 1 km spatially-resolved flux measurements. Therefore in order to further reduce these uncertainties, each flight track is divided into four parts, each approximately 11 km in length. The first two segments represent outer regions of SW London (zones 1 and 2) and the second set (zones 3 and 4) for central and NE London. The flux data are then averaged for each segment to give a single NO_x emission flux. The uncertainty associated for these fluxes is greatly reduced compared to the more spatially resolved flux data, as shown in Fig. 5, now being in the range 30–45%.

Fig. 6 depicts the average ratios of the measured to emission inventory-estimated fluxes from all the flight legs. The error bars denote the standard error of the 17 legs. Zones 1 and 2 in outer London show ratios of around 1.5, with the ratio increasing to over 2 for zones 3 and 4. These discrepancies are similar in magnitude to those reported from measurements made by Lee *et al.* (2015)[12] from 180 m above ground level on the BT Tower in central London.

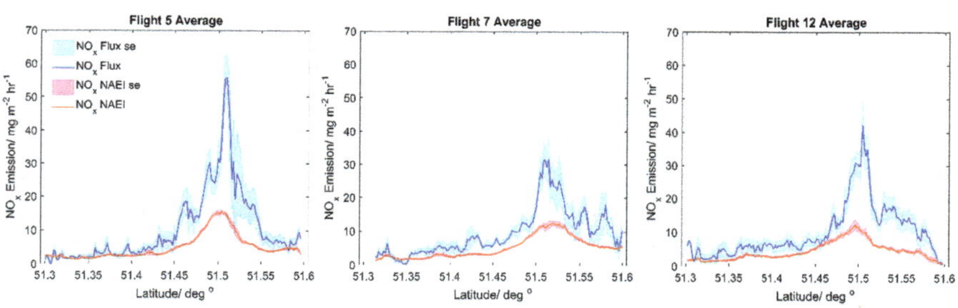

Fig. 5 Average for Flights 5, 7 and 12, measured NO_x flux plotted in blue with standard error as the shaded blue area. Calculated and scaled NAEI estimates of NO_x emission are plotted in red, with standard error as the shaded red area.

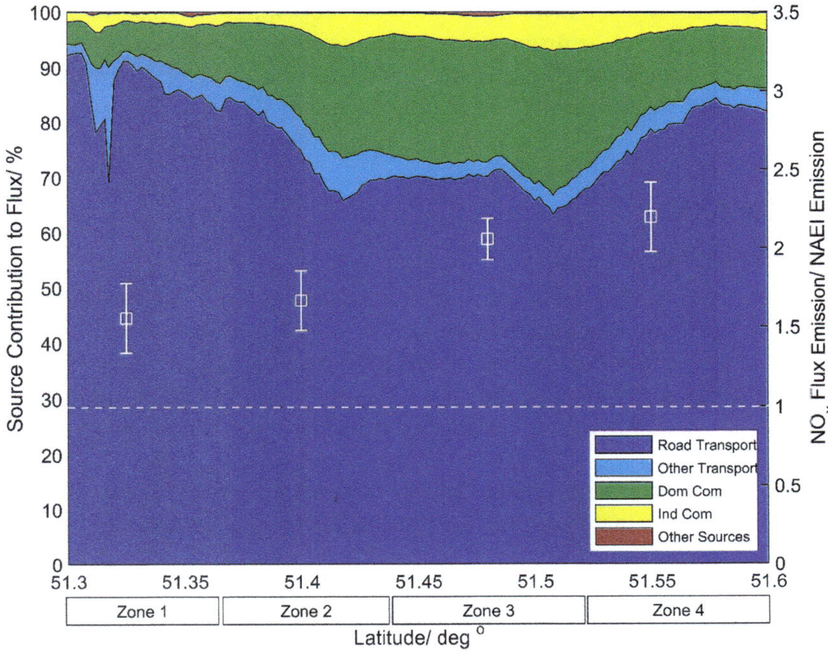

Fig. 6 Calculated emission source contributions from the NAEI plotted on the first *y* axis, with sources being broken down into road transport; other transport (such as rail), domestic and commercial combustion, industrial combustion and other sources. Also plotted on the second *y* axis as white squares is the ratio of measured to NAEI estimated NO_x flux, with the error bars denoting the standard error of all points included. The flight is broken into 4 zones, with 1 and 2 being SW outer London, 3 central London and 4 NE outer London.

It is clear that the NAEI is unable to accurately represent emissions of NO_x from London, with particularly large discrepancies in the central London area. One possible explanation for this is in the way in which the inventory accounts for road transport emissions. Fig. 6 shows that across the whole of the area surveyed by our flights, road transport is the dominant source of NO_x. Traffic contributes up to 90% of NO_x emissions in outer London (zones 1 and 2), strongly suggesting that significant error in the road traffic source in the inventory is responsible for the underestimation of NO_x emissions by the inventory relative to our flux observations. In central and NE London (zones 3 and 4), where the underestimation of the inventory is highest, road transport, while still the largest source, is responsible for 65–70% of the total NO_x emissions in the inventory, with most of the remainder due to domestic and commercial combustion of gas for space and water heating. The relatively low use of gas for heating during July is taken account of *via* the emission scaling factors applied to the NAEI, but the large underestimation in the inventory in the central London area suggests that as well as underestimation of traffic-derived NO_x, there is also some error in the NAEI treatment of domestic and commercial combustion, or there is a major missing source of NO_x in the inventory.

The results described above are potentially important as the NAEI is used to provide emissions for air quality forecasting models, which inform potential future air quality abatement strategies. The following sections will discuss potential reasons for the discrepancy, including measurement errors and investigate the advantages of using a more London-specific emissions inventory with a more explicit treatment of the road traffic source.

Discussion

Measurement errors

Using the method of Karl *et al.* (2013)[24] we are able to quantify a limit of detection (LOD) for our flux measurements from the calculated covariance function, above which the measured flux can be distinguished from the combined effects of instrument noise.[24,37] We find the LOD varies from flight to flight, being in the range 0.075–0.180 mg m^{-2} h^{-1}, with our measured data almost always significantly above this.

Some NO_x emitted at the ground surface may be lost before it is detected at the flight altitude, causing an error in our flux estimates. Using measured meteorological parameters and the wavelet transform, we calculate Deardorff velocities (w^*) for each flight. w^* gives a turn-over time for all the influencing eddies which account for the measured flux. We found w^* to be in the range of 0.5–3.0 m s^{-1}, which gives a vertical ascent time of 2–8 minutes from the surface to the measurement altitude of 300 m.

Loss of emitted substances can occur *via* three mechanisms: chemical reaction, weakening vertical transport and storage. To account for chemical losses in the atmosphere, we use the vertical flux profiles obtained in the Karl *et al.* (2013) study.[24] They were able to calculate the loss of isoprene emission during transport through the boundary layer. To use this method for NO_x, we substitute NO_x chemistry into the model in place of isoprene chemistry, so taking into account the lifetime of NO_x in the troposphere with respect to chemical reaction of NO_x with the hydroxyl radical (OH). We use the middle-of-the-day OH abundances measured in London during summer 2012, as reported in the Lee *et al.* (2015) study,[38] of 2.0×10^6 molecule per cm^3. This gives a lifetime of NO_x above London of ~11 hours. Using this calculated NO_x lifetime, we estimate that the loss of NO_x flux between the ground surface and the flight altitude is only between 1 and 2%. We have not considered other chemical NO_x loss processes (*e.g.* PAN formation) as we believe these will be small compared to the reaction of NO_2 with OH. Loss due to weakening vertical transport accounts for the loss due to vertical momentum decreasing as altitude increases and is estimated to be 25–30%. The final loss process involves storage of some of the emission within the urban structure, such as street canyons.[39,40] However we do not consider this to be an important loss process here. Significant storage of NO_x within the street canyon in London would result in a steady build up of concentrations, something that is not typically observed. Thus, if deposition processes are ignored, the transfer of NO_x out of the street canyon is essentially equal to the emissions, with the only result being a potential time lag from emission to measurement. Future work in this area could gain from using the Lenschow *et al.* (2015) study to account for all three loss terms in unison.[41] All of these loss processes have the potential to increase our measured fluxes, which would further increase the discrepancy between measured and inventory-estimated emissions.

As discussed above, our study shows that, even in outer London where traffic sources are estimated to contribute 90% of the emitted NO_x, there is observed inventory under-estimation by a factor of around 150%, adding further evidence that it is the traffic source sector that contains the major error. There are several sources of disagreement between the flux emission estimates and those from the emission inventories. The emission inventories focus on longer term (annual) emission totals rather than providing estimates by hour of the year. For this reason the emissions for a particular hour of the year need to be estimated through the scaling factors described in the previous section. It is however difficult to quantify the additional uncertainties introduced when scaling factors are used in emission inventories in this way.

The London atmospheric emissions inventory

The London Atmospheric Emissions Inventory (LAEI) provides emission estimates for eight key air pollutants (including NO_x) at 1 km^2 resolution across all London boroughs, to the outer M25 motorway boundary that encompasses Greater London. The inventory reflects the geography of the roads in London, enabling an accurate assessment of population exposure and health impacts. The inventory is broken down into source sectors, contributing to the total annual estimates. Sources within the LAEI include: road transport (exhaust and non-exhaust), large regulated industrial processes, small regulated industrial processes, large boiler plants, gas heating (domestic and industrial-commercial), oil combustion sources (domestic and commercial), coal combustion sources (domestic and commercial), agricultural and natural sources, rail, ships, airports and others such as sewage plants. The LAEI also contains a non-road mobile machinery source (*e.g.* cranes, small electricity generators and other construction machinery), which has been shown previously to be a significant source of NO_x in central London.[13]

The LAEI uses a 'bottom up' road traffic inventory taking vehicle flow and speed on each road and combining these with national and London-specific vehicle stock data (including buses and taxis) to calculate emissions for each of the 11 vehicle types and combining these to create emissions at 1 km^2 resolution. The 'enhanced LAEI' results used here benefited from roadside emissions measurements, obtained using the University of Denver Fuel Efficiency Automobile Test (FEAT) system (IR/UV absorption spectroscopy), deployed for a 6 week campaign and taking measurements from 70 000 vehicles, including cars, buses and taxis, at four locations across London.[42] Vehicle number plates were recorded for each vehicle and these were cross-referenced against vehicle databases to obtain relevant vehicle details including their Euro emissions classification. The roadside emissions measurements quantified exhaust emissions of both total NO_x and NO_2 as a ratio to carbon dioxide (CO_2), with the results then combined with CO_2 estimates from the LAEI to create NO_x and NO_2 emissions in g km^{-1}. The emissions from vehicle types that were not measured during the campaign, most notably articulated heavy goods vehicles, were taken from the published LAEI results.[13]

In order to try to produce an improved agreement between the inventory measured emissions, we have produced an 'enhanced' version of the NAEI. We use all sources from the NAEI except its road emission source to form the base of the inventory. For the road emission source we use the enhanced LAEI road emission estimates scaled using the described roadside emissions measurements data, which increases the road traffic source of NO_x in the inventory by an average of ~50% across London. We also use a revised version of the LAEI's non-road mobile machinery (NRMM) source, which gives an improved estimate to that in the NAEI.

Comparing the enhanced NAEI with our measurements allows us to better assess the effect of using a more explicit treatment of traffic emissions in the national inventory. The same methodologies, as used for NAEI comparison, were used in analysis of the enhanced NAEI. Each enhanced NAEI source sector was analysed to give emission estimates along the flight track at 1 km resolution. Each sector was scaled individually to allow for time of day comparison using the scaling factors described earlier, then added together to give overall NO_x emission estimates.

Fig. 7 shows how enhanced NAEI emission estimates in red (standard error as light red shading) compare with the measured NO_x fluxes in blue (standard error as light blue shading) for each flight average. It is apparent that the agreement between the measurement and the inventory is better than for the standard NAEI. Outer regions of London seem to compare very well, with all flight legs showing only a small degree of discrepancy. However, despite the agreement being improved and almost always falling within the error of the 1 km spatially resolved flux measurement, there is still a significant discrepancy observed in central London. As with the standard NAEI data, the flight track was also divided into four parts, each approximately 11 km in length, with the flux data then averaged for each segment to give a single NO_x emission flux with reduced error (again in the range 30–45%). Fig. 8 depicts the average ratio of the measured to enhanced emission inventory flux from all the flight legs, with the error bars denoting the standard error of the 17 legs included, along with the source sector contribution as before. The contribution from the different source sectors is similar for the enhanced NAEI as for the standard inventory. The main difference is the addition of the non-road mobile machinery source, which now contributes up to 9% of total NO_x emissions in central London. The ratio of measurement to enhanced emissions inventory for all zones shows considerable improvement compared to the standard NAEI. For the SW outer London zones, the ratio is close to one (average of 1.1), which is within the standard error of the measurements. However, in central and NE London, there is still some significant underestimation (average measurement to inventory ratio of 1.48), which is outside the (30–

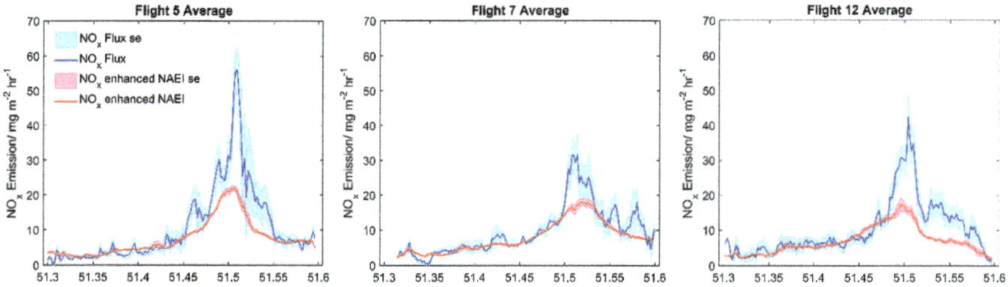

Fig. 7 Average for Flights 5, 7 and 12, measured NO_x flux plotted in blue with standard error as the shaded blue area. Calculated and scaled enhanced NAEI estimates of NO_x emission are plotted in red, with standard error as the shaded red area.

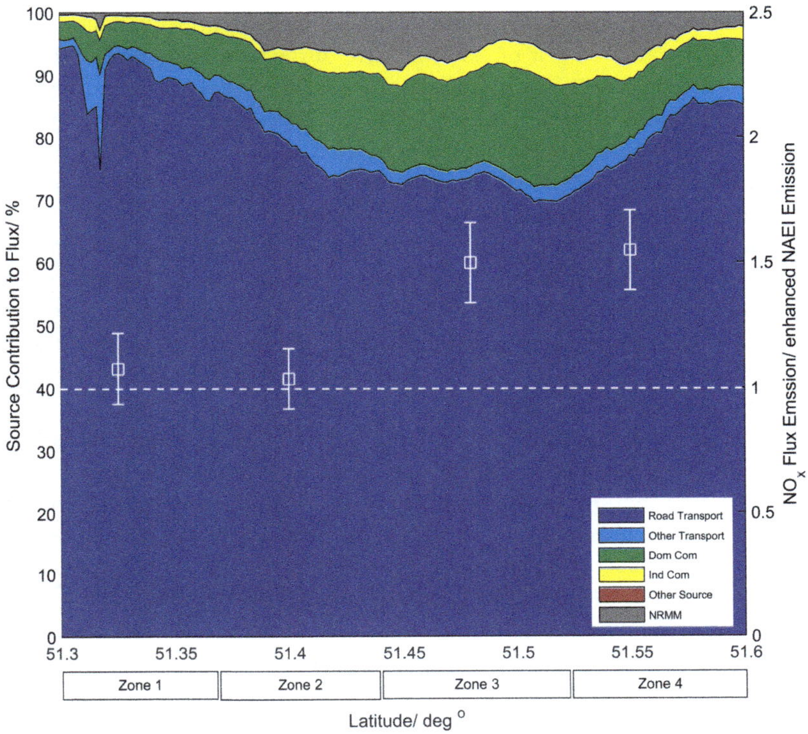

Fig. 8 Calculated emission source contributions from the enhanced NAEI (see text for details) plotted on the first *y* axis, with sources being broken down into road transport; other transport (such as rail), domestic combustion, industrial combustion, other sources and non-road mobile machinery (NRMM). Also plotted on the second *y* axis as white squares is the ratio of measured to enhanced NAEI estimated NO_x flux, with the error bars denoting the standard error of all points included. The flight is broken into 4 zones, with 1 and 2 being SW outer London, 3 central London and 4 NE outer London.

45%) flux measurement uncertainty. The observed improvements in the enhanced emission inventory estimation of NO_x can be directly related to the improved road transport source increasing the magnitude of the road transport emissions and the addition of the NMRR as a key source not currently accounted for in the standard NAEI.

Summary and conclusions

In this study, NO_x fluxes have been measured from an aircraft flying low over London using continuous wavelet transform, allowing for spatial analysis of NO_x emissions. NO_x fluxes were observed to be largest in central London, with maximum emissions of ~80 mg m^{-2} h^{-1} observed on flights during the morning rush hour. After calculating flux footprints along the flight tracks, measurements were compared to scaled NAEI estimates, providing a top-down analysis of the inventory. A significant discrepancy was identified between NAEI emission estimates and actual flux measurements, with the highest underestimation being a factor of two in central London. In outer London, where the source of NO_x is almost exclusively from road transport (~90%), there is an underestimation in the inventory of around a factor of 1.5, indicative of the poor treatment of the road traffic source in the NAEI. Due to the potential loss of some NO_x emitted at ground level during its transport to the measurement altitude, the calculated inventory underestimation can be considered as conservative.

We also compare our measurements to an enhanced version of the NAEI, containing both non-road mobile machinery emission sources from the LAEI and real world traffic emissions from the enhanced LAEI, which provides a much better treatment of NO_x coming from these sources. The comparison yielded better agreement between the two, especially in outer London, where the measured to inventory ratio was found to be around 1.1. In central London however there is still an average underestimation of around a factor of 1.5 compared to the measurements.

The current work has provided important information on the spatial variation in NO_x emissions over a large and complex urban environment. There are several areas where further research would be beneficial. First, the inhomogeneous terrain in London presents a challenge for flux measurements, both in terms of the inhomogeneous nature of the roughness and distributions of emissions (horizontally and vertically). Further measurements over different city areas would help better understand these effects. For example, measurements over a city such as Paris that has fewer high-rise buildings than London would provide a useful contrast. Second, the measurement of the fluxes of other species such as CO and CO_2 would help provide more comprehensive analysis of the NO_x flux emission estimates. More work is also required on understanding and quantifying the short-term accuracy of emission inventories. Emission inventories tend to focus on providing longer term (annual) emission estimates and are less able to provide information on the temporal nature of emissions. Improved information on the temporal characteristics of emissions would improve the reliability of short-term flux measurements when compared with emission inventories. In particular, our measurements highlight the critical importance of obtaining independent measurements of pollutant emission rates from vehicles during on-road driving conditions and using these data in emission inventories, rather than relying on emissions data obtained during artificial test driving conditions or provided by vehicle manufacturers.

Author contribution statement

CNH, ACL, JL and RP conceived the study and obtained funding. AV and JL wrote the paper. AV, SM, PM, AG and TK carried out the flux data analysis and footprint modelling. JL, MS, ACL, CNH, RP and BD made the measurements. DC and SB provided the emissions inventory and scaling data, as well as help in interpreting the results. All authors contributed to the discussion and commented on the manuscript.

Acknowledgements

We would like to thank Captain Carl Joseph and co-pilot James Johnson of the Natural Environment Research Council (NERC) Airborne Research and Survey

Facility for their contribution to the OPFUE campaign during 2013, through flight planning and flying of the Dornier 228 research aircraft. We would also like to thank NERC (grant NE/J00779X/1) and the UK Department of Environment, Food and Rural Affairs for funding. The National Ecological Observatory Network is a project solely sponsored by the National Science Foundation and managed under cooperative agreement by NEON, Inc. This material is based upon work supported by the National Science Foundation under Cooperative Service Agreement EF-1029808. Any opinions, findings, and conclusions or recommendations expressed in this material are those of the author(s) and do not necessarily reflect the views of the National Science Foundation.

References

1 V. Strand, M. Svartengren, S. Rak, C. Barck and G. Bylin, *Eur. Respir. J.*, 1998, **12**, 6–12.

2 R. W. Atkinson, I. M. Carey, A. J. Kent, T. P. van Staa, H. R. Anderson and D. G. Cook, *Epidemiology*, 2013, **24**, 44–53.

3 WHO, *Review of evidence on health aspects of air pollution – REVIHAAP project: final technical report*, 2013.

4 WHO, *Health risks of air pollution in Europe – HRAPIE project, Recommendations for concentration–response functions for cost–benefit analysis of particulate matter, ozone and nitrogen dioxide*, 2014.

5 S. Sillman, *Atmos. Environ.*, 1999, **33**, 1821–1845.

6 D. C. Carslaw, S. D. Beevers, J. E. Tate, E. J. Westmoreland and M. L. Williams, *Atmos. Environ.*, 2011, **45**, 7053–7063.

7 D. C. Carslaw, *Atmos. Environ.*, 2005, **39**, 4793–4802.

8 M. Weiss, P. Bonnel, R. Hummel, A. Provenza and U. Manfredi, *Environ. Sci. Technol.*, 2011, **45**, 8575–8581.

9 E. A. Mazzi and H. Dowlatabadi, *Environ. Sci. Technol.*, 2007, **41**, 387–392.

10 T. Bush, I. Tsagatakis, K. King and N. Passant, *NAEI UK emission mapping methodology 2006*, 2008.

11 J. Brioude, S. W. Kim, W. M. Angevine, G. J. Frost, S. H. Lee, S. A. McKeen, M. Trainer, F. C. Fehsenfeld, J. S. Holloway, T. B. Ryerson, E. J. Williams, G. Petron and J. D. Fast, *J. Geophys. Res.: Atmos.*, 2011, **116**, 19.

12 J. D. Lee, C. Helfter, R. M. Purvis, S. D. Beevers, D. C. Carslaw, A. C. Lewis, S. J. Moller, A. Tremper, A. Vaughan and E. G. Nemitz, *Environ. Sci. Technol.*, 2015, **49**, 1025–1034.

13 Greater London Authority, London Datastore, http://data.london.gov.uk/.

14 M. D. Shaw, J. D. Lee, B. Davison, A. Vaughan, R. M. Purvis, A. Harvey, A. C. Lewis and C. N. Hewitt, *Atmos. Chem. Phys.*, 2015, **15**, 5083–5097.

15 J. D. Lee, S. J. Moller, K. A. Read, A. C. Lewis, L. Mendes and L. J. Carpenter, *J. Geophys. Res.: Atmos.*, 2009, **114**, DOI: 10.1029/2009jd011879.

16 I. B. Pollack, B. M. Lerner and T. B. Ryerson, *J. Atmos. Chem.*, 2010, **65**, 111–125.

17 K. M. Beswick, M. W. Gallagher, A. R. Webb, E. G. Norton and F. Perry, *Atmos. Chem. Phys.*, 2008, **8**, 5449–5463.

18 C. Torrence and G. P. Compo, *Bull. Am. Meteorol. Soc.*, 1998, **79**, 61–78.

19 C. Thomas and T. Foken, *Boundary-Layer Meteorology*, 2007, **123**, 317–337.

20 B. Yuan, L. Kaser, T. Karl, M. Graus, J. Peischl, T. L. Campos, S. Shertz, E. C. Apel, R. S. Hornbrook, A. Hills, J. B. Gilman, B. M. Lerner, C. Warneke,

F. M. Flocke, T. B. Ryerson, A. B. Guenther and J. A. de Gouw, *J. Geophys. Res.: Atmos.*, 2015, **120**, 6271–6289.
21 S. Metzger, W. Junkermann, M. Mauder, K. Butterbach-Bahl, B. T. Y. Widemann, F. Neidl, K. Schafer, S. Wieneke, X. H. Zheng, H. P. Schmid and T. Foken, *Biogeosciences*, 2013, **10**, 2193–2217.
22 A. Held, *Boundary-Layer Meteorology*, 2014, **151**, 79–94.
23 T. Karl, E. Apel, A. Hodzic, D. D. Riemer, D. R. Blake and C. Wiedinmyer, *Atmos. Chem. Phys.*, 2009, **9**, 271–285.
24 T. Karl, P. K. Misztal, H. H. Jonsson, S. Shertz, A. H. Goldstein and A. B. Guenther, *J. Atmos. Sci.*, 2013, **70**, 3277–3287.
25 P. K. Misztal, T. Karl, R. Weber, H. H. Jonsson, A. B. Guenther and A. H. Goldstein, *Atmos. Chem. Phys.*, 2014, **14**, 10631–10647.
26 R. Atkinson, *Atmos. Environ.*, 2000, **34**, 2063–2101.
27 J. Mann and D. H. Lenschow, *J. Geophys. Res.: Atmos.*, 1994, **99**, 14519–14526.
28 H. P. Schmid, *Agr. Forest Meteorol.*, 2002, **113**, 159–183.
29 T. Vesala, U. Rannik, M. Leclerc, T. Foken and K. Sabelfeld, *Agr. Forest Meteorol.*, 2004, **127**, 111–116.
30 T. Vesala, N. Kljun, U. Rannik, J. Rinne, A. Sogachev, T. Markkanen, K. Sabelfeld, T. Foken and M. Y. Leclerc, *Environ. Pollut.*, 2008, **152**, 653–666.
31 M. Y. Leclerc and T. Foken, *Footprints in micrometeorology and ecology*, Springer, 2014.
32 S. Metzger, W. Junkermann, M. Mauder, F. Beyrich, K. Butterbach-Bahl, H. P. Schmid and T. Foken, *Atmos. Meas. Tech.*, 2012, **5**, 1699–1717.
33 D. R. Drew, J. F. Barlow and S. E. Lane, *J. Wind Eng. Ind. Aerod.*, 2013, **121**, 98–105.
34 D. C. Carslaw and S. D. Beevers, *Atmos. Environ.*, 2004, **38**, 3585–3594.
35 Defra and DECC, NAEI data, http://naei.defra.gov.uk/.
36 EEA, *EMEP/EEA air pollutant emission inventory guidebook 2013, Technical guidance to prepare national emission inventories*, Report 12/2013, 2013.
37 D. P. Billesbach, *Agr. Forest Meteorol.*, 2011, **151**, 394–405.
38 J. D. Lee, L. K. Whalley, D. E. Heard, D. Stone, R. E. Dunmore, J. F. Hamilton, D. E. Young, J. D. Allan, S. Laufs and J. Kleffmann, *Atmos. Chem. Phys. Discuss.*, 2015, **15**, 22097–22139.
39 A. Bjorkegren, C. Grimmond, S. Kotthaus and B. Malamud, *Atmos. Environ.*, 2015, **122**, 775–790.
40 A. R. Desai, K. Xu, H. Tian, P. Weishampel, J. Thom, D. Baumann, A. E. Andrews, B. D. Cook, J. Y. King and R. Kolka, *Agr. Forest Meteorol.*, 2015, **201**, 61–75.
41 D. H. Lenschow, D. Gurarie and E. G. Patton, *Geoscientific Model Development Discussions*, 2015, **8**, 9323–9372.
42 D. C. Carslaw and G. Rhys-Tyler, *Atmos. Environ.*, 2013, **81**, 339–347.

Faraday Discussions

DISCUSSIONS

Urban case studies: general discussion

William Brune, William Bloss, Zongbo Shi, Francis Pope, Gary Fuller, Paul S. Monks, Alison Tomlin, Thomas Karl, Matthew Hort, Claudia Mohr, Rob MacKenzie, Athanasia Vlachou, Zhe Tian, Louisa J. Kramer, Dwayne Heard, Ruth Purvis, Xavier Querol, Urs Baltensperger, Rachel Dunmore, Roy Harrison, Tim Murrells, Jose L. Jimenez, Eben Cross, Gordon McFiggans, Astrid Kiendler-Scharr, Tzer-Ren Ho, Aurélie Charron, Timothy Wallington, Nivedita Krishna Kumar, Simone Pieber, Franz Geiger, Andreas Wahner, Edward Mitchell, André Prévôt, Andreas Skouloudis, Markus Kalberer, Brian McDonald, C. N. Hewitt, Costas Sioutas, Neil. M. Donahue, James Lee, Dominik van Pinxteren, Sarah Moller, María Cruz Minguillón, Martin Shafer, David Carslaw, Christian Ehlers and Spyros Pandis

DOI: 10.1039/C6FD90021F

Simone Pieber opened a general discussion of the paper by Spyros Pandis: from your presentation together with recent source apportionment studies such as *e.g.* Crippa *et al.* for Europe,[1] the relative importance of other emissions from the residential heating and cooking sector, especially wood combustion, becomes more and more evident. Despite this, big efforts are still being made to clean up emissions from the transport sector. Do you think this is still relevant, or should we focus our attention differently, and what do you think is currently hindering us from tackling the heating sector?

1 M. Crippa *et al.*, *Atmos. Chem. Phys.*, 2014, **14**, 6159–6176.

Spyros Pandis answered: This is an excellent observation and it is one of the main points that our paper tries to make in the "Age of Foolishness" section. While transportation remains a significant source sector in urban areas, its importance has been decreasing. Despite this change, the required rebalancing of our control efforts is not happening in most cities. This is partially due to inertia: we have all been taught that transportation is the main cause of air quality problems in big cities and we have a hard time realizing that all our efforts to control it are actually having an impact and what was true in the 1980s and 90s may not be true any more in the 2010s. The second problem is the challenge of dealing with a lot of small individual sources and a lot of people responsible for these heating and cooking emissions. These two obstacles together have made progress in controlling these "non-traditional" urban air pollution sources difficult.

Athanasia Vlachou asked: In your paper you suggest that primary OA can be categorized as fresh and aged POA by performing tests on cooking and biomass burning. What about traffic? Could one make such a separation for this primary source as well?

Spyros Pandis responded: In the present paper we discuss evidence of the conversion of cooking and biomass burning OA to more oxidized forms that then appear in the PMF analysis as the corresponding AMS source-specific factors. When the chemical conversion is fast, as is the case for the COA, we could find only one of the factors: the fresh one in the winter and the more oxidized one in the summer. For wood burning, two factors could be found in the same period, at least in some cases.[1] This suggests a more gradual conversion of the fresh biomass burning OA, to its more aged form, and finally to the oxidized OA (OOA). To the best of our knowledge, there has not been a report of an aged form of HOA in a field study. One could speculate that HOA, as it ages chemically, is converted to one of the reported forms of OOA (SV-OOA and/or LV-OOA) without maintaining its source identity, at least from the point of view of the AMS.

1 A. Bougiatioti, I. Stavroulas, E. Kostenidou, P. Zarmpas, C. Theodosi, G. Kouvarakis, F. Canonaco, A. S. H. Prévôt, A. Nenes, S. N. Pandis, and N. Mihalopoulos, *Atmos. Chem. Phys.*, 2014, **14**, 4793–4807.

Astrid Kiendler-Scharr said: You point out the strong contribution of SOA to fine PM. How would SOA be regulated? Does this call for stronger VOC emission controls? What would your take-home message be?

As a brief comment, I would also like to point out that cooking organic aerosol (COA) is not observed in sites like Cabauw,[1] which is categorized as a rural site. If the reason for the lack of COA is atmospheric aging during transport, which patterns should we look for in AMS or ACSM data sets to assign this source?

1 P. Schlag, A. Kiendler-Scharr, M. J. Blom, F. Canonaco, J. S. Henzing, M. M. Moerman, A. S. H. Prévôt, and R. Holzinger, *Atmos. Chem. Phys. Discuss.* 2015, **15**, 35117–35155.

Spyros Pandis replied: The regulation of SOA clearly requires regulation of its precursor vapors (volatile, intermediate volatility and semi-volatile organic compounds). Reductions of sulphur dioxide, oxides of nitrogen, and primary OA could also help, but there is a lot of debate and uncertainty regarding the effectiveness of these measures in reducing OA in specific urban areas. Dilution and chemical aging are expected to reduce the contribution of fresh COA to total OA in rural sites to a few percent. At the same time the COA temporal signal will be smeared out to a large extent, so it will be difficult to separate from sources like the HOA. So I am not surprised that PMF does not find a COA factor in Cabauw. Unfortunately, the features of the COA spectrum do depend on the cooking habits in each area. Therefore one would probably need to obtain a representative Dutch COA spectrum in Amsterdam or another big city in the area, and then use it in ME-2 as a way to get a rough estimate of the probably small COA contribution to the total OA.

Neil Donahue opened a general discussion of the papers by Dominik van Pinxteren and Spyros Pandis: I have a question for both speakers and for Astrid

Kiendler-Scharr as well. The cooking factor COA is associated with distinct maxima at mealtimes, especially lunch. This feature will be in the data for local sources but may well be smeared out for upwind sources by the time they arrive at the receptor site. Does this mean that the PMF methods will be biased toward finding the COA factor in urban settings but will fail to find it in the background?

Spyros Pandis replied: Indeed the ability of PMF to identify and separate sources relies heavily on the temporal evolution of the contribution of each source. As one moves away from the source, this temporal information gets weaker and weaker and eventually the regular PMF algorithm cannot separate the specific source from the rest. However, the contribution of the fresh OA from the source to the total will also be getting lower and for COA it will probably be just a few percent. Techniques like the ME-2 algorithm may help in this case as long as there is reliable information about the COA AMS spectrum for the specific area. In general, all these source apportionment techniques have a hard time with the identification and accurate quantification of sources that contribute only a few percent to the total and do not have unique features in their fingerprint.

Dominik van Pinxteren said: I cannot really add to the AMS COA factor interpretation, but generally I would agree that if there is not enough variance in contributions from a given source (e.g. because concentration peaks are smeared out) PMF will not be able to resolve it. In our study, I think the (tentatively identified) cooking source was only resolved because we pooled different sites into one dataset which included the residential site, where factor concentrations were significant, and other sites, where its concentrations were negligible.

Astrid Kiendler-Scharr added: A fundamental assumption that is made in factor analysis is that factors feature invariant mass spectral profiles and show variable concentration over time. Therefore an additional level of complexity in source identification is added when the primary aerosol changes composition - and thereby mass spectral fingerprint - during transport. As we heard before, COA changes mass spectral fingerprint after exposure to oxidation. This and the diversity of mass spectral signatures from different COA will make the source identification difficult away from direct sources.

Roy Harrison asked: On the specific issue of AMS/PMF analysis of cooking aerosols, I was a collaborator in a study at a rural site in Northern Italy. This was well away from any source of cooking emissions, but picked up signals with mass spectral characteristics and temporal variation that matched previously published cooking emissions. We came to the conclusion that this factor was largely unrelated to cooking, and some other studies refer to cooking-like organic aerosol. I think we need to be extremely careful with this.

Spyros Pandis answered: I agree that one needs to be careful with the attribution of any AMS/PMF factor to cooking emissions and that the temporal profile of the factor may not be sufficient. In the case of Athens as discussed in our paper, we performed similar measurements in another Greek city (Patras), analyzed the data separately and came up with practically the same AMS COA spectra in the two cities in each season. We then performed laboratory experiments by cooking meat (greek souvlakia) and confirmed that the AMS spectrum of the fresh emissions was the same as the one that the PMF analysis determined during the winter, in both Athens and Patras. Finally we exposed the fresh COA to ambient levels of ozone (around 60 ppb) or sunlight and showed that the AMS spectrum changed rapidly, so it now matched the one determined by the PMF in the two cities during the summertime. Additional evidence like this can help improve our confidence regarding the assignment of the corresponding PMF factor to cooking activities in urban areas.

Roy Harrison addressed Dominik van Pinxteren and Spyros Pandis: As part of the ClearfLo programme in London, we measured a range of organic molecular markers. It was during the winter campaign so hopefully there was not much photochemistry at the time. We carried out source apportionment with a chemical mass balance model using organic molecular markers and compared our results to simultaneous estimates of aerosol sources, including cooking aerosol, measured by James Allan at Manchester University using an AMS. For the cooking organic aerosol (COA), the AMS results were about double those from the mass balance model. There was a later report in the literature that showed that the collection efficiency of the AMS for cooking aerosol is close to 100%, whereas James had used 50%, which reconciles the difference in the two measurements. Do you think there are often over-estimations in measurements due to this? Regarding the Leipzig measurements, I do not believe that the PMF chemical profile can be very convincingly associated with cooking emissions. It could be due to a number of things, including vegetative detritus.

Spyros Pandis answered: The AMS collection efficiency issue does introduce additional uncertainty in the AMS/PMF results. In our ambient measurements we did not dry the ambient particles, something that in general leads to higher collection efficiencies and lower uncertainties. We then estimated the AMS CE, combining the corresponding AMS size/composition distributions with the SMPS ones using the volume distribution matching algorithm of Kostenidou *et al.*[1] Our laboratory measurements of COA are consistent with a CE of unity. I do agree with your assessment that the AMS CE (especially if it is different for different OA particles) introduces additional uncertainty in the corresponding source apportionment results. The authors of the Leipzig study are clearly better qualified to address your comment regarding the results of their study, so I will defer to them.

1 E. Kostenidou, R. K. Pathak and S. N. Pandis, *Aerosol Sci. Technol.*, 2007, **41**, 1002–1010.

Dominik van Pinxteren added: Regarding our Leipzig measurements, I agree that the identification of the cooking factor is only tentative, and we also state this in the paper. Beyond the (indeed ambiguous) chemical profile, the association was mainly done based on the pattern of factor concentrations for the different sites and seasons (no seasonal trend, highest concentrations at a residential site with a number of restaurants) which we thought fit better to cooking emissions rather than *e.g.* plant detritus. It is clear, however, that the inclusion of cooking-specific molecular tracers, such as cholesterol or others, would be needed to increase confidence in this factor, and we are planning to do this in future studies.

André Prévôt commented: In an urban area, what you show and what we have also shown, is that where you have lots of restaurants, cooking is a factor. But if you have externally mixed oil droplets then we might overestimate by a factor two. Any unmixed organics and secondaries might also be overestimated. An additional problem is that PMF is tough. Primary organic aerosol contributions are very low – 10 to 15% - and cooking etc. and often the emissions happen at the same time and thus correlate. One issue concerning cooking is probably that organics from cooking change their mass spectra a bit faster, as their mass spectra are less stable, so you have partially oxidised organic aerosols.

Gordon McFiggans addressed Spyros Pandis: In the source apportionment of the various contributions of the short-, medium- and long-range sources to PM, there will inevitably be dependencies on the prescribed volatility distributions of the organic components, their ageing rates and the assumptions used in matching the VBS boundary condition and emissions to the reported "OC" emissions from, for example, the TNO database. Could you comment on the magnitude of these uncertainties and robustness of the attributions? I would recommend inclusion of error bars to represent these uncertainties in the paper.

Spyros Pandis answered: The uncertainty of the contributions of different source areas, shown for example in Figure 3, is different for different PM components and depends on the uncertainty of their emissions (or the emissions of their precursors), production mechanisms, atmospheric processing, and removal mechanisms. These uncertainties are lower for the elemental carbon and sulfate shown in the first two graphs of this figure. For SOA the uncertainties are higher, but much lower than the uncertainty of the overall VBS scheme used in this study. The source area attribution for SOA is determined to a large extent by the corresponding timescales for the oxidation of its precursors to SOA. These timescales (*e.g.*, reaction rates of VOCs and IVOCs with oxidants) are relatively well known and this makes the attribution of the SOA to source areas (with the coarse resolution of short-, medium- and long-range used here) quite robust. For example, the conclusion that most of the SOA in Paris during the summer MEGAPOLI campaign was due to medium-range sources does not change for reasonable choices of the volatility distributions, aging ranges, emissions, *etc.* A detailed calculation of the error bars of these estimates in a 3D chemical transport model is clearly challenging.

Nivedita Krishna Kumar asked: Did you not identify a cooking organic aerosol factor in your PMF results from Athens, like Crippa *et al.* did in Paris?[1]

1 M. Crippa *et al.*, *Atmos. Chem. Phys.*, 2014, **14**, 6159–6176.

Spyros Pandis answered: In our measurements in both Athens and Patras during both winter and summer fresh cooking organic aerosol was as important as hydrocarbon-like organic aerosol which is attributed mostly to transportation emissions. These finding were similar to the results of Crippa *et al.* in the MEGAPOLI campaigns in Paris despite the differences in size and practices among the three cities.

Zhe Tian continued the discussion of the paper by Dominik van Pinxteren: I have noticed that the traffic contributions are much higher than traffic exhaust in your paper. Could you give us more details about how you define "traffic" and "traffic exhaust" and your choice of tracers? And have you tried any organic markers in your PMF model?

Dominik van Pinxteren answered: Traffic exhaust was mainly identified from high mass contributions of water-insoluble carbon (*i.e.* EC and hydrophobic organics), high species contributions of hopanes, and a homohopane index close to mineral oil derived emissions (rather than coal combustion, which can be another source of hopanes). This source was, however, only resolved in the ultrafine and the coarse particle size ranges. In accumulation mode particles it was mixed into a "general traffic" factor, which in addition to exhaust emissions also includes other traffic-related sources such as brake wear, tire wear, dust resuspension *etc.* and was characterised by large contributions of several metals (see our paper for details). With our PMF results we can thus not fully differentiate exhaust and non-exhaust across all particle size ranges and "traffic exhaust" contributions in total PM_{10} are therefore missing the accumulation mode fraction, which is at least part of the reason why they are smaller than "general traffic". With regards to organic markers – yes, we included many of them, and you can find all the details in the paper.

Spyros Pandis added: Our paper focuses on traffic emissions including both exhaust and non-exhaust emissions. In the analysis presented here only AMS spectra have been used. There have been a number of efforts in the literature (for Mexico City, Los Angeles, and Pittsburgh) in which separate source apportionment analysis has been performed using AMS data and organic tracer measurements; however, the comparison of the corresponding results were outside the scope of the current work.

Edward Mitchell said: Both studies have identified residential solid fuel combustion as an increasingly important source of ambient pollution. In comparison to other sources such as traffic, it can be argued that regulations and abatement technologies for heating stoves are decades behind. In Europe we now have the Ecodesign regulations limiting efficiency, PM, CO, NO_x and OGC emissions, but these will not apply until 2022. Do the speakers believe that policies such as this will be sufficient to reduce emissions from this source, or do we need additional abatement technologies such as filters and catalysts as already used in the traffic sector?

Spyros Pandis responded: I think that the Ecodesign regulations will help, but I am afraid that it will be too little too late. I personally believe that it does not make any sense to allow wood burning in fireplaces or wood stoves in large urban areas that already have air pollution problems. The same applies also to agricultural and open burning in areas around big urban centers.

Dominik van Pinxteren said: In Germany, emissions from small wood stoves have been regulated already to some extent, leading to the replacement of very old stoves within a transition period. Given that these old and very inefficient stoves have much higher pollutant emissions than more modern ones, I would expect a decrease of emissions. Additional technological developments would certainly help to reduce emissions even more. Whether this will be enough, however, to observe substantial reductions of emissions at a given site is another question. In our study in Leipzig we have seen that during the most intense pollution episodes, more than 80% of solid fuel combustion aerosol was transported into the city, likely due to long-range transport from Eastern European areas. A modernisation of house heating infrastructure in these countries might thus have a large positive influence on the air quality in Leipzig, and many other cities as well.

Tim Wallington returned to the discussion of the paper by Spyros Pandis: Figure 1 in your paper reminded me of the Kuznets curve. Is there scope for an economic analysis looking at level of urban air pollutant emissions in the 5 cities as a function of economic activity to compare with the Kuznets curve?

Spyros Pandis answered: This is a very interesting observation. The curve in Figure 1 does indeed look like the Kuznets curve after replacement of the economic inequality axis of the former with the pollution level here. There is clearly room for interesting economic analysis here, trying first to add actual data points in the graph. One of our observations for Athens is that the path along this curve may not be reversible, but may actually include a hysteresis if a city moves from right to left. A modest (but rapid) decline in development due to a financial crisis may result in significant increase in air pollution levels if practices like wood burning are widely adopted.

Zongbo Shi remarked: We often see that when people describe different PMF factors they give a percentage of, for example, COA, HOA, or SOA. What are the uncertainties in these estimates? Should all papers be clear that these are estimates which have relatively large uncertainties?

Spyros Pandis replied: There are a number of formal methods for the estimation of the uncertainty of PMF solutions. These include classical bootstrap, displacement of factor elements, and bootstrap enhanced by displacement.[1] These approaches capture the uncertainty of the PMF results due to random errors and rotational ambiguity. These approaches have been used in a number of AMS/PMF studies. For example, Lanz *et al.* estimated moderate uncertainties for the OA components in Zurich during summertime: OOA-I 39.8–50%, OOA-II 18.9–22.3%, HOA 5.6–8.6%, COA 10.1–14.7%, wood burning 7.4–10.3%, and other minor sources 5.2–7.4%.[2] Even lower uncertainties have been reported by Ulbrich *et al.* for Pittsburgh.[3] These ranges could be optimistic, because they probably capture only some of the sources of uncertainty. Another approach is to view these source apportionment results as revealing the relative magnitude of the corresponding sources: small, medium and large. For example, if PMF suggests that a component represents 20% of the total OA in a polluted area, it is probably quite significant and its sources should be controlled. I refer you also to my response to the previous comment by Roy Harrison.

1 S. G. Brown, S. Eberly, P. Paatero and G. A. Norris, *Sci. Total Environ.*, 2015, **518–519**, 625-635.

2 V. A. Lanz, M. R. Alfarra, U. Baltensperger, B. Buchmann, C. Hueglin and A. S. H. Prévôt, *Atmos. Chem. Phys.*, 2007, **7**, 1503–1522.
3 I. M. Ulbrich, M. R. Canagaratna, Q. Zhang, D. R. Worsnop and J. L. Jimenez, *Atmos. Chem. Phys.*, 2009, **9**, 2891–2918.

Xavier Querol returned to the discussion of Spyros Pandis' and Dominik van Pinxteren's papers: Thanks a lot for your excellent presentations and papers. Two comments; firstly, I missed in "A tale of five cities" the inclusion of the non-exhaust vehicle emissions and their contributions to ambient PM. I think these are relevant for air quality science in the urban atmosphere, and yet there hasn't been much discussion about them in the meeting either. There are very interesting points, such as phase transformations and toxicity of brake and tyre wear (Mo, Cu, Zn and Mn compounds) – also how to apportion the brake, tire, pavement and resuspension contributions – that should be taken into account when dealing with urban PM. Secondly, Dominik's paper attributed the reduction in elemental carbon (EC) to the efficiency of the Low Emission Zone (LEZ) in Leipzig. There is a lot of scientific debate on that. Can you identify effects due exclusively to the LEZ implementation rather than effects due to other actions, such as the progressive renewal of the fleet by Euro 5 diesel (with filter traps for PM) since 2009, or other simultaneous actions? Other cities such as Barcelona, without LEZ being implemented, did reduce EC levels in a similar proportion. This does not contradict at all the potential efficiency of the LEZ on EC levels, it is meant only as a consideration to take into account.

Dominik van Pinxteren responded: In our study we compare the urban EC increment in the years 1999–2000 (way before the implementation of the LEZ) with the ones observed in 2013–2015 (a few years after its implementation) and we therefore cannot exactly distinguish between the impacts of the LEZ and other actions. However, we know from another study we have done at the same traffic site in Leipzig[1] that BC concentrations sharply decreased immediately after the implementation of the LEZ in 2011 due to a reduction of heavy-duty traffic. It has also been shown that the Leipzig LEZ accelerated the modernisation of the car fleet as compared to other Saxonian cities.[2] This is why we believe that the LEZ did indeed have a strong impact on the observed EC reduction in Leipzig.

1 F. Rasch, W. Birmili, K. Weinhold, S. Nordmann, A. Sonntag, G. Spindler, H. Herrmann, A. Wiedensohler and G. Löschau, *Gefahrstoffe - Reinhalt. Luft*, 2013, **73**, 483–489.
2 G. Löschau *et al.*, *Umweltzone Leipzig, Teil 3 Immissionssituation 2010-2013,* Dresden, Germany, 1st edn, 2015, https://publikationen.sachsen.de/bdb/artikel/23885 (German only).

Spyros Pandis replied: The non-exhaust vehicle emissions are clearly important for the urban atmosphere - refer also to my response to Zhe Tian's earlier question. However, a significant fraction of these particulate emissions are in the coarse size range, so they are a lot more important for the regulation of PM_{10} compared to the fine PM ($PM_{2.5}$ or PM_1) that is the focus of our work. These components (metal oxides, crustal material, *etc.*) are included in the primary fine PM in Figures 2 and 4 of our paper. I do agree with your statement though, that these emissions do not receive the attention that they deserve. The examples that you mention do underline the challenge of attributing a reduction in the concentration of a pollutant (EC in this case) to a specific measure when multiple changes are taking place at the same time.

Claudia Mohr addressed Spyros Pandis: In your presentation and paper you make the statement that SOA is becoming relatively more important. In absolute numbers, does it show a decreasing trend as well? Do we know how it is evolving globally? Does better POA control also lead to SOA reduction? Are there studies on this?

Spyros Pandis responded: The available data in specific areas like Los Angeles with long fine particulate matter data sets suggest that SOA is decreasing too in areas where significant reductions of pollutant emissions have taken place. Unfortunately, generalizing to other urban areas or even globally is a lot more challenging because older PM chemical composition data sets have little information about the chemical composition of OA. AMS data sets go only back to the early 2000s. The effect of POA control on SOA levels depends a lot on the ability of these compounds to form a common solution (see also Neil Donahue's paper (DOI: 10.1039/C5FD00214A). This ability appears to be limited initially but it increases as the POA ages chemically and is converted to SOA. Based on this, one would expect a small effect near the sources when POA is still fresh and a modest effect further away when the POA has been oxidized but has also been diluted. However, the reduction of SOA precursors (volatile, intermediate volatility, and semivolatile organic compounds) that often accompanies the reduction of POA will cause a reduction in SOA. There have been a number of modeling efforts to quantify these effects (for example Megaritis *et al.*[1]) but their results do depend a lot on the assumptions about the mixing state of the OA components.

1 A. G. Megaritis, C. Fountoukis, P. E. Charalampidis, C. Pilinis, and S. N. Pandis, *Atmos. Phys. Chem.*, 2013, **13**, 3423-3443.

Jose-Luis Jimenez said to Spyros Pandis: Following up on Gordon McFiggans' comment that there is a lot of uncertainty about SOA results with the VBS due to different yield and aging parameters available: when one fits urban SOA results to fast OA mass observations, you can only extract 2 pieces of information, an amount and a timescale of SOA formation (see, for example, Hodzic and Jimenez[1] and Hayes *et al.*[2]). Then one can simulate regional SOA with many VBS versions, which have from dozens to over 100 parameters. Many of those parameters may give similar results for the urban SOA over short timescales, although sometimes those results are still inconsistent with observations when analyzed trying to isolate the urban component (*e.g.* Woody *et al.*[3]) - especially, the different versions can give very different answers over longer timescales (see *e.g.* Hayes *et al.*[2]), and also for biogenic SOA (*e.g.* your own work, and several other papers). So I would argue that the VBS versions in use are very under-constrained, and we don't really know whether often we are getting agreement for a combination of wrong reasons.

1 A. Hodzic and J. L. Jimenez, *Geosci. Model Dev.*, 2011, **4**, 901–917.
2 P. L. Hayes *et al. Atmos. Chem. Phys.*, 2015, **15**, 5773–5801.
3 M. C. Woody, K. R. Baker, P. L. Hayes, J. L. Jimenez, B. Koo and H. O. T. Pye, *Atmos. Chem. Phys.*, 2016, **16**, 4081–4100.

Spyros Pandis answered: I agree that while the VBS is a robust modeling framework, it requires a number of parameters that are not well constrained right now. There is a lot of ongoing work around the world providing constraints with measurements of volatility distributions, emissions of IVOCs, chemical aging rates, *etc.*, but there are still a lot of knowledge gaps. If a set of such VBS parameters captures the SOA formation, it probably means that it is capturing the two important SOA formation characteristics: the SOA formed and the formation rate. This can still result in the right answer for the wrong reasons. However, returning to the comment by Gordon McFiggans, a model that can capture these two characteristics is probably providing a decent answer to the source attribution question, at least for the coarse resolution used here.

Simone Pieber continued the discussion of the paper by Dominik van Pinxteren: Contribution of coal burning emissions were observed dominantly during the winter period and when air masses arrived from the east. Does this imply that the role of emissions from industrial coal burning and coal burning in power plants for electricity generation, which should also contribute during summer, is negligible? Could you please comment on the relative importance of the residential *vs.* the industry sector for those fuels?

Dominik van Pinxteren answered: Yes, it would be also my interpretation that coal burning emissions are dominated by house heating, and I think it is consistent with industrial coal burning emissions being much more regulated than emissions from house heating. Also, a recent review on PM_{10} apportionment in cities[1] shows that domestic fuel burning explains a much larger share of PM_{10} in Central and Eastern Europe than industrial emission as a whole (including oil and coal combustion, but also other emissions).

1 F. Karagulian, C. A. Belis, C. F. C. Dora, A. M. Prüss-Ustün, S. Bonjour, H. Adair-Rohani and M. Amann, *Atmos. Environ.*, 2015, **120**, 475–483.

Gary Fuller addressed Spyros Pandis: There was no mention in your paper about the progressive dieselisation in your five cities. There are contrasting trajectories between the European and non-European cities, and within Europe, Athens is following a different trajectory to other countries. Where would you put dieselisation into your matrix of foolishness, wisdom, hope or despair? Secondly, we should be careful how we present the concept of medium range contributors. Policy makers often think they can't do anything about this, but it really depends on where you are standing; someone's medium range is someone else's local source. Uniform policies across multiple cities could be a helpful approach.

Spyros Pandis responded: This is indeed an interesting potential additional chapter in the "Tale of Five Cities", with diesel cars dominating the new additions to the corresponding fleet in Paris (as well as in a lot of the EU countries), while gasoline-powered ones can be mainly found in the other four cities. Paris is currently rethinking its strategy and considering a move away from diesel, while Athens has recently allowed the use of diesel cars, but appears to have had second thoughts and may change its mind again and ban them once more. As a first approximation, the modern cars (either gasoline or diesel) emit a lot less than the corresponding cars in the late 20th century. As a result, all five cities have managed to reduce their fine particulate matter concentrations, even if they have followed different trajectories. I think that the ultimate answer to your question depends on the city, and varies from hope for some European cities to foolishness. I would use foolishness for Athens, due to the low quality of the diesel fuel often found in the Greek market that can create major problems for the control systems of modern diesel cars. I do agree with your comment about the need for uniform policies across multiple cities and even better large areas. I think that the US approach, with the definition of Transport Regions and then the development of policies for all sources in the states in the Transport Region together (including emissions trading), has proven to be quite effective. It could be a good policy model for groups of European countries (or even all of Europe) that face more or less the same fine particulate problem, exchanging particles and their precursors with each other.

William Bloss opened a general discussion of the paper by Kebin He:† Does the change in the NOR and SOR ratios imply a change in the oxidation capacity of the atmosphere over the period of the emissions reductions – *i.e.* a reduction in OH. Is there any evidence for this in terms of changes in *e.g.* O_3?

Zongbo Shi responded: As shown in Fig. 7, the SOR and NOR decreased substantially during both day and night, implying that the secondary oxidation processes of SO_2 and NO_2 were suppressed due to emission reductions. This was consistent with the changes of O_3, which reduced from 0.065 ppm before the parade to 0.040 ppm during the parade.

Gary Fuller said: You said in the summary presentation and in the paper that COA also went down during the parade period. I assume that there wouldn't have been any cooking restrictions. Was this reduction because there were fewer people in the city, or was this because what we think of as cooking emissions is actually something else – an industrial emission for example?

Zongbo Shi replied: As described in line 34 of page 9, open charcoal grills were strictly banned during the emission control period, which may have led to the reduction of COA. Our result was consistent with that observed during the 2014 APEC control period, which also showed a reduction of COA from 7.8 to 5.2 $\mu g\ m^{-3}$. No further evidence showed that cooking emissions may be actually an industrial emission.

Roy Harrison addressed Zongbo Shi : I am very concerned with the interpretation of sulphur oxidation and nitrogen oxidation ratios. Taking sulphur as an example, sulphate is likely to be generated from SO_2 emissions originating at a considerable distance from the point of measurement, whereas SO_2 will probably be emitted rather locally. As a consequence, I am not convinced that we can infer anything about the efficiency of reduction of sulphate concentrations in response to reduced emissions of SO_2 using the ratio, as it is comparing two measurements which are responding to emissions into different locations. Do you think that this is a fair interpretation?

† Kebin He's paper was presented by Zongbo Shi, *University of Birmingham, UK*

Zongbo Shi answered: I agree that sulphate could be generated from both locally and regionally emitted SO_2. But the lifetime of SO_2 could also be from several hours to several days, leading to the regional transport of SO_2. In addition, the emission controls during the parade were not only implemented in Beijing but also the surrounding areas, as described in the introduction of our paper. Thus I think it is reasonable to use sulphur oxidation and nitrogen oxidation ratios here.

Claudia Mohr opened a general discussion of the paper by María Cruz Minguillón: What were the numbers of vehicles on the road at the "low" and "high traffic" periods, respectively? What was the difference in the numbers of diesel and non-diesel vehicles? Were diesel and non-diesel car numbers affected equally by holiday and non-holiday season?

María Cruz Minguillón replied: The traffic intensity increased 37% from the low to the high traffic periods (from about 40 000 vehicles per day to about 55 000 vehicles per day on average, including weekends). The % of diesel light duty vehicles of the fleet is up to 50%, according to statistics from the DGT (the traffic authority in Spain). Nevertheless, there is no information available about the % of diesel and non-diesel vehicles for specific periods such as the ones in our study. We can assume that the % of diesel vehicles was similar between both low and high traffic periods. The average daily patterns of the traffic intensity for the low and high traffic periods are shown in Fig. S5.

Athanasia Vlachou queried: Barcelona is characterized by high dust concentrations. Did you account for carbonates, especially in the radiocarbon analysis, to avoid any overestimation of the organic carbon?

María Cruz Minguillón responded: The ^{14}C analyses were performed for $PM_{2.5}$ samples. The content of carbonates in these samples can be estimated based on the Ca concentrations, which were experimentally determined. The estimated average carbonate concentrations for the five days with radiocarbon analyses ranged between 0.3 and 0.5 $\mu g\ m^{-3}$. These concentrations are considered negligible compared to the OA concentrations (ranging between 7 and 14 $\mu g\ m^{-3}$).

Andreas Wahner remarked: you showed in your Barcelona study that you had a higher biogenic component to SOA formed in high traffic conditions. Is your conclusion based on the assumption that the photochemical state of the atmosphere was the same under both conditions? Or could this be the reason, and not the different emissions?

María Cruz Minguillón answered: The conclusion is based on the assumption that the photochemical state of the atmosphere was similar, since the toluene/benzene ratio is not significantly different between the low traffic and the high traffic periods (average daily ratios ranged from 3.1 to 4.3 for the days when ^{14}C measurements were carried out), although the toluene/benzene ratios were quite noisy on an hourly basis. If no changes occurred in terms of the nature of sources between the two periods (additional sources with a higher toluene/benzene ratio during the HT period, for example), OH concentrations in the atmosphere of

Barcelona cannot be considered as significantly different from one period to another. Nevertheless, some other sources of toluene may exist, as indicated by the PMF analysis applied to the PTR-MS dataset (results not shown in the paper).

Astrid Kiendler-Scharr asked: Is the strong increase in particulate nitrate during the second part of your measurements significant? Is the higher non-fossil SOA contribution a consequence of higher NO_x concentrations and gas phase precursor chemistry or do you have indications for other sources of the non-fossil SOA?

María Cruz Minguillón responded: The nitrate concentrations increased from 0.5 μm^{-3} on average during the low traffic period to 1.0 $\mu g\ m^{-3}$ during the high traffic period. These concentrations are quite low and close to the detection limit of the instrument, and hence the reported increase in percentage (+99%) should be considered with caution, and should not be strictly compared to the increase in OA concentrations, for instance, which were well above the detection limit. Nitrate only represents 5% of the PM_1 mass concentration on average.

Our hypothesis is that the increase in the % of non-fossil SOA during the high traffic period is related to the increase in NO_x concentrations, by favoring the formation of SOA from the available biogenic precursors. Nevertheless, the presence of other non-fossil sources, such as SOA from cooking, should not be discarded, as this contribution has been found to be significant at other places; for example, Hayes *et al.* found that SOA from cooking was 19–35% of total SOA in Los Angeles.[1] Hence, considering that the COA contribution increased by about 60% from the low to the high traffic period (due to the higher activity as people is back from holidays), the associated SOA-cooking could well have increased as well.

1 P. L. Hayes *et al.*, *Atmos. Chem. Phys.*, 2015, **15**, 5773–5801.

Jose-Luis Jimenez said: My question concerns the possible interpretations for the higher non-fossil carbon during the period with the higher vehicle and NO_x emissions, which appears surprising. I think you propose that the increase of NO_x leads to higher regional levels of NO_3 radicals at night, which then oxidize more biogenic VOCs at night regionally. That is certainly a plausible explanation, and it would be very useful to try to evaluate it with a 3D model with the relevant chemistry. However, I wonder about an alternative or complementary explanation. At high NO_x levels such as those in Barcelona, it is expected that OH radical levels are substantially reduced as NO_x increases. This was clearly shown, for example, in the recent CalNex 2010 study for the Los Angeles area.[1] Thus it is plausible that the lower OH during the high NO_x period reduces SOA formation from anthropogenic precursors, thus reducing the fossil SOA. One would need to normalize the fossil SOA per excess CO in order to properly account for this possibility. Could you comment on this hypothesis?

1 I. B. Pollack *et al.*, *J. Geophys. Res.*, 2012, **117**, D00V05.

María Cruz Minguillón replied: The suggested explanation is well received; it may well be that during the high traffic period, when higher NO_x concentrations are recorded and hence there are lower OH radical levels, the SOA formation from anthropogenic precursors is reduced.

Two different calculations have been made to further investigate this possibility: normalized fossil SOA per excess CO, and percentage of fossil SOA with respect to total fossil OA (in terms of OC : OOCf/OCf). It should be noted that the precision of the available CO measurements is 0.1 mg m^{-3} (equivalent to about 85 ppb) and the hourly concentrations ranged from 0.2–1.2 mg m^{-3}, and hence the interpretation of the CO and excess CO (CO concentrations after subtracting 100 ppb as the CO background) should be made with caution. The results obtained support the suggested hypothesis, since the fossilSOA/excessCO ratios were found to be higher for the samples during the low traffic period and the first sample for the high traffic period (17-21 μg m^{-3} of fossilSOA per ppm CO), whereas they were lower for the second and third samples of the high traffic period (13 and 10 μg m^{-3} of fossilSOA per ppm CO, respectively), pointing to lower fossil SOA formation under high NO_x conditions (the NO_x average concentrations were 23, 21, 33, 40, and 59 μg m^{-3} for the five samples analyzed, in chronological order). The difference between the low traffic and the high traffic period, on average, is 17%. Regarding the percentage of fossil SOA with respect to total fossil OA, it remains relatively constant for all the samples (OOCf/OCf is between 0.72–0.76) except for the last one, with a ratio of only 0.51. This last sample corresponds to the day with a higher impact of road traffic based on the BC, NO_x and HOA concentrations. The difference between the OOCf/OCf ratios in the low traffic and the high traffic period, on average, is 10%, which agrees reasonably well with the 17% difference found for the fossilSOA/excessCO considering all the uncertainties associated with these calculations. However the toluene/benzene ratio is not significantly different between the low traffic and the high traffic period (average daily ratios ranging from 3.1 to 4.3). If no changes occurred in terms of the nature of sources between the two periods (additional sources with a higher toluene/benzene ratio during the HT period, for example) OH concentrations in the atmosphere of Barcelona cannot be considered as significantly different from one period to another.

Considering all of the above, we acknowledge the comment and certainly accept the hypothesis as a possible complementary explanation of the observed measurements.

Andreas Wahner remarked: The conclusion that enhanced BSOA formation during the HT period compared to the LT period could be due to enhanced NO_x concentration during HT depends on the correlation of NO_x to OH radical, which is not linear but has a maximum at a specific NO_x concentration. Did you analyze the consequences of this dependence?

María Cruz Minguillón responded: As you say, the NO_x influence on SOA formation is not linear. We may not have enough information to completely attribute the enhanced BSOA formation to the NO_x difference, but it is a plausible explanation for the observations/measurements made.

Dwayne Heard said: Normally if you plot the concentration of OH *versus* the concentration of NO_x, there is initially an increase in OH (owing to the HO_2 + NO reaction increasing in importance as an OH source), but once a certain level of

NO_x is reached, the OH concentration peaks and then starts to decrease again (owing to the importance of the OH + NO_2 reaction in forming nitric acid removing OH). What concentration of NO and NO_2 do you have here?

María Cruz Minguillón responded: Concentrations of NO and NO_2 were 2.5 and 15.8 $\mu g\ m^{-3}$ on average, respectively, during the low traffic period and 4.6 and 28.8 $\mu g\ m^{-3}$ during the high traffic period. This means an increase of about 80% in both NO and NO_2 concentrations from the low to the high traffic period. Maximum hourly concentrations reached 120 and 130 $\mu g\ m^{-3}$ of NO and NO_2, respectively. Unfortunately we do not have OH information available.

André Prévôt said: I was wondering if you could use the relative VOC composition to say something about the OH levels, or even NO_3 levels. Toluene/benzene ratios will change during the day. How much oxidation of OH is there and in the evening, how prominent is the decrease in alkanes due to VOC chemistry? The ratios of VOCs may indicate how much oxidation is going on.

María Cruz Minguillón answered: The toluene to benzene ratio (on an hourly basis) was relatively noisy during the first period (from 1 to >10), and more stable (from 2 to 6) during the second period. There was not a clear daily pattern that repeated every day, so the average daily pattern does not give very valuable information.

It is possible that there is an additional source of toluene, as indicated by the PMF analysis carried out on the PTR-MS dataset (results not shown in the paper). Anyway, considering an OH concentration of 5.106 molecule cm^{-3} and following the equation described by El Haddad *et al.*,[1] we obtain 0–0.5 days of aging during the campaign, but it is quite noisy, maybe due to the low concentration measured.

For the alkanes point, the PTR-MS data does not allow dissociation of the alkanes from the alkenes, as they give the same fragments (43, 57, 71…).

1 I. El Haddad, *Atmos. Chem. Phys.*, 2013, **13**, 7875–7894.

Simone Pieber commented: On the question of a changed fleet composition during the high traffic and low traffic period in the data-set, as raised by Dr Claudia Mohr: if those data are available, it might be useful to look, in addition to BC/NO_x traces, into changing mass spectra for the VOCs as identified with the PTR-MS, in order to obtain more information on relative compositions of diesel and gasoline fuelled vehicles; these should have different VOC compositions based on the differences in the raw fuel and after-treatment systems.

María Cruz Minguillón replied: The BC/NO_x ratios varied between 39 and 61 (ng m^{-3} of BC per $\mu g\ m^{-3}$ of NO_x) during the low traffic period and between 38 and 60 (ng m^{-3} of BC per $\mu g\ m^{-3}$ of NO_x) during the high traffic period, calculated on an hourly basis for the average daily pattern for each of the periods. Regarding the differences in VOCs, Fig. 2 shows the average daily pattern for some selected VOCs. The toluene and benzene concentrations in that plot show a similar variation. To better check this, the same data treatment (average daily pattern values) for the toluene/benzene ratio has been applied. The toluene/benzene ratio is higher during the night for both the low and high traffic periods (around 4), and lower during the middle hours of the day (12–16 UTC) (toluene/benzene ratio around 3). More intra-day variation is observed for the low traffic than for the high traffic period.

Sarah Moller asked: In your paper you state that similar meteorological patterns were observed during both the high and low traffic periods. You then go on to say that during the high traffic period, lower wind speed and higher pressure were recorded as a consequence of an anticyclonic episode, leading to atmospheric stagnation. You also say that there were differences in air mass origin between the two periods. Do you believe that the observed changes in atmospheric concentrations and composition are due predominantly to the change in traffic flow rather than meteorological variation? Did you attempt to quantify the impact, or relative importance, of the meteorological changes?

María Cruz Minguillón responded: The changes are mainly attributed to the increase in traffic intensity, although an effect of other factors is also present, of course. The traffic intensity increased by 37% from the low to the high traffic period on average, whereas the BC concentrations increased by 60%, and the NO and NO_2 concentrations increased by 80%. Hence, there are two things here: a) the variations in traffic intensity do not have a linear relationship with the BC, NO and NO_2 concentrations, and/or b) there are other factors favoring the accumulation or low dispersion of pollutants during the high traffic period with respect to the low traffic period. We did not attempt to quantify the impact of these meteorological changes. Maybe to do so we'd need a longer sampling period, so that we can really identify similar patterns in a significant number of days so that the differences are meaningful. Alternatively, and most likely a better option, modelling the formation of SOA could help in the interpretation.

Jose-Luis Jimenez commented: Our results from Los Angeles, as well as results from several other groups, show that biogenic VOCs emitted within the urban area make too little SOA to affect the urban concentrations (see *e.g.* Hayes *et al.*[1] or Heo *et al.*[2]). It is the biogenic VOCs emitted over much larger regional scales that produce the large majority of the biogenic SOA observed at the urban sites. Do you agree with this result for Barcelona?

1 P. L. Hayes *et al.*, *Atmos. Chem. Phys.*, 2015, **15**, 5773–5801.
2 J. Heo, B. de Foy, M. R. Olson, P. Pakbin, C. Sioutas and J. J. Schauer, *Atmos. Environ.*, 2015, **103**, 171–179.

María Cruz Minguillón answered: We do not have evidence that the precursors for the biogenic SOA present in Barcelona are emitted within the Barcelona metropolitan area, as we do not have evidence to the contrary. Considering the VOC emissions information, it makes sense to conclude that the biogenic SOA precursors are coming from outside the city. This would make a lot of sense considering the wind regime, dominated by NW flow during the night and SE-S flow during the day. This wind regime facilitates both the transport of pollutants from the city inland during the day and the transport of emissions from outside the city (such as biogenic precursors) towards the city during the night. It also agrees with the relatively higher monoterpene concentrations during the night (as shown in Fig. 2), and the relatively higher SV-OOA concentrations during the night (Fig. 8).

Jose-Luis Jimenez returned to the discussion of Kebin He's paper: From previous work on which I collaborated, on air quality during the Beijing Olympic Games, I understand that the emission area that affects Beijing is very large – at least a radius of 500km/1000km, – since fine aerosols are dominated by secondary species that are formed over such length scales. Thus a very large radius for the emission controls would have needed to obtain the reported results. What was the radius of the area where emissions control was applied around Beijing? Was it big enough to have an effect?

Zongbo Shi answered: As the question described, the emission area that affects Beijing is very large. During the 2008 Beijing Olympic Games, Beijing and its five neighboring administrative regions (*i.e.*, Tianjin, Hebei, Shanxi, Shandong and Inner Mongolia) enacted strict emission control measures to improve the air quality. And many previous studies have indicated significant reduction of air pollutants during the Beijing Olympic Games. For this study, as we elucidated in the paper, Henan province was also involved in the emission control action during the V-day Parade, besides the six regions we mentioned above. The radius of the area for emission controls is almost 500km–800km. We thought it was big enough to have an effect.

Urs Baltensperger inquired: I would expect that primary OA would react more rapidly to a change in emissions than the secondary OA, but your study shows the opposite – what are your thoughts on this?

Zongbo Shi answered: Thanks for the question. In this study, our monitoring site was located in the urban area of Beijing. While primary OA was mainly influenced by local emission sources, secondary OA could react to emission changes on a large regional scale. In the paper, we have taken the cluster 1 case as an example to explain this. Consistent with the significantly reduced VOC emissions of 58%–66% of southern Hebei, the concentration of secondary OA under cluster 1 decreased by 48% as a response to the regional emission control. For primary OA, it was also reduced by 48% in the cluster 1 case, mainly influenced by local emissions.

Andreas Skouloudis said: In the emission control period for the Parade Blue, this work focused on the effects before and during this parade. However, it would be interesting to generalise this. Did you observe similar effects afterwards, and if this happened several years running, will draw the same conclusions?

Zongbo Shi answered: Thanks for the question. The periods after this parade were much influenced by high precipitation. Thus it may be not reasonable to explore the effects of emission control by comparing aerosol characteristics during and after the parade. As the same emission control would not be continuously implemented for several years at this stage, I am not sure if we would draw the same conclusions. But the results concluded from this study are consistent with those from other two similar emission control periods, during the

2008 Beijing Olympic Games and the 2014 Asia-Pacific Economic Cooperation summit in Beijing.

Aurélie Charron addressed María Cruz Minguillón: Have you considered the contribution of biofuels included in fuels to the amount of non-fossil carbon? As an example, I give you the average percentage of biofuels in French fuels – that is, 7% – although I do not know the figures for Spain.

María Cruz Minguillón replied: The biofuel use in Spain is very low (below 2%, according to the statistics of the DGT, the traffic authority in Spain), and this is true also for the study region. Hence we did not consider biofuels when doing the calculations and interpretations in our study. In case the use of biofuel increases, future studies should consider it. Some studies highlight the formation of SOA from biofuel vehicles.[1] To better understand how the use of biofuel affects the mixture of ambient OA in terms of fossil and nonfossil fractions, specific studies on the biofuel combustion emissions and aging are required. The primary emissions from biofuel vehicles are expected to be still partially fossil, given that the biofuel engines use lubricating oils, responsible for primary OA emissions. On the other hand, the SOA produced from the precursors emitted from biofuel vehicles would be non-fossil (or mainly non-fossil), considering that the precursors are compounds from the biofuel that have been released to the atmosphere by evaporation. On the other hand, the % of biofuel use, even if it is very low as stated, is not expected to change between the first and the second periods of our study.

1 R. Suarez-Bertoa *et al.*, *Atmos. Environ.*, 2015, **117**, 200–211.

Matthew Hort returned to the discussion of Kebin He's paper: In the paper, you talk about the fact that the analysis was grouped by similar meteorological conditions. This is a very good thing to do and I am pleased to see it. However, it was not clear to me what time period had been used in defining the similar meteorology.

I also wondered what meteorological data had been used to draw the comparison. It seemed that a single point had been used. Depending on the relevant transport and possibly reaction times, it might mean that meteorology of the wider synoptic picture would be a better method for characterisation (*i.e.* as a single point can be the same under different wider conditions that may affect the trajectory of the air mass). Though this might be harder to define, one possibly useful technique is the use of Lagrangian back trajectories.

Zongbo Shi replied: The procedure that we followed to select data populations with similar meteorological conditions has been described in detail in the second part of our results. Not only a single point was used for comparison. Briefly speaking, local temperature, RH, wind speed and PBL height were very similar before and during the emission control period, as presented on page 10 of the paper. However, local wind direction was very different. Then we further used the back trajectory cluster analysis to select comparable conditions of wind direction and regional air mass transport. Finally, the cases of cluster 1 and cluster 3 following back trajectory analysis were selected. As shown in Fig. 4, data populations under these two cases had similar weather conditions.

Neil Donahue asked Costas Sioutas and Martin Shafer: Can you comment on the DTT assay and biogenic SOA?

Costas Sioutas responded: in LA we don't have much SOA derived from biogenic photo-oxidation; however, the DTT studies in Atlanta by Weber and Verma (and others) do indicate a relationship between biogenic SOA and DTT activity.[1]

1 T. Fang *et al.*, *Atmos. Chem. Phys.*, 2016, **16**, 3865–3879.

Martin Shafer added: I don't believe our study can contribute substantially to furthering our understanding of the role of biogenic SOA versus anthropogenic SOA on the oxidative potential of the SOA as estimated with the DTT assay. Studies have clearly shown the role of anthropogenic SOA precursors (SO_2, NO_x) on biogenic SOA formation, and thus the oxidative potential of these PM/SOA components are intimately linked. That said, it is reasonable to expect that the oxidation products of biogenic precursors (isoprene, terpenes, etc.) may be more active in the DTT assay than the SOA from major anthropogenic sources. I would refer readers to Kramer *et al.*[1] and Chen *et al.*,[2] and references cited within.

1 A. J. Kramer, W. Rattanavaraha, Z. Zhang, A. Gold, J. D. Surratt, and Y. H. Lin, *Atmos. Environ.*, 2016, **130**, 211–218.
2 X. Chen, P. K. Hopke and W. P. L. Carter, *Environ Sci Technol.*, 2011, **45**, 276–282.

Gary Fuller addressed Costas Sioutas: I'm really intrigued about the sub-story in the paper; the increase in abrasion products. These are traced by copper, barium, manganese in your study, so by abrasion do you mean brake wear? Can you account for it? We have found in some London roads that there is an increase in PM_{10} despite a decrease in $PM_{2.5}$ (as shown in Fig. 1) – largely these are locations where we have an increase in heavy duty vehicles.

1 A. Font and G. Fuller, Roadside air quality in London – identifying the outliers: Part 1, King's College London, 2015.

Costas Sioutas answered: We do mean mostly tire wear – please read the related discussion in our paper.

Urs Baltensperger asked: I presented in my talk yesterday that the brake wear size distribution is such that most of its mass is found in the coarse or medium mode, and only a small fraction is found in the fine mode (see Visser *et al.*[1]). So, could the brake wear effects that you see in the fine mode be due to a cross correlation with the traffic exhaust from exhaust pipes?

1 S. Visser *et al.*, *Atmos. Chem. Phys.*, 2015, **15**, 11291–11309.

Costas Sioutas responded: We don't necessarily agree with the statement that brake wear is in the coarse PM mode only – Ba as a tracer for brake wear almost exclusively comes from brake wear abrasion.[1]. Brake wear abrasion contributes to both fine and coarse PM. Several studies have shown that particles generated from brake wear abrasion in fact maximize at around 4 μm – for instance, the study of Ijima *et al.*[2] on particle size and composition distribution of automotive

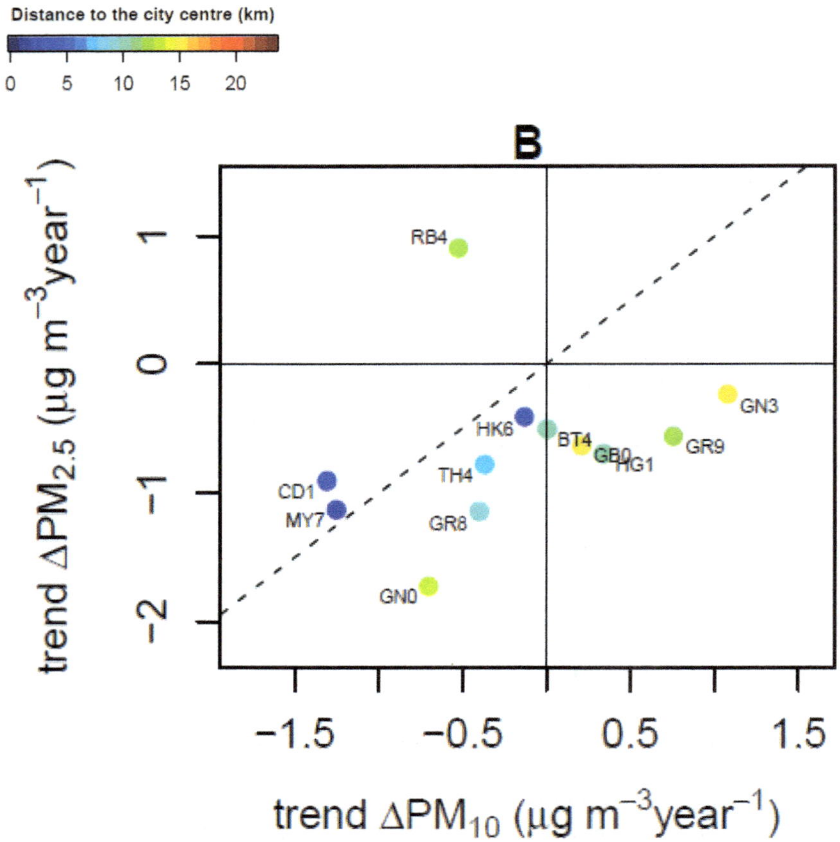

Fig. 1 Trends in roadside increment of $PM_{2.5}$ and PM_{10} from London 2010–2014. Reproduced with permission from Font and Fuller[1]

brake abrasion, where the shapes of the distributions of five metallic elements (Fe, Cu, Zn, Sb and Ba) were very similar to the total mass distribution of the brake abrasion dusts and had a mode of around 3–4 µm. This is also in line with the study of Harrison *et al.*[3] in which they took Ba as a tracer of brake dust - their scaled Ba representative of brake abrasion also showed a peak at around 4 µm and a smaller peak in the $PM_{2.5}$ range. A study by Ntziachristos *et al.* also showed that "Most elements fall into two distinct concentration profiles with particle size. First, several elements present a monotonic increase in their profile with size, such as Na, S, Ti, V, Mn, Fe, Sn, and Ba. The highest concentrations appear in the $PM_{0.18-2.5}$ mode."[4] In their study of urban air, Birmili *et al.* found similar profiles for Na, Fe, and Ba in the size range 0.5–7.2 µm,[5] while Lough *et al.* also presented similar monotonic profiles up to 2.5 µm for Ba in a road-tunnel.[6]

One should also take into account the effect of brake pad materials itself in contributing to size-fractionated PM emission. The brake dust profiles which Schauer *et al.* reported[7] were for semimetallic and low-metallic brake pads. The semimetallic and low-metallic designations are types of brake pads discussed in the literature, but the range of actual brake pad formulations on the market is vast. They reported their profiles for PM_{10} and $PM_{2.5}$. Interestingly, the contribution of low metallic brake pads to $PM_{2.5}$ (13.2 mg g^{-1} brake dust PM) was about 2.3 times higher than for PM_{10} (5.8 13.2 mg g^{-1} brake dust PM). However, for semimetallic ones they were more or less comparable (13.4 and 14.2 for $PM_{2.5}$ and PM_{10}, respectively).

In our own work on source apportionment of metals in the ultrafine size range in LA,[8] we state that, "While road dust has been identified as a major source of coarse particle emissions (Amato *et al.*;[9] Pakbin *et al.*[10]), other studies have consistently been reporting road dust emissions originating from tire–pavement interactions to be present in the smaller size fractions as well, with mass mean particle diameters between 15 and 50 nm (Dahl *et al.*;[11] Gustafsson *et al.*[12,13])."

1 P. Sanderson, J. Maria Delgado-Saborit and R. M. Harrison, *Atmos. Environ.*, 2014, **94**, 353–365.
2 A. Iijima, K.I Sato, K. Yano, H. Tago, M. Kato, H. Kimura and N. Furuta, *Atmos. Environ.*, 2007, **41**, 4908–4919.
3 R. M. Harrison, A. M. Jones, J. Gietl, J. Yin and D. C. Green, *Environ. Sci. Technol.*, 2012, **46**, 6523–6529.
4 L. Ntziachristos, Z. Ning, M. D. Geller, R. J. Sheesley, J. J. Schauer and C. Sioutas, *Atmos. Environ.*, 2007, **41**, 5684–5696.
5 W. Birmili, A. G. Allen, F. Bary and R. M. Harrison, *Environ. Sci. Technol.*, 2006, **40**, 1144–1153.
6 G. C. Lough, J. J. Schauer, J.-S. Park, M. M. Shafer, J. T. DeMinter and J. P. Weinstein, *Environ. Sci. Technol.*, 2005, **39**, 826–836.
7 J. J. Schauer, G. C. Lough, M. M. Shafer, W. F. Christensen, M. F. Arndt, J. T. DeMinter and J.-S. Park, *Res. Rep. Health Eff. Inst.*, 2006, **133**, 1–76, discussion 77–88.
8 A. Saffari, N. Daher, M. M. Shafer, J. J. Schauer and C. Sioutas, *Environ. Pollut.*, 2013, **181**, 14–23.
9 F. Amato, M. Pandolfi, A. Escrig, X. Querol, A. Alastuey, J. Pey, N. Perez, and P. K. Hopke, *Atmos. Environ.*, 2009. **43**, 2770–2780.
10 P. Pakbin, Z. Ning, M. M. Shafer, J. J. Schauer and C. Sioutas, *Aerosol Sci. Technol.*, 2011, **45**, 949–963.
11 A. Dahl, A. Gharibi, E. Swietlicki, A. Gudmundsson, M. Bohgard, A. Ljungman, G. R. Blomqvist and M. Gustafsson, *Atmos. Environ.*, 2006, **40**, 1314–1323.
12 M. Gustafsson, G. R. Blomqvist, A. Gudmundsson, A. Dahl, E. Swietlicki, M. Bohgard, J. Lindbom and A. Ljungman, *Sci. Total Environ.*, 2008, **393**, 226–240.
13 M. Gustafsson, G. R. Blomqvist, A. Gudmundsson, A. Dahl, P. Jonsson and E. Swietlicki, *Atmos. Environ.*, 2009, **43**, 4699–4702.

Eben Cross asked: The composition of fuel and lubricant additives include metal species that overlap with the non-tailpipe chemical emission profile characterized here. Can the authors comment on how potential exhaust-derived metal contributions are separated from non-tailpipe sources?

Costas Sioutas responded: That's a good point when it comes to species such as Zn, CU *etc,*; however, Ba is a very reliable tracer of brake wear.

Markus Kalberer inquired of Costas Sioutas and Martin Shafer: I have a general question to both of you – you have DCFH measurements on one hand, and DTT assays on the other, while some groups use biological measurements instead. So there are many different methods to use in trying to come up with a metric for oxidative potential. What's your opinion on which is best?

Costas Sioutas answered: I am in no position to offer an informed response, since I have not evaluated carefully each of these methods individually and using evaluation metrics that are yet to be defined; I can at least speak for DTT, for which we have evidence of correlations with health effects observed in human panel studies.

Martin Shafer added: Our ROS measurements are biologically based (using alveolar macrophages) and in our opinion incorporating "biology" in increasingly relevant models is key to advancing oxidative potential metrics. The issue really is less the specific ROS probe, and more the complexity of the biological model and how close a surrogate it is for the functioning of human systems. Multi-cellular *in-vitro* models likely provide a more robust predictive capability. That said, in going after the integrated oxidative potential (in whatever model), a general probe like DCFH is most useful. If pathway analysis is needed, then less effort should be placed on characterizing specific ROS species, but instead focus should be on expression/production of pathway characteristic signaling molecules (cytokines, chemokines, *etc.*).

Urs Baltensperger commented: I do not believe that we will be able to describe biological effects on human beings with just a single assay. This means we need to run a variety of tests in order to see which combination provides us with the best possible answer. Here I would love to see closer collaboration between chemists and biologists to come up with a combined and more complete picture of what is going on, and then in the future possibly try to reduce the number of assays.

Xavier Querol addressed Costas Sioutas and Martin Shafer: Thanks a lot for your excellent presentations and papers. Two questions; firstly, did you evaluate if the higher ROS of the fine *versus* coarse fraction is softened if you normalise the ROS by the surface area of these fractions? Or do you think that the higher ROS is due to the differences in chemical composition and oxidation state, of the two fractions? Secondly, I observed in tables of the paper relating to European cities that ROS was highly correlated for the coarser fraction with a group of elements of typically crustal origin (Ca, Ti, Mg, Al, and specially K, the latter in the coarser fraction being usually associated with minerals more than biomass burning). The correlation is better for these than for other typically ROS-associated trace metals. However, as we move to finer fractions the correlation decreases. What do you interpret as being the causes for this?

Costas Sioutas replied: We did not reach conclusive results by normalising ROS *vs.* surface area, but this can be done with specific surface area measurements.

Martin Shafer responded: Thank you for your insightful comments. We did not examine the potential influence of surface area normalization on the specific activity of the PM size classes – primarily because, as noted in your comments, a large fraction of the PM mass (and an even larger fraction of the ROS activity) in the PM_3 size class (and a somewhat smaller fraction in the larger size classes) was water soluble. This is fundamentally driven by a contrast in chemical composition of the size-classes. Additionally, particle phagocytosis is a primary initiating step in ROS generation in our macrophage model, and this process is relatively insensitive to particle size (within some limits).

The contrast in ROS activity across PM size classes reflects the net effect of several factors, including major differences in chemical composition, water solubility and mechanisms of uptake/interaction with the macrophages. The specific activity (*i.e.* mass normalized) of the PM_3 fraction is actually significantly greater (more than 2-fold) than PM_{7-3} and over 4-fold greater than the super-coarse fraction. This likely reflects a greater contribution from soluble higher activity organic species such as SOA, some contribution from soluble transition metals, and "dilution" of the specific activity of larger PM by relatively inactive components. The chemical species correlations in the larger size fractions likely reflect only secondarily the chemical composition but more importantly the greater insoluble fraction (which effectively means crustal material –thus the enhanced correlation with Ca, Ti, Mg, Al, K). Macrophages are particularly sensitive to particulates "themselves", which are strong initiators of the respiratory burst upon phagocytosis. The chemical components with which Ca, Mg and K are associated in the fine PM are different to those in the coarser fractions and are significantly more soluble. These elements are inherently non-reactive, thus the ROS correlation with these elements in the fine fraction "dissolves".

With regards to the finding of a higher specific activity of PM at the background sites, we hypothesize a greater contribution of higher activity organic SOA components at these sites.

Xavier Querol replied to Martin Shafer: I agree with this, but this does not answer my question, because what you have found is a correlation between ROS and Ca in the coarse fraction, and no correlation or even negative correlation of these two parameters in the fine fraction. And why is ROS more correlated in the coarse fraction with K than with classical ROS-related metals?

Roy Harrison remarked: I have been interested in these *in vitro* tests of particle toxicity for many years. I convened a meeting in London almost 10 years ago which was attended by leading international workers, including Professor Sioutas. This left me deeply sceptical over the value of the tests and I have not had reason to change my view since. Martin's paper raises a number of important issues. Firstly, the tests using a measure of exposure (*e.g.* the DTT assay) do not correlate with those giving a measure of effect (*e.g.* the TNF alpha measurement), hence I wonder if the exposure tests are telling us anything useful. Work that I have done in this field collaborating with Professor Frank Kelly of Kings College London uses a simulated lung fluid and looks at depletion of anti-oxidant in the lung fluid. This shows the greatest activity in the coarse particle fraction at roadside which is in complete contrast to Martin's results. Clearly, the Kings College test is being driven by very different factors from the DTT assay. These kinds of inconsistencies come up quite widely. Also, the spatial distribution of activity in Martin's results is most unexpected. Since the urban background aerosol is typically a mixture of the regional background with urban emissions, one would expect its activity to lie between the rural and street samples, but in Martin's data it shows the highest oxidative potential per unit mass, which is highly counter-intuitive. My other concern beyond the lack of correlation of the results of different tests of oxidative potential is that there is a scarcity of evidence of clinically significant effects in human populations. This is critical to establishing the value of such tests.

Costas Sioutas replied: I will send you one of our papers[1] in which we clearly demonstrate the strong association between these assays and biomarkers of systemic inflammation in human subjects in LA.

1 R. J. Delfino, N. Staimer, T. Tjoa, M. Arhami, A. Polidori, D. L. Gillen, S. C. George, M. M. Shafer, J. J. Schauer and C. Sioutas, *Epidemiology*, 2010, **21**, 892–902.

Xavier Querol added: I remember a paper by the Sioutas team where they found more ROS in the LA pollution plume than in the city, and they attributed this to ROS of SOA.

Urs Baltensperger remarked: We know that a large fraction of the atmospheric aerosol is water soluble, so for anything water soluble the surface area is not relevant. The same is true for the number concentration. This needs to be taken into account when searching for additional metrics that could describe the health effects of aerosols.

Xavier Querol responded: I agree, but I am still convinced that surface area might have a role here, because when we applied a DNA array assay to fine and coarse fractions and we normalised these by surface area, differences were reduced. Many PM components having a high ROS effect, such as metallic species and PAH for example, are not water-soluble.

Neil Donahue commented: I agree with Urs that surface area may be irrelevant when most of the material on fine particles is water soluble. However, I can easily hypothesize a situation where the urban background could show the highest toxicity. The organic fraction evolves dynamically, and if relatively short-lived first- or second-generation oxidation products are the most toxic, we would see the greatest toxicity some time after emission. That could be the urban background.

Tzer-Ren Ho asked Martin Shafer: did you measure the cellular viability of the macrophage when exposed to the aerosols? Exposure time is only two hours, so cells may have undergone apoptosis but not be dead - this might explain why the TNF alpha results don't correlate to the ROS results. In my results, when I exposed cells to diesel exhaust particles some of the cells did die, which might have affected your results in this case.

Martin Shafer responded: We have measured plasma membrane integrity with either PI or Sytox green fluorescent probes. These probes of cell viability (not necessarily apoptosis) show no indication of NR8383 cell death at a 2 h exposure time in most of the samples that we have examined. In addition, our detailed dose-response exposure protocol is designed to detect sub-lethal toxicity, and if such non-linearities are observed data at those concentrations are not included in the slope analysis. If necessary, exposures are repeated at more dilute concentrations to ensure that the dose-response line is adequately predicted. The generally poor correlation between the TNF alpha data and ROS in this data set is likely the result of several factors, including (a) a contrast in the kinetic profiles of ROS and cytokine induction, (b) a much stronger preference of particulate phases over soluble phases for TNF alpha expression in comparison with ROS generation, and (c) a smaller cytokine dataset in comparison with the ROS dataset in this study.

Tzer-Ren Ho remarked: One of the ways to be sure is to use annexin V staining during flow cytometry, to ensure cells are not undergoing apoptosis

Martin Shafer answered: Yes, that is a good suggestion. We have considered this protocol, but it would require some modification to our exposure media (and added uncertainty) as a result of the sensitivity of this stain to calcium concentrations. While we believe apoptosis is probably not a significant issue in our cell model and exposure protocols (as mentioned in my response to your previous question), it is something we will consider evaluating in on-going studies.

Markus Kalberer said: I have a comment regarding Roy Harrison's remark on the seeming inconsistencies of the different assays. We have tried the DCFH assays as well as determining ascorbic acid consumption of aerosol extracts. There will be different reactivity for different assays depending on what you throw at them – for example, ascorbic acid is easily oxidised by transition metals, but you can add milligrams of OAs to it and it won't react. Secondly, I have a comment to Neil Donahue regarding short lived components – this relates to the poster which we have presented at this Discussion, where we try to convert the DCFH methodology to an online method so we can measure within a couple of seconds. This could be a strategy to follow up, to get a more consistent picture.

Andreas Wahner returned to the discussion of Costas Sioutas' paper: you showed in your studies that you identified certain fractions of aerosols to be reactive in your test on oxidative potential. Did you also confirm with artificial aerosols that they give the same results and in similar quantities? For example, could you use just aerosols from brakes?

Costas Sioutas replied: We haven't done tests with artificial brake aerosols – for more information I refer you to Shuster-Meiseles *et al.*[1] and Zhao *et al.*[2]

1 T. Shuster-Meiseles, M. M. Shafer, J. Heo, M. Pardo, D. S. Antkiewicz, J. J. Schauer, A. Rudich and Y. Rudich, *Environ. Res.*, 2016, **146**, 252–262.
2 J. Zhao, N. Lewinski and M.l Riediker, *Aerosol Sci. Technol.*, 2015, **49**, 65–74.

Urs Baltensperger addressed Martin Shafer: You showed in your talk that you found spatially quite a uniform oxidative activity across Europe, while other people see substantially higher activity in street canyons than at urban background sites. I guess you must have come up with some kind of hypothesis to explain this?

Martin Shafer answered: It was uniform within a general factor of 5 when expressed in terms of air concentrations (activity per volume of air), and a bit more uniform when normalized to PM mass. And yes, in general the "background" sites exhibited comparable activity (or higher when normalized to PM mass) to the urban canyon sites. Our working hypothesis is that the PM at the background sites are reflecting a significantly higher fraction of intrinsically higher activity PM components. SOA from urban emissions formed on time-scales of multiple hours and SOA transported into the urban air-shed from more rural regions is a likely candidate. This hypothesis is supported by our observation of the generally poor correlation of the PM_3 elemental species typically considered as vehicle emission tracers with ROS and DTT activities.

Urs Baltensperger remarked: We have shown at PSI in our smog chamber that with the formation of secondary organic aerosol the ROS concentration increases substantially, regardless of the precursor type. So secondary aerosol will definitely contribute to the toxicity.

Roy Harrison addressed Costas Sioutas: One big unsolved question in this field is the fact that assays of this kind, and human challenge studies, attribute low toxicity to sulphate. But many epidemiological studies show sulphate as showing the strongest correlation with mortality and adverse outcomes. Could it be that actually sulphate is a surrogate for low volatility SOA, and it's actually SOA that's driving this toxicity?

Costas Sioutas replied: Indeed, sulfate could be a surrogate of SOA.

Martin Shafer added: I agree with Prof. Harrison's observation/analysis that assays of this type generally indicate that sulfate has little or no ROS activity, and that is not unexpected from a biochemistry perspective. In fact, in the large number of field studies in which we have performed these ROS assays, sulfate at best shows no positive correlation with ROS, and in several studies is anti-correlated. So while the interpretation of epidemiological data in terms of chemical drivers is complex, I do not disagree with the hypothesis that sulfate may indeed be acting as a surrogate for SOA in these models, particularly for large spatial scale studies with a mix of rural and urban samples.

William Brune commented: There are a few studies of air pollution effects on humans that are being done using human subjects, including studies of inhalation of diesel exhaust. These studies are much more expensive than studies with cells and it can be hard to find volunteers, but these studies are being done.

Tzer-Ren Ho said to Martin Shafer: If you look at the sputum obtained from sputum challenge, you can find macrophages and dendritic cells in the upper airways as well, so you can definitely find them there. Henceforth, particles that deposit in the upper airways cannot be disregarded.

Martin Shafer replied: I don't believe there is much disagreement on this fact, though mucus can be escalated up the respiratory tract from quite deep in the lungs. The toxicity of coarse PM may also speak to the fact that upper airway processes cannot be ignored.

William Brune asked: Following up on Neil Donahue's comment about oxygenated hydrocarbons in particles being important for air pollution impacts on human health, do you think that the way that you collect the particles for analysis can affect their chemical composition and thus their impact on human health? If the particle chemical composition is degraded, how do you deal with this?

Martin Shafer replied: Most likely – that is the case with nearly all commonly implemented field techniques for aerosol sampling. Quantifying that potential bias is very challenging and requires both new approaches for PM collection and new tools that are sensitive to the speciation relevant to human health. On the inorganic side, chemical speciation (*e.g.* oxidation state), not chemical composition, is the biggest concern and we (and others) are developing and implementing techniques that will address this for the primary elements of concern. On the organic compound side, our analytical tools have advanced to the point where relevant species quantification is likely not the limiting factor in addressing collection biases – one must build these tools into studies comparing field collection methods and proxies for human health impacts. Obviously we take great care to minimize post-collection biases by freezing and in many recent studies (including the European Aerosol Study) sealing the filter in air-impermeable, nitrogen purged bags. Significant efforts are being dedicated to developing more sensitive assays that can be used with much shorter time-integrated samples (1–2 hours, where potential speciation biases are likely to be much smaller). Novel tools for collection of aerosol PM for health studies are being developed that collect and concentrate PM directly into aqueous media (or even culture media) on time scales of an hour, and couple this with near-real time chemical characterization. Our group has several publications where these novel approaches are detailed. On an encouraging note, the limited toxicological data from these novel methods appears to confirm the integrity and validity of 24-hour integrated filter samples. But clearly more directed study is essential.

Costas Sioutas added: This is an important issue and obviously we thought about it – we have conducted a detailed characterization of our PM in all of our publications and have shown that the chemical composition is very similar to that measured by reference samplers (USEPA frm) sampling in parallel.

Neil Donahue said: My mental model is that there are many toxicants in fine particles, with different dose responses, and that the overall toxicity is some sort of combination of these effects. There is no magic bullet. However, I have always been puzzled why particles seem to be especially toxic. Is there any evidence that particles deposited on cells more or less overwhelm individual cells or groups in ways that the same amount of vapor, distributed across the lung, could not?

Costas Sioutas responded: Neil, as you know we have discussed this and indeed it could be the foundation of an interesting series of inhalation experiments whereby we provide an atmosphere containing SVOC in their particle phase and the same amount/concentration of SVOC in the gas/vapor phase to see how different the results might be. We developed/modified our VACES concentrators to do just this experiment, but never got a chance to carry it out – see our paper discussing VACES modification.[1]

1 P. Pakbina, Z. Ning, A. Eiguren-Fernandez and C. Sioutas, *J. Aerosol Sci.*, 2011, **42**, 555–566.

Martin Shafer said: I totally agree that ambient PM toxicology is very complex, both in the chemical/physical drivers of responses and the biochemical modes of action, and clearly no single assay can reproduce that complexity – a battery of assays will be required. Direct comparisons between particles, water soluble components of those particles, and vapor forms of the particulate species are difficult to construct, let alone interpret. Such experiments would be extremely interesting and valuable and new tools may allow us to do this with environmentally relevant PM/SVOC species. But, yes there is literature indicating that alveolar response is especially sensitive to particulates. This may reflect the mode in which invading particulates are dealt with in the airways – usually by recruitment of macrophages to the PM. Phagocytosis of the particles is a powerful initiator of the oxidative stress cascade, a pathway that would not be expressed with a similar mass of vapor.

André Prévôt inquired: It's intriguing that the ROS activity per mass is so much lower at the urban canyon than the urban background. Could you look at the mass balance, especially for the small particle size, to see whether the road emissions contribute anything? Can you work this out from the data?

Martin Shafer responded: It would be conceptually possible to source apportion the ROS (*e.g.* using the approach we used in our previous work[1]). And we may attempt to do so in a planned follow-up manuscript; however, we cannot at this stage define explicitly the road emissions contribution. There is little doubt though that primary road emissions do contribute to the ROS burden as evidenced in numerous roadside studies and in studies of tailpipe, brake, tire, and roadway dust source samples.

1 Y. Zhang, J. J. Schauer, M. M. Shafer, M. P. Hannigan and S. J. Dutton, *Environ. Sci. Technol.*, 2008, **42**, 6502–7509.

William Brune addressed Martin Shafer and Urs Baltensperger :I may have missed this in the published literature, but are there scatter plots of ROS *versus* an indicator of aerosol particle oxidation, such as perhaps O:C ratio?

Martin Shafer responded: I'm not aware of published correlations between metrics of the bulk oxidation status/state of the aerosol particle and macrophage-based ROS or DTT. Oxygen is rarely if ever quantified in the PM and though theoretically derivable from detailed organics analysis, I suspect the uncertainty would be large. Several papers have examined correlations of ROS with specific organic components or fractions of OC that could potentially serve as surrogates of the oxidation status of the PM and depending upon the ROS indicator, correlations of reasonable significance are observed. However most of these studies were not specifically designed to address this issue and confounding with other variables contributes to high uncertainty. Even less work has been published examining the relationships between oxidation status of redox active metals in the PM and ROS and this is an area that we are exploring in on-going research.

Urs Baltensperger responded: We did not publish a scatter plot, but we showed the temporal evolution of the normalized ROS concentration (ROS concentration measured in moles of hydrogen peroxide equivalents normalized to the molar organic carbon concentration per m^3) and the O:C ratio.[1] There we indeed saw a correlation in the first one to two hours, but after this the relative ROS concentration stopped increasing, while the O:C ratio continued to increase.

1 S. M. Platt *et al.*, *Nat. Commun.*, 2014, **5**, 3749.

Markus Kalberer said: You discussed a possible link between O:C ratios in organic aerosols and the response of biological assays. One should be very careful in interpreting such correlations. One of these assays, DCFH, is very sensitive to hydrogen peroxide but not at all to carboxylic acids, even though these functional groups have the same number of oxygens. I would call to go back to more basic laboratory experiments to characterise these assays with different aerosols, rather than doing more field experiments where everything is extremely complex.

Martin Shafer answered: I concur (nearly) completely with your comments. Clearly the utility of bulk O:C ratios in the context of ROS/DCFH sensitivity must be interpreted with appropriate circumspection. And much can/should be done in the lab with well-defined aerosols (and controlled "aged" aerosols) to understand the response (and utility) of a broad range of *in vitro* assays. Field studies do have a place, if only to provide context and limits to lab studies. Also deconstructing (in the lab) actual field aerosol to provide information on the activity of specific PM components (using targeted physical/chemical "separation" tools) can actually be quite powerful and relevant.

Rachel Dunmore opened a general discussion of the paper by Christian Ehlers: You state that a GC instrument was specifically designed to be able to measure C8–C16 VOCs; could you provide a justification for this statement? Specifically, details of the sampling and analysis method for the larger carbon number VOCs. Also, what was the sensitivity of this instrument for C11 VOCs (*i.e.* n-undecane) in your calibration standard *versus* that of a smaller carbon number VOC (such as pentane)? Fig. 2, which is taken from our own work, shows a reconstituted 1D GC chromatogram from a winter ClearfLo GCxGC chromatogram (using Zoex GC Image software) with three major aromatics and the C8–C12 n-alkanes identified. The majority of the material in between these is made up of other aliphatic compounds, with a few aromatics and oxygenated compounds. Without the addition of GCXGC the additional aliphatic material, made up of many low concentration isomers, is hard to quantify and after 45 min appears as an increasing baseline with a small number of peaks. With the GCxGC chromatogram, it was possible to quantify this other VOC loading and separate the contribution from grouped aliphatic, aromatic and oxygenated compounds. For example, in winter in London, the average mixing ratio for n-undecane was 397 pptv (*cf.* toluene 635 ppt) while the average for the grouped C11 aliphatics was 459 pptv.[1] This additional aliphatic material would be impossible to quantify using the 1D chromatogram alone. Did you make any estimation of the other aliphatic VOCs in addition to the n-alkane?

1 R. E. Dunmore, J. R. Hopkins, R. T. Lidster, J. D. Lee, M. J. Evans, A. R. Rickard, A. C. Lewis and J. F. Hamilton, *Atmos. Chem. Phys.*, 2015, **15**, 9983–9996.

Christian Ehlers replied: The GC instrument was originally developed to analyze sesquiterpenes and monoterpene compounds. It consists of a thermodesorption system combined with a cold injection system where the components are refocused (TDS, Gerstel, Mülheim). The compounds are separated on a 60 DB1-column before they are detected by FID and MS (Agilent, MSD 5975c). The adsorption tubes were desorbed at 320°C. The transfer lines were held at 300°C and the column was ramped to its maximum temperature of 320°C to prevent losses of high boiling compounds. Compounds lighter than octane are not completely retained in the focusing unit. When sampling diesel exhaust a large number of peaks between decane and tetradecane were observed which appear as a baseline increase in the FID signal. By using the single ion traces in the mass spectrometer it was possible to deconvolute the compound mix which consists of straight chained alkanes, and single and double methyl substituted alkanes as well as substituted cyclohexanes. Also, substituted benzenes, naphthalenes and decalins were found. Still, the total FID signal of compounds larger than dodecane was low compared to the lighter compounds.

Alison Tomlin opened a general discussion of the paper by James Lee: Even in the interpretation of the measurements it seems that you are using a model: *e.g.* to calculate the relevant footprint. Can you be sure that the remaining discrepancy (following the update of the emissions inventory) is due to an inventory problem and is not due to sensitivities to other model parameters? One parameter which may be important is the assumed roughness length for each city neighbourhood. The method of Drew *et al.* that you cited in your paper does not account for heterogeneous building heights and therefore is likely to underestimate

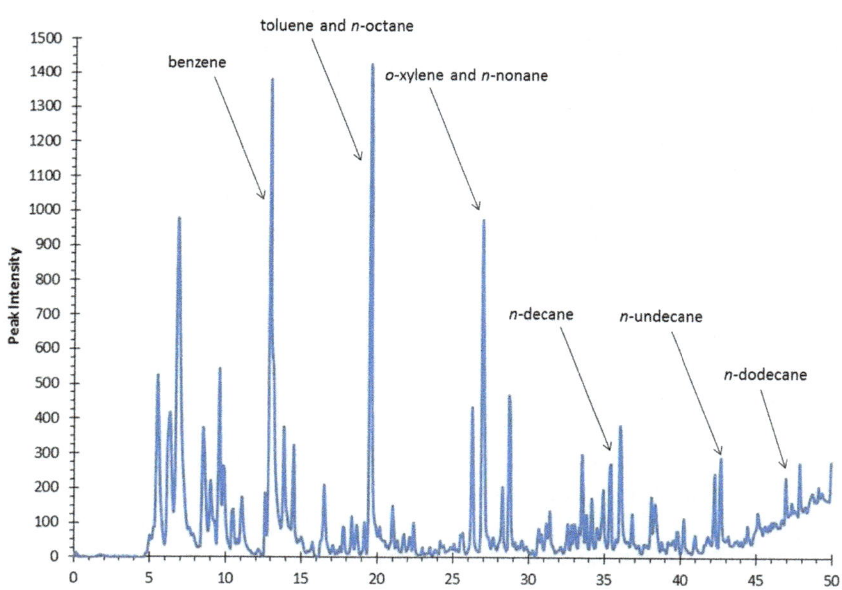

Fig. 2 1D GC chromatogram from a winter ClearfLo GCxGC chromatogram (using Zoex GC Image software) with three major aromatics and the C8–C12 n-alkanes identified. Reproduced with permission from Dunmore *et al.*[1]

roughness lengths and displacement heights for London's central region, based on the work of Millward Hopkins *et al.*[1] Can you comment on the sensitivity to these surface aerodynamic parameters?

1 J. T. Millward-Hopkins, A. S. Tomlin, L. Ma, D. Ingham and M. Pourkashanian, *Boundary-Layer Meteorology*, 2011, **141** 443–465.

James Lee responded: This is a good point. Our footprint model is dependent on a variety of factors, including the roughness length, which as is pointed out we have taken from the Drew *et al.* 2011 paper.[1] We appreciate the comment that there are other measures of roughness length for urban areas and that the measurement we use may be an underestimate. We have now carried out a sensitivity study of our footprint model to roughness length by doubling the value we use. This does indeed cause a reduction in our footprint area and thus an increase in the NO_x emission from the inventory for central London. This data is shown in Fig. 3. In central London the ratio is reduced from about 1.5 to 1.35, although in the outer London zones the ratio is actually increased from around 1.05 to 1.15. This shows the importance of having the correct roughness length for the interpretation of our results, and we will work on using the best possible data in our future work.

1 D. R. Drew, J. F. Barlow and S. E. Lane, *J. Wind Eng. Ind. Aerod.*, 2013, **121** ,98–105.

Thomas Karl addressed Christian Ehlers: In your paper (page 12, line 30) you state that the total decrease of NO_x emissions in your Fig 13 is quite comparable to the observed trend of NO_x concentrations (Fig 11), based on the updated HBEFA3.2 database that now also includes RDE. Yet, a closer look at these figures suggests that the NO_x emission inventory drops about 60% faster than NO_x concentrations during the discussed time period. Your statement therefore implies (1) either significant emissions of NO_x from other sectors in German cities, that have not declined or increased, (2) a significant increase of advected NO_x over German cities (*e.g.* due to agricultural activities) or (3) a significant change of other processes leading to lower atmospheric loss rates of NO_x in German cities (*e.g.* atmospheric oxidation, deposition, conversion to HNO3, alkylnitrates, PANs *etc.*). I would think that (1) is the most realistic; but in this case I was wondering whether you can really assume that a major fraction of NO_x in German cities really originates from sources other than traffic. Another possibility would be that even updated bottom-up emission inventories still largely under-predict urban NO_x from traffic related emission sources, questioning the reliability of these bottom-up models due to inherent uncertainties of the upscaling approach.

Christian Ehlers answered: For urban areas we would not expect other important sources of NO_x apart from traffic emissions. We would rather expect that even the updated emission inventories are still underestimating the emission factors.

Tim Wallington asked James Lee: Wouldn't the simplest explanation for the discrepancy in downtown London be that you're underestimating "other" emissions aside from traffic?

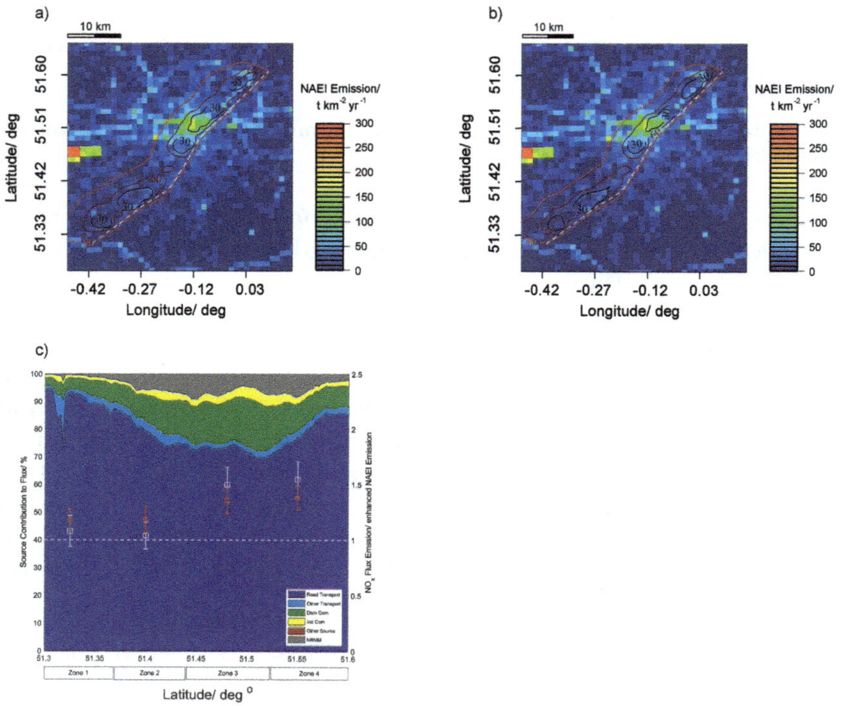

Fig. 3 a) The original footprint calculated using the Drew *et al.* roughness length as in our paper, b) the footprint calculated when we double the roughness length and c) the effect of this on the measured/NAEI emission ratio (white points are the original data, red points are the new data)

James Lee replied: This is a possible explanation. However, as even in central London traffic is estimated to account for around 80% of the NO_x source, it would need the other sources to be severely underestimated (by a factor of several) to bring the inventory in agreement with the measurement. We believe this is unlikely to be the case and therefore think traffic emissions are likely to be the main source of error in the inventory.

Marliyyah Mahmood asked Christian Ehlers: Did you measure and if so what are, the minimum and maximum concentrations of the local ozone production?

Christian Ehlers replied: The comparison to field observations was not done explicitly for the data shown in our paper. In principle, the local ozone production rate can be determined by measurements of RO_x and NO concentrations. Such observations were done from our institute in urban and sub-urban regimes in China and in the vicinity of Berlin and Freiburg. We know from past campaigns using the RO_x LIF technique that in environments with large R_VOC (above 10 s^{-1}) and small NO_x (in the NO_x limited regime) OH is underestimated by our models.[1] In these environments the concentration of HO_2 seems to be correctly calculated by our models (see Hofzumahaus *et al*[2]). In the R_VOC regime below 5 s^{-1}, the ozone production rate was calculated from direct measurements of peroxy radical and NO concentrations using the MIESR technique.[3] The observed

O3 production rate showed a strong increase with a NO mixing ratio between 0.1 and 0.3 ppb, where $P(O_3)$ reached a maximum value of 8 ppb h^{-1}. The dependence of $P(O_3)$ on NO at mixing ratios below 0.3 ppb had a positive slope of 53 ppb h^{-1} per ppb NO. Above 0.5 ppb NO the rate of ozone production decreased linearly with a slope of 3.3 ppb h^{-1} per ppb NO. These experimental data were explicitly compared to model calculations (MCM 3.1) showing good agreement. That gives us reason to believe that our current models can calculate the local ozone production rate from RO_x + NO, but not the degradation rate of VOCs by OH in the NO_x limited regime.

1 F. Rohrer *et al.*, *Nature Geoscience*, 2014, 7, 559–563.
2 A. Hofzumahaus *et al.*, *Science*, 2009, **324**, 1702–1704.
3 D. Mihelcic *et al.*, *J. Geophys. Res.*, 2003, **108**, 8254.

Franz Geiger opened a general discussion of the paper by David Carslaw: The comment about diesel traffic being unique in central London made me think about the ongoing efforts to increase the share of biodiesel in transportation fuels. How do you think transitioning into biodiesel might affect air quality in London?

David Carslaw answered: For the issues covered in our paper, *i.e.* NO_x and NO2 emissions and concentrations, I do not think that a move to using more biodiesel would make much difference. There could, however, be greater effects on the concentrations of other species such as acetaldehyde.

Francis Pope remarked: Are the vehicle age dependent changes in the NO2:NO_x ratio accompanied by changes in the total NO_x emissions? Also, are the changing NO2:NO_x ratios accompanied by changes in the CO and particulate matter emissions?

David Carslaw responded: We find there has been essentially no change in total NO_x emissions for Euro 4/5 diesel passenger cars with vehicle age, as shown in Fig. 4 below. The CO data for diesel vehicles do not show obvious variation with vehicle age, although the data are noisier. We do not have data for particle emissions.

Paul Monks said to Christian Ehlers: Given that the VOC to NO_x reactivity has changed so markedly over the last 10 years, from about 70% reacting with VOC and 30% with NO_2 to about the opposite situation, would you imagine there would be more HNO_3 formed?

Christian Ehlers responded: The changes in the reactivity contributions of VOC and NO_x which were observed over the past years would lead to the assumption that more HNO_3 can be formed. Up to now we have not measured HNO_3 with our mobile laboratory, but we are considering investigating this topic in the future.

Astrid Kiendler-Scharr added: It has to be pointed out in the first place, that while the relative contribution of NO_x to OH reactivity in urban environments has increased, both the VOC- and NO_x-reactivity have decreased. It seems unlikely that this would result in an increase in HNO_3 formation.

Andreas Skouloudis addressed James Lee: It's interesting that you have this measuring campaign using an aircraft. These campaigns are very expensive but I wish we could have this. My question is: what made you select July as the date for this campaign and not winter? What would your conclusions mean for the annual emissions?

James Lee responded: We chose July as it was a month where we were most likely to get good weather. Conducting low flights over land requires no low cloud and so we decided that this was the best time of year to go for. We would also have found it difficult to fly within the boundary layer in winter, making flux measurements impossible. Funding restrictions did not allow us to conduct two separate campaigns at different times of year. We agree that ideally sampling should take place at different times of year in order to compare measurements with the annual estimate of inventories, however as we were not able to do this we used scaling factors to scale the inventory output to the time of year, day of week and time of day that the measurements were taken (as referenced in our paper).

Tim Murrells said: I value these discussions on the emissions inventory. One thing that is quite clear is that it's crucial to indicate what version of the national inventory you're comparing against. There clearly is an inconsistency with the national inventory's estimates for London, which should be addressed. Some of the differences are due to the fact that the NAEI is an inventory based on national data and trends, which is spatially distributed using some local data. The fleet in London is quite unique. We have been treating London in some more detail in

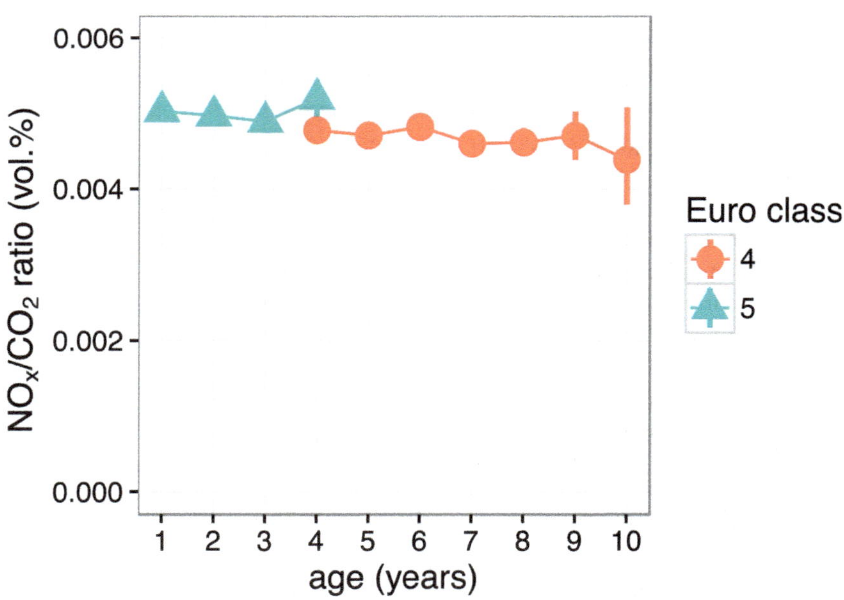

Fig. 4 NO_x/CO_2 ratio for Euro 4 and Euro 5 diesel passenger cars as a function of vehicle age

more recent inventory versions using fleet data from TfL, although probably still not as much detail as used in the London Atmospheric Emissions Inventory. Most national inventories use the same source of emission factors, although databases of emission factors can be parameterised in different ways. I believe that the German inventory uses factors from HBEFA which are of the same origin as the COPERT factors used in the UK inventory, namely from the European ERMES source. These are published sources of factors specifically designed for national inventories. Both sources represent real-world emission factors, but factors for Euro 5 and later vehicles are based on tests done on a relatively small number of vehicles and none of the factors take into account recent remote sensing data. I note that the enhanced LAEI uses factors derived from remote sensing as well as more detailed traffic data which would explain the better agreement with your observations. Variability in NO_x and primary NO_2 emissions between vehicles with different technologies, but made to the same Euro standard, can be large and this makes it difficult to represent in an inventory without detailed knowledge of the local vehicle fleet. Another key point when making comparisons between measurements and inventories is how emission factors were parameterised. The NAEI uses simple relationships between average vehicle speed and emission factor which works alright on a national scale, but is not an ideal approach for modelling emissions on a more local scale and on specific roads. Here, a more dynamic vehicle emission simulation model to estimate emissions at a greater time and spatial resolution along lengths of a road would be better, but requires more detailed traffic data.

Gary Fuller addressed David Carslaw: Thanks, David, for the excellent paper. You present your findings as a change in primary NO_2 emissions leading to a repartitioning of NO and NO_2. Looking at London roadside measurements (2010–2014 inclusive) we have seen that about 2/3 of roads have decreasing NO_2, in agreement with your findings (see Fig. 5). However, in these places there's an approximate 1 : 1 correspondence with decreasing NO_x suggesting that NO_2 is being preferentially abated rather than simply emitted as NO. Can you offer insights on this from your review of the technologies? Additionally, in about 1/4 of roads we see increases in both NO_x and NO_2 which suggests that NO_2 might not have peaked everywhere.

1 A. Font and G. Fuller, Roadside air quality trends in London – identifying the outliers: Part 1, King's College London, 2015.

David Carslaw answered: Thank you for your interesting question. Overall I think there is considerable site-to-site variation in both the trends and the NO_2/NO_x ratios from vehicles, as your work nicely shows. One of the key influences for London I believe is the bus fleet, which is both large and highly variable with respect to NO_x – and particularly NO_2 emissions. Our earlier paper[1] shows that there is a considerable variation in NO_2 emissions from the London bus fleet; from almost zero direct NO_2 up to about 40% of the total NO_x. Because bus technologies tend to be associated with specific bus routes, this has the effect of adding additional variation to the estimated NO_2/NO_x ratios for ambient measurements close to London roads. It is challenging, however, to robustly link vehicle flow and type information on individual road links with observed changes and variations in NO_x and NO_2 concentrations. The approach in our paper has therefore focused on determining the overall effects, while acknowledging there is site-to-site variation that is captured well in your work. Finally, it has become increasingly challenging to make these linkages due to the much wider range of engine and aftertreatment technologies used on vehicles.

1 D. C. Carslaw and G. Rhys-Tyler, *Atmos. Environ.*, **81**, 339–347.

Francis Pope inquired of James Lee: Have you compared your NO_x observations with leaf area index or other proxies of plant density? Is there any evidence that plants increase the dry deposition of NO_x?

James Lee replied: We have not specifically looked at leaf area index. It is possible that there could be net deposition of NO_x in areas of London such as the parks, where there is little anthropogenic emission. However, our flux footprint is significantly larger than the area of a single park and thus we believe that any deposition of NO_x to plant surfaces within this will be insignificant compared to the emissions.

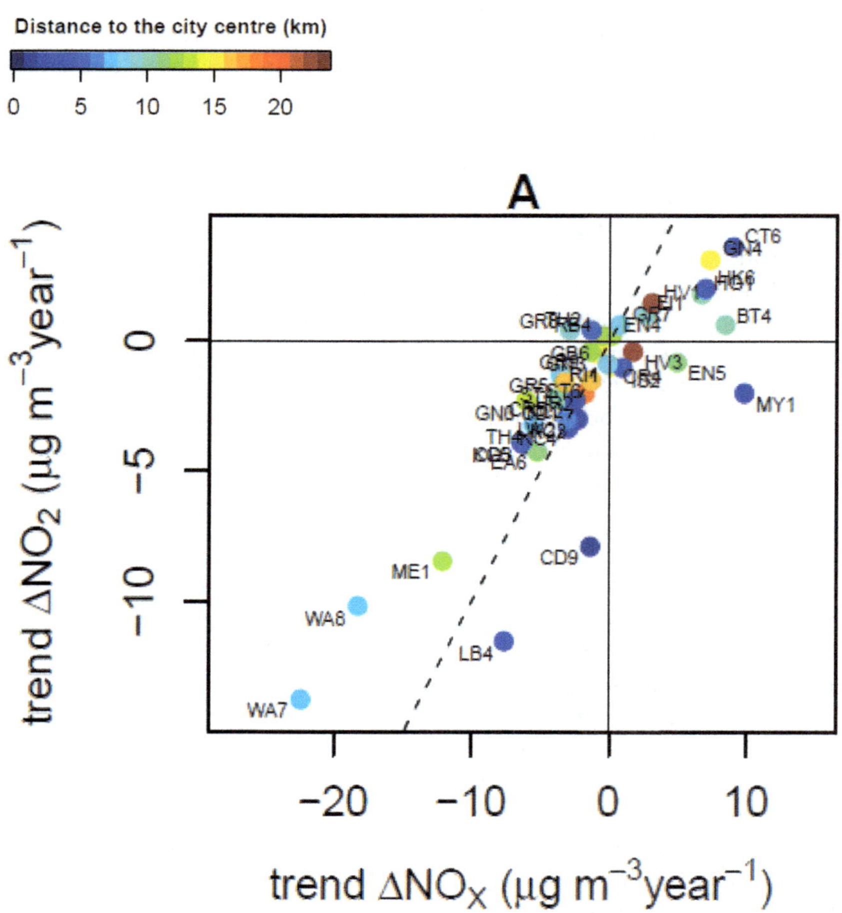

Fig. 5 Trends in roadside increment of NO_2 and NO_x from London 2010–2014. Reproduced with permission from Font and Fuller[1]

William Brune addressed Christian Ehlers: You showed the ozone isopleth diagram with NO_x and VOCs axes and the paths that a pollution plume might take through it. Can you think of a way to do an experiment to test if these paths are really what happens in the atmosphere?

Christian Ehlers responded: The path shown in the paper is a hypothetical pathway for an air-parcel originating from a pollution hotspot in an urban area and ending up in a forest. The starting point as well as the end-point in the forest can be related to measurements in the corresponding regions. The path in-between is based on two assumptions: starting from a pollution hotspot, a plume will be diluted while passing over an urban background region. Dilution here means mixing-in of air containing very much lower concentrations from the background (we have chosen RVOC = 3 s^{-1} and RNO_2 = 0.2 s^{-1}). This will influence both the VOC and the NO_2 concentrations in an equal manner. So for the first part the trajectory is following a dilution path almost approaching background conditions. Entering a region with biogenic influence, the VOC reactivity will rise due to the biogenic emissions. A further clarification of the assumed path could be done by either installing a network of measuring stations or using a mobile laboratory. In both cases the effort would be high, as not only NO_x measurements but also specified VOC measurements from a mobile-system would be needed to obtain the required time resolution. In the future our mobile laboratory could be used for such a study, following the wind from a pollution source to the endpoint in the forest, because we are working towards such a mobile GC system. It is proposed to complement this type of investigation by releasing an appropriate tracer to track the specific air parcels on their way.

Dwayne Heard commented: When you calculated the ozone production you allowed the model to run for 10 min to allow NO, NO_2 and OH radicals to reach a *quasi*-steady state. Did you vary this at all to investigate any influence of the value on the results? Also did you compare the calculated OH concentrations for the various combinations of VOC and NO_x OH reactivities with any previous field measurements of OH at suitable locations?

Christian Ehlers answered: We tried two different pathways: a) we used a traffic-VOC-mix and a biogenic-VOC-mix dominated by isoprene with 10 min time scale, and b) we used time scales between 5 and 15 min to test the attainability of the *quasi*-steady state. When all important boundary conditions (HCHO/R_VOC ratio, CO/R_VOC ratio, O_3, H_2O, temperature, H_2O_2, HONO, J values, k(NO_2+OH), k(O_3+NO)) were matched, the results were very similar. Regarding your second question on comparison to field observations – please refer to my response to the question from Marliyyah Mahmood.

Louisa Kramer commented: You mention in the paper that ambient NO_2 was detected after conversion to NO using either a molybdenum converter or a photolytic converter. Were the photolytic converters used to replace the molybdenum converters in the ambient monitors? Or were they used in different monitors? Given that molybdenum converters are known to convert other nitrogen oxide species, resulting in an overestimation of NO_2, would this have a large impact on your results?

Christian Ehlers responded: For all the measurements, photolytic converters were used. For the measurements located close to the emission sources, one would not expect the molybdenum converters to cause a substantial artifact. Substances causing such artifacts are not emitted from traffic in large amounts but produced photochemically. As a consequence the artifact will be largest in rural regions and not in urban areas.

Ruth Purvis asked: Please could you explain your techniques for measuring the heavier weight hydrocarbons more fully, especially the sampling systems? Have the losses due to sampling systems (WAS *vs.* tubes *vs.* online) been investigated? As there is discussion as to whether these heavier weight hydrocarbons are actually seen in diesel emissions, are there any plans for a comparison with other measurement systems to rule out instrument error?

Christian Ehlers replied: The exhaust was sampled onto carbotrap cartridges to analyse heavier compounds. The cartridges were heated up to 320°C for desorption in a thermodesorber (Gerstel, Mülheim, Germany). Before separation the compounds where refocussed on cooled empty glass liners which were heated to 350°C for desorption. Transfer lines were held at a temperature of 300°C to avoid compound losses. Liners and transfer lines were passivated with SilcoNert 2000 (SilTek, Bad Homburg, Germany). The separation was performed on a 60 m x 0.25 mm DB1 capillary column which was heated to 320°C. One sample was analysed by both GCs and compared to direct OH loss measurement using laser induced fluorescence (LIF). Here the observed OH loss rate could be explained by the total contribution of the individual compounds measured.

Tim Wallington said: You showed a large reduction of VOCs from traffic over the past 20 years. Do you have an opinion on what the next 20 years will bring regarding NO_x and VOC emission trends from vehicles?

Christian Ehlers responded: Regarding petrol vehicles, the VOC emissions can be attributed to cold start conditions and conditions where the catalytic converter is not working at full efficiency. Here a heating system for the catalyst could increase the efficiency and reduce the VOC emissions. The VOC emissions from new diesel vehicles are quite low and possible further means of reduction will have less effect on atmospheric VOC concentration. For the NO_x emissions, selective catalytic reduction systems as they are used for EURO6 diesel vehicles could be further improved to reduce emissions. For petrol cars these systems could reduce the NO_x emissions in the same way. One point which should be regarded for all the possible emission-reduction systems is the energy consumption of these systems which will in consequence increase the CO_2e-missions of such a vehicle.

Roy Harrison asked: Does this work inform us as to what the best strategy would be for the EU to adopt to control tropospheric ozone?

Christian Ehlers responded: In this study we calculated ozone production rates for air-masses close to the pollution sources and the pathway of the dilution to a biogenically dominated environment. The results of our calculation show that the decrease in the VOC/NO_x ratio resulted in decreasing ozone production rates for the regions close to the source and, as a consequence, a decrease of the highest peak concentrations of ozone. The results of this study showed a decreasing ozone production rate related to lower VOC/NO_x ratios for the urban areas. In rural regions where VOC emissions are biogenically dominated the effect could be opposite.

William Bloss said: Is there any significant influence from trends in background ozone levels on the modelled NO_2 abundance (NO_x speciation) for a given emission over the 1–2 decade timescale being explored?

Christian Ehlers replied: The background ozone trend seems to be small (see Fig. 1 in our paper) even on an 20-year time scale. But the NO_x speciation near traffic emissions is strongly influenced by the NO/NO_2 ratio from the car fleet and by the total concentration of NO_x reached at the hot-spots. Close to those hot-spots ozone is typically small and NO very large, sometimes larger than NO_2. As we explained in the paper, Fig. 26 was calculated without taking these effects into consideration. The ozone production rate at these locations would be altered *via* less OH production by ozone photolysis but at the same time affected positively by larger RO_2–HO_2 recycling by NO. We would expect the second effect to be larger, so we would expect larger ozone production rates than anticipated in these hot-spots. Since this effect depends on the concentrations of NO_x relative to background ozone, this problem would be restricted to NO_2 reactivities above 10 s^{-1} (equal to 44 ppb). Below that threshold, NO_x speciation and ozone concentration would approach normal conditions.

William Bloss addressed David Carslaw: The changing (reducing) urban decrement or titration of O_3 in urban environments also provides information on NO_x emissions and speciation - does this offer any additional constraint on trends in NO_x emissions over, for example, the past decade, or is this too spatially broad and imprecise to be useful?

David Carslaw answered: The more general role of O_3 on urban NO_2 trends is an important aspect of understanding the trends over the past decade. Any analysis needs to understand the competing influences of ozone and primary NO_2 emissions. The analysis in our paper using a simple constrained chemistry model accounts for the changes in ozone concentrations over time and the influence on concentrations of NO_2.

William Brune remarked: An experiment has been done on the East Coast of the US where coal-fired power plants have been replaced by natural gas power plants and the levels of NO_x have dropped dramatically. So has the ozone, even in the cities. So I think that the correct strategy to reduce ozone would be to get rid of NO_x.

Paul Monks responded: To add to the European context, it is clear that there is a distinction between the urban and "regional" contexts. In the urban context, reduction in NO_x raises ozone (due to a reduction in NO–O_3 titration) levels, while in the regional context, ozone is clearly mainly under NO_x control. Much of the current situation was summarised in our recent ozone review.[1]

1 P. S. Monks *et al.*, *Atmos. Chem. Phys.*, 2015, **15**, 8889–8973.

Brian McDonald said: In the U.S., emissions of nitrogen oxides (NO_x) have been coming down from motor vehicles along with volatile organic compounds (VOCs). In general, ozone has decreased in different regions of the country. More stringent VOC control relative to NO_x appears to have been particularly effective in reducing ozone in California. However, we're now starting to see faster NO_x control over VOCs, due to emission controls implemented on power plants, and fuel-switching from coal to natural gas, and new regulations requiring selective catalytic reduction (SCR) systems on heavy-duty trucks. Consequently, the Eastern U.S. appears to be having greater sensitivity to these NO_x reductions. Overall, for all regions to benefit, it is important to just control both NO_x and VOCs.

Tim Murrells commented: Going back to Tim Wallington's question about future vehicle emissions: we have made some emission projections for the UK and it's hard to see how we can do much more for reducing VOCs from road traffic sources, at least for the lighter ones that are regulated. Our projections show that there will be much greater VOC emissions from other sources that need to be targeted. The big issue for road transport is NO_x and whether the new Euro 6 regulations based on Real Driving Emission test procedures (RDE), to be introduced from 2018, will deliver the anticipated reductions from diesel cars and vans. Our NO_x emission projections are critically dependent on this.

André Prévôt commented: I disagree with the statement that we have done everything we could regarding VOCs in the transport sector. There is more room to improve VOCs from gasoline cars, especially during cold starts.

Ruth Doherty said: I would suggest caution with the usage of "region" in seeing ozone reductions with precursor emissions controls. For the Task Force on Hemispheric Transport of Air Pollution[1] we have performed experiments to assess the ozone response to substantial (20%) emission reductions. In many mid-latitude regions, notably considerable parts of China and the Middle East as well as areas in Europe, we find ozone increases in response to emissions reductions all year round (due to reduced NO_x titration). Only when we move to the sub-tropics do we see year round regional ozone decreases due to 20% emission reductions, possibly due to less limitation on photochemistry that would reproduce ozone from NO_2.

1 Hemispheric Transport of Air Pollution, www.htap.org

Rob MacKenzie asked Christian Ehlers: How will you communicate Fig. 26 of your paper to the transport sector and to policy-makers? The trajectories depicted on the figure seem to imply that the partial adherence to policy between 1994 and 2014 has had the "happy accidental consequence" of moving urban air to part of the "response space" in which ozone production is lower than it would have been if policy had been rigorously adhered to.

Christian Ehlers responded: In the 1990s, policy makers in Germany and elsewhere tried to decrease the concentration of ozone near the surface. At that time, ozone was known to be an important risk to human health and for the production of crops. For that reason, the catalytic reduction technique for VOC and NO_x emissions of petrol cars was introduced to meet the more and more stringent legislation for car emissions in Europe. Diesel cars were rare at that time and were believed to be environmentally preferable because they had fewer engine-emissions and lower fuel consumption compared to petrol cars. The subsequent shift in the passenger car fleet from petrol to diesel cars which took place in the last 20 years was therefore politically supported. The principle technical difficulty for diesel cars in reducing NO_x emissions was somehow overlooked. Although the photochemical principles shown in our Fig. 26 were known 20 years ago, we do not think that the strong reduction of VOC, and the very small reduction of NO_x, traffic emissions occurring in Germany were intended. Nevertheless, the original political intention to reduce the ozone problem was achieved. You may call that accidental. A good deal later than the first efforts for the reduction of ozone, there was a certain change in the line of action. Besides the regulation of car emissions, ambient concentrations of pollutants like NO_2 were recognized to be health risks on their own account and were therefore limited by the European legislation. And that is the situation nowadays.

Nick Hewitt remarked: There is a question as to whether or not street trees are beneficial to urban air quality – it is a highly topical and important question, but is difficult to answer. Trees in cities have at least four effects on air quality: 1) trees can enhance deposition of pollutants by increasing surface area and enhancing deposition velocities: modelling shows that enhanced deposition to trees can only have an appreciable effect on air quality when acting on a relatively small volume of air within a street canyon. 2) Trees are an important source of VOCs and can then contribute to SOA and ozone formation: whether this is significant for air quality depends on the local chemistry, but in any case any effects will probably occur downwind of the source region. 3) Trees planted in an urban street canyon can reduce ventilation and so reduce mixing of background air from above into the canyon, which can potentially dramatically increase street level concentrations of pollutants. This decrease in mixing is possibly the most important effect. 4) CFD modelling has shown that street trees can increase dispersion, hence reducing street level concentrations in some places and increasing them in others. It is therefore highly likely that planting significant numbers of trees in city centres may be beneficial to air quality in some circumstances, but detrimental to air quality in others. There is more information on this in work by Pugh *et al.*,[1] McDonald *et al.*[2] and Donovan *et al.*[3]

1 T. A. M. Pugh, A. Robert MacKenzie, J. Duncan Whyatt and C. Nicholas Hewitt, *Environ. Sci. Technol.*, 2012**46**, 7692–7699.
2 A. G. McDonald, W. J. Bealey, D. Fowler, U. Dragosits, U. Skiba, R. I. Smith, R. G. Donovan, H. E. Brett, C. N. Hewitt and E. Nemitz, *Atmos. Environ.*, 2007, **41**, 8455–8467.
3 R. G. Donovan , H. E. Stewart , S. M. Owen, A. R. MacKenzie and C. N. Hewitt, *Environ. Sci. Technol.*, 2005, **39**, 6730–6738.

María Cruz Minguillón responded: This is clearly an interesting point. To properly quantify the different effects caused by the presence of trees on the air quality, which may be opposite in terms of increasing or decreasing ambient PM concentrations, further research is needed. I would be interested in Neil Donahue's opinion on this.

Paul Monks commented: To add to the comments on the role that trees play in the urban atmosphere – a further role is illustrated in our recent paper[1] in that trees across an urban landscape can act to disperse pollution. In this modelling study it was shown that trees caused a net dispersion of 7% for a pollutant across urban Leicester.

1 A. P. R. Jeanjean, G. Hinchliffe, W. A. McMullan, P. S. Monks and R. J. Leigh, *Atmos. Environ.*, 2015, **120**, 1–14.

Nick Hewitt replied: It is important to recognise that placing trees in an urban street canyon can inhibit downward mixing of background air into the canyon, thus elevating street level pollutant concentrations, but can also cause greater turbulence within the canyon, aiding dispersion, as shown in the paper you mentioned. This latter effect will cause street level concentrations to fall in some places but to rise in others as the pollutant emissions become more effectively mixed at street level. Whether trees are good for urban air quality is obviously a difficult, and site specific, question to answer.

Andreas Skouloudis remarked: I would like to reassure you that we have used Ehlers' graphs, both during the first and second AutoOil programmes, and the industrial partners were fully aware of their importance for the emission reductions of ozone precursors. What is interesting is that this paper is the realisation of actual emission reductions over the years, and what these reductions meant for regional ozone.

Faraday Discussions

PAPER

Non-linear partitioning and organic volatility distributions of urban aerosols

S. Madronich,*[a] A. J. Conley,[a] J. Lee-Taylor,[a] L. I. Kleinman,[b] A. Hodzic[a] and B. Aumont[c]

Received 3rd December 2015, Accepted 17th December 2015

DOI: 10.1039/c5fd00209e

Gas-to-particle partitioning of organic aerosols (OA) is represented in most models by Raoult's law, and depends on the existing mass of particles into which organic gases can dissolve. This raises the possibility of non-linear response of particle-phase OA mass to the emissions of precursor volatile organic compounds (VOCs) that contribute to this partitioning mass. Implications for air quality management are evident: a strong non-linear dependence would suggest that reductions in VOC emission would have a more-than-proportionate benefit in lowering ambient OA concentrations. Chamber measurements on simple VOC mixtures generally confirm the non-linear scaling between OA and VOCs, usually stated as a mass-dependence of the measured OA yields. However, for realistic ambient conditions including urban settings, no single component dominates the composition of the organic particles, and deviations from linearity are presumed to be small. Here we re-examine the linearity question using volatility spectra from several sources: (1) chamber studies of selected aerosols, (2) volatility inferred for aerosols sampled in two megacities, Mexico City and Paris, and (3) an explicit chemistry model (GECKO-A). These few available volatility distributions suggest that urban OA may be only slightly super-linear, with most values of the normalized sensitivity exponent in the range 1.1–1.3, also substantially lower than seen in chambers for some specific aerosols. The rather low exponents suggest that OA concentrations in megacities are not an inevitable convergence of non-linear effects, but can be addressed (much like in smaller urban areas) by proportionate reductions in emissions.

1. Introduction

Organic aerosols (OA) are ubiquitous in the atmosphere and especially abundant in urban areas where they are a major part of air pollution.[1–4] They are associated with volatile organic compounds (VOCs), both those emitted directly and those

[a]National Center for Atmospheric Research, Boulder, Colorado, USA. E-mail: sasha@ucar.edu

[b]Brookhaven National Laboratory, Upton, New York, USA

[c]Laboratoire Interuniversitaire des Systèmes Atmospheriques, UMR 7583, CNRS, Université Paris Est Créteil et Université, Paris Diderot, 94010 Créteil, France

produced by atmospheric reactions of other VOCs. Great advances have been made over the past decade in understanding OA formation, removal, and properties through synergies of laboratory and chamber measurements, observations from networks, field campaigns, and satellite platforms, and modeling on multiple scales.[3–13] Still, the relationship between OA and VOCs is not fully understood, particularly for realistic urban mixtures. One of the simplest questions to an air quality model could be: if VOC emissions are changed by one percent, by what percent will OA change? The answer depends of course on so many factors that it is seldom forthcoming.

The possibility of a non-linear dependence of OA on VOCs appears within the absorptive partitioning model formulated by Pankow,[14,15] in which OA forms a quasi-ideal solution that further enhances gas-to-particle absorption. A corresponding non-linearity is found in smog chamber experiments, under conditions where the yield of OA is seen to increase with OA itself,[4,16] and often stated as a mass-dependent yield but is in fact a result from the same partitioning considerations. The question arises whether such non-linearities occur also in real urban atmospheres, and if they could be exacerbating already high pollution levels in megacities like Mexico City[17] or Beijing.[18] Even in less polluted environments, understanding and quantifying this non-linearity would seem a useful step in any regulatory cost-benefit analysis.

There are potentially other sources of non-linearity between OA and VOCs that can be identified and distinguished from those due to absorptive partitioning. For example, variations in VOCs could cause changes in OH concentrations, with a broad array of consequences for chemical reactions leading to the condensable species that ultimately form OA. These could include changes in ozone (O_3) and NO_x concentrations which in turn could affect both yields and product identities. Although there have been many modeling studies looking at the effect of reducing VOC emissions on OA concentrations [*e.g.*, ref. 19–22] it remains unclear how much of the response arises from non-linear partitioning *vs.* other chemical changes. Most 3D models[11] have fairly rudimentary parameterizations for OA formation from VOCs, with largely untested dependencies.

Within the partitioning model framework, the relationship between OA and VOCs is described by the volatility distribution, which is the amount of material (gas + particle) present across a spectrum of saturation concentrations (or vapor pressures, often divided in decadal bins). The volatility distribution is a defining characteristic of an aerosol composed of a mixture of many organic compounds. According to absorptive partitioning theory, mixture components with low volatility will condense nearly completely to the particle phase, while those with intermediate volatility will partition to the particle phase in proportion to the total particle mass, which includes the OA itself. Remarkably, very few volatility distributions for realistic aerosol mixtures have been reported. Here we examine a few available determinations of these volatility spectra, including two urban settings (Mexico City and Paris), several chamber studies, and estimates based on the chemical composition predicted by an explicit chemical box model. We use the volatility distributions to estimate the scaling of organic aerosols on VOC loading, $OA \sim VOC^S$, where S is a non-linear scaling or sensitivity exponent. The results are shown to be consistent with previously reported field-campaign observations of the rate of growth of OA downwind of intense pollution sources (Mexico City), compared to growth rates in less polluted regions (Eastern U.S.).

2. Modeling framework

Consider a mixture of organic species, indexed i, that may be present both in the gas and particle phases with respective amounts g_i and p_i totaling t_i,

$$g_i + p_i \equiv t_i \tag{1}$$

with all amounts expressed in identical units, *e.g.*, μg m^{-3}. According to quasi-ideal absorptive partitioning theory,[4,15] the gas to particle ratio is given by Raoult's law,

$$g_i/p_i = c_i\Big/\sum_i p_i \tag{2}$$

where c_i is the sub-cooled saturation amount of i. Combining these equations gives a system of non-linear equations,

$$p_i = t_i\Bigg/\left(1 + c_i\Big/\sum_i p_i\right) \tag{3}$$

coupled through the total particle phase, $\sum_i p_i$. For a single component system, eqn (3) reduces to

$$p_1 = t_1 - c_1 \tag{4}$$

showing that the particle mass increases linearly with the excess above threshold ($t_1 > c_1$), rather than being directly proportional to the total, t_1. For mixtures of multiple components, eqn (3) must be solved numerically for each p_i as a function of c_i and t_i; threshold effects then become less pronounced, but may still manifest as a stronger-than-linear increase in particle amount p_i for locally small increases in t_i.

A typical urban aerosol mixture may contain thousands of different compounds, and in principle the summations in eqn (2) and (3) extend over all species. Following Donahue *et al.*,[23] and noting that the sub-cooled saturation amount of a species, c_i, is a quasi-intensive property, the individual species can be grouped together according to similar c_i values. Here we also use decadal c_i spacing centered on integer values of $\log_{10} c_i$, with the understanding that the subscript i now refers to these decadal volatility bins, rather than to individual molecular species. The volatility distribution, t_i, is the total amount of organic material (gas + particle) present in each volatility bin c_i, before consideration of partitioning, and is determined by prior processes, *e.g.* direct emissions, or as a myriad of complex intermediates from the gas-phase oxidation of VOCs. Depending on the functional form $t_i(c_i)$, gas–particle partitioning may be linear or non-linear to various extents.

The potential non-linearity is illustrated in Fig. 1 by solving eqn (3) for the simple case of two components, a non-volatile seed ($c_1 = 0$, $p_1 = t_1 = 1$) and a partly volatile compound with saturation concentration c_2. The figure shows how the partly volatile compound will partition between the gas and particle phases, as a function of the total available material (t_2 = gas + particle), with each curve representing a possible compound having a different value of c_2. Two distinct regimes are evident: if seed is abundant (specifically if $c_2 \lesssim p_1 = 1$), the

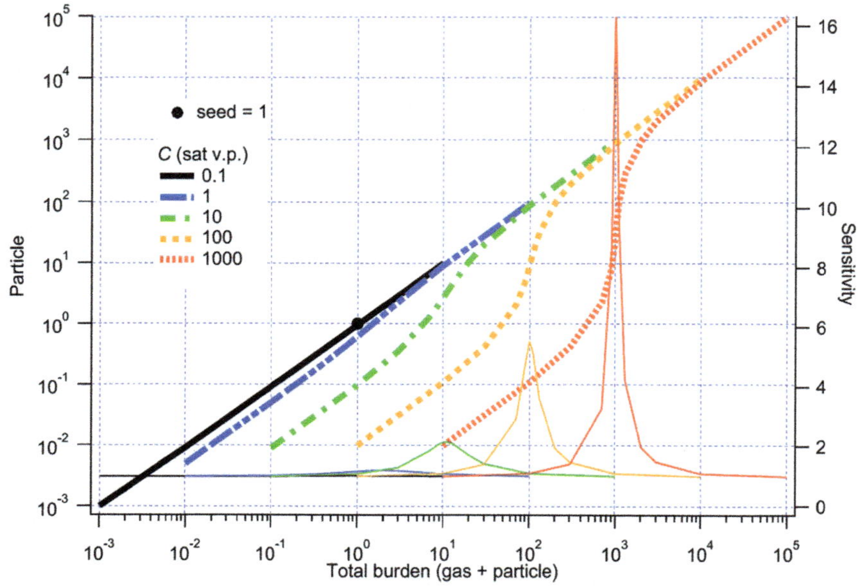

Fig. 1 Idealized gas–particle partitioning in a two-component system where one component is not volatile (seed), and the volatility c of the other component is varied. The amount partitioned to the particle is shown on the left scale, while the thin curves and right scale shows local sensitivity (log–log slope) to changes in the total burden t. Units are arbitrary if consistent but for definiteness can be taken as µg m^{-3} for all quantities.

growth will be linear with essentially complete partitioning to the particle phase, over a wide range of t_2. On the other hand, if c_2 is significantly larger than the seed, non-linear growth can occur as t_2 approaches c_2. In this case, for small burdens ($t_2 < c_2$) the particle amount is still sensitive to the amount of seed, but becomes saturated and therefore simply linear when $t_2 > c_2$.

We define a normalized sensitivity coefficient S as the logarithmic derivative

$$S = \mathrm{d}\ \ln P/\mathrm{d}\ \ln T \tag{5}$$

where $P = \sum_i p_i$ and $T = \sum_i t_i$. For simplicity, variations in T were applied uniformly across all species and therefore vapor pressure bins, specifically by multiplying each t_i by the same factor, X, so that $S = \mathrm{d}\ \ln P/\mathrm{d}\ \ln X$. This is also illustrated in Fig. 1 (right axis), where for the case $c_2 = 1000$ the sensitivity coefficient reaches a value of 16 (such large near-threshold values are apparently not realized in the atmosphere according to our analyses below).

The Generator of Explicit Chemistry and Kinetics of Organics – Atmosphere (GECKO-A) model was used to simulate the gas-phase chemistry and gas–particle partitioning using a zero-dimensional (box) framework.[24,25] Two cases are considered representing (i) anthropogenic VOCs in Mexico City and its surroundings (MILAGRO field campaign),[17] and (ii) a terpene-dominated forest in Colorado (BEACHON-ROCS field campaign);[26] both simulations have been described in detail by Lee-Taylor *et al.*[27] The Mexico City simulation consisted of two phases, an urban "Eulerian" period lasting 2 days that included emissions, dry deposition, diurnal variation of temperature and PBL height, as well as entrainment (dilution) in both vertical and horizontal dimensions, followed by a 4 day "Lagrangian" outflow that included only horizontal dilution and diurnally invariant temperature. GECKO-A calculates the chemical transformations of gases and particles, tracking a large number ($\approx 10^5$) of explicit species for which saturation concentrations are known or estimated, allowing their assignment to any volatility binning scheme. For the BEACHON simulations, the biogenic precursor mix was dominated by α- and β-pinene, limonene, and carene, with trace amounts of isoprene and methyl butenol, based on observations.[26]

3. Results and discussion

3.1. Volatility distributions

Volatility distributions have been inferred in controlled chambers using several techniques, including secondary OA (SOA) growth as a function of reagent consumed, from evaporated fractions remaining through thermal denuders, or from dilution experiments.[28] All of these methods have complications, *e.g.*, need to estimate enthalpy of vaporization for thermal denuders, changes in chemical composition in all cases, and wall losses. To our knowledge, only two studies reported volatility distributions of OA from ambient urban air: from the Mexico City T0 MILAGRO site,[29] and from a suburban MEGAPOLI site near Paris.[30] From models, the only reported urban volatility distributions appear to be those based on our explicit chemical mechanism GECKO-A for MILAGRO and BEACHON.[27]

Reported volatility distributions for various organic aerosols are shown in Table 1, and are essentially the values of t_i in the notation used here. Measurements from different studies are not always comparable directly because they differ in their total amount, $T = \sum_i t_i$, as shown in the table under the heading "Sum". Volatility distributions were normalized to $T = 1.0\ \mu g\ m^{-3}$ by Presto and Donahue,[31] Grieshop *et al.*,[32] and Paciga *et al.*[30] but to $T = 2.5\ \mu g\ m^{-3}$ by Robinson *et al.*[8] The precise normalization is unimportant, as T can contain large contributions from highly volatile compounds that make only small contributions to the particle amount, $P = \sum_i p_i$. What is important, however, is that T be scaled to give (*via* eqn (3)) particle amounts that are typical of urban atmospheres, *e.g.* approximately in the typical range 1–10 $\mu g\ m^{-3}$ of OA. Therefore, the t_i values from Table 1 were multiplied by various scale factors, then used in eqn (3) to calculate the corresponding p_i values, which were finally summed to obtain the total particle-phase amount, P. The results, shown in Fig. 2, are curves describing the growth of particle amount P as a function of the total amount T. Some curvature is seen even on logarithmic scales, so the slopes were evaluated over the typical range 1–10 $\mu g\ m^{-3}$, as also indicated by the placement of the slope values in the figure. These log–log slopes are the sensitivity factors, S, giving the (%) response of particle amount to variations in precursors, and are seen to fall in the range 1.3–1.8. These are rather high values that could have some regulatory relevance if typical of urban environments.

The volatility distribution for OA collected in Mexico City during MILAGRO was normalized by Cappa and Jimenez[29] to a particle mass of 17 $\mu g\ m^{-3}$, the campaign-average OA value. Values of t_i were obtained directly from their Fig. 5

Table 1 Volatility distributions for organic aerosols, μg m^{-3}

		$\log_{10} c_i$															
	Ref.	−7	−6	−5	−4	−3	−2	−1	0	1	2	3	4	5	6	Sum	Sensitivity
Chamber, alpha-pinene + ozone. SOA formation *vs.* VOC consumed	31						0.005	0.01	0.04	0.1	0.2	0.4				0.71	1.75
Dilution of diesel POA	8						0.03	0.06	0.1	0.15	0.2	0.3	0.4	0.5	0.8	2.54	1.34
Wood smoke, thermal denuder	32						0.05	0	0.05	0.25	0.15	0.5	0	0.1		1.00	1.64
Diesel exhaust and lubricating oil, dilution and thermal denuder	32						0.01	0.01	0.04	0.21	0.18	0.5	0.1			1.00	1.79
Mexico City (MILAGRO) total organic carbon (OC)	29		1.5	1.6	1.6	1.7	1.8	2	2.5	4.1	8.3	20				45.10	1.17
HOA	29		0.7	0.8	0.9	1	1.3	2.1	3.5	6.5	12	22				50.80	1.28
BBOA	29			0.1	0.3	0.6	1.3	2.4	4.3	7.5	14	24				54.50	1.35
OOA	29		1.8	1.8	1.9	1.9	2	2.1	2.2	2.9	4.3	7.2				28.10	1.10
LV-OOA	29	4.5	3.5	2.5	1.7	1	0.6	0.3	0.2	0.1						14.40	1.01
SV-OOA	29		1.2	1.5	1.8	2.1	2.8	3.4	4.4	5.8	7.5					30.50	1.14
Paris (MEGAPOLI) HOA-summer	30				0.12	0.13	0.08	0.03	0.05	0.58						0.99	1.08
SVOOA-summer	30				0.06	0.14	0.15	0.13	0.18	0.34						1.00	1.19
LVOOA-Summer	30	0.2	0.25	0.28	0.26	0.04										1.03	1.00
COA-summer	30			0.13	0.15	0.08	0.2	0.08	0.37							1.01	1.14
MOA-summer	30				0.03	0.03	0.05	0.28	0.41	0.2						1.00	1.52
HOA-winter	30				0.11	0.09	0.07	0.12	0.11	0.5						1.00	1.20
COA-winter	30		0.12	0.11	0.14	0.43	0.12	0.1								1.02	1.02
BBOA-winter	30				0.2	0.09	0.08	0.14	0.1	0.4						1.01	1.13
OOA-winter	30				0.3	0.09	0.07	0.09	0.1	0.35						1.00	1.14
MILAGRO day 1 GECKO-A model	27				0.2	0.5	0.693	1.6	1.87	3.1	6.38	5	9.4	15	20	63.98	1.27
BEACHON day 1 GECKO-A model	27				0.05	0.1	0.178	0.21	0.27	0.28	0.35	0.5	0.8	1.4	1.4	5.52	1.21

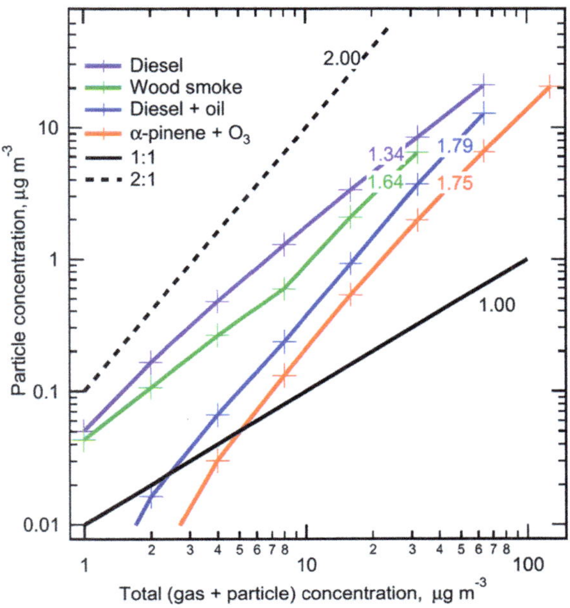

Fig. 2 Growth of particle mass computed for the volatility distributions measured in chambers, including secondary organic aerosols formed by ozonolysis of α-pinene,[31] diesel exhaust,[8] and wood smoke and diesel/oil mixtures.[32] Crosses show where gas–particle partitioning was calculated. Lines with log–log slopes of 1 and 2 are also shown, with arbitrary vertical offsets.

and used to compute p_i (by solving eqn (3)) and P (which was confirmed as ≈17 μg m^{-3}). Superimposed variations of t_i (10% increments) were used to evaluate the sensitivity $S = \mathrm{d}\ln P/\mathrm{d}\ln T$, with values shown in Table 1 to range from 1.00 for low-volatility oxygenated SOA, to 1.35 for biomass burning aerosols. On the basis of these sensitivity factors, the non-linearity in Mexico City is judged to be fairly small and decreasingly important (approaching unity) in the order BBOA, HOA, OA, SV-OOA, OOA, LV-OOA (respectively biomass burning, hydrocarbon-like, all, semivolatile, oxygenated, and low-volatility OA).

The volatility distribution in and near Mexico City was also computed with the GECKO-A model.[27] The saturation vapor pressure for each non-radical molecule was calculated using the method of Nannoolal *et al.*,[33] followed by grouping of individual molecule concentrations into volatility bins. This procedure was repeated at each time step, allowing us to follow the time evolution of the volatility spectrum. Table 1 shows the volatility distribution at the end of two days of urban processing for Mexico City, and after one day in the terpene-dominated Colorado forest. Both cases show rather low sensitivity exponents, 1.27 for Mexico City and 1.21 for the forest, due to the substantial amounts of low volatility organics, *cf.* with c_i values of 0.1 μg m^{-3} or lower. Fig. 3 shows how the Mexico City volatility evolves over time, for urban air (day 1) and in the outflow after several days of chemical processing. The plume-integrated mass is shown to account for dilution. On the first day, there is relatively little material of very low saturation concentration c_i, but this fraction increases progressively during several days due to ongoing oxidation leading to molecules that are generally more functionalized,

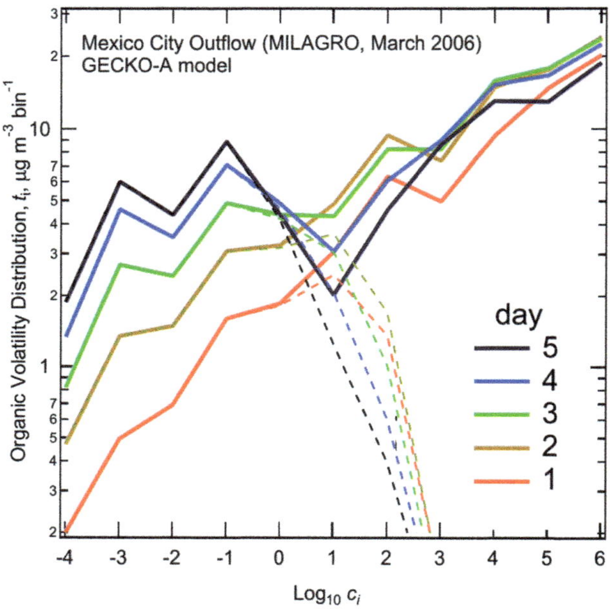

Fig. 3 Volatility distribution of organic aerosols in Mexico City outflow predicted by the GECKO-A model,[27] shown as the total amount available t_i (particle + gas) in each decadal saturation range (bin) c_i. The distribution is shown for five days, the first of which represents the urban environment while the later ones represent multiday outflow. Dashed curves show particle phase.

with higher O/C ratios,[9] higher oxidation states,[10] higher polarizability,[34] higher Henry's law coefficients,[35] and lower vapor pressures.[36] Growth in the lower volatility bins is evident and is due to photochemical transformations from higher volatility bins, *e.g.* from 0–2 on the $\log_{10} c_i$ scale, which are seen to decrease from day 2 to day 5.

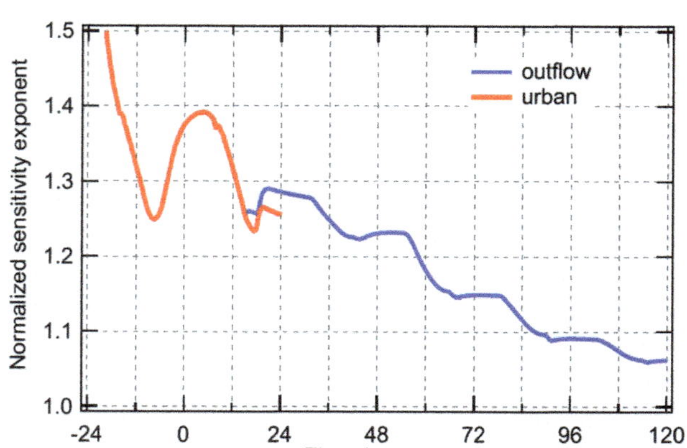

Fig. 4 Normalized sensitivity coefficient of the production of organic aerosols by VOCs, within Mexico City (red) and its outflow (blue), calculated using volatility distributions from the GECKO-A model.[27] Negative times represent model spin-up.

The effect of the aging on the sensitivity coefficient is seen in Fig. 4. As might be expected, the increasing fraction of particles with very low volatility acts effectively as a seed for the more volatile compounds, and moves the partitioning further into the seed-dominated nearly linear regime. The highest sensitivities are found in the urban setting (the first day) and may be enhanced by the colder night-time temperatures. On the other hand, the high initial sensitivities may be related to model spin-up with low seed during the first few hours of the simulations.

Organic volatility measurements have also been recently reported for a suburban site 20 km SW of Paris center.[30] Values given in Table 1 (estimated from their Fig. 5) are normalized to unity, but were then scaled until the average observed particulate mass concentrations (given in their Table 1) were reached, individually for HOA, OOA (including SV-OOA and LV-OOA), BBOA, OA from cooking (COA), and occasionally of marine origin (MOA). Thus, the sensitivity calculation was carried out for values of t_i scaled to give the observed particle masses, with additionally a 10% perturbation to obtain S. With exception of MOA, all exhibit near-linear sensitivities, in the range 1.0–1.2. Among these, the lowest values are associated with LV-OOA, and somewhat higher values for HOA, but also with some unexpected trends *e.g.*, the sensitivity of SV-OOA exceeding that of HOA in summer, that are probably in any case within the range of uncertainties. The case of the marine aerosol is interesting because of its generally more volatile nature, but it should be noted that this represented only a small fraction (a tenth or less) of the total average particulate mass in that study.

3.2. Other evidence and uncertainties

To summarize Table 1, the few available volatility distributions for Mexico City and Paris imply that OA may be only slightly super-linear, with most values of the sensitivity exponent in the range 1.1–1.3, notably lower than seen in chambers for some specific aerosols. There are a couple of other lines of evidence that are consistent with this result.

Kleinman *et al.*[37] used measurements of OA and carbon monoxide (CO) from the DOE G1 aircraft, to compare the rate of production of OA in and near Mexico City to that in the less-polluted Eastern U.S. With observed OA growth rates normalized by ΔCO (plume CO – background CO) concentrations to account for differences in emissions and dilution, the ratio OA/ΔCO is seen to increase over the course of several days downwind from sources, due to ongoing chemistry providing condensable gases. Kleinman *et al.*[37] estimated the first-day growth rate in the OA/ΔCO ratio was 62 $\mu g\ m^{-3}\ ppm^{-1}$ for Mexico City, essentially identical to the values estimated for the NEAQS2002 (66 $\mu g\ m^{-3}\ ppm^{-1}$)[38] and NEAQS/ITCT2004 (60 $\mu g\ m^{-3}\ ppm^{-1}$)[39,40] field campaigns in the U.S., where CO levels were on average lower by more than a factor of two. Thus, their analysis does not support the notion that OA production is disproportionately strong in a megacity, at least to the extent that CO may be considered a surrogate for VOCs. Large uncertainties in these estimates were recognized and discussed by Kleinman *et al.*,[37] and include, for example, the need to use the NO_x/NO_y ratio to estimate when photochemical age reaches one day. Furthermore, emission ratios of many VOCs relative to CO are significantly larger in Mexico City than in the U.S.,[41] complicating direct comparison between the different locations. The apparently good numerical agreement between the three campaigns may thus be fortuitous. Nevertheless, their tentative conclusion, that Mexico City does not display a disproportionate rate of OA production due to non-linear partitioning, is consistent with the conclusions drawn from the volatility distributions, that non-linearity if any is minor.

It might be possible to detect non-linear partitioning using other observations as surrogates, *e.g.* correlations between particulate organic carbon (OC, the carbon content of OA) and elemental carbon (EC, typically as soot). Because EC is emitted directly, it may bear a simple proportionality to the emitted VOC pool, while OC is subject to non-linear partitioning similarly to OA. A major advantage of considering this ratio is that the OC and EC measurements are made rather frequently, and usually as a pair, due to relatively simple instrumentation such as thermo-optical analysis.[42,43] Thus, OC/EC correlations could provide some indication of OA/VOC scaling. A brief literature review did not show any obvious non-linear relationship, with large variability with OC/EC ratios in the range 1–10.[44–46] However, many factors are known to affect this ratio and measurements from different locations may not necessarily be comparable.[44,47]

There are several reasons why the above volatility-based sensitivity exponents, small as they are, could be over-estimates. Most notably, variations were made to all t_i simultaneously, as if they were all from the same pool of precursors. In reality, one often regulates only a subset, *e.g.* aromatics but not biogenics, so the overall sensitivity may be smaller.[4] But there are also reasons why these could be under-estimates. One problem can be seen in Fig. 4, near the beginning of the simulation. Non-linearities are stronger during these first few hours despite the imposition of initial seed (2 $\mu g\ m^{-3}$), and may be sensitive to it until enough semi-volatile mass condenses to overwhelm any seed effect. Some caution in this early regime is warranted to ensure that model representations are not overly sensitive to poorly constrained parameters such as initial seed. How this initiation occurs is not completely clear and a number of additional complexities are possible, *e.g.* from new particle formation processes.

More generally, the sensitivity exponents computed here from volatility distributions are local, in the sense that they predict the response of gas–particle partitioning to small changes in their total amount without regard for the history of the air parcel. A more global sensitivity analysis could consider responses of OA to changes in initial conditions (or emissions), or changes in the nature of the seed. While there may be advantages to each approach, the local sensitivities used here lend themselves to easy interpretation in terms of local partitioning.

4. Conclusions

The urban volatility distributions derived from measurements in Mexico City and Paris, and from simulations with the GECKO-A explicit chemical model, all indicate that in complex urban mixtures the organic particle mass increases nearly linearly with the VOC burden, with a sensitivity exponent $\approx 1.2 \pm 0.2$, considerably lower than chamber systems. Such linearity implies that there are no disproportionate benefits of emission reductions. Rather, megacity OA pollution is not an inevitable convergence of non-linear effects, but can be addressed (much like in smaller urban areas) by rational and proportionate reductions in emissions.

Photochemistry is the main driving force for the change in volatility distributions, pushing organic material from higher volatility to lower volatility, at least in the urban environment. In Table 1, aerosols that have undergone intense photochemical processing (OOA and its variants) show sensitivities very close to unity, while HOA, BBOA, and MOA – all presumably less processed by tropospheric chemistry – show higher sensitivities. Ultimately, the role of photochemistry may reverse, leading to net fragmentation rather than further functionalization.[48,49] Recent studies suggest that OA chemistry may be more dynamic than previously believed, on both the formation and removal sides,[50] and it is likely that volatility distributions also evolve over the lifetime of the aerosol.

However, there are still large uncertainties in all methods used, whether based on volatility spectra from thermal denuders, OA/CO from aircraft observations, or models. Very few measurements of OA volatility distributions in urban (or any) environments have been reported. Given that most 3D models treat gas–particle partitioning using the equations described here (or slight variants thereof), additional comparisons between modeled and observed volatility distribution may be useful to evaluate whether such models can reasonably represent the sensitivity of OA to changing emissions of VOCs.

Acknowledgements

The National Center for Atmospheric Research is sponsored by the National Science Foundation. SM and JLT were partly supported by a grant from the US Department of Energy, Office of Science BER/DE-SC0006780.

References

1 G. M. Hidy, B. R. Appel, R. J. Charlson, W. E. Clark, S. K. Friedlander and D. H. Hutchinson, *J. Air Pollut. Control Assoc.*, 1975, **25**, 1106–1114.
2 S. I. Lamb, C. Petrowski, I. R. Kaplan and B. R. T. Simoneit, *J. Air Pollut. Control Assoc.*, 1980, **30**, 1098–1115.
3 B. Finlayson-Pitts and J. N. Pitts, *Chemistry of the Upper and Lower Atmosphere*, Academic Press, 1999.
4 J. H. Seinfeld and S. N. Pandis, *Atmospheric Chemistry and Physics: From Air Pollution to Climate Change*, John Wiley and Sons, Hoboken, New Jersey, 2nd edn, 2006.
5 M. C. Jacobson, H.-C. Hansson, K. J. Noone and R. J. Charlson, *Rev. Geophys.*, 2000, **38**, 267–294.
6 M. Kanakidou, J. H. Seinfeld, S. N. Pandis, I. Barnes, F. J. Dentener, M. C. Facchini, R. van Dingenen, B. Ervens, A. Nenes, C. J. Nielsen, E. Swietlicki, J. P. Putaud, Y. Balkanski, S. Fuzzi, J. Horth, G. K. Moortgat, R. Winterhalter, C. E. L. Myhre, K. Tsigaridis, E. Vignati, E. G. Stephanou and J. Wilson, *Atmos. Chem. Phys.*, 2005, **5**, 1053–1123.
7 S. Fuzzi, M. O. Andreae, B. J. Huebert, M. Kulmala, T. C. Bond, M. Boy, S. J. Doherty, A. Guenther, M. Kanakidou, K. Kawamura, V.-M. Kerminen, U. Lohmann, L. M. Russell and U. Pöschl, *Atmos. Chem. Phys.*, 2006, **6**, 2017–2038, DOI: 10.5194/acp-6-2017-2006.

8 A. L. Robinson, N. M. Donahue, M. K. Shrivastava, E. A. Weitkamp, A. M. Sage, A. P. Grieshop, T. E. Lane, S. N. Pandis and J. R. Pierce, *Science*, 2007, **315**, 1259–1262.
9 J. L. Jimenez, *et al.*, *Science*, 2009, **326**, 1525–1529.
10 J. H. Kroll, N. M. Donahue, J. L. Jimenez, S. H. Kessler, M. R. Canagaratna, K. R. Wilson, K. E. Altieri, L. R. Mazzoleni, A. S. Wozniak, H. Bluhm, E. R. Mysak, J. D. Smith, C. E. Kolb and D. R. Worsnop, *Nat. Chem.*, 2011, **3**, 133–139, DOI: 10.1038/nchem.948.
11 K. Tsigaridis, *et al.*, *Atmos. Chem. Phys.*, 2014, **14**, 10845–10895, DOI: 10.5194/acp-14-10845-2014.
12 S. N. Pandis, N. M. Donahue, B. N. Murphy, I. Riipinen, C. Fountoukis, E. Karnezi, D. Patouliasa and K. Skyllakoua, *Faraday Discuss.*, 2013, **165**, 9–24, DOI: 10.1039/c3fd00108c.
13 R. Zhang, G. Wang, S. Guo, M. L. Zamora, Q. Ying, Y. Lin, W. Wang, M. Hu and Y. Wang, *Chem. Rev.*, 2015, **115**, 3803–3855.
14 J. F. Pankow, *Atmos. Environ.*, 1987, **21**, 2275–2283.
15 J. F. Pankow, *Atmos. Environ.*, 1994, **28**, 185–188.
16 J. R. Odum, T. Hoffmann, F. Bowman, D. Collins, R. C. Flagan and J. H. Seinfeld, *Environ. Sci. Technol.*, 1996, **30**, 2580–2585.
17 L. T. Molina, S. Madronich, J. S. Gaffney, E. Apel, B. de Foy, J. Fast, R. Ferrare, S. Herndon, J. L. Jimenez, B. Lamb, A. R. Osornio-Vargas, P. Russell, J. J. Schauer, P. S. Stevens and M. Zavala, *Atmos. Chem. Phys.*, 2010, **10**, 8697–8760.
18 S. Guo, *et al.*, *Proc. Natl. Acad. Sci. U. S. A.*, 2014, **111**, 17373–17378, DOI: 10.1073/6 pnas.1419604111.
19 K. Nguyen and D. Dabdub, *Aerosol Sci. Technol.*, 2002, **36**, 560–572.
20 J. Chen, J. Lu, J. C. Avise, J. A. DaMassa, M. J. Kleeman and A. P. Kaduwela, *Atmos. Environ.*, 2014, **92**, 182–190.
21 C. L. Blanchard, G. M. Hidy, S. Shaw, K. Baumann and E. S. Edgerton, *Atmos. Chem. Phys.*, 2016, **16**, 215–238, DOI: 10.5194/acp-16-215-2016.
22 S. Aksoyoglu, J. Keller, D. C. Oderbolz, I. Barmpadimos, A. S. H. Prévôt and U. Baltensperger, *Int. J. Environ. Pollut.*, 2012, **50**, 451–459, DOI: 10.1504/ijep.2012.051215.
23 N. M. Donahue, A. L. Robinson, C. O. Stanier and S. N. Pandis, *Environ. Sci. Technol.*, 2006, **40**, 2635–2643.
24 B. Aumont, S. Szopa and S. Madronich, *Atmos. Chem. Phys.*, 2005, **5**, 2497–2517.
25 B. Aumont, M. Camredon, C. Mouchel-Vallon, S. La, F. Ouzebidour, R. Valorso, J. Lee-Taylor and S. Madronich, *Faraday Discuss.*, 2013, **165**, 105–122, DOI: 10.1039/c3fd00029j.
26 J. Ortega, A. Turnipseed, A. B. Guenther, T. G. Karl, D. A. Day, D. Gochis, J. A. Huffman, A. J. Prenni, E. J. T. Levin, S. M. Kreidenweis, P. J. DeMott, Y. Tobo, E. G. Patton, A. Hodzic, Y. Y. Cui, P. C. Harley, R. S. Hornbrook, E. C. Apel, R. K. Monson, A. S. D. Eller, J. P. Greenberg, M. C. Barth, P. Campuzano-Jost, B. B. Palm, J. L. Jimenez, A. C. Aiken, M. K. Dubey, C. Geron, J. Offenberg, M. G. Ryan, P. J. Fornwalt, S. C. Pryor, F. N. Keutsch, J. P. DiGangi, A. W. H. Chan, A. H. Goldstein, G. M. Wolfe, S. Kim, L. Kaser, R. Schnitzhofer, A. Hansel, C. A. Cantrell, R. L. Mauldin and J. N. Smith, *Atmos. Chem. Phys.*, 2014, **14**, 6345–6367.

27 J. Lee-Taylor, J. A. Hodzic, S. Madronich, B. Aumont, M. Camredon and R. Valorso, *Atmos. Chem. Phys.*, 2015, **15**, 595–615, DOI: 10.5194/acpd-14-17999-2014.
28 M. Bilde, *Chem. Rev.*, 2015, **115**, 4115–4156.
29 C. D. Cappa and J. L. Jimenez, *Atmos. Chem. Phys.*, 2010, **10**, 5409–5424.
30 A. Paciga, E. Karnezil, E. Kostenidou, L. Hildebrandt, M. Psichoudaki, G. J. Engelhart, B.-H. Lee, M. Crippa, A. S. H. Prévôt, U. Baltensperger and S. N. Pandis, *Atmos. Chem. Phys.*, 2016, **16**, 2013–2023, DOI: 10.5194/acp-16-2013-2016.
31 A. A. Presto and N. M. Donahue, *Environ. Sci. Technol.*, 2006, **40**, 3536–3543.
32 A. P. Grieshop, M. A. Miracolo, N. M. Donahue and A. L. Robinson, *Environ. Sci. Technol.*, 2009, **43**, 4750–4756.
33 Y. Nannoolal, J. Rarey and D. Ramjugernath, *Fluid Phase Equilib.*, 2008, **269**(1–2), 117–133.
34 J. F. Pankow and K. C. Barsanti, *Atmos. Environ.*, 2009, **43**, 2829–2835.
35 A. Hodzic, B. Aumont, C. Knote, J. Lee-Taylor, S. Madronich and G. Tyndall, *Geophys. Res. Lett.*, 2014, **41**, 4795–4804, DOI: 10.1002/2014gl060649.
36 R. Valorso, B. Aumont, M. Camredon, T. Raventos-Duran, C. Mouchel-Vallon, N. L. Ng, J. H. Seinfeld, J. Lee-Taylor and S. Madronich, *Atmos. Chem. Phys.*, 2011, **11**, 6895–6910.
37 L. I. Kleinman, S. R. Springston, P. H. Daum, Y.-N. Lee, L. J. Nunnermacker, G. I. Senum, J. Wang, J. Weinstein-Lloyd, M. L. Alexander, J. Hubbe, J. Ortega, M. R. Canagaratna and J. Jayne, *Atmos. Chem. Phys.*, 2008, **8**, 1559–1575.
38 L. I. Kleinman, P. H. Daum, Y.-N. Lee, G. I. Senum, S. R. Springston, J. Wang, C. Berkowitz, J. Hubbe, R. A. Zaveri, F. J. Brechtel, J. Jayne, T. B. Onasch and D. Worsnop, *J. Geophys. Res.*, 2007, **112**, D09310, DOI: 10.1029/2006jd007786.
39 A. P. Sullivan, R. E. Peltier, C. A. Brock, J. A. de Gouw, J. S. Holloway, C. Warneke, A. G. Wollny and R. J. Weber, *J. Geophys. Res.*, 2006, **111**, D23S46, DOI: 10.1029/2006jd007072.
40 R. J. Weber, A. P. Sullivan, R. C. Peltier, A. Russell, B. Yan, M. Zheng, J. de Gouw, C. Warnke, C. Brock, J. S. Holloway, E. L. Atlas and E. Edgerton, *J. Geophys. Res.*, 2007, **112**, D13302, DOI: 10.1029/2007jd008408.
41 J. A. de Gouw, D. Welsh-Bon, C. Warneke, W. C. Kuster, L. Alexander, A. K. Baker, A. J. Beyersdorf, D. R. Blake, M. Canagaratna, A. T. Celada, L. G. Huey, W. Junkermann, T. B. Onasch, A. Salcido, S. J. Sjostedt, A. P. Sullivan, D. J. Tanner, O. Vargas, R. J. Weber, D. R. Worsnop, X. Y. Yu and R. Zaveri, *Atmos. Chem. Phys.*, 2009, **9**, 3425–3442.
42 T. Novakov and C. E. Corrigan, *Microchim. Acta*, 1995, **119**, 157–166.
43 A. Karanasiou, M. C. Minguillón, M. Viana, A. Alastuey, J.-P. Putaud, W. Maenhaut, P. Panteliadis, G. Močnik, O. Favez and T. A. J. Kuhlbusch, *Atmos. Meas. Tech. Discuss.*, 2015, **8**, 9649–9712, DOI: 10.5194/amtd-8-9649-2015.
44 C. Pio, M. Cerqueira, R. M. Harrison, T. Nunes, F. Mirante, C. Alves, C. Oliveira, A. Sanches de la Campa, B. Artiñano and M. Matos, *Atmos. Environ.*, 2011, **45**, 6121–6132.
45 Z. Niu, F. Zhang, Z. Kong, J. Chen, L. Yin and L. Xu, *J. Environ. Monit.*, 2012, **14**, 2961–2967.

46 Z. Pusheng, D. Fan, Y. Yadong, H. Di, Z. Xiujuan, Z. Wenong, Y. Qing and L. Huaiyu, *Atmos. Environ.*, 2013, **71**, 389–398.
47 M. C. Day, M. Zhang and S. N. Pandis, *Atmos. Environ.*, 2015, **12**, 317–325.
48 A. Hodzic, S. Madronich, P. S. Kasibhatla, G. Tyndall, B. Aumont, J. L. Jimenez, J. Lee-Taylor and J. Orlando, *Atmos. Chem. Phys.*, 2015, **15**, 9253–9269.
49 K. Dzepina, C. Mazzoleni, P. Fialho, S. China, B. Zhang, R. C. Owen, D. Helmig, J. Hueber, S. Kumar, J. A. Perlinger, L. J. Kramer, M. P. Dziobak, M. T. Ampadu, S. Olsen, D. J. Wuebbles and L. R. Mazzoleni, *Atmos. Chem. Phys.*, 2015, **15**, 5047–5068, DOI: 10.5194/acp-15-5047-2015.
50 A. Hodzic, P. S. Kasibhatla, D. S. Jo, C. Cappa, J. L. Jimenez, S. Madronich and R. J. Park, *Atmos. Chem. Phys. Discuss.*, 2015, **15**, 32413–32468.

Faraday Discussions

PAPER

Modelling component evaporation and composition change of traffic-induced ultrafine particles during travel from street canyon to urban background

Irina Nikolova,* A. Rob MacKenzie,† Xiaoming Cai, Mohammed S. Alam and Roy M. Harrison‡

Received 13th November 2015, Accepted 14th December 2015
DOI: 10.1039/c5fd00164a

We developed a model (CiTTy-Street-UFP) of traffic-related particle behaviour in a street canyon and in the nearby downwind urban background that accounts for aerosol dynamics and the variable vapour pressure of component organics. The model simulates the evolution and fate of traffic generated multicomponent ultrafine particles (UFP) composed of a non-volatile core and 17 Semi-Volatile Organic Compounds (SVOC, modelled as *n*-alkane proxies). A two-stage modelling approach is adopted: (1) a steady state simulation inside the street canyon is achieved, in which there exists a balance between traffic emissions, condensation/evaporation, deposition, coagulation and exchange with the air above roof-level; and (2) a continuing simulation of the above-roof air parcel advected to the nearby urban park during which evaporation is dominant. We evaluate the component evaporation and associated composition changes of multicomponent organic particles in realistic atmospheric conditions and compare our results with observations from London (UK) in a street canyon and an urban park. With plausible input conditions and parameter settings, the model can reproduce, with reasonable fidelity, size distributions in central London in 2007. The modelled nucleation-mode peak diameter, which is 23 nm in the steady-state street canyon, decreases to 9 nm in a travel time of just 120 s. All modelled SVOC in the sub-10 nm particle size range have evaporated leaving behind only non-volatile material, whereas modelled particle composition in the Aitken mode contains SVOC between $C_{26}H_{54}$ and $C_{32}H_{66}$. No data on particle composition are available in the study used for validation, or elsewhere. Measurements addressing in detail the size resolved composition of the traffic emitted UFP in the atmosphere are a high priority for future research. Such data would improve the representation of these particles in dispersion models and provide the data essential for model validation. Enhanced knowledge of the

University of Birmingham, School of Geography, Earth and Environmental Sciences, Edgbaston, B152TT, Birmingham, UK. E-mail: i.nikolova@bham.ac.uk

† Also at Birmingham Institute of Forest Research, Edgbaston, B152TT, Birmingham, UK.

‡ Also at Department of Environmental Sciences/Center of Excellence in Environmental Studies, King Abdulaziz University, PO Box 80203, Jeddah, 21589, Saudi Arabia.

chemical composition of nucleation-mode particles from diesel engine exhaust is needed to predict both their atmospheric behaviour and their implications for human health.

Introduction

Current air quality regulations are set in terms of $PM_{2.5}$ or PM_{10} mass concentrations. However, those mass measurements reflect primarily the accumulation mode and coarse mode particles. Regulatory measures for ultrafine particles (UFP, diameter $D_p < 100$ nm) have not been introduced, mainly due to insufficient knowledge of the human toxicity of these particles. Recent study suggests that ultrafine particles, by number count, play an important role in the toxicity of airborne particulate matter[1] showing a link between cardiovascular health outcomes with UFP exposure. The majority of UFP in the urban atmosphere arise from road traffic emissions, and road traffic is also recognised to be the largest source of ultrafine particles in the UK national emissions inventory.[2] At a London roadside site, 71.9% of particles by number arise from traffic, with 38% of those in the nucleation mode ($D_p < 30$ nm) and 53% in the exhaust solid mode (Aitken, $30 < D_p < 100$ nm).[3] Despite large scientific interest, and many contributions to the topic of UFP over the last few decades, the behaviour of ultrafine particles still remains an area with open questions and even contradictory opinions. Specifically open questions remain as to what extent the composition affects the evolution, properties and atmospheric behaviour of UFPs. The chemical composition of particles from engine exhaust is complex, but traffic emitted particles can be approximated as multicomponent organic solutions, coated around a refractory graphitic core. Rapid new particle formation and condensation, with a characteristic time-scale of seconds, is responsible for the formation of UFP in the near-wake region of the vehicle. Studies in the wake of vehicles[4–6] show that primary UFP can grow substantially due to condensation of semi-volatile organic compounds (SVOC). Nonetheless, those SVOC are largely uncharacterised because they are unresolved by traditional gas chromatography and often referred to as an unresolved complex mixture (UCM). New methods have been developed to investigate which compounds are encapsulated in the UCM, such as two dimensional gas chromatography, a promising method to resolve this complex mixture.[7] Worton and co-authors[8] have reported that traffic induced UFP at a tailpipe are rich in semi-volatile organic compounds, predominantly *n*-alkanes, alkyl-cyclohexanes and aromatics. Our own measurements of diesel exhaust composition, analysed with two dimensional gas chromatography, point as well to the wealth of SVOC.[9] The presence of semi-volatile components within vehicle exhaust particles is responsible for their growth and also has implications for their behaviour in the atmosphere. During the process of dilution downstream from the source, a reduction in vapour and particle concentrations will occur, leading to a shift in favour of evaporation of the SVOC.[10]

The evaporation process reduces the diameter of the UFP, with implications for the efficiency with which they are taken up by inhalation.[11] A study of Dall'Osto and co-authors[12] reports particle size spectra at three sites (street canyon, urban park and tower) in London, UK. They demonstrated a very significant shift in particle size distributions towards smaller sizes during advection from a traffic source to a nearby urban park, at a distance of about 665 metres. The sub-10 nm mode at Regents Park (urban park in London) could potentially arise from localised nucleation processes, but the diurnal pattern and relationship to gaseous pollutants did not show any evidence to support this concept. Particle concentrations correlated with NO_x and black carbon point to a traffic source.[12] These findings merit further investigation, by means of numerical simulations, to characterise the evaporative potential and composition changes of the UFP during the advection from the street canyon to the urban park. Typically, in urban areas, high volumes of traffic in street canyons result in some of the highest concentrations of UFP due to the limited dilution of the traffic exhaust particles. These hot spots are of great interest due to the possible adverse health effects being attributed to such exposure. Therefore an investigation of the dispersion of UFP and their dynamics within such environments is of growing importance.

A number of studies have reported the modelling of ultrafine particle concentrations in street canyons,[13–17] showing great variations in spatial and temporal scales. Other studies have focused on aerosol dynamics (mainly deposition and coagulation), showing the effect of each process on the size distribution.[18–20] A general consensus is that removal processes such as dilution and dry deposition should be considered at street scale.[21] However these studies did not study the volatility of the particles and their complex size-dependent composition given the wealth of traffic-emitted SVOC. There is an apparent gap in the realistic representation of composition of traffic-generated UFP among the street canyon studies. Our aim is to address this gap and develop a model of traffic particle behaviour in a street canyon and in the near vicinity that accounts for the dynamics of the nucleation-mode particles composed predominantly of organic compounds with a small core of metallic ash or sulphuric acid, with robust insight into the evolution and fate of traffic induced multicomponent UFP in the presence of vapour-phase SVOC. We simulate the evaporation and composition change of multicomponent organic particles in realistic atmospheric conditions and put forward recommendations for further studies.

Methodology

Field measurements: area and meteorology

Data were taken from a measurement campaign which was part of the REPARTEE multi-faceted study of aerosols and gases in the atmosphere of London, UK.[22] Briefly, air sampling was carried out in 2007 between October and November at Marylebone Road and Regents Park, London. The sampling site at Marylebone Road (a street canyon of near unity height/width ratio running approximately WSW-ENE) is on the kerbside of a heavily trafficked London road carrying about 80 000 vehicles per day. Traffic count data were provided from inductive loops in the road. On average the share of Heavy Duty Vehicles was 13.7%. UFP counts and size distributions in the street canyon were determined using Condensation Particle Counters (model TSI 3022A) and a Scanning Mobility Particle Sizer (SMPS, Model 3080 Electrostatic Classifier). The sampling site at Regents Park is about 665 m (Euclidean distance) to the north of the street canyon. It is in the centre of a large area of park, and air sampling was from an elevated inlet with a Differential Mobility Particle Sizer consisting of two "Vienna"-style DMAs – an ultrafine DMA for particles of 3.4–34 nm and a standard DMA for particles of

30–830 nm, with TSI model 3025 and 3010 condensation particle counters respectively.

Meteorological data, such as synoptic wind speed and wind direction, were obtained from the Heathrow airport site (UK). Days with southerly wind direction, approximately perpendicular to the street canyon axis, were selected in order to evaluate the advection and behaviour of UFP from Marylebone Road to Regent's Park. Mean wind speed for the selected days was 3 m s^{-1}, but the lowest wind speed was 1.5 m s^{-1}. These wind speeds correspond to minimum travel times from the street canyon to the nearby park of between 222 s and 443 s, respectively. The importance of above-roof-top wind speed and traffic-induced turbulence is addressed in several studies.[23–26] These studies show that for low wind speeds, typically below 1.5 m s^{-1}, traffic induced turbulence plays a more prominent role in the dilution process. Traffic-induced turbulence was not considered in the present study and its effect on the dilution is neglected. More details on the UFP number counts, size distributions and meteorological conditions can be found elsewhere.[12]

Model description

The model system consists of the street canyon model CiTTy-Street,[27] which is the urban version of the CiTTyCAT trajectory model,[28] coupled with a UFP aerosol dynamics module. CiTTy-Street is an urban canopy chemistry model designed to capture the exchange across the street/rooftop concentration gradients in pollutants and in-canyon pollutant deposition of gases, $PM_{2.5}$ and PM_{10}. CiTTy-Street consists of two boxes: the lower box represents either a single canyon or a range of street canyons of height H and width W within a city. The upper box represents the urban boundary layer, typically with a depth of 1000 m. Mixing between the boxes is parameterised using a dimensionless air exchange rate[29] for an above-roof wind perpendicular to the along-canyon axis, which produces in-canyon concentrations matching long-term observations (http://www2.dmu.dk/atmosphericenvironment/Trapos/datadoc.htm;[27]). We have further developed the model to include size-resolved mass and number of UFP. The size range and size resolution considered by the UFP module is user-defined; currently, the UFP module consists of 15 discrete uniform size bins in a logarithmic scale in the size range of 5.8–578 nm. Processes such as dry deposition, coagulation and condensation/evaporation in the presence of SVOC are considered. Dry deposition is based on the resistance scheme described by Seinfeld and Pandis[30] and used by Nikolova and co-authors.[20] The deposition velocity contains a variety of physical processes and is a function of the aerodynamic resistance r_a, the quasi-laminar resistance r_b and the sedimentation speed. Particles deposit on different surfaces like ground, building walls and roofs; the effect of green infrastructure on deposition[28] can be accommodated by varying the deposition resistances. Polydisperse coagulation is also taken into consideration, following the mathematical description in discrete form provided by Seinfeld and Pandis[30] and briefly described by Nikolova and co-authors.[20]

The UFP module also accounts for the multicomponent nature of the particles. The driving force for condensation/evaporation is the difference between the partial pressure of each representative SVOC component and its saturation vapour pressure. SVOC measurements of vapour and particulate phases were not conducted during the REPARTEE campaign. However, Harrad and co-authors[31] reported SVOC measurements from Birmingham, which is the second largest city in the UK. Vapour and particulate phases were measured at two sites: Bristol road (BROS), next to a busy traffic road; and at the Birmingham university campus (EROS), an urban background site. The range of SVOC varies from C_{16} to C_{34}, and we have considered SVOC up to C_{32} (Table 1). Harrison and co-authors[32] show that these SVOC are actively partitioning between vapour and particle and are representative of the volatility range of compounds in both ambient airborne and diesel exhaust particles. BROS and EROS SVOC vapour and particulate measurements were used as an initial (in the street canyon) and inlet background (above roof-top) concentration, respectively. Particles in the model are composed of 17 representative SVOC, however, one more component was added to the system to represent a non-volatile component in the nucleation and Aitken modes in and above the street canyon. The composition of the particles in the nucleation mode is initialised with 1% non-volatile material (by volume), while Aitken-mode particles are mainly non-volatile, having 90% non-volatile material (Table 2), which is discussed in detail in the Results and discussion. These non-volatile particles have grown because of the condensation of semi-volatile material, mainly hydrocarbons, during the dilution stage. Typically, if the pre-existing aerosol concentration is low, homogeneous nucleation is favoured and new particle formation occurs. However, high concentrations of pre-existing particles promote the condensation of semi-volatile compounds onto these pre-existing particles. Close to the tailpipe, 1–3 s after emission, supersaturation may be large enough to trigger nucleation. In addition, in the presence of a large surface area of particles, the SVOC and sulphuric acid will condense quickly on the particles, as discussed by Zhang and Wexler.[36] In our study, we have assumed that all processes related to the formation of new particles and subsequent growth and/or

Table 1 Mean vapour and particulate phases of SVOC measured at a busy traffic road (BROS) and urban background (EROS) sites in Birmingham, year 1999/2000. In parenthesis is the standard deviation (+/−) of the mean. Unit: ng m^{-3}

	BROS vapour	BROS particulate	EROS vapour	EROS particulate
C16	6.42 (3.62)	0.65 (0.55)	2.47 (1.46)	0.45 (0.41)
C17	9.32 (4.31)	0.85 (0.54)	3.67 (2.36)	0.58 (0.41)
C18	9.36 (4.41)	0.82 (0.43)	3.61 (2.12)	0.47 (0.28)
C19	9.74 (4.09)	1.32 (1.02)	4.07 (1.96)	0.57 (0.38)
C20	8.35 (3.59)	1.93 (1.53)	3.13 (1.41)	0.69 (0.59)
C21	6.20 (3.06)	2.63 (1.84)	2.96 (1.61)	1.31 (1.23)
C22	3.97 (1.98)	3.25 (1.64)	1.64 (0.86)	1.51 (1.54)
C23	2.32 (1.25)	3.73 (1.63)	1.49 (1.35)	2.67 (2.15)
C24	1.59 (1.79)	3.34 (1.70)	0.88 (1.25)	2.51 (2.29)
C25	1.20 (0.52)	4.46 (2.41)	0.78 (0.69)	4.05 (3.42)
C26	1.00 (0.50)	2.37 (1.45)	0.74 (0.43)	3.01 (3.74)
C27	1.28 (0.53)	3.61 (2.46)	0.84 (0.45)	4.49 (4.16)
C28	1.03 (0.53)	1.89 (1.32)	0.76 (0.48)	2.51 (3.42)
C29	1.24 (0.53)	4.89 (3.84)	0.79 (0.46)	5.19 (5.06)
C30	0.80 (0.38)	2.03 (1.26)	0.54 (0.37)	1.94 (2.21)
C31	0.77 (0.30)	3.60 (3.23)	0.50 (0.32)	3.81 (4.78)
C32	0.42 (0.18)	1.32 (0.89)	0.28 (0.21)	1.07 (1.31)

Table 2 Particles composition for nucleation mode (NM) and Aitken mode (AiM) for all input organic components expressed as % of the volume concentration. Above roof top composition is used for initial and inlet background concentrations

	Particulate mass fractions NM		Particulate mass fractions AiM	
Component	Street canyon	Above roof top	Street canyon	Above roof top
Non-volatile	1.0	1.0	90.0	90.0
C16	1.5	1.2	0.3	0.3
C17	2.0	1.6	0.8	1.0
C18	1.9	1.3	0.5	0.5
C19	3.1	1.5	1.1	1.4
C20	4.5	1.8	0.4	0.7
C21	6.1	3.5	0.8	1.2
C22	7.5	4.1	0.6	0.8
C23	8.7	7.2	1.0	1.1
C24	7.8	6.7	0.8	0.7
C25	10.4	10.9	0.9	0.7
C26	5.5	8.1	0.8	0.4
C27	8.4	12.1	0.6	0.4
C28	4.4	6.7	0.5	0.2
C29	11.4	13.9	0.3	0.2
C30	4.7	5.2	0.2	0.1
C31	8.3	10.3	0.2	0.2
C32	3.1	2.9	0.2	0.1

condensation onto pre-existing particles have completed before particles reach the kerbside where size distributions are measured. Condensation/evaporation with mass accommodation coefficient $\alpha = 1$,[33] and accounting for the Kelvin effect is treated with a fully-moving-diameter scheme[34] and redistribution onto fixed size bins in a mass and number conserving approach.[35] Steady-state simulations are performed using the CiTTy-Street-UFP model in and above the street canyon and results for the UFP number size distributions, size resolved composition of the particles and aerosol dynamics are presented in the Results and discussion. In order to mimic the advection of particles from above roof-top (above Marylebone Road) to the nearby urban park (Regent's Park), we have allowed particles from the upper box (above roof-top) to evolve, using a fully Lagrangian microphysics scheme, for a period equal to the travel time between the two sites. The Lagrangian microphysics scheme allows particles of UFP to grow/evaporate to their exact size.

Results and discussion

Stage 1: modelling in-canyon processes

The steady state model output was analysed in the street canyon (Marylebone Road) and the size distribution is plotted in Fig. 1. In addition, on the same figure we present the 'inlet' background size distribution (*i.e.*, the background size distribution that is advected into the above-roof box by the mean wind) and the size-resolved emission distribution, in order to illustrate the relative influences on the steady state size distribution from the background and the emissions,

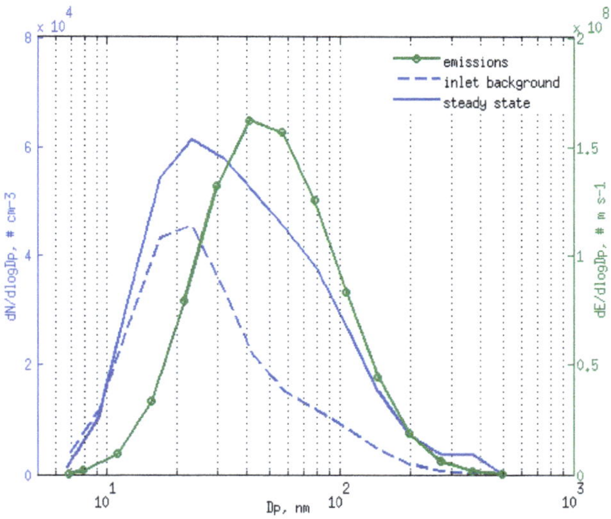

Fig. 1 Steady state number size distribution in the street canyon (Marylebone Road) for a cross-canyon wind speed of 1.5 m s^{-1}. On the same figure, the inlet background size distribution and the size resolved emission distribution are plotted to illustrate the relative influences on the steady state size distribution from the background and the emissions, respectively.

respectively. The size resolved emissions are initialised with a bi-modal log normal distribution, with geometric mean diameters in nucleation and Aitken modes at 35 nm and 65 nm, respectively (Fig. 1). The modelled size distribution peaks in the nucleation mode with a geometric mean diameter at 23 nm and is followed by a small shoulder in the Aitken mode. The nucleation mode, which is semi-volatile, makes an important contribution to the total particle number close to traffic sources.[3] The total number of particles in the street canyon at steady state for a wind speed of 1.5 m s^{-1} is 54 500 # cm^{-3}.

Role of deposition and coagulation in the street canyon

Deposition and coagulation are the processes that reduce the number of particles per size bin and therefore the total number of particles. Their overall effect depends on factors such as particle number concentration, particle diameter, state of the atmosphere, wind speed, the nature of the deposition surface, *etc.* For $PM_{2.5}$ and chemical tracers, Pugh and co-authors[27] describe the relative importance of in-canyon deposition as a function of canyon retention-time, defined and discussed below. We have examined the effect of deposition and coagulation on the steady state particle size distribution inside the street canyon (*cf.* Fig. 1) for a wind speed of 1.5 m s^{-1} and 3.0 m s^{-1}, respectively. The steady state total number concentration without deposition and coagulation is 59 400 # cm^{-3} and 35 200 # cm^{-3}, respectively. Deposition reduces the total number concentration by 4.1% (for wind speed 1.5 m s^{-1}) and 3.4% (for wind speed 3 m s^{-1}) - see Table 3. The rate of coagulation is found to be faster than that of deposition for a wind speed of 1.5 m s^{-1}, with an overall 4.8% reduction in the total number concentration at steady-state. However, the role of coagulation in the well-mixed street

Table 3 Reduction in total number of particles in the street canyon (%) for different wind speeds and accounting for different processes. All comparisons are against a base case steady-state solution without coagulation or deposition

Wind speed, m s^{-1}	Coagulation + deposition	Coagulation	Deposition	Total number UFP, # cm^{-3}
1.5	8.2	4.8	4.1	59 400
2	6	2.7	3.5	47 200
3	4.7	1.7	3.4	35 200

canyon is diminished with an increase in the wind speed. Particle concentration decreases only by 1.7% for a wind speed of 3 m s^{-1}. The overall reduction of particle numbers due to the combined effect of deposition and coagulation is 8.2% for 1.5 m s^{-1} and 4.7% for 3 m s^{-1} wind speeds. Deposition is found to be faster than coagulation for the steady state size distribution inside the street canyon for wind speeds above 2 m s^{-1} (Table 3). In terms of reduction of particle concentration per size bin, losses due to deposition at low wind speed (1.5 m s^{-1}) for particles with a diameter of 6.7 nm can reach up to 19%, while losses due to coagulation are higher, approximately 25%. With the increase of particle diameter, losses per size bin decrease as shown in Fig. 2. In the case of a wind speed of 3 m s^{-1}, deposition losses in the smallest size bin (6.7 nm) are approximately 20%, while the loss due to coagulation is about 15% (Fig. 3). Deposition remains faster in reducing the number of particles per size bin than coagulation for particles with a diameter below 50 nm.

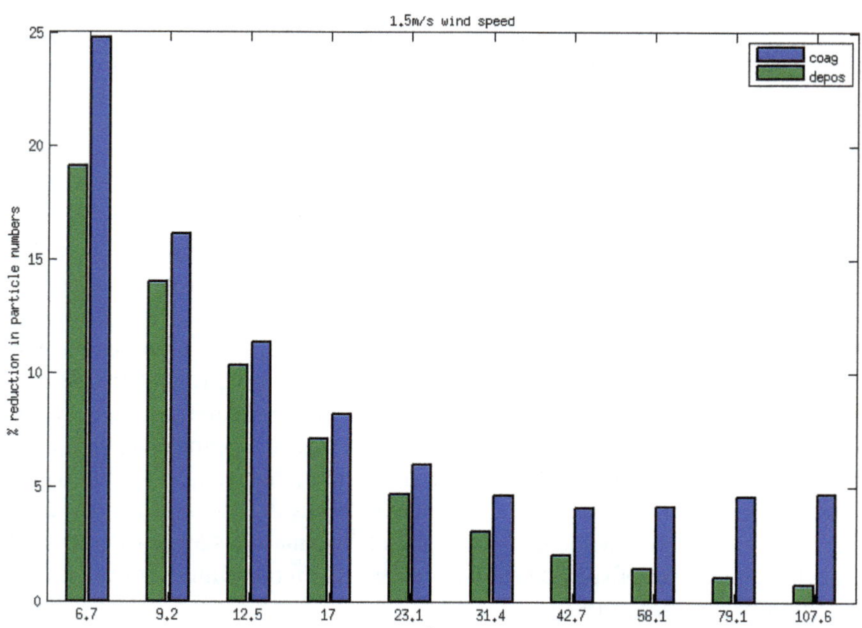

Fig. 2 Reduction in particle number concentration in the street canyon (%) due to coagulation and deposition per size bin at a wind speed of 1.5 m s^{-1}.

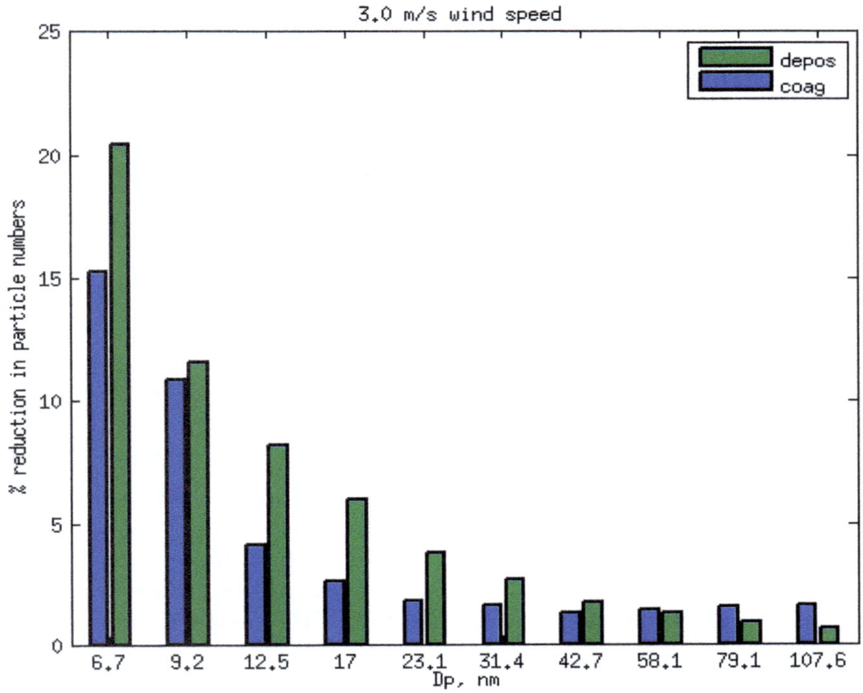

Fig. 3 Reduction in particle number concentration in the street canyon (%) due to coagulation and deposition per size bin at a wind speed of 3 m s^{-1}.

There are different opinions on the importance of deposition and coagulation in the literature. For example Gidhagen and co-authors[13] found that losses at kerbside due to deposition and coagulation are overall small and typically less than 10% with deposition being faster for higher wind speeds than coagulation. Deposition and coagulation were most effective for the smallest ultrafine particles, which is in line with our findings. Nikolova and co-authors[16] show that during the 'road-to-ambient' dilution stage, dilution is the major mechanisms in altering the aerosol size distribution, while coagulation and deposition play minor roles. Kumar and co-authors[37] also support the view that coagulation at street scale could be neglected especially when particles with a diameter of less than 10 nm are not considered. Based on the urban scale modelling study of Ketzel and Berkowicz,[21] deposition plays a more pronounced role in reducing the total particle number concentration than coagulation. Several studies[20,38–40] agree on the negligible role of coagulation in the street environment. However, Li and co-authors[41] have measured particle size distributions at different heights in a street canyon and have argued that the shift in the peak diameter in the Aitken mode to larger diameters is attributed to the effect of coagulation. Stagnant air flows favour an increased particle number concentration through reduced vertical dispersion that may promote particle to particle coagulation. The role of deposition in Li and co-authors[41] in reducing particle number concentration was not addressed and the measured decrease in particle number concentration with height was mainly attributed to dilution and coagulation. Zhu and co-authors[42] reported that atmospheric dilution and coagulation played substantial roles in the rapid decrease of the ultrafine particle concentration and transformation of the size distribution; however, their study was near to a highway. Overall, the removal of particles due to deposition and coagulation depends on a number of factors such as particle number concentration, particle diameter, ventilation in and out of a street canyon, the complexity of the urban area which influences the air flow between buildings, *etc.*[43,44] and so cannot, in general, be neglected. Nonetheless, all studies agree that the prime mechanism of reducing particle number concentration is dilution. Time-scale analysis elucidates the relative importance of the removal processes, as suggested in Ketzel and Berkowicz.[19] We have evaluated the characteristic timescales for dilution, deposition, and coagulation. Dilution is the fastest process with a time scale of 147 s for a wind speed of 1.5 m s^{-1} and canyon height of 22 m, followed by coagulation (250 s), and then deposition (288 s). Dilution remains the fastest process in reducing the particle number concentration for a wind speed of 3 m s^{-1}, with a time scale of 73 s, while deposition and coagulation have much slower time scales of 320 s and 500 s, respectively. That is, as above-roof wind speed increases, the scale separation between dilution and the microphysical processes also increases, leading to an increasing dominance of dilution over other processes in determining the steady-state. Another useful parameter in the diagnosis of the model dynamic system is the so-called retention time $\tau = H/\mathrm{we}$, where H is the building height and we is the exchange velocity, which is linearly proportional to the background reference wind speed above the roof level. Cai[45] shows that the exchange coefficient (*i.e.*, the constant of proportionality between we and the above-roof mean wind), based on large eddy simulation (LES) for a unity canyon aspect ratio, is about 0.01 for a neutrally stratified atmosphere. For $H = 22$ m and a wind speed at roof-top of 1.5 m s^{-1}, τ is approximately 1467 s. For a wind speed at roof-top of 3 m s^{-1}, τ is about 733 s. In both cases we see a large retention time in the street canyon and, hence, the opportunity for the ultrafine particles to reach a steady state as well as for in-canyon microphysics to influence the size distribution within and above the canyon. Pugh and co-authors[27] (see especially their Fig. 3) find that in-canyon steady-state $PM_{2.5}$ mass concentrations are a strong function of deposition velocity for $\tau > 100$ s. Given the strong dependence of deposition efficiency on particle size in the ultra-fine size range (Fig. 2 and 3), we can deduce that increasing the deposition surface without increasing τ, *e.g.* by judicious use of green infrastructure, will greatly reduce nucleation-mode number concentrations.

Stage 2: above roof-top and downwind advection: the role of SVOC on particle behaviour

Above roof-top, the steady-state size distribution is plotted in Fig. 4 along with the inlet background size distribution. The modelled total number of particles above roof-top is ~30 500 # cm^{-3}, *i.e.*, 56% of the in-canyon steady-state concentration. Particles from above roof-top are subject to condensation/evaporation for a time similar to that for advection between the street canyon (Marylebone Road) and urban park using a fully Lagrangian size-distribution solver. For the lowest wind speed of 1.5 m s^{-1} and distance between the two sites of 665 m, the travel time is at least 443 s. The distribution after 443 s is plotted in Fig. 4. A remarkable shift in

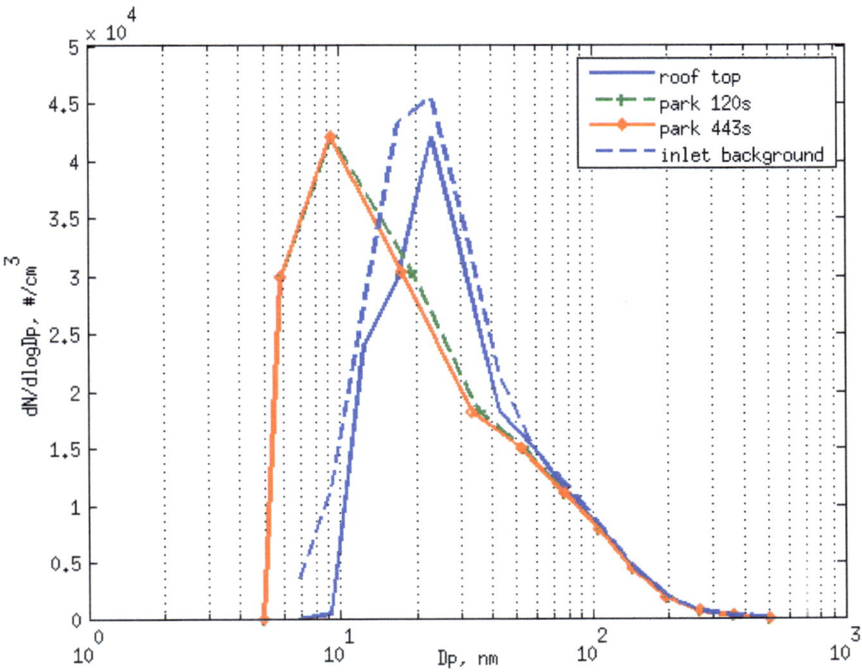

Fig. 4 Particle size distributions due to the mixing of particles between the street canyon air and above roof-top (labelled 'roof-top'), the inlet background size distribution ('inlet background') and during advection from above roof-top to the urban park plotted at two selected times.

the nucleation-mode particles is modelled, with very little effect on particle diameter in the Aitken mode. The modelled 23 nm geometric mean diameter in the nucleation mode has shrunk to ~9 nm, consistent with the measurements in the urban park, which showed a peak at $D_p \sim$ 7–8 nm.[12] Moreover, the remarkable shift in particle diameter in the nucleation mode happens in ~120 s, which is much faster than the travel time between the street canyon and the park. Vertical mixing in the urban boundary layer is neglected due to the relatively large time scale involved, which is *ca.* 10 min under conditions of high turbulence and more than 20 min under stable conditions.[46] For a wind speed of 3 m s^{-1} at roof-top, the travel time to the urban park would be ~222 s, nearly twice as long as the modelled time of ~120 s for the detection of a major diameter shift. In the sub-10 nm-diameter particle size range, the *n*-alkanes in the range $C_{16}H_{34}$–$C_{22}H_{46}$ have entirely evaporated and the particles are solely composed of the non-volatile material (see Table 4). Particles with an initial $D_p = 31$ nm have shrunk to $D_p = 18$ nm and are composed of a non-volatile material and the slowly evaporating $C_{31}H_{64}$ and $C_{32}H_{66}$. Aitken-mode particles continue to contain $C_{26}H_{54}$–$C_{32}H_{66}$ SVOCs, although in relatively low concentrations compared to the concentration of non-volatile material. *n*-Alkanes in the range $C_{24}H_{50}$–$C_{32}H_{66}$ can be found in all particles in the accumulation mode ($D_p > 100$ nm), while $C_{23}H_{48}$ is present in particles with $D_p \geq 143$ nm at the end of the simulation.

Table 4 Component mass-size distributions, dM/dlog D_p (ng m^{-3}), after advection from the street canyon to the urban park. D_p is particle diameter (in nm) after evaporation to exact size. 'D_p < 5 nm' indicates that the distribution goes beyond the lower bin bound and therefore concentration for all components is set to 0. Initial D_p (in nm): particle diameter at roof-top at steady state. SVOC concentration for particles with D_p > 5 nm are set to 0 in the table when the concentration for that component is less than 0.005 ng m^{-3}. dlog D_p = 0.1337

Non-volatile mass, μg m^{-3}	dM/dlog D_p, individual component, ng m^{-3}																	Total mass, μg m^{-3}	D_p, nm	Initial D_p, nm
	C32	C31	C30	C29	C28	C27	C26	C25	C24	C23	C22	C21	C20	C19	C18	C17	C16			
0	0.00	0.00	0.00	0.00	0.00	0.00	0.00	0.00	0.00	0.00	0.00	0.00	0.00	0.00	0.00	0.00	0.00	0	<5	7
0	0.00	0.00	0.00	0.00	0.00	0.00	0.00	0.00	0.00	0.00	0.00	0.00	0.00	0.00	0.00	0.00	0.00	0	<5	9
0	0.00	0.00	0.00	0.00	0.00	0.00	0.00	0.00	0.00	0.00	0.00	0.00	0.00	0.00	0.00	0.00	0.00	0	<5	12
0.00309	0.00	0.00	0.00	0.00	0.00	0.00	0.00	0.00	0.00	0.00	0.00	0.00	0.00	0.00	0.00	0.00	0.00	0.00309	6	17
0.0173	0.00	0.00	0.00	0.00	0.00	0.00	0.00	0.00	0.00	0.00	0.00	0.00	0.00	0.00	0.00	0.00	0.00	0.0173	9	23
0.0847	0.42	0.14	0.00	0.00	0.00	0.00	0.00	0.00	0.00	0.00	0.00	0.00	0.00	0.00	0.00	0.00	0.00	0.0852	18	31
0.3407	4.51	6.46	0.91	0.27	0.00	0.30	0.00	0.00	0.00	0.00	0.00	0.00	0.00	0.00	0.00	0.00	0.00	0.3531	33	43
1.074	6.34	12.73	3.36	2.30	0.35	2.34	0.00	0.00	0.00	0.00	0.00	0.00	0.00	0.00	0.00	0.00	0.00	1.101	52	58
2.597	6.10	11.92	4.84	4.95	1.31	5.60	0.16	0.00	0.00	0.00	0.00	0.00	0.00	0.00	0.00	0.00	0.00	2.632	77	79
4.783	6.79	10.04	5.80	5.91	2.87	9.02	0.33	0.18	0.11	0.00	0.00	0.00	0.00	0.00	0.00	0.00	0.00	4.824	105	108
6.700	8.56	11.03	7.69	7.68	5.54	14.64	0.87	0.27	0.16	0.13	0.00	0.00	0.00	0.00	0.00	0.00	0.00	6.757	143	146
7.136	9.15	11.70	8.75	9.21	8.14	18.99	2.80	0.71	0.19	0.14	0.00	0.00	0.00	0.00	0.00	0.00	0.00	7.206	195	199
5.780	7.52	9.75	7.55	8.41	8.36	17.83	5.77	2.72	0.38	0.12	0.00	0.00	0.00	0.00	0.00	0.00	0.00	5.848	265	271
3.563	4.68	6.14	4.85	5.63	6.01	12.09	6.61	5.30	1.33	0.13	0.00	0.00	0.00	0.00	0.00	0.00	0.00	3.616	360	369

Discussion

The entrained lubricating oil from cylinder walls is partially burned and together with unburned fuel is responsible for the organic carbon associated with the diesel exhaust particles. The formation of nucleation-mode particles is driven by the concentration of the nucleating species, mainly sulphuric acid and hydrocarbons. During dilution and cooling of the exhaust plume, volatile precursors may become sufficiently supersaturated to nucleate new particles and promote growth of existing particles. These vapour precursors may also be subject to gas-to-particle conversion by adsorption onto the accumulation mode, thus reducing the concentration of precursors and, hence, the driving force for nucleation and subsequent growth. Therefore, for a given concentration of precursors, nucleation and growth will be suppressed by existing accumulation-mode particles. Current legislation standards aim at reducing the mass concentration of the accumulation-mode particles, hence unintentionally promoting the formation of nucleation-mode particles. The studies of Sakurai and co-authors,[47] Tobias and co-authors,[48] and Ziemann and co-authors[49] suggest that the nucleation-mode particles consist primarily of heavy hydrocarbons mainly associated with lubrication oil. However, there is still debate regarding the existence of non-volatile material in particles in the nucleation mode, while studies broadly agree on the predominance of non-volatile material in the Aitken mode. Biswas and co-authors[50] point out that the smallest particles ($\sim$20 nm) in the vicinity of a highway are predominantly volatile, shrinking below the lower cut-off size of their SMPS ($\sim$7 nm), while the remaining particles in the upper nucleation mode shrink down to $\sim$10 nm. Aitken-mode particles ($D_p > 40$ nm) show evidence of external mixing with a non-volatile fraction that increases with particle diameter. The 80 nm particles, which are a good indicator of soot aerosol emissions, had the highest non-volatile fraction of all particle sizes.[50] Similar findings are reported by Tiitta and co-authors[51] from a thermodenuder study of roadside ultrafine particles. They show a major fraction of nucleation-mode particles that have partly evaporated with a non-volatile core of 10 nm. They also found that for the non-volatile constituents that are dominant in the Aitken-mode particles[4] more than 97% of the volume of 12 and 30 nm particles evaporated when heated to 400 °C, which is consistent with the volatility of $C_{24}H_{50}$–$C_{32}H_{66}$. Such heavy alkanes are more prevalent in lubricating oil than in fuel.[49,52] According to Ronkko and co-authors,[53] particle size distributions measured after a thermodenuder consist of two modes with geometric mean diameters smaller than 9 nm and 37–47 nm, indicating that both the nucleation-mode and Aitken-mode particles contain non-volatile material. The non-volatile material (core) of the particles is formed under high temperature conditions, most likely in the combustion chamber. Nonetheless, the formation mechanism and composition of these non-volatile fractions are unresolved due to the limited experimental data and sampling challenges. Ronkko and co-authors[54] discuss the existence of a non-volatile fraction in nucleation-mode particles with or without after-treatment equipment and point out that the non-volatile diameter is typically below 7 nm. Kirchner and co-authors[55] show that during cold start and idle not all nucleation-mode particles are removed after being heated in the thermodenuder at 280 °C. Particles with a diameter of 30–40 nm shrunk to 10–20 nm having smaller spots (1–3 nm) with very high contrast when examined under TEM (transmission electron microscopy). Birmili and co-authors[56] express the view that all particles above 12 nm contain a non-volatile material. On the other hand, Wehner and co-authors[57] did not detect any non-volatile material in their thermodenuder measurements of traffic influenced particles near a highway. They found that the nucleation-mode particles appear to be completely volatile, while 30 nm particles have non-volatile material above the instrument cut-off size of 10 nm. Nearly all the particles in the upper Aitken (~80 nm) and lower accumulation mode (~150 nm) contain a non-volatile core with a thickness of the volatile layer below 10% of particle radius.

Our approach of considering a non-volatile fraction in both the nucleation and Aitken modes is in line with the studies discussed above. We show that under meteorological conditions at the time, in order to explain the evaporative diameter shrinkage in the nucleation-mode particles measured by Dall'Osto and co-authors[12] in London, particles in the nucleation mode should have 99% by volume of volatile material in the 23 nm particles. In addition, we have performed sensitivity analysis to assess the importance of the non-volatile fractions on the size distribution and the overall role of evaporation. By increasing the non-volatile material from 1% to 3% in the nucleation mode in both model stages, we simulate a diameter shrinkage from 23 nm to 11.6 nm, *i.e.* ~97% of the 23 nm particle's volume is volatile. Measurements of the composition of UFP in the street canyon and the urban park are not currently available, which hinders our model validation for the composition of UFP at this stage. Moreover, the health effects of particles containing non-volatile material are still unknown.[54] The non-volatile material contains mainly carbon and possibly some metal oxides from the engine lubricant and sulphate from oxidation of S in the fuel. Li and co-authors[58] show that an emitting diesel truck during idle induces a high level of oxidative stress in human aortic endothelial cells, mainly related to the type of metals and trace elements. However, the emissions from the same diesel truck under an urban dynamometer driving cycle induce a pro-inflammatory response, attributed to the enriched content of organic species. Xia and co-authors[59] show that traffic-related UFP, enriched in polycyclic and other semivolatile hydrocarbons, act to promote allergic airway inflammation. The relative importance of these particles in creating a health risk from exposure warrant further investigations.

Conclusions

We have developed a model (CiTTy-Street-UFP) of traffic-emitted particle behaviour in a street canyon and in the near vicinity that accounts for particle dynamics and the complex mixing between particles with different compositions, and considers the multicomponent nature of the UFP in the presence of 17 SVOC (modelled as *n*-alkane proxies). We evaluated the evaporative potential of UFP and the composition changes of multicomponent organic particles under realistic atmospheric conditions. Steady state simulations were performed in a street canyon and above roof-top. The above-roof-top size distribution of UFP was advected to the nearby urban park during which particles were allowed to grow/shrink to their exact size. We found that during the advection, particles in the nucleation mode shrunk remarkably in agreement with the measurements of Dall'Osto and co-authors.[12] The modelled nucleation-mode peak diameter at 23 nm decreased to 9 nm in 120 s, much less than the straight-line travel time between sites even at roof-top wind speeds of 3 m s^{-1}. No SVOC remained in the sub-10 nm range; only non-volatile material remained. In order to simulate the nucleation-mode UFP observed, therefore, we require either involatile cores in the nucleation-mode particles emitted from the street canyon, or a composition consisting in part of organic compounds with volatilities much lower than those of $C_{32}H_{66}$ *n*-alkanes. Particles in the Aitken mode had SVOC present between $C_{26}H_{54}$ and $C_{32}H_{66}$. Although we found a very good agreement between modelled and measured results for the overall shape of the size distribution, there is no data on particle composition in the urban park with which to validate the composition aspect of our results. Moreover, robust measurements of the composition of UFP at street scale are lacking. Measurements devoted to addressing in detail the composition of the traffic emitted UFP in the atmosphere are required. This would improve the representation of these particles in dispersion models and provide data for model validation. In addition, the robust representation of the chemical composition would provide more insights into the mechanisms of mixing in the atmosphere, the typical lifetime and overall behaviour of the traffic-emitted ultrafine particles. Data on the composition of UFP will allow an assessment of the complex spatial and temporal composition variations at or close to traffic sites. Such data would provide information on possible trends and help identify the non-volatile and volatile materials of relevance for future epidemiological and toxicological studies investigating the health effects of UFP.

Acknowledgements

This work is part of the FASTER project, sponsored by the European Research Council (ERC).

References

1 R. W. Atkinson, G. W. Fuller, H. R. Anderson, R. M. Harrison and B. Armstrong, *Epidemiology*, 2010, **21**(4), 501–511.
2 AQEG, Air Quality Expert Group, *Department for Environment*, Food and Rural Affairs Defra, London, 2005.
3 R. M. Harrison, D. C. S. Beddows and M. Dall'Osto, *Environ. Sci. Technol.*, 2011, **45**, 5522–5528.
4 D. B. Kittelson, W. F. Watts and J. P. Johnson, *J. Aerosol Sci.*, 2006, **37**, 913–930.
5 U. Uhrner, S. von Lowis, H. Vehkamaki, B. Wehner, S. Brasel, M. Hermann, F. Stratmann, M. Kulmala and A. Wiedensohler, *Atmos. Environ.*, 2007, **41**, 7440–7461.
6 B. Albriet, K. N. Sartelet, S. Lacour, B. Carissimo and C. Seigneur, *Atmos. Environ.*, 2010, **44**, 1126–1137.
7 A. W. H. Chan, G. Isaacman, K. R. Wilson, D. R. Worton, C. R. Ruehl, T. Nah, D. R. Gentner, T. R. Dallmann, T. W. Kirchstetter, R. A. Harley, J. B. Gilman, W. C. Kuster, J. A. de Gouw, J. H. Offenberg, T. E. Kleindienst, Y. H. Lin, C. L. Rubitschun, J. D. Surrat, P. L. Hayes, J. L. Jimenez and A. H. Goldstein, *J. Geophys. Res.*, 2013, **118**, 6783–6796.

8 D. R. Worton, G. Isaacman, D. R. Gentner, T. R. Dallmann, A. W. H. Chan, C. Ruehl, W. Kirchstetter, K. R. Wilson, R. A. Harley and A. H. Goldstein, *Environ. Sci. Technol.*, 2014, **48**(7), 3698–3706.
9 M. S. Alam, S. Z. Rezaei, C. P. Stark, Z. Liang, H. M. Xu and R. M. Harrison. *Faraday Discussion, Chemistry of Urban Atmosphere*, 2016.
10 A. L. Robinson, N. M. Donahue, M. K. Shrivastava, E. A. Weitkamp, A. M. Sage, A. P. Grieshop, T. E. Lane, J. R. Pierce and S. Pandis, *Science*, 2007, **315**, 1259–1262.
11 R. M. Harrison, C. Giorio, D. C. Beddows and M. Dall'Osto, *Sci. Total Environ.*, 2010, **409**, 289–293.
12 M. Dall'Osto, A. Thorpe, D. C. S. Beddows, R. M. Harrison, J. F. Barlow, T. Dunbar, P. I. Williams and H. Coe, *Atmos. Chem. Phys.*, 2011, **11**, 6623–6637.
13 L. Gidhagen, C. Johansson, J. Langner and V. Foltescu, *Atmos. Environ.*, 2005, **39**, 1711–1725.
14 A. Abdulsaheb and P. Kumar, *18th International Symposium on Transport and Air Pollution*, 2010, pp. 1–6.
15 P. Kumar, A. Garmory, M. Ketzel, R. Berkowicz and R. Britter, *Atmos. Environ.*, 2009, **43**, 949–958.
16 I. Nikolova, S. Janssen, K. Vrancken, P. Vos, V. Mishra and P. Berghmans, *Sci. Total Environ.*, 2011, **412**, 336–343.
17 G. Habilomatis and A. Chaloulakou, *Sci. Total Environ.*, 2015, **530–531**, 227–232.
18 L. Gidhagen, C. Johansson, J. Langner and G. Olivares, *Atmos. Environ.*, 2004, **38**, 2029–2044.
19 M. Ketzel and R. Berkowicz, *Atmos. Environ.*, 2004, **38**, 2639–2652.
20 I. Nikolova, S. Janssen, P. Vos and P. Berghmans, *Aerosol Air Qual. Res.*, 2014, **14**, 145–155.
21 M. Ketzel and R. Berkowicz, *Atmos. Environ.*, 2005, **39**, 3407–3420.
22 R. M. Harrison, M. Dall'Osto, D. C. S. Beddows, A. J. Thorpe, W. J. Bloss, J. D. Allan, H. Coe, J. R. Dorsey, M. Gallagher, C. Martin, J. Whitehead, P. I. Williams, R. L. Jones, J. M. Langridge, A. K. Benton, S. M. Ball, B. Langford, C. N. Hewitt, B. Davison, D. Martin, K. F. Petersson, S. J. Henshaw, I. R. White, D. E. Shallcross, J. F. Barlow, T. Dunbar, F. Davies, E. Nemitz, G. J. Phillips, C. Helfter, C. F. Di Marco and S. Smith, *Atmos. Chem. Phys.*, 2012, **12**, 3065–3114.
23 N. A. Mazzeo and L. E. Venegas, *Int. J. Environ. Pollut.*, 2005, **25**, 1–4.
24 S. Di Sabatino, P. Kastner-Klein, R. Berkowicz, R. E. Britter and E. Fedorovich, *Environ. Fluid Mech.*, 2003, **3**(2), 129–143.
25 P. Kastner-Klein, E. Fedorovich, M. Ketzel, R. Berkowicz and R. Britter, *Environ. Fluid Mech.*, 2003, **3**, 145–172.
26 E. Solazzo, S. Vardoulakis and X. Cai, *Atmos. Environ.*, 2007, **41**, 5357–5370.
27 T. A. M. Pugh, A. R. MacKenzie, *et al.*, *Environ. Sci. Technol.*, 2012a, **46**, 7692–7699.
28 T. A. M. Pugh, M. Cain, J. Methven, O. Wild, S. R. Arnold, E. Real, K. S. Law, K. M. Emmerson, S. M. Owen, J. A. Pyle, C. N. Hewitt and A. R. MacKenzie, *Geosci. Model Dev.*, 2012b, **5**, 193–221.
29 C.-H. Liu, D. Y. C. Leung and M. C. Barth, *Atmos. Environ.*, 2005, **39**, 1567–1574.
30 J. H. Seinfeld and S. N. Pandis, *Atmospheric Chemistry and Physics*, Wiley, New York, 1998.

31 S. Harrad, S. Hassoun, M. S. C. Romero and R. M. Harrison, *Atmos. Environ.*, 2003, **37**, 4985–4991.
32 R. M. Harrison, A. M. Jones, D. C. Beddows, M. Dall'Osto and I. Nikolova, *Atmos. Environ.*, 2016, **125**, 1–7.
33 J. Julin, P. M. Winkler, N. M. Donahue and P. E. Wagner, *Environ. Sci. Technol.*, 2014, **48**, 12083–12089.
34 M. Z. Jacobson, *Aerosol Sci. Technol.*, 1997, **27**, 491–498.
35 E. Debry, K. Fahey, K. Sartelet, B. Sportisse and M. Tombette, *Atmos. Chem. Phys.*, 2007, **7**, 1537–1547.
36 K. M. Zhang and A. S. Wexler, *Atmos. Environ.*, 2004, **38**(38), 6643–6653.
37 P. Kumar, M. Ketzel, S. Vardoulakis, L. Pirjola and R. Britter, *J. Aerosol Sci.*, 2011, **42**, 580–603.
38 E. Vignati, R. Berkowicz, F. Palmgren, E. Lyck and P. Hummelshoj, *Sci. Total Environ.*, 1999, **235**, 37–49.
39 M. Pohjola, L. Pirjola, J. Kukkonen and M. Kulmala, *Atmos. Environ.*, 2003, **37**, 339–351.
40 L. Pirjolaa, T. Lähdea, J. V. Niemic, A. Kousac, T. Rönkköe, P. Karjalainene, J. Keskinene, A. Freyf and A. Hillamof, *Atmos. Environ.*, 2012, **63**, 156–167.
41 X. L. Li, J. S. Wang, X. D. Tu, W. Liu and Z. Huang, *Sci. Total Environ.*, 2007, **378**, 306–316.
42 Y. Zhu, W. Hinds, S. Kim, S. Shen and C. Sioutas, *Atmos. Environ.*, 2002, **36**, 4323–4335.
43 B. Wehner, W. Birmili, T. Gnauk and A. Wiedensohler, *Atmos. Environ.*, 2002, **36**, 2215–2223.
44 A. Charron and R. M. Harrison, *Atmos. Environ.*, 2003, **37**, 4109–4119.
45 X. Cai, *Atmos. Environ.*, 2012, **51**, 268–277.
46 J. F. Barlow, T. M. Dunbar, E. G. Nemitz, C. R. Wood, M. W. Gallagher, F. Davies, E. O'Connor and R. M. Harrison, *Atmos. Chem. Phys.*, 2011, **11**, 2111–2125.
47 H. Sakurai, P. Park, P. H. McMurry, D. D. Zarling, D. B. Kittelson and P. J. Ziemann, *Environ. Sci. Technol.*, 2003b, **37**, 5487–5495.
48 H. Tobias, D. E. Beving and P. J. Ziemann, *Environ. Sci. Technol.*, 2001, **35**(11), 2233–2243.
49 P. Ziemann, H. Sakurai and P. H. McMurry, Final report, CRC Project no. E-43-44, 2002.
50 S. Biswas, L. Ntziachristos, K. Moore and C. Sioutas, *Atmos. Environ.*, 2007, **41**(16), 3479–3493.
51 P. Tiitta, P. Miettinen, P. Vaattovaara, J. Joutsensaari, T. Petaja, A. Virtanen, T. Raatikainen, P. Aalto, H. Portin, S. Romakkaniemi, H. Kokkola, K. E. J. Lehtinen, M. Kulmala and A. Laaksonen, *Atmos. Environ.*, 2010, **44**(7), 976–986.
52 H. Sakurai, H. J. Tobais, K. Park, D. Zarling, K. S. Docherty, D. B. Kittelson, P. H. McMurry and P. J. Ziemann, *Atmos. Environ.*, 2003a, **37**, 1199–1210.
53 T. Ronkko, A. Virtanen, J. Kannosto, J. Keskinen, M. Lappi and L. Pirjola, *Environ. Sci. Technol.*, 2007, **41**, 6384–6389.
54 T. Ronkko, T. Lahde, J. Heikkila, L. Pirjola, U. Bauschke, F. Arnold, H. Schlager, D. Rothe, J. Yli-Ojanpera and J. Keskinen, *Environ. Sci. Technol.*, 2013, **47**, 11882–11889.
55 U. Kirchner, V. Scheer, R. Vogt and R. Kägi, *J. Aerosol Sci.*, 2009, **40**(1), 55–64.

56 W. Birmili, K. Heinke, M. Pitz, J. Matschullat, A. Wiedensohler, J. Cyrys, H.-E. Wichmann and A. Peters, *Atmos. Chem. Phys.*, 2010, **10**, 4643–4660.
57 B. Wehner, S. Philippin, A. Wiedensohler, V. Scheer and R. Vogt, *Atmos. Environ.*, 2004, **38**(36), 6081–6090.
58 R. Li, Z. Ning, R. Majumdar, J. Cui, W. Takabe, N. Jen, C. Sioutas and T. Hsiai, *Part. Fibre Toxicol.*, 2010, **7**(6), 1–12.
59 M. Xia, L. Viera-Hutchins, M. Garcia-Lloret, M. N. Rivas, P. Wise, S. A. McGhee, Z. K. Chatila, N. Daher, C. Sioutas and T. A. Chatila, *J. Allergy Clin. Immunol.*, 2015, **136**, 441–453.

Faraday Discussions

PAPER

Simulation of particle diversity and mixing state over Greater Paris: a model–measurement inter-comparison†

Shupeng Zhu,[a] Karine N. Sartelet,*[a] Robert M. Healy[b] and John C. Wenger[c]

Received 17th November 2015, Accepted 4th December 2015
DOI: 10.1039/c5fd00175g

Air quality models are used to simulate and forecast pollutant concentrations, from continental scales to regional and urban scales. These models usually assume that particles are internally mixed, *i.e.* particles of the same size have the same chemical composition, which may vary in space and time. Although this assumption may be realistic for continental-scale simulations, where particles originating from different sources have undergone sufficient mixing to achieve a common chemical composition for a given model grid cell and time, it may not be valid for urban-scale simulations, where particles from different sources interact on shorter time scales. To investigate the role of the mixing state assumption on the formation of particles, a size-composition resolved aerosol model (SCRAM) was developed and coupled to the Polyphemus air quality platform. Two simulations, one with the internal mixing hypothesis and another with the external mixing hypothesis, have been carried out for the period 15 January to 11 February 2010, when the MEGAPOLI winter field measurement campaign took place in Paris. The simulated bulk concentrations of chemical species and the concentrations of individual particle classes are compared with the observations of Healy *et al.* (*Atmos. Chem. Phys.*, 2013, **13**, 9479–9496) for the same period. The single particle diversity and the mixing-state index are computed based on the approach developed by Riemer *et al.* (*Atmos. Chem. Phys.*, 2013, **13**, 11423–11439), and they are compared to the measurement-based analyses of Healy *et al.* (*Atmos. Chem. Phys.*, 2014, **14**, 6289–6299). The average value of the single particle diversity, which represents the average number of species within each particle, is consistent between simulation and measurement (2.91 and 2.79 respectively). Furthermore, the average value of the mixing-state index is also well represented in the simulation (69% against 59% from the measurements). The spatial distribution of the mixing-state index shows that the

[a]*CEREA (Joint Laboratory Ecole des Ponts ParisTech – EdF R& D), Université Paris-Est, Champs-sur-Marne, France. E-mail: sartelet@cerea.enpc.fr; Fax: +33 1 64 15 21 70; Tel: +33 1 64 15 21 57*

[b]*Environmental Monitoring and Reporting Branch, Ontario Ministry of the Environment and Climate Change, Toronto, Canada*

[c]*Department of Chemistry & Environmental Research Institute, University College Cork, Cork, Ireland*

† Electronic supplementary information (ESI) available. See DOI: 10.1039/c5fd00175g

particles are not mixed in urban areas, while they are well mixed in rural areas. This indicates that the assumption of internal mixing traditionally used in transport chemistry models is well suited to rural areas, but this assumption is less realistic for urban areas close to emission sources.

1 Introduction

Atmospheric particles possess a large diversity of chemical compositions, which are constantly evolving as particles mix with each other and interact with surrounding gases. The diversity of particle compositions in a population of a given size range is often referred to as the particle mixing state.[1] Although most measurements are performed for "bulk" chemical compounds, *i.e.* without considering the particle mixing state (*e.g.*, Aerodyne high-resolution time-of-flight aerosol mass spectrometer (HR-ToF-AMS),[2] and multi-angle absorption photometer (MAAP)[3]), other field studies have focused on measuring individual particle compositions using single particle mass spectrometers such as the aerosol time-of-flight mass spectrometer (ATOFMS). Particle mixing state information can aid identification of particle sources,[4–7] as well as determine climate-relevant properties such as cloud condensation nuclei (CCN) activity, hygroscopicity, optical absorption and scattering.[8–11] Furthermore, information on the particle mixing state can be a powerful tool to assess ageing processes and the relative impact of local and regional sources of ambient particles in urban environments.[12]

Aside from *in situ* field observations, the modelling of composition-resolved particles remains challenging. In most air quality and climate models, for computational reasons, the particle diversity is not considered, and it is assumed that all particles within the same size bin[13] or within the same mode[14] have homogeneous composition. These models are often referred to as internal-mixing models. An external-mixing model provides additional complexity by allowing for multiple particle compositions within a given size range. Limited by complexity and computational resources, most of the developed external-mixing models are 0-D box models.[15–20] Few attempts have been made to simulate externally-mixed particles in three dimensions, usually neglecting coagulation, and introducing assumptions. In Kleeman and Cass,[21] different particle distributions were associated with different emission sources. However, particles were not allowed to exchange freely between different populations, and only transportation and interaction with gas phase species was simulated. In the works of Stier *et al.*[22] and Bauer *et al.*,[23] the particle distribution was represented by mixed and unmixed modes of predefined compositions; in Oshima *et al.*,[24] the compositions of particles were discretised based on the mass fraction of black carbon in the particles. Riemer *et al.*[25] modeled externally-mixed particles using an accurate stochastic approach, which might be computationally expensive when the number of particles is high. Zhu *et al.*[20] developed a Size-Composition Resolved Aerosol Model (SCRAM), where both the size and mass fractions of chemical components of particles are discretised. For a given particle size bin, particles have distinct chemical compositions. When discretising mass fractions, chemical components may be grouped into several aggregates to reduce the computational cost. SCRAM takes into account the processes of coagulation, condensation/evaporation and nucleation.

The SCRAM model has already been integrated into the Polyphemus air quality platform[26] and used to evaluate the particle properties in the Greater Paris region from 28 June to 5 July 2009.[27] This evaluation showed that SCRAM is able to give satisfactory results for both $PM_{2.5}/PM_{10}$ mass concentrations and aerosol optical depths, as assessed from comparisons with observations. Furthermore, the model has the ability to analyse the particle mixing state, as well as the impact of the mixing-state assumption on particle formation and properties. However, as no field measurements of single particle composition and mixing state are available for this period, the model performance could not be evaluated.[27] In this work, a simulation is conducted for January/February 2010 when the composition of individual particles was measured[1] during the winter campaign of MEGAPOLI (Megacities: emissions, urban, regional and Global Atmospheric POLlution and climate effects, and Integrated tools for assessment and mitigation). As a result, the model performance in simulating the particle mixing state can be evaluated for the first time through a model–measurement inter-comparison.

2 Model description

The Polair3D air quality model[28] of the Polyphemus air quality platform[26] is used to simulate air quality over the Paris area. The Carbon Bond 05 model (CB05)[29] is used for gas-phase chemistry and the VSRM model is applied for the in-cloud processing of aerosols.[30] The SCRAM model[27] is used to simulate the dynamics of the aerosol size distribution.

In SCRAM, the mass fraction of a given chemical component or aggregate of chemical components within each particle is discretised into bins. Compared to previous work,[20,27] SCRAM now allows the user to independently discretise the mass-fraction bins for each chemical component or aggregate of chemical components. In other words, the discretisation and the number of mass-fraction bins can be different for each chemical component or aggregate of chemical components. A comprehensive particle composition list is generated by selecting the possible combinations of the mass-fraction bins of each chemical component or aggregate of chemical components. A combination of mass fractions is defined as acceptable only if the sum of the lower bounds of each fraction bin within the combination is lower than 100%.

Three aerosol dynamic processes are taken into account in SCRAM: coagulation, condensation/evaporation and the homogeneous binary nucleation of sulfate and water. In this work, the condensation/evaporation of inorganic aerosols is determined using ISORROPIA[31] with a fully dynamic approach to compute dynamically the mass transfer rate between gas and particles for each particle size and composition bin. For secondary organic aerosol (SOA) formation, the H_2O model[32] is used, and bulk equilibrium is always assumed for organic species due to limitations of the H_2O model. After condensation/evaporation, the moving-center algorithm is used for mass-number redistribution among fixed size bins and composition bins. More details about the discretisation method, the mathematical derivation, the different condensation/evaporation approaches and model validations can be found in Zhu *et al.*[20]

Although SCRAM has the ability to simulate externally-mixed (EM) particles, it may also be used for internally-mixed (IM) particles, *i.e.* using the internal-mixing assumption by having only one mass-fraction bin (between 0 and 100%) for each chemical component or aggregate of chemical components.

3 Simulation setup and measurement database

Simulations were conducted over Greater Paris between 15 January and 11 February 2010. The domain covers the whole Greater Paris region ([1.35°E, 3.55°E] × [48.00°N, 49.50°N]) with a horizontal resolution of 0.02° × 0.02°, and 9 vertical layers from ground level to 12 000 m.

In order to generate initial and boundary conditions for the simulation over Greater Paris, two nested simulations over Europe and France were conducted (see Fig. 1 for details of the domains). Because of the limitation of computational resources, these simulations were conducted with the internal-mixing assumption and the bulk-equilibrium approach for condensation/evaporation. The initial and boundary conditions for the Europe simulation were obtained from the Model for OZone And Related chemical Tracers (Mozart v2.0),[33] and those for the France simulation were obtained from the Europe simulation. For both the France and Europe simulations, anthropogenic emissions of gases and particles were taken from the EMEP inventory,[34] while the biogenic emissions were computed using the Model of Emissions of Gases and Aerosols from Nature (MEGAN).[35] Meteorology was from reanalysis of the European Centre for Medium-Range Weather Forecasts (ECMWF) model.

For the simulations over Greater Paris, the anthropogenic emissions of gases and particles were obtained from the Airparif (the Paris air quality agency, http://www.airparif.asso.fr) inventory for the year 2005 over Ile de France and from the EMEP inventory outside Ile de France. Following Couvidat *et al.*,[36] gas-phase semi-volatile organic compound (SVOC) emissions were estimated from primary

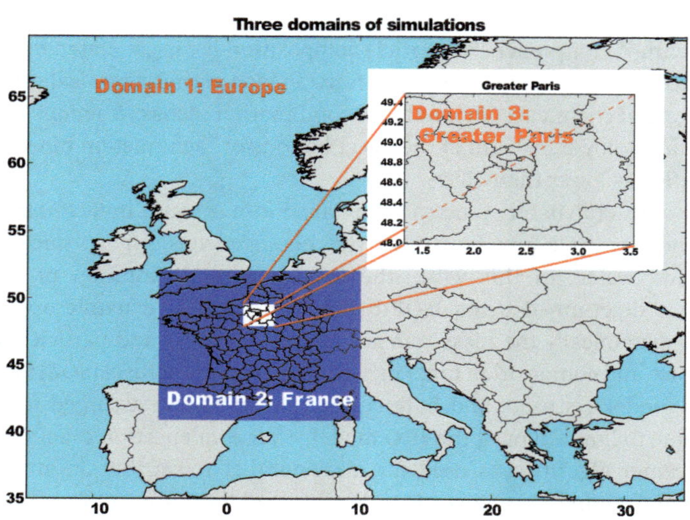

Fig. 1 Map of the three domains used for simulations: Europe ([−14.75°E, 34.75°E] × [35.25°N, 69.75°N], resolution: 0.5° × 0.5°), France ([−5.0°E, 10.0°E] × [41°N, 52°N], resolution: 0.1° × 0.1°) and Greater Paris ([1.35°E, 3.55°E] × [48.00°N, 49.50°N], resolution: 0.02° × 0.02°).

organic aerosol (POA) emissions. Although a ratio (SVOC Emissions/POA Emissions) of 5 was used in Couvidat *et al.*,[36] this ratio was set to 2.5 here, as derived from measurement data contained in a recent traffic emission report.[37] Meteorology was simulated with the Weather Research & Forecasting (WRF) version 3.6 model[38] using the urban canopy model and the Corine land-use database[39] with the YSU parametrization[40] for the planetary boundary layer. The WSM6 scheme[41] was used for the microphysics option of WRF, and the Kain–Fritsch convective parametrization[42] was used for cumulus physics.

Two simulations were conducted over Greater Paris: one with the internal-mixing (IM) assumption and one with the external-mixing (EM) assumption. The size distribution ranging from 0.01 to 10 μm was discretised into five size bins with bounds at 0.01, 0.1585, 0.4, 1.0, 2.5119 and 10 μm. As detailed in Couvidat *et al.*,[32] 31 particulate species were included in the simulations. In order to establish an appropriate comparison with the chemical species available from the observations, the 31 species were divided into 5 groups in the EM simulation: Elemental Carbon (EC), sulfate (SO_4), nitrate (NO_3), organic aerosol group (OA), and OThers (OT), which includes ammonium and sea salt. For the external-mixing (EM) simulation, the mass fraction of the first group, EC, was discretised into three mass-fraction bins ([0.0–0.1], [0.1–0.9], [0.9–1.0]), and the SO_4, NO_3 and OA groups were discretised into two mass-fraction bins ([0.0–0.1], [0.1–1.0]). If the mass fraction of one chemical group is located within the range [0.1–1.0], then it is considered as one of the main chemical groups of the bin, otherwise it is regarded as insignificant. EC was discretised with one more mass-fraction bin than the other groups ([0.9–1.0]), to be able to distinguish freshly emitted EC particles from aged particles. The last group OT was not discretised and its mass fraction was obtained by mass conservation. A total of 17 possible particle compositions were generated, as presented in Table 1. The names assigned to the

Table 1 17 Externally-mixed particle compositions

		Mass fraction of each group			
Index	Composition names	EC	SO_4	NO_3	OA
1	OT	0–0.1	0–0.1	0–0.1	0–0.1
2	OA	0–0.1	0–0.1	0–0.1	0.1–1
3	NO_3	0–0.1	0–0.1	0.1–1	0–0.1
4	OA–NO_3	0–0.1	0–0.1	0.1–1	0.1–1
5	SO_4	0–0.1	0.1–1	0–0.1	0–0.1
6	OA–SO_4	0–0.1	0.1–1	0–0.1	0.1–1
7	SO_4–NO_3	0–0.1	0.1–1	0.1–1	0–0.1
8	OA–SO_4–NO_3	0–0.1	0.1–1	0.1–1	0.1–1
9	EC+	0.1–0.9	0–0.1	0–0.1	0–0.1
10	EC–OA	0.1–0.9	0–0.1	0–0.1	0.1–1
11	EC–NO_3	0.1–0.9	0–0.1	0.1–1	0–0.1
12	EC–OA–NO_3	0.1–0.9	0–0.1	0.1–1	0.1–1
13	EC–SO_4	0.1–0.9	0.1–1	0–0.1	0–0.1
14	EC–OA–SO_4	0.1–0.9	0.1–1	0–0.1	0.1–1
15	EC–SO_4–NO_3	0.1–0.9	0.1–1	0.1–1	0–0.1
16	EC–OA–SO_4–NO_3	0.1–0.9	0.1–1	0.1–1	0.1–1
17	EC	0.9–1	0–0.1	0–0.1	0–0.1

particle compositions were chosen depending on their main chemical groups. The water content within each particle was also computed and tracked with SCRAM, while it was excluded when computing the mass fraction of each chemical group.

The measurement data of O_3, PM_{10} and $PM_{2.5}$ from the BDQA database ("Base de Données de la Qualité de l'Air": the French Database for Air Quality that covers France) are used to evaluate the model performance. More detailed observation data is obtained during the winter MEGAPOLI (Megacities: Emissions, urban, regional and Global Atmospheric POLlution and climate effects, and Integrated tools for assessment and mitigation) campaign in 2010,[12] where chemical composition at the single particle level was measured using an Aerosol Time-of-Flight Mass Spectrometer (ATOFMS) for particles within the size range 150–1067 (nm) at the urban background site of the Laboratoire d'Hygiène de la Ville de Paris (LHVP), Paris (48.75°N, 2.36°E) between 15 January and 11 February. Ten carbonaceous classes were identified from approximately 1.50 million detected mass spectra using a *K*-means algorithm.[43] First, the number concentration of each ATOFMS class was calculated for each hour of the campaign, then the corresponding mass concentration of each particle class was estimated based on particle diameter and an assumed density value of 1.5 g cm^{-3}. The hourly-resolved mass fraction of each chemical species was also measured for each size bin, which helped determine the bulk mass concentration of each species during the measured period. Finally, the ATOFMS-derived total mass concentrations for each species were multiplied by a factor of 1.24 to account for the low bias of the ATOFMS.[12]

4 Results and discussion

4.1 Bulk mass concentrations

Concentrations of $PM_{2.5}$, PM_{10} and O_3 from both the IM and EM simulations are first compared to the observations from the BDQA database to evaluate the model performance. Table 2 shows the definitions of the statistical indicators used in this study, while Table 3 compares the results from the IM and EM simulations with the measurements over the entire simulation period. Both the EM and IM simulations have good model performance, as the hourly O_3 concentration satisfies the recommended performance criteria[44] ($|MNGB| \leq 15\%$ and $MNGE \leq 30\%$), and both PM_{10} and $PM_{2.5}$ meet the model performance criteria proposed by Boylan and Russell[45] with $MFE \leq 75\%$ and $|MFB| \leq 60\%$. The statistics for $PM_{2.5}$ are even closer to the model performance goal ($MFE \leq 50\%$ and $|MFB| \leq 30\%$) proposed by Boylan and Russell,[45] while PM_{10} is rather under-estimated, which is consistent with previous simulations over the same region.[27,36,46] Possible reasons for this under-estimation of coarse particles may be that re-suspension is not modelled and that boundary conditions (*i.e.*, pollution transported into Greater Paris from other regions) may be under-estimated.

The concentrations of PM_{10} and $PM_{2.5}$ from the IM and EM simulations are very similar, which is consistent with the findings of the summer simulations.[27] However, the similarity between PM_{10} and $PM_{2.5}$ from the IM and EM simulations may be artificial, as it may be caused by the bulk equilibrium approach used to compute the condensation/evaporation of organics. Organic mass is found to be the largest component of the total aerosol mass here. As shown in Zhu *et al.*[27] for

Table 2 Definitions of the statistics used in this work. $(o_i)_i$ and $(c_i)_i$ are the observed and the simulated concentrations at time and location i, respectively. n is the number of data

Statistic indicator	Definition
Root mean square error (RMSE)	$\sqrt{\frac{1}{n}\sum_{i=1}^{n}(c_i - o_i)^2}$
Correlation	$\frac{\sum_{i=1}^{n}(c_i - \bar{c})(o_i - \bar{o})}{\sqrt{\sum_{i=1}^{n}(c_i - \bar{c})^2}\sqrt{\sum_{i=1}^{n}(o_i - \bar{o})^2}}$
Mean normalised gross bias (MNGB)	$\frac{1}{n}\sum_{i=1}^{n}\frac{o_i - c_i}{c_i}$
Mean normalised gross error (MNGE)	$\frac{1}{n}\sum_{i=1}^{n}\frac{\lvert o_i - c_i\rvert}{c_i}$
Normalised mean bias (NMB)	$\frac{\sum_{i=1}^{n} o_i - c_i}{\sum_{i=1}^{n} c_i}$
Normalised mean error (NME)	$\frac{\sum_{i=1}^{n}\lvert o_i - c_i\rvert}{\sum_{i=1}^{n} c_i}$
Mean fractional bias (MFB)	$\frac{1}{n}\sum_{i=1}^{n}\frac{c_i - o_i}{(c_i + o_i)/2}$
Mean fractional error (MFE)	$\frac{1}{n}\sum_{i=1}^{n}\frac{\lvert c_i - o_i\rvert}{(c_i + o_i)/2}$

inorganics, the mixing assumption strongly impacts the concentrations of semi-volatile components when the condensation/evaporation is computed dynamically.

More detailed analyses are made for PM_1 by comparing individual chemical components in the model simulations with the observation data from Healy *et al.*[12] Because dust and sea-salt are not taken into account in the measurements

Table 3 Statistics for the IM and EM simulation results (15 January to 11 February) and measurements of the BDQA network during the MEGAPOLI winter campaign. (Obs. stands for observation. Sim. stands for simulation. Corr. stands for correlation)

Species	Sim.	Obs. mean μg m^{-3}	Sim. Mean μg m^{-3}	RMSE μg m^{-3}	Corr.%	MNGB%	MNGE%	MFB%	MFE%
O_3	IM	31.59	37.07	18.72	61.23	−2.78	20.44	−8.13	23.71
	EM	31.59	37.07	18.72	61.23	−2.78	20.44	−8.13	23.71
PM_{10}	IM	30.68	19.63	22.49	38.26	−24.33	51.06	−46.64	63.34
	EM	30.68	19.44	22.65	37.86	−25.10	51.41	−47.69	64.08
$PM_{2.5}$	IM	24.95	21.65	19.02	39.58	19.53	66.86	−9.39	55.57
	EM	24.95	21.53	19.21	38.65	19.55	67.73	−10.08	56.28

of Healy *et al.*,[12] they are not included in the computation of simulated PM_1 concentrations for the comparison. In other words, PM_1 is computed by summing the concentrations of the first three size bins (between 0.01 μm and 1 μm) and all composition bins, excluding the concentrations of dust and sea salt. The concentrations of elemental carbon (EC), sulfate (SO_4), nitrate (NO_3), ammonium (NH_4), organic aerosols (OA) in PM_1 are available in the measurements and are compared to the simulation results. The statistical performance of the IM and EM simulations is presented in Table 4 with the measurements for the entire simulation period. The temporal evolution of the different chemical components is plotted in Fig. 2.

Over the entire simulation period, Table 4 and Fig. 2 show that both the IM and EM results compare well to observations. The statistics for comparisons of PM_1 and EC meet both the model performance goal and criterion proposed by Boylan and Russell,[45] while the statistics of the other components meet the model performance criterion. EC, SO_4 and NO_3 are slightly under-estimated, while OA and NH_4 are slightly over-estimated. SO_4 tends to be more under-estimated during the first two weeks because of the under-estimation of some events, as shown in Fig. 2. On 18 January, an increase in SO_4 was caused by a strong fog event. The model is not able to reproduce this peak, as fog events may not be properly modelled in our simulations, where cloud chemistry occurs depending on the liquid water content of the grid cell. This also explains the under-estimation of the peak concentrations of NO_3 and NH_4 during the same period. The under-estimation of sulfate may also be linked to the formation of organo-sulfate, as Healy *et al.*[12] observed internally-mixed organics and sulfate during that fog event. Another peak of SO_4 occurs between 25 and 28 January, where highly polluted continental air masses, which most likely originated from north-western and eastern Europe, are transported over Greater Paris.[12] The under-estimation of this peak may be due to uncertainties in the larger-scale simulations (France and Europe), leading to an under-estimation of the amount of sulfate transported into

Table 4 Statistics for the IM and EM simulation results (15 January to 11 February) and measurements obtained[12] at the LHVP site (48.75°N, 2.36°E) during the MEGAPOLI winter campaign (Obs. stands for observation. Sim. stands for simulation. Corr. stands for correlation)

Species	Sim.	Obs. mean μg m^{-3}	Sim. Mean μg m^{-3}	RMSE μg m^{-3}	Corr.%	NMB%	NME%	MFB%	MFE%
PM_1	IM	14.21	12.16	9.53	59.47	−14.42	43.45	−8.83	45.84
	EM	14.21	11.99	9.62	59.06	−15.60	43.82	−10.62	46.52
EC	IM	2.08	1.46	1.42	46.82	−29.59	46.56	−26.33	49.35
	EM	2.08	1.47	1.41	46.90	−29.45	46.54	−26.16	49.30
SO_4	IM	1.62	0.84	1.82	54.52	−48.48	65.34	−31.36	70.14
	EM	1.62	0.83	1.83	53.55	−49.04	65.91	−33.02	71.30
NO_3	IM	3.59	2.20	2.85	64.79	−38.73	49.15	−39.47	61.79
	EM	3.59	2.23	2.77	67.57	−37.90	49.06	−40.67	64.49
NH_4	IM	1.15	1.54	1.10	61.47	33.42	70.29	38.98	69.07
	EM	1.15	1.34	0.95	61.88	15.74	59.62	32.64	65.12
OA	IM	4.65	6.20	6.41	38.72	33.19	77.86	10.28	60.59
	EM	4.65	6.21	6.42	38.79	33.44	77.99	10.42	60.64

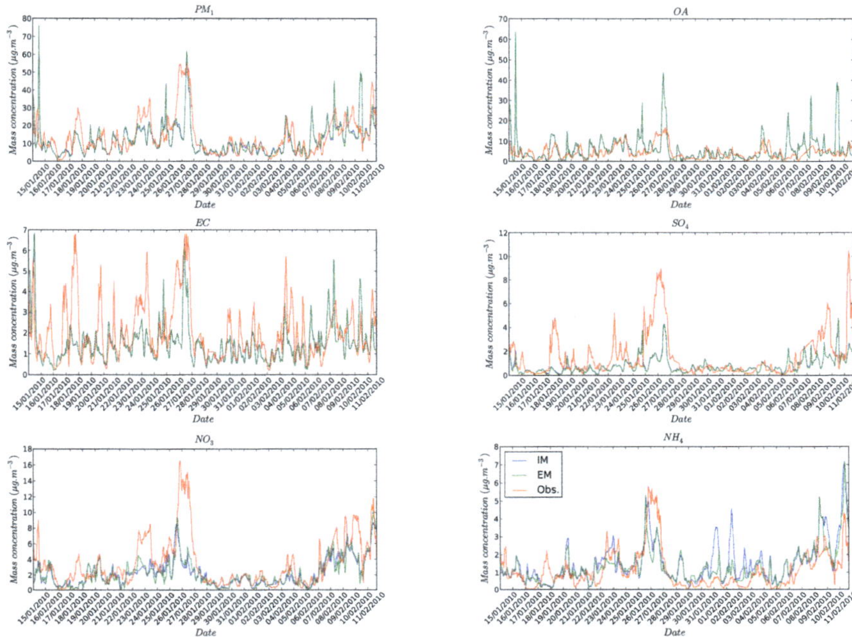

Fig. 2 Comparisons of the IM (blue line) and the EM (green line) simulation results with ATOFMS-derived mass concentrations for PM_1, EC, OA and inorganic ions (Obs. stands for observation – red line), when only the green line is visible the difference between the runs (IM *vs.* EM) is less than the thickness of the line.

Greater Paris through boundary conditions. Such under-estimation of sulfate has been observed in previous simulations over Europe during the winter period of 2001,[28] and likely originates from uncertainties in sulfur aqueous chemistry. As emphasized by Bessagnet *et al.*,[47] sulphur aqueous chemistry, which is predominant in winter, is very difficult to simulate, because it is very sensitive to temperature and pH. Note that during the same event, EC is relatively well modelled, while OA is over-estimated. There is no peak of SO_4 during the period 28 January–7 February, which is dominated by local emissions and is well modelled.

As for PM_{10} and $PM_{2.5}$, the differences in PM_1 concentrations between the IM and EM simulations are not significant, although noticeable differences occur in the NO_3 and NH_4 peaks, as shown in the time evolution curves of Fig. 2. As expected, both the IM and EM simulations lead to similar concentrations of non-volatile components (EC, SO_4) and also of OA (because of the bulk-equilibrium assumption made when computing condensation/evaporation of organics). Because EC, SO_4 and OA represent more than 71% of the PM_1 concentration, the variations of the PM_1 concentration with the mixing-state hypothesis are limited. For volatile inorganics (NO_3 and NH_4), the EM assumption results in higher NO_3 concentration and lower NH_4, as noted in the summer simulations.[27] The modelled mass concentrations for NH_4 are lower in the EM scenario, between 30 January and 3 February (see Fig. 2), which is more consistent with observations.

Fig. 3 represents the size distribution of mass concentration and mass fraction for each chemical species. Compared to Healy *et al.*,[12] a similar distribution is found for EC with higher concentrations at low diameters (<200 nm), and for secondary

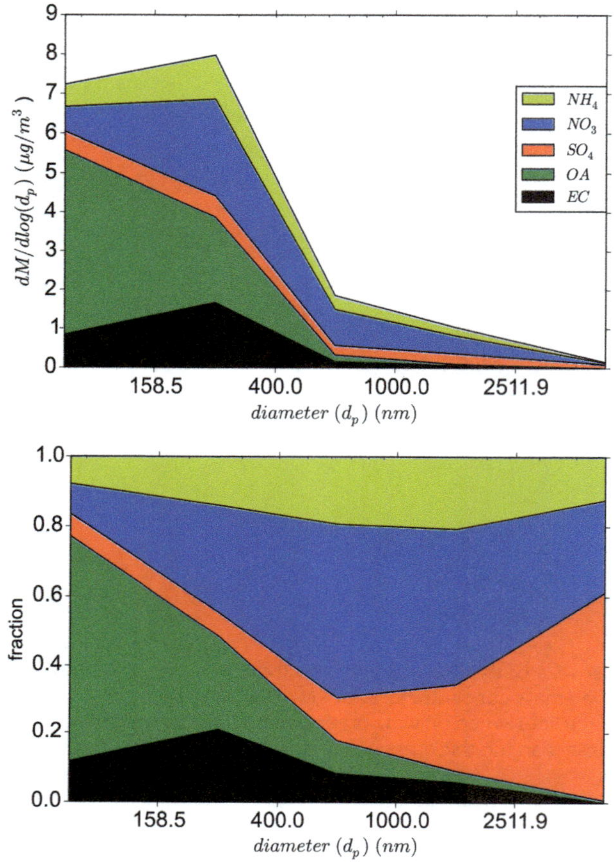

Fig. 3 Mass-size distribution (up, stacked) and size-resolved average mass fractions (down, stacked) for each chemical species from the external mixing simulation.

inorganic components such as NO_3, SO_4 and NH_4, larger mass fractions are found at higher diameters (>400 nm). However, in contrast to the measurements, most of the OA mass is found in the smallest size bins, probably because of the redistribution algorithm used in the condensation/evaporation of organics, which neglects the Kelvin effect. The condensation/evaporation of organics is performed using a bulk-equilibrium approach, which determines, for each organic chemical component, the gas-phase concentration and the particle-phase concentration summed over all size and composition bins. The condensed OA concentration is redistributed onto particle bins depending on the number concentrations of the bins and their mean diameters. In order to improve the size distribution of OA, the redistribution of the OA mass should be improved by taking the Kelvin effect into account in the redistribution algorithm (see ESI† for more detailed discussions), or by computing dynamically the condensation/evaporation of OA.

4.2 Concentrations of particle composition classes

In the EM simulation, 17 particle classes are discretised based on the combinations of the mass-fraction bins of the different chemical components (or groups of chemical components), as defined in Table 1. The chemical group OT contains NH_4, dust and sea salt. However, as dust and sea salt are not included in the measurements of Healy *et al.*,[12] they are not considered when computing particle total mass, and the composition of each particle class is therefore recomputed after the simulation. In the updated particle classes, the only component considered in the group OT is NH_4. Table S1† shows the average mass fraction of each chemical component at the LHVP site for PM_1 particles, as well as the mass fraction of each particle class within the PM_1 concentration. It is clear that most of the particle mass is concentrated in a few particle classes. OA dominated particles represent more than half of the total particle mass (60.1%), while the rest of the particle mass is represented by the classes OA–NO_3 (19.8%), EC–OA (13.1%), EC (2.9%), OA–SO_4 (1.3%) and EC–OA–NO_3 (1.0%). The particle composition classes from the simulation and measurements are classified so that they can be easily compared. The details of the classification are presented in the ESI,† together with the mass fraction of the chemical components of each particle class (Tables S2 and S3†).

The mixing state of EC is well reproduced in the simulation (Table 5). As shown in Tables S2 and S3,† the simulation results estimate 18.0% of EC-rich particles and 82.0% of EC-poor particles, in excellent agreement with the ratio from the measurements: 16.8% EC-rich particles and 83.2% EC-poor particles. Furthermore, 10.3% of EC particles are found to be mixed with OA (EC–OA class) in the measurements against 13.1% in the simulation. The percentage of EC–OA–NO_3 particles is under-estimated in the simulation (1.1%) compared to the measurements (3.8%), as well as the percentage of EC–OA–SO_4 particles (0.2% in the simulation compared to 2.7% in the measurements). In the measurements, EC–OA–SO_4 particles have the highest concentrations during the fog event, indicating that the under-estimation of EC–OA–SO_4 particles in the simulation is probably linked to the fog event when sulfate concentrations are strongly under-estimated in the simulation.

As for the total EC, the mixing state of EC-rich particles is relatively well simulated. In the simulation (see Table S1†), more than 84% of EC-rich particles are mixed (with a mass fraction of EC lower than 90%). In the measurements, all EC-rich particles are found mixed with other species. This is slightly higher than determined in the simulation, partly because in the measurements the classification was originally based on sources rather than mass fractions of the different chemical components. The average mass fraction of EC is around 0.45 for all EC-rich particles in the simulation, which is close to the value from the

Table 5 Mass fractions of each chemical group for EC-rich and EC-poor particles from both observation (Obs.) and simulation (Sim.)

			Mass fraction of each group				
Cases	Particle class	Mass ratio%	EC	SO_4	NO_3	OA	NH_4
Obs.	EC-rich	16.8	0.48	0.10	0.08	0.31	0.03
	EC-poor	83.2	0.07	0.14	0.32	0.36	0.11
Sim.	EC-rich	18.04	0.45	0.02	0.03	0.48	0.02
	EC-poor	81.96	0.01	0.03	0.08	0.83	0.05

measurements (0.48). This indicates that the average degree of mixing of EC-rich particles is similar between simulation and measurements. However, the mixing state of EC-poor particles is not as well simulated as the mixing state of EC-rich particles. Most of the EC-poor particles are nearly unmixed (73%) in the simulation, while all of the particles are found to be well mixed (with OA mass fraction between 0.27 and 0.52) from the measurements. Most of the mass of EC-poor particles is represented by OA, as indicated by the high value (0.83) of the OA mass fraction for the EC-poor particles from the simulation. These EC-poor particles tend to have low diameters (92% below 0.4 μm). The over-estimation of the percentage of OA in these EC-poor particles is a consequence of the redistribution algorithm used in the condensation/evaporation of organics, as discussed in the previous section.

Table 6 presents the statistics for EC-rich and EC-poor particles where a relatively good correlation can be found for both EC-rich and EC-poor particles (43.1% and 60.5% respectively). The statistics of EC-poor particles even satisfy the model performance goal of Boylan and Russell[45] for PM simulation. The statistics of EC-rich particles are not as good as those of EC-poor particles, because the particle mass is under-estimated due to the underestimation of EC, NO_3 and SO_4 (EC-rich particles not only contain EC mass, but also the mass of the other species mixed with EC). Nevertheless, these results show that the SCRAM model can simulate the concentrations of EC-rich and EC-poor particles reasonably well.

4.3 Mixing state analysis

Healy *et al.*[1] compute a quantitative mixing-state index (χ) of measured particles based on the information-theoretic entropy approach proposed by Riemer and West.[48] The same method is used in this study to derive the mixing-state index of simulated particles. A detailed description of this method can be found in Riemer and West.[48] For a population of N particle compositions ($N = 17$ in this study) and A distinct chemical components (or species, $A = 5$ in this study), the mass of species a in particle i is denoted as μ_i^a for $i = 1, \ldots, N$ and $a = 1, \ldots, A$; the total mass of particle i is μ_i; the total mass of species a in the total particle population is μ^a, and the total mass of the entire population is μ. These masses are computed as follows:

$$\mu_i = \sum_{a=1}^{A} \mu_i^a \quad \mu^a = \sum_{i=1}^{N} \mu_i^a \quad \mu = \sum_{i=1}^{N} \mu_i^a = \sum_{a=1}^{A} \mu^a \tag{1}$$

Table 6 Statistics for EC-rich and EC-poor particles obtained in simulations (15–28 January) and measurements obtained[12] at the LHVP site (48.75°N, 2.36°E) (Obs. stands for observation. Sim. stands for simulation. Corr. stands for correlation)

Particle class	Obs. mean μg m^{-3}	Sim. mean μg m^{-3}	RMSE μg m^{-3}	Corr.%	NMB%	NME%	MFB%	MFE%
EC rich	4.77	2.14	3.71	43.11	−55.13	63.11	−83.68	89.96
EC poor	11.82	9.85	8.46	60.50	−16.68	45.56	−7.61	48.98

The mass fraction of species a in particle i (p_i^a), the mass fraction of particle i in the population (p_i), and the mass fraction of species a in the population (p^a) can be calculated as follows:

$$p_i^a = \frac{\mu_i^a}{\mu_i} \quad p_i = \frac{\mu_i}{\mu} \quad p^a = \frac{\mu^a}{\mu} \tag{2}$$

This information can be computed either for particles within each size bin or for particles merged from several size bins. In this study, for comparison with measurements of PM_1, the analysis is performed for particles in the first three size bins (10–1000 nm).

The information-theoretic Shannon entropy is a measure of the uncertainty associated with a random variable. According to Riemer and West,[48] the Shannon entropy associated with each single particle (H_i) can be computed based on the mass fraction of each species within that particle:

$$H_i = \sum_{a=1}^{A} -p_i^a \ln p_i^a \tag{3}$$

While the average per-particle Shannon entropy (H_α) is given by:

$$H_\alpha = \sum_{i=1}^{N} -p_i H_i \tag{4}$$

Finally, the Shannon entropy of the entire bulk population (H_Υ) can be expressed as:

$$H_\Upsilon = \sum_{a=1}^{A} -p^a \ln p^a \tag{5}$$

Based on these Shannon entropies, the corresponding diversity values can be derived:

$$D_i = e^{H_i} \quad D_\alpha = e^{H_\alpha} \quad D_\Upsilon = e^{H_\Upsilon} \tag{6}$$

where D_i is the particle diversity, or the effective number of species in particle i, D_α is the average single particle diversity, and D_Υ is the bulk population diversity.

Finally, the mixing-state index χ can be derived from the following equation:

$$\chi = \frac{D_\alpha - 1}{D_\Upsilon - 1} \times 100\% \tag{7}$$

To summarize, D_i represents the number of chemical components in a particle, D_α represents the average value of D_i over the entire particle population, and D_Υ shows the bulk population diversity which is consistent under both IM and EM assumptions. Eventually, χ represents the degree of similarity between an arbitrary mixing case and a pure internal mixture. As a result, $\chi = 0\%$ for a pure external mixture, because $D_i = D_\alpha = 1$, while $\chi = 100\%$ for a pure internal mixture as $D_\alpha = D_\Upsilon$.

The single particle diversity (D_i) is computed from the simulation results and compared to the measurements at the LHVP site. Most of the simulated particles (64.5%) are within the D_i range 1.8 to 3.0, which is very close to the value of 71% obtained by Healy *et al.*[1] for the same D_i range. As most particles are concentrated in the smallest size bins, the D_i values are normalised by the number of particles in each size bin in order to get a better representation of the variation of D_i with particle size. Fig. 4 shows the dependence of the time-averaged number concentration on diversity and particle diameter, as well as the dependence of time-averaged particle composition on particle diameter. As noted by Healy *et al.*,[1] smaller particles have higher mass fractions of EC and OA, while larger particles have higher inorganic mass fractions. Although the particle number concentrations tend to be under-estimated in the simulation, high number concentrations are observed at low diameters (below 200 nm) for low diversities (between 1.8 and 2.3 in the measurements and between 1.8 and 3 in the simulation) and at high diameters for high diversity (between 3.6 and 4.2 in the measurements and between 3.4 and 4.2 in the simulation). In agreement with the measurements, at low diameters, most of the particles are composed of 2 or 3 species, mainly OA and EC. At high diameters, particles are composed of inorganics (80% in the simulation and 60% in the measurements). As the particle diameter grows, the mass fraction of inorganics also grows in the measurements, indicating that the condensation/evaporation of inorganic species is correctly modelled. However, the OA mass is under-estimated at high diameters and over-estimated at low diameters, indicating as noted before that the redistribution of OA after condensation/evaporation needs to be improved or to be computed dynamically.

As in the measurements, the mass fraction of EC decreases from 26% for the bin of diameters between 158.6 nm and 400 nm to a value lower than 10% at higher diameters. The variation of EC with particle diameter is largely linked to

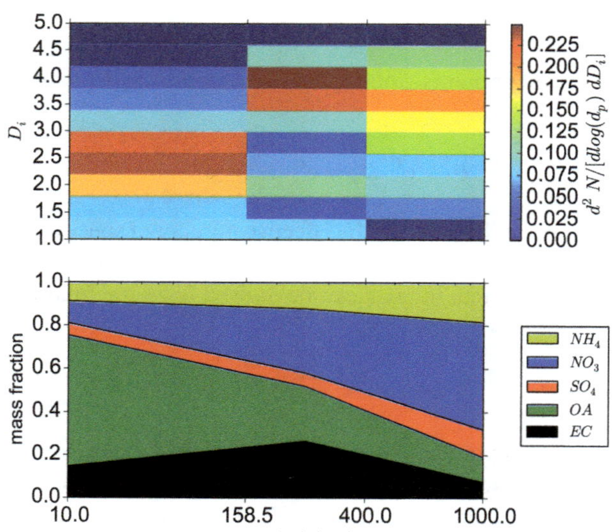

Fig. 4 Dependence of the time-averaged number concentration on single particle diversity (diversity normalised by the number of particles in each bin) and particle diameter (upper panel), and dependence of time-averaged particle composition on particle diameter (lower panel).

the size distribution used in the emissions: 54% was emitted in the bin [10–158.5 nm], 42% in the bin [158.5–400 nm] and 3% in the bin [400–1000 nm]. Although the mass fraction of EC increases to about 52% in the measurements, it remains below 30% in the simulation, because of the over-estimation of OA at low diameters, as well as the assumed size-distribution of emissions.

The average single particle diversity (D_α) and the species diversity of the bulk population (D_γ) are also derived from the simulation results. Unlike D_i, which represents the characteristics of a single particle, D_α and D_γ describe the overall attributes of the entire particle population. The mixing-state index χ can potentially range from 0% (fully externally mixed) to 100% (fully internally mixed). The averaged value of D_α from the simulation (2.91) is close to that obtained from the measurements (2.79), while that averaged value of D_γ tends to be under-estimated (3.76 from the simulation and 4.04 from the measurements), because of the bulk over-estimation of OA and the bulk under-estimation of SO_4. As a result, the mixing-state index χ is slightly over-estimated (69% in the simulation and 59% in the measurements), although the average single particle diversity is well represented. These values of the mixing-state index indicate that the particle population at LHVP is not internally mixed.

The relationship between D_α, D_γ and χ at the LHVP site is displayed in Fig. 5. The values of the mixing-state index range from 23% to 90%, with a mean value of 69%. This result is generally consistent with the measurements of Healy *et al.*,[1] where the mixing-state index is found to vary between 37% and 72%, with a mean value of 59%. Another phenomenon observed from the measurements, and also apparent in Fig. 5 is that the average single particle diversity (D_α) is never higher than 4 (3.5 in the measurements), even when the bulk population diversity (D_α) approaches the maximum value of 5. This indicates that a variety of different chemical mixing states exist for particles of the same size at all times.

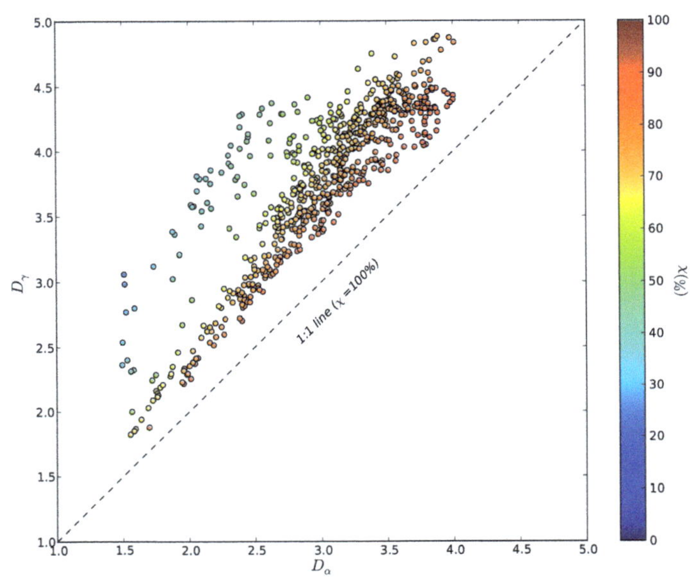

Fig. 5 Scatter plot of the hourly mixing-state index (χ) as a function of average particle diversity (D_α) and average bulk population diversity (D_γ).

The diurnal variations of D_α, D_γ and χ are computed from both simulation and observation results, and the dependence of their mean value upon time of day is shown in Fig. 6, as well as the corresponding variation of the mass fraction of each chemical component. The averaged value of D_α from the simulation (2.91) is very close to the one from the measurement (2.79). However, D_γ is relatively underestimated from the simulation due to the overestimation of OA and underestimation of SO_4. As a result, the mixing-state index χ is overestimated. A decrease of both D_α and χ can be observed between 5:00 and 8:00 in the simulation, due to the increase of traffic emissions, which introduce a large amount of particles with high EC mass fractions and low D_i values. However, D_γ does not increase during this period, as suggested by the measurements because of the increase of the mass fraction of EC and OA. In both the simulation and the measurements, the value of χ increased between 8:00–14:00 due to the combined effect of increasing D_α and decreasing D_γ. The increase of D_α during this period is due to the decrease of traffic emissions and ageing of existing particles, while the decrease of D_γ is largely caused by the increase of the OA fraction and the decrease of the EC and inorganic fractions. In the simulation, χ drops between 14:00 and 20:00 due to the increase of D_γ as a result of slowly increasing EC mass fraction (evening traffic). However, this decrease is limited to 14:00–16:00 in the measurements. An increase of D_α can be observed between 20:00 and 22:00 in both the simulation and the measurements, as a result of enhanced ammonium nitrate during night

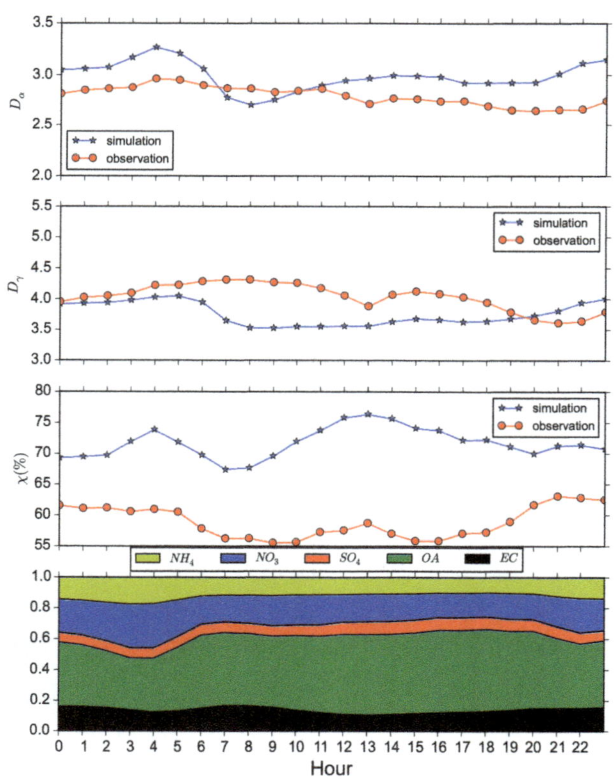

Fig. 6 Diurnal variations of the mixing-state index (χ) and bulk population mass fraction.

time, which is reflected by the increase of inorganic mass fraction during that period, as well as an increase of D_γ.

So far, the discussion was based on the measurements and simulation results at the LHVP site. Because our modelling study provides comprehensive data for the study of the spatial distribution of particle properties, Fig. 7 presents the spatial distribution of D_α, D_γ and χ over the entire simulated domain, averaged over the simulation period. The values of D_α are lower close to the centre of Paris and in areas with significant traffic but higher in rural areas. Diversity is lower where emissions are high, because freshly emitted particles mostly consist of one chemical component with a single particle diversity close to 1. The spatial distribution of D_γ is relatively uniform as it represents the number of bulk chemical components. However, there are regions where D_γ can suddenly increase or decrease, and these variations may not be similar to those of D_α. For example, in the region within the black rectangle in Fig. 7 (strong dust emissions from sand and stone mining industries along the Seine river), some parts have low D_γ and low D_α. This type of region (low D_α and low D_γ) may imply a relatively homogeneous emission source, where only one type of particle is emitted. Other regions, such as the south west of Paris and Roissy airport (denoted by the black circle in Fig. 7) have low D_α values and high D_γ values. In such regions, there may be heterogeneous emission sources, where particles with distinct compositions may be emitted from the same place or transported to that place. Considering the spatial distribution of the mixing-state index χ, regions with a larger difference between D_α and D_γ usually have lower χ, which means they are less internally mixed. Usually, the regions of low χ correspond to regions of low D_α (high traffic emissions). Note that for regions of low D_α but low D_γ, such as within the black rectangle in Fig. 7, the mixing-state index can be high, although the region may

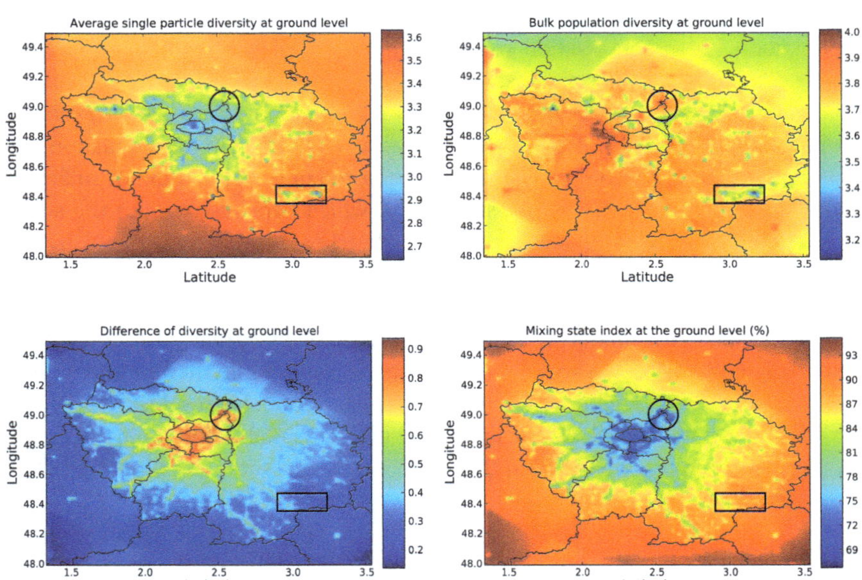

Fig. 7 Spatial distribution of time-averaged average single particle diversity (D_α), average bulk population diversity (D_γ), difference between D_α and D_γ, and mixing-state index (χ) over Greater Paris.

correspond to a location characterised by strong local emissions. Thus, χ also reflects the homogeneity of the particle population, and a group of particles with a heterogeneous composition has a low mixing-state index and is considered as externally mixed. On the other hand, a group of particles with a homogeneous composition has a high mixing-state index and is considered as internally mixed, even if the dominant particles consist of only one species (from one source).

5 Conclusions

Particle diversity and mixing state was studied using the newly developed size-composition resolved aerosol model (SCRAM) coupled to the Polyphemus air-quality platform. Two simulations were conducted over Greater Paris with different mixing-state assumptions (internal mixing IM or external mixing EM).

Both simulations reproduce well the concentration of O_3, PM_{10} and $PM_{2.5}$, as assessed from comparisons to observations from the BDQA network. The simulation results of both bulk concentrations of chemical species and the concentrations of individual particle classes are compared with the observations of Healy *et al.*[12] Good correlations are found between simulation results and measurements for both PM_1 and bulk species concentrations, and the statistical indicators for most species satisfy the model performance criteria proposed by Boylan and Russell,[45] although the concentration of SO_4 is under-estimated during periods where significant chemical processing occurs, *e.g.* during a strong fog event and periods dominated by continental (aged) air masses.

For the concentrations of individual particles obtained from the EM simulation, a very good agreement is found for the mass fractions of EC-rich and EC-poor particles between simulation results and observations. The single particle diversity and mixing-state index are computed from the results of the EM simulation based on a new quantification approach developed by Riemer and West,[48] and are compared with the observation based analyses of Healy *et al.*[1] at the urban site LHVP. The average value of the single particle diversity is consistent between the simulation and the measurements (2.91 and 2.79, respectively), while the averaged bulk population diversity is slightly under-estimated (3.76 from the simulation and 4.04 from the observation), probably because of the under-estimation of sulfate or nitrate in the model. The mixing-state index, which depends on both the single particle diversity and the bulk population diversity, is well represented by the simulation, while a mean value of 69% from the simulation and 59% from the measurements indicates that the particles are not internally mixed. Finally, the time-averaged spatial distribution of the mixing-state index shows that particles are more externally mixed in urban regions in and around Paris, while they are more internally mixed in rural regions. This indicates that traditional aerosol models, which assume that particles are internally mixed, may be suitable for simulating rural regions, while in urbanised areas, the internal-mixing assumption does not hold.

References

1 R. M. Healy, N. Riemer, J. Wenger, M. Murphy, M. West, L. Poulain, A. Wiedensohler, I. O'Connor, E. McGillicuddy, J. Sodeau, *et al.*, *Atmos. Chem. Phys.*, 2014, **14**, 6289–6299.

2 P. F. DeCarlo, J. R. Kimmel, A. Trimborn, M. J. Northway, J. T. Jayne, A. C. Aiken, M. Gonin, K. Fuhrer, T. Horvath, K. S. Docherty, *et al.*, *Anal. Chem.*, 2006, **78**, 8281–8289.

3 A. Petzold and M. Schönlinner, *J. Aerosol Sci.*, 2004, **35**, 421–441.

4 D. Murphy, D. Cziczo, K. Froyd, P. Hudson, B. Matthew, A. Middlebrook, R. E. Peltier, A. Sullivan, D. Thomson and R. Weber, *J. Geophys. Res.: Atmos.*, 2006, **111**, DOI: 10.1029/2006JD007340.

5 A. P. Ault, C. J. Gaston, Y. Wang, G. Dominguez, M. H. Thiemens and K. A. Prather, *Environ. Sci. Technol.*, 2010, **44**, 1954–1961.

6 R. M. Healy, J. Sciare, L. Poulain, K. Kamili, M. Merkel, T. Müller, A. Wiedensohler, S. Eckhardt, A. Stohl, R. Sarda-Estève, *et al.*, *Atmos. Chem. Phys.*, 2012, **12**, 1681–1700.

7 M. Dall'Osto, J. Ovadnevaite, D. Ceburnis, D. Martin, R. M. Healy, I. O'Connor, I. Kourtchev, J. Sodeau, J. Wenger and C. O'Dowd, *Atmos. Chem. Phys.*, 2013, **13**, 4997–5015.

8 M. Mallet, J. Roger, S. Despiau, J. Putaud and O. Dubovik, *J. Geophys. Res.: Atmos.*, 2004, **109**, DOI: 10.1029/2008JD011574.

9 H. Furutani, M. Dall'osto, G. C. Roberts and K. A. Prather, *Atmos. Environ.*, 2008, **42**, 3130–3142.

10 H. Herich, L. Kammermann, B. Friedman, D. S. Gross, E. Weingartner, U. Lohmann, P. Spichtinger, M. Gysel, U. Baltensperger and D. J. Cziczo, *J. Geophys. Res.: Atmos.*, 2009, **114**, DOI: 10.1029/2003JD003940.

11 E. Crosbie, J.-S. Youn, B. Balch, A. Wonaschütz, T. Shingler, Z. Wang, W. Conant, E. Betterton and A. Sorooshian, *Atmos. Chem. Phys.*, 2015, **15**, 6943–6958.

12 R. M. Healy, J. Sciare, L. Poulain, M. Crippa, A. Wiedensohler, A. S. Prévôt, U. Baltensperger, R. Sarda-Estève, M. L. McGuire, C.-H. Jeong, *et al.*, *Atmos. Chem. Phys.*, 2013, **13**, 9479–9496.

13 E. Debry, K. Fahey, K. Sartelet, B. Sportisse and M. Tombette, *Atmos. Chem. Phys.*, 2007, **7**, 1537–1547.

14 K. Sartelet, H. Hayami, B. Albriet and B. Sportisse, *Aerosol Sci. Technol.*, 2006, **40**, 118–127.

15 M. Z. Jacobson, R. P. Turco, E. J. Jensen and O. B. Toon, *Atmos. Environ.*, 1994, **28**, 1327–1338.

16 L. M. Russell and J. H. Seinfeld, *Aerosol Sci. Technol.*, 1998, **28**, 403–416.

17 M. Jacobson, *J. Geophys. Res.*, 2002, **107**, 1327–1338.

18 J. Lu and F. M. Bowman, *Atmos. Chem. Phys.*, 2010, **10**, 4033–4046.

19 H. Dergaoui, K. N. Sartelet, É. Debry and C. Seigneur, *J. Aerosol Sci.*, 2013, **58**, 17–32.

20 S. Zhu, K. Sartelet and C. Seigneur, *Geosci. Model Dev.*, 2015, **8**, 1595–1612.

21 M. J. Kleeman and G. R. Cass, *Environ. Sci. Technol.*, 2001, **35**, 4834–4848.

22 P. Stier, J. Feichter, S. Kinne, S. Kloster, E. Vignati, J. Wilson, L. Ganzeveld, I. Tegen, M. Werner, Y. Balkanski, *et al.*, *Atmos. Chem. Phys.*, 2005, **5**, 1125–1156.

23 S. Bauer, D. Wright, D. Koch, E. Lewis, R. McGraw, L.-S. Chang, S. Schwartz and R. Ruedy, *Atmos. Chem. Phys.*, 2008, **8**, 6003–6035.

24 N. Oshima, M. Koike, Y. Zhang, Y. Kondo, N. Moteki, N. Takegawa and Y. Miyazaki, *J. Geophys. Res.*, 2009, **114**, 1–17.

25 N. Riemer, M. West, R. A. Zaveri and R. C. Easter, *J. Geophys. Res.*, 2009, **114**, D09202.
26 V. Mallet, D. Quélo, B. Sportisse, M. Ahmed de Biasi, É. Debry, I. Korsakissok, L. Wu, Y. Roustan, K. Sartelet, M. Tombette, *et al.*, *Atmos. Chem. Phys.*, 2007, **7**, 5479–5487.
27 S. Zhu, N. K. Sartelet, Y. Zhang and A. Nenes, *J. Geophys. Res.*, 2016, accepted.
28 K. Sartelet, E. Debry, K. Fahey, Y. Roustan, M. Tombette and B. Sportisse, *Atmos. Environ.*, 2007, **41**, 6116–6131.
29 G. Sarwar, D. Luecken, G. Yarwood, G. Z. Whitten and W. P. Carter, *Journal of Applied Meteorology and Climatology*, 2008, **47**, 3–14.
30 K. M. Fahey and S. N. Pandis, *Atmos. Environ.*, 2001, **35**, 4471–4478.
31 A. Nenes, S. N. Pandis and C. Pilinis, *Aquat. Geochem.*, 1998, **4**, 123–152.
32 F. Couvidat, É. Debry, K. Sartelet and C. Seigneur, *J. Geophys. Res.: Atmos.*, 2012, **117**, DOI: 10.1029/2011JD017214.
33 L. W. Horowitz, S. Walters, D. L. Mauzerall, L. K. Emmons, P. J. Rasch, C. Granier, X. Tie, J.-F. Lamarque, M. G. Schultz, G. S. Tyndall, *et al.*, *J. Geophys. Res.: Atmos.*, 2003, **108**, DOI: 10.1029/2002JD002853.
34 V. Vestreng, *Review and Revision. Emission data reported to CLRTAP. EMEP MSC-W Status Report 2003*, Meteorological synthesizing centre-west, norwegian meteorological institute, oslo technical report, 2003.
35 A. Guenther, T. Karl, P. Harley, C. Wiedinmyer, P. Palmer and C. Geron, *Atmos. Chem. Phys.*, 2006, **6**, 3181–3210.
36 F. Couvidat, Y. Kim, K. Sartelet, C. Seigneur, N. Marchand and J. Sciare, *Atmos. Chem. Phys.*, 2013, **13**, 983–996.
37 A. Charron and G. Aymoz, *PM-DRIVE: Particulate DiRect and Indirect On-Road Vehicular Emissions*, L'agence de l'environnement et de la maîtrise de l'energie (ademe) technical report, 2015.
38 W. Skamarock, J. Klemp, J. Dudhia, D. Gill, D. Barker, M. Duda, X. Huang, W. Wang and J. Powers, *A description of the advanced research WRF version 3*, NCAR Technical Note, NCAR/TN\ u2013475? STR, 2008, p. 123.
39 Y. Kim, K. Sartelet, J.-C. Raut and P. Chazette, *Boundary-Layer Meteorology*, 2013, **149**, 105–132.
40 S.-Y. Hong, Y. Noh and J. Dudhia, *Mon. Weather Rev.*, 2006, **134**, 2318–2341.
41 S.-Y. Hong and J.-O. J. Lim, *Asia-Pacific Journal of Atmospheric Sciences*, 2006, **42**, 129–151.
42 J. S. Kain, *J. Appl. Meteorol.*, 2004, **43**, 170–181.
43 D. S. Gross, R. Atlas, J. Rzeszotarski, E. Turetsky, J. Christensen, S. Benzaid, J. Olson, T. Smith, L. Steinberg, J. Sulman, *et al.*, *Environ. Model. Software*, 2010, **25**, 760–769.
44 A. Russell and R. Dennis, *Atmos. Environ.*, 2000, **34**, 2283–2324.
45 J. W. Boylan and A. G. Russell, *Atmos. Environ.*, 2006, **40**, 4946–4959.
46 Y. Wang, K. N. Sartelet, M. Bocquet and P. Chazette, *Atmos. Chem. Phys.*, 2014, **14**, 3511–3532.
47 B. Bessagnet, A. Hodzic, R. Vautard, M. Beekmann, S. Cheinet, C. Honoré, C. Liousse and L. Rouil, *Atmos. Environ.*, 2004, **38**, 2803–2817.
48 N. Riemer and M. West, *Atmos. Chem. Phys.*, 2013, **13**, 11423–11439.

Faraday Discussions

PAPER

The treatment of uncertainties in reactive pollution dispersion models at urban scales

A. S. Tomlin,*[a] T. Ziehn,†[a] P. Goodman,‡[b] J. E. Tate[b] and N. S. Dixon[c]

Received 10th November 2015, Accepted 27th November 2015
DOI: 10.1039/c5fd00159e

The ability to predict NO_2 concentrations ($[NO_2]$) within urban street networks is important for the evaluation of strategies to reduce exposure to NO_2. However, models aiming to make such predictions involve the coupling of several complex processes: traffic emissions under different levels of congestion; dispersion *via* turbulent mixing; chemical processes of relevance at the street-scale. Parameterisations of these processes are challenging to quantify with precision. Predictions are therefore subject to uncertainties which should be taken into account when using models within decision making. This paper presents an analysis of mean $[NO_2]$ predictions from such a complex modelling system applied to a street canyon within the city of York, UK including the treatment of model uncertainties and their causes. The model system consists of a micro-scale traffic simulation and emissions model, and a Reynolds averaged turbulent flow model coupled to a reactive Lagrangian particle dispersion model. The analysis focuses on the sensitivity of predicted in-street increments of $[NO_2]$ at different locations in the street to uncertainties in the model inputs. These include physical characteristics such as background wind direction, temperature and background ozone concentrations; traffic parameters such as overall demand and primary NO_2 fraction; as well as model parameterisations such as roughness lengths, turbulent time- and length-scales and chemical reaction rate coefficients. Predicted $[NO_2]$ is shown to be relatively robust with respect to model parameterisations, although there are significant sensitivities to the activation energy for the reaction NO + O_3 as well as the canyon wall roughness length. Under off-peak traffic conditions, demand is the key traffic parameter. Under peak conditions where the network saturates, road-side $[NO_2]$ is relatively insensitive to changes in demand and more sensitive to the primary NO_2 fraction. The most important physical parameter was found to be the background wind direction. The study highlights the key parameters required for reliable $[NO_2]$ estimations suggesting that accurate reference measurements for wind direction should be a critical part of air quality assessments for in-street locations. It also highlights the importance of street

[a]*Energy Research Institute, School of Chemical and Process Engineering, University of Leeds, Leeds, UK. E-mail: A.S.Tomlin@leeds.ac.uk*
[b]*Institute for Transport Studies, University of Leeds, Leeds, UK*
[c]*School of Earth and Environment, University of Leeds, Leeds, UK*
† Current address: CSIRO, Oceans and Atmosphere, Aspendale, Australia.
‡ Current address: School of Civil Engineering and Geosciences, Newcastle University, UK.

scale chemical processes in forming road-side [NO_2], particularly for regions of high NO_x emissions such as close to traffic queues.

A Introduction

Although European directives to reduce NO_x emissions from vehicles have been in operation for well over a decade, many urban areas across Europe are still failing to meet the NO_2 air quality standards set by the EU Directive 2008/50/EC. Within the UK there are a large number of Air Quality Management Areas (AQMAs) which have been declared on the basis of NO_2, a large proportion of which are in highly trafficked urban areas. Strategies must therefore be put in place to address potential reductions in NO_2 concentrations, particularly focussed on traffic sources. To aid in this AQM process, models can be developed which aim to predict road-side NO_2 as a function of important traffic characteristics and can therefore be used to assess the potential impact of traffic management or emissions based intervention measures. Such models should also include relevant features of the urban environment that affect pollution formation and dispersion such as wind speed and direction, rapid chemical transformations and street topologies. The latter feature should be included since it is well understood that urban buildings interact with background winds to modify the turbulent flow structures within the streets, restricting the dispersion of traffic related pollutants out of the street network,[1–5] and potentially allowing time for in-street chemical processing of emissions. Models aiming to achieve road-side predictions therefore should contain representation of this potentially circulating flow occurring within the street canyons, as well as vertical and horizontal fluxes into and out of the street network.[3] Representations of chemical transformations are also required, in particular those relevant to the formation of secondary NO_2. A number of modelling approaches have been suggested to address the dispersion part of the problem including Gaussian based models such as OSPM (Operational Street Pollution Model),[6] network models such as SIRANE,[7] and compartment based models where pollutant exchange rates are parameterised according to canyon aspect ratios,[8] as well as high resolution computational fluid dynamics (CFD) approaches.[9,10] Several studies have attempted to couple models of the complex turbulent flow with chemical sub-models, albeit for single or small networks of streets.[11–15] These studies have highlighted the influence of incomplete mixing on the formation of secondary pollutants in regions of high primary NO_x emissions. This suggests that coarse resolution models (*e.g.* urban air shed models) are unlikely to be of direct relevance for the study of road-side concentrations and exposure, although they may provide boundary conditions for higher resolution studies at the street scale. However, detailed representations of the interplay between emissions, chemical transformations and dispersion within the street network will be critical for near field exposure. Few models, however, have attempted to couple this modelling of dispersion at the micro-scale with both chemical transformation processes and high resolution traffic emissions models. The current work aims to present such an integrated system which couples a micro-scale traffic emissions model with a turbulent reactive dispersion model based on a combined CFD and reactive Lagrangian particle dispersion approach.[16,17]

The use of such models within the AQM framework requires an understanding of the confidence that can be placed in their predictions. Lack of confidence, or uncertainty, can result from a lack of detailed knowledge of the model parameterisations. It follows that model evaluation will benefit from the inclusion of sensitivity studies that highlight the impact of uncertain input parameters on predicted output concentrations. As part of an uncertainty/sensitivity study we might ask questions such as: (i) how robust are predicted road-side NO_2 concentrations, *i.e.* how wide are the predicted distributions if we properly account for uncertainties within the model input parameters, and (ii) taking into account uncertainties in the model parameters, can we trace the effects of potential traffic management strategies such as reductions in demand, reductions in emissions, and changes in emissions profiles, or are these effects swamped by uncertainties due to the model itself?

Within CFD based approaches, the use of Reynolds Averaged Navier Stokes (RANS) models in particular has raised questions as to their suitability for accurately describing turbulent chemical interactions when they contain only averaged representations of turbulent length and time-scales. On the plus side, such models allow the representation of reactive dispersion within complex street topologies at lower computational cost than, for example, Large Eddy Simulations (LES). On the negative side, they require parameterisation of mixing lengths rather than resolving eddies in an explicit way. They do however, provide much more realistic descriptions of in-street processing of pollutants than urban air shed type models which are too coarse to capture complex street geometries and the effects of buildings on in-street flow profiles. Given the potential utility of such RANS models, it is worth considering how robust their simulations are to the parameterisations chosen. We attempt to address some of these questions here and present an approach for the assessment of sensitivities for a complex multi-component model aiming to predict time-averaged road-side concentrations of NO_2 as a function of street topologies, background meteorology, traffic characteristics and temperature dependent parameterisations of chemical reactions.

B Methodology

(i) Case study and dispersion model structure

The location modelled in this study is that of Gillygate, York, UK, the site of an extensive measurement campaign[2,18] that has provided observations used in previous evaluations of some of the model components used here.[10] Gillygate is a relatively narrow street with an aspect ratio (building height to street width) of approximately 0.8, leading to cross-street recirculating flow under a range of background wind directions, restricting the dispersion of pollutants out of the street. The traffic flow along Gillygate is quite high with significant periods of congestion, and it therefore represents a potential pollution hot spot. It was included within the stated AQMAs by the City of York Council in 2014.[19] 19 exceedances of the hourly limit value for NO_2 of 200 $\mu g\ m^{-3}$ were noted in 2011, and annual average [NO_2] values varied between 30 and 50 $\mu g\ m^{-3}$ from (2010–2015) based on data from the air quality monitoring site which is situated away from the congested junction at the Northern end of the street. Diffusion tube data from various points within Gillygate[19] showed significant variation in concentrations along the street, with annual average values exceeding 50 $\mu g\ m^{-3}$ in some locations. It therefore provides a useful polluted road-side case study on which to demonstrate the approach adopted for model uncertainty analysis in this work.

Fig. 1 (top) shows the grid and the building configuration of Gillygate and the surrounding area that were used for the simulations in this study. The building heights in metres are indicated in the legend. Fig. 1 (bottom) provides an aerial view of the main part of the modelling domain. The basis for the underlying flow and turbulence model under consideration is the k–ε RANS model MISKAM.[20] This model was chosen on the basis that it is commonly used as an operational model[9] and has undergone previous evaluation for street canyon case studies.[10] It has been shown to substantially improve the representation of dispersion in built up areas due to improved representation of turbulent flow patterns compared to more general dispersion models such as those used in compliance assessment.[21] In this study, a non-equidistant grid was used to enable a higher resolution within the area of interest.

Fig. 1 shows the two locations G3 and G4 that were used in the original measurement campaign as well as the mast location that was used to obtain reference wind speed and direction. The experimental measurements were of concentrations of the non-reactive tracer carbon monoxide. Unfortunately no measurements of [NO_2] are available from the campaign but previous evaluations with respect to [CO] highlighted the ability of the model to capture the main flow and dispersion characteristics within the street.[10] The uncertainty analysis here allows additional investigation into the robustness of the model with respect to the chemical parameterisations. We use the same locations here as in the previous study for investigating [NO_2] predictions, as well as three other sites on each side of the street at 20 m intervals to the South of G3 and G4. A wind direction of 0° represents channelled flow from North to South along the street canyon. The wind directions sampled in the case study (110–130°) represent oblique flow over the building adjacent to G3 towards the North of the domain and lead to a helical in-street recirculating flow with a northerly channelled component.[2] We focus on a single wind sector here due to the computational cost of the random sampling approach required for the global sensitivity calculations. In reality of course, many different wind directions would be used within an air quality assessment. Since model validation is usually performed using concentrations normalised by a reference wind speed, the inflow wind speed is kept to a constant value of 5 m s^{-1} at a height of 50 m, with the vertical profile then determined using a log-law which is based on the sampled values of the roughness length for the incoming flow.

The output from MISKAM is used as the underlying turbulent flow structure for a dispersion model based on the Lagrangian stochastic particle dispersion approach with micro-mixing and chemical sub-models (for a full description and evaluation see ref. 16 and 17). The complex dispersion modelling system was used previously to investigate a reactive plume of nitrogen oxides (NO_x) released into an approximately homogeneous turbulent grid flow doped with ozone (O_3) for comparison against wind tunnel experiments.[17] The chemical and micro-mixing sub-models used here are the same as those specified in the photolysis extended case described in Ziehn *et al.*[17] The chemical reactions included are detailed in Table 1. In summary, only simple reactions between NO, NO_2 and O_3 are included in the chemical model but these include the photolysis of NO_2 and O_3. Quite broad ranges have been included for photolysis rates reflecting variation in daytime conditions. No organic reactions are included in the analysis. Rather, any reactions

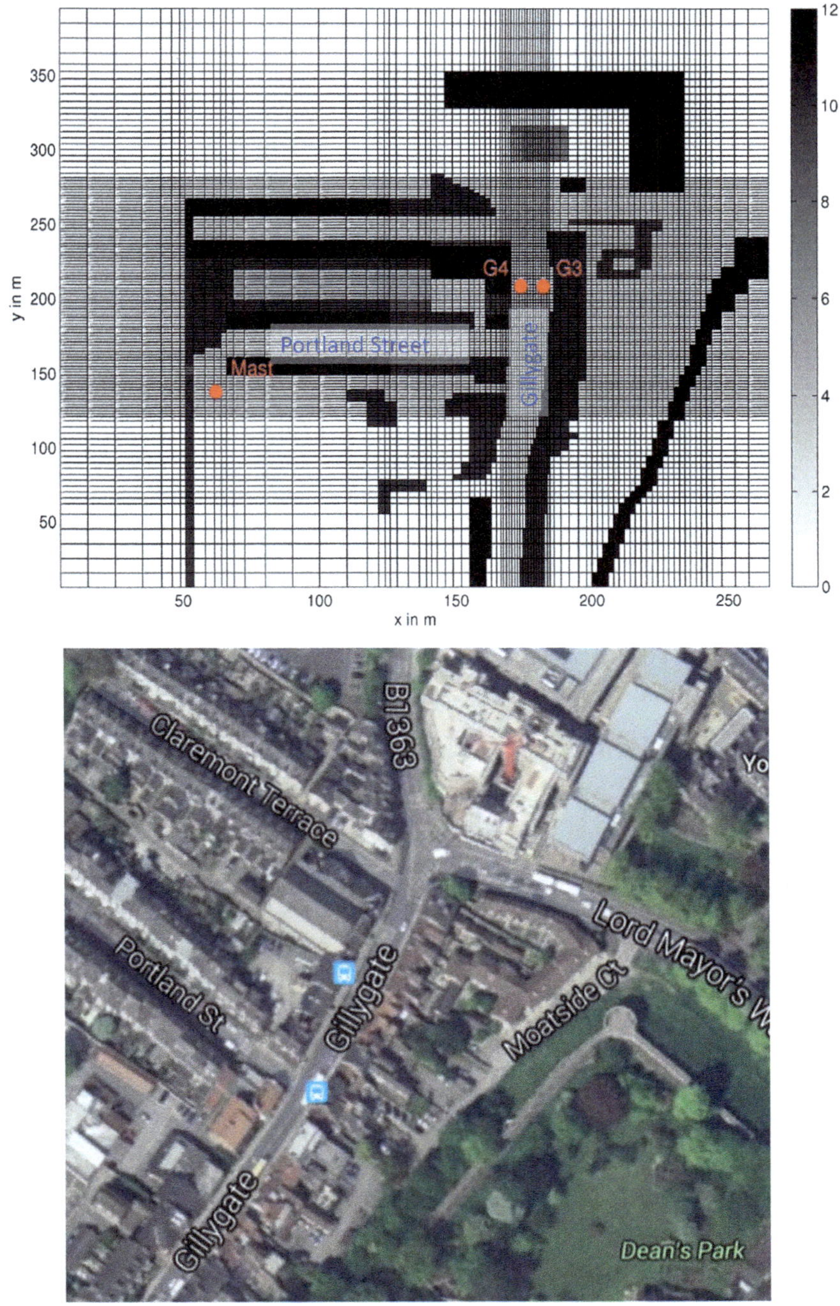

Fig. 1 Top: Site schematic for the York Gillygate site showing the grid and building configuration as used in MISKAM. Bottom: Aerial view of the site. ©2015 Infoterra Ltd. & Bluesky, ©2015 Google.

occurring at longer time-scales are represented implicitly within the description of above-roof O_3 concentration which is considered as an uncertain parameter. A high sensitivity with respect to this parameter would indicate the importance of long

Table 1 Chemical scheme parameter ranges used within the uncertainty analysis. Concentrations of $[N_2] = 2.03 \times 10^{19}$ molecule cm^{-3} and $[O_2] = 5.45 \times 10^{18}$ molecule cm^{-3} have been used

Parameter	Nominal value	Minimum	Maximum	Unit	Source
A factor, R1 $O + O_2 + N_2 = O_3 + N_2$	5.60×10^{-34}	4.991×10^{-34}	6.283×10^{-34}	cm^6 per molecule^2 per s	33
A factor, R2 $O + O_2 + O_2 = O_3 + O_2$	6.00×10^{-34}	5.348×10^{-34}	6.732×10^{-34}	cm^6 per molecule^2 per s	33
A factor, R3 $O + O_3 = 2O_2$	8.00×10^{-12}	6.654×10^{-12}	9.618×10^{-12}	cm^3 per molecule per s	33
A factor (k_0), R4 $O + NO + M = NO_2 + M$	1.00×10^{-31}	0.794×10^{-31}	1.259×10^{-31}	cm^6 per molecule^2 per s	33
A factor R4 (k_1)	3.00×10^{-11}	1.504×10^{-11}	5.986×10^{-11}	cm^3 per molecule per s	33
A factor, R5 $O + NO_2 = NO + O_2$	5.50×10^{-12}	4.790×10^{-12}	6.315×10^{-12}	cm^3 per molecule per s	33
A factor, R6 $NO + O_3 = NO_2 + O_2$	1.40×10^{-12}	1.165×10^{-12}	1.683×10^{-12}	cm^3 per molecule per s	33
Photolysis rate *J* O_3, R7 $O_3 = O + O_2$	2.75×10^{-5}	1.0×10^{-5}	4.5×10^{-5}	s^{-1}	34
Photolysis rate *J* NO_2, R8 $NO_2 = NO + O$	0.0075	0.004	0.011	s^{-1}	35
n for reaction R1	2.6	2.1	3.1	—	33
n for reaction R2	2.6	2.1	3.1	—	33
E/*R* for reaction R3	2060	1860	2260	K^{-1}	33
n_0 for reaction R4	1.6	1.3	1.9	—	33
n_∞ for reaction R4	−0.3	−0.6	0	—	33
E/*R* for reaction R5	−188	−268	−108	K^{-1}	33
E/*R* for reaction R6	1310	1110	1510	K^{-1}	33

range chemical processes for the net formation of NO_2. For the current study the coupled dispersion model is further linked to a traffic micro-simulation model and a zero background concentration of $[NO_2]$ is assumed so that the modelled concentrations represent road-side increments above background.

(ii) Traffic micro-simulation modelling

Vehicle flows within the study area were modelled using an established, commercial traffic micro-simulation package AIMSUN 5.1.10 (ref. 22) which represents the movement of individual vehicles through a road network using discrete time intervals of the order of one second. Individual components within the micro-simulation govern the interaction of vehicles with one another, the interaction of vehicles with traffic signals, how vehicles make lane-changing manoeuvres and how vehicles accept gaps in traffic streams. Within each time-step individual components are called, using information from the previous time step, to assign new kinematic information (acceleration, speed and position) to every vehicle. Vehicle and driver parameters which are considered static within a given run (*e.g.* maximum vehicle acceleration rates) are generated on vehicle entry to the network, and sampled from appropriate distributions. Given the fine spatial and temporal resolution of traffic micro-simulation, output statistics may be aggregated over a wide variety of scales, for use within appropriate environmental models. A substantial body of literature already exists on methodologies detailing such approaches, *e.g.* ref. 23 and 24.

(iii) Traffic emissions modelling

Using the AIMSUN model, total NO_x emissions were calculated using the polynomial emissions functions proposed by Int Panis *et al.*[25] based on vehicle type, instantaneous speed, and acceleration parameters. Instantaneous emissions rates (g s^{-1}) were calculated for each vehicle at each model time-step (~1 s) using polynomial functions. These were converted to mass values, based on initial vehicle velocity, and the assumption of linear acceleration over the time-step. Given the position of the vehicle at the start and end of a time-step, contributions were then apportioned to the individual sections, through which the vehicle had passed. The final output from the bespoke software was the emission rates (g h^{-1}) for each 10 m road section, based on the integrated total of all contributions from vehicles passing through the sections in that period. These were then converted to g m^{-1} s^{-1} for input into the dispersion model.

The specific fraction of NO_2 within the total NO_x was treated as an uncertain parameter as discussed later. Calculated emissions were then linked by vehicle position to a particular 10 m section of road giving spatial-profiles of emissions along Gillygate *via* bespoke software external to AIMSUN.[26] The overall traffic network consisted of 4 km of roads surrounding Gillygate and 8 intersections, including 2 which were signalised. Four categories of vehicles were considered: cars, vans, HGVs and buses, for compatibility with Int Panis *et al.*[25] The dynamic demand in the network (the number of vehicles desiring to travel through the network within the simulated hour) was varied over two sets of normalised ranges. The first is an "off-peak" case from 0.8–1.2 with the mean value of 1.0 representing 'typical' inter-peak demand. The second was a "peak" case with demand varying from 1.2–1.6. Each simulation run therefore represented 1 h at a particular level of demand using a random sampling approach within the specified ranges.

The normalised demand level was derived from a year of traffic flow data obtained from York's urban traffic control system equating to a two-way flow of ~880 veh per h along Gillygate. Additional to the dynamic demand was a fixed level of demand from buses based on timetable information. At the base demand level, the network is considered as busy, but in an 'under-saturated' state, *i.e.* able to cope with the level of demand, with only transient queues forming at junctions. At demand levels above 1.1, modelled speeds begin to decline rapidly from ~20 km h^{-1} to ~10 km h^{-1} at a demand of 1.4. At these higher demand levels, substantial over-saturated queues form as vehicles are unable to clear signalised junctions within a single signal period. For off-peak, under capacity periods, total emissions increase in a slightly non-linear fashion with the volume of traffic as shown for a typical section of Gillygate in Fig. 2a. Some of the non-linearity may be explained by the increasing

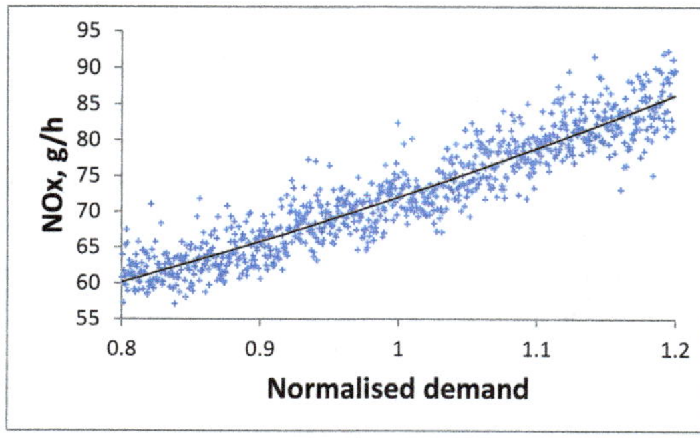

a)

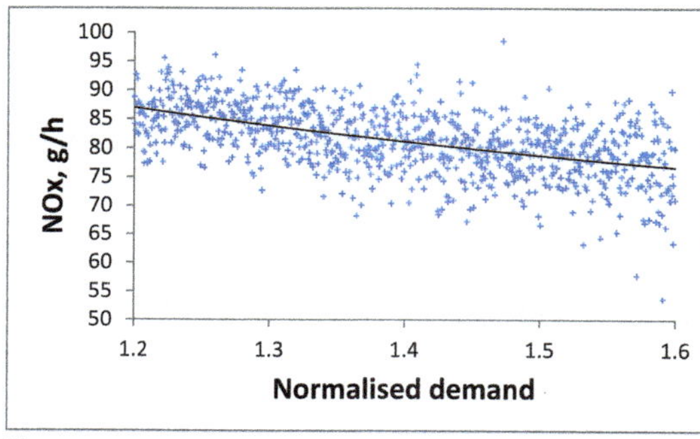

b)

Fig. 2 Total exhaust emissions of NO_x for a sample section of Gillygate for (a) off-peak and (b) peak demand scenarios.

relative fraction of HGVs present, whose contribution to NO_x emissions starts to dominate those of passenger cars. This phase is followed by a transitional period as demand approaches and exceeds network capacity, where emissions stabilise at a high overall level as shown in Fig. 2b. The influence of these characteristics on modelled road-side mean NO_2 levels is discussed in the next section.

(iv) Model parameterisations and ranges of uncertainty

Close to a surface the effects of turbulence in a RANS model need to be modelled using boundary conditions that reflect the surface roughness. MISKAM represents these boundary conditions using an idealised log-law based on three surface roughness lengths z_o for the incoming boundary flow, the urban surface and wall surfaces, respectively. It follows that parameterisations of these roughness lengths will be a possible source of uncertainty in the final output predictions since they determine the near surface velocity profiles and therefore influence both dispersion and deposition[27] processes. A study by Benson *et al.*,[28] based on the application of MISKAM to a street canyon scenario, showed that overall, the uncertainty in the predicted mean and turbulent flow fields due to roughness lengths was small in comparison to the mean outputs. The study concluded that the model was well defined even with large ranges of input parameter uncertainty. In general, at a particular receptor, the closest surface was found to be most influential on the model output except when a recirculating canyon flow pattern was strongly present. In this latter case background wind angle was found to dominate.

The turbulent flow field is used as an input to the Lagrangian particle dispersion model where it is assumed that the velocity of a fluid element, for times larger than the Kolmogorov time-scale, is a Markov process that is a continuous function of time.[29] Within the Lagrangian particle model framework, the two important parameterisations are the Lagrangian structure coefficient c_0, and the mixing time-scale coefficient α. The Lagrangian structure function is defined as the ensemble average of the square of the change in Lagrangian velocity and the definition of c_0 is therefore important in determining the effective turbulent diffusion in velocity space. There is some debate within the literature as to whether its value can be universally defined for all types of turbulent flows with a range of values between 2 and 10 quoted from different studies.[30–32] It is interesting to establish therefore how sensitive concentration predictions are to the chosen value. Within the model tested, a simple particle mixing model is adopted, that of interaction by exchange with the mean (IEM) concentration.[29] In order to provide generality, the mixing model uses a coefficient α which defines the relationship between the turbulent time-scales (total turbulent kinetic energy and its dissipation rate) and the mixing time-scale at every point in the flow as defined by the following equation:

$$t_{\mathrm{m}} = \alpha \frac{k}{\varepsilon}$$

The specification of α should also be considered to be uncertain and the relative time-scales of the chemical processes compared to the mixing time-scale could have an important influence on the formation of secondary species such as NO_2.

Uncertainties in the traffic emissions model have been adopted for the level of traffic demand as discussed above, and the $NO : NO_x$ ratio for the emissions source which determines the fraction of NO *vs.* primary NO_2 assumed at source. The range adopted was chosen to reflect levels of primary NO_2 estimated for UK vehicle fleets,[36] taking into account a range of possible fuel types as well as engine exhaust treatment methodologies (15–25% primary NO_2). The 26 model parameters varied within the sensitivity/uncertainty analysis can therefore be summarised as:

- Velocity structure function coefficient c_0 [4–6].
- Mixing time-scale coefficient α [0.6–3].
- Surface roughness length z_o for inflow, surface and wall [5–50, 0.5–50, 0.5–10 cm].
- Temperature dependent Arrhenius rate parameters for $NO/NO_2/O_3$ reactions, photolysis rate parameters for J O_3 and J NO_2 [see Table 1 for details].
- Background wind direction θ [110–130°].
- Temperature [273–298 K].
- Background ozone concentration [7.35×10^{11} to 1.23×10^{12}, molecules cm^{-3} or 30–50 ppb].
- $NO : NO_x$ ratio for traffic emissions [0.75–0.85].
- Normalised traffic demand [off peak 0.8–1.2, peak 1.2–1.6].

Where the ranges used are shown in the square brackets except for the Arrhenius and photolysis parameters which were detailed in Table 1. Of these parameters, most relate to uncertainties in the model formulation. For a given urban morphology, it is mainly the last three, background O_3 concentration, $NO : NO_x$

ratio and traffic demand, that could be affected by the implementation of pollution mitigation strategies. Therefore if the model were to be highly robust, we would like to see the overall sensitivities of the model dominated by these last three parameters. In a wider study it would of course be possible to assess the impact of changes to urban form including "urban greening[8]" which may impact on roughness lengths and therefore dispersion and deposition.

(v) Global uncertainty and sensitivity analysis methods

Within a global uncertainty analysis the aim is to determine the possible range of predicted outputs (in this case time-averaged road-side NO_2 concentration increments) given the ranges of uncertainty within the model input parameters. Often a Monte Carlo type approach is adopted where random or quasi-random samples of the inputs are used within the adopted ranges of uncertainty, and many model simulations are performed; one for each random sample. The result is a predicted distribution of outputs instead of a single value. The width of this distribution gives the modeller an idea of the robustness of the model given the model input uncertainties. Statistical parameters such as standard deviation can be used to provide simple quantified measures of the output uncertainty in cases where the distribution does not deviate too far from being Gaussian. Otherwise higher moments may also be useful. The aim of improving model robustness is to provide better and better quantification of the model inputs such that the predicted output distribution becomes as narrow as possible *i.e.* the predicted output uncertainty is low.

In order to assist in the model improvement processes it is useful to determine which of the uncertain model inputs have the largest influence of predicted output uncertainties. This is achieved here through a global sensitivity analysis based on the ANalysis Of VAriance (ANOVA) approach.[37] In such methods, the output variance is decomposed into component functions representing the effects of individual and groups of parameters whose importance can then be ranked according to global sensitivity indices.

The global sensitivity analysis has been achieved using the RS-HDMR (Random Sampling High Dimensional Model Representation) method introduced by Rabitz *et al.*[38] to express the input–output relationship of complex models with large numbers of input parameters, and further developed into a user friendly Matlab package GUI-HDMR by Ziehn and Tomlin.[39] The mapping between input parameters $x_1, \ldots, x_n$ and output variables $f(x) = f(x_1, \ldots, x_n)$ in the domain R^n is written in the form:

$$f(x) = f_0 + \sum_{i=1}^{n} f_i(x_i) + \sum_{1 \le i < j \le n} f_{ij}(x_i, x_j) + \ldots + f_{12\ldots n}(x_1, x_2, \ldots, x_n) \tag{1}$$

where f_0 denotes the mean effect (or zeroth-order term), which is a constant. The function $f_i(x_i)$ is a first-order term (or first-order component function) giving the effect of parameter x_i acting independently (although generally nonlinearly) upon the output $f(x)$. What this component function shows is the independent influence of parameter x_i on the output right across the domain of uncertainty for x_i. The function $f_{ij}(x_i, x_j)$ is a second-order term describing the co-operative effects of the parameters x_i and x_j upon the output $f(x)$. The higher-order terms reflect the co-operative effects of increasing numbers of input parameters acting together to influence the output $f(x)$.

Due to its formulation as a set of hierarchical component functions, the HDMR expansion provides the possibility to determine sensitivity indices for each of the input parameters in an automatic way for selected model outputs. For given input parameter ranges, these indices indicate the relative contribution of each parameter to the predicted output variance. Thus they can be directly used to rank the importance of each individual parameter in determining the model output variance and to explore parameter interactions. The HDMR expansion is computationally very efficient if higher order input parameter interactions are weak and can therefore be neglected. For many systems a HDMR expression up to second-order already provides satisfactory results and a good approximation of $f(x)$ (*e.g.* ref. 28).

In RS-HDMR, a number of model simulations are performed using a quasi-random set of input samples. This set of model simulations is then used to fit polynomial expressions for each component function in eqn (1). The sensitivity coefficients for individual parameters or for interaction terms can then be easily calculated from the coefficients of the polynomial expansion (see Ziehn and Tomlin[39] for details). For the current studies, the 26 dimensional input space is sampled 512 times using a quasi-random approach from uniform distributions within the parameter ranges specified. The RS-HDMR meta-model fit is then generated where the output function $f(x)$ represents the NO_2 concentration at the 8 in-street locations discussed above. In practice, the larger the sample size, the better the fitted representation of the component functions and therefore the more accurate the sensitivity indices will be. In reality sample size is often limited by computational resources, particularly for complex high resolution models with substantial individual run times. When using a limited sample size such as used here, it is very important to assess the accuracy of the functional fit to eqn (1) and the GUI-HDMR code provides the facility to do this. Previous applications of the method in chemical kinetics problems[40] has shown that large sample sizes (>1000) are usually only needed where significant second-order terms are present (*i.e.* important parameter interactions exist).

C Results and discussion

(i) Overall uncertainty of predicted $[NO_2]$ and accuracy of HDMR fits

The predicted $[NO_2]$ distributions for sites G3 and G4 are shown in Fig. 3 for the off-peak case study. The most frequently predicted $[NO_2]$ at G3 is around 1×10^{12} molecules cm^{-3} which equates to $\sim$76 $\mu g\ m^{-3}$ and is twice that at G4 confirming the influence of the in-street recirculation on the concentration profiles across the canyon. Direct comparison with high time resolution measured data is not possible but data from the diffusion tube network operated by the City of York Council offers a qualitative comparison.[19] In 2013, their measurement site in Gillygate opposite Portland street (close to G3, see Fig. 1) gave an annual average for $[NO_2]$ of 48.4 $\mu g\ m^{-3}$ compared to only 21 $\mu g\ m^{-3}$ at various sites along Portland street itself. It is clear therefore that due to the in-street dispersion processes large variations in NO_2 concentration can be seen over quite short distances. The distributions show that, given the uncertainties adopted for the input parameters, the predicted $[NO_2]$ on both sides of the street can vary by up to a factor of 2. This suggests that more accurate parameterisations of the inputs are necessary to improve the robustness of the model predictions.

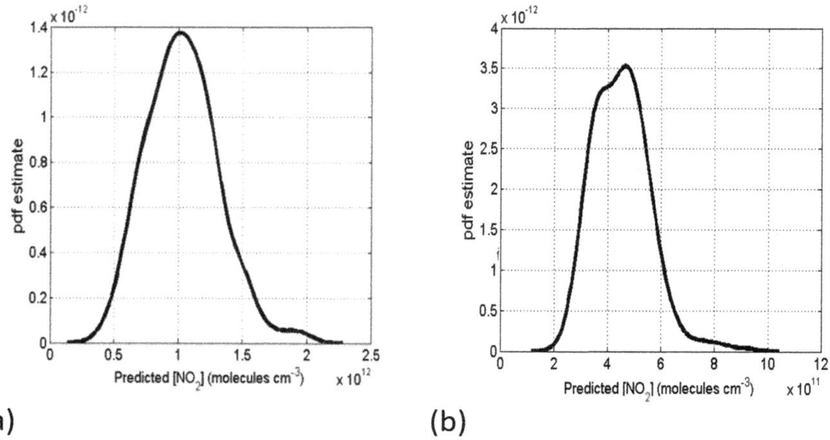

Fig. 3 Predicted [NO_2] distributions for sites (a) G3 and (b) G4 based on a quasi-random sample of 512 runs and the uncertainty limits described in Section B.

By decomposing the predictive variance into contributions from each of the uncertain inputs, the global sensitivity methodology allows us to target such model improvement strategies on the most important parameters. However, in order to exploit the HDMR component functions for sensitivity analysis purposes, it is first important to establish that the HDMR meta-model gives a reasonable fit to the outputs from the full model runs. This test is especially important for the current example since the combined model simulation time was of the order of an hour and therefore the sample size of 512 was limited by available computer resources. The coefficient of determination or R^2 values comparing the fitted second-order HDMR meta-model with the data from the full model simulations vary between 0.905 and 0.951 illustrating that the second-order meta-model gives a good fit despite the limited sample size. This provides confidence in the accuracy of the HDMR component functions and the sensitivity results derived from them. The percentage of total variance accounted for by first-order effects ranges from 79–94% for the different sites indicating that the variance is dominated by sensitivities to individual parameters, and that the effects of parameter interactions are quite small. Where second- or higher-order effects do exist they tend to lead to tails in the predicted output distributions which can be seen to some extent in Fig. 3.

(ii) Sensitivity coefficients – overall parameter importance

Turbulent flow parameterisations. The sensitivity coefficients calculated using HDMR provide the relative influence of each parameter on the variance within distributions such as that shown in Fig. 3. They are scaled between 0 and 1, where 1 represents 100% contribution to the output variance. Fig. 4 presents the average sensitivity coefficients across the 8 sample locations for predicted road-side time-averaged [NO_2]. The mixing time-scale coefficient α and other parameters with low importance are not shown on the plot. The lack of sensitivity to α is important and shows that the mean concentration predictions are not greatly affected by the mixing time-scale and that the simple IEM mixing model is valid in this context. The Lagrangian structure function coefficient c_0 also exhibits a low sensitivity. This

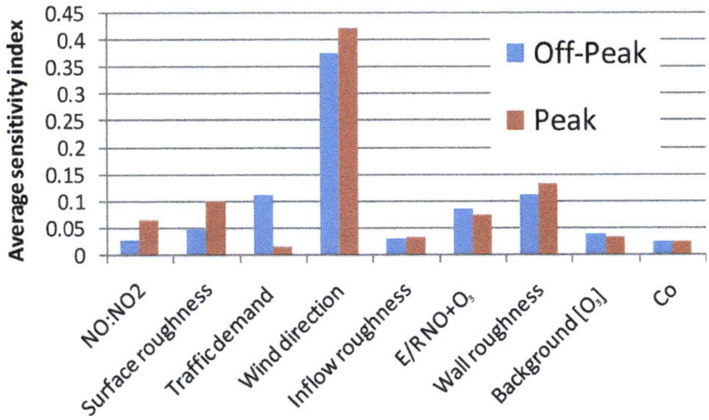

Fig. 4 The average first-order sensitivity coefficients across all 8 sites for the peak and off-peak normalised traffic demand studies.

is an encouraging result and suggests that the simulated concentrations are not highly sensitive to the chosen turbulence model parameterisations. There are however, sensitivities to the parameterisation of the velocity profiles at the model surfaces (*i.e.* to the chosen values of roughness lengths z_0). The lowest sensitivity is to the inflow roughness, which is used to define the incoming logarithmic velocity profile at the boundary of the domain. The low sensitivity indicates that the computational domain was large enough to reduce the influence of the inflow boundary conditions on the overall dispersion profiles. The wall roughness however, exhibits a mean sensitivity of over 10%. Detailed calculations show that for site G3, this can be as high as 30%. This suggests that predicted [NO_2] close to the street canyon walls (within 2 m in the case of G3) can be highly sensitive to the near wall flow parameterisations as was previously suggested to be the case by Benson *et al.*[28] for velocity and turbulence fields in these locations. We will return to this point in the next section where sensitivity coefficients for individual sites are discussed in more detail.

Physical and chemical parameterisations. Wind direction is the major physical parameter which dominates the prediction of [NO_2] at all locations. On average it accounts for around 40% of the variance in predicted [NO_2]. This implies that a reliable reference measurement of background wind conditions is an essential input for air quality modelling systems, particularly those aiming to represent recirculating flows within urban street canyons and to assess road-side exposure. The sensitivity of predicted mean [NO_2] to background [O_3] is on average quite low, which may be surprising given that the main formation route for NO_2 on short time-scales that are comparable with the in-street recirculation times, is *via* the reaction of NO with O_3. However, a higher sensitivity to the activation energy for the reaction of NO with O_3 is seen. This suggests that there is a higher sensitivity to the rate of this short time-scale reaction occurring within the street than to the amount of ozone mixing into the canyon due to longer range processes. This, coupled with the importance of wind direction, indicates that if we wish models to capture exposures at the road-side, then detailed models of flow, reaction and dispersion on very short spatial scales are necessary. More accurate parameterisation of the temperature dependence of the reaction NO + O_3, using for example, *ab initio* calculations, would be critical to improving model robustness.

In terms of the influence of traffic characteristics, there are clear differences between the two modelled demand scenarios. For off-peak conditions, there is clearly a response to the levels of traffic demand with an average contribution of ~11% to the predicted [NO_2]. Under peak conditions where the network saturates, road-side [NO_2] is relatively insensitive to changes in demand and more sensitive to the primary NO_2 fraction (see Fig. 4). The results show that in order to have a substantial effect on road-side NO_2 through traffic demand reduction measures, it would be necessary to reduce demand to the lower end of the sensitive region, *i.e.* by 60% or more.

(iii) Sensitivity coefficients – site to site variation

In this section we focus on the off-peak traffic demand scenario and investigate the responses to changes in selected parameters at individual sites. First we compare sites G3 and G4 which lie on opposite sides of the canyon close to the signalled junction on the leeward and windward sides of the canyon, respectively. Fig. 3 shows that G3 displayed the higher NO_2 concentrations due to the in-canyon helical circulation present. At G3 the highest sensitivity ($S_i = 0.3$) is to the wall roughness length z_o with the response shown in Fig. 5a. The greater the assumed roughness length, the higher the predicted [NO_2], since a higher roughness leads to lower in-canyon flow velocities as the boundary is approached. In these figures the red line displays the first-order component function for the selected parameter (see eqn (1)). Were the response to depend only on a single parameter with an S_i of 1, then all points would lie on the red line. In reality, several parameters contribute to the output variance and hence the scatter points in the plots represent the effects of the other important parameters. For the wall z_o, the range spanned by the component function for this parameter (shown in red) is about 5×10^{11} molecules cm^{-3} *i.e.* almost 40 μg m^{-3} and suggests that the near wall flow is highly influential on pollutant exposure within street canyon type geometries. Depending on the type of wall, windows, street furniture *etc.* the z_o

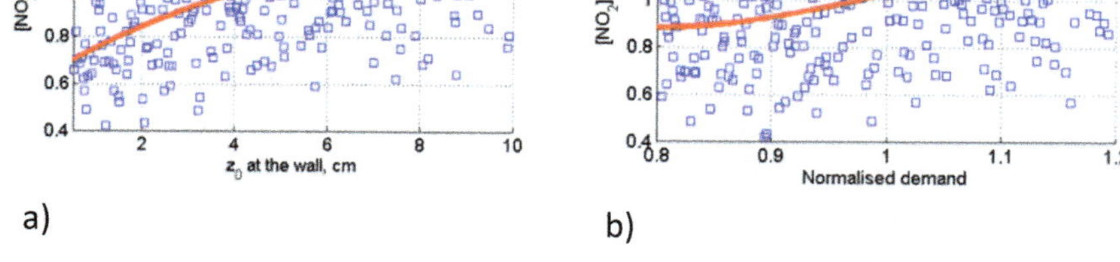

Fig. 5 Scatter plots and first-order component functions highlighting the response to changes in individual parameters on time-averaged [NO_2] at receptor G3 (a) wall roughness length z_o (b) normalised traffic demand. The scatter points show the overall response based on a random sample size of 512. The red lines show the first-order component functions *i.e.* the individual response to each parameter.

value may vary significantly from site to site and is a difficult parameter to estimate causing problems for improving model robustness.

The second most influential parameter is traffic demand ($S_i = 0.15$) as shown in Fig. 5b. Changes in demand of 40% lead to a range of NO_2 concentrations of a width of about 3×10^{11} molecules cm^{-3} *i.e.* 23 μg m^{-3}. This illustrates that within the off-peak flow regime, reductions in demand could lead to substantial reductions in road-side mean [NO_2] levels. The figure shows that as the demand levels increase, the slope of the sensitivity decreases which is consistent with the fact that overall emissions flatten off under highly congested conditions. The pay-off is therefore greater for the lower demand scenarios.

Lower [NO_2] concentrations are seen at G4 since the flow has circulated around the canyon before reaching this receptor point and hence primary emissions of NO_x have dispersed to a certain extent. At G4 the most dominant parameter is now the normalised demand ($S_i = 0.25$) which suggests that the model is relatively robust to the flow and turbulence parameterisations at this receptor. Chemical processes are, however, more dominant at this windward location than they were at G3. The activation energy for reaction $NO + O_3 = NO_2 + O_2$ has a high sensitivity ($S_i = 0.24$) at this site, as does the background ozone concentration ($S_i = 0.08$). The component functions and scatter plots for these two parameters are shown in Fig. 6a and b respectively. As the activation energy is lowered, the production of NO_2 across the canyon increases and assumed uncertainties in this parameter can account for differences in predicted [NO_2] of around 15 μg m^{-3} based on the predicted range of the component function (shown in red). The influence of background [O_3] is about half as strong based on the assumed uncertainties.

Wind direction was not highly influential at sites G3 and G4 since the most important dispersion process is the recirculating flow which is driven by a strong cross-street wind component for all reference directions tested. However, as we move down the canyon away from the congested junction, then it begins to play a more and more dominant role as it affects the along street dispersion of emissions from the heavily trafficked junction. At sites A and B, 60 and 40 m to the

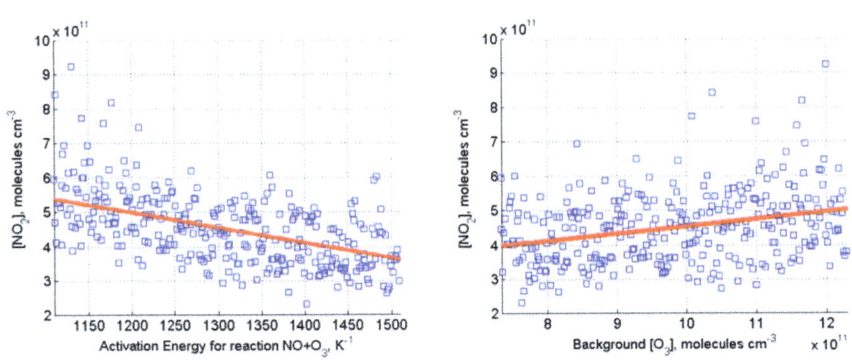

Fig. 6 Scatter plots and first-order component functions highlighting the response to changes in individual parameters on time-averaged [NO_2] at receptor G4, (a) activation energy for $NO + O_3$ reaction and (b) background ozone concentration. The scatter points show the overall response based on a random sample size of 512. The red lines show the first-order component functions *i.e.* the individual response to each parameter.

South of G4 respectively, the sensitivity coefficients with respect to wind angle are >0.7, and the response to background wind direction is highly non-linear as shown in Fig. 7 for site A. The complex interplay between the topology of the street network and the flow patterns established means that the concentrations of NO_2 at site A are mostly dominated by primary emissions, and therefore as the wind direction takes on an increasingly southerly component, the concentration decreases. The relative importance of in-street chemical processes will be discussed further in the next section.

(iv) Discussion of chemical processes

In-street NO_2 concentration increments above background will be affected by both primary NO_2 emissions being dispersed from vehicle exhausts, and the formation of secondary NO_2 from primary NO through its reaction with ozone which is mixing into the street from the background. The relative importance of these processes will vary with location within the street and can be explored *via* analysis of the sensitivity coefficients. In this section we therefore also consider the sensitivity of in-street time-averaged ozone concentrations ($[O_3]$) to the model parameterisations.

The main first-order sensitivity indices for sites G3 (leeward close to junction), G4 (windward close to junction), A (windward 60 m South of G4), and D (leeward 60 m South of G3) are shown in Fig. 8 for both ozone and NO_2 concentrations. In-street ozone is mainly controlled by the background concentration except for at site G3 where high concentrations of NO_x are expected due to its close proximity to queueing traffic and the in-street recirculation patterns. However, this sensitivity to background ozone is not seen for $[NO_2]$ which suggests that concentration increments of NO_2 above background within congested street networks of the street canyon type are unlikely to be dominated by long range transport of ozone. Instead, in-street NO_2 tends to be dominated by the effects of wind direction, traffic demand, local secondary NO_2 formation rates, and near wall velocity

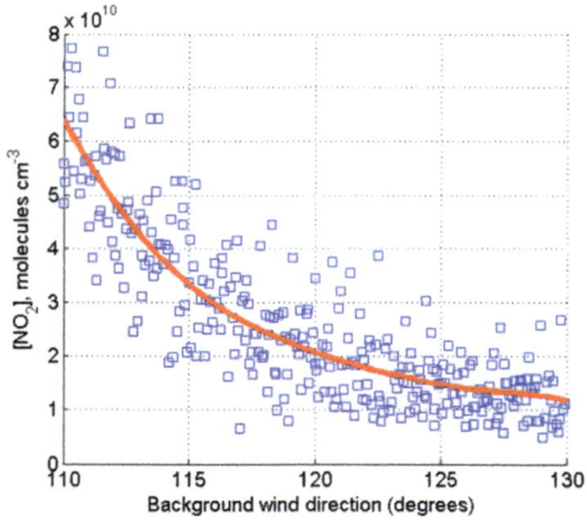

Fig. 7 Scatter plot and first-order component function highlighting the response to changes in reference wind direction on time-averaged $[NO_2]$ at a receptor 60 m to the South of G4. The scatter points show the overall response based on a random sample size of 512. The red lines show the first-order component functions *i.e.* the individual response to each parameter.

profiles governed by roughness parameterisations. The importance of local NO_2 formation is indicated by the importance of the rate parameters for the inorganic chemical reactions with the activation energy for NO + O_3 being the dominant parameter. This parameter is particularly influential at sites G3 and G4 close to the junction where NO_x concentrations are high. It is also the dominant parameter affecting in-street $[O_3]$ at site G3 since a higher rate for this reaction increases the titration of ozone by NO.

(v) Uncertainties in model structure and issues of "uniqueness of place"

The approach for uncertainty assessment used here is parameter based and therefore assumes that the chosen model structure is appropriate for the intended application. It assesses uncertainties in prediction due to lack of knowledge of parameter values but not due to missing processes within the model. There are a number of challenges in choosing an appropriate model structure. The first is computational cost. It is unlikely that a high resolution CFD model that resolves all eddy structures formed within the complex geometry (*e.g.* using LES or direct numerical simulation), as well as all possible chemical transformations, can be used within the context of compliance assessment due to computational requirements. Even such highly detailed models require accurate boundary and initial conditions. At the other end of the spectrum, semi-empirical models such as those based on a Gaussian or multi-compartment approaches are based on parameterisations informed by street topologies, rather than detailed representations of in-street flow characteristics. The use of parameters based on geometric features such as canyon width to height ratios has provided a way of generalising the use of such models within different city geometries. However, site specific

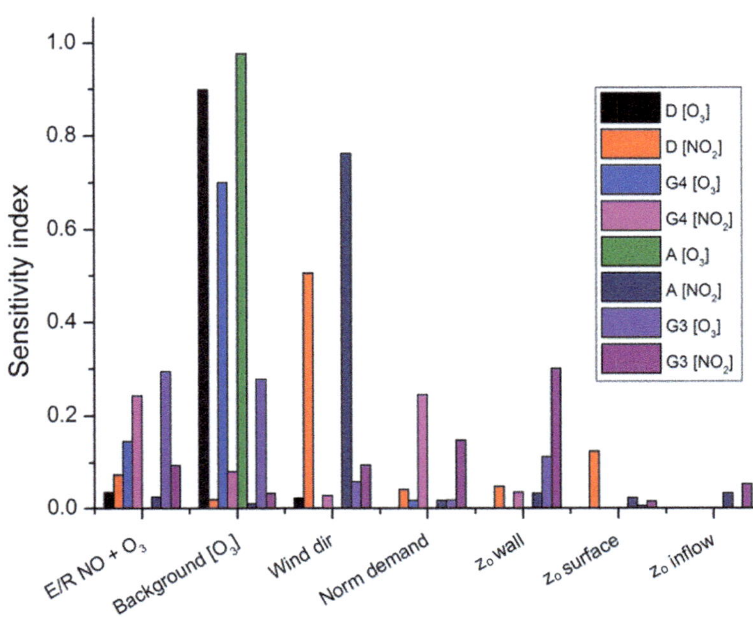

Fig. 8 Main first-order sensitivity indices for sites G3, G4, A and D. An index of 1 implies that a single parameter totally determines the variance in predicted concentrations.

features such as complex intersections and roof shapes cannot be captured by such models and yet the DAPPLE (Dispersion of Air Pollution and its Penetration into the Local Environment) field campaigns highlighted the importance of a number of these features in determining in-street flow and dispersion.[41,42] Such models then may contain structural uncertainties that limit their generality. This raises the issue of "uniqueness of place" for models that has been previously highlighted within the field of hydrological modelling.[43] LES models may require less site specific parameterisation, but may be unaffordable from a computational perspective.

The RANs CFD approach adopted in this work attempts to seek the middle ground, but each simulation still takes of the order of 10 minutes to an hour on a standard PC, and hence the use of such models is difficult within the context of rapid screening. These models, however, can capture the average effects of flows within complex geometries that lead to spatial variation in pollutant concentrations over small scales. A key question that remains is whether uncertainties within their parameterisations can be reduced significantly or whether they still suffer from issues of "uniqueness of place", where parameterisations are highly site specific and difficult to quantify. Background wind direction ranked highly amongst the sensitive parameters affecting in-street [NO_2]. In principle, above-roof measurements could be obtained for all sites of interest but in reality this would be practically difficult. The need for a reference site free of interference from local flow features was discussed by Barlow *et al.* for the case of London.[44] Roughness lengths were also important, in particular those of the nearest surface to the receptor site studied. Roughness lengths cannot be measured directly and whilst approaches have been recently suggested to estimate these for complex city-scale surfaces,[45] in reality canyon wall roughness lengths will always have to be estimated and may be site specific.

D Conclusions

A global uncertainty and sensitivity study was used to investigate the robustness of predictions of road-side time averaged NO_2 concentrations based on a Reynolds averaged flow model coupled with a reactive Lagrangian particle dispersion model using NO_x emission rates calculated from a traffic micro-simulation model. Overall, predicted [NO_2] at a number of in-street sites was uncertain to within approximately a factor of 2. The predictions were fairly robust with respect to the parameterisations of turbulent dispersion and mixing time-scales. Uncertainties in the parameterisation of these processes did not greatly affect the time-averaged predictions. At near wall sites close to the high emission traffic junction, however, predictions were strongly influenced by the parameterisation of near boundary flows determined by the wall roughness length z_o. This is a difficult parameter to estimate but could strongly influence the prediction of in-street exposure within narrow street canyons such as the one studied here. Further away from the traffic junction, reference wind direction was the most dominant parameter affecting predicted [NO_2]. Whilst this parameter could be measured above-roof, it is not a routine measurement that is carried out within UK urban areas. Rather, reference meteorological measurements are commonly obtained from nearby rural sites or local airports. The direction of above-roof winds has been shown to strongly affect in-street flow and dispersion processes[5,44] and thus it is not surprising that it influences the transport of NO_x emissions from the congested junction to other in-street receptors in this case. However, obtaining long term appropriate reference measurements is a difficult challenge for most cities.

With respect to chemical processes, it is the short time-scale, in-street processes that seem to most strongly affect road-side [NO_2]. In particular the activation energy for the reaction of NO with O_3 is the most critical parameter. Better quantification of the rate of this reaction would help to improve model robustness.

In terms of mitigation strategies, two patterns emerge. Under peak traffic conditions the model suggests that even moderate reductions in traffic demand are unlikely to reduce in-street [NO_2]. For such congested conditions, the fraction of NO_2 in total NO_x was shown to be more influential than demand, indicating that reducing primary NO_2 could be a key factor in reducing NO_2 in congested street canyon situations. For lower demand scenarios, there were demonstrable benefits to reducing traffic demand, mainly at sites close to the traffic queue. Overall, despite the scatter in the data due to uncertainties within the model parameterisations, there were clearly discernible effects of possible demand or emissions management measures suggesting that the model set up was robust enough to be useful within the context of exposure mitigation.

Acknowledgements

This work was supported by EPSRC grant GR/S90881/01.

References

1 S. K. Park, S. D. Kim and H. Lee, *Sci. Total Environ.*, 2004, **323**, 263–271.
2 J. W. D. Boddy, R. J. Smalley, N. S. Dixon, J. E. Tate and A. S. Tomlin, *Atmos. Environ.*, 2005, **39**, 3147–3161.
3 P. Louka, S. E. Belcher and R. G. Harrison, *Atmos. Environ.*, 2000, **34**, 2613–2621.
4 A. S. Tomlin, R. J. Smalley, J. E. Tate, J. F. Barlow, S. E. Belcher, S. J. Arnold, A. Dobre and A. Robins, *Atmos. Environ.*, 2009, **43**, 5027–5037.
5 A. Dobre, S. J. Arnold, R. J. Smalley, J. W. D. Boddy, J. F. Barlow, A. S. Tomlin and S. E. Belcher, *Atmos. Environ.*, 2005, **39**, 4647–4657.
6 R. Berkowicz, *Environ. Monit. Assess.*, 2000, **65**, 323–331.
7 M. Carpentieri, P. Salizzoni, A. Robins and L. Soulhac, *Environ. Modell. Softw.*, 2012, **37**, 110–124.
8 T. A. M. Pugh, A. R. MacKenzie, J. D. Whyatt and C. N. Hewitt, *Environ. Sci. Technol.*, 2012, **46**, 7692–7699.
9 M. Ketzel, R. Berkowicz and A. Lohmeyer, *Environ. Monit. Assess.*, 2000, **65**, 363–370.
10 N. S. Dixon, J. W. D. Boddy, R. J. Smalley and A. S. Tomlin, *Atmos. Environ.*, 2006, **40**, 958–972.
11 J. J. Baik, Y. S. Kang and J. J. Kim, *Atmos. Environ.*, 2007, **41**, 934–949.
12 A. Garmory, I. S. Kim, R. E. Britter and E. Mastorakos, *Atmos. Environ.*, 2009, **43**, 4670–4680.
13 V. B. Bright, W. J. Bloss and X. M. Cai, *Atmos. Environ.*, 2013, **68**, 127–142.

14 J. Baker, H. L. Walker and X. M. Cai, *Atmos. Environ.*, 2004, **38**, 6883–6892.
15 J. Zhong, X. M. Cai and W. J. Bloss, *Environ. Pollut.*, 2015, **200**, 42–52.
16 N. S. Dixon and A. S. Tomlin, *Atmos. Environ.*, 2007, **41**, 8114–8127.
17 T. Ziehn, N. S. Dixon and A. S. Tomlin, *Atmos. Environ.*, 2009, **43**, 5978–5988.
18 J. W. D. Boddy, R. J. Smalley, P. S. Goodman, J. E. Tate, M. C. Bell and A. S. Tomlin, *Atmos. Environ.*, 2005, **39**, 3163–3176.
19 City of York Council, *2014 Air Quality Progress Report for City of York Council*, 2014.
20 J. Eichhorn and A. Kniffka, *Meteorol. Z.*, 2010, **19**, 81–90.
21 M. O. Letzel, T. Flassak and D. Angel, *Gefahrstoffe - Reinhalt. Luft*, 2012, **72**, 329–334.
22 TSS, *AIMSUN – The Integrated Transport Modelling Software*, http://www.aimsun.com/wp/.
23 F. An, M. Barth, J. Norbeck and M. Ross, *Transp. Res. Rec.*, 1997, **1587**, 52–62.
24 M. Zallinger, J. Tate and S. Hausberger, *16th International Transport and Air Pollution Conference*, Graz, Austria, 2008.
25 L. I. Panis, S. Broekx and R. H. Liu, *Sci. Total Environ.*, 2006, **371**, 270–285.
26 P. S. Goodman and G. Rhys Tyler, *6th AIMSUN User Group Meeting*, Capri, Italy, 2008.
27 M. W. Gallagher, E. Nemitz, J. R. Dorsey, D. Fowler, M. A. Sutton, M. Flynn and J. Duyzer, *J. Geophys. Res.: Atmos.*, 2002, 107.
28 J. Benson, T. Ziehn, N. S. Dixon and A. S. Tomlin, *Atmos. Environ.*, 2008, **42**, 1874–1891.
29 B. L. Sawford, *Flow, Turbulence and Combustion*, 2004, **72**, 133–160.
30 D. Anfossi, G. Degrazia, E. Ferrero, S. E. Gryning, M. G. Morselli and S. T. Castelli, *Bound.-Lay. Meteorol.*, 2000, **95**, 249–270.
31 S. M. Du, *Bound.-Lay. Meteorol.*, 1997, **83**, 207–219.
32 A. M. Reynolds, *Bound.-Lay. Meteorol.*, 1998, **89**, 161–170.
33 R. Atkinson, D. L. Baulch, R. A. Cox, J. N. Crowley, R. F. Hampson, R. G. Hynes, M. E. Jenkin, M. J. Rossi and J. Troe, *Atmos. Chem. Phys.*, 2004, **4**, 1461–1738.
34 R. E. Shetter, C. A. Cantrell, K. O. Lantz, S. J. Flocke, J. J. Orlando, G. S. Tyndall, T. M. Gilpin, C. A. Fischer, S. Madronich, J. G. Calvert and W. Junkermann, *J. Geophys. Res.: Atmos.*, 1996, **101**, 14631–14641.
35 R. E. Shetter, A. H. McDaniel, C. A. Cantrell, S. Madronich and J. G. Calvert, *J. Geophys. Res.: Atmos.*, 1992, **97**, 10349–10359.
36 D. C. Carslaw, *Atmos. Environ.*, 2005, **39**, 4793–4802.
37 I. M. Sobol, *Math. Comput. Simulat.*, 2001, **55**, 271–280.
38 H. Rabitz, O. F. Alis, J. Shorter and K. Shim, *Comput. Phys. Commun.*, 1999, **117**, 11–20.
39 T. Ziehn and A. S. Tomlin, *Environ. Modell. Softw.*, 2009, **24**, 775–785.
40 T. Turányi and A. S. Tomlin, *Analysis of Kinetic Reaction Mechanisms*, Springer, 2014.
41 A. A. Balogun, A. S. Tomlin, C. R. Wood, J. F. Barlow, S. E. Belcher, R. J. Smalley, J. J. N. Lingard, S. J. Arnold, A. Dobre, A. G. Robins, D. Martin and D. E. Shallcross, *Bound.-Lay. Meteorol.*, 2010, **136**, 489–513.
42 C. R. Wood, S. J. Arnold, A. A. Balogun, J. F. Barlow, S. E. Belcher, R. E. Britter, H. Cheng, A. Dobre, J. J. N. Lingard, D. Martin, M. K. Neophytou, F. K. Petersson, A. G. Robins, D. E. Shallcross, R. J. Smalley, J. E. Tate, A. S. Tomlin and I. R. White, *Bull. Am. Meteorol. Soc.*, 2009, **90**, 955–970.

43 K. J. Beven, *Hydrol. Earth Syst. Sci.*, 2000, **4**, 203–213.
44 J. F. Barlow, A. Dobre, R. J. Smalley, S. J. Arnold, A. S. Tomlin and S. E. Belcher, *Atmos. Environ.*, 2009, **43**, 5536–5544.
45 J. T. Millward-Hopkins, A. S. Tomlin, L. Ma, D. B. Ingham and M. Pourkashanian, *Bound.-Lay. Meteorol.*, 2013, **146**, 447–468.

Faraday Discussions

PAPER

Assessing chemistry schemes and constraints in air quality models used to predict ozone in London against the detailed Master Chemical Mechanism

Tamsin L. Malkin,[a] Dwayne E. Heard,[ab] Christina Hood,[c] Jenny Stocker,[c] David Carruthers,[c] Ian A. MacKenzie,[d] Ruth M. Doherty,[d] Massimo Vieno,[e] James Lee,[fg] Jörg Kleffmann,[h] Sebastian Laufs[h] and Lisa K. Whalley*[ab]

Received 11th December 2015, Accepted 8th February 2016
DOI: 10.1039/c5fd00218d

Air pollution is the environmental factor with the greatest impact on human health in Europe. Understanding the key processes driving air quality across the relevant spatial scales, especially during pollution exceedances and episodes, is essential to provide effective predictions for both policymakers and the public. It is particularly important for policy regulators to understand the drivers of local air quality that can be regulated by national policies *versus* the contribution from regional pollution transported from mainland Europe or elsewhere. One of the main objectives of the Coupled Urban and Regional processes: Effects on AIR quality (CUREAIR) project is to determine local and regional contributions to ozone events. A detailed zero-dimensional (0-D) box model run with the Master Chemical Mechanism (MCM*v3.2*) is used as the benchmark model against which the less explicit chemistry mechanisms of the Generic Reaction Set (GRS) and the Common Representative Intermediates (CRI*v2-R5*) schemes are evaluated. GRS and CRI are used by the Atmospheric Dispersion Modelling System (ADMS-Urban) and the regional chemistry transport model EMEP4UK, respectively. The MCM model uses a near-explicit chemical scheme for the oxidation of volatile organic compounds (VOCs) and is constrained to observations of VOCs, NO_x, CO, HONO (nitrous acid), photolysis frequencies and meteorological parameters measured during the ClearfLo (Clean Air for London) campaign. The sensitivity of the less explicit chemistry schemes

[a]School of Chemistry, University of Leeds, Leeds, UK. E-mail: L.K.Whalley@leeds.ac.uk
[b]National Centre for Atmospheric Science, University of Leeds, Leeds, UK
[c]Cambridge Environmental Research Consultants, 3 King's Parade, Cambridge, UK
[d]School of GeoSciences, University of Edinburgh, Edinburgh, UK
[e]Centre for Ecology & Hydrology, Bush Estate, Penicuik, UK
[f]Department of Chemistry, University of York, York, UK
[g]National Centre for Atmospheric Science, Wolfson Atmospheric Chemistry Laboratory, York, UK
[h]Physikalische und Theoretische Chemie/Fakultät Mathematik und Naturwissenschaften, Bergische Universität Wuppertal (BUW), Wuppertal, Germany

to different model inputs has been investigated: Constraining GRS to the total VOC observed during ClearfLo as opposed to VOC derived from ADMS-Urban dispersion calculations, including emissions and background concentrations, led to a significant increase (674% during winter) in modelled ozone. The inclusion of HONO chemistry in this mechanism, particularly during wintertime when other radical sources are limited, led to substantial increases in the ozone levels predicted (223%). When the GRS and CRI*v2-R5* schemes are run with the equivalent model constraints to the MCM, they are able to reproduce the level of ozone predicted by the near-explicit MCM to within 40% and 20% respectively for the majority of the time. An exception to this trend was observed during pollution episodes experienced in the summer, when anticyclonic conditions favoured increased temperatures and elevated O_3. The *in situ* O_3 predicted by the MCM was heavily influenced by biogenic VOCs during these conditions and the low GRS [O_3] : MCM [O_3] ratio (and low CRI*v2-R5* [O_3] : MCM [O_3] ratio) demonstrates that these less explicit schemes under-represent the full O_3 creation potential of these VOCs. To fully assess the influence of the *in situ* O_3 generated from local emissions *versus* O_3 generated upwind of London and advected in, the time since emission (and, hence, how far the real atmosphere is from steady state) must be determined. From estimates of the mean transport time determined from the NO_x : NO_y ratio observed at North Kensington during the summer and comparison of the O_3 predicted by the MCM model after this time, ~60% of the median observed [O_3] could be generated from local emissions. During the warmer conditions experienced during the easterly flows, however, the observed [O_3] may be even more heavily influenced by London's emissions.

1. Introduction

Air pollution is the environmental factor with the greatest impact on human health in Europe[1]. More than 65% of the population in Europe live in cities where levels of particulate matter (PM) exceed the WHO Air Quality Guidelines, and when considering exceedances in the secondary pollutant ozone (O_3), this figure rises to 95%.[1,2] $PM_{2.5}$ (particulate matter less than 2.5 micrometres in diameter) and O_3 are estimated to contribute to 458 000 and 17 400 premature deaths each year across Europe, respectively.[1] Both PM and O_3 concentrations are influenced by many factors including local emissions, chemistry and regional advection.[3] Understanding the key processes driving air quality across the relevant spatial and temporal scales, especially during pollution episodes, is necessary to inform both policymakers and the public. For the development of effective mitigation policies, it is important to isolate the contribution of local emissions and production during peak events from that due to long-range transport of regional pollution.

Air pollution events in the UK are generally associated with stagnation events, which can occur at any time, but which in summer may be coincident with heatwaves.[4] Large contributions of European pollution to the UK are often associated with slow-moving easterly air masses.[3,5] Several studies have highlighted the downward entrainment of O_3-rich air from Europe to ground level in the UK during the heatwave of 2003.[6–8] Interactions between the weather conditions and chemistry processes can also contribute to elevated O_3, for example, the role of enhanced biogenic emissions in the production of ozone in urban plumes in the UK was first considered by MacKenzie *et al.*[9] Other processes including reduced dry deposition, and extensive forest fires in mainland Europe during heatwaves, can also influence UK O_3.[5–8]

There remains some debate in the analysis of surface ozone levels as to the origin of the ozone and in particular the role of long-range transport.[10] Coupled modelling studies allow regional and local processes affecting air quality across the UK and, specifically, in London to be examined. Recent intensive measurements from the NERC ClearfLo (Clean Air for London) and REPARTEE (Regent's Park and Tower Environmental Experiment) campaigns provide an opportunity to combine box models and measurements to gain an insight into the relative contributions of locally generated and transported O_3. During the winter campaigns of REPARTEE (October 2006 and October/November 2007),[11] O_3 concentrations recorded at the BT Tower were always higher than at surface sites, consistent with the city acting as an efficient chemical sink for regional O_3, which was confirmed by strong downward fluxes of O_3 measured on the BT Tower.[11]

The ClearfLo project was a measurement programme in and around London lasting 2 years (2011–2012) including two month-long intensive observation periods (IOPs) in the summer and wintertime.[3] The summer IOP, which coincided with the London 2012 Olympics, provides the opportunity to investigate factors controlling O_3 when photochemical activity is high, in contrast to the earlier REPARTEE campaign which took place in winter. Understanding O_3 production in summer is critical as this is when the exceedances typically occur. During the summer IOP, a number of high pollution events were observed where meteorological conditions favoured sustained, elevated O_3 levels (peaking at ~100 ppb[3]). In addition to the high O_3 levels recorded during the summer IOP, much lower O_3 concentrations were recorded during the winter IOP (January–February, 2012), with a peak concentration of ~40 ppb recorded.

One of the main objectives of the Coupled Urban and Regional processes: Effects on AIR quality (CUREAIR) project is to determine local and regional contributions to ozone (O_3) events. This paper presents a comparison of O_3 concentrations predicted to be produced locally in London (both during the summertime and wintertime) using three different chemistry schemes. The level of chemical complexity between the three schemes investigated varies greatly from (1) a simple 7 step reaction scheme (the generic reaction scheme – GRS) which is used in the Atmospheric Dispersion Modelling System (ADMS-Urban)[12] to (2) a lumped, 609 reaction, chemistry scheme (Common Representative Intermediates – CRI) used in some versions of the EMEP4UK regional transport model to (3) a near-explicit, 16 940 reaction, chemistry scheme (the Master Chemical Mechanism – MCM*v3.2*) used in this paper as the benchmark scheme against which the smaller chemistry schemes are assessed.

The sensitivity of simulated O_3 produced by the chemistry schemes to a variety of influences including the effects of VOC and NO_x concentrations, and the photolysis frequency of NO_2, as well as the impact of HONO are presented. Comparison of the O_3 simulated by the three schemes when equivalent constraints are used is also presented.

This paper first describes the model methodology and is then divided into 4 results sections: (1) investigating the impact of VOCs, NO_x, NO_2 photolysis frequency and HONO during the summer ClearfLo IOP on the O_3 concentration predicted by the GRS scheme, (2) investigating the influence of HONO and NO_x during the summer ClearfLo IOP on the O_3 concentration predicted by the CRI*v2*-

R5 scheme, and (3) and (4) investigating the same impacts as (1) and (2) respectively but for the winter ClearfLo IOP. The paper then concludes with a summary of the results and discussion of the influence of chemistry and transport on O_3 production in London.

2. Methodology

Models of atmospheric composition typically fall into one of several categories according to the spatial coverage (global, regional, urban, street canyon, single point (box)). Urban-scale dispersion models allow simulation at roadsides and hotspots but rely on background observations for their initialisation conditions whereas regional chemistry-climate models can be used for future climate simulations but have insufficient spatial resolution for direct comparison with urban monitoring sites. A combination of these two modelling strategies can be used to study regional and local drivers of urban air quality. In addition, chemical box models can provide a detailed representation of the chemical environment and may be used for evaluation of simpler chemical oxidation schemes represented in regional and urban-scale models.

2.1 Master Chemical Mechanism

The Master Chemical Mechanism (MCM*v3.2*)[13,14] is a near-explicit chemical mechanism which describes the detailed gas-phase chemical processes involved in the tropospheric degradation of a series of primary emitted VOCs. The entire MCM treats the degradation of methane and 142 non-methane VOCs and considers photolysis and oxidation by OH, Cl, O_3 and NO_3. In its entirety, the MCM (*v*3.2) contains 5734 species and 16 940 gas-phase reactions. The MCM has been utilised using input data from numerous field campaigns to investigate the chemistry of the polluted urban boundary layer,[15,16] marine boundary layer,[17,18] continental low-NO_x regions influenced by biogenic emissions[19,20] and polar regions;[21,22] a comprehensive review of these and more can be found in Stone *et al.*, 2012.[23]

In this work, a 0-D box model was run using the MCM*v3.2* and constrained to measurements from the ClearfLo project of NO, CO, CH_4, HONO, 62 individual VOC species measured by GC-FID and also 2D-GC,[24,25] PAN, HCHO, HNO_3, water vapour, temperature, pressure and measured photolysis frequencies (including j(O^1D), j(NO_2), j(HONO), j(HCHO), j(CH_3COCH_3) and j(CH_3CHO)) calculated using wavelength-resolved actinic flux measurements by a spectral radiometer. A constant H_2 concentration of 500 ppb was assumed.[26] The model constraints were updated hourly. For species measured more regularly, data were averaged to hourly intervals. The model was run unconstrained to O_3 and NO_2 and the predicted *in situ* (or locally produced) O_3 was output once a steady state concentration had been reached (further details on how the box model was run is provided in Section 2.4).

For all model generated species (including O_3 and NO_2) a first order loss rate (k) was included. This loss rate varied as a function of the measured boundary layer height (h) ($k = V_d/h$), where V_d represents a deposition velocity which was taken to be equal to 1 cm s^{-1} for all of the model-generated species.

In addition to the MCM, two smaller chemistry schemes are used in this study and run within the same box model framework: the Generic Reaction Set (GRS) as used in the urban dispersion model, ADMS-Urban, and the Common Representative Intermediates (CRI) mechanism as used in the regional chemistry transport model, EMEP4UK. Both these models are described below. The scheme sizes and capabilities are summarised in Table 1.

2.2 ADMS-Urban dispersion model and GRS mechanism

ADMS-Urban[27] is a quasi-Gaussian air dispersion modelling tool able to resolve the details of concentration fields within an urban area at high resolution (tens of metres) by explicitly representing the near-field features of the dispersion of emissions from all source types; specifically: point, line, area, volume, road and airport runway source types. ADMS-Urban performs well in comparison to measured data for a wide range of pollutants including NO_2, PM_{10} and O_3 and by explicitly modelling complex urban features such as street canyons, tunnels and noise barriers and including a relatively simple photolytic NO_x chemistry scheme, the model is able to predict concentrations at kerbside locations as well as at urban background sites; consequently, the model is used worldwide for air quality management and assessment studies. ADMS-Urban can be used as a stand-alone system, where upwind measurements of meteorology and pollutants are used to drive the model. Alternatively, the model can be coupled to regional modelling systems, for both assessment[12,28] and forecasting applications.[29] Fig. 1a shows an ADMS-Urban dispersion model contour plot of the annual mean O_3 concentration in London for 2010.

Within the ADMS-Urban domain the local NO_x and O_3 chemistry is represented by the GRS, initially developed at the Commonwealth Scientific and Industrial Research Organisation (CSIRO)[30,31] in Australia. The GRS was

Table 1 Key attributes of the various chemistry schemes used

	Species	Reactions	Chemistry includes oxidation of	Models used in
[a]MCM*v3.2*	5734 (3789)	16 940 (11 410)	143 VOCs including 22 ≤C12 alkanes, 16 ≤C12 alkenes, 9 aldehydes, 18 aromatics, isoprene, α- and β-pinene	Benchmark scheme
[b]GRS	7	7	Lumped VOCs	[c]ADMS-Urban
[d]CRI*v2-R5*	220	609; 4% The size of the MCM	22 VOCs including ≤C4 alkanes and alkenes, 9 oxygenated compounds, benzene, toluene, isoprene, *o*-xylene, α- and β-pinene	[e]EMEP4UK

[a] MCM (Master Chemical Mechanism), subset sizes in brackets. [b] GRS (Generic Reaction Set). [c] ADMS-Urban (Atmospheric Dispersion Modelling System). [d] CRI (Common Representative Intermediates). [e] EMEP4UK (European Monitoring and Evaluation Programme (for the UK)).

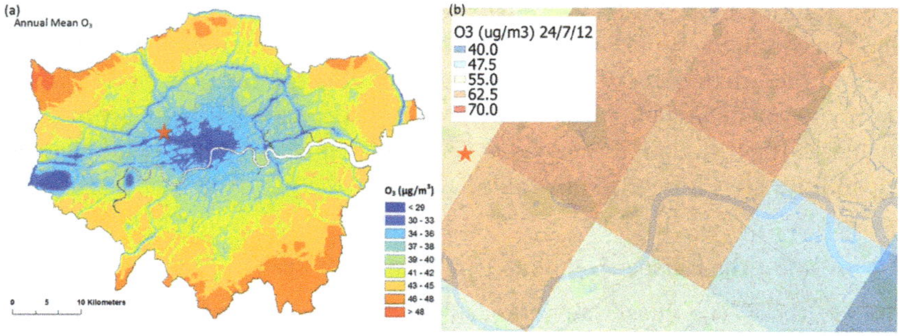

Fig. 1 (a) Contour plot of London showing the annual mean O_3 concentrations predicted by ADMS-Urban for 2010 constrained to upwind observations. (b) EMEP4UK hourly average surface O_3 concentrations at 5 km resolution over central London at 12:00, 24/07/2012 (during ClearfLo summer IOP). The star represents the North Kensington site in each case.

developed by fitting model estimates of O_3 concentrations to data obtained from an outdoor smog chamber using VOC and NO concentrations typical of urban areas.[30,31] It is a semi-empirical photochemical mechanism which reduces the many thousands of chemical reactions involving NO, NO_2, O_3 and many hydrocarbons (which are treated explicitly in the MCM) to the following seven reactions involving a number of surrogate species:

$$\mathrm{ROC} + h\nu \xrightarrow{\mathrm{O_2}} \mathrm{RP} + \mathrm{ROC} \tag{R1}$$

$$\mathrm{RP} + \mathrm{NO} \rightarrow \mathrm{NO_2} \tag{R2}$$

$$\mathrm{NO_2} + h\nu \xrightarrow{\mathrm{O_2}} \mathrm{NO} + \mathrm{O_3} \tag{R3}$$

$$\mathrm{NO} + \mathrm{O_3} \rightarrow \mathrm{NO_2} \tag{R4}$$

$$\mathrm{RP} + \mathrm{RP} \rightarrow \mathrm{RP} \tag{R5}$$

$$\mathrm{RP} + \mathrm{NO_2} \rightarrow \mathrm{SGN} \tag{R6}$$

$$\mathrm{RP} + \mathrm{NO_2} \rightarrow \mathrm{SNGN} \tag{R7}$$

where ROC refers to Reactive Organic Compounds, RP is the Radical Pool, SGN is the Stable Gaseous Nitrogen product and SNGN is the Stable Non-Gaseous Nitrogen product.

Reactions (3) and (4) represent chemically exact mechanisms, while the rest of the reactions are only approximate representations of their chemical counterparts. Reaction (1) is a semi-empirical representation of all the processes that lead to radical production from VOCs through photo-oxidation. ROC is conserved in the reaction; thus, becomes a surrogate for the products of the initial oxidation of the emitted VOCs. Reaction (2) represents the conversion of NO to NO_2 by reaction with radicals and leads to the termination of the radical, RP. Reaction (5) represents another sink for the radical pool. Reactions (6) and (7) lead to the formation of organic and inorganic nitrates. The rate of (R1) is a function of the photolysis frequency for NO_2 (further details below), an empirically determined weighted mean reactivity coefficient of the ROC, and the temperature. By normalising the emitted VOCs by their rates of radical production, a composite ROC concentration is derived; in effect, the ROC concentration determined is weighted by the ability of the different VOCs used in the calculation to produce radicals.

Typically, the GRS chemistry scheme used in ADMS-Urban is run constrained to background measurements of NO, NO_2, O_3 and, if available, ROC. These background measurements are from a location approximately 10 km upstream from the point of interest and are representative of rural background concentrations. The reactions (R1)–(R7) are applied in the model in two steps: initially the rural background concentrations and sources undergo reaction, and then secondly the contribution of sources nearest to the receptor point (typically 72 s upwind, but dependent on local meteorology) undergo reaction (integrated for 72 s on average) generating the final modelled O_3 concentration.

A 0-D box model was run using the GRS mechanism and (in the base model run) constrained to concentrations of ROC and NO which were also used as the near-field constraints in ADMS-Urban and which derive from ADMS-Urban dispersion calculations, including emissions and background concentrations,[32] (further discussion on how ROC is determined is provided in Section 3.1.1.). Analogous to the MCM box model described above, the GRS box model was run unconstrained to O_3 and NO_2 and the predicted *in situ* (or locally produced) O_3 was output once a steady state concentration had been reached. As for the MCM box model, the GRS model constraints were updated hourly. In the case of the GRS box model a constant loss rate of O_3 and NO_2, equal to 3×10^{-5} s^{-1}, was used to represent deposition or loss of these species from the model box. The GRS predicted *in situ* O_3 is particularly sensitive to this loss rate (owing to the lack of competing reactions in this mechanism). Despite this, the influence of the various model parameters on O_3 reported in Sections 3.1.1 and 3.2.1 does not change even if this loss rate is varied and so it is still possible to assess the sensitivity of the GRS mechanism to various parameters.

2.3 EMEP4UK regional model and CRI*v2-R5* chemistry mechanism

The EMEP MSC-W (Meteorological Synthesizing Centre-West) model[33] is used by the United Nations Economic Commission for Europe Convention on Long Range Transboundary Air Pollution, at a grid resolution of 50 km to assess transboundary air pollution in Europe. EMEP4UK[7,34] is a version of the EMEP4UK model targeted specifically at UK air quality and incorporating a nested sub domain with 5 km resolution over the British Isles. Fig. 1b shows the resolution and capability of the EMEP4UK model predicting the hourly average surface O_3 concentration for London at 12:00 on 24/07/2012. As part of the CUREAIR project, EMEP4UK will be used to provide chemical boundary conditions for high resolution city-scale chemical dispersion modelling with ADMS-Urban.

Within the CUREAIR project, the EMEP4UK regional model will be run with the CRI*v2-R5* reduced chemistry mechanism as in a number of previous studies.[7,34]

Reduced chemistry mechanisms such as CRI are achieved either by (i) reducing the complexity of the chemistry for the considered suite of VOCs and/or (ii) by lumping emissions so that the chemistry for one VOC can be used to represent that of a number of VOCs. Method (i) was used to generate the CRI*v2* mechanism which reduced the number of species and reactions used in the MCM*v3.1* by 90% (*i.e.* to 434 species and 1183 reactions), whilst still adequately describing the degradation of methane and 115 non-methane VOCs.[35] The non-methane VOC groups are defined by the Global Emissions Inventory Activity[36] to impose a more severe level of reduction through emissions lumping. In this case, more limited selections of VOCs were considered to represent each VOC group, with the choice of species taking account of their photochemical ozone creation potentials (POCP) value, abundance in the detailed speciation, and the simplicity of the associated CRI*v2* degradation mechanism. This allowed reductions of up to 55% in the numbers of reactions and species relative to CRI*v2*. The mechanism retains CRI*v2* chemistry for isoprene, α-pinene and β-pinene and this contributes to the number of VOCs, species and reactions given. These reduced mechanisms display a degree of compromise in the O_3-forming ability of the VOC sub-categories, but retain a good level of overall performance.[37]

By considering a series of emission lumping options for anthropogenic VOCs (method (ii)), a set of further reduced CRI mechanisms were developed by Watson *et al.*[37] The smallest of these CRI schemes (known as CRI*v2-R5*) has 220 species and 609 reactions, with the suite of emitted non-methane VOCs represented by 22 compounds. This was a further 49% reduction in the numbers of reactions and species relative to CRI*v2*. The performances of CRI*v2* and its reduced variants (including CRI*v2-R5*) were tested during development against that of MCM*v3.1* for a wide range of ambient conditions, using box model scenarios and simulations of a major field campaign.[35,37]

A 0-D box model was run using CRI*v2-R5* and (similar to the MCM box model) run constrained to measurements from the ClearfLo project of NO, CO, CH_4, VOCs, PAN, HCHO, HNO_3, water vapour, temperature, pressure and measured photolysis frequencies. The model was run unconstrained to O_3 and NO_2 (and unconstrained to observed HONO in the base run) and the predicted *in situ* (or locally produced) O_3 was output once steady state conditions were reached. For all model generated species (including O_3 and NO_2) a first order loss rate (k) which varied with boundary depth was included as in the MCM box model. The CRI*v2-R5* model constraints were updated hourly.

2.4 Running the box models to steady state conditions

In the three box models constrained to chemical mechanisms of varying complexity, each model point was initialised with zero $[O_3]$ and zero $[NO_2]$ and then run forward with O_3 and NO_2 concentrations allowed to vary until steady state concentrations were reached. This was found to take up to 168 hours (seven days) for the MCM model, but was quicker for the simpler GRS mechanism (<3 days typically). The O_3 predicted by each of the three chemical mechanisms (run with equivalent NO and VOC constraints) for one model point on the 2nd August is presented in Fig. 2. All constraints including photolysis rates were held constant during each model point. By running to steady state in this way, the full O_3 creation potential of the VOCs present under a particular NO_x and radiation loading is determined. As can be seen in Fig. 2, this O_3 creation potential is specific to the chemical mechanism employed. Given the length of time it can take before O_3 reaches a steady state concentration, it is unlikely that the real

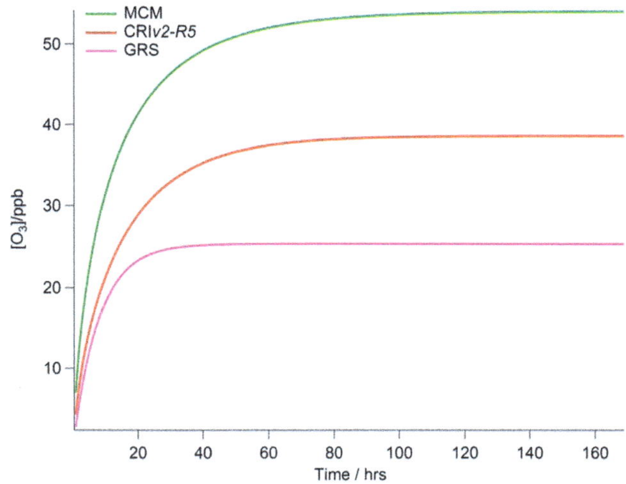

Fig. 2 Example of the modelled O_3 concentration determined by the three chemistry schemes for a single model point run forward until steady state conditions are reached; $[O_3]$ output hourly for 7 model days.

atmosphere is in steady state; Lee *et al.*[38] calculate a mean transport time to the North Kensington site since emission of ~40 to 50 minutes during the summer from the NO_x/NO_y ratio. Furthermore, in the ADMS-Urban dispersion model the GRS scheme is run for just 72 s on average. Nevertheless, to evaluate the ability of different chemical mechanisms to produce O_3 it is necessary to compare each under steady state conditions to prevent any bias caused by variability in the length of time it may take for $[O_3]$ to reach steady state when the complexity of the chemical mechanisms varies.

3. Results and discussion

3.1 Modelled O_3 concentrations during the ClearfLo summer IOP

The majority of the summer measurement period was characterised by south-westerly winds, with the wind speed showing a diurnal cycle of less than 1 m s^{-1} at night to 4–6 m s^{-1} in late afternoon. The exception to this are two periods from 24–27 July and 8–10 August, during which the site was subjected to an easterly flow, with lower wind speed. Due to central London being to the east of the site, these periods are characterised by higher levels of NO_x (up to 60 ppb of NO and 50 ppb of NO_2), which has its source mainly from traffic.

The measured O_3 concentrations during the ClearfLo summer IOP are compared to the modelled O_3 concentrations from the ADMS-Urban dispersion model and EMEP4UK regional model (Fig. 3a). Reasonable agreement is achieved, with both models able to reproduce the variability observed day to day. When compared as a median IOP diurnal cycle (Fig. 3c), the ADMS-Urban and EMEP4UK models under-predict the peak O_3 concentrations by 30–40% during the day and this under-prediction is most evident on the days when the observed O_3 was most elevated (*e.g.* on the 24th and 25th July). Fig. 3b compares the *in situ* O_3 predicted by the three chemistry mechanisms. The MCM box model run to steady state

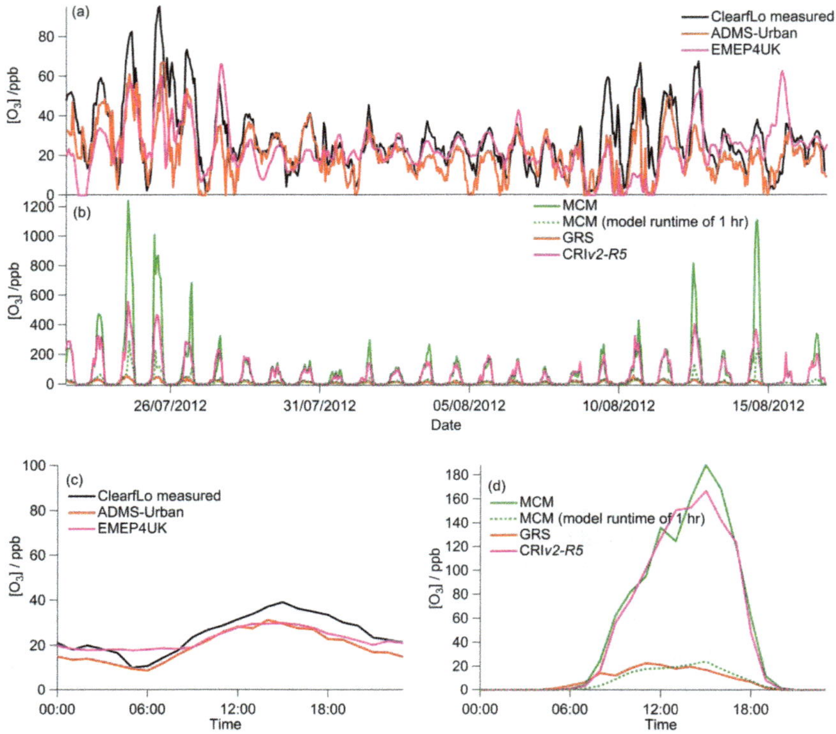

Fig. 3 (a) Full time series of ADMS-Urban, EMEP4UK modelled and measured O_3 concentrations from ClearfLo summer IOP. (b) Time series of the ozone predicted by the chemistry schemes GRS, CRI*v2-R5* and MCM*v3.2* run in box models to steady state (MCM [O_3] predicted after 1 h model run time also shown) from ClearfLo summer IOP (c) corresponding median diurnal cycle profile from the data in panel (a). (d) Corresponding median diurnal cycle from the data in panel (b).

generates significant concentrations of O_3 each day (substantially more than was observed), with a peak concentration of 1240 ppb predicted on the 24th July during the polluted easterly flow. For reference, the [O_3] predicted by the MCM after a model run time of 1 hour (which is similar to the estimated mean transport time to the North Kensington site since emission[38]) is also presented in Fig. 3b and d. The median predicted [O_3] by the MCM model run for 1 hour peaks during the afternoon at 24 ppb (Fig. 3d) suggesting that ~60% of the observed [O_3] may be generated from local emissions. There is considerable variability day to day in the [O_3] predicted by the MCM after a run time of 1 hour, however, with ~290 ppb generated on the 24th July suggesting, under certain conditions, that the observed [O_3] may be even more heavily influenced by London's emissions.

Elevated O_3 concentrations are also predicted by the CRI*v2-R5* model although the peak O_3 predicted on the 24th and 25th July is roughly half that predicted by the MCM. During the south-westerly flows, when the daily maximum [O_3] observed was lower, the CRI*v2-R5* model is in reasonable agreement with the MCM. The GRS base model predicts significantly lower *in situ* O_3, relative to the MCM model, with a peak concentration of 59 ppb simulated on the 24th July. It is evident from Fig. 3 that the GRS base model which is constrained with the same input parameters as ADMS-Urban is unable to generate as much O_3 as the MCM model which is constrained to the observations made during ClearfLo. These differences may be due to differences in the model constraints or differences in the mechanisms themselves. In the following sections the sensitivity of the modelled O_3 concentration from the GRS and CRI*v2-R5* models to different variables (including HONO) is investigated. The ability of the GRS and CRI*v2-R5* mechanisms to reproduce the level of O_3 predicted by the explicit MCM when all models are constrained with equivalent inputs is also evaluated.

3.1.1 Sensitivity analysis of the GRS-modelled O_3 to various model constraints and comparison to the MCM-modelled O_3. The impact of NO and ROC concentrations, measured photolysis frequencies and the introduction of HONO chemistry on the predicted O_3 concentration using the GRS chemistry scheme is investigated below. The effect of NO and ROC concentrations were investigated by either altering their background concentration by a factor of 2, 10 or 0.1 or by utilising the measured ClearfLo concentrations. The ClearfLo NO and ROC concentrations were approximately (on average) a factor of 8.5 times lower and 3.5 times higher than concentrations used in the base GRS run (and used in ADMS-Urban) respectively. The comparison between ClearfLo IOP measured concentrations and those from the ADMS-Urban dispersion calculations, including emissions and background concentrations, can be seen in Fig. 4. The difference between the base ROC concentration (used in the GRS model run presented in Fig. 3) and the ClearfLo derived ROC is in part due to the number of individual VOC species used to derive ROC in each case. The base run ROC (which represents the ROC concentration used in ADMS-Urban) utilises inputs from the ADMS-Urban dispersion calculations, including emissions of (lumped) VOCs, NO_x and PM and measured O_3 and NO_x background concentrations. These lumped VOC emissions are based primarily on road transport (exhaust emissions). Speciation of the lumped VOC emission comes from the Standard Automatic Hydrocarbon network dataset, which currently monitors the emission factors for 32 separate NMVOCs.[39] This is significantly less than the 78 VOCs (36 aliphatics, 19 monoaromatics, 21 oxygenated and 2 halogenated) measured during ClearfLo.[24] It should be noted that the MCM model was only constrained to the 62 VOCs for which degradation mechanisms explicitly exist. Nevertheless the contribution of a significant fraction of VOCs present in London's air is not considered in the base GRS run.

Further under-representation of ROC in the base run likely derives from the use of a single factor to convert VOC concentrations to ppb and the subsequent conversion to ROC. The ADMS-Urban model currently uses a factor of 0.31 (the value appropriate for benzene) to convert VOC concentrations from $\mu g\ m^{-3}$ to ppb. This is likely to be too low in practice, hence underestimating the ppb concentrations of VOC. VOCs are converted to the composite ROC by their ability to produce radicals (RO_2 or HO_2). The concentration of ROC is defined as a reactivity coefficient multiplied by VOC concentration. For example, Johnson[31] used $[ROC] = 0.0067[VOC]$ for typical 1980s Australian urban air dominated by motor vehicles; ADMS-Urban currently uses 0.1. If measured rural concentrations of VOC are unavailable for input to ADMS-Urban, the increased reactivity coefficient partially compensates for the reduced concentration. Empirically determined reactivity coefficients for individual VOC species are available from smog chamber experiments, whilst numerically determined reactivity coefficients have

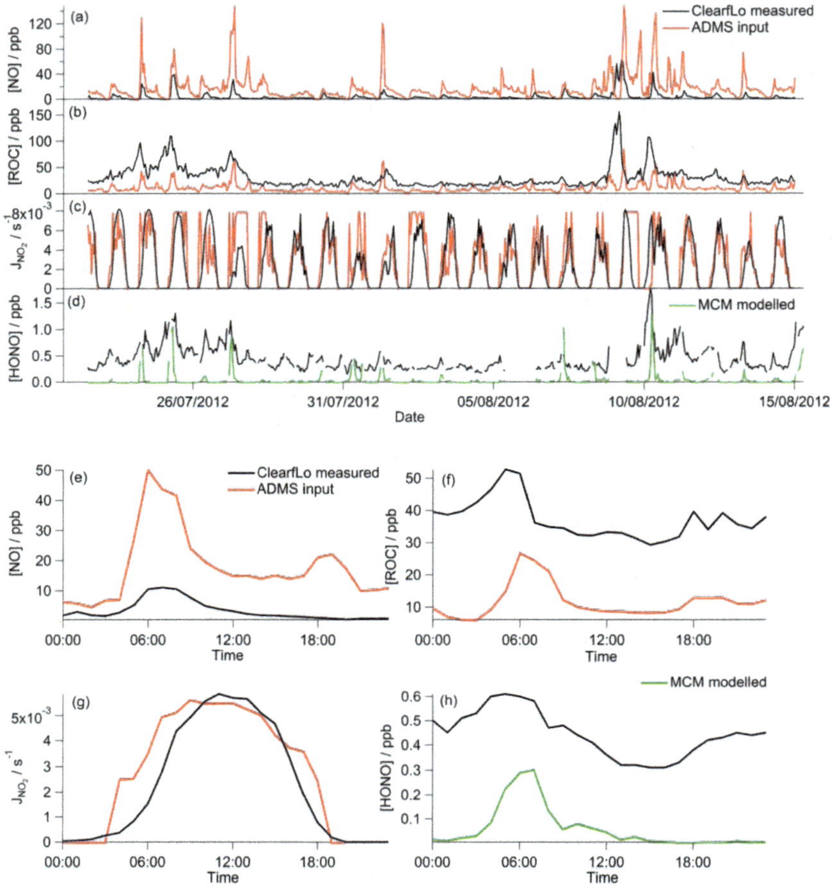

Fig. 4 (a–c) Full time series of ADMS-Urban inputs and measured NO and VOC concentrations and NO_2 photolysis frequency from ClearfLo summer IOP. (d) Full time series of MCM modelled HONO concentrations and ClearfLo measured HONO. (e–h) Average diurnal cycle of the time series in panels a–d.

been calculated using a comparison of the GRS mechanism with more complex mechanisms.[31,40] A full description of this methodology for conversion can be found in Johnson[31] or Venkatram *et al.*[45] The latter numerical approach is used to convert the comprehensive measurement suite of VOC concentrations from the ClearfLo campaigns to ROC for the GRS chemistry scheme.

Owing to a lack of long-term monitoring of solar actinic flux from which photolysis frequencies may be determined, the rate of photolysis of NO_2 (ρ_3) used in the GRS scheme is calculated from the background concentrations of O_3 ($C_{BGD}(O_3)$), NO ($C_{BGD}(NO)$) and NO_2 ($C_{BGD}(NO_2)$):

$$\rho_3 = \min(\rho_3(Q_{\max}), \overline{\rho_3})\ \ s^{-1} \tag{E1}$$

$$\rho_3 = 8 \times 10^{-4} \exp\left(\frac{-10}{Q}\right) + 7.4 \times 10^{-6} Q \tag{E2}$$

$$\overline{\rho_3} = \frac{C_{BGD}(O_3) \times C_{BGD}(NO) \times \rho_4}{C_{BGD}(NO_2)} \tag{E3}$$

$$\rho_4 = 4.405 \times 10^{-2} \exp\left(\frac{-1370}{T_0}\right) \tag{E4}$$

where Q is solar radiation in W m^{-2}, Q_{max} is the maximum possible value of Q (assuming a solar elevation angle of 90 degrees and zero cloud cover), and T_0 is temperature in Kelvin. In case of erroneous background data, a minimum of $\rho_3(Q_{max})$ and $\overline{\rho_3}$ is taken to ensure that the reactions occur at a realistic rate. By utilising the measured j(NO_2) from the ClearfLo IOP, the sensitivity of O_3 to this parameterisation is investigated.

HONO can represent a major source of OH particularly in urban regions[38] and has been shown to influence O_3 and PM.[41,42] The importance of HONO on the oxidation capacity during the ClearfLo campaign has recently been highlighted by Lee *et al.*[38] From simultaneous measurements of HONO and OH during the summer ClearfLo IOP and subsequent modelling activities, Lee and co-authors determined an average 40% under-prediction in the daytime [OH] when the model was unconstrained to the measured [HONO] as opposed to constrained. HONO as a source of radicals (and NO) was included in the GRS scheme by the addition of the following reaction and using the measured photolysis frequency for HONO:

$$\text{HONO} \rightarrow \text{RP} + \text{NO} \tag{R8}$$

This reaction assumes that the OH generated from HONO photolysis instantaneously converts to a peroxy radical (RP).

Fig. 5 presents the correlation of the MCM modelled O_3 concentration from the summer ClearfLo campaign with the GRS-scheme modelled O_3 for each scenario discussed above. As discussed above, the base GRS run predicts significantly less *in situ* O_3 than the MCM model with a ratio of GRS O_3 : MCM O_3 of 0.05. This ratio increases to 0.11 when only MCM-modelled O_3 below 400 ppb is considered indicating that the base GRS model more closely reproduces the [O_3] predicted by the MCM in the cleaner south-westerly flows than in the polluted easterlies. Nevertheless, the base GRS model predicts significantly lower O_3 than the MCM model.

The sensitivity of O_3 to j(NO_2) (Fig. 5a and b) was investigated by replacing the calculated photolysis frequency (ρ_3 and eqn (E1)) with the observed photolysis frequency. The level of agreement between the GRS model and the MCM remained very similar as did R^2 (Table 2). It is apparent from Fig. 4c and g that the calculated NO_2 photolysis frequency replicates the actual photolysis frequency on average over the time series (Fig. 4g) and was able to replicate the peak rate of NO_2 photolysis. However, on a number of days (*e.g.* 25th, 27th) the calculated photolysis frequency is determined from $\rho_3(Q_{max})$ and, as a consequence of this, on the 27th the correlation with the measured photolysis frequencies is poor. This difference likely contributes to the small decrease in GRS-predicted O_3 that is seen (9%) when only [O_3] <400 ppb is considered.

Both decreasing the NO concentration by a factor of 10 (Fig. 5g) or using the measured NO concentration (Fig. 5c) improves the agreement between the GRS-

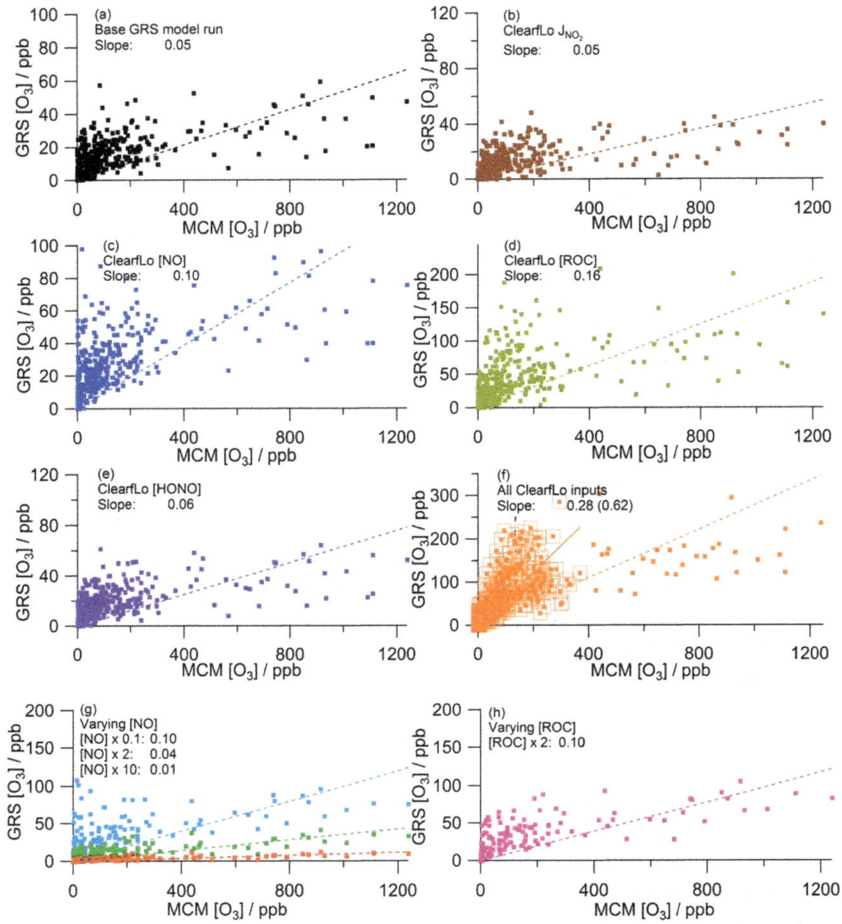

Fig. 5 Correlation of MCM modelled [O_3] from summer ClearfLo IOP against GRS modelled [O_3]. (a) Base GRS scheme, (b) ClearfLo measured j(NO_2), (c) measured ClearfLo [NO], (d) measured ClearfLo [ROC], (e) measured ClearfLo [HONO], (f) all ClearfLo measured data (value in parentheses = slope when MCM modelled [O_3] is <400 ppb), (g) & (h) varying [NO] and [ROC]. All data can be found in Table 2.

predicted O_3 and that of the MCM (GRS O_3 : MCM O_3 ratio = 0.10). Increasing NO concentration from the base GRS run by a factor of 2 or 10 leads to reductions in the GRS-predicted O_3 indicative of a NO_x-saturated regime (Fig. 5g). The GRS O_3 : MCM O_3 ratio falls to 0.04 when NO is doubled and to 0.01 when NO is increased by a factor of 10.

The sensitivity of the GRS-predicted O_3 to the ROC concentration is significant (Fig. 5a, d and h): increasing the total ROC concentration by a factor of 2 leads to a 100% increase in [O_3] relative to the base GRS run. The ClearfLo ROC, which is ~3.5 times higher than the base ROC, improves the agreement between GRS and MCM modelled O_3 concentrations further, with the GRS O_3 : MCM O_3 ratio now equal to 0.16 (or 0.34 when MCM-modelled O_3 is less than 400 ppb). A major cause of the improved agreement between the GRS and MCM modelled O_3 concentrations is due to the significant concentration of VOCs that are missing from the

Table 2 Summary of parameters obtained from the correlation plots shown in Fig. 5

	Model	Modelled GRS $[O_3]$ / Modelled MCM$[O_3]$	R^2	% change from base run[a]
a	Base GRS[a]	0.05 (0.11)[b]	0.25 (0.36)[b]	—
b	Measured j(NO_2)	0.05 (0.10)[b]	0.23 (0.42)[b]	0 (−9)[b]
c	ClearfLo [NO]	0.10 (0.20)[b]	0.17 (0.28)[b]	+100 (+82)[b]
d	ClearfLo [VOC]	0.16 (0.34)[b]	0.23 (0.36)[b]	+220 (+209)[b]
e	ClearfLo [HONO]	0.06 (0.13)[b]	0.20 (0.37)[b]	+20 (+18)[b]
f	All ClearfLo inputs	0.28 (0.62)[b]	0.33 (0.57)[b]	+460 (+464)[b]
g	Varying [NO] × 0.1	0.10 (0.21)[b]	0.07 (0.20)[b]	+100 (+91)[b]
	Varying [NO] × 2	0.04 (0.07)[b]	0.28 (0.38)[b]	−20 (−36)[b]
	Varying [NO] × 10	0.01 (0.02)[b]	0.30 (0.38)[b]	−80 (−82)[b]
h	Varying [ROC] × 2	0.10 (0.21)[b]	0.22 (0.35)[b]	+100 (+91)[b]

[a] Base GRS is the model run with the GRS emission inputs. [b] Values in parentheses derive from correlation of MCM modelled $[O_3]$ <400 ppb against GRS modelled $[O_3]$.

ADMS-Urban dispersion calculations, which include VOCs that are based primarily on road transport (exhaust emissions).[32] These lumped transport VOCs may account for a reasonable proportion of the total VOCs, but omits any biogenics *e.g.* isoprene, monoterpenes and also substituted aromatics such as the xylenes which are known to have a very high potential for O_3 formation.[43] Many biogenic VOCs are highly reactive and are readily oxidised by the OH radical to form peroxy radicals. It has been previously identified that the production of O_3 in urban areas (including London) is highly sensitive to biogenic emissions.[9,25,44]

The sensitivity of GRS modelled O_3 to HONO chemistry was investigated by including reaction (R8) in the GRS mechanism and constraining to the measured HONO concentration and HONO photolysis frequency from the ClearfLo IOP. The introduction of this additional radical (RP) source in the GRS chemistry scheme increased the GRS O_3 concentration (+20% from the base run) and agreement with the MCM-predicted O_3 improved marginally (Fig. 5a and e).

When the GRS model was constrained to all the measured ClearfLo parameters (j(NO_2), ClearfLo ROC, NO and HONO) (Fig. 5f) agreement between the GRS-predicted O_3 and MCM-predicted O_3 increased to a ratio of 0.28 for the whole

Table 3 Summary of the correlation plots in Fig. 6

	Model	Modelled CRI $[O_3]$ / Modelled MCM $[O_3]$	R^2	% increase from chemistry only[a]
a	Base CRI	0.51 (0.85)[b]	0.75 (0.83)[b]	—
b	ClearfLo HONO	0.53 (0.91)[b]	0.73 (0.83)[b]	+4 (+7)[b]
c	2 × [NO]	0.40 (0.57)[b]	0.79 (0.81)[b]	−22 (−33)[b]
d	10 × [NO]	0.14 (0.17)[b]	0.73 (0.71)[b]	−73 (−80)[b]

[a] Base CRI, panel a: CRI*v2-R5* run unconstrained to HONO. [b] Values in parentheses derive from correlation of MCM modelled $[O_3]$ <400 ppb against GRS modelled $[O_3]$.

campaign and to 0.62 when only MCM-modelled O_3 <400 ppb was considered. The correlation coefficient also increases ($R^2 = 0.33$ or 0.57). Interestingly, the percentage increase from the base run (+460%) when all the ClearfLo parameters are included is greater than when the parameters are added into the GRS model individually, highlighting the non-linear dependence of the O_3 chemistry on NO_x and VOC concentrations.

Overall, when equivalent inputs are used as constraints in both the MCM and GRS schemes, the predicted O_3 is significantly lower if the lumped, eight reaction GRS scheme (including HONO reaction) is used. This trend is most evident, however, on the days when the observed O_3 was most elevated, *i.e.* during the easterly flows, typified by warm, stagnant conditions. If the modelled O_3 is even more heavily filtered and only MCM-modelled O_3 below 150 ppb is considered, the ratio GRS $[O_3]$: MCM $[O_3]$ increases to 0.96 ($R^2 = 0.7$) demonstrating that under certain conditions, the heavily lumped GRS scheme has the ability to reproduce the O_3 predicted by the more explicit MCM scheme. Given that the ADMS-Urban dispersion model has a tendency to under-predict the peak $[O_3]$ observed during the easterly flows experienced at the start of the IOP (Fig. 3a), it becomes relevant to consider differences caused by the simplifications in the chemistry scheme employed. The MCM model generates a significant concentration of model-generated intermediate species which contribute ~9 s^{-1} to the total OH reactivity during the polluted easterly flows.[25] These intermediates (deriving largely from the biogenic species of alpha-pinene and limonene) increase the MCM-modelled peroxy radical concentrations and drive up the predicted O_3.[25] [ROC] is conserved in the GRS by reaction (R1) and this effectively increases the O_3 creation potential of the GRS. It is apparent, however, that this single reaction under-represents the influence of the secondary chemistry of the MCM model intermediates at the start of the IOP which is dominated by the oxidation of the biogenics. Although biogenic species contribute only ~1% to the total VOC loading (excluding methane), the contribution they make to [ROC] is ~20% demonstrating their high radical (and O_3) creation potential. From the comparison of the GRS constrained to ClearfLo ROC with the near-explicit MCM, however, it is evident that the conversion of these VOC types to ROC (*i.e.* the ability of the biogenics to generate radicals) is under-estimated by the methodology employed. In the [ROC] constraint used within the ADMS-Urban model the impact of biogenics is not included at all, yet in a warming atmosphere, in the presence of NO_x, the impact of these species on $[O_3]$ is likely to become increasingly relevant.

3.1.2 Sensitivity analysis of the CRI*v2-R5*-modelled O_3 to various model constraints and comparison to the MCM-modelled O_3. The CRI*v2-R5* box model is constrained using the same ClearfLo measurements as the benchmark MCM model but the anthropogenic VOC species are subject to systematic lumping as discussed in Section 2.3.

The regional chemistry transport model EMEP4UK can be run with the CRI*v2-R5* chemistry scheme and, hence, here the ability of this simplified (but still somewhat complex relative to GRS) chemistry scheme to predict O_3, as compared to that simulated by the near-explicit MCM model, is assessed. Sensitivity tests investigating the impact of constraining to measured concentrations of HONO are presented. The impact of NO is also investigated (by doubling and increasing the NO concentration ten-fold) to contrast to the impact of NO in the GRS box model.

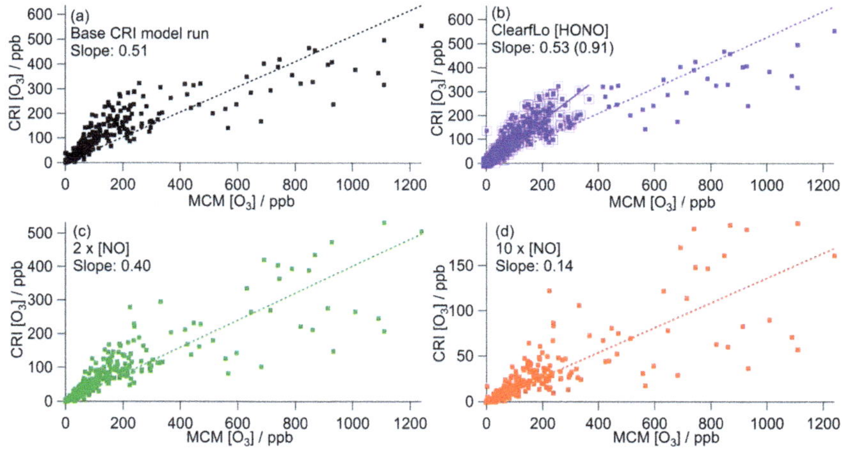

Fig. 6 Correlation of CRI modelled $[O_3]$ from summer ClearfLo IOP against MCM modelled $[O_3]$. (a) Base CRI model run (value in parentheses = slope when MCM modelled $[O_3]$ is <400 ppb). (b) CRI constrained to ClearfLo measured HONO. (c) Base CRI model run, but with [NO] doubled. (d) Base CRI model run, but with [NO] increased ten-fold. All data can be found in Table 3.

Fig. 6 shows the correlation of modelled CRI*v2-R5* O_3 concentrations with those from the MCM benchmark model. As discussed in Section 3.1, the base CRI*v2-R5* run, unconstrained to HONO, predicts less O_3 than the MCM box model with a ratio of CRI O_3 : MCM O_3 = 0.51, although this increases to 0.85 if only MCM-modelled O_3 which is less than 400 ppb is included. The correlation coefficient is significantly better ($R^2 = 0.75$ (or 0.83)) than all the GRS–MCM correlations, likely reflecting the origin of the CRI scheme. The ratio CRI O_3 : MCM O_3 increases (to 0.53 or 0.91) when the CRI*v2-R5* model is run constrained to the observed HONO demonstrating the impact of HONO as a radical source and ultimately as a source of O_3.

Increasing the [NO] by a factor of 10 (to concentrations similar to those used in the GRS base run) decreases the CRI*v2-R5* modelled O_3 by 73% relative to the base CRI*v2-R5* run, demonstrating a slightly weaker NO dependence than observed in the GRS box model as NO concentrations were varied. This weaker NO dependence likely reflects the lower VOC : NO_x ratios of the base GRS box model relative to the CRI*v2-R5* box model.

The CRI*v2-R5* when constrained to equivalent inputs to the MCM reproduces the MCM-predicted O_3 well (ratio of CRI*v2-R5* : MCM = 0.91) at all times apart from during the polluted easterly conditions when elevated $[O_3]$ prevailed. As discussed in Section 3.1.1, the oxidation of monoterpenes significantly increases the MCM model intermediates which enhance modelled peroxy radical concentrations and O_3 [25] at these times. Although the concentration of individual monoterpenes can be represented explicitly in the CRI*v2-R5* scheme (rather than as a single lumped biogenic) the oxidation scheme for these species differs from the MCM somewhat. Considering the ozonolysis of alpha pinene and the subsequent RO_2 + NO reactions (proceeding along the dominant reaction pathways in both the MCM and CRI*v2-R5*), 10 molecules of NO_2 can be generated using the MCM scheme relative to 8 molecules in CRI*v2-R5*. Furthermore, 3 OH radicals are consumed in the CRI*v2-R5* scheme relative to 1 OH radical in the MCM, although this is offset somewhat by 4 HO_2 radicals being generated along the CRI*v2-R5* reaction pathway relative to 2 HO_2 radicals in the MCM. Overall these differences lead to CRI*v2-R5* predicting around twice as much HO_2 as the MCM on the afternoon of the 25th July; however, the MCM predicts up to 4 times as much RO_2 as CRI*v2-R5* at this time and ultimately, as a consequence, more O_3 is generated by the MCM scheme. EMEP4UK, run with CRI*v2-R5*, has a tendency to under-predict O_3 under these conditions and this analysis suggests that the differences in the *in situ* chemistry of biogenic species in CRI*v2-R5* could, in part, contribute to this under-prediction at these times.

3.2 Modelled O_3 concentrations during the ClearfLo winter IOP (January–February 2012)

During the winter, the majority of the measurement period was characterised by south-westerly winds. Wind speeds were variable; early in the campaign (13–19 January) wind speeds dropped and extremely elevated levels of NO_x and VOCs were observed (Fig. 8a and b). During this period O_3 was almost entirely titrated away by NO. Similar stagnation events were also observed for shorter periods later

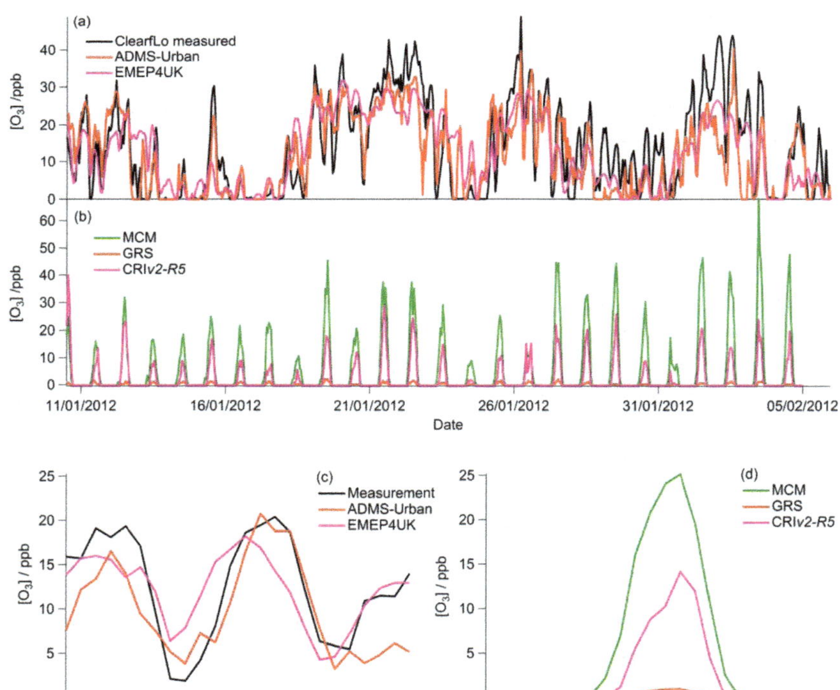

Fig. 7 (a) Full time series of ADMS-Urban, EMEP4UK modelled and measured O_3 concentrations from ClearfLo winter IOP. (b) Time series of the GRS, CRI*v2-R5* and MCM*v3.2* box model O_3 concentrations from ClearfLo winter IOP. (c) Corresponding IOP median diurnal cycle profile from the data in panel (a). (d) Corresponding median diurnal cycle from the data in panel (b).

in the campaign, *e.g.* on the 24th January. The ClearfLo measured O_3 concentrations are compared to the modelled O_3 concentrations from ADMS-Urban and EMEP4UK for the winter IOP (Fig. 7a). The ADMS-Urban and EMEP4UK models reproduce the measured winter O_3 concentration time-series well, although the EMEP4UK model over-predicts the median O_3 during the morning and under-predicts O_3 concentrations during the afternoon (Fig. 7c). Both models capture the nighttime O_3 trend reasonably well, with only a slight under-prediction observed from midnight to 6 am. The MCM box model (Fig. 7b) run to steady state predicts more O_3 than the base CRI*v2-R5* and the GRS models. The chemical O_3 creation potential determined (by running the box models to steady state with respect to O_3) is substantially lower than in summer and the daytime O_3 concentrations predicted by the MCM are only slightly higher than observed, suggesting that the maximum O_3 creation potential of the air-mass is close to being realised during the winter. It also takes less time to reach a steady state $[O_3]$ when running the winter MCM model than running the summer MCM model owing to the higher $[NO_x]$ in winter; this is most evident during the stagnation events, when wind speeds were extremely low. When wind speeds picked up (20–23 January), the MCM under-predicts the daytime O_3 that was observed, highlighting that a fraction of the O_3 must have been transported from local/regional/continental sources at these times, in agreement with results from REPARTEE.[11] The MCM box model is unable to reproduce the observed O_3 at night (Fig. 7b and d), again suggesting that local chemistry (even when the full O_3 creation potential is determined by running to steady state conditions) is not controlling the observed O_3 at these times.

The time series and median diurnal cycles for the box models can be seen in Fig. 7b and d. The GRS base model significantly under-predicts O_3 relative to the MCM box model (and observations). The CRI*v2-R5* model (unconstrained to observations of HONO) also under-predicts O_3 compared to the MCM model albeit to a lesser extent. The sensitivity of the modelled O_3 concentration to different variables in the GRS and CRI*v2-R5* chemistry schemes is investigated below using a comparison with the MCM box model in Sections 3.2.1 and 3.2.2. The sensitivity of the GRS and CRI*v2-R5* chemistry schemes during the winter IOP were subject to the same investigations as for summer.

3.2.1 Sensitivity analysis of the GRS-modelled O_3 to various model constraints and comparison to the MCM-modelled O_3. The impact of NO and ROC concentrations, measured NO_2 photolysis frequency and the introduction of HONO chemistry on the predicted O_3 concentration using the GRS chemistry scheme is investigated. The comparison between ClearfLo IOP measured concentrations and those from the ADMS-Urban dispersion calculations, including emissions and background concentrations, can be seen in Fig. 8(a–c) and (e–g).

As discussed in Section 3.2, the GRS box model (Fig. 9a) significantly underestimates O_3 relative to the MCM box model, predicting ~30 times less O_3. In contrast to summer, the agreement with the MCM model improves modestly (6% increase relative to the base GRS run) when the calculated NO_2 photolysis frequency is replaced with the observed NO_2 photolysis frequency (Table 4, Fig. 9a and b), likely reflecting the faster photolysis frequency observed (on average) than calculated during the winter project (Fig. 8g).

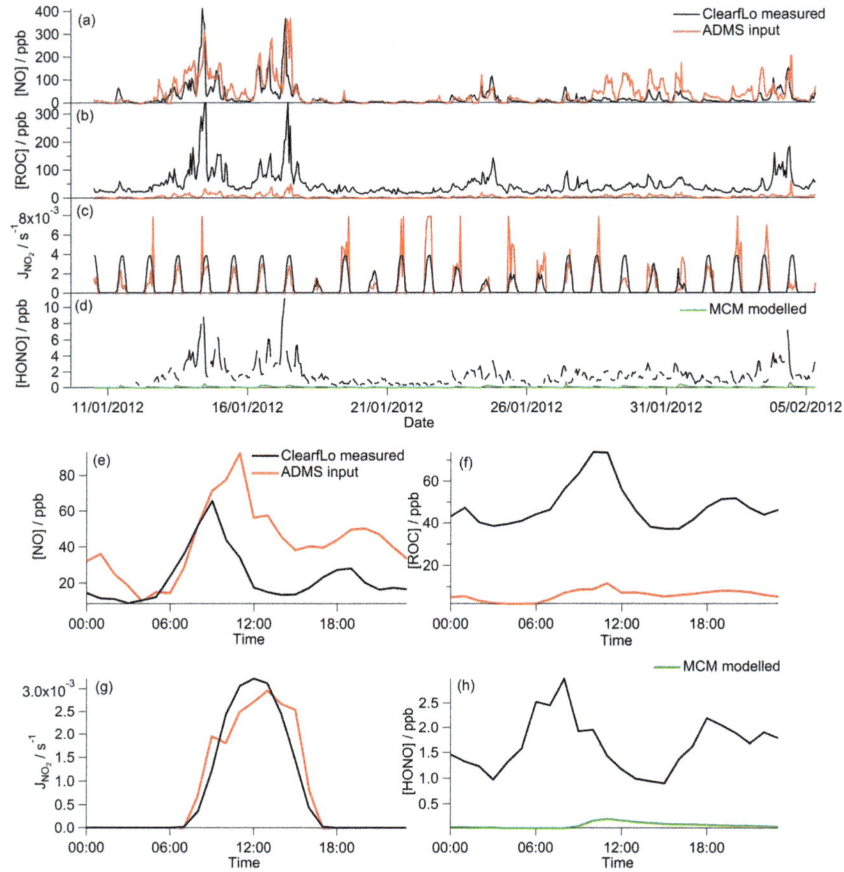

Fig. 8 (a–c) Full time series of ADMS-Urban inputs and measured NO and VOC concentrations and NO_2 photolysis frequency from ClearfLo winter IOP. (d) Full time series of MCM modelled HONO concentrations and ClearfLo measured HONO. (e–h) Average diurnal cycle of the time series in panels a–d.

On average, the ClearfLo observed NO concentration was a factor of 2 lower than the NO constraint used in the GRS base run. This difference is less than observed during the summer, where the ClearfLo observations were a factor of ~8.5 lower than the GRS base model constraint. Fig. 8a highlights reasonable agreement between the GRS base model NO constraint and the NO observed during the stagnation events and so the difference in the model constraints between the two IOPs likely reflects the ability of the ADMS-Urban dispersion calculations, including emissions and background concentrations, to capture these high NO episodes. Both decreasing the NO concentration by a factor of 10 (Fig. 9g) or using the measured NO concentration (Fig. 9c) improves the agreement between the GRS-predicted O_3 and that of the MCM, with a 226% or 123% increase respectively in the GRS modelled O_3 relative to the base run. Increasing NO concentration from the base GRS run by a factor of 2 or 10 leads to reductions in the GRS-predicted O_3 indicative of a NO_x-saturated regime (Fig. 9g); this is analogous to the summertime. The GRS O_3 : MCM O_3 ratio falls to 0.019 when NO is doubled and to 0.005 when NO is increased by a factor of 10.

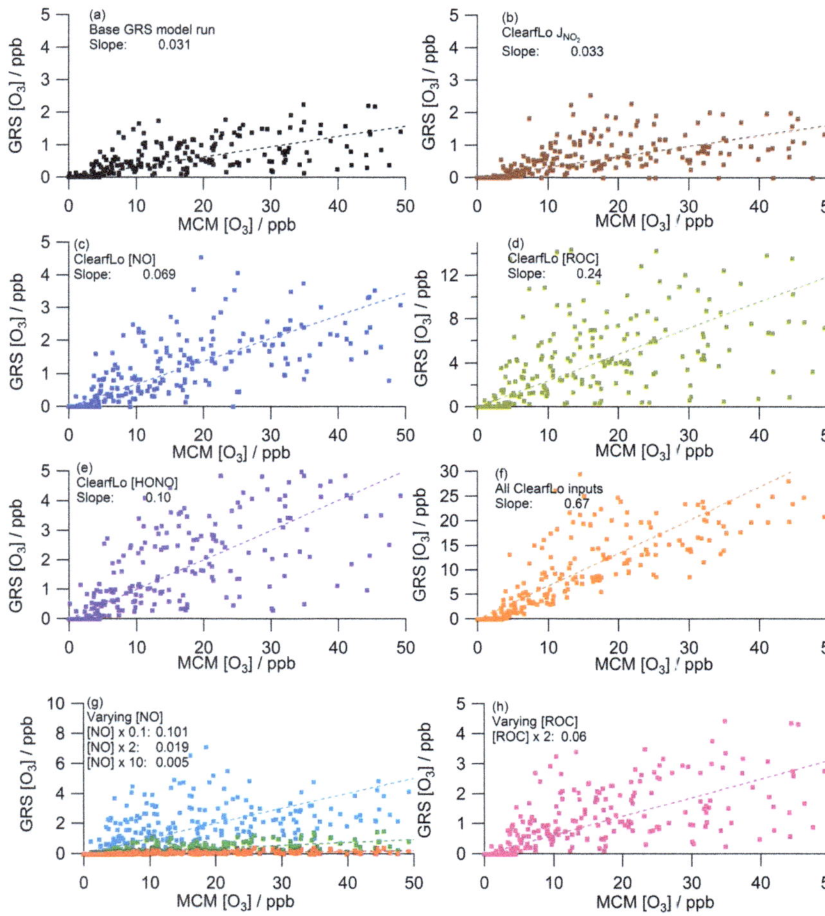

Fig. 9 Correlation of MCM modelled [O_3] from winter ClearfLo IOP against GRS modelled [O_3], (a) Base GRS scheme, (b) ClearfLo measured j(NO_2), (c) measured ClearfLo [NO], (d) measured ClearfLo [ROC], (e) measured ClearfLo [HONO], (f) all ClearfLo measured data, (g) & (h) varying [NO] and [ROC].

The ClearfLo-derived ROC concentration was approximately 9 times greater than the concentration used in the base GRS run (and used in ADMS-Urban). This difference is even greater than that observed during the summer comparisons where, in summer, the ClearfLo-derived ROC was ~3.5 times greater than the concentration used to constrain the base GRS run. In winter, the concentration of substituted aromatics which are not considered in the base ROC constraint was ~6 to 8 times greater than during the summer months and these species contribute to the ClearfLo-derived ROC significantly. Constraining the GRS box model to the ClearfLo-derived ROC increases the predicted O_3 concentration by 674% relative to the base case and the GRS O_3 : MCM O_3 ratio increases to 0.24.

The introduction of the additional gas-phase HONO chemistry reactions to the GRS chemistry scheme improved the correlation between the GRS and MCM modelled O_3 concentrations (Fig. 9a and e). Inclusion of the observed HONO as a GRS-model constraint and RP source increases the GRS predicted O_3 by 223%

Table 4 Summary of parameters derived from the correlation plots shown in Fig. 9

	Model	Modelled GRS $[O_3]$ / Modelled MCM $[O_3]$	R^2	% change from base run[a]
a	Base GRS[a]	0.031	0.62	—
b	Measured j(NO_2)	0.033	0.56	+6
c	ClearfLo [NO]	0.069	0.61	+123
d	ClearfLo [VOC]	0.24	0.52	+674
e	ClearfLo [HONO]	0.10	0.67	+223
f	All ClearfLo inputs	0.67	0.78	+2061
g	Varying [NO] × 0.1	0.101	0.50	+226
	Varying [NO] × 2	0.019	0.59	−39
	Varying [NO] × 10	0.005	0.54	−84
h	Varying [ROC] × 2	0.06	0.62	+94

[a] Base GRS is the model run with the GRS emission inputs.

relative to the base run. The impact of HONO on O_3 is much more significant during the wintertime than summertime, where, in summer, predicted O_3 increased by just 20% relative to the base run. HONO concentrations were more elevated during the winter IOP, with peak HONO concentrations of 11.1 ppb observed during the stagnation event; in summer the peak HONO concentration observed was 1.77 ppb.

When the model is constrained to all the measured ClearfLo parameters (Fig. 9f), significantly more O_3 is predicted relative to the base run and the GRS $O_3 : MCM\ O_3$ ratio increases to 0.67. Analogous to summer, the percentage increase from the base run (+2061%) when all the ClearfLo parameters are included is greater than when the parameters are added into the GRS model

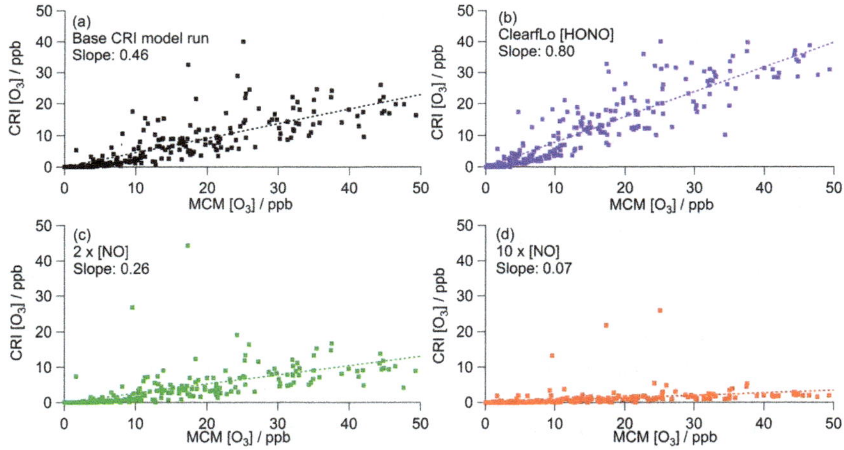

Fig. 10 Correlation of CRI modelled $[O_3]$ from winter ClearfLo IOP against MCM modelled $[O_3]$, (a) Base CRI model, (b) CRI constrained to ClearfLo measured HONO, (c) Base CRI model run, but with [NO] doubled, (d) Base CRI model run, but with [NO] increased ten-fold. All data can be found in Table 5.

Table 5 Summary of the parameters derived from the correlation plots shown in Fig. 10

	Model	Modelled CRI $[O_3]$ / Modelled MCM $[O_3]$	R^2	% difference from chemistry only[a]
a	Base CRI	0.46	0.77	—
b	ClearfLo HONO	0.80	0.89	+74
c	2 × [NO]	0.26	0.47	−43
d	10 × [NO]	0.07	0.20	−85

[a] Base CRI, panel a: CRI*v2-R5* run unconstrained to HONO.

individually, highlighting again the non-linear dependence of the O_3 chemistry on NO_x and VOC concentrations.

The GRS $[O_3]$: MCM $[O_3]$ ratio in winter when each scheme is constrained with equivalent inputs is similar to the summer ratio when only MCM O_3 <400 ppb is considered (winter ratio = 0.67, summer ratio = 0.62) demonstrating the ability of this simple scheme to replicate the near-explicit MCM reasonably well under a range of conditions including the higher NO_x conditions experienced during the winter. The ADMS-Urban dispersion model is able to predict the observed O_3 well throughout the winter IOP despite the very limited ability of the GRS, when constrained to the ADMS-Urban input parameters (Fig. 9a) to generate O_3, indicating that the *in situ* chemistry is not controlling the $[O_3]$ observed in winter.

3.2.2 Sensitivity analysis of CRI*v2-R5*-modelled O_3 to various model constraints and comparison to the MCM-modelled O_3. Analogous to the GRS–MCM model comparison in the winter (Section 3.2.1), constraining the CRI*v2-R5* box model to the observed HONO concentrations significantly increases the concentration of O_3 predicted by the CRI*v2-R5* box model (by 74%) (Fig. 10a and b; Table 5); in summer the CRI*v2-R5* box model constrained to HONO predicted 4% (or 7% for MCM-modelled O_3 <400 ppb) more O_3 relative to the base run. This result further demonstrates the role of HONO as an important O_3 source during the winter when other sources of radicals (which drive the *in situ* O_3 production) are limited. The influence of NO on the predicted O_3 (Fig. 10c and d) again highlights a NO_x-saturated regime.

Constrained to the observed HONO, the CRI scheme is able to reproduce the MCM-predicted O_3 for the winter ClearfLo IOP to within 20% and the R^2 of 0.89 demonstrates the similarity of the O_3 predicted by the two schemes throughout the IOP, during which a variety of chemical conditions were experienced.

Table 6 Chemistry scheme run times

Chemistry scheme	Run time[a]
MCM*v3.2*	1105
CRI*v2-R5*	1.00
GRS	0.015

[a] Run time relative to that of CRI*v2-R5*.

4. Conclusions

This paper has analysed the ability of the reduced chemistry schemes of the GRS and CRI*v2-R5* mechanisms to predict O_3 compared to the near-explicit MCM*v3.2* scheme constrained to measurements made in London, and has looked at the impact different model constraints have upon modelled O_3. The main conclusions of these model comparisons and sensitivity analyses are as follows:

1. The GRS semi-empirical chemistry mechanism utilised in the ADMS-urban dispersion model predicts lower O_3 concentrations both during the summer and winter months compared to the MCM box model when constrained to the NO and ROC derived from dispersion calculations. The GRS scheme reproduces the *in situ* O_3 predicted by the MCM box model to within 40% on most days in the summer and winter when the ROC constraint is increased to reflect the VOC observations during ClearfLo and the NO constraint is decreased to reflect observations. However, during the elevated O_3 episodes experienced during the easterly flows in the summer, when temperatures increased and wind speeds dropped, the agreement between the GRS scheme and the MCM decreased. Similarly, agreement between CRI*v2-R5*- and MCM-predicted O_3 also worsened during these periods. The *in situ* O_3 predicted by the MCM was heavily influenced by biogenic VOCs during these conditions and the low GRS $[O_3] : MCM\ [O_3]$ ratio (and low CRI*v2-R5* $[O_3] : MCM\ [O_3]$ ratio) demonstrates that the lumped schemes under-represent the full O_3 creation potential of these species. Biogenics are not considered in the total [ROC] used to constrain the ADMS-Urban model, and even though they are only a small proportion of the total measured VOCs (1%), they have a high O_3 creation potential. In a warming atmosphere, in the presence of NO_x, the impact of these species on $[O_3]$ is likely to become increasingly relevant.

In winter, omission of substituted aromatic species such as xylenes in the total [ROC] constraint used by ADMS-Urban, which have high O_3 creation potentials similar to the biogenic VOCs, leads to an under-representation of the total ROC and, similar to summer, lowers the O_3 concentration predicted by the GRS scheme.

2. Despite the very limited ability of the GRS to generate O_3, when constrained to ADMS-Urban input parameters in winter, the ADMS-Urban dispersion model is able to predict the observed O_3 well indicating that the *in situ* chemistry is not controlling the $[O_3]$ in London in the winter. This is further reflected by the inability of the near-explicit MCM run to steady state (reflecting the full chemical O_3 creation potential) to predict the observed $[O_3]$ on most days of the winter IOP other than during the stagnation episodes.

During the summer IOP, the near-explicit MCM run to steady state predicts extremely high O_3 concentrations, up to 1240 ppb. The O_3 observed was significantly lower, indicating that the real London atmosphere was far from steady state. To fully assess the influence of the *in situ* O_3 generated from local emissions *versus* O_3 generated upwind of London and advected in, knowledge of the time since emission (and, hence, how far the real atmosphere is from steady state) is needed. From estimates of the mean transport time determined from the $NO_x : NO_y$ ratio observed at North Kensington during the summer and comparison of the O_3 predicted by the MCM model after this time, ~60% of the median observed $[O_3]$ could be generated from local emissions. During the warmer conditions experienced during the easterly flows, however, the observed [O_3] may be even more heavily influenced by London's emissions.

3. The O_3 predicted by the reduced chemistry mechanisms of GRS and CRI*v2-R5* is highly sensitive to HONO chemistry in the wintertime; less so in the summer. An inclusion of a radical source from HONO and an accurate representation of the HONO concentrations observed in urban centres could improve the *in situ* O_3 predicted by these reduced chemistry schemes. Long-term HONO measurements are not performed as standard and the contribution of different HONO sources, many of which are heterogeneous in nature, remains uncertain. HONO has been shown to correlate with NO_2 at urban sites and so its concentration may be parameterised from NO_2 observations as a first step to representing this radical source. HONO can be an important source of OH radicals in the urban environment so any underestimation has a significant impact on the oxidising capacity of the atmosphere and on secondary organic aerosol (SOA) production as well as influencing *in situ* O_3 as demonstrated here.

4. One of the advantages of using reduced chemistry mechanisms is that computer processing time and hence the required computer power can be minimised. It can be seen from Table 6 that by replacing the explicit MCM with the chemistry mechanisms of CRI*v2-R5* and GRS the run times can be significantly reduced by a factor of 1000 and 60 000 respectively. This is an important consideration when investigating large time series (*e.g.* multiple years) or running chemistry in conjunction with other computationally intensive simulations, *e.g.* of transport and dispersion. The inclusion of the additional HONO chemistry caused little impact on these run times (Table 6).

ADMS-Urban and EMEP4UK contain simple chemical schemes and are constrained by incomplete predictions of source/sink species, *e.g.* volatile organic compounds and nitrous acid, and contain parameterisations for some inputs, *e.g.* photolysis frequencies. Despite these simplifications, these schemes are used widely to provide forecasts and assessments of O_3 and other air quality markers for the community. In this paper we have provided an analysis of the conditions under which the simple schemes appear able to deliver reliable forecasts of a key pollution marker, O_3, with the simple chemical schemes performing better (with respect to the explicit MCM) in winter and in summer under lower O_3 conditions. In addition, we show the sensitivity towards O_3 production of these simple schemes for variations of NO_x and VOC loading/speciation. We also demonstrate the impact of the inclusion of nitrous acid (HONO) chemistry which is overlooked in these types of mechanism. HONO can play a dominant role in radical production particularly in winter when other sources of radicals which lead to the production of O_3 are limited. Both the ADMS-Urban and EMEP4UK models have a tendency to under-predict peak daytime O_3 concentrations in London, particularly in the summer during high O_3 episodes. Increasing the contribution of the local chemistry, with ROC (VOC) and HONO concentrations representative of the levels observed during ClearfLo (and enhancing the full O_3 creation potential of biogenic VOC in line with the near-explicit MCM) may help to reduce this under-prediction.

Acknowledgements

This work was funded by the UK NERC under grant NE/M002381/1. The authors would like to thank all participants of ClearfLo.

References

1 *EEA Report No 5/2014*, 2014.

2 WHO final technical report, *Review of evidence on health aspects of air pollution – REVIHAAP project*, 2013.

3 S. I. Bohnenstengel, S. E. Belcher, A. Aiken, J. D. Allan, G. Allen, A. Bacak, T. J. Bannan, J. F. Barlow, D. C. S. Beddows, W. J. Bloss, A. M. Booth, C. Chemel, O. Coceal, C. F. Di Marco, M. K. Dubey, K. H. Faloon, Z. L. Fleming, M. Furger, J. K. Gietl, R. R. Graves, D. C. Green, C. S. B. Grimmond, C. H. Halios, J. F. Hamilton, R. M. Harrison, M. R. Heal, D. E. Heard, C. Helfter, S. C. Herndon, R. E. Holmes, J. R. Hopkins, A. M. Jones, F. J. Kelly, S. Kotthaus, B. Langford, J. D. Lee, R. J. Leigh, A. C. Lewis, R. T. Lidster, F. D. Lopez-Hilfiker, J. B. McQuaid, C. Mohr, P. S. Monks, E. Nemitz, N. L. Ng, C. J. Percival, A. S. H. Prévôt, H. M. A. Ricketts, R. Sokhi, D. Stone, J. A. Thornton, A. H. Tremper, A. C. Valach, S. Visser, L. K. Whalley, L. R. Williams, L. Xu, D. E. Young and P. Zotter, *Bull. Am. Meteorol. Soc.*, 2014, **96**, 779–804.

4 A. M. Fiore, V. Naik, D. V. Spracklen, A. Steiner, N. Unger, M. Prather, D. Bergmann, P. J. Cameron-Smith, I. Cionni and W. J. Collins, *Chem. Soc. Rev.*, 2012, **41**, 6663–6683.

5 S. Solberg, Ø. Hov, A. Søvde, I. S. A. Isaksen, P. Coddeville, H. De Backer, C. Forster, Y. Orsolini and K. Uhse, *J. Geophys. Res.: Atmos.*, 2008, **113**, D07307.

6 X. V. Francis, C. Chemel, R. S. Sokhi, E. G. Norton, H. M. A. Ricketts and B. E. A. Fisher, *Atmos. Environ.*, 2011, **45**, 6880–6890.

7 M. Vieno, A. J. Dore, D. S. Stevenson, R. Doherty, M. R. Heal, S. Reis, S. Hallsworth, L. Tarrason, P. Wind, D. Fowler, D. Simpson and M. A. Sutton, *Atmos. Chem. Phys.*, 2010, **10**, 7963–7978.

8 J. D. Lee, A. C. Lewis, P. S. Monks, M. Jacob, J. F. Hamilton, J. R. Hopkins, N. M. Watson, J. E. Saxton, C. Ennis, L. J. Carpenter, N. Carslaw, Z. Fleming, B. J. Bandy, D. E. Oram, S. A. Penkett, J. Slemr, E. Norton, A. R. Rickard, L. K Whalley, D. E. Heard, W. J. Bloss, T. Gravestock, S. C. Smith, J. Stanton, M. J. Pilling and M. E. Jenkin, *Atmos. Environ.*, 2006, **40**, 7598–7613.

9 A. R. MacKenzie, R. M. Harrison, I. Colbeck and C. N. Hewitt, *Atmos. Environ., Part A*, 1991, **25**, 351–359.

10 J. M. Harris and S. J. Oltmans, *J. Geophys. Res.: Atmos.*, 1997, **102**, 8781–8791.

11 R. M. Harrison, M. Dall'Osto, D. C. S. Beddows, A. J. Thorpe, W. J. Bloss, J. D. Allan, H. Coe, J. R. Dorsey, M. Gallagher, C. Martin, J. Whitehead, P. I. Williams, R. L. Jones, J. M. Langridge, A. K. Benton, S. M. Ball, B. Langford, C. N. Hewitt, B. Davison, D. Martin, K. F. Petersson, S. J. Henshaw, I. R. White, D. E. Shallcross, J. F. Barlow, T. Dunbar, F. Davies, E. Nemitz, G. J. Phillips, C. Helfter, C. F. Di Marco and S. Smith, *Atmos. Chem. Phys.*, 2012, **12**, 3065–3114.

12 J. Stocker, C. Hood, D. Carruthers and C. McHugh, *Int. J. Environ. Pollut.*, 2012, **50**, 308–316.

13 M. E. Jenkin, S. M. Saunders, V. Wagner and M. J. Pilling, *Atmos. Chem. Phys.*, 2003, **3**, 181–193.

14 S. M. Saunders, M. E. Jenkin, R. G. Derwent and M. J. Pilling, *Atmos. Chem. Phys.*, 2003, **3**, 161–180.

15 K. M. Emmerson, N. Carslaw, D. C. Carslaw, J. D. Lee, G. McFiggans, W. J. Bloss, T. Gravestock, D. E. Heard, J. Hopkins, T. Ingham, M. J. Pilling, S. C. Smith, M. Jacob and P. S. Monks, *Atmos. Chem. Phys.*, 2007, **7**, 167–181.
16 R. M. Harrison, J. Yin, R. M. Tilling, X. Cai, P. W. Seakins, J. R. Hopkins, D. L. Lansley, A. C. Lewis, M. C. Hunter, D. E. Heard, L. J. Carpenter, D. J. Creasey, J. D. Lee, M. J. Pilling, N. Carslaw, K. M. Emmerson, A. Redington, R. G. Derwent, D. Ryall, G. Mills and S. A. Penkett, *Sci. Total Environ.*, 2006, **360**, 5–25.
17 R. Sommariva, A. L. Haggerstone, L. J. Carpenter, N. Carslaw, D. J. Creasey, D. E. Heard, J. D. Lee, A. C. Lewis, M. J. Pilling and J. Zádor, *Atmos. Chem. Phys.*, 2004, **4**, 839–856.
18 R. Sommariva, W. J. Bloss, N. Brough, N. Carslaw, M. Flynn, A. L. Haggerstone, D. E. Heard, J. R. Hopkins, J. D. Lee, A. C. Lewis, G. McFiggans, P. S. Monks, S. A. Penkett, M. J. Pilling, J. M. C. Plane, K. A. Read, A. Saiz-Lopez, A. R. Rickard and P. I. Williams, *Atmos. Chem. Phys.*, 2006, **6**, 1135–1153.
19 P. M. Edwards, M. J. Evans, K. L. Furneaux, J. Hopkins, T. Ingham, C. Jones, J. D. Lee, A. C. Lewis, S. J. Moller, D. Stone, L. K. Whalley and D. E. Heard, *Atmos. Chem. Phys.*, 2013, **13**, 9497–9514.
20 D. Stone, M. J. Evans, R. Commane, T. Ingham, C. F. A. Floquet, J. B. McQuaid, D. M. Brookes, P. S. Monks, R. Purvis, J. F. Hamilton, J. Hopkins, J. Lee, A. C. Lewis, D. Stewart, J. G. Murphy, G. Mills, D. Oram, C. E. Reeves and D. E. Heard, *Atmos. Chem. Phys.*, 2010, **10**, 9415–9429.
21 W. J. Bloss, M. Camredon, J. D. Lee, D. E. Heard, J. M. C. Plane, A. Saiz-Lopez, S. J. B. Bauguitte, R. A. Salmon and A. E. Jones, *Atmos. Chem. Phys.*, 2010, **10**, 10187–10209.
22 P. Edwards, M. J. Evans, R. Commane, T. Ingham, D. Stone, A. S. Mahajan, H. Oetjen, J. R. Dorsey, J. R. Hopkins, J. D. Lee, S. J. Moller, R. Leigh, J. M. C. Plane, L. J. Carpenter and D. E. Heard, *J. Geophys. Res.: Atmos.*, 2011, **116**, D22306.
23 D. Stone, L. K. Whalley and D. E. Heard, *Chem. Soc. Rev.*, 2012, **41**, 6348–6404.
24 R. E. Dunmore, J. R. Hopkins, R. T. Lidster, J. D. Lee, M. J. Evans, A. R. Rickard, A. C. Lewis and J. F. Hamilton, *Atmos. Chem. Phys.*, 2015, **15**, 9983–9996.
25 L. K. Whalley, D. Stone, B. Bandy, R. Dunmore, J. F. Hamilton, J. Hopkins, J. D. Lee, A. C. Lewis and D. E. Heard, *Atmos. Chem. Phys.*, 2016, **16**, 2109–2122.
26 G. L. Forster, W. T. Sturges, Z. L. Fleming, B. J. Bandy and S. Emeis, *Tellus B*, 2012, **64**, 17771.
27 D. J. Carruthers, H. A. Edmunds, A. E. Lester, C. A. McHugh and R. J. Singles, *Int. J. Environ. Pollut.*, 2000, **14**, 364–374.
28 J. Stocker, C. Hood, D. Carruthers, M. Seaton, K. Johnson and J. Fung, 13th Annual CMAS Conference, Chapel Hill, NC, USA, 2014.
29 http://www.airtext.info/.
30 M. Azzi, G. Johnson and M. Cope, *Presented in part at the 11th International Conference of the Clean Air Society of Australia and New Zealand*, 1992.
31 G. Johnson, *presented in part at the 8th International Conference of the Clean Air Society of Australia and New Zealand*, 1984.
32 *Defra's Emission Factor Toolkit, EFTv6.0.2*, 2014, http://laqm.defra.gov.uk/review-and-assessment/tools/emissions-factors-toolkit.html.
33 D. Simpson, A. Benedictow, H. Berge, R. Bergström, L. D. Emberson, H. Fagerli, C. R. Flechard, G. D. Hayman, M. Gauss, J. E. Jonson,

M. E. Jenkin, A. Nyíri, C. Richter, V. S. Semeena, S. Tsyro, J. P. Tuovinen, Á. Valdebenito and P. Wind, *Atmos. Chem. Phys.*, 2012, **12**, 7825–7865.

34 M. Vieno, M. R. Heal, S. Hallsworth, D. Famulari, R. M. Doherty, A. J. Dore, Y. S. Tang, C. F. Braban, D. Leaver, M. A. Sutton and S. Reis, *Atmos. Chem. Phys.*, 2014, **14**, 8435–8447.

35 M. E. Jenkin, L. A. Watson, S. R. Utembe and D. E. Shallcross, *Atmos. Environ.*, 2008, **42**, 7185–7195.

36 http://www3.epa.gov/ttnchie1/conference/ei19/session5/granier2.pdf.

37 L. A. Watson, D. E. Shallcross, S. R. Utembe and M. E. Jenkin, *Atmos. Environ.*, 2008, **42**, 7196–7204.

38 J. D. Lee, L. K. Whalley, D. E. Heard, D. Stone, R. E. Dunmore, J. F. Hamilton, D. E. Young, J. D. Allan, S. Laufs and J. Kleffmann, *Atmos. Chem. Phys.*, 2016, **16**, 2747–2764.

39 http://uk-air.defra.gov.uk/networks/network-info?view=hc.

40 P. Hurley, *TAPM V4. Part 1: technical description. CSIRO Marine and Atmospheric Research Internal Report No. 5*, 2008.

41 M. Gonçalves, D. Dabdub, W. L. Chang, O. Jorba and J. M. Baldasano, *Atmos. Environ.*, 2012, **54**, 168–176.

42 Y. F. Elshorbany, B. Steil, C. Brühl and J. Lelieveld, *Atmos. Chem. Phys.*, 2012, **12**, 9977–10000.

43 E. von Schneidemesser, P. S. Monks and C. Plass-Duelmer, *Atmos. Environ.*, 2010, **44**, 5053–5064.

44 B. K. Pun, S.-Y. Wu and C. Seigneur, *Environ. Sci. Technol.*, 2002, **36**, 3586–3596.

45 A. Venkatram, P. Karamchandani, P. Pai and R. Goldstein, *Atmos. Environ.*, 1994, **28**, 3665–3678.

Faraday Discussions

PAPER

Verifiable emission reductions in European urban areas with air-quality models

A. N. Skouloudis*[a] and D. G. Rickerby[b]

Received 27th November 2015, Accepted 5th January 2016
DOI: 10.1039/c5fd00189g

The first and second AutoOil programmes were conducted since 1992 as a partnership between the European Commission and the automobile and oil industries. These have introduced emission reductions in Europe based on numerical modelling for a target year. They aimed to identify the most cost-effective way to meet desired future air quality over the whole European Union. In their time, these regulatory efforts were considered an important step towards a new approach for establishing European emission limits. With this work, we review the effectiveness of forecasts carried out with numerical modelling and compare these with the actual measurements at the target year, which was the year 2010. Based on these comparisons and new technological innovations these methodologies can incorporate new sectorial assessments for improving the accuracy of the modelling forecasts and for examining the representativeness of emissions reductions, as well as for the simultaneous assessment of population exposure to cocktails of toxic substances under realistic climatological conditions. We also examined at the ten AutoOil domains the geographical generalisation of the forecasts for CO and NO_2 at 1065 European urban areas on the basis of their population and the local population density.

Introduction

The key elements of utilising modelling for demonstrating compliance with the requirements of European Environmental Policy are not new.[1] The use of these models for regulatory purposes requires the coherent coupling of many disciplines relating to anthropogenic and natural processes. For policy purposes, these are then linked to economic data for identifying cost-optimised solutions achieving maximum pollutant reductions over regional and urban domains. This has been the case when industrial partners participated in common actions with regulatory authorities for achieving the best air-quality outcomes over large regional territories. In the past, vehicle emission limits have been set using the

[a]*Institute for Environment & Sustainability, European Commission, Joint Research Centre, via E. Fermi 2749, Ispra 21027, Italy. E-mail: andreas.skouloudis@jrc.ec.europa.eu*
[b]*Institute for Health and Consumer Protection, European Commission, Joint Research Centre, via E. Fermi 2749, Ispra 21027, Italy*

'best available technology' approach. As standards have become increasingly stringent, it will be essential to adopt new approaches for future emission reductions. During the first two AutoOil programmes,[2,3] extensive analysis of emissions reductions based on automotive fuel quality and engine technologies have resulted in several legislative efforts like the so-called Daughter Directives, the Ozone and PM directives and eventually the Clean Air for Europe Programme. All of these had 2010 as the target year and were based on several forecasts for emission reductions.

In their time, these regulatory efforts were considered an important step towards a new approach for establishing European emission limits. Despite various emission reduction efforts, air pollution remains in many areas of Europe above acceptable limits. Since it is not possible to assess if the forecasted emissions were correct at the regulatory base year, we can verify the air-quality concentrations, which is a result of the emission reductions at the target year. In this work, we are reviewing the options that were available at the time of the forecasts and we compare these with the actual measurements for the target year 2010.

Since then, we have incorporated several innovations into these methodologies for sectorial assessment and for examining the representativeness of abatement strategies, as well as for the simultaneous assessment of population exposure to cocktails of toxic substances under different climatological conditions. We examine also, why air quality issues have become more complex at the regional level with a wider impact of oxidants on population and rural ecosystems. The spatial resolution of an ambient air quality model (that is, the area over which the predicted concentrations are averaged) may vary from several meters to several thousand kilometres.

Evidence is also presented for why integration in regulatory applications is necessary across a range of scales (local, regional), sectors (transport, energy) and issues (ozone, particulate matter) together with health effects as identified in the European Commission Action Plan for Environment and Health (COM-2004-416).

We also examine what types of numerical modelling innovations can be carried out after incorporating recent advances in distributed data architectures and sensor technologies. We also show what type of integrated new applications can be used to understand the hazards at specific urban hot spots and to reduce the burden of human exposure for the local population.

The successes of new regulatory approaches depend on the implementation of new technologies, for setting realistic targets and for monitoring at intermediate steps on how these targets are achieved, without the need to look in the long-term future with unverifiable parameters that lead to emission reductions that are beyond present capabilities.

The AutoOil approaches

In planning for the abatement and control of air pollution in regional and urban domains several complex issues might arise. The aim of the first AutoOil programme was to achieve compliance with the EU and WHO air-quality standards by looking into classical urban pollutants (such as CO, SO_2, NO_2 and C_6H_6) over seven urban domains and by looking into regional photochemical pollution over the whole of Europe. The seven cities studied were Athens, Cologne, the Hague,

London, Lyon, Madrid and Milan. These cities were considered representative in terms of air quality, location and size. Since measurements are not possible at all locations and at all times, simulations were carried out by three models using 3D Eulerian models for meteorology coupled also with dispersion models without allowing for chemical reactions.[2,4] This assumption was necessary because of the large number of emission scenarios that were required. In addition, the period of simulation in each city domain was restricted to four days. Several urban background locations were used for identifying the four day interval when the measured concentrations were close to the annual averages simultaneously for all four pollutants. For establishing this period the measured concentrations were analysed over several years.[4] In each city, the size of each urban domain was 100 × 100 km^2 and the size of each grid was 2 × 2 km^2.

In order to establish the emission reduction targets for ozone another approach was utilised. It was considered necessary to look into regional simulations over the whole of Europe. The simulations were carried out over a year but by assuming that ozone concentrations are highest during afternoon hours; the 8 hour averaged concentrations were estimated by taking the mean of the calculated concentrations at 12:00 and 18:00. The calculated ozone concentrations over Europe are shown in Fig. 1. Simulations were carried out with a single layer Lagrangian model.[5] The advantage of this model is its simplicity for calculating over long time-periods, nevertheless there were still computational difficulties so the model was applied every six hours on a scale of each grid being 150 × 150 km^2.

Ideally, it is essential to relate the ozone concentrations to emissions on a finer grid. In the past, this problem was prohibitive in terms of computational times and for this reason today the EMEP grids have become three times finer.[6] Nevertheless, whenever a large geographical area is projected in a grid based on km the spacing will change with latitude. This is only due to the mathematical projection of the spherical coordinates over a flat Eulerian map. As such, it means that in northern latitudes, there is 27% more area averaging than in southern grids. While this is perhaps acceptable for emissions due to small spatial

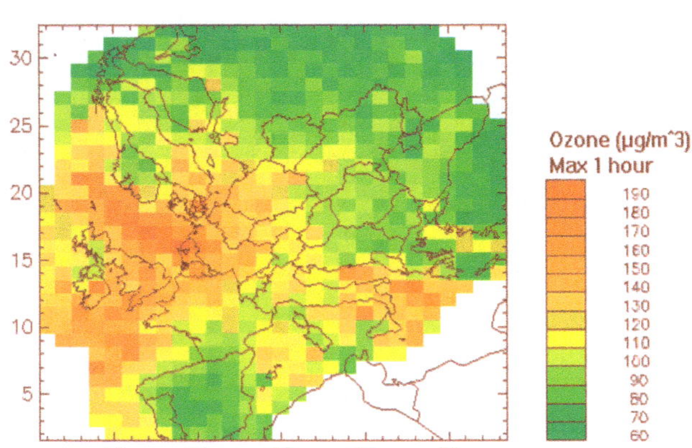

Fig. 1 The grid used by the EMEP model with the calculated maximum one hour ozone concentrations in Europe plotted over the 150 × 150 km^2 polar-stereographic projection.[5,6]

variations in rural uninhabited areas, it introduces significant underestimation of concentrations emitted and transported due to large area averaging for each grid cell.

Recent advancements in computational speeds coupled with the reduced costs of electronic storage have resolved some of these limitations in the second AutoOil programme. The spatial resolution of an ambient air quality model (that is, the area over which the predicted concentrations are averaged) may vary from one to several km. However, the projection of the spherical geographical coordinates in a flat Eulerian grid expressed in km was abandoned. Due to the vastness of the overall European domain a grid based on geographical degrees was utilised. The partial differential equations are solved numerically for the ten domains shown in Fig. 2 as well as in the whole European domain. Twenty levels were considered in the vertical domain up to 13 km for the meteorological calculations and up to 4 km for the dispersion model. For the latter, calculations were carried out with an appropriate chemical scheme. Apart from the improved treatment of the boundary, the other obvious advantage of these models is that the coarser domains allow simulations to be carried out over long modelling periods (several months).

The choice of the spatial grid on which the equations are solved is governed by the degree of spatial detail in the emissions inventory and the accuracy of the meteorological variables. Sometimes it is desired to predict pollutant concentrations in the immediate vicinity of sources, in a city centre as well as the greater downwind area. In such cases the spatial resolution of the concentrations might be as little as a few km at the centre but coarser in the outer regions. The coupling

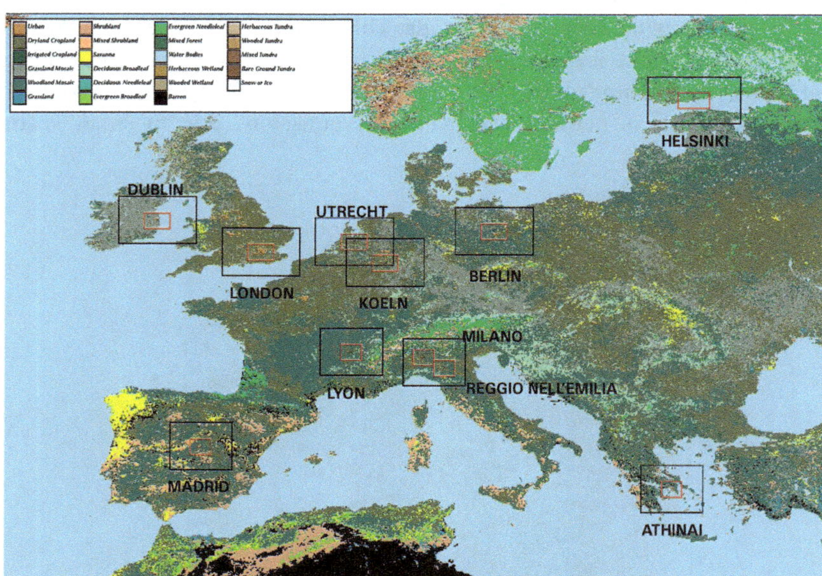

Fig. 2 The three levels of two-way nested domains used for the second AutoOil programme. Red is the inner urban domain, black is the outer domain and the overall image is the European regional domain. The colours in the legend show the corresponding CORINE land cover.

of these resolutions in multi-nested layers and the online interaction of the solution schemes (two-way nesting) provides the basis for accurately examining the impact of man-made emissions and their impact in larger geographical areas. A typical outer domain with two inner domains is depicted in Fig. 3 with information about the urban and suburban territories aggregated from the detailed CORINE categories of land-cover shown in Fig. 2.

In regulatory applications, spatial resolution is frequently considered of secondary importance in favour of providing simulations over long modelling periods. However, for verifying the effectiveness of regulatory assessments, the spatial resolution in the horizontal and vertical directions is a key area that needs to be properly addressed. Assuming that parametrisation of physical and chemical processes is the same, models with the finer resolution in the horizontal direction allow a more accurate spatial representation of the overall processes and are expected to characterise more accurately the size of hot-spot areas. On the other hand, emission of pollutants at different heights from industrial and traffic sources coupled with vertical atmospheric transport requires that the models have a reasonable number of levels in the vertical direction. The proximity of maximum ozone concentrations near large urban agglomerations in South Europe and the existence of significantly varying conditions in densely populated areas means that average representation in coarse regional grid sizes (more than 50×50 km^2) is rather inappropriate and modelling windows with an adequate size need to be utilised.

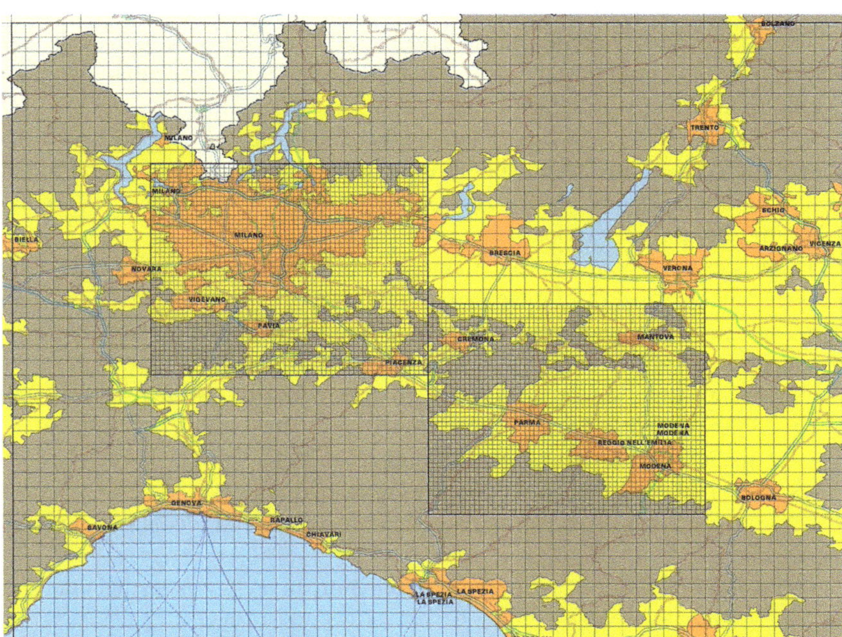

Fig. 3 Detail of the three levels of grid in Milano and Reggio Nel Emilia with the urban (orange) and suburban (yellow) landscape areas on the background. These areas are identified from the CORINE categories for land-cover categorisation.

Yet there remains the problem of the accurate estimation of emissions, which needs to be given at each grid cell for the corresponding temporal resolution, especially for natural sources. The set of emission categories that were taken into consideration were for traffic sources (passenger cars gasoline and LPG, passenger cars diesel, light duty commercial vehicles, heavy duty commercial vehicles, buses, two wheeled vehicles), medium area sources (non-industrial combustion plants, combustion in manufacturing industry, solvent and other product use, extraction and distribution of fossil & other fuels, combustion in energy & transformation industries <50 MW, gas turbines, stationary engines), large area sources (combustion in energy & transformation industries 50–300 MW, combustion in energy & transformation industries >300 MW, production process industries, waste treatment & disposal) and small area sources (other mobile sources and machinery, agriculture, natural emissions).

For the aforementioned emission categories inventories were constructed for NO_x, VOC, CO, SO_2 and PM_{10}. The requirements for implementing a model are strongly governed by its temporal resolution. The temporal resolution of emissions (that is, the time period over which these are assumed constant in the air-quality simulations) may vary from several minutes to one year. Starting from the annual emissions for these inventories were disaggregated to daily and hourly values according to chapter 4.2 of the air-quality report.[3] For assessing the impact of emission reduction under annual mean and the most severe episodic conditions in each urban domain the steps that were followed were:

• The monitoring sites were characterised according to their relevance in representing the domain and the exposure significance.

• Measured data for monitoring sites were used for at least one to five calendar years previously (rolling data, one year before the current calendar date) for examining the temporal trends.

• Statistical analysis was carried out to identify representative periods with air-quality similar to annual mean conditions and for identifying a similar period of severe conditions (98 percentile). This step was carried out once for conventional pollutants and once for ozone.

• Forecasts for meteorological parameters were calculated for a grid cell on the basis of data obtained from meteorological offices.

• The current air-quality concentrations were calculated and compared with measurements for the base year.

• Emissions were weighted according to anthropogenic and natural estimate for the target year 2010.

• Air-quality concentrations were calculated for 2010.

• Source apportionment simulations were carried out.

• Spatial synthesis of all appropriate parameters from all modelling domains was carried out.

• Spatial generalisation of the results in other urban areas was carried out from the intermediate domain.

These steps, although simplified, required data handling, retrieval and harmonisation as well as computational capacities that represent big-data and distributed network architectures that only recently are becoming available.

Forecasted concentrations for the target year

The aim of the legislative efforts in the AutoOil programmes was to identify suitable measures that could be introduced in the coming years that would lead to achieving compliance with the desired air-quality standards by 2010. This compliance should be "ideally" achieved for all pollutants and should have an impact both on regional and urban scales.

A certain aspect of the forecasts is the evolution of emissions for the target year and how these will be translated into air-quality concentration especially for chemically reacting pollutants. The evolution of the emissions used in these two programmes are shown in Fig. 4 by all of the then member states of EU. Despite the fact that the projection of emissions inventories is uncertain due to the complexity of anthropogenic activities and the statistical nature of forecasts, the updating of emission inventories can be described as a purely logistical process. From this figure it is clear that after 2010 will start the period of less possible reductions with all, by then, existing technologies.

The application of regulatory models of this type for long-term forecasts requires, firstly, the suitable updating of emission inventories and, secondly, the identification of suitable emission reductions in order to achieve compliance with specific air-quality targets.

Regional scale forecasts

A typical example is ozone. Taking into consideration the already agreed technological measures until 2010 and photochemical calculations carried out in the first AutoOil, maps were generated with EMEP like those shown in Fig. 1. From these maps the percentage of the EU15 area exposed to each ozone concentration was deduced with GIS processing as in Fig. 5; by this time there were only 15 member states in EU. The reduction of emissions will improve the percentage of the European area which achieves compliance, but with varying impact as shown in the same figure. The straight line in the figure shows the limit values.

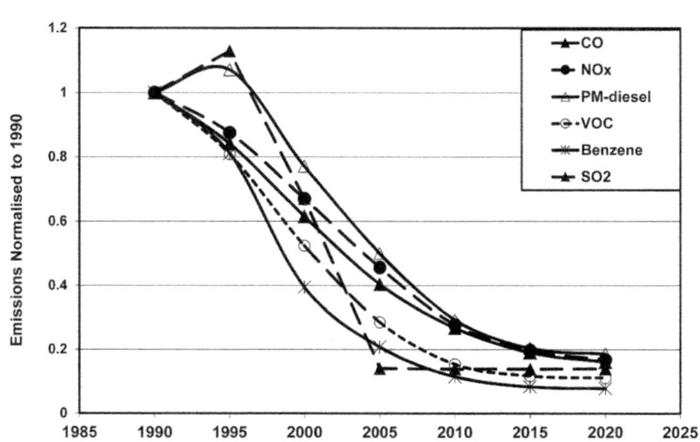

Fig. 4 The forecasted emission for the EU15 member states normalised to 1990.

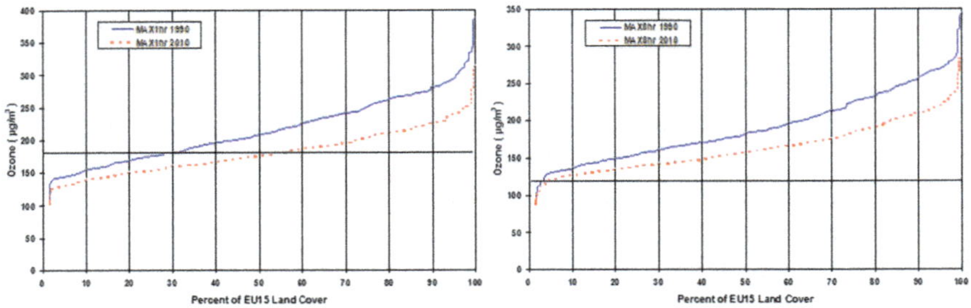

Fig. 5 Ozone concentrations in the EU15 states from EMEP simulations after the impact of agreed measures on emissions until 2010. The horizontal axis shows the percent of EU15 land area effected from these concentrations for the 1 hour maximum values (left), and the 8 hours maximum values (right). The desirable limit values are also indicated with the horizontal lines.

The intersection of the distribution curve with the limit value indicates the percentage of the area above which the limit value is exceeded. It is evident that the limit values of the 1 hour maximum values of ozone and the 8 hours average values are different. The latter is practically exceeded all over Europe. The same figure shows that 73% of the European territory was exposed to 1 hour average concentrations of 180 μg m^{-3} in 1990 but this will be reduced to 47% by 2010.

In order to improve the compliance area further reductions in NO_x and VOC emissions are needed. Sensitivity calculations for 2010 as shown in Fig. 6 show that, despite significant improvements in the area, in the European territory experiencing high one hour concentrations, there are still areas where the targets are exceeded even for reductions of NO_x emission reductions by 60%, 70% and 80%. Of course these are results from simulations with a coarse geographical grid and a single vertical layer. Even, ignoring the uncertainty of numerical simulations, it is clear that uniform reductions are necessary and these reductions might

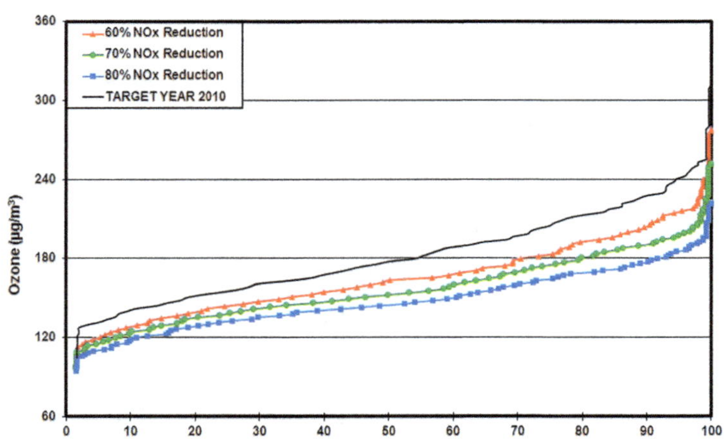

Fig. 6 Ozone concentrations and the percent of EU15 area effected. Impact of 60%, 70% and 80% NO_x emission reductions on maximum 1 hour values.

not be translated with the same level of success in other target limits for the same air pollutant.

Forecasts for urban areas

Article 4 of the European Directive on passenger car emissions is designed to produce effects to meet the requirements of the EU air-quality standards and related objectives at the least cost. However, the major challenge of the AutoOil-2 programme was to take account of all emission sources in an integrated manner. For this reason the Working Group for air-quality decided to proceed without the use of attenuation factors (deduced from emission ratios) for projecting the results of air-quality to other pollutants. For assessing the impact of emission reduction in metropolitan areas the existing and projected air-quality conditions were analysed with 3D photochemical models.

Following the emission projections for the target year, the photochemical simulations produced spatial contours maps showing the detailed areas of exceedance over the target year. On the basis of these, in the second AutoOil programme were calculated: (a) the maximum and (b) mean concentrations in each domain, (c) the area of exposure for the population and the (d) percentage of the urban area above the WHO target values. The simulations were repeated to examine the source apportionment for different emission categories so the last option offered by this programme was (e) break down of pollutant concentrations at each grid cell according to the real emission sources.

Two of these terms (a and d) are shown for all ten domains in Fig. 7. GIS was used for the evaluation of (d) on the basis of maps similar to those shown in Fig. 3. A similar process was used for the population exposure (c) whereas maps similar to those of Fig. 3 were used but with the population density in each grid cell.

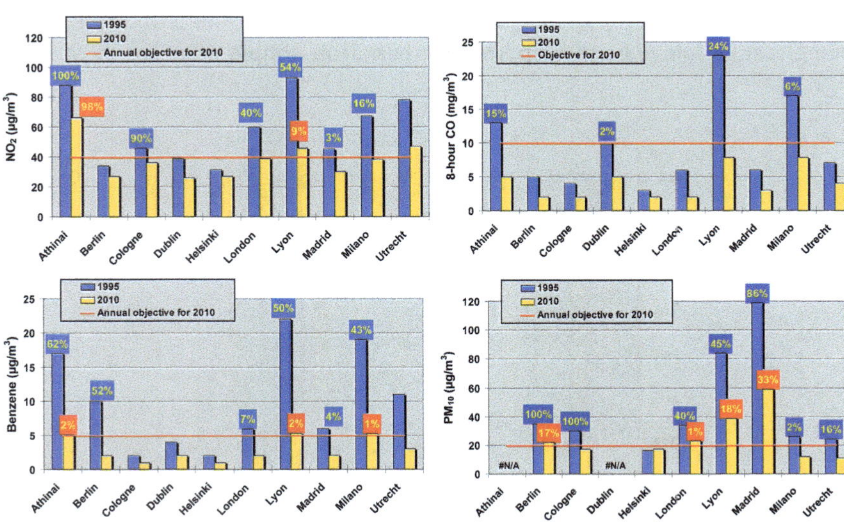

Fig. 7 Maximum predicted values at each of the ten AutoOil-2 domains for the annual mean concentrations of NO_2, CO, C_6H_6 and PM_{10}. The urban area above the 2010 target values is indicated by the percentages calculated from the GIS maps shown in Fig. 3. For PM_{10} the calculations are for limited measurements only where available for 1995.

These population density maps were unique at the time because they were based on census data at high spatial resolution and disaggregated to local grid cells according to land-cover and light intensity consumption from satellite images. This process resulted in annual mean projections that were closer to physical and chemical reality than the projections forecasted by the extrapolations of only the emission as was done in the first AutoOil.

The mean concentrations (b) for three pollutants in the inner urban domains are shown at Fig. 8. The vertical axis shows the values normalised against the values obtained for the base year 1995 at the same domain. It is worth noticing the increase of the ozone values in view of the local emission inventories provided by the local authorities.

For assessing pollutants that depend on the dispersion of several precursors and that are subject to chemical transformations it is necessary to establish an accurate relationship between air-quality concentration and emissions (e). Despite the complexity of establishing this link (especially for photochemical processes), additional simulations were performed for source apportionments in characterising the "reality" in each grid cell. Naturally, the emission inventories need to be sufficiently disaggregated into the same source categories.

Because in the second AutoOil simulations were carried out separately for 18 categories, it was possible to identify directly the attribution of Traffic Sources (CAT_20), the Elevated Area Sources (CAT_21), the Large Point Industrial Sources (CAT_22), the Other Area Sources & Nature (CAT_23) as well as the Non-Attributable Secondary Concentrations (NASC) part of primary emissions, which was due to the chemical reaction potential of the different portions of the concentration originating from natural background and the portion originating from the boundaries (*i.e.* the imposed lateral boundary concentrations).

Fig. 9 shows the breakdown of the 40 highest pollutant concentrations at the inner domains for Athens, Lyon and Milan. This figure also contains a portion of "non-attributable" concentrations with the green colour which is not directly attributable to anthropogenic sources. In reality this portion is the reason why simple emission reduction approaches based on simple attenuation of "base

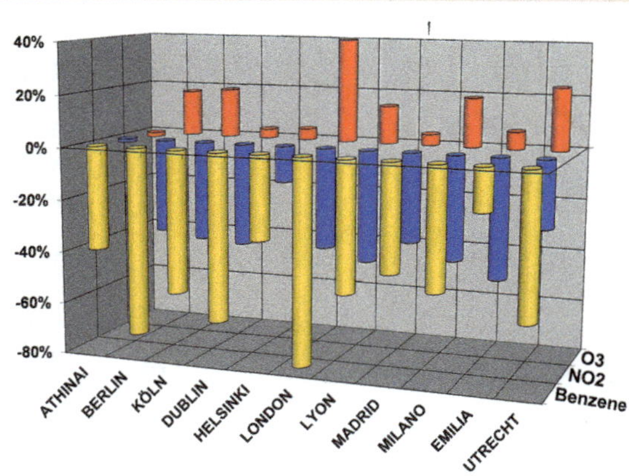

Fig. 8 The percentage of change between 1995 and 2010 for the mean air-quality concentrations in each of the urban inner domains of the second AutoOil on the basis of numerically calculated values.

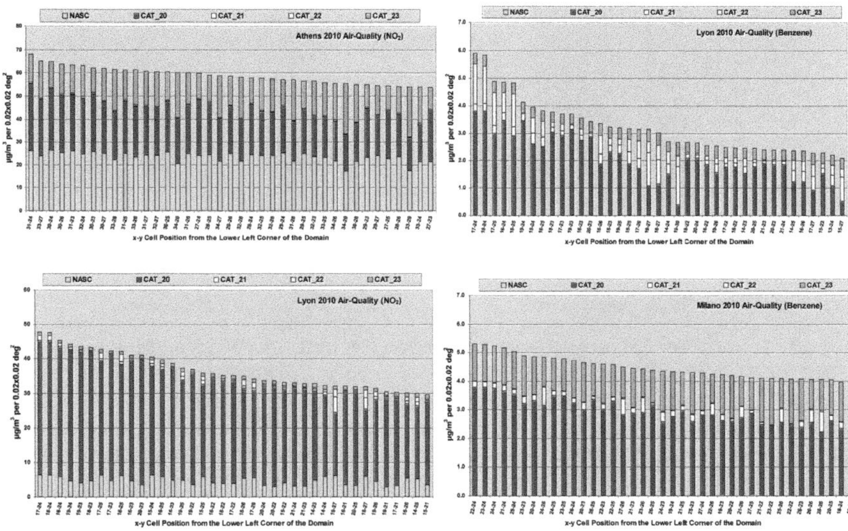

Fig. 9 Emission source attribution of air-quality for 40 grid-cells with the largest concentrations of NO_2 and C_6H_6 in the city domains of Athens, Lyon and Milan.

year" air-quality from factors derived on emission have unpredictable consequences. This is also the case with cost-effective models that utilise emissions only and air-quality concentrations are utilised off-line after optimising on emission files only.

Although for many years in scientific literature, comparisons with measurements have been carried out, these usually are not focused or are not general. Hence, they do not help in increasing the confidence on modelling data, especially if these conveniently examine only one component (*e.g.* ozone) and ignore evaluations for the precursors or the meteorology. The next section shows comparisons with measurements at the target year and generalisations over large geographical areas.

Actual measurements at the target year

Most urban domains contain a number of air-monitoring stations, operated under the auspices of the local authorities, at which average pollutant concentrations levels are reported at times ranging from minutes to hours. By looking into the actual measurements at the target year we can see how much of this process has been successful.

Between 1995 and 2010 the number of monitoring stations operating at the ten AutoOil domains changed significantly. In 1995, 242 monitoring stations reported ozone in the black rectangles shown in Fig. 2 (outer domains). Out of these 242 stations, 73 operated at the red rectangles (inner domains). In 2010, the corresponding number of stations were 529 and 195 respectively. Consequently, we are able to look relatively accurately what was the effectiveness of the emission reductions introduced between the 1995 and 2010. Fig. 10 shows the comparison of measured concentrations between the base and the target year. The comparisons are shown for O_3 and NO_2, but similar figures have been generated for CO,

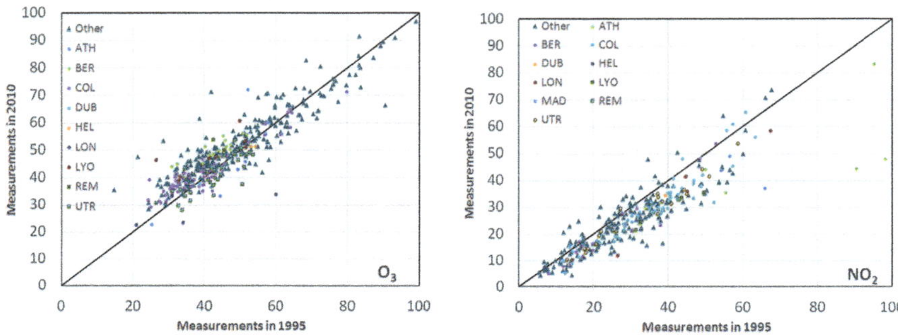

Fig. 10 Comparison of measurements between the base year 1995 and the target year 2010 for O_3 (left) and NO_2 (right). The units are in μg m^{-3}. The diagonal line indicates no variation in concentrations between 1995 and 2010. The area above this line represents deterioration and the area below the line improvements in the measured concentrations at the target year. With triangles are represented the monitoring stations outside the ten outer AutoOil domains.

C_6H_6 and PM_{10}. The stations reporting within the AutoOil domains (both inner and outer) are shown separately for each modelling domain. With triangles are shown the values for all other European stations which are in the European grid but outside the AutoOil domains. The diagonal line divides the graph in two parts. The top-left area represents the monitoring stations with deterioration in 2010 of the annual mean values of 1 hour maximum concentration for O_3 and the 1 hour averaged concentration for NO_2. The lower right area below the diagonal line shows the stations with improvements in 2010.

As seen by the two parts of Fig. 10 for O_3 and NO_2 respectively, the expectations of the modelling simulation of Fig. 8 are confirmed. Similar comparisons with the other pollutants confirmed the same trend. These comparisons with measurements implicitly means that the detailed setup with the urban and regional representation of the atmosphere was correct despite being numerically laborious with the ICT tools that were available when these modelling simulations were carried out.

Experimental comparisons were also made between measurements and modelled results for the target year. The purposes of these comparisons were to examine at the target year:

- How well the annual average conditions are represented.
- Examine the frequency distribution during the modelled annual-mean period and demonstrate that this is similar to the distribution at the target year.
- Demonstrate spatial variability by looking into several monitoring stations at least for the annual mean and the 98 percentile values.

Unfortunately, the verification of modelling simulations with time-series comparisons for pollutant concentrations and meteorology during the target year were not possible due to the limitations of the modelling methodology in working with surrogate periods. This is because the surrogate annual-mean period and the period for the 98 percentile episodes occurred at different dates in 1995 and in 2010.

For future regulatory approaches, it is worth mentioning that in the modelling area there have been significant improvements in numerical tools and in the information available from monitoring and emission databases. The former can already benefit from big data and utilise assimilated techniques for reducing the uncertainty of initial conditions and improve the spatial representativeness. The following areas are now technologically feasible and will contribute in the development of a new generation of modelling methodologies.

First, the increase in the global deployment of compact weather monitoring devices is now capable to provide in real time 10 min averaged values over a dense grid as shown by Fig. 11. This will not only improve the spatial initialisation, but because of lowering the temporal interval it will make meteorological tools more sensitive to acute episodes.

Secondly, supplementary to static networks, mobile sensor systems have also become available. These have the advantage of reducing the number of sensor nodes required, at the expense of increasing the complexity of the data processing. Data from vehicle mounted sensors in conjunction with GPS can be used to generate a map of pollution distribution in a specific area; wireless sensor networks could potentially fulfil this function by supplementing the data from existing monitoring stations to enable extremely localised mapping of air pollution to analyse peak events and identify hot spots.[7] Their wide scale deployment in densely populated urban areas would allow the spatial and temporal variation of air pollutants.

Thirdly, algorithms have been developed to optimise the sampling rate to obtain the best accuracy while reducing data transmission costs and verified by simulations and tests with a prototype mobile sensor system.[8] Mobile sensor platforms have been designed to be mounted on both public transport and private vehicles. In the latter case individual drivers would have the option of

Fig. 11 Meteorological measurements for the national network of weather stations in the Athens domain of AutoOil and the surrounding area (the values in each blue box are in degrees C).

measuring air quality data for private use only or to participate in community based sensing.[9]

Stricter emission regulations will in future also require more efficient on-board diagnosis systems. Aside from monitoring air pollution, metal oxide sensors can also be used to control the levels of pollutants present in exhaust gases.[10] Because of their low cost and the ease with which they can be integrated into mobile devices and networks, these sensor technologies may also find application in variable engine emission monitoring as well as personal exposure to air pollutants.[11]

Geographical generalisation

An important remaining issue for regulatory comparisons remains the geographical representativeness over the domain of interest. Ideally, the model simulations should be repeated in verification runs with the actual meteorology and emission inventories during the target year. Comparison of the forecasted conditions with actual up-to-date conditions from verification runs could be very useful in further regulatory efforts for 2020 and 2050. This process is laborious even with modern ICT resources due to heterogeneous data structures and data gaps. So it might be useful on the basis of inner and outer modelling windows to propose a geographical generalisation of the simulated concentrations.

Within the ten modelling domains shown in Fig. 2 were situated 1065 urban settlements. These represented 46.5% of EU15 urban population. Out of these settlements 281 are within the 11 inner modelling domains and 784 within the ten outer domains. These settlements are located primarily at the states of Belgium, France, Finland, Germany, Greece, Ireland, Italy, Luxembourg, the Netherlands, Spain and United Kingdom, although there are few additional urban settlements from Estonia, Switzerland and Poland. This is a sufficiently large base on which to establish a generalised relationship between air quality concentrations and the local population density.

By normalising local concentrations for each pollutant against the population density and the percentage of EU15 urban population the results for the data set of 1065 settlements are shown in Fig. 12. The local concentrations are extracted from the modelling results in the grid cell where each settlement is situated. The city population density is according to the Eurostat census data in the respective urban areas. The upper part is for the maximum 8 hour averaged CO and the lower for hourly NO_2. The line in these figures represents the following function:

$$c \approx a p_{\mathrm{d}} {u_{\mathrm{p}}}^{b+1}$$

where, "c" is the pollutant concentration (expressed in μg m^{-3}), "p_{d}" is the population density (expressed as inhabitants per km^2) and "u_{p}" is the percentage of population against the total EU15 urban population and "a" and "b" are empirical constants. These take different values for the base year and the target year on the basis of emission reductions and change of the climate. For NO_2 the values of $a = 1.63 \times 10^{-4}$ and $b = -2.2124$ in 1995 and $a = 1.03 \times 10^{-4}$ and $b = -2.2163$ for 2010. For max-8H CO the values of $a = 3.38 \times 10^{-2}$ and $b = -2.0255$ in 1995 and $a = 2.10 \times 10^{-2}$ and $b = -2.0695$ for 2010.

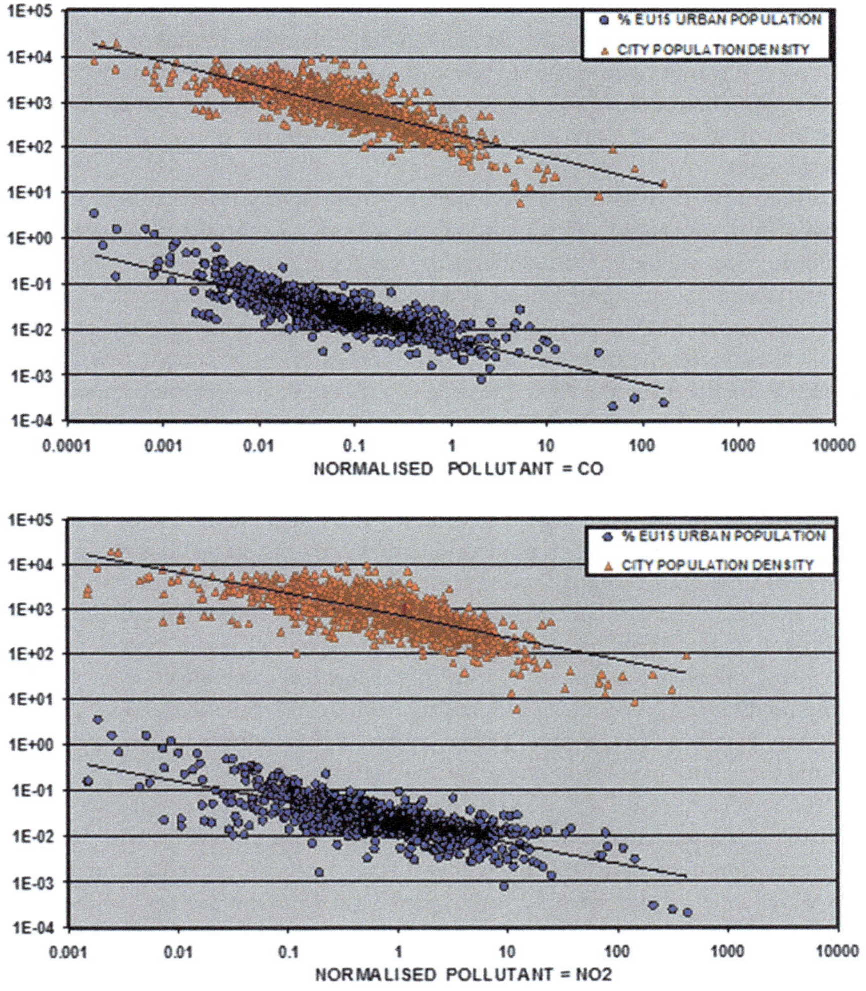

Fig. 12 Normalised maximum 8 hours CO (upper) and the hourly NO_2 concentrations in 1995 (base year). These are expressed in the *y*-axis against the percentage of EU urban population (circles) and the local population density (triangles) for all settlements in the AutoOil domains.

The advantages of this approach are that the sample is large (more than 1000 settlements), with a consistent geographical grid of at worst 6×6 km^2 as in Fig. 3, which is by far better than most regional models. The emissions are not directly utilised which might introduce uncertainty or *ad-hoc* assumptions about the representation of monitoring stations.

Conclusions

In the past, the perception of sustainable urban air-quality conditions led to emission control strategies that were fundamentally linked only to the magnitude of the emission sources and all atmospheric processes were assumed to be static.

Subsequently these strategies were coupled with measurements of air-quality through networks of measuring stations, which also had limitations in representing reality over large areas. However, even in the latter case, serious problems might arise when such measurements are extrapolated to locations where data are missing or when interpreting concentration in the proximity of measuring instruments.

With this work we presented a modelling methodology for describing the air-quality in a target year after analysing the current conditions in a base year. Following the validation simulations on the base year's emission reductions several scenarios were introduced aiming to achieve compliance to air-quality limit values. Several such regulatory efforts have been carried out in the past but their outcome on the planned "target year" has rarely been analysed. With this work we analysed for the first time measurements from over more than 1000 monitoring sites both during the "base year" as well as the "target year". These demonstrate the effectiveness of emission reductions in Europe in achieving compliance to well-known air-quality standards. Concerning modelling, since air-quality problems have become more complex at the regional and local level with a wider impact of oxidants on population and rural ecosystems, it is essential to carry out simulations with sufficient physical, geographical and temporal details.

It is evident that not all emission reductions will work everywhere and at all temporal scales. Even more, compliance to targets with different temporal weighting might not necessarily mean that it has the same effect, even for the same pollutant. As technology progresses, it might be realistic to adopt targets that vary locally and with time. Hence, technological infrastructures and algorithms that allow such variations might be useful.

The verification with actual air-quality measurements in the target year shows that the scenarios used until now have achieved maximum results with reasonable emission reductions. It is likely that further emission reductions might not lower the concentrations further but these might shift the problem in another source category. Without new vehicle technologies and fuels, scenarios of further reducing emissions in one sector might shift the problem in another sector or simply to another geographical area.

The integrated use of urban and regional models in this study has already forecast some reductions and some deterioration in ozone concentrations. However, these were more significant outside the urban areas where the mechanism of ozone formation is more closely related to the background concentrations and to the improvement due to the climatological conditions.

New technologies for monitoring can further enhance all steps of regulatory simulations and introduce significant advancement in the actual verification of abatement strategies. Conventional air quality monitoring systems are expensive and consist of a limited number of stations installed in fixed locations, with consequently coarse spatial resolution. In order to increase the areal coverage and usefulness of air quality data it is necessary to acquire real-time information and provide this in a dynamic way to both regulatory authorities and the general public.

References

1 *Proc. of the workshop on the Objectives for Next Generation of Practical Short-Range Atmospheric Dispersion Models*, ed. H. R. Olesen and T. Mikkelsen,

National Environmental Research Institute, Risø, P.O. Box 358, DK-4000 Roskilde, Denmark, 1992.

2 A. N. Skouloudis, *Issues in Environmental Science and Technology*, Royal Society of Chemistry, 1997, Issue 8, pp. 67–93.

3 A. N. Skouloudis, *EUR Report No EUR19725EN*, European Commission, EU Bookshop ISSN: 1018-5593, 2000, pp. 1–264.

4 AutoOil-I, *Sub Group-2, DG-XI/D3 Report on 'Air quality report of the AutoOil programme'*, European Commission, Belgium, 1996.

5 D. Simpson, *Atmos. Environ.*, 1992, (9), 1609–1634.

6 C. Dore and S. Vidič, EMEP Centres, Note 3, 2012, http://www.unece.org/fileadmin/DAM/env/documents/2012/air/EMEP_36th/n_3_EMEP_note_on_grid_scale__projection_and_reporting.pdf.

7 P. Völgyesi, A. Nádas, X. Koutsoukos and A. Lédeczi, *Proc. of the 7th International Conference on Information Processing in Sensor Networks*, IEEE Computer Society Washington, DC, 2008, pp. 529–530.

8 S. C. Hu, Y.-C. Wang, C.-Y. Huang and Y. C. Tseng, *Journal of Systems and Software*, 2011, **84**, 2005–2012.

9 S. Devarakonda, P. Sevusu, H. Liu, R. Liu, L. Iftode and B. Nath, *Proc. of the 2nd ACM SIGKDD International Workshop on Urban Computing*, ACM, New York, 2013, pp. 1–22.

10 M. Miura, Novel Carbon Resource Sciences, *Newsletter*, 2010, **2**, 10–16.

11 A. de Nazelle, E. Seto, D. Donaire-Gonzalez, M. Mendez, J. Matamala, M. J. Nieuwenhuijsen and M. Jerrett, *Environ. Pollut.*, 2013, **176**, 92–99.

Faraday Discussions

DISCUSSIONS

Numerical modelling strategies for the urban atmosphere: general discussion

Rob MacKenzie, Alison Tomlin, Jörg Kleffmann, Thomas Karl, C. N. Hewitt, Dwayne Heard, Karine Sartelet, Roberto Sommariva, Urs Baltensperger, Roy Harrison, Sasha Madronich, Gordon McFiggans, Spyros Pandis, John Wenger, Astrid Kiendler-Scharr, Neil M. Donahue, Rachel Dunmore, Ruth Doherty, Sarah Moller, Iarla Kilbane-Dawe, Brian McDonald, Andreas Wahner, Shupeng Zhu, Albert Presto, Markus Kalberer, Matthew Hort, James Lee, Irina Nikolova, Jose L. Jimenez, Lisa Whalley, Mohammed Salim Alam and Andreas Skouloudis

DOI: 10.1039/C6FD90022D

Astrid Kiendler-Scharr opened a general discussion of the paper by Sasha Madronich: I understand that it's convenient to work with volatility distributions – why is it that we would assume that the chemistry actually leads to a population of each of the volatility bins? It seems that recent findings of auto-catalytic reactions in the atmosphere and oxidation reactions leading to highly oxidized molecules point to mechanisms where the consecutive formation of oxidation products with ever lower volatility during the course of a reaction sequence is not happening. Are these contradicting the use of the volatility basis set? Second, it seems that vapour pressures of compounds and compound classes are among the largest uncertainties of models describing partitioning - could you guide the experimentalists on what needs to be done to provide more precise vapor pressures?

Sasha Madronich responded: In explicit chemical models like GECKO-A, vapor pressures of all species (reagents and products) are calculated from molecular structures and can be mapped onto any discretized volatility basis set (VBS). However, decadal VBS bins may be inefficient if using several bins to represent species of very low volatility that are essentially all condensed, or may have insufficient resolution for species having saturation concentrations (vapor pressures converted to C*, $\mu g\ m^{-3}$) near ambient aerosol loading. If for some systems the vapor pressures drop more rapidly (*e.g.*, by autooxidation), these changes can still be represented within GECKO-A, but cannot be described by a VBS multigenerational stepping algorithm in which molecular identities have been anonymized. As for the determination of vapor pressures, it may be useful to note that the existing thermodynamic estimation methods agree quite well for pure hydrocarbons and mono-functional oxygenated species, but diverge from one another for increasingly functionalized molecules.[1]

1 R. Valorso, B. Aumont, M. Camredon, T. Raventos-Duran, C. Mouchel-Vallon, N. L. Ng, J. H. Seinfeld, J. Lee-Taylor and S. Madronich, *Atmos. Chem. Phys.*, 2011, **11**, 6895–6910.

Gordon McFiggans asked: We have previously carried out investigations using different models (*e.g.* MCM and GECKO/A) and found significant differences in how the volatility bins are filled. Does the multigenerational approach, giving a continuous distribution, inevitably dictate that there will be very low sensitivity to the volatility distribution? How dependent is this continuity in the distribution on the SARs used in the model? Will the volatility distribution in the real atmosphere truly be as continuous as predicted by such a complex multigenerational model, or will the mixing of airmasses of different ages give more variability and dictate that a simple trajectory box modelling framework will always underestimate the sensitivity?

Sasha Madronich answered: We made a comparison of 3 vapor pressure SARs[1] and found very similar continuity in the distribution of the vapor pressure. This continuity is dictated by the multigenerational oxidation process producing a myriad of species with a large (and so *quasi*-continuous) spectrum of properties including vapor pressure. In our simulations of Mexico City, photochemical processing tends to generate a substantial amount of low-volatility compounds fairly quickly (*e.g.* see Fig. 3 of the paper), although of course this fraction grows over the course of several days. These early low-volatility compounds provide sufficient self-seeding to the mixture that the sensitivity exponent is only slightly above unity, *i.e.*, nearly linear. Any external sources of, *e.g.*, aged low-volatility particles would also contribute to the seeding and could result in even more linear sensitivity.

Neil Donahue said: To be provocative, I don't feel we really need to know vapor pressures to answer this. It is very important for explicit, predictive models to be able to predict partitioning, but within the VBS framework the partitioning behavior is empirical; the specific compounds are interesting but not essential to know. Furthermore, if we consider the whole system, other than very close to sources, such as roadways (which are very important of course in the urban context) I would not expect the system to be very sensitive to volatility (just as Sasha has shown).

Spyros Pandis asked: One of the limitations of the estimated organic aerosol volatility distributions based on thermodenuder measurements is that they cover only the material that is in the particulate phase. Therefore, they do not include organic material with higher volatilities that is in the gas phase during the period of the measurements. Could this effect explain the calculated increase of the nonlinear response of the organic aerosol system reported in this work, as one moves from the relatively clean conditions of the Paris campaign, to the more polluted conditions in Mexico City, and finally to the higher levels of organic aerosol in the smog chamber experiments?

Sasha Madronich answered: Estimation of volatility distributions from thermodenuder (TD) measurements is challenging for many reasons,[1] including ambiguity as to whether gas-particle equilibrium has been achieved during the experiment. For example, Cappa and Jimenez[2] used a TD residence time of 16 s for the Mexico City samples, and developed a complex model to simulate evaporation during this limited time, which presumably accounted for the initial absence of the gas phase from collected filters, but the experimental situation is obviously complex. Certainly, the effects would be largest in more polluted situations, consistent with the non-linearity rankings found in our study (smog chambers > Mexico City > Paris). However, the issues in interpreting TD observations are not relevant for the GECKO-A model results presented here (for the MILAGRO and BEACHON field campaigns), which are based on a theoretical estimate of vapor pressures.

1 M. Bilde, *Chem. Rev.*, 2015, **115**, 4115–4156.
2 C. D. Cappa and J. L. Jimenez, *Atmos. Chem. Phys.*, 2010, **10**, 5409–5424.

Spyros Pandis commented: If one is concerned not only about the local sensitivity but rather the response of the system to a change of the order of 50%, is the response similar for both an increase of the organic material and for a decrease? How does the addition of an extra volatility bin with a reasonable amount of material (not included in the thermodenuder measurements) affect this response?

Sasha Madronich replied: The relatively small non-linearities found here would indicate roughly proportional increases or decreases, even with respect to moderately large variations (*e.g.* a factor of 2) in available organic material. Local sensitivity could, however, be distinguished spatially from its regional values, or temporally between instantaneous or multi-day values, for example, noting that regional background seed aerosol on one day may often be the result of urban aerosol exported on previous days.

Urs Baltensperger opened a general discussion of the paper by Irina Nikolova: You conclude that there is probably a non-volatile core to the particles, otherwise there would be complete evaporation. In addition, the constant particle number concentration seems to indicate that every particle contains a non-volatile core. Can you speculate as to what that might be, specifically in the smallest fraction of 10 nm particles?

Irina Nikolova replied: Currently the composition of the non-volatile core is not fully understood. The relatively high sulphur content (less than 50 ppm) in the fuel may contribute to the constituents of this core. In addition, trace metals from the lubricating oil (such as Ca, Fe) may contribute to the composition of this core as well. Most of the hydrocarbon species evaporate from the unburned lubricating oil droplets heated at high temperatures. Such a metallic core could be coated with volatile species upon leaving the tailpipe. Metallic ash residues could also contribute. In our model, we do not specify the content of the non-volatile core; however, the health-related response to that core may be different depending on its chemical composition, and merits further investigation.

Alison Tomlin asked: It is maybe a bit surprising that the particle number concentrations don't change very much overall. Are coagulation and deposition processes so slow that they do not lead to overall particle losses or are new particles being formed?

Irina Nikolova responded: In our study we have examined the role of coagulation and deposition for a typical traffic induced particle number size distribution for three wind speeds (see Table 3 in our paper). All comparisons are against a base case steady-state solution without coagulation or deposition, but accounting for ultrafine particle emissions, condensation/evaporation, mixing with background air and air exchange between the street canyon and the rooftop. There is no new particle formation. For the ambient conditions and the street canyon geometry, we have estimated the time scales for dilution, coagulation and deposition as well. Coagulation and deposition are found to be slow processes, with a maximum reduction in the total number of particles of 8.2% for a wind speed of 1.5 m s^{-1} at rooftop and both processes acting on the particle size distribution. The reduction in particle number concentration decreases to 4.7% for a wind speed of 3 m s^{-1} at rooftop.

Sasha Madronich asked: Did you consider dry deposition of organic gases, particularly SVOCs, that are partitioning dynamically with the particle phase? Deposition velocities for some of these gases (especially those already partly oxidized) could be quite large, and could result in a significant loss of particle mass by evaporative readjustments.

Irina Nikolova replied: In our study we have considered the deposition of both gases and particulates. However, our model does not account for oxidation of SVOC gases due to the relatively large time scales involved in that process compared to the 20-minute model run-time we typically deploy. The emitted particles have only a very small component of partially oxidised compounds.

Neil Donahue said: I have a visceral reaction to the term non-volatile. While it is true that there may be a non-volatile core, I don't believe it is necessary to explain what you observe. The vapor pressure of a 0.01 μg m^{-3} constituent is just 1 part per trillion. Furthermore, the system should equilibrate on a timescale roughly equal to the condensation sink, meaning pretty quickly in this case I should think. Have you considered an urban background concentration during your mixing calculations?

Irina Nikolova answered: Urban background concentration was considered during the simulations. In our calculations particle composition and vapour-phase concentration data were used based on measurements at traffic and background sites (please refer to Table 1 in our paper). Figure 1 in our paper provides the inlet background size distribution as well as the emission size distribution profile. Mixing between background particles and in-street canyon particles as well as vapour-phase mixing are parameterised using a dimensionless air exchange rate.[1]

1 C.-H. Liu, D. Y. C. Leung and M. C. Barth, *Atmos. Environ.*, 2005, **39**, 1567–1574.

Neil Donahue said: At $d_p = 5$nm the Kelvin effect should take the saturation vapour pressure from 1 part per trillion to 10 parts per trillion. So I think things will go into steady state.

Albert Presto asked: I disagree with Neil as I think there are non-volatile cores present. We've done sampling in a traffic tunnel and when you heat to 250 °C you end up with particles remaining at 10 nm, which are definitely not organic. Other exhaust studies have used catalytic strippers and end up with material left around at 10 nm, so maybe they are metal because these cores don't burn in the stripper. They are at least functionally non-volatile and don't interact chemically.

Neil Donahue replied: I'm not saying that the particles do not have refractory cores, I'm saying that they don't need to. I agree that there is plenty of evidence that they do.

Gordon McFiggans addressed Irina Nikolova: The persistence of the finest particles, whilst being consistent with every particle containing a solid core, is also consistent with a very low concentration of very low volatility vapours being present in the background atmosphere. Why is it more likely that all particles have a solid involatile core? On the contrary, is it not more likely that the air that is mixing with the evolving and diluting plume contains a continuous distribution of volatility, including small mixing ratios of very low volatility components, sufficient to stop evaporation of the finest particles? This would appear to be consistent with the multigenerational modelling approach such as presented in the last talk. This calls into question the assumptions in the model boundary conditions in the present study. In order to draw any conclusions about the composition of the finest particles, I would suggest that it is necessary to conduct sensitivity simulations. These should include a range of initialisations with dilution air containing a continuous distribution of organic components down to low volatilities, to establish the required magnitude of the lowest volatility components that would be consistent with the persistence of the finest particles. This could be used to quantify the fraction of fine particles that need to have a solid involatile core for a given amount of low volatility components, hence establishing the likelihood of both explanations.

Irina Nikolova replied: Thank you for the valuable suggestions. Our current laboratory study with the GCxGC-ToF-MS (DOI: 10.1039/C5FD00185D) shows evidence of SVOCs originating from diesel fuel and lubricating oil contributing to the particle phase compounds. The largest concentration detected in the particle phase is found in the C23–C29 region, corresponding to the contribution of the lubricant oil. The diesel fuel contributes to the particulate composition with low molecular weight SVOCs (less than or equal to C20). A recent study by Gkatzelis *et al.*[1] shows that 40% of the traffic related particles evaporate in a thermodenuder at 400 °C, while the remaining 60% have a non-volatile core. In support of a non-volatile core are the studies referred to in our paper.[2–7] We currently use vapour phase SVOC concentrations measured in Birmingham in the model, and we have recently collected new data from London.

1 G. I. Gkatzelis, D. K. Papanastasiou, K. Florou, C. Kaltsonoudis, E. Louvaris and S. N. Pandis, *Atmos. Meas. Tech.*, 2016, **9**, 103–114.
2 S. Biswas, L. Ntziachristos, K. Moore and C. Sioutas, *Atmos. Environ.*, 2007, **41**, 3479–3493.
3 P. Tiitta *et al.*, *Atmos. Environ.*, 2010, **44**, 976–986.
4 T. Ronkko, A. Virtanen, J. Kannosto, J. Keskinen, M. Lappi and L. Pirjola, *Environ. Sci. Technol.*, 2007, **41**, 6384–6389.
5 T. Ronkko, T. Lahde, J. Heikkila, L. Pirjola, U. Bauschke, F. Arnold, H. Schlager, D. Rothe, J. Yli-Ojanpera and J. Keskinen, *Environ. Sci. Technol.*, 2013, **47**, 11882–11889.
6 U. Kirchner, V. Scheer, R. Vogt and R. Kägi, *J. Aerosol Sci.*, 2009, **40**, 55–64.
7 W. Birmili, K. Heinke, M. Pitz, J. Matschullat, A. Wiedensohler, J. Cyrys, H.-E. Wichmann and A. Peters, *Atmos. Chem. Phys.*, 2010, **10**, 4643–4660.

Sasha Madronich replied: The multigenerational GECKO-A model does indeed predict the presence of very low volatility components, initially with small mixing ratios but increasing as photochemical processing of gases from higher volatility bins continues. However, particle-phase processes such as oligomerization (not currently in GECKO-A) could contribute, and the model is not able to distinguish between its model construct of a *quasi*-ideal solution, and a solid involatile core.

Karine Sartelet asked: I think your study is very interesting and shows the impact of the Kelvin effect on organics. I was wondering what parameters (*e.g.* surface tension) you used to model the Kelvin effect?

Irina Nikolova responded: Surface tension was considered following the work of Queimada *et al.*[1]

1 A. J. Queimada, Isabel M. Marrucho, J. A. P. Coutinho, *Fluid Phase Equilib.*, 2001, **4639**, 1–10.

Karine Sartelet asked: Did you investigate the sensitivity of your results to the number of size sections used?

Irina Nikolova replied: Our goal is to evaluate in detail the physical behaviour and chemical composition of the ultrafine particles (with diameters of less than 100 nm). We have performed tests varying the number of the size bins (from 5 to 20). However, during the redistribution step after condensation/evaporation with a low number of size bins, the redistribution scheme becomes quite diffusive. This problem can be overcome by increasing the number of size bins and finding the balance between the computational load and the reduction in the numerical diffusivity. In our study, that balance occurred when using 15 discrete size bins.

Karine Sartelet asked: I wonder what the sensitivity to the number of size sections is, because processes such as deposition are known to be sensitive to the size distribution. Because 15 sections are used in this work, the sensitivity to the number of sections may be low.

Irina Nikolova replied: We have used the scheme of resistances to estimate the deposition velocities. Deposition velocities are higher in the nucleation mode particles (especially in the sub-10 nm size range) than in the Aitken mode particles due to the Brownian diffusion of the particles. Due to the evaporation of SVOCs, the particle size distribution shifts significantly to the lower end of the nucleation mode. A lower number of size bins (*e.g.* 5) on a logarithmic scale is too coarse in the nucleation mode and does not provide the detail we are interested in to investigate.

Rob MacKenzie opened a general discussion of the paper by Karine Sartelet and addressed Shupeng Zhu: Please could you comment on what, other than runtime constraints, determined your choice of size bins?

Shupeng Zhu responded: Yes, the number of size sections is limited by the computation cost. The choice of the size bounds was initially made based on a log size discretization (0.01, 0.0398, 0.1585, 0.6310, 2.5119 and 10.0) μm. However, as the measurement data provide information on PM_1 (diameter < 1 μm) and most of the BC mass observed in the measurements is less than 0.4 μm, we manually inserted two size bounds 0.4 μm and 1.0μm between 0.1585 μm and 2.5119 μm. We removed size bounds 0.0398 μm and 0.6310 μm in order to keep 5 size sections.

Spyros Pandis said: What was assumed regarding the mixing state and the size/composition distributions of the boundary conditions?

Shupeng Zhu replied: In the simulations of this paper, boundary conditions come from a larger-scale simulation, in which the internal mixing assumption is used. Therefore, we do not have information on the mixing state of the boundary conditions. Therefore, all particles coming from the boundary conditions are assumed to be aged and internally mixed. So for each size section, there is only one composition for particles for the boundary conditions.

Spyros Pandis commented: Can you please explain what was assumed regarding the size/composition distribution of the transportation emissions? Did all emitted particles have the same composition (*e.g*, the same black carbon, organic carbon, *etc.*)?

Shupeng Zhu replied: All emitted particles are considered as particles with pure substance. In this study only elemental carbon (*i.e.*, black carbon) and dust are directly emitted in particle phase. So each emitted particle is either made of pure elemental carbon or pure dust. All organic carbon (POA) is considered as semi-volatile and put into gas phase then allowed to condense into the particle phase within the SOA treatment.

Karine Sartelet added: For transportation, we assume that black carbon and dust are emitted from exhaust and resuspension respectively. We assume that there are two distinct populations. For SOA we assume that everything is in the gas phase, then let them partition.

Shupeng Zhu said: So all emissions in particle phase are considered as fully unmixed and only contain one chemical species. For particles originating from initial and boundary conditions, we assumed that they are internally mixed: within the same size section, they all have the same composition.

Roy Harrison remarked: I agree that the ATOFMS is an imperfect instrument with very different sensitivities to different elements. With careful calibrations against independent measurements, this could be compensated for to some degree, but you still cannot compensate for the fact that hit rates (the ratio of ionised particles to those entering the instrument) is different for particles of differing composition. There are also significant matrix effects for chemically-mixed particles, so there are major weaknesses in the technique and I am not entirely comfortable with using it as a basis for a simulation. However, having said that, I am not sure that there is any other technique available giving better information on single particles.

John Wenger commented: I agree with Prof. Harrison that the aerosol time of flight mass spectrometer (ATOFMS) has limitations. But if we're addressing the diversity of particles at an individual level, then the ATOFMS is the best technique for doing this. I also think that characterisation of the chemical diversity of particles is a useful approach which allows us to compare and contrast ambient particle mixing states at different locations around the world.

Gordon McFiggans addressed Shupeng Zhu and Karine Sartelet: A statement is made in the paper that a dynamic inorganic component treatment is used, with organics treated in equilibrium followed by redistribution across the particle sizes, due to the limitations of the H_2O model. That redistribution will put the organics into discrete size fractions of particles and hence dictate the mixing-state of all particles containing organics. Can you elaborate more on the redistribution? How confident are you that this redistribution puts them in the right place? Most importantly, how is the robustness of your conclusions dependent on the skill with which the organics are distributed? I would suggest that a sensitivity study is used to attribute error bars in the mixing state of organics to this redistribution.

Shupeng Zhu answered: Details about the redistribution method of organics species are given in the electronic supporting information (ESI) (see DOI: 10.1039/c5fd00175g). Basically, the mass redistributed after the equilibrium is related to both the number of particles and the condensation/evaporation kernel within each bin. This method tends to redistribute more organic aerosol (OA) mass into the smallest size bins due to their high number concentrations, and it leads to an overestimation of OA in those sizes. It is true that this redistribution method needs to be improved, and we also discussed one possible improvement by introducing the Kelvin effect into the weighting factor of the redistribution in the ESI. A sensitivity study is a good suggestion, but the best way would be to directly use a dynamic approach to compute the organics partitioning. Such approach already exists and we are planing to implant it into our model in future work.

Neil Donahue asked: I still don't understand how you have a large Kelvin influence for relatively large particles. How does this work out?

Shupeng Zhu responded: The particle diameter studied here is between 0.01 and 10 μm. For particles in the first size section, which have a mean diameter of around 0.029 μm, the Kelvin effect is not negligible. If the Kelvin influence is not taken into account, then more mass will be redistributed into the smallest size section during each time step. Even if this amount of mass is small for each single time step, the accumulated effect will be very significant as the over-estimated mass will feed back and produce a even larger weighting factor for the small size section in the next time step. Another factor is that we find the size distribution of inorganic species to be more reasonable as this is computed dynamically and includes the Kelvin effect. So, the ideal solution would be to use a dynamic computation for the organic species as well in our future work.

Astrid Kiendler-Scharr addressed Shupeng Zhu, Sasha Madronich and Irina Nikolova: I have a question to all three of you, coming from an experimental point of view. If you were to choose measurements to most efficiently improve your models and constrain the input to models, which measurements would these be? What is the largest gap in observational knowledge and which parameters should be measured?

Shupeng Zhu replied: I think a more precise measurement on a single particle level from some specifically designed gas chamber experiment would be very helpful for the improvement of our understanding of particle mixing, as well as the model design. The largest gap in observational knowledge is that we lack the means to monitor how the chemical composition evolved within individual particles during the mixing process. The parameter we are interested in is the mass fraction of each chemical component within each particle.

Sasha Madronich responded: One useful bit of information would be the distribution of specific functional groups per condensed molecule (*e.g.* various -OH, -CO, -COOH, -ONO_2, and -OOH moieties) and in particular the occurrence of multiple functional groups per molecule. FTIR quantification of some functional groups (*e.g.* Maria *et al.*[1]) has already shown potential for sorting particles of different origins (*e.g.* Russell *et al.*[2]). Among other things, such measurements could help assess to what extent the molecular identity of predicted gas phase photochemical products is retained in the condensed phase, and whether the particle bulk O/C, N/C, and H/C ratios can be reconstructed from the molecular scale information.

1 S. F. Maria, L. M. Russell, B. J. Turpin and R. J. Porcja , *Atmos. Environ.*, 2002, **36**, 5185–5196.
2 L. M. Russell, R. Bahadur and P. J. Ziemann, *Proc. Nat. Acad. Sci.*, 2012, **108**, 3516–3521.

Irina Nikolova responded: I agree with what has been said so far, that measurements of the mass fraction of each individual component and the distribution of specific functional groups per molecule are valuable inputs for modelling work. In order to understand the complex behaviour and interactions between particles in the atmosphere with multiple sources, there is a need for a clearer understanding on how individual particle composition changes per source. High spatial and temporal resolution measurements of particle composition in the ambient atmosphere are also highly needed.

Urs Baltensperger returned to a discussion of the paper by Sasha Madronich: I read from your modelling framework that you consider a mixture of organic species. Do you allow for inorganics and water as well, or do you just avoid mixing?

Sasha Madronich replied: Particle-phase inorganics and water were not considered in this study, but certainly should be in future work. Most current parameterizations of organic aerosols are still relatively simple, particularly those implemented in 3D chemistry transport models, and are largely based on smog chamber yields, where organic mass tends to overwhelm particle-phase inorganics (*e.g.*, those used as seeds), or on algorithms such as the multi-generational volatility basis set (VBS) that also assume a purely organic phase. Current thermodynamic models (*e.g.*, ISORROPIA, MADRID, MOSAIC and AIM) have limited representation of organics, and certainly not in the molecular detail provided by explicit gas-phase models such as MCM or GECKO-A. Neil Donahue's paper (DOI: 10.1039/C5FD00214A) showed that even different organic aerosols may or may not be miscible depending on their precise nature. As we gain better understanding of this complexity from laboratory and field measurements, improved parameterizations can hopefully be developed to represent mixing appropriately, rather than avoiding it.

Roy Harrison commented: I would like to question whether the Pankow model of absorption–adsorption into an organic medium is the correct paradigm. The atmosphere typically contains some liquid water associated with particles and our work on the mass closure of atmosphere aerosol clearly shows the presence of water. Since the more oxidised organic aerosol is likely to be hydrophilic, would a Henry's Law type model be more appropriate for the oxidised species?

Sasha Madronich responded: The Pankow paradigm (which for absorption into a liquid is essentially Raoult's law) has been used extensively to represent gas-particle partitioning of organics, especially in 3D models as it can be parameterized efficiently, *e.g.* in terms of chamber-measured yields, or with algorithms such as the multi-generational Volatility Basis Set (VBS). You are correct that this may not be always the best approach, given that water-soluble organic aerosols are at least as abundant as hydrophobic aerosol (especially after photochemical ageing), and that water is frequently a major constituent of the particles. The Henry's law limit is obviously relevant to organics diluted in cloud drops, (*e.g.*, Graedel and Goldberg[1] and Hermann *et al.*[2]) but it is also likely to be dominant for many aerosol systems, especially in high humidity environments such as the South Eastern U.S. and the Amazon.[3,4] Liquid–liquid phase separation within the aerosol is also possible (*e.g.*, Zuend and Seinfeld[5]). Unified models that encompass both Raoult's and Henry limits have been developed (*e.g.*, Pun *et al.*[6] and Bowman and Eskelson[7]) but have not been used widely in 3D models due to computational complexity and scarcity of experimental constraints. Better parameterizations of these processes could be a real opportunity to improve the predictive ability of the models.

1 T. E. Graedel, and K. I. Goldberg, *J. Geophys. Res.*, 1983, **88**, 10865–10882.
2 H. Herrmann, T. Schaefer, A. Tilgner, S. A. Styler, C. Weller, M. Teich and T. Otto, *Chem. Rev.*, 2015, **115**, 4259–4334.
3 B. Aumont, S. Madronich, I. Bey and G. S. Tyndall, *J. Atmos. Chem.*, 2000, **35**, 59–75.
4 B. Ervens, B. J. Turpin and R. J. Weber, *Atmos. Chem. Phys.*, 2011, **11**, 11069–11102.

5 A. Zuend and J. H. Seinfeld, *Atmos. Chem. Phys.*, 2012, **12**, 3857–3882.
6 B. K. Pun, R. J. Griffin, C. Seigneur and J. H. Seinfeld, *J. Geophys. Res.*, 2002, **107**, 4333.

Karine Sartelet remarked: In our model, for organics we distinguish hydrophobic from hydrophilic components. Each component is a surrogate for a class of organics. We do have water-soluble organics.

Gordon McFiggans addressed Karine Sartelet: There's a whole range of different treatments available for multicomponent thermodynamics and partitioning, with ion interaction models for aqueous electrolytes at one end, and absorptive partitioning for organic components at the other. Most widely used recent treatments include the AIOMFAC group contribution approaches to drive multicomponent inorganic and organic partitioning. It is readily possible to cover a wide range of different levels of complexity, dependent on the questions to be answered. I am not sure that it is the case that all large scale models use "Pankow" partitioning, and it's quite straightforward, if the activity coefficient reference state is correct, to include water soluble particle components. I think it is important to keep in mind these capabilities when choosing the appropriate tools to address the question being asked. Internally consistent frameworks do exist and may be more appropriate in investigations of the mixing of inorganic components with insoluble and water soluble organics.

Karine Sartelet replied: Although the activity coefficients were not modified to take into account the interactions between organics and inorganics in this work (only organic–organic interactions), the mixing state of organics is influenced by inorganics in our model H_2O, as soluble organics condense onto inorganics and insoluble organics do not. Separately from this work, we have also developed a model, SOAP,[1] to take into account the dynamics of condensation/evaporation of organics and organic–inorganic interactions using AIOMFAC. We did not use SOAP in this work, but we are planning to couple it to the mixing-state resolved model SCRAM to investigate the mixing state of inorganics with organics.

1 F. Couvidat, and K. Sartelet, *Geosci. Mod. Dev.*, 2015, **8**, 1111–1138.

Shupeng Zhu added: Our organic model differentiates between soluble and insoluble organics. In this work, activity coefficients for organic–organic interactions were taken into account using UNIFAC, but inorganic–organic interactions were not considered.

Rob MacKenzie opened a general discussion of the paper by Andreas Skouloudis: Please could you comment on whether you consider your results – showing no change in urban ozone over time – are consistent with the data interpretation and modelling of the paper presented by Christian Ehlers earlier (especially their Fig. 1, 2, and 26)?

Andreas Skouloudis answered: Although ozone precursors have been falling between the base year and the target year, the ratio of NO and NO_2 gases in the atmosphere has changed. We should also consider that Christian's manuscript examined only two sites (one at Mace Head in Ireland and one at

Hohenpeissenberg in Germany). The measurements we presented are in all monitoring sites at the ten modelling domains of the second AutoOil, and we show the actual situation all over Europe as was available at that time. Obviously, we did not examine the frequency of exceedances of the limit values because these limits depend on the climate conditions between the base and the target year, and because the number of hours of exceedance is subjective for different limit values, as you can easily see from Fig. 5 in our work. However, in constant conditions, there is a neat cycle in which nitrogen dioxide helps to form ozone and nitric oxide helps to break it apart. This means, basically, that we are taking away some of the ozone suppression and, as shown in Fig. 26 of Christian's work, this might result in increases of ozone concentrations. This indeed is in line with what we have observed in Fig. 10 of our work. This cycle appears to have been perturbed by control measures that were actually intended to remove the fine particles and soot in vehicle exhausts. It remains to be seen if these levels continue to stay high or if eventually we might see a reduction of ozone in future years. This is one of the reasons why models should become operational over larger scales, in order to account for what is happening in other regions of the world and identify properly the import of ozone and its precursors from outside. In the meantime, the ground-level of ozone might remain persistently high even in regions with restricted sunlight and not-so-hot temperatures.

Andreas Wahner commented: What was presented in the paper by C. Ehlers were measurements by us and statistics from the German federal agencies for German cities. In the paper, Fig. 1 shows measurements of background ozone at GAW Hohenpeisenberg, and there you hardly see any change in background ozone, except for maybe a small decline over the last 20 years. You were referring to many stations throughout Europe, and you can show that the long term trends at different stations are not the same. There seems to be no clear trend in ozone.

In the past years there have been advances in detailed model descriptions of aerosol concentration and formation, and in techniques for particle characterization. Do you see these advancements being used in the models you use in developing policy?

Andreas Skouloudis replied: Indeed, there have been many advancements focusing on aerosol science, their association with atmospheric chemistry and their size and number characterisation. There have been also been advancements in numerical techniques, as you can see from my other publication.[1] Yet, for policy development implementation there is still a lot of empiricism and the models used have incorporated parametrisations that hinder the generalisation of size and number distributions and create doubts on proposed solutions. Unfortunately, apart from inter-comparison of schemes, more research publications are needed for characterising nucleation, coagulation condensation deposition and other loss mechanisms of aero-disperse aerosols and the emissions from relevant anthropogenic sources as seen in M. Prank *et al.*[2], so far in coarse spatial geographical resolution.

1 A. N. Skouloudis and W. Eifler, Paper 7-46, Eurotrac Symposium Garmisch-Partenkirchen Germany, SPB Academic Publishing 11-15, 1994.
2 M. Prank *et al.*, *Atmos. Chem. Phys.*, 2016, **16**, 6041–6070.

Andreas Wahner remarked: In many cases of model and measurement comparison, simplified or parameterized models and limited datasets are used and models/ parameters are sometimes optimized to give better agreement. Should it be a task for the future, with growing accessibility to much observational data, to compare the full complexity of data with as detailed models as we know of, using inverse modelling techniques to find the weak point in our mechanism in an algorithmic optimized way? This may be possible in future if the advancements in computer power in the next 5 to 10 years are increasing as expected. Should our community not be prepared for these opportunities?

Matthew Hort said: I would agree with the need for high resolution modelling in a general sense but would urge caution in thinking that high resolution is the answer to everything. The resolution needs to be relevant for the question being "answered" and also needs to be considered in context of the other model inputs (*e.g.* emissions), parameters (*e.g.* chemical and physical) and available observation data for initiation, verification and validation. Also, in thinking about high resolution we have to think about the genuine predictability of the relevant phenomena, and if the only valid approach might be probabilistic. The exact scope will indeed depend on the application/task. I am coming from a large scale *i.e.* city and regional, suitable for air quality assessments and forecasts. At these scales the only viable approach is similar to that followed by Meteorological Services where measurement data are used to constrain the model initial state and verify, but this requires a sound model, daily detailed measurements and sound methods of comparison.

I think computers are decades away from being powerful enough to run models for real world applications, in the context I have stated above, at the kind of complexity Andreas has asked about (street resolving), for anything but highly idealised cases *i.e.* single street canyons or road junctions.

Alison Tomlin responded: I certainly agree with your first point. The topic of the paper is probabilistic modelling. For national scale forecasting, 1 km seems like a realistic objective in terms of scale. However, many pollution hot-spots might occur because of small scale issues such street topographies, particular traffic hot spots, localised emissions from combustion for heat applications *etc.* There is therefore a role for nested higher resolution, and building resolving models for such applications – assuming, as you say, that emissions data are available at such scales and that the models can be suitably validated. Even a CFD model such as that used in the present study can be run in less than 1 h on a standard PC so perhaps we are not decades away from using such models for particular case studies of importance. Models such as ADMS and OSPM, of course, can be run much faster than this and also have a role to play in nesting models for city applications. It will be important to carry out the types of validation and sensitivity studies such as the one described in our paper, to establish whether such models are fit for purpose for particular applications. If they can be shown to be relevant, then they could be used to answer some questions of local relevance that could not be answered by larger scale forecasting models.

Andreas Skouloudis replied: Although I agree with the necessity for better model inputs regarding emissions, sound physical and chemical models and initial conditions that are based on detailed hourly measurements, I do not think that computational capacity is the main bottleneck for the proliferation of advanced modelling applications, even nowadays. For this purpose, I would like to draw to your attention the article by Overpeck *et al.*[1] which shows that modelling potential will be steadily increasing until 2030 in parallel to initial measurements with satellite and sensor technologies. These are already, in 2015, resolving a lot of initialisation demands and eliminating the constrains of empirical assumptions. The main challenge ahead is not to increase the resolution of the modelling domain all the way down to street canyon or road junction level, but to demonstrate the success of modelling in resolving topics that otherwise could not be solved in various interlinked geographical scales (ideally two-way nested). The aim of our present work was to show the success of realistic regulatory forecasts on the basis of real measurements in the target year. This is perhaps a small effort in the direction of genuine predictability of the evolution of urban pollution. As for the uncertainty initial conditions, taking into consideration the revolution in meteorological services with the use of compact sensor technologies in association with satellite observations, and in association with sensors and RFIDs implemented in everyday human applications, I think that there is no need for caution. In reality we need, today, innovative modelling deployments that will substitute for the empiricism of parametrisations and the uncertainty of initial conditions, taking into account the advantages of actual real-time measurements. Certainly, we can do this even today in certain domains in Europe, otherwise all these increases of modelling potential will be wasted.

1 J. T. Overpeck, G. A. Meehl, S. Bony and D. R. Easterling, *Science* , 2011, **331**, 700–702.

Ruth Doherty asked: Going back to the previous questions regarding the idea to include all the chemical complexity we know about in air quality models – if we did this with the ADMS-Urban and other such models then the model wouldn't be able to run. This is because these urban models, which are typically local dispersion and chemistry models, were designed to be used by local authorities and planners and hence typically run on a single Windows or other PC. So as suggested previously by Dr Hort, we need to go back to the question/s asked and how we can best answer that. For example, if we are interested in the effect of long-term exposure to $PM_{2.5}$ concentrations, how would we deal with chemical complexity and what level of complexity would be needed to calculate chronic health impacts?

Alison Tomlin replied: There is a large body of work on methods for chemical model reduction (see Turanyi and Tomlin[1] for an extensive review). These methods have been very successfully used in combustion modelling but have been perhaps under-utilised in atmospheric chemistry models. Perhaps they would have something to offer in terms of providing a consistent way of reducing the size of reaction mechanisms or models whilst retaining the important chemical complexity.

1 T. Turanyi and A. S. Tomlin, *Analysis of Kinetic Reaction Mechanisms*, Springer-Verlag, Berlin, Heidelberg, 2014.

Nick Hewitt addressed Andreas Skouloudis: Concerning the geographical aggregation of data – Fig. 12 in your paper shows the log of concentration against the log of a population parameter. You define the functions in the text, but don't discuss the implication of the gradient of these lines for exposure and regulation across Europe. Could you elaborate?

Andreas Skouloudis responded: The details about how the local air quality concentrations were generalised all over the ten modelling domains, and can be found in our corresponding report[1]. These relationships were developed originally for the generalisation of the modelling results over urban domains in Europe and certainly link pollutant concentrations to human exposure. They have not yet been verified with individual exposure assessments since they refer to integrated values over large urban areas, but such verifications could certainly become feasible in the near future on the basis of real-time measurements from mobile monitoring devices. We are willing to discuss the implication of these relationships further, if suitable exposure data become available, and always on the basis of quantifiable parameters based on measurements.

1 EUR Report No EUR19725EN, European Commission, EU Bookshop ISSN: 1018-5593, 2000, pp 1–264.

Jörg Kleffmann opened a general discussion of the paper by Alison Tomlin: Did you include in your MISKAM model the turbulences which were induced by cars (the "piston effect")? How would this affect the high variability of NO_2 modelled inside the street canyon? We recently measured NO_x in a double lane traffic tunnel in Brussels,[1] showing homogenous mixing throughout the whole cross-section, which is also reasonable at least for the lower section (<5 m altitude) of a street canyon. Potential spatial variability (or not) is of high importance for defining locations for NO_2 measurements in urban environments.

1 M. Gallus *et al.*, *Build. Environ.*, 2015, **84**, 125–133.

Alison Tomlin responded: We did not include traffic induced turbulence in our simulations. Under low wind speeds and high traffic speeds this might be important. However, typical average vehicle velocities within the street varied between 10 and 20 km h^{-1} due to the levels of congestion experienced. These are unlikely to produce enough turbulence to make the canyon well mixed. The model was validated against observations for the low reactivity tracer carbon monoxide within an earlier study by Dixon *et al.*[1] That study showed a clear difference (up to a factor of 3) between concentrations in the leeward *vs.* windward canyon locations. The turbulent kinetic energy was also seen to vary between leeward and windward locations. For one particular wind direction where a low velocity convergence zone was shown to be formed by the model, there was a significant discrepancy between the model and observations. Traffic-produced turbulence could have been affecting the dispersion for these particular conditions.

1 N. S. Dixon, J. W. D Boddy, R. J. Smalley and A. S. Tomlin, *Atmos. Environ.*, 2006, **40**, 958–972.

Rob MacKenzie asked: It is interesting and very challenging to see the dependence of results on roof-top wind direction. What are the prospects of generating well-characterised roof-top wind directions for studies that do not measure it directly?

Alison Tomlin responded: This is a difficult challenge for many urban areas, since above roof meteorological measurements are not routinely obtained within most cities. In addition, in order to serve as a reliable reference measurement, the site needs to be far enough above the roof to reduce the influence of localised roof-top structures. In a study by Barlow *et al.*,[1] conducted during the DAPPLE campaign, a comparison was made between a near roof-top site and a site on the top of the BT Tower in terms of the ability of the reference data to be able to explain the in-street variability in channelling and recirculating flow components. In fact the higher measurement (190 m) explained more of the in-street flow than the site that was only 2 m above the roof-top; the latter showed interference from local roof-top structures. Long term measurements from the BT tower could form a suitable reference for London based studies and it would be very useful to perform a comparison between the use of data from the BT tower and that from nearby reference sites such as Heathrow. Some other cities may have equivalent sites; Leeds City Council, for example, has collected meteorological observations from an above-roof reference tower within the city centre for several years. These measurements are, however, piecemeal and there is no guarantee of their longevity due to funding challenges. A second option within a nested modelling scenario would be to use a coarser resolution model to obtain the boundary conditions for the street scale model. For this to be successful, however, the coarser model would need an accurate representation of the surface aerodynamic characteristics. This is a challenging area and there is a need for comparison between different types of reference measurements/models in terms of their ability to explain in-street flow characteristics for a wider range of case studies.

1 J. F. Barlow, A. Dobre, R. J. Smalley, S. J. Arnold, A. S. Tomlin and S. E. Belcher, *Atmos. Environ.*, 2009, **43**, 5536–5544.

Brian McDonald remarked: With regards to high resolution atmospheric modelling, we've been talking about chemical and meteorological complexity. What about emissions complexity? In order to run high-resolution models, high-resolution emission inventories are needed as input. To develop emission maps finer than 1 km is a big effort, and potentially fraught with large uncertainties. Let's not forget about the need to develop bottom-up emission inventories as well. We need better maps of sources.

Rob MacKenzie said to Alison Tomlin: Several comments from the floor – in response to a number of papers – have commented on the need to move to dynamic simulation of emissions. Please could you comment on this aspect of your own paper? Does this part of the model contribute much to variabilities in NO_2 concentration at street level?

Alison Tomlin responded: A dynamic traffic micro-simulation model was used in the work with the intention of assessing the impact of traffic demand on roadside increments in NO_2. What the paper showed was that for the lower demand scenarios the predicted concentrations were quite sensitive to the level of demand. Under these scenarios, transient queues built up at the signal controlled intersections, but they cleared between signals. For higher demands congestion could build up within the street, with permanent queues forming. Under these scenarios, predicted road-side increments in NO_2 were not very sensitive to changes in demand; larger reductions in demand were required in order to reduce permanent queue build up. This type of dynamic model therefore allows different traffic scenarios to be investigated. The challenge is to link the outputs of the traffic model to relevant real world emissions factors that accurately represent the effects of vehicle speed, acceleration, age, condition, mitigation technology *etc.* Such factors need to be representative of urban driving conditions that include a large degree of stop–start behaviour close to intersections.

Spyros Pandis asked: Our models are used both for improving our understanding of the present (often for scientific reasons), but also for predicting future changes due to changes in emissions (mostly for regulatory use). Are their sensitivities to different parameters, for example wind speed and direction, the same in these two cases? Clearly, the predicted concentration of a pollutant is quite sensitive to wind speed. Is, however, the change in this concentration for a change in emissions as sensitive to the wind? This issue is obviously relevant if one focuses on chemical mechanisms and specific reactions in there.

Alison Tomlin answered: This is an interesting question and there are perhaps different ways of looking at the issue. Firstly, one of the main conclusions of the paper was that despite the impacts of uncertainties in the model parameters on the width of predicted concentration distributions, it was still possible to assess the impact of potential traffic management measures, such as reductions in demand, using the HDMR method for variance decomposition. ANOVA type methods allow for the effects of individual parameters to be clearly seen within the scatter caused by other uncertainties in the model. Secondly, the treatment of model parameter uncertainty helps to avoid coming to the wrong conclusions in assessing the causes of concentration trends *i.e.* attributing changes in concentrations to the wrong processes. For example, it is common in the interpretation of pollution time series to factor out the effects of long term trends in wind speeds. If this is not done, then trends in concentration could be wrongly attributed to changes in emissions when they could be a result of differing climatology between different parts of the time series. If the site in question was within a street canyon, then changes in reference wind direction could have an even more dramatic effect on changes in measured concentrations than wind speeds. Therefore, if one time period had substantially differing background wind directions to another, these effects would need to be accounted for in the same way as wind speeds. For the assessment of the causes of short term episodes and intervention measures, this may be especially important. Finally, models are often used to test emissions inventories, which in turn are used in the assessment of pollution trends. Differences between model predictions and measurements are sometimes used to suggest errors within the inventory. However, they may also be due to a range of other model uncertainties, and these need to be taken into account before conclusions can be drawn about inventory problems. The treatment of uncertainties is therefore very important in inverse modelling.

Sasha Madronich remarked: You showed the sensitivity of model results to many parameters, including some arising from uncertainties in chemical reaction kinetics. The sensitivity to Arrhenius parameters (A-factor and activation energy, or E/R) is interesting in its own right, but should be distinguished from a sensitivity to the rate coefficient, $k(\mathrm{T}) = A\ \exp[-E/(RT)]$. When fitting plots of log $k(T)$ *vs.* $1/T$, covariance between intercept A and slope E/R makes their estimation less certain than the k data from which they are derived (*e.g.*, Cvetanovic *et al.*[1]). The IUPAC (your ref. 33) recommendation for the reaction NO + O_3 is $k(298\ \mathrm{K}) = 1.8 \times 10^{-14}\ \mathrm{cm^3\ s^{-1}}$ with an uncertainty of 8% ($\Delta\log = \pm 0.08$), and $k(T) = 1.4 \times 10^{-12}\ \mathrm{cm^3\ s^{-1}}$ (no uncertainty given) $\exp[-(1310 \pm 200\ \mathrm{K})/T]$.

Propagating just the E/R range (1110–1510 K) through the Arrhenius equation gives a range of $k(298\ \mathrm{K}) = (0.9\ \text{to}\ 3.3) \times 10^{-14}\ \mathrm{cm^3\ s^{-1}}$, indicating a factor of 2 uncertainty, much larger than the recommended 8% uncertainty of k near 298 K. It is the latter that controls the evolution of the chemical system – though the direct sensitivity to E/R certainly matters where temperature variability is an issue.

1 R. J. Cvetanovic, D. L. Singleton and G. Paraskevopoulos, *J. Phys. Chem.* 1979, **83**, 50–60.

Alison Tomlin replied: Thanks for the comment. I think the uncertainty factor of 0.08 was for log(k) which leads to a 20% uncertainty in k at 298 K. That aside, I agree with you that treating correlations between Arrhenius parameters is an important topic. It was recently addressed in a very interesting paper by Nagy and Turanyi.[1] The problem is that these correlations are very rarely presented in the literature. They have to be derived from a wide variety of data that constrains the rate constant rather than individual studies. This requires the application of some kind of optimisation strategy and whilst such studies are beginning to emerge, correlation information is available for only a very limited number of systems. For combustion systems, older evaluations such as those of Atkinson and Baulch do often provide uncertainties in k at different temperatures, but these were not presented for the reaction under debate. Therefore we were left with using the uncertainty in E/R as presented. Treating correlations is a very interesting topic that should be considered in much more detail in the future.

1 T. Nagy and T. Turányi, *Int. J. Chem. Kinet.*, 2011, **43**, 359–378.

Dwayne Heard commented: It is interesting to see that your sensitivity studies show that the model results are sensitive to the value of the rate coefficient for the reaction of NO with ozone, and also its temperature dependence. Coincidentally, I was recently asked about the kinetics of this reaction in connection with another modelling study, in particular the behaviour at lower temperatures. The kinetics of this reaction were studied in some detail in the 1970s and early 1980s, so more than 30 years ago. A more recent study of this reaction was published in 1998 by Moonen *et al.*[1], where it was stated that previous measurements had been made at low total pressures and/or with very large mixing ratios of NO and O_3 relative to those used in the atmosphere. Moonen *et al.* used 10 s of ppbs of the reagents, and studied the reaction at tropospheric temperatures and pressures. Although they found some differences in the activation energy and pre-exponential factors compared with the earlier studies, the rate coefficient at 298 K was similar to the recommended value, and the lack of a pressure dependence was confirmed. Bedjanian *et al.*[2] also published a rate constant in the same year, but it was only obtained at 298 K and at a low pressure of 1 Torr (in helium), agreeing with the recommended value. Although there is no reason to suspect that the kinetic data are incorrect, the kinetics of this reaction were established before the development of some of the modern techniques used for studying kinetics. Given the importance of this reaction as highlighted in this paper, and for atmospheric chemistry in general (it is difficult to think of a more important reaction for urban areas!), it may be prudent to embark on a fresh study of the kinetics of this reaction, particularly over a wide range of temperatures. I would like to thank Professor Neil Donahue for useful discussions regarding the kinetics of this reaction.

1 P. C. Moonen, J. N. Cape, R. L. Storeton-West and R. McColm, *J. Atmos. Chem.*, 1998, **29**, 299–314.
2 Y. Bedjanian, G. Le Bras and G. Poulet, *J. Phys. Chem.*, 1998, **102**, 10501–10511.

Alison Tomlin replied: I would agree with the comment that it may be useful to perform further detailed kinetic studies on this reaction. It would seem that there have been no recent studies, and none since the evaluation of Atkinson *et al.*[1] that determined the uncertainty in activation energy used in the sensitivity analysis of our work.

1 R. Atkinson, D. L. Baulch, R. A. Cox, J. N. Crowley, R. F. Hampson, R. G. Hynes, M. E. Jenkin, M. J. Rossi, and J. Troe, *Atmos. Chem. Phys.*, 2004, **4**, 1461–1738.

Thomas Karl communicated to Alison Tomlin: We have recently shown experimentally[1] that significant bias for treating chemical reactions can occur in RANS depending on Dahmköhler numbers and the underlying degree of heterogeneity of the reactant surface emissions. These complex and non-linear effects depend on spatial model resolution, chemical reactivity and scales of surface heterogeneity. Our experimental findings support recent LES modelling studies[2] performed over dense plant canopies, which are characterized by non-linear heterogeneity driven mixing effects similar to what we find in "urban canopies". Since the spatial scale of the presented modelling domain in your paper varies, the question arises as to how heterogeneity driven segregation of the reactants is treated as a function of Dahmköhler numbers in your modelling study.

1 L. Kaser *et al.*, *Geophys. Res. Lett.*, 2015, **42**, 10894–10903.
2 H. G. Ouwersloot, J. Vilà-Guerau de Arellano, C. C. van Heerwaarden, L. N. Ganzeveld, M. C. Krol and J. Lelieveld, *Atmos. Chem. Phys.*, 2011, **11**, 10681–10704.

Alison Tomlin replied: As explained in the paper, the model describes mixing by the Interaction by Exchange with the Mean (IEM) model where the mixing time-scale is determined by the mixing-constant and the ratio of turbulent kinetic energy to its dissipation. Therefore it is not possible to relate this directly to the Dahmköhler number. We accept that this is a simplification compared to considering the high temporal resolution intermittency that may be present in such situations. However, a compromise was sought since it was felt that LES would be too computationally expensive to study anything other than simple street canyon situations. And even with LES, a sub-grid scale turbulence model would be required below the 1 m grid resolution used. The model set-up was tested against experimental wind tunnel data (including a full global uncertainty analysis) for a NO_x plume dispersing in background ozone from an earlier paper by Ziehn *et al.*[1] What we found in this study was that the model gave good agreement with the measurements of mean NO_x and NO_2 concentrations but that the errors were higher for predictions of the concentration variances. We therefore feel that for the high model resolution used, the IEM model is a suitable compromise for predicting mean concentrations.

1 T. Ziehn, N. S. Dixon and A. S. Tomlin, *Atmos. Environ.*, 2009, **43**, 5978–5988.

Roberto Sommariva opened a general discussion of the paper by Lisa Whalley: Your models are run constrained to all measured species and parameters, except NO_2 and O_3. This approach is fine for short lived species, but not so much for long lived species, such as ozone. It takes a very long time for ozone to reach steady-state (7 days in the case of the MCM model!) and as a consequence, the concentrations of all the non-measured intermediates are very different at the end of the calculation of a single data point compared to at the beginning. This may lead to significant errors in the calculation of the ozone concentration and complicate the comparison with the O_3 observations, which is shown in your Fig. 3.

Lisa Whalley answered: In Fig. 3 we do not attempt to compare the steady state ozone concentration predicted by the box models to the observed ozone. We have avoided doing this, as it is unlikely in an urban centre such as London (with a multitude of NO_x and VOC emission points) that ozone will have reached steady state conditions. Instead we are comparing the steady state ozone concentrations predicted by the different mechanisms to each other, and assessing whether the differences observed are due to the model inputs used to constrain the mechanisms or due to the differences in the complexity of the chemistry mechanisms. We have chosen to run all the box models to steady state before comparison, so as to prevent any bias caused by variability in the length of time it may take for ozone concentrations to reach steady state when the complexity of the chemical mechanism differs. We have estimated *in situ* ozone production using the MCM box model after 1 h, which is the average time (since emission) it took an air-mass to reach the measurement site, and have compared this to the observed ozone and see that *in situ* production could account for ~ 60% of the ozone observed on average. During the transit from emission to the observation site oxidised VOC intermediates will have been generated within the air-mass. These oxidised products may not be detectable or identifiable by the GC systems used to measure the VOCs; nevertheless, the influence of these intermediate species should be considered when estimating the *in situ* ozone production.

Iarla Kilbane-Dawe remarked: I would like to come back to emissions modelling, rather than dynamics or concentrations. Given the residence time of the species in urban environments, the emissions are an important factor. It is really important to acknowledge how poor our understanding and characterisation of the emissions that are occurring in the urban environments are. Tim Murrels touched on some of these in a previous discussion session, but it is clear that many of the models and assumptions we use to implement spatially and temporally varying emissions, especially at high resolution in urban sites, are wrong – sometimes by as much as an entire order of magnitude for important species and partitioning of species within families. In many cases, important physical, chemical and socioeconomic effects are simply ignored.

The implication is that, since important emissions are wrong, then all of the adjustments being made to a model to make it agree with the verified concentration measurements are likely to be wrong. And by implication, much of the scientific effort we've seen today is being put in the wrong place.

Also, on the density point that Matthew Hort mentioned – I think it's timely to do high resolution modelling now, as I think we're about to see a great improvement in the quality of data available across cities.

Karine Sartelet commented: With regards to model uncertainty – we did an internal comparison of our model. We found that the internal uncertainties in the model were lower. Structural uncertainties were larger than parametric uncertainties. The uncertainty comes from the input data as well as the model itself.

Alison Tomlin replied: I agree that uncertainties originating from differences in model structures are also important. They are perhaps more difficult to quantify than parametric uncertainties since in principle an infinite number of model structures is possible (using, for example, different resolutions, turbulence models, sizes of chemical mechanisms, *etc.*). One approach taken in the climate change community is to perform regular model inter-comparison exercises. By including all available models run as ensembles, then an envelope of predictions can be given which at least traces uncertainties due to differences in model set-ups. The difficult next step is to trace the origins of these uncertainties back to specific parts of the models. For parametric uncertainty this is time consuming, but conceptually relatively straightforward using the ANOVA type approaches discussed in our paper. For structural uncertainties this is a bigger challenge.

Andreas Skouloudis said to James Lee: It would be interesting, also, to talk about the certainties of emissions inventories. For the London emissions inventory it would be very interesting for the younger generations, to carry out comparisons of inventories that were constructed with bottom-up synthesis, as well as with the disaggregation of top-down constructed inventories. But this work requires the synthesis of data from citizen monitoring, which has only been emerging in recent years together with the evolution of sensor technologies. This could inspire future research in this area and facilitate similar conclusions as we have seen in this important paper, and improve the accuracy of inventories from real-time data. I wonder if you would consider examining such types of comparisons in the future?

James Lee responded: I do not believe that, at present, data available from citizen monitoring (which will largely be based around small sensor technologies) is of sufficient quality to allow meaningful comparison with inventories. Even if the measurements prove that the concentrations measured by such systems are of good quality, this still does not give a direct comparison with emissions inventories. A model would be required to make the step from emission to concentration. I do think that personal monitors may have a role to play in the future in improving air pollutant data coverage across a city (especially one without such a large traditional network as London), something which could prove extremely useful for mitigation strategies.

Sarah Moller returned to the discussion of the papers by Andreas N. Skouloudis: You suggested that inventories at the local/city scale should be developed. That presents a major challenge to many areas in the UK. While the data available on sources and activity in London are reasonable, the same data are not collected for other cities and regions, for example the drive cycles developed for London and used in the London emission inventory. What are your thoughts on this? There is good data for some sources that could be used in regional/city scale inventories but there are also major pollution sources for which the geographical distribution in the UK national emission inventory is a poor representation of the actual situation. In these cases the activity data is often not available to give a better representation for inclusion in local inventories. Do you think it is feasible to expect cities to be able to produce reasonable inventories that are an improvement on what the national inventory can do? Are there good examples of this elsewhere in Europe that we could learn from?

Andreas Skouloudis answered: It is true that the emissions inventory data are not sufficiently good and there is always room for improvements. These improvements can be either directed towards improving the spatial resolution or even becoming more realistic on the temporal scale. It is true that in London human activities, transport, energy consumption *etc.* are detailed and very accurate. However, for the regulatory applications I have described in this work, it is necessary to have the same detail and temporary accuracy everywhere in Europe. Unfortunately, at the time when this work was carried out, the activity data were not all sufficiently reliable. In such cases, the comparison of emission inventories constructed from disaggregating top-down data with emission data from aggregating bottom-up data from different activities and sources can help in identifying the level of uncertainty. However, this is not going to resolve the actual situation of the modelling needs. We will need a process that calculates the emissions on the basis of citizens' actions in terms of real-time assessment of transport needs (*e.g.* from satellite and *in situ* images) and in terms of accurately monitoring the energy and telematic consumption everywhere. Since 2005 we have been working with big-data in automating this process and for this purpose it might be useful to consult our recent review publication[1.] This work shows some of the instruments that we could potentially use for improving the local activities and make it feasible to produce reasonable inventories. This may also be relevant to the comment made by Brian McDonald earlier.

1 P. Kumar, A. N. Skouloudis, M., M. Viana, M. Cristina Carotta, G. Biskos and L. Morawsk, *Sci. Total Environ.*, 2016, **560–561**, 150–159.

Brian McDonald remarked: On the question of bottom-up inventories and what we can do with them in the future: I think there's some hope here. I'm always amazed at how much data is available on human activity, from Google, our smartphones, *etc.* – "big data". It's not easy because some data may be proprietary, but there are opportunities here, and this kind of activity data could be useful in developing better maps of critical emission sources.

Gordon McFiggans commented: I disagree with Andreas Wahner that we should be attempting to include all known chemistry with full complexity in numerical models of the urban atmosphere. There are very many chemical, physical and dynamic processes competing for computational resources in numerical simulations. It will simply not be possible in a reasonable timeframe to explicitly treat all processes at the highest level of detail and complexity to answer all the questions of importance to our understanding of the urban atmosphere. Interpretation of kerbside and roadside exposure and street canyon variability of pollutant concentrations requires metre-scale resolution. A two-way coupled nested simulation covering the 5 domains shown in the first plot down to a 1 km resolution, including the intermediate chemical scheme in Lisa Whalley's paper (DOI: 10.1039/C5FD00218D) with detailed inorganic and organic size-resolved aerosol, runs at 5 times real time on the UK National Supercomputing Facility. The second plot, running at the 10 cm resolution required to capture steady-state CFD predictions of particle number concentration, including the semi-volatile redistribution of components in a street canyon, required 3 simulation months. These two very different approaches to answer very different questions of importance to urban air quality could not possibly include full complexity treatments of all processes. For the foreseeable future, it will be necessary to reduce the complexity of treatments in a manner appropriate to the questions being addressed. We need to choose the right tools for the job.

Andreas Skouloudis answered: I think that simulations with a CPU time of five times real-time can be still very useful. Especially if we identify periods of less than 4–5 days that are suitable for representing the annual mean of episodes with different percentile severity according to annual meteorology. We have been very successful in identifying such periods in several parts of the world and we are willing to help if asked.

Also, from your comment, it is not also clear if the experience of the modelling system is based on several processors and models that are designed, compiled and optimised to run on high performance parallel machines. Perhaps a scale between 1 km and 10 cm might be more rational, in particular if some of the meteorological/photochemical parameters become numerically unstable in spatial resolutions beyond 1 km. This is certainly an issue to be addressed elsewhere. Nevertheless, sometimes it might be better to wait between 20–25 days rather than taking regulatory decisions based on coarse models which in the end will lead to wrong abatement solutions or damaging regulatory interventions, as we have seen several times in the past two decades.

Iarla Kilbane-Dawe commented: Bottom-up inventories do make sense – there are quite rich data available from transport models, but there are difficulties in getting hold of them. We need to build relationships with local transport modellers. The scientific community needs to consider what the uncertainties are in the emissions factors tied to the activity data – and what the right models are to build these emissions factors. This is an effort where much greater effort is required.

Mohammed Salim Alam said: So what we have spoken about over the last few days is how there are too many unknowns in the characterisation of the emissions from diesel vehicles. The issue is that there is not much information in the literature on high molecular weights in particular, which is what we presented during the first session. A lot of what we currently "know" and have talked about is focused on low molecular weight species. There are far too many gaps in determining the composition of diesel exhaust emissions, particularly for high molecular weights, and I think that this is an area that should be focused on in order to determine its potential in SOA formation.

Matthew Hort remarked: Some of the challenge with mass sensors is that they can be of varying specification and also less than ideally located, resulting in varying quality of the return measurement. The meteorological community has been looking at tackling this issue, as high quality observations won't proliferate to the levels needed for high resolution modelling. One interesting development is in what is sometimes called opportunistic data (*i.e.* information from existing equipment that was built for another purpose). This is cheap, as it is an extra use of something already built, but it is often of high quality. This, I think, highlights an important additional dimension to cheap data, in that the sensors do not also need to be cheap. While this is a new area there may be lessons to be learnt from the Meteorological community and its extensive use of measurement networks.

The Met Office have a community data collection initiative and web site called WOW (Weather Observation Website):[1]Anyone can upload measurement data and information. The site now has thousands of additional measurement sites, over and above the official Met Office and other agency sites. This data, while from varying instruments and of varying quality and representivity, has proved very valuable in understanding the evolution of bad and intense weather events. The site is open to other sorts of measurements and in combining a range of environmental measurements it might offer a sensible and useful tool in expanding air quality observations.

1 Weather Observation Website, http://wow.metoffice.gov.uk.

James Lee commented: We are talking about two ways of using these cheap sensors. One way is to use a network of sensors to measure things like compliance, or to look at health effects of air pollution, and the other is using them for validation of models. The confidence of the scientific community in these sensors isn't high at the moment. I think people won't want to run computationally expensive models with measurements they don't have faith in.

Andreas Skouloudis answered: For the first aim, in order to show compliance, perhaps it will be sufficient to examine concentrations averaged over long time periods. For such purposes the passive sampler approach might not sufficient and certainly cheap. These can also be used for assessing chronic health effects.

However, for acute health effects there are no such cheap sensor products. Nanocrystalline sensors are expensive but sufficiently accurate. Their advantage is that they are mobile, and certainly more geographically representative, than expensive conventional monitoring.

Ruth Doherty opened a general discussion of the Concluding remarks by Jose-Luis Jimenez: Regarding your last slide on indoor air pollution, are the uncertainties large because we don't know enough about the indoor sources and their emissions?

Jose-Luis Jimenez replied: There has been very little study of indoor air chemistry, compared to outdoor air chemistry. If one searches, for example, Web of Science, the ratio of outdoor/indoor air chemistry publications is around 100. Many key variables (*e.g.* photolysis rates, OH, NO_3 radicals, HONO *etc.*) have never been measured indoors, or only once or twice. Many of the newer instruments developed for outdoor air research have never, or almost never, been used indoors either. Combined with the high diversity of indoor environments, materials, ways in which humans use these, *etc.*, and the importance of processes that are not relevant for outdoor air (emissions from materials, surface chemistry on materials and humans, impact of human breath) there are many questions about indoor chemistry that we know very little about. The Sloan Foundation in the US recognized these gaps and has started a funding program for indoor air chemistry,[1] which is how we have been able to do the initial measurements I discussed in my talk, but a much larger international effort is needed in this area in my opinion.

1 Alfred P. Sloan Foundation, STEM Research - Chemistry of Indoor Environments, http://www.sloan.org/major-program-areas/stem-research/chemistry-of-indoor-environments/.

Rachel Dunmore said: The graph comparing indoor and outdoor is normalised, can you comment on the scale of difference between the two when not normalised?

Jose-Luis Jimenez replied: The absolute total concentrations of the 115 measured acids were higher indoors, although the time periods were a few days apart, due to limitations on the location of the ToF-CIMS instrument for sampling indoor *vs.* outdoor air. This is consistent with prior literature results for formic and acetic acids, which are typically higher indoors. These results are being finalized for submission to ES&T, where the full details will be available.

Roberto Sommariva commented: You have talked about the problem of doing too many field campaigns and not spending enough time thinking about the data. This is in part due to human nature, and in part driven by the way the funding system works. Do you think there is something we – as individual scientists – can do to deal with this issue, or is it a structural problem of modern science that needs to be addressed at a higher level?

Jose-Luis Jimenez replied: This is a very good question, one that I have spent a considerable amount of time wondering about.

In terms of the causes: it seems to me that there are two main factors that drive the lack of sufficient funding for data analysis and modeling of field campaign data: one is that funding programs and the program managers who run them need "news" to justify funding of their particular program to their superiors, the broader government, and society (and to advance their own careers, in the case of the program managers). News of a new field campaign with an airplane, or ship, or containers and people in the field seem to sell a lot better than some additional papers from an "old" study done 4 years ago. A second factor is that for an individual scientist, leading a proposal for a new field campaign is a good way to advance one's own career, and also to increase one's chances of funding. The fact that things have been done this way for a long time has created a "culture" in the field, that almost all program managers and scientists seem to share, that, for example, "a field proposal gets you 3 years of funding". Of course one can write a proposal focused on data analysis, but the funding rate of those is typically much lower than of proposals to acquire new data. For example, for my own group, we have a 100% success rate on proposals to acquire new field measurements, but only a 25% success rate on (equally good, if not better) proposals to analyze data from past field studies. I have talked to multiple program managers about this problem, and they are always sympathetic – except when they call you and ask to cut your budget massively for the next field study (effectively greatly reducing analysis time), or risk being left out of the campaign.

This is a huge problem that leads to a lot of wasted effort and prevents some of the best science from being followed up on and published, as researchers are distracted into going into the field and producing new data, and end up abandoning all data. Even if one manages to get some extra funding for analysis of a given campaign, the fact that most other participants did not is a serious problem, as their responses to queries about their data and model drag, as they long ago ran out of funding for that campaign and are busy with other projects. In that situation, any analyses that would require significant time from other groups become unfeasible. As science becomes ever more complex and specialized, some of the most interesting findings happen on the 2nd or 3rd round of analysis, after the initial papers are out and people have had a chance to digest their own measurements and think about how they relate to findings from others in the same campaign, and the broader state of the field and other fields. The advent of very powerful but very complex instruments in our field (*e.g.* ToF-CIMS, FIGAERO-ToF-CIMS, TOF-PTRMS, SV-TAG *etc.*) will make this worse, as 3 years is woefully inadequate for datasets from such instruments during complex field (or lab) campaigns. Many such instruments produce data at the same rate (MB h^{-1}) as satellite instruments.

In terms of what can be done, I do believe that the funding agencies and the culture of our scientific field can be changed, even if incrementally. But many of us need to keep making the case many times, with specific examples of how the short timeline of funding is damaging to scientific output. This has happened in some meetings I have attended, but much more is needed. In the medium term, I think we will make some progress and lessen the problem, as long as enough people join in with these efforts. But the problem will probably remain serious by the time I retire.

Faraday Discussions

PAPER

Concluding remarks: Faraday Discussion on chemistry in the urban atmosphere

Jose L. Jimenez

Received 2nd June 2016, Accepted 2nd June 2016
DOI: 10.1039/c6fd90019d

This article summarises the Concluding remarks from the Faraday Discussion on Chemistry in the Urban Atmosphere. The following themes are addressed: (a) new results that inform our understanding of the evolving sources and composition of the urban atmosphere ("News"); (b) results that identify gaps in our understanding that necessitate further work ("Gaps"); (c) the emerging instrumentation revolution and some of the challenges that it brings; (d) the structural issues of insufficient support for the analysis of field campaigns; and (e) some important areas that were missing from this Faraday Discussion and that should receive an increasing focus in the future.

Introduction

Urban atmospheric chemistry has been and continues to be a very active research area, and thus this Faraday Discussion was "sold-out," and led to many intense discussions at a high level. Baltensperger (DOI: 10.1039/c6fd00065g) has done an excellent job in discussing the main areas of interest and their scientific background in the introductory article for this issue, so I will not review that further here. Two important "boundary conditions" of this field should be mentioned at the outset. First, a key long-term objective of research in this area is to gain the ability to use numerical models to predict concentrations of key species accurately, over a wide range of conditions and in response to perturbations, and with confidence that predictions will work for the right reasons. Only with such confidence will we be able to design effective and efficient policies to control pollution in ever-evolving urban areas under a changing global climate. That has not always been the case in the past, as funding agencies have sometimes decided to mostly move on from some problems before such confidence was established. Second, urban areas do not exist in isolation, and many of the chemical species present in their atmosphere were emitted or formed outside of them (see *e.g.* Fig. 1 from Baltensperger, DOI: 10.1039/c6fd00065g). The longer-lived a species is, the higher the contribution of the "regional background" to urban concentrations.

Dept. of Chemistry and Biochemistry, and Cooperative Institute for Research in the Environmental Sciences, University of Colorado, Boulder, CO 80309, USA. E-mail: jose.jimenez@colorado.edu

Many advances were presented and discussed at the meeting. Rather than trying to review every paper presented, the following sections highlight key examples while grouping together several items of the same type: (a) new results that inform our understanding of the evolving sources and composition of the urban atmosphere ("News"); (b) results that identify gaps in our understanding that necessitate further work ("Gaps"); (c) the emerging instrumentation revolution and some of the challenges that it brings; (d) the structural issues of insufficient support for the analysis of field campaigns; and (e) some important areas that were missing from this Faraday Discussion and that should receive an increasing focus in the future.

News

The enormity of the emissions control problem that China is facing was made obvious from the work of Li *et al.* (DOI: 10.1039/c6fd00004e), who reported on the reduction of particulate matter (PM) concentrations following the imposition of strict emission controls in a very large area around Beijing, including controls that led to reduced economic output such as shutting down factories and construction sites, in preparation for a major WWII anniversary parade. Concentrations of fine PM were reduced by about a factor of two by these controls. Given that PM concentrations in Beijing and elsewhere in China often exceed WHO health guidelines by an order-of-magnitude, the challenge of developing efficient pollution controls that do not severely impact economic output appears striking.

Several papers concerned nitrogen oxides (NO_x) emissions and concentrations. Diesel vehicles are important sources of NO_x, especially in Europe. While most NO_x emissions from transportation sources are in the form of NO, the fraction of the emissions occurring as NO_2 has been increasing due to the impacts of some exhaust control technologies (Euro 3–5 cars with diesel oxidation catalysts). Carslaw *et al.* (DOI: 10.1039/c5fd00162e) presented results from cross-road remote sensing measurements in London that indicate that the NO_2/NO_x fraction peaked around 2010 and has decreased substantially since. These decreases are thought to result from the increasing penetration fractions of different control technologies (Euro-6 diesels), as well as the aging of some older technologies, and have helped in reducing air quality exceedances for NO_2.

The formation of tropospheric ozone (O_3) is a complex and highly nonlinear process that remains insufficiently understood. Ehlers *et al.* (DOI: 10.1039/c5fd00180c) presented a paper on twenty years of ambient monitoring of the key O_3 precursors, volatile organic compounds (VOC) and NO_x, at near-roadway sites in Germany. A large decrease of the VOC/NO_x ratio was observed, which was larger than expected based on emission inventories. This decrease was predicted to lead to a large decrease in local O_3 production efficiency. This was the paper that elicited the most intense discussion (as part of an already very intense meeting), in particular in terms of the best way to reduce O_3 in the future (NO_x *vs.* VOC controls), highlighting the variety of views within the community and the need for further research on the topic.

Gaps

The use of ethanol in vehicle fuels has been shown to lead to increasing emissions of it, as well as of acetaldehyde, a reactive and toxic gas. Such species also affect the formation of O_3. Dunmore *et al.* (DOI: 10.1039/c5fd00190k) presented measurements of both species during 2012 and showed that ethanol was the most abundant VOC in London air. They used the Master Chemical Mechanism (MCM) to interpret the observed acetaldehyde, and concluded that only 16% of the observed concentrations could be explained by the model. A large missing primary source was the most likely candidate to explain the observations.

A conceptually related closure exercise was presented by Kim *et al.* (DOI: 10.1039/c5fd00230c), who compared measured OH reactivity with that calculated from all the measured OH reactants at two sites in South Korea. At the urban site there was good agreement between both quantities, but at a nearby forested site 50–70% of the OH reactivity could not be accounted for by the measured species. The discrepancy is attributed to unmeasured oxidation products of biogenic VOC.

The sources of PM are not well understood and have been the subject of intense research over the last decade. Pandis *et al.* (DOI: 10.1039/c5fd00212e) presented a comparison of five recent large urban studies, and concluded that secondary PM (formed from chemistry in the atmosphere) now dominates over primary PM (that enters the atmosphere already in the particle phase) at most times and locations. As vehicle exhaust emissions are continually being reduced, the impact of other "non-traditional sources" is becoming more important. In particular cooking emissions have been identified in multiple recent studies as rivaling emissions from traffic, but their identification remains controversial, and their atmospheric reactivity and secondary aerosol formation potential are practically unknown. The substantial impact of cooking emissions on the non-fossil carbon budget of urban aerosols is also frequently ignored. Non-exhaust traffic emissions (*e.g.* from brake wear and dust resuspension) are another example of a poorly characterized source whose relative importance is increasing, also discussed by Shirmohammadi *et al.* (DOI: 10.1039/c5fd00166h).

Organic aerosols (OA) are an important component of fine PM, and their sources are particularly unclear. Minguillón *et al.* (DOI: 10.1039/c5fd00182j) performed a field study in Barcelona that contrasted periods with similar meteorology but very different vehicle traffic, just before and after the end of the August summer vacation period. Surprisingly the secondary OA (SOA) from non-fossil sources was larger in the periods with increasing traffic, when the opposite would have been expected due to the larger fossil OA and precursor emissions from traffic. While the causes for this observation are not completely clear, they include potentially enhanced biogenic SOA formation due to higher levels of NO_3 radicals at night and/or slower urban photochemistry due to higher ambient NO_x in the high-traffic period.

Reactive oxygen species (ROS) are a poorly characterized fraction of PM that are thought to play a role in PM health effects by leading to oxidative stress after deposition in the lungs. Shafer *et al.* (DOI: 10.1039/c5fd00196j) presented results from ROS surrogates from samples taken in six European cities in 2012. A remarkable uniformity of ROS concentrations across the continent was observed, which together with higher specific activity for background and rural sites suggests the importance of regional secondary sources for ROS. However, insoluble (and thus primary) species were increasingly important for ROS in coarse particles. An important topic for further research is the relationship between assays such as those used in this study *vs.* human health outcomes.

Low-cost electrochemical sensors have been recently gaining popularity for pollution monitoring, as a potential replacement for more expensive traditional monitors. The availability of inexpensive portable sensors capable of reasonably accurate measurements could indeed open many doors for investigation in urban atmospheric chemistry, such as the study of fine-scale pollutant variability, "personal monitoring," *etc.* Lewis *et al.* (DOI: 10.1039/c5fd00201j) evaluated multiple low-cost sensors in laboratory and field settings, and reported large interferences from *e.g.* CO_2 as well as numerous other problems with sensor performance, which varied by pollutant. These authors proposed some strategies to mitigate the observed problems, but much more high-quality testing by our community needs to be performed to gain (or not) confidence in this approach. At present this type of low-cost sensor strikes this author as a potential danger for our field, especially if the scientific community is not closely involved in their use. Public administrators could be convinced to install networks of these sensors instead of the more accurate and traceable (and expensive) traditional measurements. This risk may be especially large in developing countries, where little data exist and cost considerations are paramount, and could lead to a substantial waste of resources and a situation in which only very unreliable data is available for some locations in the future.

Instrumentation revolution

Atmospheric chemical instrumentation is constantly being improved, and it can be argued that the history of atmospheric chemistry closely follows improvements in its instrumentation: as more species could be monitored at increasingly lower concentrations, insights naturally followed. We are in a period of especially rapid technological improvements at the high-end of the available instrumentation (in contrast with the concern about poor performance of low-end sensors just discussed). A particularly impressive driver is the jump in field-portable mass spectrometry technology applied in our field, from 1960s technology (quadrupoles) to 1990s technology (high-resolution time-of-flight mass spectrometers). For example, data from the CLOUD collaboration shown by Baltensperger (DOI: 10.1039/c6fd00065g) convincingly demonstrated for the first time that organic molecules were often involved in the nucleation of new particles in the atmosphere. Data from single-particle quantitative aerosol mass spectrometry allowed Robinson *et al.* (DOI: 10.1039/c5fd00214a) to obtain unique insights about organic gas–particle partitioning. Although not discussed at the meeting, recent developments for simultaneous real-time gas and particle analyses using advanced mass spectrometers[1,2] are starting to revolutionize several subfields.

The application of commercial offline techniques to our field is also keeping pace. Alam *et al.* (DOI: 10.1039/c5fd00185d) presented results from the analysis of vehicle exhaust using two-dimensional gas chromatography with variable energy ionization and time-of-flight mass spectrometry, resulting in a very detailed characterization of this source. Tong *et al.* (DOI: 10.1039/c5fd00206k) applied ultrahigh-resolution mass spectrometry to ambient samples, resulting in the characterization of hundreds of different polycyclic aromatic hydrocarbons (PAHs), a type of species of high health concern.

Another measurement capability undergoing impressive development is the aircraft measurement of species fluxes. Vaughan *et al.* (DOI: 10.1039/c5fd00170f) presented results for fluxes of NO_x over Central London, and concluded that the emissions inventory may be underestimated by as much as 40%. This also qualifies as a significant piece of "news", revealing important gaps in even carefully-constructed inventories. Many advanced mass spectrometers have the time resolution and signal-to-noise to report data for hundreds of species at the sub-1 second data rates needed, and their application to airborne flux measurements is starting to yield unique insights.[3]

Lack of sufficient analysis support for field studies

Field studies are critical for understanding atmospheric chemistry. Unfortunately field studies in our field, at least in the US and the UK, appear to suffer from a chronic structural problem: too many studies are carried out, and too little support is allocated to analysis and interpretation of the data. Informal conversations with dozens of field scientists indicate that the problem is pervasive and of a large magnitude. Reducing the number of field studies performed by approximately a factor of two and doubling the personnel time allocated to data analysis and publication strikes most of the scientists I have talked to as reasonable. The instrumentation revolution just described is only making the situation worse, as advanced instruments have much larger and more complex outputs and larger calibration/characterization/interpretation needs than earlier instrument generations. For example the data rates (many MB min^{-1}) of some *in situ* mass spectrometers are comparable to those of atmospheric chemistry sensors deployed aboard NASA satellites.

As our understanding becomes ever more complex and specialized, some of the most interesting findings do not arise until a 2nd or 3rd round of analysis, or the detailed and careful comparison of complex experimental results with complex models. Performing those analyses and modeling confidently requires that the initial papers have been published, so that their authors have had a chance to digest their own measurements (with feedback from the larger community as part of conferences and peer-review) and think about how they relate to findings from others in the same campaign, the broader state of the field, and other fields. Also synthesis efforts across multiple field campaigns and laboratory studies suffer in this environment, limiting the creation of a detailed "big picture view" of the issues. The lack of sufficient analysis support is a major problem that reduces the return on the research investment and prevents some of the best science from being followed up on and published, as researchers are distracted into going into the field and producing new data, and end up mostly abandoning old data. This problem is widely recognized and frequently discussed, yet little progress appears to have been made in addressing it in the 20 years that this scientist has been active in research. Of course one can write a proposal focused on data analysis, but the funding rate of those is typically much lower than of proposals to acquire new data. Even if an individual group manages to obtain some extra funding for analysis of a past campaign, the fact that most other participants did not is a serious problem, as "their heads are no longer in it". *E.g.* any analyses that would require significant time from other groups often become unfeasible.

In terms of what can be done to lessen this problem, I do believe that the funding agencies and the culture of our scientific field could be changed, even if incrementally. But many of us need to keep making the case many times, with specific examples of how the short timeline of funding is damaging to scientific output. Funding programs should be receptive to a change that increases their scientific return-on-investment. This has happened recently in some forums, *e.g.* the report from a 2015 US atmospheric chemistry community meeting states "There was overwhelming support by the community [...] for re-analysis of previous field campaigns [...] and a desire to convey this message to [the funding agency]."[4]

Missing areas

Some topics of importance for urban atmospheric chemistry were not covered by any of the papers presented at this Faraday Discussion. This probably reflects the interests and funding available to the researchers that participated in the meeting, but it is worth highlighting for future reference.

First, there was very limited discussion of the impact of urban emissions on larger regional and global scales, and very little use of satellite data. The scientific communities working on urban *vs.* larger scales are often pretty disjointed and narrowly focused, which can lead to a lack of sufficient bi-directional communication. For example, discussions about OA in the Southeast US in the summer, a hotspot of biogenic hydrocarbon emissions, often implicitly take for granted that almost all the OA present must be of biogenic origin. However, recent results show that ~28% of the summer OA in that region arises directly from urban sources.[5]

Second, almost all the research presented focused on the current situation in developed countries, mostly the UK and Western Europe. Most of the human population lives in urban areas at present,[6] but the economic level of most of those areas is significantly lower than those that were the focus of this meeting. The air quality problems in the developing world are often more acute,[7,8] and some distinct emission sources such as the burning of trash[9] are major in those areas and minor in the developed world. Although it can be difficult to perform studies in the developing world for a variety of reasons, developed world researchers should commit more effort to this area.

Finally, there has been very little study of indoor air chemistry, compared to outdoor air chemistry. If one searches *e.g.* Web of Science, the ratio of outdoor/indoor air chemistry publications is around 100. Many key variables (*e.g.* photolysis rates, OH, NO_3 radical, HONO...) have never been measured indoors, or only once or twice. Many of the newer instruments developed for outdoor air research have never or almost never been used indoors. Combined with the high diversity of indoor environments and ways in which humans use them, *etc.*, and the importance of processes that are minor or negligible for outdoor air (emissions from materials, surface chemistry on materials and humans, impact of human breath) there are many questions about indoor chemistry about which we know very little. The Sloan Foundation in the US recognized these gaps and has started a small funding program for indoor air chemistry[10] and a few groups including ours have started using advanced techniques in indoor environments recently as a result,[11,12] but a much larger effort is needed in this area. It is important that such an effort be international, as there are very large differences in building designs, materials, ventilation practices, and indoor sources in different parts of the world that are likely to greatly impact indoor air chemistry.

Acknowledgements

The author is grateful for many in-person and email discussions with Faraday Discussion participants, and for support from a College Scholar Award from the University of Colorado.

References

1 F. D. Lopez-Hilfiker, C. Mohr, M. Ehn, *et al.*, A novel method for online analysis of gas and particle composition: description and evaluation of a Filter Inlet for Gases and AEROsols (FIGAERO), *Atmos. Meas. Tech.*, 2014, **7**, 983–1001.

2 A. W. H. Chan, N. M. Kreisberg, T. Hohaus, *et al.*, Speciated measurements of semivolatile and intermediate volatility organic compounds (S/IVOCs) in a pine forest during BEACHON-RoMBAS 2011, *Atmos. Chem. Phys.*, 2016, **16**, 1187–1205.

3 G. M. Wolfe, T. F. Hanisco, H. L. Arkinson, *et al.*, Quantifying sources and sinks of reactive gases in the lower atmosphere using airborne flux observations, *Geophys. Res. Lett.*, 2015, **42**, 8231–8240.

4 Report from the 2015 ACCORD (Atmospheric Chemistry Center for Observational Research and Data) Workshop, US National Center for Atmospheric Research. Accessed 12 May 2016. https://www2.acom.ucar.edu/accord/accord-2015-workshop.

5 P. S. Kim, D. J. Jacob, J. A. Fisher, *et al.*, Sources, seasonality, and trends of southeast US aerosol: an integrated analysis of surface, aircraft, and satellite observations with the GEOS-Chem chemical transport model, *Atmos. Chem. Phys.*, 2015, **15**, 10411–10433.

6 *World Urbanization Prospects, 2014 Revision*, United Nations DESA's Population Division, 2014.

7 World Health Organization, *WHO's Urban Ambient Air Pollution Database–Update 2016*, WHO, 2016.

8 M. Subramanian, Can Delhi save itself from its toxic air?, *Nature*, 2016, **534**, 166–169.

9 C. Wiedinmyer, R. J. Yokelson and B. K. Gullett, Global emissions of trace gases, particulate matter, and hazardous air pollutants from open burning of domestic waste, *Environ. Sci. Technol.*, 2014, **48**, 9523–9530.

10 Research Program on the Chemistry of Indoor Environments, Alfred P. Sloan Foundation, http://www.sloan.org/major-program-areas/stem-research/chemistry-of-indoor-environments/, Accessed 15 May 2016.

11 X. Tang, P. K. Misztal, W. W. Nazaroff and A. H. Goldstein, Siloxanes Are the Most Abundant Volatile Organic Compound Emitted from Engineering Students in a Classroom, *Environ. Sci. Technol. Lett.*, 2015, **2**, 303–307.

12 S. Liu, R. Li, R. J. Wild, C. Warneke, J. de Gouw, S. S. Brown, S. L. Miller, J. C. Luongo, J. L. Jimenez and P. J. Ziemann, Contribution of human-related sources to indoor volatile organic compounds in a university classroom, *Indoor Air*, 2016, DOI: 10.1111/ina.12272.

Poster titles

What effect does VOC sampling time have on derived OH reactivity?, **P. Monks, H. Sonderfeld, I. White, I. Goodall, A. Lewis, J. Hopkins and R. Koppmann,** *Department of Chemistry, United Kingdom*

A portable, automated on-line instrument to quantify PMI-bound reactive oxygen species (ROS): with an example of urban ambient measurement, **F. Wragg, S. Fuller, R. Freshwater, M. Kalberer and D. Green,** *University of Cambridge Department of Chemistry, United Kingdom*

Urban flux measurements in an Alpine valley, **T. Karl, M. Graus, A. Held, M. Striednig, L. von der Heyden, M. J. Deventer, C. Lamprecht, A. Hammerle, and G.Wohlfahrt,** *University of Innsbruck, Austria*

Using a large population of urban monitoring sites to investigate the outcome of policies to manage traffic emissions, **G. Fuller and A. Font,** *King's College London, United Kingdom*

Posttranslational modification of Birch protein by common gas phase pollutants, NO_2 and O_3, **M. Mahmood, F. Pope and W. Bloss,** *University of Birmingham, United Kingdom*

VOCs in the urban atmospheric chemistry: Impacts of large-scale biofuel burn, **P. A. Dominutti, T. Nogueira, A. Borbon and A. Fornaro,** *University of São Paulo, Brazil*

Investigating wood burning and fossil fuel contribution to black carbon and carbon monoxide concentrations in Athens, during wintertime, **A.-C. Kalogridis, S. Vratolis, N. Mihalopoulos, and K. Eleftheriadis,** *ERL, Institute of Nuclear and Radiological Science & Technology, Energy & Safety, NCSR Demokritos, Greece*

Effect of highway proximity on CO_2, CH_4 and NO_2 emissions from an agricultural field near Jefferson City, Missouri, U.S.A, **N. Nkongolo,** *Lincoln University of Missouri, U.S.A*

Using dispersion calculations for the quantification of the contribution of biomass burning in heating systems to ambient PM levels in Thessaloniki, Greece, **N. Moussiopoulos, G. Tsegas, I. Douros and C. Sioutas,** *Laboratory of Heat Transfer and Environmental Engineering, Greece*

Organic aerosol in the urban environment: spatial variations in concentration and gas/particle partitioning, **A. Presto and Z. Li,** *Carnegie Mellon University, U.S.A*

Impact of transboundary peatland burning smoke influx on carbonaceous profile of urban aerosols in Singapore, **S. Sarkar, S. Jia, J. Reid, S. Mishra, S. Swarup, C. N. Ong and L. Yu,** *NUS Environmental Research Institute, Singapore*

Phase separation of organics and sulphate in individual aerosol particles, **W. Li and Z. Shi,** *Shandong University, China*

Reduction in local ozone levels in urban São Paulo due to a shift from ethanol to gasoline use, **F. Geiger and A. Salvo,** *Northwestern University, U.S.A*

Characterization of organic molecular markers in $PM_{2.5}$ aerosol at an urban site in Sydney, **C. Meng, and P. Nelson,** *Macquarie University, Australia*

SIFT-MS: a new technique for comprehensive air analysis, **M. Perkins, V. Langford, D. Padayachee, J. Buckland, D. Hera and M. McEwan,** *Anatune Ltd, United Kingdom*

Atmospheric reaction of Cl atoms, OH radicals and ozone with guaiacol: rate constants and secondary organic aerosol yields, **A. Soto, B. Ballesteros, E. Jiménez, T. Fagniez, C. Coeur-Tourneur and J. Albaladejo,** *University of Castilla la Mancha, Spain*

A high spatio-temporal resolution agricultural residue burning smoke emissions product to support air pollution studies in China, **T. Zhang, M. Wooster, D. Green and B. Main,** *King's College London, United Kingdom*

Combination of SMPS and 4λ-PAS opens up novel perspectives in the characterisation of diesel exhaust particles, **T. Ajtai, N. Utry, M. Pintér, G.Kiss-Albert, G. Szabó and Z. Bozóki,** *University of Szeged, Hungary*

Particulate OH-PAH emissions from a residential wood log stove, **R. Avagyań, R.Nyström, R. Lindgren, C. Boman and R. Westerholm,** *Stockholm University, Sweden*

Evaluation of the behaviour of polycyclic aromatic hydrocarbons (PAH) in urban plumes, **J. Dang, M. Alam, and R. Harrison,** *University of Birmingham, United Kingdom*

A multi-scale modelling system for studying urban chemistry and transport, **R. Doherty, C. Hood, J. Stocker, I. MacKenzie, D. Carruthers, M. Vieno, C. Lin, T. Malkin, L. Whalley, D. Heard and O. Wild,** *University of Edinburgh, United Kingdom*

Using observations to constrain volatile organic compound emissions for urban air quality modeling, **B. McDonald, J. de Gouw, J. Gilman, S.-W. Kim, S. McKeen, G. Frost and M. Trainer,** *NOAA ESRL, U.S.A*

Size-differentiated chemical composition of organic components of particles arising from road traffic, **R. Xu, M. Alam and R. Harrison,** *University of Birmingham, United Kingdom*

O and NO_2 low-cost sensor network in Zurich, **M. Mueller, P. Graf, B. Schwarzenbach and C. Hueglin,** *EMPA, Swiss Federal Laboratories for Materials Science and Technology, Switzerland*

Unsaturated ethers in the atmosphere: Reactivity towards OH radicals, response under the sunlight and contribution to the global warming, **E. Jiménez, M. Antiñolo, A. Ocaña, B. Ballesteros, J. Aranguren, S. Lane and J. Albaladejo,** *University of Castilla-La Mancha, Spain*

Spatial distribution and annual trend of $PM_{2.5}$ in Spain, **A. Santurtún, M. Santurtún, A. Villar and M. T. Zarrabeitia,** *Unit of Legal Medicine, Department of Physiology and Pharmacology, University of Cantabria, Spain*

Real time source apportionment and air quality monitoring using multi-wavelength photoacoustic spectroscopy, **T. Ajtai, N. Utry, M. Pintér, G. Kiss-Albert, P. Orvos, G. Szabó and Z. Bozóki,** *University of Szeged, Hungary*

Surface ozone concentration trend differences between urban and rural environments in Spain, **A. Santurtún, A. Villar and M. T. Zarrabeitia,** *Unit of Legal Medicine, Department of Physiology and Pharmacology, University of Cantabria, Spain*

Influence of synoptic weather on UK surface and tropospheric ozone, **R. Pope and M. Chipperfield,** *University of Leeds, United Kingdom*

Relationship between NO_2 and stroke in the north of Spain, **A. Santurtún, A. Villar, P. Bolivar Ruiz and M. T. Zarrabeitia,** *Unit of Legal Medicine, Department of Physiology and Pharmacology, University of Cantabria, Spain*

Geographical distribution and trend of particulate matter (PM_{10}) in urban areas of Spain, **A. Santurtún, A. Villar, Alejandro, J. Lopez Gutierrez M. T. Zarrabeitia,** *Unit of Legal Medicine, Department of Physiology and Pharmacology, University of Cantabria, Spain*

Suitability of acetonitrile as a tracer for residential wood burning in an urban environment in Europe, **S. Reimann, T. Fröhlich, C. Hueglin and M. Hill,** *EMPA, Switzerland*

Source apportionment of brown carbon from urban ambient aerosols, **N. Krishna Kumar, K. Dällenbach, J. Slowik, U. Baltensperger, A. Prévôt and I. El Haddad,** *Paul Scherrer Institute, Switzerland*

Yearly cycle source apportionment of organic aerosol in urban and rural environment: Zurich *vs* Magadino, **A. Vlachou, K. Dällenbach, U. Baltensperger, C. Bozzetti, J. Slowik, A. Prévôt and I. El Haddad,** *Paul Scherrer Institut, Switzerland*

Air pollution study in coastal area of North Sea, **A. Sokolov, E. Dmitriev, H. Delbarre, P. Augustin and C. Gengembre,** *Laboratory for Physico-Chemistry of the Atmosphere, University of Littoral Coted'Opale, France*

Contribution of solid fuel burning to $PM_{2.5}$ in residential areas of Ireland, **J. Wenger, I. O'Connor, E. McGillicuddy, J. Arndt, S. Hellebust, P. Buckley, R. Healey and J. Sodeau,** *University College Cork, Ireland*

Analysis of atmospheric particles at a low-income high population density location of Old Naledi in the capital city of Gaborone, Botswana, **B. Mmereki and T.Khumoetsile,** *University of Botswana, Botswana*

Assessing the variability of pollution along roads in Birmingham using a mobile monitoring laboratory, **L. Kramer, L. Crilley, T. Adams, S. Ball, F. Pope and W. Bloss,** *University of Birmingham, United Kingdom*

How important are local and regional sources of fine and coarse particles in London? Results from Clearflo winter and summer measurements, **L. Crilley,**

W. Bloss, J. Yin, D. Beddows, R. Harrison, P. Zotter, A. Prévôt, J. Allan, S. Szidat, F. Lucarelli, G. Calzolai, S. Nava, G. Valli, V. Bernardoni and R. Vecchi, *University of Birmingham, United Kingdom*

Inorganic chlorine in an urban environment: Measurements and seasonal patterns, **R. Sommariva, L. Hollis, S. Ball, R. Cordell, T. Brown, M. Evans, Z. Fleming and P. Monks,** *University of Leicester, United Kingdom*

Air pollution particulate matter promotes DC maturation and enhances their stimulation of CD8 lymphocytes, **T.-R. Ho, P. Pfeffer, E. Mann, N. Matthews, I. Mudway, F. Kelly and C. Hawrylowicz,** *MRC & Asthma UK Centre in Allergic Mechanisms of Asthma, United Kingdom*

KEY-VOCs project: metrology for VOC indicators in air pollution and climate change, **M. Corbel, A. Baldan, S. Persijn and L. Spinelle,** *NPL, United Kingdom*

Chemical characterization of $PM_{2.5}$ during haze episodes in Beijing, **M. Zheng, C. Yan and X. Li,** *Peking University, China*

City-sonde science: street level NO_2 from bicycle measurements using the KNMI NO_2 -sonde, **D. Stein Zweers, M. den Hoed and E. Tenner,** *Royal Netherlands Meteorological Institute, Netherlands*

Carbon in aerosol particulate matter: an innovative way to standardise measurements, **A. Santoro, T. Carey, A. Held, J.-P. Putaud, E. Grandesso, F. Cavalli and F. Torrisi,** *Joint Research Centre Institute for Reference Materials and Measurements, Belgium*

Ozonolysis of ultrasonically levitated droplets containing unsaturated carboxylic acids, **E. R. Cabrera-Martinez, H. J. Ancill, K. Rastogi and C. Pfrang,** *University of Reading, United Kingdom*

Distributed low-cost air quality sensing: challenges and potential of neighborhood-scale AQ monitoring, **E. Cross, D. Hagan, L. Williams, G. Adamkiewicz, A.Backus, D. Worsnop, J. Kroll and J. Jayne,** *Aerodyne Research, Inc., U.S.A*

Investigation of the spatio-temporal variation of air pollutants in the city of Zurich using mobile measurements, **C.Hueglin, M. Hundt, M. Mueller and L. Emmenegger,** *EMPA, Switzerland*

Source apportionment of $PM_{2.5}$ in Jinan, China: emerging trend, **Z. Tian, Z. Shi, R. Harrison, L. Xin, C. Shurui, J. Zhang, J. Sun, J. Chi, W. Li and P. Pant,** *University of Birmingham, United Kingdom*

The formation of highly oxidised multifunctional organic compounds (HOM) in oxidation of benzene, **O. Garmash, M. Rissanen, I. Pullinen, S. Schmitt, T. Mentel, J. Wildt, E. Kleist, M. Kulmala and M. Ehn,** *University of Helsinki, Finland*

Mass spectrometer *vs* human nose: a comparison of organic aerosol characterization by AMS-PMF with city smells in Barcelona, **C. Mohr, R. Schifanella, D.Quercia, L. M. Aiello, A. Prévôt and X. Querol,** *Karlsruhe Institute of Technology, Germany*

Highly oxidized multifunctional organic compounds from anthropogenic precursors, **U. Molteni, F. Bianchi, C. Frege, I. El Haddad, F. Klein, M. Rossi, J. Dommen and U. Baltensperger,** *Paul Scherrer Institut, Switzerland*

Estimation of NO_2 emissions from Lahore using car MAX-DOAS observations and comparison with OMI satellite data, **M. Razi, R. Shaiganfar, M. Khokhar, N. Ahmad, S. Dörner, S. Beirle and T. Wagner,** *Max Planck Institute for Chemistry, Mainz, Germany, Germany*

Our take on the state-of-the-science for SOA from urban VOC sources, **J.-L. Jimenez, P. Hayes *et al*,** *University of Colorado at Boulder, U.S.A*

Atmospheric chemistry implications of residential solid fuel combustion, **E. Mitchell, A. Williams, J. Jones and A. Lea-Langton,** *University of Leeds, United Kingdom*

Black and organic carbon from a light-duty diesel engine: contrasting the behaviour of particulates emitted under cold-start and running conditions, **G. McFiggans, S. Alam, M. R. Alfarra, J. Allan, R. Dunmore, J. Hamilton, R. Harrison, S. Kong, D. Liu, K. Pereira, J. Whitehead, D. Beddows and H. Coe,** *University of Manchester, United Kingdom*

Volatility of shipping fuel combustion emissions, **S. Pieber, J. Orasche, B. Stengel, Y. Zhao, F. Klein, D. Kilic, I. El Haddad, J. Corbin, J. Slowik, A. Robinson, N. Donahue, R. Zimmermann, U. Baltensperger, H. Czechand A. Prévôt,** *Paul Scherrer Institute, Switzerland*

The Skinner Prize for the best poster was jointly awarded to Miss Chuanping Meng of Macquarie University, Australia, for her poster on characterization of organic molecular markers in $PM_{2.5}$ aerosol at an urban site in Sydney, and Miss Edna Cabrera- Martinez of the University of Reading, United Kingdom, for her poster on ozonolysis of ultrasonically levitated droplets containing unsaturated carboxylic acids.

List of participants

Dr Tibor Ajtai, *University of Szeged*, *Hungary*
Dr Mohammed Salim Alam, *University of Birmingham*, *United Kingdom*
Professor Paulo Artaxo, *University of Sao Paulo, Institute of Physics*, *Brazil*
Professor Dr Urs Baltensperger, *Paul Scherrer Institute*, *Switzerland*
Dr William Bloss, *University of Birmingham*, *United Kingdom*
Professor William Brune, *Pennsylvania State University*, *USA*
Professor Lucy Carpenter, *University of York*, *United Kingdom*
Dr Antonia Carruthers, *Newcastle University*, *United Kingdom*
Dr David Carslaw, *Ricardo Energy & Environment*, *United Kingdom*
Dr Aurélie Charron, *IFSTTAR*, *France*
Ms Marivon Corbel, *NPL*, *United Kingdom*
Dr Leigh Crilley, *University of Birmingham*, *United Kingdom*
Dr Eben Cross, *Aerodyne Research, Inc.*, *USA*
Miss Juan Dang, *University of Birmingham*, *United Kingdom*
Ms Mirjam den Hoed, *Royal Netherlands Meteorological Institute*, *Netherlands*
Dr Ruth Doherty, *University of Edinburgh*, *United Kingdom*
Professor Neil Donahue, *Carnegie Mellon University*, *USA*
Dr Rachel Dunmore, *University of York*, *United Kingdom*
Dr Christian Ehlers, *Forschungszentrum Jülich GmbH*, *Germany*
Dr Anna Font, *King's College London*, *United Kingdom*
Professor David Fowler, *NERC Centre for Ecology & Hydrology*, *United Kingdom*
Dr Gary Fuller, *King's College London*, *United Kingdom*
Professor Dr Franz Geiger, *Northwestern University*, *USA*
Dr Jacqueline Hamilton, *University of York*, *United Kingdom*
Professor Roy Harrison, *University of Birmingham*, *United Kingdom*
Professor Dwayne Heard, *University of Leeds*, *United Kingdom*
Professor Hartmut Herrmann, *TROPOS Leipzig*, *Germany*
Professor Nick Hewitt, *Lancaster University*, *United Kingdom*
Dr Matthew Hort, *Met Office*, *United Kingdom*
Dr Christoph Hueglin, *Empa*, *Switzerland*
Professor Jose-Luis Jimenez, *University of Colorado at Boulder*, *USA*
Dr Elena Jiménez, *University of Castilla-La Mancha*, *Spain*
Dr Markus Kalberer, *University of Cambridge*, *United Kingdom*
Dr Athina-Cerise Kalogridis, *ERL, Institute of Nuclear and Radiological Science & Technology, Energy & Safety, NCSR. Demokritos*, *Greece*
Professor Dr Thomas Karl, *University of Innsbruck*, *Austria*
Professor Dr Astrid Kiendler-Scharr, *Forschungszentrum Jülich GmbH*, *Germany*
Professor Saewung Kim, *Department of Earth System Science, University of California, Irvine*, *USA*
Dr Jörg Kleffmann, *University of Wuppertal*, *Germany*
Dr Louisa J. Kramer, *University of Birmingham*, *United Kingdom*
Dr James Lee, *University of York*, *United Kingdom*

Professor Alastair Lewis, *National Centre for Atmospheric Science, University of York, United Kingdom*
Professor Angus MacKenzie, *University of Birmingham, United Kingdom*
Dr Sasha Madronich, *NCAR, USA*
Dr Brian McDonald, *NOAA ESRL, USA*
Professor Gordon McFiggans, *University of Manchester, United Kingdom*
Dr María Cruz Minguillón, *Institute of Environmental Assessment and Water Research (IDAEA), CSIC, Spain*
Dr Baagi Mmereki, *University of Botswana, Botswana*
Dr Claudia Mohr, *Karlsruhe Institute of Technology, Germany*
Dr Victor Moll, *Givaudan UK Ltd, United Kingdom*
Dr Sarah Moller, *NCAS, University of York, United Kingdom*
Professor Paul Monks, *Department of Chemistry, United Kingdom*
Dr Juliane Mossinger, *Nature, United Kingdom*
Professor Andrew Mount, *The University of Edinburgh, United Kingdom*
Dr Michael Mueller, *Empa, Switzerland*
Dr Tim Murrells, *Ricardo Energy & Environment, United Kingdom*
Dr Irina Nikolova, *University of Birmingham, United Kingdom*
Dr Micheal O'Dwyer, *Environmental Protection Agency Ireland, Ireland*
Professor Spyros Pandis, *University of Patras, Greece*
Mr Mark Perkins, *Anatune Ltd, United Kingdom*
Dr Francis Pope, *University of Birmingham, United Kingdom*
Dr Richard Pope, *University of Leeds, United Kingdom*
Dr Albert Presto, *Carnegie Mellon University, USA*
Dr André Prévôt, *Paul Scherrer Institut, Switzerland*
Dr Ruth Purvis, *NCAS, University of York, United Kingdom*
Professor Dr Xavier Querol, *IDAEA-CSIC, Spain*
Dr Stefan Reimann, *Empa, Switzerland*
Dr Andrew Rickard, *NCAS, University of York, United Kingdom*
Dr Anke Roiger, *DLR-IPA, Germany*
Dr Harald Saathoff, *Karlsruhe Institute of Technology, Germany*
Dr Anna Santoro, *Joint Research Centre Institute for Reference Materials and Measurements, Belgium*
Dr Ana Santurtún, *Unit of Legal Medicine, Department of Physiology and Pharmacology, University of Cantabria, Spain*
Dr Sayantan Sarkar, *NUS Environmental Research Institute, National University of Singapore, Singapore*
Dr Karine Sartelet, *CEREA - Ecole des Ponts, France*
Dr Martin Shafer, *University of Wisconsin-Madison, USA*
Dr Marvin Shaw, *University of York, United Kingdom*
Dr Zongbo Shi, *University of Birmingham, United Kingdom*
Professor Costas Sioutas, *University of Southern California, USA*
Dr Andreas Skouloudis, *European Commission, Italy*
Dr Anton Sokolov, *Laboratory for Physico-Chemistry of the Atmosphere, University of Littoral Cote d'Opale, France*
Dr Roberto Sommariva, *University of Leicester, United Kingdom*
Dr Hannah Sonderfeld, *University of Leicester, United Kingdom*
Miss Amparo S, *University of Castilla la Mancha, Spain*
Professor Alison Tomlin, *University of Leeds, United Kingdom*
Dr Shengrui Tong, *Institute of Chemistry, Chinese Academy of Sciences, China*
Dr Dominik van Pinxteren, *Leibniz-Institute for Tropospheric Research, Germany*
Dr Mar Viana, *IDAEA-CSIC, Spain*
Professor Dr Andreas Wahner, *Forschungszentrum Jülich GmbH, Germany*

Dr Timothy Wallington, *Ford Motor Company*, *USA*
Professor John Wenger, *University College Cork*, *Ireland*
Professor Roger Westerholm, *Stockholm University*, *Sweden*
Dr Lisa Whalley, *University of Leeds*, *United Kingdom*
Miss Ruixin Xu, *University of Birmingham*, *United Kingdom*
Ms Mei Zheng, *Peking University*, *China*
Dr Shupeng Zhu, *ENPC*, *France*